Quantum Effects, Heavy Doping, and the Effective Mass

Series on the Foundations of Natural Science and Technology

Published:

Vol. 1: Space and Time, Matter and Mind: The Relationship between Reality and Space-Time
by W. Schommers

Vol. 2: Symbols, Pictures and Quantum Reality: On the Theoretical Foundations of the Physical Universe
by W. Schommers

Vol. 3: The Visible and the Invisible: Matter and Mind in Physics
by W. Schommers

Vol. 4: What is Life? Scientific Approaches and Philosophical Positions
by H.-P. Dürr, F.-A. Popp and W. Schommers

Vol. 5: Grasping Reality: An Interpretation-Realistic Epistemology
by H. Lenk

Vol. 6: Nano-Engineering in Science and Technology: An Introduction to the World of Nano-Design
by M. Rieth

Vol. 7 Magneto Thermoelectric Power in Heavily Doped Quantized Structures
by K. P. Ghatak

Vol. 8 Quantum Effects, Heavy Doping, and the Effective Mass
by K. P. Ghatak

Vol. 9 The Timeless Approach
Frontier Perspectives in 21st Century Physics
by D. Fiscaletti

Series on the Foundations of Natural Science and Technology — Vol. 8

Quantum Effects, Heavy Doping, and the Effective Mass

Kamakhya Prasad Ghatak

University of Engineering and Management, India

NEW JERSEY • LONDON • SINGAPORE • BEIJING • SHANGHAI • HONG KONG • TAIPEI • CHENNAI • TOKYO

Published by

World Scientific Publishing Co. Pte. Ltd.
5 Toh Tuck Link, Singapore 596224
USA office: 27 Warren Street, Suite 401-402, Hackensack, NJ 07601
UK office: 57 Shelton Street, Covent Garden, London WC2H 9HE

Library of Congress Cataloging-in-Publication Data
Names: Ghatak, Kamakhya Prasad, author.
Title: Quantum effects, heavy doping and the effective mass / Kamakhya Prasad Ghatak, Institute of Engineering and Management, India.
Other titles: Series on the foundations of natural science and technology ; v. 8.
Description: Hackensack, NJ : World Scientific, [2017] | Series: Series on the foundations of natural science and technology ; vol. 8 | Includes bibliographical references and index.
Identifiers: LCCN 2016025416| ISBN 9789813146518 (hardcover ; alk. paper) | ISBN 9813146516 (hardcover ; alk. paper)
Subjects: LCSH: Doped semiconductors. | Effective mass (Physics) | Nanostructures. | Nanostructured materials.
Classification: LCC QC611.8.D66 G44 2017 | DDC 537.6/221--dc23
LC record available at https://lccn.loc.gov/2016025416

British Library Cataloguing-in-Publication Data
A catalogue record for this book is available from the British Library.

Desk Editor: Christopher Teo

Typeset by Stallion Press
Email: enquiries@stallionpress.com

Printed in Singapore

Our pains occur from contact-born pleasures

Dedication

Lovable MAA,

This is the record of not only placing all the books and the research papers which bear my name as a co-author at your Lotus feet for your kind acceptance (although I suspect deeply that you are the real hidden first author and the main Driving force) but also my eternal gratitude to you. My Dear MAA, when I came to this globe called Earth through you I was alone, not consulted and obviously I registered my arrival by mere crying. After then, being my real friend you did not forget to place enough number of trials in my life so that I may not be negatively critical of others who are forced to face trials in their life and fail consequently after tremendous positive action in their own ways. You are constantly wiping each and every error of mine in Your own unseen ways. You are constantly teaching me that my education is nothing but heducation and the real education is the permanent creative stage of mind which if at all arrives must stay till Death and my real ***I****, is not at all equal to the sum of the body I wear and mind which Drives the body.*

MAA, this is my first and last letter to You in printed form. I do not know Your address since you are living in a wonderful world from where the return visa is not possible although I feel that Your radiant existence is being reflected in each and every aspect of mine. MAA, You are constantly teaching me that nobody is to be held responsible for any aspect of my life. My ego is the basic cause of my bondage and if I can transform this ego into something which goes eternally then and only then I am FREE in the real sense of the term. Before mailing the manuscript of this book to C. Teo, In-house Editor, World Scientific Publishing Co. ***the solo Driving force behind all our***

***books from World Scientific**, in this morning, I saw in a dream and heard that MAA, you are telling me the following sentences which I wrote before they vanished in the sea my mind:*

"Hear my son as you pass by,
Where you are now, so was once, I.
Where I am now, so you must be,
Delete everything to follow me."

*MAA, I am now eagerly waiting for the kiss of GOD, known as death since I treat GOD and Death on something like equal terms. Dear GOD, please grant my very Death as the last Death and place the probability of new arrival in **the infinitely deep multi-dimensional quantum wells in Hilbert space and for that thought, my Dear MAA, no tears are enough to wash Your Lotus feet**.*

With unconditional surrender at Your lotus feet,
Yours and only Yours,
Kamakhya Prasad Ghatak

Let us dream more while we are AWAKE

Preface

The unification of the concept of asymmetry of the wave vector space of the charge carriers in semiconductors with the modern fabricational techniques such as Molecular Beam Epitaxy, Metallo-Organic-Chemical-Vapor Deposition and Fine Line Lithography in one, two and three dimensions (such as ultrathin films, inversion and accumulation layers, quantum well super-lattices, nipi structures, carbon nano-tubes, quantum wires, quantum wire super-lattices, quantum dots, magneto inversion and accumulation layers, quantum dot super-lattices, etc.) spawns not only useful quantum effect devices but also unearth new concepts in the realm of Nano-Electronics and related disciplines. It is worth remarking that these semiconductor nanostructures occupy a paramount position in the entire arena of nano science and technology by their own right and find extensive applications in quantum registers, resonant tunneling diodes and transistors, quantum switches, quantum sensors, quantum logic gates, hetero-junction field-effect, quantum well and quantum wire transistors, high-speed digital networks, high-frequency microwave circuits, quantum cascade lasers, high-resolution terahertz spectroscopy, super-lattice photo-oscillator, advanced integrated circuits, super-lattice photo-cathodes, thermoelectric devices, super-lattice coolers, thin film transistors, intermediate-band solar cells, micro-optical systems, high performance infrared imaging systems, band-pass filters, thermal sensors, optical modulators, optical switching systems, single electron/molecule electronics, nano-tube based diodes, quantum switches, quantum sensors and other nano-electronic devices. Knowledge regarding these quantized structures may be gained from original research contributions in scientific journals, proceedings of international conferences and different review articles respectively [1–14].

Mathematician Simmons rightfully tells us [15] that the mathematical knowledge is said to be doubling in every 10 years and in this context we can also envision the extrapolation of the Moore's law by projecting it in the perspective of the advancement of new research and analyses, in turn, generating novel concepts particularly in the area of nano-science and technology [16]. In this context, it may be noted that the available reports on the said areas cannot afford to cover even an entire chapter regarding the EFFECTIVE MASS (EM) in heavily doped (HD) quantized structures and after thirty years of continuous effort, we see that the complete investigations of the EM comprising of the whole set of materials and allied sciences is really a sea and is a permanent member of the domain of impossibility theorems.

It is well-known that the concept of the EM in quantum materials is one of the basic pillars in the realm of modern low dimensional quantum science and technology [17]. Among the various definitions of the effective mass (e.g. effective acceleration mass, density-of-state effective mass, concentration effective mass, conductivity effective mass, Faraday rotation effective mass, etc.) [18], it is the effective momentum mass that should be regarded as the basic quantity [19], since it is this mass which appears in the description of transport phenomena and all other properties of the carriers in materials having arbitrary band structures [19] ***together with the fact that only the basic definition of mass generates the Einstein's*** $\boldsymbol{E = mc^2}$ ***where E, m and c are the energy, mass and velocity of light respectively equation for photons and under the condition of stationary frame of reference*** [20]. It can be shown that it is the effective momentum mass which enters in various transport coefficients and plays the most dominant role in explaining the experimental results of different scattering mechanisms through Boltzmann's transport equation [21]. The carrier degeneracy in materials influences the effective mass when it is energy dependent. Under degenerate conditions, only the electrons at the Fermi surface of n-type materials participate in the conduction process and hence, the effective momentum mass of the electrons corresponding to the Fermi level would be of interest in electron transport under such conditions. The Fermi energy is again determined by the carrier energy spectrum and the electron statistics and therefore, these two features would determine the dependence of the EM in degenerate n-type materials under the degree of carrier degeneracy. In recent years, various energy wave vector dispersion relations have been proposed [19-32] which have created the interest in studying the EM in such materials under external conditions. Besides the physical properties of various quantum materials have been studied intensively [33]. The nature of variations of the EM has widely been studied in the literature [34–49]. Some of the significant features, which have emerged from these investigations, are:

1. The EM (effective mass) increases monotonically with increasing carrier concentration.
2. The EM increases with doping in heavily doped materials in the presence of band tails.
3. The nature of variations is significantly influenced by the energy band constants of various materials having different band structures.
4. The EM oscillates with inverse quantizing magnetic field due to SdH effect. The EM in Bismuth under magnetic quantization depends both on the Fermi energy and the magnetic quantum number due to the presence of band nonparabolicity only.
5. The EM increases with the magnitude of the quantizing electric field in n-channel inversion layers of III–V materials and depends on the sub-band index for both weak and strong electric field limits.
6. The EM in ultrathin films of nonlinear optical materials depends on the Fermi energy and size quantum numbers due to the specific dispersion relations.
7. The EM has significantly different values in superlattices and also in the presence of quantum confined superlattices of small gap

materials with graded interfaces together with effective mass of lattices.

The EM in GaN_xAs_{1-x}/GaAs quantum wells (QWs) has been investigated, and detected by cyclotron resonance technique by P. N. Hai *et al.* [50]. The values of EM are $0.12m_0$ and $0.19m_0$ which are directly determined for the 70-Å-thick QWs with N composition of 1.2% and 2.0%, respectively. This sizable increase in the EM is consistent with the earlier theoretical predictions based on the strong interaction of the lowest conduction band states with the upper lying band states or impurity band induced by the incorporation of N. DiVincenzo *et al.* [51] formulated the self consistent effective mass theory for intra layer screening in graphite intercalation compounds. The effective mass approximation differential equations appropriate for impurities in a graphite host are constructed and are used to solve self-consistently for the screening response surrounding a single intercalant atom. The screening cloud is found to have a very slow algebraic decay with a characteristic length of 0.38 nm which is due to both the semi metallic and two dimensional character of graphite. The transferred charge in alkali-metal-graphite intercalation compounds is distributed nearly homogeneously on a carbon plane.

Perlin *et al.* [52] performed infrared reflectivity and Hall effect measurements on highly conducting n-type GaN ($n \approx 6 \times 10^{19}$ cm^{-3}) bulk crystals grown by the high-pressure high-temperature method for the purpose of the experimental determination of the EM of GaN. Values of electron-plasma frequency and free-electron concentration were determined for each sample of the set of seven crystals. It enabled them to calculate the perpendicular EM in the wurtzite structure of GaN as $m^* = 0.22 \pm 0.02\ m_0$ and the effects of non-parabolicity together with the difference between parallel and perpendicular components of the effective mass are small and do not exceed the experimental error. The EM has been determined by magneto-photoluminescence in as-grown and hydrogenated $GaAs_{1-x}N_x$ samples for a wide range of nitrogen concentrations (from $x < 0.01\%$ to $x = 1.78\%$) by the group of

Masia *et al.* [53]. A modified **k·p** model, which takes into account hybridization effects between N cluster states and the conduction band edge, reproduces quantitatively the experimental values up to $x \leq 0.6\%$. Experimental and theoretical evidence is provided for the N complexes responsible for the nonmonotonic and initially puzzling compositional dependence of the EM. Sewall *et al.* [54] investigated the experimental tests of EM and atomistic approaches to quantum dot electronic structure. The overall symmetry of the envelope functions for the four lowest energy excitonic states in colloidal CdSe quantum dots are assigned using excitonic state-resolved pump/probe spectroscopy.

G.E Smith [55] performed the experimental determination of the EM's in Bismuth-Antimony alloy. It was found that the EM's in $Bi_{95}Sb_5$ are smaller by about a factor of two than that of pure Bi and the hole masses are essentially unchanged. SdH investigations on n-InP are presented by the group of Schneider *et al.* [56] and EM as a function of carrier concentration has been determined. The experiments were carried out with bulk and liquid phase epitaxically grown material and carrier concentrations between $n_0 = 10^{20}$ m^{-3} and 10^{22} m^{-3} within the ranges of temperature between 2 K to 77 K and magnetic field B = 22 Tesla. The experimental result agrees very well with the theoretical relations. The values of the EM in Cd_3As_2 were obtained from low temperature SdH, magneto-Seebeck and Hall measurements by Caron *et al.* [57]. The theoretical estimation of the variation of the energy gap at Γ as a function of temperature and pressure has been obtained. There is a band reversal in $Cd_{3-x}Zn_xAs_2$ and $Cd_3As_xP_{2-x}$alloys.Bhattacharya *et al.* [58] evaluated the EM in compound semiconductor films of CdS_xTe_{1-x} and CdS_xSe_{1-x} which showed bowing phenomena similar to those for optical band gaps for the above alloy films.

B. Slomski *et al.* [59] has investigated the in-plane EM (effective mass) of quantum well states in thin Pb films on a Bi reconstructed Si (111) surface by angle-resolved photoemission spectroscopy. It is found that this EM is a factor of 3 lower than the unusually high values reported for Pb films grown on a Pb reconstructed Si (111) surface. Through a quantitative low-energy electron diffraction analysis the change in EM as a function of coverage and for the different interfaces is linked to a change of about 2% in the in-plane lattice constant. To corroborate this correlation, density functional theory calculations are performed on freestanding Pb slabs with different in-plane lattice constants. These calculations show an anomalous dependence of the EM on the lattice constant including a change of sign for values close to the lattice constant of Si (111). This unexpected relation is due to a combination of reduced orbital overlap of the $6pz$ states and altered hybridization between the $6pz$ and the $6pxy$ derived quantum well states. Furthermore, it is shown by core-level spectroscopy that the Pb films are structurally and temporally stable at temperatures below 100 K.

The EM's for spin-up and spin-down electrons of a partially spin-polarized Fermi liquid are theoretically different as proposed by L.M. Wei *et al.* [60]. They extracted the spin-up and spin-down EM's from magnetotransport measurements at different temperatures for a 2D electron gas in an $In_{0.65}Ga_{0.35}As/In_{0.52}Al_{0.48}As$ quantum well exhibiting zero-field spin splitting. Two analytical methods are used, one involving the simultaneous fitting of fast Fourier transform (FFT) spectra and the other involving inverse FFT analysis. Both methods confirm that the EM's for spin-up and spin-down are different, consistent with theoretical expectations. The group of Karra *et al.* [61] performed Cyclotron-resonance measurements for wide (100–300 nm) modulation-doped $Al_xGa_{1-x}As$ graded parabolic quantum wells for electron areal densities $10^9/cm^2$–$2.5\times10^{11}/cm^2$. A clear dependence of the cyclotron frequency on N_s is observed in the extreme quantum limit which is understood in terms of alloy effects. Self-consistent calculations that include the x dependence of the local effective mass and exchange-correlation effects in a local density approximation are in quantitative agreement with the measurements for high densities. At low densities a pinning of the cyclotron frequency is observed that is not predicted by the model.

Rößner *et al.* [62] reported the dependence of the effective masses on hole density in remotely doped strained Ge layers on relaxed $Si_{0.3}Ge_{0.7}$ buffers with sheet densities from 2.9×10^{11} cm^{-2} to

1.9×10^{12} cm^{-2}. The masses have been determined using temperature dependent Shubnikov–de Haas oscillations. No noticeable dependence of the mass on the magnetic field has been found. The extrapolated G point effective mass has been found to be 0.080 times the free electron mass. From the measured data the variation of the mass with kinetic energy and the shape of the topmost heavy hole sub-band have been calculated. The results are in good agreement with theoretical predictions. The determination of the EM of the two-dimensional electron gas (2DEG) and nonparabolicity effects in modulation-doped $In_{0.65}Ga_{0.35}As/In_{0.52}Al_{0.48}As$ single quantum well were investigated by T. M. Kim *et al.* [63] by performing temperature-dependent Shubnikov–de Haas (SdH) measurements and fast Fourier transformation (FFT) and the inverse FFT (IFFT) analyses. The result of the angular dependent SdH measurements clearly demonstrated the occupation of two sub-bands in the quantum wells by the 2DEGs. The EM's determined from temperature-dependent S±dH measurements and the FFT and IFFT analyses were 0.05869 and 0.05385m_e for the first and zeroth sub-bands, respectively. The EM's obtained from the S-dH measurements and the FFT and IFFT analyses measurements qualitatively satisfy the nonparabolicity behavior in the $In_{0.65}Ga_{0.35}As$ single quantum well.

The non-parabolic EM's in InGaAs quantum wells (QWs), sandwiched by thick InAlAs barriers of 0.52-eV band offset, were studied by N. Kotera *et al.* [64] in normal and parallel directions to the QW plane. The normal mass was experimentally obtained by observing interband photocurrent spectra of undoped InGaAs multi-QW structures. The mass increased by more than 50% from the bulk band edge mass, 0.041m_0. Electron eigenenergies were calculated in QWs based on Kane's three-level band theory. The calculated 'apparent' normal mass as a function of kinetic energy up to 0.5 eV agreed well with experiments. The parallel mass in n-type modulation-doped InGaAs QWs was experimentally obtained by pulse cyclotron resonance up to 100 T. The analysis in quantizing magnetic fields, modified for two-dimensional QWs, fits well with cyclotron energy. The 'apparent' parallel mass as a function of energy was obtained consistently. Interband optical transitions of $In_{0.53}Ga_{0.47}As/In_{0.52}Al_{0.48}As$

multi-quantum wells have been observed by Tanaka *et al.* [65] in photocurrent spectra. Interband transitions were assigned from the spectral structures. Eigen-energies of conduction band were not proportional to the square of quantum numbers. An EM normal to the quantum well plane was 50%-heavier than the bulk band-edge mass of InGaAs.

The electronic structures of Bi_2Te_3 and Sb_2Te_3 were computed and related to the thermoelectric properties of Bi_2Te_3 and Sb_2Te_3 superlattices by Wang and Cagin [66]. They found that the similarity of the electronic structure of the two materials permits the Bi_2Te_3 and Sb_2Te_3 superlattices to inherit high band edge degeneracy, and thus have high electrical conductivity. From the calculated EM along the superlattice growth direction, they infer that presence of more Sb_2Te_3 than Bi_2Te_3 in the superlattice leads to a smaller EM and enhanced carrier mobility. Furthermore, their results suggest that external tensile strain parallel to the interface may further improve the thermoelectric performance of the Bi_2Te_3 and Sb_2Te_3 superlattices. Engineered energy-wave-vector dispersion relations of either electrons or holes hold great promise for realizing fundamental oscillators at terahertz frequencies if they contain sections with a negative EM at appropriate energy levels as suggested by Gribnikov *et al.* [67], although, neither bulk semiconductor materials nor quantum wells or quantum wires exhibit such negative EM sections in the dispersion relations at favorable energy levels. Therefore, the novel use of a nanostructure is proposed to create an NEM section of electrons at suitable energy levels. This structure utilizes a heterojunction with a QW channel grown perpendicular to a superlattice. At small values of the wave-vector k; the electron wave function ψ resides mostly in the QW channel and, as k increases, ψ extends further into the superlattice. This spread of ψ induces a negative EM section in the energy dispersion relation and several combinations of suitable material systems are considered by them.

It is well known that heavy doping and carrier degeneracy are the keys to unlock the important properties of semiconductors and they are especially instrumental in dictating the characteristics of Ohmic contacts and Schottky contacts respectively [68–78]. It is an amazing fact that although the heavily doped semiconductors (HDS) have been investigated in the literature but the study of the corresponding EMs of HDS ***is still one of the open research problems***. **This first monograph solely investigates the EM** in HD (Heavily Doped) non-linear optical, III–V, II–VI, Gallium Phosphide, Germanium, Platinum Antimonide, stressed, IV–VI, Lead Germanium Telluride, Tellurium, II–V, Zinc and Cadmium diphosphides, Bismuth Telluride, III–V, II–VI, IV–VI and HgTe/CdTe quantum well HD super-lattices with graded interfaces under magnetic quantization, III–V, II–VI, IV–VI and HgTe/CdTe HD effective mass super-lattices under magnetic quantization, quantum confined effective mass super-lattices and super-lattices of HD optoelectronic materials with graded interfaces respectively. Our method is not at all related with the DOS technique as used in the aforementioned works. From the electron energy spectrum, one can obtain the DOS but the DOS technique, as used in the literature cannot provide the $E-k$ dispersion relation. ***Therefore, our study is more fundamental than those in the existing literature, because the Boltzmann transport equation, which controls the study of the charge transport properties of the semiconductor devices, can be solved if and only if the E-k*** **dispersion relation** ***is known.***

This book, containing eight chapters and seven Appendices is partially based on our on-going research on the EM in HDS from 1994 and an attempt has been made to present a cross section preview of the EM for wide range of HDS and their quantized counter parts with varying carrier energy spectra under various physical conditions.

It is well known that the band tails are being formed in the forbidden zone of the HDS and can be explained by the overlapping of the impurity band with the conduction and valence bands [79]. Kane [80] and Bonch Bruevich [81] have independently derived the theory of band tailing for semiconductors having unperturbed parabolic energy bands. Kane's model [80] was used to explain the experimental results on tunneling [82] and the optical absorption edges [83, 84] in this context. Halperin and Lax [85] developed a model for band tailing applicable only to the deep tailing states. Although Kane's concept is often used in the literature for the investigation of band tailing [86, 87], it may be noted that this model [80, 88] suffers from serious assumptions in the sense that the local impurity potential is assumed to be small and slowly varying in space coordinates [87]. In this respect, the local impurity potential may be assumed to be a constant. In order to avoid these approximations, we have developed in this book, the electron energy spectra for HDS for studying the EM based on the concept of the variation of the kinetic energy [79, 87] of the electron with the local point in space coordinates. This kinetic energy is then averaged over the entire region of variation using a Gaussian type potential energy. On the basis of the E–k dispersion relation, we have obtained the electron statistics for different HDS for the purpose of numerical computation of the respective EMs. It may be noted that, a more general treatment of many-body theory for the DOS of HDS merges with one-electron theory under macroscopic conditions [79]. Also, the experimental results for the Fermi energy and others are the average effect of this macroscopic case. So, the present treatment of the one-electron system is more applicable to the experimental point of view and it is also easy to understand the overall effect in such a case [78]. In a HDS, each impurity atom is surrounded by the electrons, assuming a regular distribution of atoms, and it is screened independently [86, 89, 90]. The interaction energy between electrons and impurities is known as the impurity screening potential. This energy is determined by the inter-impurity distance and the screening radius (popularly known as the Debye screening length). The screening length changes with the band structure. Furthermore, these entities are important for HDS in characterizing the semiconductor properties [91–94] and the modern electronic devices [86, 93]. The works on Fermi energy and the screening length in an n-type GaAs have already been initiated in the literature [94], based on Kane's model. Incidentally, the limitations of Kane's model [80, 87], as mentioned above, are also present in their studies.

In Chapter 1 we study the EM in QWs of HD nonlinear optical materials on the basis of a generalized electron dispersion law introducing the anisotropies of the effective masses and the spin orbit splitting constants respectively together with the inclusion of the crystal field splitting within the framework of the ***k.p*** formalism. ***We will observe that the complex electron dispersion law in HDS, instead of real one, occurs from the existence of the essential poles in the corresponding electron energy spectrum in the absence of band tails.*** The physical picture behind the existence of the complex energy spectrum in HD non-linear optical semiconductors is the interaction of the impurity atoms in the tails with the splitting constants of the valance bands. The more is the interaction, the more the prominence of the complex part than the other case. In the absence of band tails, there is no interaction of the impurity atoms in the tails with the spin orbit constants and consequently, the complex part vanishes. One important consequence of the HDS forming band tails is that ***the EM exists in the forbidden zone, which is impossible without the effect of band tailing. In the absence of band tails, the effective mass in the band gap of semiconductors is infinity. Besides, depending on the type of the unperturbed carrier energy spectrum, the new forbidden zone will appear within the normal energy band gap for HDS.***

The results of HD III-V (e.g. InAs, InSb, GaAs etc.), ternary (e.g. $Hg_{1-x}Cd_xTe$), quaternary (e.g. $In_{1-x}Ga_xAs_{1-y}P_y$ lattice matched to InP) compounds form a special case of our generalized analysis under certain limiting conditions. The EM in HD QWs of II–VI, IV–VI, stressed Kane type semiconductors, Te, GaP, $PtSb_2$, Bi_2Te_3, Ge, GaSb, II–V, Lead Germanium Telluride, Zinc and Cadmium Diphosphides has also been investigated in the appropriate sections. The importance of the aforementioned semiconductors has also been described in the same chapter. In the absence of band tails and under the condition of extreme carrier degeneracy together with certain limiting conditions, all the results for all the EMs for all the HD QWs of Chapter 1 get simplified into the form of isotropic parabolic energy bands exhibiting the necessary mathematical compatibility test.

With the advent of modern experimental techniques of fabricating nano-materials, it is possible to grow semiconductor super-lattices (SLs) composed of alternative layers of two different degenerate layers with controlled thickness [95]. These structures have found wide applications in many new devices such as photodiodes [96], photoresistors [97], transistors [98], light emitters [99], tunneling devices [100], etc [101–112]. The investigations of the physical properties of narrow gap SLs have increased extensively, since they are important for optoelectronic devices and also since the quality of heterostructures involving narrow gap materials has been greatly improved. It may be noted that the nipi structures, also called the doping super-lattices, are crystals with a periodic sequence of ultrathin film layers [113, 114] of the same semiconductor with the intrinsic layer in between together with the opposite sign of doping. All the donors will be positively charged and all the acceptors negatively. This periodic space charge causes a periodic space charge potential which quantizes the motions of the carriers in the z-direction together with the formation of the subband energies. The electronic structures of the nipis differ radically from the corresponding bulk semiconductors as stated below:

a. Each band is split into mini-bands
b. The magnitude and the spacing of these minibands may be designed by the choice of the super-lattices parameters and
c. The electron energy spectrum of the nipi crystal becomes two-dimensional leading to the step functional dependence of the density-of-states function.

In Chapter 2 we shall study the EM in NIPIS of HD nonlinear optical, III–V, II–VI, IV–VI and stressed Kane type semiconductors.

It is well known that the electrons in bulk semiconductors in general, have three dimensional freedom of motion. When, these electrons are confined in a one dimensional potential well whose width is of the order of the carrier wavelength, the motion in that particular direction gets quantized while that along the other two directions remains as free. Thus, the energy spectrum appears in the shape of discrete levels for the one dimensional quantization, each of which has a continuum for the two dimensional free motion. The transport phenomena of such one dimensional confined carriers have recently been studied [115–135] with great interest. For the metal-oxide-semiconductor (MOS) structures, the work functions of the metal and the semiconductor substrate are different and the application of an external voltage at the metal-gate causes the change in the charge density at the oxide semiconductor interface leading to a bending of the energy bands of the semiconductor near the surface. As a result, a one dimensional potential well is formed at the semiconductor interface. The spatial variation of the potential profile is so sharp that for considerable large values of the electric field, the width of the potential well becomes of the order of the de Broglie wavelength of the carriers. The Fermi energy, which is near the edge of the conduction band in the bulk, becomes nearer to the edge of the valance band at the surface creating accumulation layers. The energy levels of the carriers bound within the potential well get quantized and form electric subbands. Each of the subband corresponds to a quantized level in a plane perpendicular to the surface leading to a quasi two dimensional electron gas. Thus, the extreme band bending at low temperature allows us to observe the quantum effects at the surface. Though considerable work has already been done, nevertheless it appears from the literature that the EM in accumulation layers of non-parabolic semiconductors has yet to be investigated in detail. For the purpose of comparison we shall also study the EM for inversion layers of non-parabolic compounds in addition to inversion layers of the said materials in Chapter 3.

It is well-known that in nano wires (NWs), the restriction of the motion of the carriers along two directions may be viewed as carrier confinement by two infinitely deep 1D rectangular potential wells, along any two orthogonal directions leading to quantization of the wave vectors along the said directions, allowing 1D carrier transport [136]. With the help of modern fabricational techniques, such one dimensional quantized structures have been experimentally realized and enjoy an enormous range of important applications in the realm of nanoscience in the quantum regime. They have generated much interest in the analysis of nanostructured devices for investigating their electronic, optical and allied properties [137–139]. Examples of such new applications are based on the different transport properties of ballistic charge carriers which include quantum resistors [140, 144, 145], resonant tunneling diodes and band filters [146, 147], quantum switches [148], quantum sensors [149–151], quantum logic gates [152–153], quantum transistors and sub tuners [154–156], heterojunction FETs [157], high-speed digital networks [158], high-frequency microwave circuits [159], optical modulators [160], optical switching systems [161, 162], and other devices. In Chapter 4 we shall study the EM in NWs of HD nonlinear optical, III–V, II–VI, IV–VI, *GaP*, $PtSb_2$, Bi_2Te_3, *Ge*, *GaSb*, stressed Kane type, Te, II–V, Lead Germanium Telluride, Zinc and Cadmium Diphosphides, respectively.

It is well known that the band structure of semiconductors can be dramatically changed by applying the external fields. The effects of the quantizing magnetic field on the band structure of compound semiconductors are more striking and can be observed easily in experiments [163–165]. Under magnetic quantization, the motion of the electron parallel to the magnetic field remains unaltered while the area of the wave vector space perpendicular to the direction of the magnetic field gets quantized in accordance with the Landau's rule of area quantization in the wave-vector space [165]. The energy levels of the carriers in a magnetic field (with the component of the wave-vector parallel to the direction of magnetic field be equated with zero) are termed as the Landau levels and the quantized energies are known as the Landau sub-bands. It is important to note that the same conclusion may be arrived either by solving the single-particle time independent Schrödinger differential equation in the presence of a quantizing magnetic field or by using the operator method. The quantizing magnetic field tends to remove the degeneracy and increases the band gap. A semiconductor, placed in a magnetic field B, can absorb radiative energy with the frequency $(\omega_0 = (|e|B/m_c))$. This phenomenon is known as cyclotron or diamagnetic resonance. The effect of energy quantization is experimentally noticeable when the separation between any two consecutive Landau levels is greater than k_BT. A number of interesting transport phenomena originate from the change in the basic band structure of the semiconductor in the presence of quantizing magnetic field. These have been widely investigated and also served as diagnostic tools for characterizing the different materials having various band structures [166–169]. The discreteness in the Landau levels leads to a whole crop of magneto-oscillatory phenomena, important among which are (i) Shubnikov-de Haas oscillations in magneto-resistance; (ii) De Haas–van Alphen oscillations in magnetic susceptibility; (iii) magneto-phonon oscillations in thermoelectric power, etc. In Chapter 5 we shall study the EM under magnetic quantization in HD nonlinear optical, III–V, II–VI, IV–VI, GaP, $PtSb_2$, Bi_2Te_3, Ge, $GaSb$, stressed Kane type, Te, II–V, Lead Germanium Telluride, Zinc and Cadmium Diphosphides, respectively.

With the advent of nano-photonics, there has been considerable interest in studying the optical processes in semiconductors and their nanostructures in the presence of intense light waves [170]. ***It appears from the literature, that the investigations in the presence of external intense photo-excitation have been carried out on the assumption that the carrier energy spectra are invariant quantities under strong external light waves, which is not fundamentally true.*** The physical properties of semiconductors in the presence of strong light waves which alter the basic dispersion relations have relatively been much less investigated in [171–173] as compared with the cases of other external fields needed for the characterization of the low dimensional semiconductors. The solo Chapter 6 investigates the EM in bulk specimens HD Kane type semiconductors under intense light waves. The same chapter studies the EM in the presence of magnetic quantization, cross-fields configuration, QWs, NWs, inversion and accumulation layers, doping superlattices, QWHD and NWHD effective mass superlattices magneto HD effective mass superlattices, QWHD and NWHD superlattices with graded interfaces and magneto HD superlattices with graded interfaces and respectively. Chapter 7 contains the applications of the content of this book in different directions. Chapter 8 contains the conclusions and the scope for future research.

It is worth remarking that the influence of crossed electric and quantizing magnetic fields on the transport properties of semiconductors having various band structures are relatively less investigated as compared with the corresponding magnetic quantization, although, the cross-fields are fundamental with respect to the addition of new physics and the related experimental findings. It is well known that in the presence of electric field (E_o) along x-axis and the quantizing magnetic field (B) along z-axis, the dispersion relations of the conduction electrons in semiconductors become modified and for which the electron moves in both the z and y directions. The motion along y-direction is purely due to the presence of E_0 along x-axis and in the absence of electric field, the effective electron mass along y-axis tends to infinity which indicates the fact that the electron motion along y-axis is forbidden. The EM of the isotropic, bulk semiconductors having parabolic energy bands exhibits mass anisotropy in the presence of cross fields and this anisotropy depends on the electron energy, the magnetic quantum number, the electric and the magnetic fields respectively, although, the EM along z-axis is a constant quantity. In 1966, Zawadzki and Lax [174] formulated the electron dispersion law for III–V semiconductors in accordance with the two band model of Kane under cross fields configuration which generates the interest to study this particular topic of solid state science in general [175–176]. ***Appendix A (Chapter 9) investigates the EM under cross-field configuration in HD nonlinear optical, III–V, II–VI, IV–VI and stressed Kane type semiconductors respectively. This chapter also tells us that the EM in all the cases is a function of the finite scattering potential, the magnetic quantum number and the Fermi energy even for HD semiconductors whose bulk electrons in the absence of band tails are defined by the parabolic energy bands.***

With the advent of nano-devices, the build-in electric field becomes so large that the electron energy spectrum changes fundamentally instead of being invariant and **Appendix B (Chapter 10)** of this book investigates the EM under intense electric field in bulk specimens of HD III–V, ternary and quaternary semiconductors. The same chapter also explores the influence of electric field on the EM in the presence of magnetic quantization, cross-fields configuration, QWs, NWs, inversion and accumulation layers, doping superlattices, QWHD and NWHD effective mass superlattices, magneto HD effective mass superlattices, QWHD and NWHD superlattices with graded interfaces magneto HD superlattices with graded interfaces under magnetic quantization, size quantization, accumulation layers, HD doping superlattices and effective mass HD superlattices under magnetic quantization respectively. **It is interesting to note that the EM depends on the strong electric field** (which is not observed elsewhere) together with the fact that the EM in the said systems depends on the respective quantum numbers in addition to the Fermi energy, the scattering potential and others system constants which are the characteristic features of such hetero-structures.

With the advent of modern experimental techniques of fabricating nano-materials, it is possible to grow semiconductor super-lattices (SLs) composed of alternative layers of two different degenerate layers with controlled thickness [177]. These structures have found wide applications in many new devices such as photodiodes [178], photo-resistors [179], transistors [180], light emitters [181], tunneling devices [182], etc. [183–194]. The investigations of the physical properties of narrow gap SLs have increased extensively, since they are important for optoelectronic devices and also since the quality of hetero-structures involving narrow gap materials has been greatly improved. It is well known that Keldysh [195] first suggested the fundamental concept of a super-lattice (SL), although it was successfully experimental realized by Esaki and Tsu [196]. The importance of SLs in the field of nano-electronics has already been described in [197–199]. The most extensively studied III–V SL is the one consisting of alternate layers of GaAs and $Ga_{1-x}Al_xAs$ owing to the relative ease of fabrication. The GaAs layers forms quantum wells and $Ga_{1-x}Al_xAs$ form potential barriers. The III–V SL's are attractive for the realization of high speed electronic and optoelectronic devices [200]. In addition to SLs with usual structure, SLs with more complex structures such as II–VI [201], IV–VI [202] and HgTe/CdTe [203] SL's have also been proposed. The IV–VI SLs exhibit quite different properties as compared to the III–V SL due to the peculiar band structure of the constituent materials [204]. The epitaxial growth of II–VI SL is a relatively recent development and the primary motivation for studying the mentioned SLs made of materials with the large band gap is in their potential for optoelectronic operation in the blue [204]. HgTe/CdTe SL's have raised a great deal of attention since 1979, promising new materials for long wavelength infrared detectors and other electro-optical applications [205]. Interest in Hg-based SL's has been further increased as new properties with potential device applications were revealed [206]. These features arise from the unique zero band gap material HgTe [206] and the direct band gap semiconductor CdTe which can be described by the three band mode of Kane [207]. The combination of the aforementioned materials with specified dispersion relation makes HgTe/CdTe SL very attractive, especially because of the possibility to tailor the material properties for various applications by varying the energy band constants of the SLs. In addition to it, for effective mass SLs, the electronic sub-bands appear continually in real space [208].

We note that all the aforementioned SLs have been proposed with the assumption that the interfaces between the layers are sharply defined, of zero thickness, i.e., devoid of any interface effects. The SL potential distribution may be then considered as a one dimensional array of rectangular potential wells. The aforementioned advanced experimental techniques may produce SLs with physical interfaces between the two materials crystallo-graphically abrupt; adjoining their interface will change at least on an atomic scale. As the potential form changes from a well (barrier) to a barrier (well), an intermediate potential region exists for the electrons. The influence of finite thickness of the interfaces on the electron dispersion law is very important, since the electron energy spectrum governs the electron transport in SLs.

In **Appendix C (Chapter 11)**, the EM in III–V, II–VI, IV–VI, HgTe/CdTe and strained layer quantum well heavily doped super-lattices (QWHDSLs) with graded interfaces will be studied. Besides the EM in III–V, II–VI, IV–VI, HgTe/CdTe and strained layer quantum well HD effective mass super-lattices respectively has also been explored in the same chapter. In **Appendix D (Chapter 12)**, the EM in quantum wire superlattices has also been investigated for all the cases of Chapter 11. The **Appendix E (Chapter 13)** contains the study of the EM in HDSLs under magnetic quantization for all the cases of Chapter 11. The EM in HD ultra thin films under cross field configuration for all the materials of Chapter 9 has been investigated in **Appendix F (Chapter 14)**. The EM in quantum wells and quantum wires of HD III-V, Ternary and Quaternary materials under different physical conditions has been studied in **Appendix G (Chapter 15).**

It is needless to say that this monograph is based on the ***'iceberg principle'*** [209] and the rest of which will be explored by the researchers of different appropriate fields. *Since, there is no existing report devoted solely to the study of EM for HD quantized structures to the best of our knowledge, we earnestly hope that the present book will a useful reference source for the present and the next generation of the readers and the researchers of materials and allied sciences in general. We have discussed enough regarding the EMs in different quantized HD materials although lots of new computer oriented numerical analysis are being left for the purpose of being computed by the readers, to generate the new graphs and the inferences from them which all together is a sea in itself.* Since the production of an error free first edition of any book from every point of view is a permanent member of impossibility theorems, therefore in spite of our joint concentrated efforts for a couple of years together with the seasoned team of World Scientific Publishing Company, the same stands very true for this monograph also. ***Various expressions and few chapters of this book are appearing for the first time in printed form.*** The suggestions from the readers for the development of the book will be highly appreciated for the purpose of inclusion in the future edition, if any. ***In this book, from Chapter 1 to till the end, we have presented 200 open research problems for the graduate students, PhD aspirants, researchers, engineers in this pinpointed research topic.*** We strongly hope that alert readers of this monograph will not only solve the said problems by removing all the mathematical approximations and establishing the appropriate uniqueness conditions, but also will generate new research problems both theoretical and experimental and, thereby, transforming this monograph into a solid book. Incidentally, our readers after reading this book will easily understand how little is presented and how much more is yet to be investigated in this exciting topic which is the signature of the coexistence between new physics, advanced mathematics combined with the inner fire for performing creative researches in this context from the young scientists since like Kikoin [210] we feel that ***"A young scientist is no good if his teacher learns nothing from him and gives his teacher nothing to be proud of".*** We emphatically write that the problems presented here form the integral part of this book and will be useful for the readers to initiate their own contributions on the EM for HDS and their quantized counterparts since like Sakurai [211] we firmly believe *"The reader who has read the book but cannot do the exercise has learned nothing".* It is nice to note that if we assign the alphabets A to Z, the positive integers from 1 to 26, chronologically, then the word "*ATTITUDE* " receives the perfect score 100 and is the vital quality needed from the readers since *attitude*is the ladder on which all the other virtues mount.

In this book, the readers will get much information regarding the influence of quantization in HD low dimensional materials having different band structures. Although the name of the book is an example of extremely high Q-factor, from the content, one can easily infer that it should be useful in graduate courses on materials science, condensed matter physics, solid states electronics, nano-science and technology and solid-state sciences and devices in many Universities and the Institutions in addition to both Ph.D. students and researchers in the aforementioned fields. Last but not the least, the author hopes that his humble effort will kindle the desire to delve deeper into this fascinating and deep topic by any one engaged in materials research and device development either in academics or in industries.

References

1. Rieth. M., Schommers. W., Handbook of computational and theoretical nanoscience, American Scientific Publishers, Los Angeles (2006), Vols. 1–10; Rieth. M., Schommers. W., (2002) *Jour. Nanosci. and Nanotechnol.* **2**,

679. Rieth. M., Nano-Engineering in Science and Technology *Series on the Foundations of Natural Science and Technology* Vol. **6** (2003).

2. Nalwa. H. S. Encyclopedia of Nanoscience and Nanotechnology, (2004/2011) American Scientific Publishers, Los Angeles, Vols. **1**–**25**; Nalwa. H. S. Soft Nanomaterials, (2009) American Scientific Publishers, Los Angeles, Vols. 1–2; Nalwa. H. S., Magnetic Nanostructures, (2002) American Scientific Publishers, Los Angeles.
3. Nalwa. H. S., Nanomaterials for Energy Storage Applications, (2009) American Scientific Publishers, Los Angeles; Nalwa. H. S., Handbook of Nanostructured Biomaterials and Their Applications in Nanobiotechnology, American Scientific Publishers, Los Angeles (2005), Vols. 2; Bandyopadhyay. S., Nalwa. H. S. Quantum Dots and Nanowires, (2003) American Scientific Publishers, Los Angeles, Vol.5.
4. Tseng. T. Y., Nalwa. H. S., Handbook of Nanoceramics and Their Based Nanodevices, American Scientific Publishers, Los Angeles (2009), Vols. 1-5; K. Ariga and H. S. Nalwa Bottom-up Nanofabrication, American Scientific Publishers, Los Angeles (2009), Vols. 1–6; Ghatak. K. P., Biswas. S. N., (1993) *Acta physica Slovaca* **43**, 425.
5. Lin. Y., Nalwa. H. S., Handbook of Electrochemical Nanotechnology, American Scientific Publishers, Los Angeles (2009), Vol. 1-2; Mohanty. A. K., M. Misra, Nalwa. H. S., Packaging Nanotechnology, (2009) American Scientific Publishers, Los Angeles.
6. Yan. S., Shi. Y., Xiao. Z., Zhou. M., Yan. W., Shen, Hu. D., (2012) *J. Nanosci. Nanotechnol.* 12, 6873; Smirnov. J. R. C., Jover E., Amade R., Gabriel. G., Villa R., Bertran. E., (2012) *J. Nanosci. Nanotechnol.* **12**, 6941.
7. Mashapa. M. G., Chetty. N., Ray. S. S. , (2012) J. Nanosci. Nanotechnol. 12, 7030; Kim. D. H., You. J. H., Kim. J. H., Yoo. K. H., Kim. T. W., (2012) *J. Nanosci. Nanotechnol.* **12**, 5687; Ooba. A., Fujimura. Y., Takahashi. K., Komine. T., Sugita. R., (2012) *J. Nanosci. Nanotechnol.* **12**, 7411; Tan. C. W., Tay. B. K., (2012) *J. Nanosci. Nanotechnol.* **12**, 707.
8. Noh. S. J., Miyamoto. Y., Okuda. M., Hayashi. N., Kim. Y. K., (2012) *J. Nanosci. Nanotechnol.* 12, 428; Padalkar. S., Schroeder. K., Won. Y. H., Jang. H. S., Stanciu. L., (2012) *J. Nanosci. Nanotechnol.* **12**, 227; Lee. K. Y., Kumar. B., Park. H. K., Choi. W. M., Choi. J. Y., Kim. S. W., (2012) *J. Nanosci. Nanotechnol.* **12**, 1551.
9. Wu. J. M., (2012) *J. Nanosci. Nanotechnol.* **12**, 1434; Zhang. Y., Su. L., Lubonja. K., Yu. C., Liu. Y., Huo. D., Hou. C., Lei. Y., (2012) *Sci. Adv. Mater.* **4**, 825.
10. Tamang. R., Varghese. B., Tok. E. S., Mhaisalkar. S., Sow. C. H., (2012) Nanosci. Nanotechnol. Lett. **4**, 716; Yue. Y., Chen. M., and Y. Clein (2012) *J. Nanosci. Nanotechnol. Lett.* 4, 414.
11. Adhikari. S. M., Sakar. A., Ghatak. K. P., (2013) Quantum Matter **2**, 455; Adhikari. S. M., Ghatak. K. P., (2013) Quantum Matter **2**, 296; Ghatak. K. P., Bose. P. K., Bhattacharya. S., Bhattacharjee. A., De. D., Ghosh. S., (2013) Quantum Matter **2**, 83; Ghatak. K. P., Bhattacharya. S., Mondal.

A., Debbarma. S., Ghorai. P., Bhattacharjee. A., (2013) *Quantum Matter* **2**, 25.

12. Bose. P. K., Paitya. N., Bhattacharya. S., De. D., Saha. S., Chatterjee. K. M., Pahari. S., Ghatak. K. P., (2012) *Quantum Matter* 1, 89; Paitya. N., Bhattacharya. S., De. D., Ghatak. K. P., (2012) *Quantum Matter* **1**, 63; Bhattacharya. S., Paitya. N., Ghatak. K. P., (2013) *Jour. Comp. Theo. Nanosci.* 10, 1999;
13. Choudhury. S., De. D., Mukherjee. S., Neogi. A., Sinha. A., Pal. M., Biswas. S. K., Pahari. S., Bhattacharya. S., Ghatak. K. P., (2008) *J. Comp. Theo. Nanosci.* **5**, 375; Bhattacharya. S., Sarkar. R., De. D., Mukherjee. S., Pahari. S., Saha. A., Roy. S., Paul. N. C., Ghosh. S., Ghatak. K. P., (2009) *J. Comput. Theo. Nanosci.* **6**, 112.
14. Bhattacharya. S., Chowdhury. S., Ghoshal. S., Biswas. S. K., De. D., Ghatak. K. P., (2006) *J. Comp. Theo. Nanosci.* **3**, 423; Bhattacharya. S., De. D., Chowdhury. S., Karmakar. S., Basu. D. K., Pahari. S., Ghatak. K. P., (2006) *J. Comp. Theo. Nanosci.* **3**, 280; Ghatak. K. P., Karmakar. S., De. D., Pahari. S., Charaborty. S. K., Biswas. S. K., Chowdhury. S., (2006) *J. Comp. Theo. Nanosci.* **3**, 153; Ghatak. K. P., Bose. P. K., Ghatak. A. R., De. D., Pahari. S., Chakraborty. S. K., Biswas. S. K., (2005) *J. Comp. Theo. Nanosci.* **2**, 423.
15. Simmons. G. E., Differential Equations with Application and Historical Notes, International Series in Pure and Applied Mathematics (McGraw-Hill, USA, 1991).
16. Huff. Howard (ed.), Into the Nano era – Moore's Law beyond Planar Silicon CMOS, Vol. 106, Springer Series in Materials Science (Springer-Verlag, Germany, 2009).
17. Ono, T., Fujimoto, Y. and Tsukamoto, S. *Quan. Matt.*, **1**, 4 (2012); H. De Raedt, F. Jin and K. Michielsen, *Quan. Matt.*, **1**, 20 (2012); R. Hofmann and D. Kaviani, *Quan. Matt.*, **1**, 41 (2012); M. Narayanan and A. J. Peter, *Quan. Matt.*, **1**, 53 (2012); S. Olszewski, *Quan. Matt.*, **1**, 59 (2012); A. Heidari, N. Heidari and M. Ghorbani, *Quan. Matt.*, **1**, 86 (2012); S. Olszewski, *Quan. Matt.* **1**, 127 (2012); V. Sajfert, P. Mali, N. Bednar, N. Pop, D. Popov and B. Tosic, *Quan. Matt.* **1**, 134 (2012); B. Tuzun and S . Erkoc, *Quan. Matt.* **1**, 136 (2012), A. Heidari, N. Heidari and M. Ghorbani, *Quan. Matt.* **1**, 149 (2012); C. Falquez, R. Hofmann and T. Baumbach, *Quan. Matt.* **1**, 153 (2012); I.S Amiri, A. Nikoukar, J. Ali and P. P. Yupapin, *Quan. Matt.* **1**, 159 (2012); N. Paitya, S. Bhattacharya, D. De and K. P. Ghatak, *Quan. Matt.* **1**, 63 (2012); S. Bhattacharya, D. De, S. Ghosh and K. P. Ghatak, *J. Comp. theo. NanoSc.* **10**, 1 (2013); S. Bhattacharya, D. De, N. Paitya, S. M. Adhikari and K. P. Ghatak, *Adv. Sc. Lett.* **16**, 348 (2012); D. De, S. Bhattacharya, S. Ghosh and K. P. Ghatak, *Adv. Sc. Engg. Medicine* **4**, 211 (2012); N. Paitya, S. Bhattacharya, D. De, S. Ghosh and K. P. Ghatak, *J. Nanoengg. Nanomanufac.* **2**, 211 (2012); N. Paitya, S. Bhattacharya, D. De and K. P. Ghatak, *Adv. Sc. Engg. and Medi.* **4**, 96 (2012); N. Paitya and K. P. Ghatak, *J. Adv. Phys.*, **1**, 1 (2012); S. Bhattacharya, D. De and K. P.

Ghatak, *J. Sc. Engg. and Medi.* **6**, 5 (2012); N. Paitya and K. P. Ghatak, *J. Nanoengg. Nanomanufac.* **2**, 1 (2012);
18. S. Adachi, *J. Appl. Phys.* **58**, R11 (1985); R. Dornhaus, and G. Nimtz, *Springer Tracts in Modern Physics*, Berlin-Heidelberg, N., Springer (1976), Vol. 78, pp-1.
19. W. Zawadzki, *Handbook of Semiconductor Physics*, (Paul, W., ed.), Amsterdam, North Holland, (1982), Vol. 1, pp-719;I. M. Tsidilkovski, Cand. *Thesis Leningrad University SSR* (1955).
20. T. N. Sen and K. P. Ghatak, *Quantum Matter* **5**, 1 (2016).
21. F. G. Bass and I. M. Tsidilkovski, *Ivz. Acad. Nauk Azerb SSR* **10**, 3 (1966).
22. B. Mitra and K. P. Ghatak, *Phys. Scrip.* **40**, 776 (1989); S. K. Biswas, A. R. Ghatak, A. Neogi, A. Sharma, S. Bhattacharya and K. P. Ghatak, *Phys. E* **36**, 163 (2007).
23. M. Mondal, S. Banik and K. P. Ghatak, *J. Low Temp. Phys.* **74**, 423 (1989).
24. K. P. Ghatak, S. Bhattacharya, H.Saikia and A. Sinha, *J. Comp. Theor. Nanoscience* **3**, 1 (2006).
25. K. P. Ghatak and S. N. Biswas, *J. Vac. Sc. Tech.* 7B, 104 (1989); P. K. Chakraborty, S. Choudhury and K. P. Ghatak, *Phys. B* **387**, 333 (2007)
26. S. Bhattacharya, S. Pahari, D. K. Basu and K. P. Ghatak, *J. Comp. Theo. Nanoscience*, **3**, 280 (2006); A. Sinha, A. K. Sharma, R. Barui, A. R. Ghatak, S. Bhattacharya, and K. P. Ghatak, *Phys. B* **391**, 141 (2007).
27. S. Choudhury, L. J. Singh and K. P. Ghatak, *Nanotechnology* **15**, 180 (2004).
28. S. Chowdhary, L. J. Singh and K. P. Ghatak, *Phys. B* **B365**, 5 (2005); P. K. Chakraborty, A. Sinha, S. Bhattacharya and K. P. Ghatak, *Phys. B* **390**, 325 (2007); K. P. Ghatak, J. P. Banerjee and D. Bhattacharyya, *Nanotechnology* **7**, 110 (1996).
29. P. K. Chakraborty, B. Nag and K. P. Ghatak. *J. Phys. Chem. Solids* **64**, 2191 (2003).
30. K. P. Ghatak, J. P. Banerjee and B. Nag, *J. Appl. Phys.* **83**, 1420 (1998)
31. B. Nag and K. P. Ghatak, *J. Phys. Chem. Solids* **59**, 713 (1998); B. Nag and K. P. Ghatak, *J. Phys. Chem. Sol.* **58**, 427 (1997).
32. K. P. Ghatak, D. K. Basu and B. Nag. *J. Phys. Chem. Solids* **58**, 133 (1997); B. Nag, K. P. Ghatak, *Nonlin. Opt. and Quan. Opt.* **19**, 1 (1998).
33. K. P. Ghatak and B. Nag, *Nanostruct. Mat.* **10**, 923 (1998).
34. P. K .Bose, N. Paitya, S. Bhattacharya, D. De, S. Saha, K. M. Chatterjee, S. Pahari and K. P. Ghatak, *Quantum Matter* **1**, 89 (2012); K. P. Ghatak, B. Mitra, *International Journal of Electronics* **72**, 541 (1992); M. Mondal, N. Chattopadhyay and K. P. Ghatak, *Journal of Low Temperature Physics* **66**, 131 (1987); P. K. Chakraborty, G. C. Datta and K. P. Ghatak, *Physica Scripta*, **68**, 368 (2003); K. P. Ghatak and M. Mondal, *Zeitschrift für Physik B Condensed Matter* **69**, 471 (1988); B. Mitra and K. P. Ghatak, *Solid-state electronics* **32**, 177 (1989).
35. S. M. Adhikari, A. Sakar and K. P. Ghatak, *Quantum Matter* **2**, 455 (2013); K. P. Ghatak, S. Bhattacharya, S. K. Biswas, A. Dey and A. K. Dasgupta, *Physica Scripta* **75**, 820 (2007); M. Mondal and K. P. Ghatak, *Physics Letters A* **131**, 529 (1988).

36. A. N. Chakravarti, K. P. Ghatak, K. K. Ghosh, S. Ghosh and A. Dhar, *Zeitschrift für Physik B Condensed Matter* **47**, 149 (1982); A. Ghoshal, B. Mitra and K. P. Ghatak, *Il Nuovo Cimento D* **12**, 891 (1990); A. N. Chakravarti, A. K. Chowdhury, K. P. Ghatak, S. Ghosh and A. Dhar, *Applied physics* **25**, 105 (1981).
37. M. Mondal and K. P. Ghatak, *Physica Status Solidi (b),* **135**, K21 (1986); M. Mondal and K. P. Ghatak, *Physica Status Solidi (b),* **123**, K143 (1984); K. P. Ghatak, A. Ghoshal and B. Mitra, *Il Nuovo Cimento D,* **13**, 867 (1991).
38. K. P. Ghatak and B. Mitra, *International Journal of Electronics Theoretical and Experimental,* **70**, 343 (1991); K. P. Ghatak, B. Mitra and A. Ghoshal, *Physica Status Solidi (b)*, 154, K121 (1989); K. P. Ghatak and M. Mondal, *Zeitschrift für Naturforschung A* **41**, 881 (1986).
39. M. Mondal and K. P. Ghatak, *Physica Status Solidi (b),* **129**, 745 (1985); K. P. Ghatak, A. Ghoshal and B. Mitra, *Il Nuovo Cimento D,* **14**, 903 (1992); B. Mitra and K. P. Ghatak, *Solid-state electronics,* **32**, 515 (1989).
40. K. P. Ghatak, N. Chatterjee and M. Mondal, *Physica Status Solidi (b)* **139**, K25 (1987); A. N. Chakravarti and K. P. Ghatak, K. K. Ghosh and H. M. Mukherjee, *Physica Status Solidi (b),* **116**, 17 (1983); A. N. Chakravarti, K. P. Ghatak, A. Dhar and K. K. Ghosh, *Czechoslovak Journal of Physics,* **33**, 65 (1983).
41. K. P. Ghatak and A. Ghosal, *Physica Status Solidi (b)* **151**, K135 (1989); S. Bhattacharya, D. De, S. M. Adhikari and K. P. Ghatak, *Superlattices and Microstructures,* **51**, 203(2012); K. P. Ghatak and M. Mondal, *Physica Status Solidi (b)* **175**, 113 (1993).
42. M. Mondal and K. P. Ghatak, *Czechoslovak Journal of Physics B* **36**, 1389 (1986); M. Mondal and K. P. Ghatak, *Physica Status Solidi (b)* **126**, K47 (1984); D.De, S. Bhattacharya, S. M. Adhikari, A. Kumar, P. K. Bose and K. P. Ghatak, *Beilstein Journal of Nanotechnology* **2**, 339 (2011).
43. M. Mondal and K. P. Ghatak, *Physica Status Solidi (b),* **139**, 185 (1987); S. Debbarma, A. Bhattacharjee, S. Bhattacharyya, A. Mondal and K. P. Ghatak, *Journal of Advanced Physics,* **1**, 84 (2012); D. Bhattacharyya and K. P. Ghatak, *Physica Status Solidi (b),* **187**, 523 (1995).
44. S. M. Adhikari, D. De, J. K. Baruah, S. Chowdhury and K. P. Ghatak, *Advanced Science Focus,* **1**, 57 (2013); P. K. Bose, S. Bhattacharya, D. De, N. Paitya and K. P. Ghatak, *Advanced Science, Engineering and Medicine,* **5**, 245 (2013).
45. S. Bhattacharya and K. P. Ghatak, *Springer Science & Business Media,* 167 (2012); K. P. Ghatak ,*Acta Physica Hungarica,* **74**, 257 (1994); M. Mondal and K. P. Ghatak, *Acta Physica Slovaca,* **36**, 325 (1986); M. Mondal, K. P. Ghatak and *Acta physica Polonica.* A, **66**, 47 (1984).
46. S. Bhattachrya, N. Paitya and K. P. Ghatak, *Journal of Computational and Theoretical Nanoscience* **10**, 1999 (2013); N. Paitya and K. P. Ghatak, *Journal of nanoscience and nanotechnology,* 12, 8985 (2012); N. Chattopadhyay, K. P. Ghatak and M. Mondal, *Acta Physica Hungarica,* **67**, 163 (1990); M. Mondal and K. P. Ghatak, *Acta physica Polonica.* A, **67**, 983 (1985);.

47. P. K. Chakraborty, G. C. Datta and K. P. Ghatak, *Physica Scripta* **68**, 368 (2003); B. Nag, P. K. Chakrabarty and K. P. Ghatak, *Journal of Wave Material Interaction,* **11**, 211 (1996).
48. K. P. Ghatak, D. K. Basu and B. Nag, *Journal of Wave Material Interaction* **10**, 29 (1995); K. P. Ghatak and S. N Banik, *FIZIKA A,* **4**, 33 (1995); K. P. Ghatak and D. Bhattacharyya, *Physica Status Solidi (b)* **179**, 383 (1993); K. P. Ghatak and D. Bhattacharyya, *Journal of Wave Material Interaction,* **8**, 233 (1993).
49. A. Ghoshal and K. P. Ghatak, *Acta Physica Slovaca, Czechoslovakia,* **42**, 370 (1992); K. P. Ghatak and S. N. Biswas, *Orlando'91, Orlando, FL,* 149 (1991); K. P. Ghatak and A. Ghoshal,*Physica Status Solidi (a)* **126**, K53 (1991); K. P. Ghatak, *Acta Physica Hungarica,* **70**, 77 (1991); M. Mondal and K. P. Ghatak, *Physica Status Solidi (b),* **128**, K133 (1985).
50. P. N. Hai, W. M. Chen, I. A. Buyanova, H. P. Xinm and C. W. Tu, *Appl. Phys. Letts.* **77**, 1843 (2000).
51. D. P. diVincenzo and E. J. Mele, *Phys. Rev. B* **29**,1685 (1984).
52. P. Perlin, E. Litwin-Staszewska, B. Suchanek, W. Knap, J. Camassel, T. Suski, R. Piotrzkowski, I. Grzegory, S. Porowski, E. Kaminska and J. C. Chervin, *Appl. Phys. Letts.* **68**, 1114 (1996).
53. D. Shiri, Y. Kong, A. Bulin, M.P Anantram, *Appl. Phys. Lett.* **93**, 073114 (2008).
54. S. L. Sewall, R. R. Cooney and P. Kambhampati, *Appl. Phys. Letts.* **94**, 243116-1 (2009).
55. G. E. Smith, *Phys. Rev. Letts.* **9**, 487 (1962).
56. D. Schneider, D. Rurup, A. Plichta, H.-U. Grubert, A. Schlachetzki and K. Hansen, *Z. Phys. B.* **95**, 281 (1994).
57. L. G. Caron, J.-P. Jay-Gerin and M. J. Aubin, *Phys. Rev. B* **15**, 3879 (1977).
58. D. Bhattacharyya, R. Pal, S. Chaudhuri and A. K. Pal, *Vaccum* **44**, 803 (1993).
59. B. Slomski, F. Meier, J. Osterwalder and J. H. Dil, *Phys. Rev B* **83**, 035409 (2011).
60. L. M. Wei, K. H. Gao, X. Z. Liu, W. Z. Zhou, L. J. Cui, Y. P. Zeng, G. Yu, R. Yang, T. Lin, L. Y. Shang, S. L. Guo, N. Dai, J. H. Chu and D. G. Austing, *J. Appl. Phys.* **110**, 063707 (2011).
61. K. Karra, M. Stopa, X. Ying, H. D. EMew and S. Das Sarma, *Phys. Rev B* **42**, 9732 (1990).
62. B. Rößner, G. Isella and H. V. Känel; *Appl. Phys. Letts.* **82**, 754 (2003).
63. T. M. Kim, M. Jung and K. H. Yoo, *J. Phys. Chem. of Solids* **61**, 1769 (2000).
64. N. Kotera, H. Arimoto, N. Miura, K. Shibata, Y. Ueki, K. Tanaka, H. Nakamura, T. Mishima, K. Aiki and M. Washima, *Phys. E* **11**, 219 (2001).
65. K. Tanaka, N. Koteraa and H. Nakamura; *Microelectron. Engg.* **47**, 309 (1999).
66. G. Wang and T. Cagin, *Appl. Phys. Letts.* **89**, 152101 (2006).
67. Z. S. Gribnikov, R. R. Bashirov, H. Eisele, V. V. Mitin and G. I. Haddad; *Phys. E* 12, 276 (2002).

68. [C. J. Hwang, *J. Appl. Phys.* **41**,2668 (1970); W. Sritrakool, H. R. Glyde, V. Sa Yakanit Can. *J. Phys.*, **60**, 373 (1982); H. Ikoma, *J. Phys. Soc. Japan*, **27**, 514 (1969).
69. A. N. Chakravarti, A. K. Chowdhury, K. P. Ghatak, S. Ghosh, A. Dhar, *Appl. Phys.* **25**, 105 (1981); A. N. Chakravarti, K. P. Ghatak, K. K. Ghosh, S. Ghosh, A. Dhar, *Zeitschrift für Physik B Conden. Matter.* **47**, 149 (1982); M. Mondal, N. Chattopadhyay, K. P. Ghatak, *Jour. of Low. Temp. Phys.* **66**, 131 (1987); M. Mondal, K. P. Ghatak, *Phys. Lett. A* **131**, 529 (1988); K. P. Ghatak, A. Ghoshal, B. Mitra, *Il Nuovo Cimento.*, D **13**, 867 (1991); K. P. Ghatak, A. Ghoshal, *Phys. Stat. Sol.* (a) **126**, K53 (1991); E. A. Aruhanov, A. F. Knyazev, A. N. Nateprov, S. I. Radautsan, *Sov. Phys. Semiconduct.* **15**, 828 (1981); P. N. Hai, W. M. Chen, I. A. Buyanova, H. P. Xin, C. W. Tu, *Appl. Phys. Letts.* **77**, 1843 (2000); D. P. diVincenzo, E. J. Mele, *Phys. Rev. B*, **29**, 1685 (1984); P. Perlin, E. Litwin-Staszewska, B. Suchanek, W. Knap, J. Camassel, T. Suski, R. Piotrzkowski, I. Grzegory, S. Porowski, E. Kaminska, J. C. Chervin, *Appl. Phys. Letts.* **68,** 1114 (1996); F. Masia, G. Pettinari, A. Polimeni, M. Felici, A. Miriametro, M. Capizzi, A. Lindsay, S. B. Healy, E. P. O'Reilly, A. Cristofoli, G. Bais, M. Piccin, S. Rubini, F. Martelli, A. Franciosi, P. J. Klar, K. Volz, W. Stolz, *Phys. Rev. B*, **73**, 073201 (2006); I. M. Tsidilkovski, *Band Structures of Semiconductors,* (Pergamon Press, London, 1982); K. P. Ghatak, M. Mondal, Z. F. *Physik B*, **B69**, 471 (1988); B. Mitra, A. Ghoshal, K. P. Ghatak, *Nouvo Cimento D* **12D**, 891 (1990); K. P. Ghatak, S. N. Biswas, *Nonlin. Opt. Quant. Opts.*, **12**, 83 (1995); K. P. Ghatak, S. N. Biswas, *Proc. SPIE* **1484**, 149 (1991); M. Mondal, K. P. Ghatak, *Graphite Intercalation Compounds: Science and Applications*, MRS Proceedings, ed. by M. Endo, M. S. Dresselhaus, G. Dresselhaus, MRS Fall Meeting, **EA 16**, 173 (1988); K. P. Ghatak, M. Mondal, Z. fur, *Nature A*, **41A**, 881 (1986); C. C. Wu, C. J. Lin, *J. Low Temp. Phys.* **57**, 469 (1984); M. H. Chen, C. C. Wu, C. J. Lin, *J. Low Temp. Phys.* **55**, 127 (1984).
70. J. N. Schulman, Y. C. Chang, *Phys. Rev. B*, **24**, 4445 (1981).
71. N. G. Anderson, W. D. Laidig, R. M. Kolbas, Y. C. Lo, *J. Appl. Phys.* **60**, (2361) (1986); N. Paitya, K. P. Ghatak *Jour. Adv. Phys.* **1**, (161) (2012); S. Bhattacharya, D. De, S. M. Adhikari, K. P. Ghatak *Superlatt. Microst.* **51**, (203) (2012); D. De, S. Bhattacharya, S. M. Adhikari, A. Kumar, P. K. Bose, K. P. Ghatak, *Beilstein Jour. Nanotech.* **2**, (339) (2012); D. De, A. Kumar, S. M. Adhikari, S. Pahari, N. Islam, P. Banerjee, S. K. Biswas, S. Bhattacharya, K. P. Ghatak, *Superlatt. and Microstruct.* **47**, (377) (2010); S. Pahari, S. Bhattacharya, S. Roy, A. Saha, D. De, K. P. Ghatak, *Superlatt. and Microstruct.* **46**, (760) (2009); S. Pahari, S. Bhattacharya, K. P. Ghatak *Jour. of Comput. and Theo. Nanosci.* **6**, (2088) (2009); L. J. Singh, S. Choudhury, D. Baruah, S. K. Biswas, S. Pahari, K. P. Ghatak, *Phys. B: Conden. Matter*, **368**, (188) (2005); S. Chowdhary, L. J. Singh, K. P. Ghatak, *Phys. B: Conden. Matter*, **365**, (5) (2005); K. P. Ghatak, J. Mukhopadhyay J. P. Banerjee, *SPIE Proceedings Series*, **4746** (1292) (2002); K. P. Ghatak, S. Dutta, D. K. Basu, B. Nag, Il Nuovo Cimento. **D**

20, (227) (1998); K. P. Ghatak, B. De, *Mat. Resc. Soc. Proc.* **300**, (513) (1993); K. P. Ghatak, B. Mitra, *Il Nuovo Cimento.* **D 15**, (97) (1993); K. P. Ghatak, *Inter. Soci.Opt. and Photon. Proc. Soc. Photo Opt. Instru. Engg.* **1626**, (115) (1992); K. P. Ghatak, A Ghoshal, S. N. Biswas, M. Mondal, *Proc. Soc. Photo Opt. Instru. Engg.* **1308**,(356) (1990); S. N. Biswas, K. P. Ghatak, Internat. *Jour. Electronics Theo. Exp.* **70**, (125) (1991); B. Mitra, K. P Ghatak, *Phys. Lett. A.* **146**, (357) (1990); B. Mitra, K. P. Ghatak, *Phys. Lett. A.* **142**, (401) (1989); K. P. Ghatak, B Mitra, A Ghoshal, *Phy. Stat. Sol.* (b) **154**, (K121) (1989); B. Mitra, K. P. Ghatak, *Phys. Stat. Sol.* (b). **149**, (K117) (1988); S. Bhattacharyya, K. P. Ghatak, S. Biswas, OE/Fibers' 87, *Inter. Soc. Opt. Photon.* (73) (1987); K. P. Ghatak, A. N. Chakravarti, *Phys. Stat. Sol.* (b). **117**, (707) (1983).

72. F. Capasso, Semiconductors, and Semimetals **22**, (2) (1985).
73. F. Capasso, K. Mohammed, A. Y. Cho, R. Hull, A. L. Hutchinson, *Appl. Phys. Letts.* **47**, (420) (1985).
74. F. Capasso, R. A. Kiehl, *J. Appl. Phys.* **58**, (1366) (1985).
75. K. Ploog, G. H. Doheler, *Adv. Phys.* **32**, (285) (1983).
76. F. Capasso, K. Mohammed, A. Y. Cho, *Appl. Phys. Lett.* (478) (1986).
77. R. Grill, C. Metzner, G. H. Döhler, *Phys. Rev. B*, **63**, (235316) (2001); *Phys. Rev. B*, **61**, (15614) (2000).
78. A. R. Kost, M. H. Jupina, T. C. Hasenberg, E. M. Garmire, *J. Appl. Phys.* **99**, (023501) (2006).
79. R. K. Willardson, ed. by A. C. Beer, **1**, *Semiconductors and Semimetals* (Academic Press, New York, 1966) p.102.
80. E. O. Kane, *Phys. Rev.* **131**,79 (1963); *Phys. Rev. B.*, **139**, 343 (1965).
81. V. L. Bonch Bruevich, *Sov. Phys. Sol. Stat.*, **4**,53 (1963).
82. R. A. Logan, A. G. Chynoweth, *Phys. Rev.*, **131**, 89 (1963).
83. C. J. Hwang, *J. Appl. Phys.* **40**, 3731 (1969).
84. J. I. Pankove, *Phys. Rev. A* **130**, 2059 (1965).
85. B. I. Halperin, M. Lax, *Phys. Rev.* **148**, 722 (1966).
86. R. A. Abram, G. J. Rees, B. L. H. Wilson, *Adv. Phys.* **27**, 799 (1978).
87. B. I. Shklovskii, A. L. Efros, *Electronic Properties of Doped Semiconductors*, Springer series in Solid State Sciences, vol. **45** (Springer, Berlin, 1984).
88. E. O. Kane, Solid State Electron, **28**, 3 (1985).
89. P. K. Chakraborty, J. C. Biswas, *J. Appl. Phys.* **82**, 3328 (1997).
90. B. R. Nag, *Electron Transport in Compound Semiconductors*, Springer Series in Solid state Sciences, vol.**11,** (Springer, Heidelberg, 1980).
91. P. E. Schmid, *Phys. Rev. B* **23**, 5531 (1981).
92. Jr. G. E. Jellison, F. A. Modine, C. W. White, R. F. Wood , R. T. Young, *Phys. Rev. Lett.* **46**,1414 (1981).
93. V. I. Fistul, *Heavily Doped Semiconductors* (Plenum, New York, 1969) ch 7.
94. C. J. Hwang, *J. Appl. Phys.* **41**, 2668 (1970); W. Sritrakool, H. R. Glyde, V. Sa Yakanit Can. *J. Phys.* **60**, 373 (1982); H. Ikoma, *J. Phys. Soc. Japan* **27**, 514 (1969); M. H. Chen, C. C. Wu, C. J. Lin, J. Low *Temp. Phys.* **55**, 127 (1984).

95. F. Capasso, *Semiconductors and Semimetals* **22**, 2 (1985).
96. F. Capasso, K. Mohammed, A. Y. Cho, R. Hull and A. L. Hutchinson, *Appl. Phys. Letts.* **47**, 420 (1985).
97. F. Capasso and R. A. Kiehl, *J. Appl. Phys.* **58**, 1366 (1985).
98. K. Ploog and G. H. Doheler, *Adv. Phys.* **32,** 285 (1983).
99. F. Capasso, K. Mohammed and A. Y. Cho, *Appl. Phys. Lett.* 478 (1986).
100. R. Grill, C. Metzner, and G. H. Döhler, *Phys. Rev. B* **63**, 235316, (2001); J. Z. Wang, Z. G. Wang, Z. M. Wang, S. L. Feng, and Z. Yang, *Phys. Rev. B* **62**, 6956 (2000); *Phys. Rev. B* **61**, 15614 (2000).
101. A. R. Kost, M. H. Jupina, T. C. Hasenberg and E. M. Garmire, *J. Appl. Phys.* **99**, 023501, (2006).
102. A. G. Smirnov, D. V. Ushakov and V. K. Kononenko, *Proc. SPIE*, **4706**, 70, (2002).
103. D. V. Ushakov, V. K. Kononenko and I. S. Manak, *Proc. SPIE*, **4358**, 171, (2001).
104. J. Z. Wang, Z. G. Wang, Z. M. Wang, S. L. Feng and Z. Yang, *Phys. Rev. B* **62**, 6956, (2000).
105. A. R. Kost, L. West, T. C. Hasenberg, J. O. White, M. Matloubian and G. C. Valley, *Appl. Phys. Lett.* **63**, 3494, (1993).
106. S. Bastola, S. J. Chua and S. J. Xu, *J. Appl. Phys.* **83**, 1476, (1998).
107. Z. J. Yang, E. M. Garmire and D. Doctor, *J. Appl. Phys.* **82**, 3874, (1997).
108. G. H. Avetisyan, V. B. Kulikov, I. D. Zalevsky and P. V. Bulaev, *Proc. SPIE*, **2694**, 216, (1996).
109. U. Pfeiffer, M. Kneissl, B. Knüpfer, N. Müller, P. Kiesel, G. H. Döhler, and J. S. Smith, *Appl. Phys. Lett.* **68**, 1838, (1996).
110. H. L. Vaghjiani, E. A. Johnson, M. J. Kane, R. Grey, and C. C. Phillips, *J. Appl. Phys.* **76**, 4407, (1994).
111. P. Kiesel, K. H. Gulden, A. Hoefler, M. Kneissl, B. Knuepfer, S. U. Dankowski, P. Riel, X. X. Wu, J. S. Smith and G. H. Doehler, *Proc. SPIE* **1985**, 278, (1993).
112. G. H. Doheler, *Phys. Script.* **24**, 430 (1981).
113. S. Mukherjee, S. N. Mitra, P. K. Bose, A. R. Ghatak, A. Neoigi, J. P. Banerjee, A. Sinha, M. Pal, S. Bhattacharya and K. P. Ghatak, *J. Compu. Theor. Nanosc.* **4**, 550, (2007).
114. T. Ando, H. Fowler and F. Stern, *Rev. Mod. Phys.* **54**, 437 (1982).
115. J. J. Quinn and P. J. Styles, (ed.) Electronic Properties of Quasi Two Dimensional Systems, North Holland, Amsterdam, (1976).
116. G. A. Antcliffe, R. T. Bate and R. A. Reynolds, Proceedings of the International Conference, Physics of Semi-metals and Narrow-Gap semiconductors (ed.) D. L. Carter and R. T. Bate, Pergamon Press, Oxford, (499) (1971).
117. Z. A. Weinberg, *Sol. Stat. Electron.* **20**, 11 (1977).
118. G. Paasch, T. Fiedler, M. Kolar and I. Bartos, *Phys. Stat. Sol.* (b) **118**, 641 (1983).
119. S. Lamari, *Phys. Rev. B* **64**, 245340 (2001).
120. T. Matsuyama, R. Kürsten, C. Meißner, and U. Merkt, *Phys. Rev. B* **61**, 15588 (2000).

121. P. V. Santos and M. Cardona, *Phys. Rev. Lett.* **72**, 432 (1994).
122. L. Bu, Y. Zhang, B. A. Mason, R. E. Doezema, and J. A. Slinkman *Phys. Rev. B* **45**, 11336 (1992).
123. P. D. EMesselhaus, C. M. Papavassiliou, R. G. Wheeler, and R. N. Sacks, *Phys. Rev. Lett.* **68**, 106 (1992).
124. U. Kunze, *Phys. Rev. B* **41**, 1707 (1990).
125. E. Yamaguchi, *Phys. Rev. B* **32**, 5280 (1985).
126. Th. Lindner and G. Paasch, *J. Appl. Phys.* **102**, 054514 (2007).
127. S. Lamari, *J. Appl. Phys.* **91**, 1698 (2002).
128. K. P. Ghatak and M. Mondal, *J. Appl. Phys.* **70**, 299 (1991).
129. K. P. Ghatak and S. N. Biswas, *J. Vac. Sc. and Tech.* **7B**, 104 (1989).
130. B. Mitra and K. P. Ghatak, *Sol. State Electron.* **32**, 177 (1989).
131. K. P. Ghatak and M. Mondal, *J. Appl. Phys.* **62**, 922 (1987).
132. M. Mondal and K. P. Ghatak, J. Magnet. *Magnetic Mat.* **62**, 115 (1986); M. Mondal and K. P. Ghatak, *Phys. Script.* **31**, 613 (1985).
133. K. P. Ghatak and M. Mondal, *Z. fur Physik B* **64**, 223, (1986); K. P. Ghatak and S. N. Biswas, *Sol. State Electron.* **37**, 1437 (1994). J. J. Quinn, P. J. Styles, (ed.) *Electronic Properties of Quasi Two Dimensional Systems*, North Holland, Amsterdam, (1976); G. A. Antcliffe, R. T. Bate, R. A. Reynolds, Proceedings of the International Conference, *Physics of Semimetals and Narrow-Gap semiconductors* (ed.) D. L. Carter, R. T. Bate, Pergamon Press, Oxford, (499) (1971); D. R. Choudhury, A. K. Chowdhury, K. P. Ghatak, A. N. Chakravarti, *Phys. Stat. Sol.* (b) **98**, (K141) (1980); A. N. Chakravarti, A. K. Chowdhury, K. P. Ghatak, *Phys. Stat. Sol.* (a) **63**, K97 (1981); M. Mondal, K. P. Ghatak, *Acta Phys. Polon. A* **67**, 983 (1985); M. Mondal, K. P. Ghatak, *Phys. Stat. Sol.* (b) **128**, K21 (1985); M. Mondal, K. P. Ghatak, *Phys. Stat. Sol.* (a) **93**, 377 (1986); K. P. Ghatak, M. Mondal, *Phys. Stat. Sol.* (b) **135**, 819 (1986); M. Mondal, K. P. Ghatak, *Phys. Stat. Sol.* (b) **139**, 185 (1987); K. P. Ghatak, N. Chattopadhyay, S. N. Biswas, OE/Fibers' **87**, 203 (1987); K. P. Ghatak, N. Chatterjee, M. Mondal, *Phys. Stat. Sol.* (b) **139**, K25 (1987); K. P. Ghatak, M. Mondal, *Phys. Stat. Sol.* **(b) 148**, 645 (1988); K. P. Ghatak, A. Ghosal *Phys. Stat. Sol.* (b) **151**, K135 (1989).
134. K. P. Ghatak, N. Chattopadhyay, M. Mondal, *Appl. Phys.* **A 48**, 365 (1989).
135. P. Harrison, *Quantum Wells, Wires and Dots*, John Wiley and Sons, Ltd, (2002); B. K. Ridley, *Electrons and Phonons in Semiconductors Multilayers*, Cambridge University Press, Cambridge (1997); G. Bastard, *Wave Mechanics Applied to Semiconductor Heterostructures*, Halsted; Les Ulis, Les Editions de Physique, New York (1988); V. V. Martin, A. A. Kochelap and M. A. Stroscio, *Quantum Heterostructures*, Cambridge University Press, Cambridge (1999).
136. C. S. Lent and D. J. Kirkner, *J. Appl. Phys.* **67**, 6353 (1990); F. Sols, M. Macucci, U. Ravaioli and K. Hess, *Appl. Phys. Lett.* **54**, 350 (1980).
137. C. S. Kim, A. M. Satanin, Y. S. Joe and R. M. Cosby, *Phys. Rev. B* **60**, 10962 (1999).
138. S. Midgley and J. B. Wang, *Phys. Rev. B* **64**, 153304 (2001).

139. T. Sugaya, J. P. Bird, M. Ogura, Y. Sugiyama, D. K. Ferry and K. Y. Jang, *Appl. Phys. Lett.* **80**, 434 (2002).
140. B. Kane, G. Facer, A. Dzurak, N. Lumpkin, R. Clark, L. PfeiKer and K. West, *Appl. Phys. Lett.* **72**, 3506 (1998).
141. C. Dekker, *Physics Today* **52**, 22 (1999).
142. A. Yacoby, H. L. Stormer, N. S. Wingreen, L. N. Pfeiffer, K. W. Baldwin and K. W. West, *Phys. Rev. Lett.* **77**, 4612 (1996).
143. Y. Hayamizu, M. Yoshita, S. Watanabe, H. A. L. PfeiKer and K. West, *Appl. Phys. Lett.* **81**, 4937 (2002).
144. S. Frank, P. Poncharal, Z. L. Wang and W. A. Heer, *Science* **280**, 1744 (1998).
145. I. Kamiya, I. I. Tanaka, K. Tanaka, F. Yamada, Y. Shinozuka and H. Sakaki, *Physica E* **13**, 131 (2002).
146. A. K. Geim, P. C. Main, N. LaScala, L. Eaves, T. J. Foster, P. H. Beton, J. W. Sakai, F. W. Sheard, M. Henini, G. Hill *et al.*, *Phys. Rev. Lett.* **72**, 2061 (1994).
147. A. S. Melinkov and V. M. Vinokur, *Nature* **415**, 60 (2002).
148. K. Schwab, E. A. henriksen, J. M. Worlock and M. L. Roukes, *Nature* **404**, 974 (2000).
149. L. Kouwenhoven, *Nature* **403**, 374 (2000).
150. S. Komiyama, O. Astafiev, V. Antonov and H. Hirai, *Nature* **403**, 405 (2000).
151. E. Paspalakis, Z. Kis, E. Voutsinas and A. F. Terziz, *Phys. Rev. B* **69**, 155316 (2004).
152. J. H. Jefferson, M. Fearn, D. L. J. Tipton and T. P. Spiller, *Phys. Rev. A* **66**, 042328 (2002).
153. J. Appenzeller, C. Schroer, T. Schapers, A. Hart, A. Froster, B. Lengeler and H. Luth, *Phys. Rev. B* **53**, 9959 (1996).
154. J. Appenzeller and C. Schroer, *J. Appl. Phys.* **87**, 31659 (2002).
155. P. Debray, O. E. Raichev, M. Rahman, R. Akis and W. C. Mitchel, *Appl. Phys. Lett.* **74**, 768 (1999).
156. P. M. Solomon, Proc. IEEE, 70, 489 (1982; T. E. Schlesinger and T. Kuech, *Appl. Phys. Lett.* **49**, 519 (1986).
157. D. Kasemet, C. S. Hong, N. B. Patel and P. D. Dapkus, *Appl. Phys. Letts.* **41**, 912 (1982); K. Woodbridge, P. Blood, E. D. Pletcher and P. J. Hulyer, *Appl. Phys. Lett.* **45**, 16 (1984); S. Tarucha and H. O. Okamoto, *Appl. Phys. Letts.* **45**, 16 (1984); H. Heiblum, D. C. Thomas, C. M. Knoedler, and M.I Nathan, *Appl. Phys. Letts.* **47**, 1105 (1985).
158. O. Aina, M. Mattingly, F. Y. Juan and P. K. Bhattacharyya, *Appl. Phys. Letts.* **50**, 43 (1987).
159. I. Suemune and L. A. Coldren, IEEE, *J. Quant. Electronic.* **24**, 1178 (1988).
160. D. A. B. Miller, D. S. Chemla, T. C. Damen, J. H. Wood, A. C. Burrus, A. C. Gossard and W. Weigmann, IEEE, *J. Quant. Electron.* **21**, 1462 (1985).
161. J. S. Blakemore, *Semiconductor Statistics*, Dover, New York (1987); K P Ghatak, S Bhattacharya, S K Biswas, A Deyand A K Dasgupta, *Phys. Scr.* **75**,820, (2007).

162. N. Miura, *Physics of Semiconductors in High Magnetic Fields*, Series on Semiconductor Science and Technology (Oxford University Press, USA, 2007); K. H. J Buschow, F. R. de Boer, *Physics of Magnetism and Magnetic Materials* (Springer, New York, 2003); D. Sellmyer (Ed.), R. Skomski (Ed.), *Advanced Magnetic Nanostructures* (Springer, New York, 2005); J. A. C. Bland (Ed.), B. Heinrich (Ed.), *Ultrathin Magnetic Structures III: Fundamentals of Nanomagnetism (Pt. 3)* (Springer-Verlag, Germany, 2005); B. K. Ridley, *Quantum Processes in Semiconductors*, Fourth Edition (Oxford publications, Oxford, 1999); J. H. Davies, *Physics of Low Dimensional Semiconductors* (Cambridge University Press, UK, 1998); S. Blundell, *Magnetism in Condensed Matter*, Oxford Master Series in Condensed Matter Physics (Oxford University Press, USA, 2001); C. Weisbuch, B. Vinter, *Quantum Semiconductor Structures: Fundamentals and Applications* (Academic Publishers, USA, 1991); D. Ferry, *Semiconductor Transport* (CRC, USA, 2000); M. Reed (Ed.), *Semiconductors and Semimetals: Nanostructured Systems* (Academic Press, USA, 1992); T. Dittrich, *Quantum Transport and Dissipation* (Wiley-VCH Verlag GmbH, Germany, 1998); A. Y. Shik, *Quantum Wells: Physics & Electronics of Two dimensional Systems* (World Scientific, USA, 1997).
163. K. P. Ghatak, M. Mondal, *Zietschrift fur Naturforschung A* **41a**, 881 (1986); K. P. Ghatak, M. Mondal, *J. Appl. Phys.* **62**, 922 (1987); K. P. Ghatak, S. N. Biswas, *Phys. Stat. Sol. (b)* **140**, K107 (1987); K. P. Ghatak, M. Mondal, *Jour. of Mag. and Mag. Mat.* **74**, 203 (1988); K. P. Ghatak, M. Mondal, *Phys. Stat. Sol. (b)* **139**, 195 (1987); K. P. Ghatak, M. Mondal, *Phys. Stat. Sol. (b)* **148**, 645 (1988); K. P. Ghatak, B. Mitra, A. Ghoshal, *Phys. Stat. Sol. (b)* **154**, K121 (1989); K. P. Ghatak, S. N. Biswas, *Jour. of Low Temp. Phys.* **78**, 219 (1990); K. P. Ghatak, M. Mondal, *Phys. Stat. Sol. (b)* **160**, 673 (1990); K. P. Ghatak, B. Mitra, *Phys. Letts. A* **156**, 233 (1991); K. P. Ghatak, A. Ghoshal, B. Mitra, *Nouvo Cimento D* **13D**, 867 (1991); K. P. Ghatak, M. Mondal, *Phys. Stat. Sol. (b)* **148**, 645 (1989); K. P. Ghatak, B. Mitra, *Internat. Jour. of Elect.* **70**, 345 (1991); K. P. Ghatak, S. N. Biswas, *J. Appl. Phys.* **70**, 299 (1991); K. P. Ghatak, A. Ghoshal, *Phys. Stat. Sol. (b)* **170**, K27 (1992); K. P. Ghatak, *Nouvo Cimento D* **13D**, 1321 (1992); K. P. Ghatak, B. Mitra, *Internat. Jour. of Elect.* **72**, 541 (1992); K. P. Ghatak, S. N. Biswas, *Nonlinear Optics* **4**, 347 (1993); K. P. Ghatak, M. Mondal, *Phys. Stat. Sol. (b)* **175**, 113 (1993); K. P. Ghatak, S. N. Biswas, *Nonlinear Optics* **4**, 39 (1993); K. P. Ghatak, B. Mitra, *Nouvo Cimento* **15D**, 97 (1993); K. P. Ghatak, S. N. Biswas, *Nanostructured Materials* **2**, 91 (1993); K. P. Ghatak, M. Mondal, *Phys. Stat. Sol. (b)* **185**, K5 (1994); K. P. Ghatak, B. Goswami, M. Mitra, B. Nag, *Nonlinear Optics* **16**, 9 (1996); K. P. Ghatak, M. Mitra, B. Goswami, B. Nag, *Nonlinear Optics* **16**, 167 (1996); K. P. Ghatak, B. Nag, *Nanostructured Materials* **10**, 923 (1998).
164. D. Roy Choudhury, A. K. Choudhury, K. P. Ghatak, A. N. Chakravarti, *Phys. Stat. Sol. (b)* **98**, K141 (1980); A. N. Chakravarti, K. P. Ghatak, A. Dhar, S. Ghosh, *Phys. Stat. Sol. (b)* **105**, K55 (1981); A. N. Chakravarti,

A. K. Choudhury, K. P. Ghatak, *Phys. Stat. Sol. (a)* **63**, K97 (1981); A. N. Chakravarti, A. K. Choudhury, K. P. Ghatak, S. Ghosh, A. Dhar, *Appl. Phys.* **25**, 105 (1981); A. N. Chakravarti, K. P. Ghatak, G. B. Rao, K. K. Ghosh, *Phys. Stat. Sol. (b)* **112**, 75 (1982); A. N. Chakravarti, K. P. Ghatak, K. K. Ghosh, H. M. Mukherjee, *Phys. Stat. Sol. (b)* **116**, 17 (1983); M. Mondal, K. P. Ghatak, *Phys. Stat. Sol. (b)* **133**, K143 (1984); M. Mondal, K. P. Ghatak, *Phys. Stat. Sol. (b)* **126**, K47 (1984); M. Mondal, K. P. Ghatak, *Phys. Stat. Sol. (b)* **126**, K41 (1984); M. Mondal, K. P. Ghatak, *Phys. Stat. Sol. (b)* **129**, K745 (1985); M. Mondal, K. P. Ghatak, *Phys. Scr.* **31**, 615 (1985); M. Mondal, K. P. Ghatak, *Phys. Stat. Sol. (b)* **135**, 239 (1986); M. Mondal, K. P. Ghatak, *Phys. Stat. Sol. (b)* **93**, 377 (1986); M. Mondal, K. P. Ghatak, *Phys. Stat. Sol. (b)* **135**, K21 (1986); M. Mondal, S. Bhattacharyya, K. P. Ghatak, *Appl. Phys. A* **42A**, 331 (1987); S. N. Biswas, N. Chattopadhyay,K. P. Ghatak,*Phys. Stat. Sol. (b)* **141**, K47 (1987); B. Mitra, K. P. Ghatak, *Phys. Stat. Sol. (b)* **149**, K117 (1988); B. Mitra, A. Ghoshal, K. P. Ghatak, *Phys. Stat. Sol. (b)* **150**, K67 (1988); M. Mondal, K. P. Ghatak, *Phys. Stat. Sol. (b)* **147**, K179 (1988); M. Mondal, K. P. Ghatak, *Phys. Stat. Sol. (b)* **146**, K97 (1988); B. Mitra, A. Ghoshal, K. P. Ghatak, *Phys. Stat. Sol. (b)* **153**, K209 (1989); B. Mitra, K. P. Ghatak, *Phys. Letts.* **142A**, 401 (1989); B. Mitra, A. Ghoshal,K. P. Ghatak, *Phys. Stat. Sol. (b)* **154**, K147 (1989); B. Mitra, K. P. Ghatak, *Sol. State Elect.* **32**, 515 (1989); B. Mitra, A. Ghoshal, K. P. Ghatak, *Phys. Stat. Sol. (b)* **155**, K23 (1989); B. Mitra, K. P. Ghatak, *Phys. Letts.* **135A**, 397 (1989); B. Mitra, K. P. Ghatak, *Phys. Letts. A* **146A**, 357 (1990); B. Mitra, K. P. Ghatak, *Phys. Stat. Sol. (b)* **164**, K13 (1991); S. N. Biswas, K. P. Ghatak, *Internat. Jour. of Elect.* **70**, 125 (1991).

165. P. R. Wallace, Phys. Stat. Sol. (b), **92**, 49 (1979).
166. B. R. Nag, *Electron Transport in Compound Semiconductors*, Springer Series in Solid-State Sciences **Vol. 11** (Springer-Verlag, Germany, 1980).
167. K. P. Ghatak, S. Bhattacharya, D. De, *Einstein Relation in Compound Semiconductors and Their Nanostructures*, Springer Series in Materials Science **Vol. 116** (Springer-Verlag, Germany, 2009).
168. C. C. Wu and C. J. Lin, *J. Low Temp. Phys.* **57**, 469 (1984); M. H. Chen, C. C. Wu, C. J. Lin, *J. LowTemp. Phys.* **55**, 127 (1984).
169. P. K. Basu, *Theory of Optical Process in Semiconductors, Bulk and Microstructures* (Oxford University Press, Oxford,1997).
170. K. P. Ghatak, S. Bhattacharya, S.Bhowmik, R. Benedictus, S. Chowdhury, *J. Appl. Phys.* **103**, 094314 (2008); K. P. Ghatak, S.Bhattacharya, *J. Appl. Phys.* **102**, 073704 (2007); K. P. Ghatak, S. Bhattacharya, S. K. Biswas, A. De, A. K. Dasgupta, *Phys. Scr.* **75**, 820 (2007); P. K. Bose, N. Paitya, S. Bhattacharya, D. De, S. Saha, K. M. Chatterjee, S. Pahari, K. P. Ghatak, *Quantum Matter* **1**, 89 (2012); K. P. Ghatak, S. Bhattacharya, A. Mondal, S. Debbarma, P. Ghorai, and A. Bhattacharjee, *Quantum Matter* **2**, 25 (2013); S. Bhattacharya, D. De, S. Ghosh, J. P. Bannerje, M. Mitra, B. Nag, S. Saha, S. K. Bishwas, M. Paul, *Jour. Comp. Theo. Nanoscience* **7**, 1066 (2010); K. P. Ghatak, S. Bhattacharya, S. Pahari, S. N. Mitra, P. K.

Bose, *Jour. Phys. Chem. Solids* **70**, 122 (2009), S. Bhattacharya, D. De, R. Sarkar, S. Pahari, A. De, A. K. Dasgupta, S. N. Biswas, K. P. Ghatak, *Jour. Comp. Theo. Nanoscience* **5**, 1345 (2008); S. Mukherjee, D. De, D. Mukherjee, S. Bhattacharya, A. Sinha, K. P. Ghatak *Physica B* **393**, 347 (2007).

171. K. Seeger, *Semiconductor Physics* (Springer-Verlag, 7th Edition. 2006).
172. B. R. Nag, *Physics of Quantum Well Devices* (Kluwer Academic Publishers, The Netherlands, 2000).
173. W. Zawadzki, B. Lax, *Phys. Rev. Lett.* **16**, (1001) (1966).
174. K. P. Ghatak, J. P. Banerjee, B. Goswami, B. Nag, *Non. Opt. and Quan. Opt.* **16**, (241) (1996); K. P. Ghatak, M. Mitra, B. Goswami, B. Nag, *Non. Opt. and Quan. Opt.* **16**, (167) (1996); B. Mitra, A. Ghoshal , K. P. Ghatak, *Phys. Stat. Sol. (b)* **154**, (K147) (1989); K. P. Ghatak, B. Goswami, M. Mitra, B. Nag, *Non. Opt. and Quan. Opt.* **16**, (9) (1996); S. Bhattaacharya, S. Choudhary, S. Ghoshal, S. K. Bishwas, D. De, K. P. Ghatak, *J. Comp. Theo. Nanoscience.* **3**, (423), (2006); M. Mondal, K. P. Ghatak, *Ann. der Physik.* **46**, (502) (1989); K. P. Ghatak, OE Fiber-DL, (435) (1991);.
175. K. P. Ghatak, B. Mitra, *Internat. Jour. Electron. Theo. Exp.* **70**, (343) (1991); K. P. Ghatak, *Proc. Soc. Photo-Optical Instru. Engg.*, USA **1280**, (53) (1990); B. Mitra, K. P. Ghatak, *Phys. Lett. A* **137**, (413) (1989); M. Mondal, K. P. Ghatak, *Annalen der Physik* **501**, (502) (1989); M. Mondal, K. P. Ghatak, *Phys. Stat. Sol. (b)* **147**, (K179) (1988); S. Biswas, N. Chattopadhyay, K. P. Ghatak, International Society for Optics and Photonics, *Proc. Soc. Photo-Optical Instru. Engg.*, USA **836**, (175) (1987).
176. K. P. Ghatak, J. Mukhopadhyay J. P. Banerjee, *SPIE Proceedings Series* **4746,** 1292 (2002); K. P. Ghatak, S. Dutta, D. K. Basu, B. Nag, *Il Nuovo Cimento.* **D 20**, 227 (1998); K. P. Ghatak, B. De, *Mat. Resc. Soc. Proc.* **300**, 513 (1993); K. P. Ghatak, B. Mitra, *Il Nuovo Cimento.* **D 15**, 97 (1993); K. P. Ghatak, A Ghoshal, S. N. Biswas, M. Mondal, *Proc. Soc. Photo Opt. Instru. Engg.* **1308**, 356 (1990); B. Mitra, K. P Ghatak, *Phys. Lett. A.* **146**, 357 (1990); B. Mitra, K. P. Ghatak, *Phys. Lett. A.* **142**, 401 (1989); B. Mitra, K. P. Ghatak, *Phys. Stat. Sol. (b).* **149**, K117 (1988); S. Bhattacharyya, K. P. Ghatak, S. Biswas, OE/Fibers' 87, *Inter. Soc. Opt. Photon.* 73 (1987); K. P. Ghatak, A. N. Chakravarti, *Phys. Stat. Sol. (b).* **117**, 707 (1983).
177. F. Capasso, *Semiconductors and Semimetals* **22**, 2 (1985).
178. F. Capasso, K. Mohammed, A. Y. Cho, R. Hull, A. L. Hutchinson, *Appl. Phys. Letts.* **47**, 420 (1985).
179. F. Capasso, R. A. Kiehl, *J. Appl. Phys.* **58**, 1366 (1985).
180. K. Ploog, G. H. Doheler, *Adv. Phys.* **32**, 285 (1983).
181. F. Capasso, K. Mohammed, A. Y. Cho, *Appl. Phys. Lett.* 478 (1986).
182. R. Grill, C. Metzner, G. H. Döhler, *Phys. Rev. B* **63**, 235316 (2001); *Phys. Rev. B* **61**, 15614 (2000).
183. A. R. Kost, M. H. Jupina, T. C. Hasenberg, E. M. Garmire, *J. Appl. Phys.* **99**, 023501 (2006).

184. A. G. Smirnov, D. V. Ushakov, V. K. Kononenko, *Proc. SPIE*, **4706**, 70 (2002).
185. D. V. Ushakov, V. K. Kononenko, I. S. Manak, *Proc. SPIE*, **4358**, 171 (2001).
186. J. Z. Wang, Z. G. Wang, Z. M. Wang, S. L. Feng, Z. Yang, *Phys. Rev. B*, **62**, 6956 (2000).
187. A. R. Kost, L. West, T. C. Hasenberg, J. O. White, M. Matloubian, G. C. Valley, *Appl. Phys. Lett.* **63**, 3494 (1993).
188. S. Bastola, S. J. Chua, S. J. Xu, *J. Appl. Phys.* **83**, 1476 (1998).
189. Z. J. Yang, E. M. Garmire, D. Doctor, *J. Appl. Phys.* **82**, 3874 (1997).
190. G. H. Avetisyan, V. B. Kulikov, I. D. Zalevsky, P. V. Bulaev, *Proc. SPIE*, **2694**, 216 (1996).
191. U. Pfeiffer, M. Kneissl, B. Knüpfer, N. Müller, P. Kiesel, G. H. Döhler, J. S. Smith, *Appl. Phys. Lett.* **68**, 1838 (1996).
192. H. L. Vaghjiani, E. A. Johnson, M. J. Kane, R. Grey, C. C. Phillips, *J. Appl. Phys.* **76**, 4407 (1994).
193. P. Kiesel, K. H. Gulden, A. Hoefler, M. Kneissl, B. Knuepfer, S. U. Dankowski, P. Riel, X. X. Wu, J. S. Smith, G. H. Doehler, *Proc. SPIE*, **85**, 278 (1993).
194. L. V. Keldysh, *Soviet Phys. Solid State* **4**, 1658 (1962).
195. L. Esaki and R. Tsu, *IBM J. Research and Develop.* **14**, 61 (1970).
196. R. Tsu, *Superlattices to Nanoelectronics*, Elsevier, (2005); E. L. Ivchenko and G. Pikus, *Superlattices and Other Heterostructures*, Springer-Berlin (1995).
197. G. Bastard, *Wave Mechanics Applied to Heterostructures*, Editions de Physique, Les Ulis, France, (1990).
198. K. P. Ghatak and B. Mitra, *Int. J. Electronics* **72**, 541 (1992); S. N. Biswas and K. P. Ghatak, *Int. J. Electronics* **70**, 125 (1991); B. Mitra and K. P. Ghatak, *Phys. Stat. Sol. (b)* **149**, K117 (1988); K. P. Ghatak, B. Mitra and A. Ghoshal, *Phys. Stat. Sol. (b)* **154**, K121 (1989).
199. K. V. Vaidyanathan, R. A. Jullens, C. L. Anderson and H. L. Dunlap, *Solid State Electron.* **26**, 717 (1983).
200. B. A. Wilson, *IEEE, J. Quantum Electron.* **24**, 1763 (1988).
201. M. Krichbaum, P. Kocevar, H. Pascher and G. Bauer, *IEEE, J. Quan. Electron.* **24**, 717 (1988).
202. J. N. Schulman and T. C. McGill, *Appl. Phys. Letts.* **34**, 663 (1979).
203. H. Kinoshita, T. Sakashita and H. Fajiyasu, *J. Appl. Phys.*, **52**, 2869 (1981).
204. L. Ghenin, R. G. Mani, J. R. Anderson and J. T. Cheung, *Phys. Rev. B* **39**, 1419 (1989).
205. C. A. Hoffman, J. R. Mayer, F. J. Bartoli, J. W. Han, J. W. Cook, J. F. Schetzina and J. M. Schubman, *Phys. Rev. B.* **39**, 5208 (1989).
206. V. A. Yakovlev, *Sov. Phys. Semicon.* **13**, 692 (1979).
207. E. O. Kane., *J. Phys. Chem. Solids* **1**, 249 (1957).
208. H. Sasaki, *Phys. Rev. B* **30**, 7016 (1984).

209. A. Pais, J. *Robert Oppenheimer* (Oxford University Press, U.K 2006) (p. xviii).
210. I. K. Kikoin, *Science for Everyone: Encounters with Physicists and Physics* (Mir Publishers, Russia, 1989), p. 154.
211. J. J. Sakurai, *Advanced Quantum Mechanics* (Addison–Wesley Publishing Company, USA, 1967)

The greatest enemy of knowledge is not ignorance, it is the illusion of Knowledge

Acknowledgments

I am grateful to A. N. Chakravarti, formerly Head of the Department of the Institute of Radio Physics and Electronics of the University of Calcutta, my mentor and an *ardent advocator of the concept of theoretical minimum of Landau*, who convinced a twenty one years old Circuit theorist that Condensed Matter Physics, in general, is the hidden dual dance of quantum mechanics, statistical mechanics together with advanced mathematical techniques, and even to appreciate the incredible beauty, he not only placed a stiff note for me to derive all the equations in the Monumental Course of Theoretical Physics, the Classics of Landau–Lifshitz together with the two-volume classics of Morse-Feshbach forty years ago but also forced me to stay in the *creative critical zone of research till date.* I express my gratitude to Late B. R. Nag, formerly Head of the Departments of Radio Physics and Electronics and Electronic Science of the University of Calcutta, to whom I am ever grateful as a student and research worker, the Semiconductor Grandmaster of India for his three books on semiconductor science in general and more than 200 research papers (many of them are absolutely in *honor's class*) which still fire my imagination. I consider myself to be rather fortunate to study Mathematics under the direct influence of Late S. C. Dasgupta, formerly Head of the Department of Mathematics of the then Bengal Engineering College, Shibpur (presently Indian Institute of Engineering Science and Technology) and M. Mitra (both of them were formidable Applied Mathematicians with deep physical insight and could solve any problem from the two-volume classics of Morse-Feshbach instantly on the blackboard) of the said

Department of Mathematics for teaching me deeply the various methods of different branches of Applied Mathematics with special emphasis to ***analytic number theory*** when I was pursuing the bachelor degree in the branch of Electronics and Telecommunication Engineering forty-five years ago. I offer my thanks to Late S. S. Boral, formerly Founding Head of the Department of Electronics and Telecommunication Engineering of the then Bengal Engineering College, Shibpur, for teaching me the Theoretical Acoustics from the Classic of Morse and Ingard by urging me to solve the problems. I am grateful to S. K. Sen, formerly Head of the Department of Electrical Engineering of Bengal Engineering College, Shibpur, Ex-Vice-Chancellor of Jadavpur University and Ex-Power Minister of West Bengal State for teaching me deeply Non-linear network analysis and synthesis and Non-linear control systems in general. I am indebted to Late C.K. Majumdar, Head of the Department of Physics of the University of Calcutta to light the fire for Theoretical Physics.

I am grateful to all my friends and colleagues for the last forty years from my school days till date for forming my inner fire to solve independently the problems from the five-volume Berkley Physics Course, three-volume Sakurai, ten volumes of Addison-Wesley Modular Series on Solid State Devices, two volumes Cohen-Tannoudji *et al.*, three volumes Edwards, three volumes Carslaw, six volumes Guillemin, three volumes Tuttle, two volumes Scott, two volumes Budak, five volumes Chen, ... It is curious to note that they insisted, in the real sense of the term, that I verify all the entries of Gradshteyn-Ryzhik and three volumes of the Bateman manuscript project for the last forty years. It may be noted that the academic output of a nice scholar is the response of a good quality RC coupled amplifier in the mid-band zone whereas the same for a fine research worker is the Dirac's delta function. Incidentally, I can safely say that I belong to neither as a consequence of the academic surroundings which this life presents to me. I express my warm gratitude to H. L. Hartnagel, D. Bimberg, W. L. Freeman and W. Schommers for various academic interactions spanning over the last three decades. I remember the sweet memories of P. N. Robson,

I. M. Stephenson, V. S. Letokhov, J. Bodnar and K. F. Brennan with true reverence for inspiring me with the ebullient idea that the publications of the research monographs from internationally reputed Publishers containing innovative concepts is of prime importance to excel in creative research activity.

In 1st January 2007, I wrote a letter to Late P. T. Landsberg (popularly known as P.T.L to his scientific friends) for writing a book on Effective Mass of nano structured materials which can at least transform me in to a better scientist but P.T. L. in turn, requested me repeatedly to complete the proposed book at the earliest and often expressed his desire to write a foreword for this book. Incidentally due to previous other heavy academic and related commitments I was unable to finish this present monograph and his wife Mrs. Sophia Landsberg in her letter of desperate sadness dated January 12, 2009 informed me that from 2008 onward due to Alzheimer's disease her husband had to give up scientific work. The disappearance of P.T.L from my research scenario was a big shock to me which I have to bear till my own physical disappearance. I still remember his power packed words that "The definition of Kamakhya Prasad Ghatak = M.Tech. + Ph.D. + D. Engg. + more than 300 research papers in nanoscience and technology in SCI Journals + more than 60 research papers and International Conferences of SPIE and MRS of USA + Ph.D. guide of more than 30 Ph.D. students + many research monographs from Springer-Verlag, Germany + HOD + Dean ..... is a **big zero to him if I cannot live in a world full with new creative concepts and not merely repeating the past successes, since past success is a dead concept at least to P.T.L.**

In 21st December, 1974, A. N. Chakravarti (an internationally recognized expert on Einstein Relation in general), my mentor in my first interaction with him emphatically energized me by making myself acquainted with the ***famous Fermi-Dirac Integrals*** and telling me that ***I must master*** "Semiconductor Statistics" (Pergamon Press, London, 1962) by J. S. Blakemore for my initiation in semiconductor physics. Later on I came in deep touch with K. Seeger, the well-known author of the book "Semiconductor

Physics" (Springer Series in Solid State Sciences, vol. 40, Springer-Verlag, Germany, 1982). The solid Mathematical Physicist Late P. N. Butcher has been a steady hidden force since 1983 before his sad demise with respect to our scripting the series in band structure dependent properties of nano-structured materials. He insisted me repeatedly regarding it and to tune with his high rigorous academic standard, myself with my prominent members of my research group wrote [1] as the first one, [2] as the second one, [3] as the third one, [4] as the fourth one, [5] as the fifth one, [6] as the sixth one, [7] as the seventh one, [8] as the eighth one, [9] as the nineth one, [10] as the tenth one and the present monograph as the eleventh one. Both P. T. L and P. N. Butcher visited the Institute of Radio Physics and Electronics of the University of Calcutta, my ALMA MATER where I started my research as an M. Tech. student and later on as a faculty member. I formed permanent covalent bonds with J. S. Blakemore, K. Seeger, P. N. Butcher and P. T. L through letters (Pre email era) and these four first class semiconductor physicists, in turn, shared with pleasure their vast creative knowledge of Semiconductors and related sciences with a novice like me.

I offer special thanks to Late N. Guhachoudhury of Jadavpur University for instilling in me the thought that the *academic output* = ((*desire X determination X dedication*) – (*false enhanced self ego pretending like a true friend although a real unrecognizable foe*)) although a thank you falls in the band gap regime for my beloved better half See who really formed the backbone of my long unperturbed research career, since in accordance with "Sanatan" Hindu Dharma, the fusion of marriage has transformed us to form a single entity, where the individuality is being lost. I am grateful to all the members of my research group (from 1983 till date) for not only quantum confining me in the infinitely deep quantum wells of ***Ramanujan and Rabindranath*** but also inspiring me to teach quantum mechanics and related topics from the eight volumes classics of Greiner *et al.*

In this context, from the flash back of my memory I wish to offer my indebtedness to M. Mondal, the first member of my research team who in 1983 joined with me to complete his Ph. D work under my guidance when R. K. Poddar, the then Vice-chancellor of the University of Calcutta selected me as a Lecturer (presently Assistant Professor) of the famous Department of Radio Physics and Electronics. In 1987, S. K. Sen, the then Vice-chancellor of Jadavpur University accepted me as the Reader (presently Associate Professor) in the Department of Electronics and Telecommunication Engineering and since then a flood of young researchers (more than 12 in number consisting of B. Mitra, A. Ghoshal, D. Bhattacharya, A. R. Ghatak,....) around me created my research team and insisted me to work with them 16 hours per day including holidays in different directions of research for the purpose of my creative conversion from an ordinary engineer to a 360° research scientist and consequently I enjoyed the high rate of research publications in different reputed International journals in various aspects of band structure dependent properties of quantized structures. It is nice to note that the said young talented researchers obtained their respective Ph. D degree under my direct supervision. Incidentally in 1994, R. K. Basu, the then Vice-chancellor of the University of Calcutta selected me as a Professor in the Department of Electronic Science and another flood of research overwhelmed me in a new direction of materials science. The persons responsible for this change include S. Datta, S. Sengupta, A. Ali....The 11th and 12th names of this golden series are S. Bhattacharya and D. De respectively who, in turn formed permanent covalent bonds with me not only with respect to research (S. Bhattacharya and D. De are respectively the co-authors of seven and two Monographs in different series of Springer) but also in all aspects of life in general.

It is curious to note that after serving 18 years as a Professor in the Department of Electronic Science, in 2012, P. K. Bose, the then Director of the National Institute of Technology, Agartala requested me to join as a Professor and Departmental Head in Electronics and Communication Engineering. Being my life-long friend, I accepted his offer and more than ten young scholars around me are again directing my research in an altogether new direction. In 2015 the respected Director Professor S. Chakrabarti of Institute of Engineering and Management in Saltlake City Kolkata, has kindly offered me the position of Research Director and Senior Professor in his famous Institute in the academic end of my life to complete my last run towards the creative knowledge temple with my new young research workers and I am grateful to him in the real sense of the term for his creative gesture. Prof. S. Chakrabarti is noted for his keen academic insight and the creation of congenial atmosphere to promote research and higher learning among the faculty members. In my 30+ years of teaching life (I have the wide experience of teaching Physics, Mathematics, Applied Mechanics (from engineering statics up to nonlinear mechanics including indeterminate structures) and 70% of the courses of Electronics and Telecommunication and Electrical Engineering respectively) and 40+ years of research life (mostly in Materials Science, Nano-science and Number Theory), I have finally realized that the teaching without research is body without brain and research without teaching is body without blood **although my all time hero, creatively prolific number theorist Godfrey Harold Hardy in his classic book entitled "A Mathematician's Apology" (Cambridge University Press, 1990) tells us "I hate teaching".**

Incidentally, one young theoretician friend of mine often tells me that many works in theoretical semiconductor science are based on the following seven principles:

1. Principles of placing the necessary and sufficient conditions of a proof in the band gap regime.
2. Principles of interchange of the summation and the integration processes and unconditioned convergences of the series and the integrals.
3. Principles of random applications of one electron theory and super-position theorem in studying the properties of semiconductors, although the many body effects are very important together with the fact that the nature is fundamentally non-linear in nature.
4. Principles of using the invariant band structure concept of semiconductors even in the presence of strong external fields (light, electric, heavy doping etc.) and the random applications

of perturbation theory, which is in a sense quantum mechanical Taylor series without considering the related consequences.
5. Principle of random applications of the binomial theorem without considering the important concept of branch cut.
6. Principle of little discussion regarding the whole set of materials science comprising of different compounds having various band structures under different physical conditions as compared with the simplified two band model of Kane for III–V semiconductors.
7. Principle of using the Fermi's golden rule, the band structure, and the related features, which are valid for non-degenerate semiconductors to materials having degenerate carrier concentrations directly.

Although my young friend is a purist in his conjecture, there are no doubt certain elements of truth inside his beautiful comments. I hope that our readers will present their intricate and advanced theories after paying due weightage of his aforementioned seven principles.

I am grateful to Professor Dr. S. Chakrabarti, the Chancellor of University of Engineering and Management in Jaipur and in Rajarhat, Kolkata and the founding Director of Institute of Engineering and Management, Kolkata for his encouragement and enthusiastic support. I must express my gratitude to Professor B. Chatterjee of Computer Science and Engineering Department of my present Institute, one of the strong members of my research group, for offering important suggestions for the condensed presentation of this monograph. I offer special thanks to T. Datta, T.N. Sen and other members of my research team for placing their combined effort towards the development of this book in the DO-LOOP of a computer and critically reading the manuscript in its last phase before sending it to C. Teo, In-house Editor World Scientific Publishing Co., Singapore. Last but not the least; I am grateful forever to our life long time tested friend S. Sanyal, Principal, Lake School for Girls, Kolkata for not only motivating me at rather turbulent moments of my academic carrier but also instilling in me the concept that ***the ratio of total accumulated knowledge in my chosen field of research to my own contribution in my topic of research***

tends to infinity at any time and is the definition of non-removable pole in the transfer function of my life.

I am grateful to Prof. Dr. Wolfram Schommers (Professor of Karlsruher Institut fűr Technologie, Germany and University of Texas at Arlington, USA), Series Editor of the Series on The Foundation of Natural Science and Technology of World Scientific, Singapore for proposing my name for being a future author in the said series and later on C. Teo, In-house Editor, for accepting our book proposal. Professor Schommers is an ardent supporter of the research work of my research group and I am also grateful to him in this regard. Naturally, the author is responsible for non-imaginative shortcomings. I firmly believe that our Mother Nature has propelled this Project in her own unseen way in spite of several insurmountable obstacles.

Kolkata, India K. P. Ghatak
19th March, 2016

References

1. Ghatak. K.P., Bhattacharya. S., De. D.,*Einstein Relation in compound semiconductors and their heterostructures*, Springer Series in Materials Science **116,** (Springer 2009).
2. Ghatak. K.P., De. D., Bhattacharya. S., *Photoemission from optoelectronic materials and their nanostructures,* Springer Series in Nanostructure Science and Technology, (Springer 2009).
3. Ghatak. K.P., Bhattacharya. S, *Thermo Electric Power In Nano structured Materials Strong Magnetic Fields,* Springer Series in Materials Science **137** (Springer 2010);.
4. Bhattacharya. S, Ghatak. K.P., *Fowler-Nordheim Field Emission: Effects in semiconductor nanostructures,* Springer Series in Solid state Sciences **170** (Springer 2012);.
5. Bhattacharya. S, Ghatak. K.P., *Effective electron mass in low dimensional semiconductors,* Springer Series in Materials Sciences **167** (Springer 2013);.
6. Ghatak. K.P., Bhattacharya. S., *Debye Screening Length: Effects of Nanostructured Materials;* Springer Tracts in Modern Physics **255** (Springer 2014);.
7. Ghatak. K.P., Bhattacharya. S., *Heavily Doped 2D Quantized Structures and the Einstein Relation;* Springer Tracts in Modern Physics **260** (Springer 2015).

8. Ghatak. K.P., *Einstein's Photo-emission: Emission fromHeavily Doped Quantized Structures*; Springer Tracts in Modern Physics **262** (Springer 2015).
9. Ghatak.K.P., Dispersion Relation in Heavily-Doped Nanostructure, Springer Tracts in Modern Physics **265** (Springer 2016).
10. Ghatak. K.P., Magneto Thermoelectric Power in Heavily doped Quantized Structures, Series on the Foundations of Natural Science and Technology, **7** (World Scientific, 2016)

Self Criticism is extremely important in life

About the Author

Born in India on 12th April, 1953, Professor Kamakhya Prasad Ghatak obtained Ph.D. (Tech) Degree from the Institute of Radio Physics and Electronics of the University of Calcutta in 1988 ON THE BASIS OF 27 PUBLISHED RESEARCH PAPERS IN INTERNATIONAL PEER-REVIEWED (IP) SCI JOURNALS WHICH IS STILL A RECORD IN THE SAID INSTITUTE. HE IS THE FIRST RECIPIENT OF THE DEGREE OF DOCTOR OF ENGINEERING OF JADAVPUR UNIVERSITY ON THE BASIS OF CIRCA 150 RESEARCH PAPERS IN IP SCI JOURNALS CONCERNED WITH TRANSPORT EFFECTS IN COMPOUND NANO-STRUCTURED MATERIALS IN 1991 SINCE THE UNIVERSITY INCEPTION IN 1955 AND IN THE SAME YEAR HE RECEIVED THE INSA VISITING FELLOWSHIP TO IIT-KHARAGPUR. He is the PRINCIPAL CO-AUTHOR of more than 500 research papers in eminent IP S.C.I. Journals, more than 60 research papers in the Proceedings of the Conferences of SPIE and MRS of USA (https://scholar.google.co.in/citations?user =q8z3CMAAAAJ&hl =en) and many of his papers are being cited many times. At present, the h-index, i10-index, total citations and the maximum citation of a research paper within three years of its publication of Professor Ghatak are 29, 139, 4104 and 325 respectively. He is the invited Speaker of SPIE, MRS, etc., the referee and Editor of different reputed Journals, the supervisor of more than two dozens of PhD candidates in various aspects of semiconductor nanoscience and many of them are working as faculty members in different Universities and Academic Institutions respectively. He is the PRINCIPAL CO-AUTHOR of NINE RESEARCH MONOGRAPHS from SPRINGER-VERLAG, GERMANY among them one from Springer series in Solid State Sciences (vol. 170), three are from Springer Series in Materials Science (vols 116, 137 and 167), one from Springer Series in Nano Structure Science and Technology and four from Springer Tracts in Modern Physics (vols. 255, 260, 262 and 265) respectively. He is the SOLO AUTHOR of one research monograph (vol. 7) in the Series on the Foundations of Natural Science and Technology of World Scientific Publishing Co. Singapore. He is the PRINCIPAL EDITOR of the two edited monographs in Series in Nanotechnology Science of NOVA , USA. He joined as Lecturer in the Institute of Radio Physics and Electronics of the University of Calcutta in 1983, Reader in the Department of Electronics and Telecommunication Engineering of Jadavpur University in 1987, Professor in the Department of Electronic Science of the University of Calcutta in 1994, Head of The Department of Electronics and Communication Engineering of National Institute of Technology, Agartala, Tripura from March 2012 and later on acted as Dean (Faculty Welfare) of the said Institute. From January 2015 he has joined as Research Director and Senior Professor in Institute of Engineering and Management, Kolkata. HE WAS AT THE TOP OF THE MERIT LIST IN ALL THE ABOVE CASES. THE ALL INDIAN COUNCIL FOR TECHNICAL EDUCATION HAS SELECTED THE FIRST RESEARCH AND DEVELOPMENT PROJECT IN HIS LIFE FOR THE BEST PROJECT AWARD IN ELECTRONICS AND SECOND BEST RESEARCH PROJECT AWARD CONSIDERING ALL THE BRANCHES OF ENGINEERING FOR THE YEAR 2006. IN 2011, THE UNIVERSITY GRANT COMMISSION AWARDED A RESEARCH PROJECT TO HIM AND PLACED HIM AT THE TOP IN THE LIST OF AWARDEES. His forty years long teaching experience include Non-Linear Circuits, Semiconductor Science, Indeterminate Structures, Non-Linear Mechanics, Collision Theory, Oscillatory Integrals and present research interests are Semiconductor Nanoscience and Analytic Number Theory respectively. His brief CV has been enlisted in many biographical references of

USA and UK. Professor Ghatak is a firm believer of the concept of theoretical minimum of Landau and being the ardent follower of GITA (well-known as the theory of KRIYA YOGA), his vision, mission and passion is to stay deeply in solitude to become a permanent member of the endless world of self-realization. RAMANUJAN AND RABINDRANATH ARE THE LIFE TIME HEROES of Professor Ghatak.

Creativity is neither brought nor taught
but can only be caught
from the master soul by the prepared mind

Contents

Chapter 1

The EM in Quantum Wells (QWs) of Heavily Doped (HD) Non-Parabolic Materials

1.1 Introduction

In recent years, with the advent of fine lithographical methods [1, 2] molecular beam epitaxy [3], organometallic vapor-phase epitaxy [4], and other experimental techniques, the restriction of the motion of the carriers of bulk materials in one (QWs, doping super-lattices, accumulation, and inversion layers), two (nanowires) and three (quantum dots, magneto-size quantized systems, magneto inversion layers, magneto accumulation layers, quantum dot super-lattices, magneto QW super-lattices, and magneto doping superlattices) dimensions have in the last few years, attracted much attention not only for their potential in uncovering new phenomena in nano-science but also for their interesting quantum device applications [5–8]. In QWs, the restriction of the motion of the carriers in the direction normal to the film (say, the z direction) may be viewed as carrier confinement in an infinitely deep 1D rectangular potential well, leading to quantization [known as quantum size effect (QSE)] of the wave vector of the carriers along the direction of the potential well, allowing 2D carrier transport parallel to the surface of the film representing new physical features not exhibited in bulk semiconductors [9–13]. The low-dimensional hetero-structures based on various materials are widely investigated because of the enhancement of carrier mobility [14]. These properties make such structures suitable for applications in QWs lasers [15], hetero-junction FETs [16,

17], high-speed digital networks [18–21], high-frequency microwave circuits [22], optical modulators [23], optical switching systems [24], and other devices. The constant energy 3D wave-vector space of bulk semiconductors becomes 2D wave-vector surface in QWs due to dimensional quantization. Thus, the concept of reduction of symmetry of the wave-vector space and its consequence can unlock the physics of low-dimensional structures. In this chapter, we study the EM in QWs of HD non-parabolic semiconductors having different band structures in the presence of Gaussian band tails. At first we shall investigate the EM in QWs of HD nonlinear optical compounds which are being used in nonlinear optics and light emitting diodes [25]. The quasi-cubic model can be used to investigate the symmetric properties of both the bands at the zone center of wave vector space of the same compound. Including the anisotropic crystal potential in the Hamiltonian, and special features of the nonlinear optical compounds, Kildal [26] formulated the electron dispersion law under the assumptions of isotropic momentum matrix element and the isotropic spin-orbit splitting constant, respectively, although the anisotropies in the two aforementioned band constants are the significant physical features of the said materials [27–29]. In Section 1.2.1, the EM in QWs of HD nonlinear optical semiconductors has been investigated on the basis of newly formulated HD dispersion relation of the said compound by considering the combined influence of the anisotropies of the said energy band constants together with the inclusion of the crystal field splitting respectively within the framework of $\vec{k} \cdot \vec{p}$ formalism. The III-V compounds find applications in infrared detectors [30], quantum dot light emitting diodes [31], quantum cascade lasers [32], QWs wires [33], optoelectronic sensors [34], high electron mobility transistors [35], etc. The electron energy spectrum of III-V semiconductors can be described by the three- and two-band models of Kane [36, 37], together with the models of Stillman *et al.* [38], Newson and Kurobe [39] and, Palik *et al.* [40] respectively. In this context it may be noted that the ternary and quaternary compounds enjoy the singular position in the entire spectrum of optoelectronic materials. The ternary alloy $Hg_{1-x}Cd_xTe$ is a classic narrow gap compound. The band gap of this ternary alloy can be varied to cover the spectral range from 0.8 to over 30 μm [41] by adjusting the alloy composition. $Hg_{1-x}Cd_xTe$ finds extensive applications in infrared detector materials and photovoltaic detector arrays in the 8–12μm wave bands [42]. The above uses have generated the $Hg_{1-x}Cd_xTe$ technology for the experimental realization of high mobility single crystal with specially prepared surfaces. The same compound has emerged to be the optimum choice for illuminating the narrow sub-band physics because the relevant material constants can easily be experimentally measured [43]. Besides, the quaternary alloy $In_{1-x}Ga_xAs_yP_{1-y}$ lattice matched to InP, also finds wide use in the fabrication of avalanche photo-detectors [44], hetero-junction lasers [45], light emitting diodes [46] and avalanche photodiodes [47], field effect transistors, detectors, switches, modulators, solar cells, filters, and new types of integrated optical devices are made from the quaternary systems [48]. It may be noted that all types of band models as discussed for III-V semiconductors are also applicable for ternary and quaternary compounds. In Section 1.2.2, the EM in QWs of HD III-V, ternary and quaternary semiconductors has been studied in accordance with the corresponding HD formulation of the band structure and the simplified results for wide gap materials having parabolic energy bands under certain limiting conditions have further been demonstrated as a special case in the absence of band-tails and thus confirming the compatibility test. The II-VI semiconductors are being used in nano-ribbons, blue green diode lasers, photosensitive thin films, infrared detectors, ultra-high-speed bipolar transistors, fiber optic communications, microwave devices, solar cells, semiconductor gamma-ray detector arrays, semiconductor detector gamma camera and allow for a greater density of data storage on optically addressed compact discs [49–56]. The carrier energy spectra in II-VI compounds are defined by the Hopfield model [57] where the splitting of the two-spin states by the spin-orbit coupling and the crystalline field has been taken into account. Section 1.2.3 contains the investigation of the EM in QWs of HD II-VI compounds.

Lead Chalcogenides (PbTe, PbSe, and PbS) are IV–VI non-parabolic semiconductors whose studies over several decades have been motivated by their importance in infrared IR detectors, lasers, light-emitting devices, photo-voltaic, and high temperature thermoelectrics [58–62]. PbTe, in particular, is the end compound of several ternary and quaternary high performance high temperature thermoelectric materials [63–67]. It has been used not only as bulk but also as films [68–71], QWs [72] super-lattices [73, 74] nanowires [75] and colloidal and embedded nano-crystals [76–79], and PbTe films doped with various impurities have also been investigated [80–87]. These studies revealed some of the interesting features that had been seen in bulk PbTe, such as Fermi level pinning and, in the case of superconductivity [88]. In Section 1.2.4, the 2D EM in QWs of HD IV–VI semiconductors has been studied taking PbTe, PbSe, and PbS as examples. The stressed semiconductors are being investigated for strained silicon transistors, quantum cascade lasers, semiconductor strain gages, thermal detectors, and strained-layer structures [89–92]. The EM in QWs of HD stressed compounds (taking stressed n-InSb as an example) has been investigated in Section 1.2.5. The vacuum deposited Tellurium (Te) has been used as the semiconductor layer in thin-film transistors (TFT) [93] which is being used in CO_2 laser detectors [94], electronic imaging, strain sensitive devices [95, 96], and multichannel Bragg cell [97]. Section 1.2.6 contains the investigation of EM in QWs of HD Tellurium. The n-Gallium Phosphide (n-GaP) is being used in quantum dot light emitting diode [98], high efficiency yellow solid state lamps, light sources, high peak current pulse for high gain tubes. The green and yellow light emitting diodes made of nitrogen-doped n-GaP possess a longer device life at high drive currents [99–101]. In Section 1.2.7, the EM in QWs of HD n-GaP has been studied. The Platinum Antimonide ($PtSb_2$) finds application in device miniaturization, colloidal nanoparticle synthesis, sensors and detector materials and thermo-photovoltaic devices [102–104]. Section 1.2.8 explores the EM in QWs of HD $PtSb_2$. Bismuth telluride (Bi_2Te_3) was first identified as a material for thermoelectric refrigeration in 1954 [105] and its physical properties were later improved by the addition of bismuth selenide and antimony telluride to form solid solutions. The alloys of Bi_2Te_3 are useful compounds for the thermoelectric industry and have been investigated in the literature [106–110]. In Section 1.2.9, the EM in QWs of HD Bi_2Te_3 has been considered. The usefulness of elemental semiconductor Germanium is already well known since the inception of transistor technology and, it is also being used in memory circuits, single photon detectors, single photon avalanche diode, ultrafast optical switch, THz lasers and THz spectrometers [111–114]. In Section 1.2.10, the EM has been studied in QWs of HD Ge. Gallium Antimonide (GaSb) finds applications in the fiber optic transmission window, hetero-junctions, and QWs. A complementary hetero-junction field effect transistor in which the channels for the p-FET device and the n-FET device forming the complementary FET are formed from GaSb. The band gap energy of GaSb makes it suitable for low power operation [115–121]. In Section 1.2.11, the EM in QWs of HD GaSb has been studied. Section 1.3 contains the result and discussions pertaining to this chapter. Section 1.4 contains a dozen open research problems.

1.2 Theoretical Background

1.2.1 *The EM in Quantum Wells (QWs) of HD nonlinear optical materials*

The form of **k·p** matrix for nonlinear optical compounds can be expressed extending Bodnar [27] as

$$H = \begin{bmatrix} H_1 & H_2 \\ H_2^+ & H_1 \end{bmatrix} \tag{1.1}$$

where,

$$H_1 \equiv \begin{bmatrix} E_{g0} & 0 & P_{\|}k_z & 0 \\ 0 & (-2\Delta_{\|}/3) & (\sqrt{2}\Delta_{\perp}/3) & 0 \\ P_{\|}k_z & (\sqrt{2}\Delta_{\perp}/3) & -\left(\delta + \dfrac{1}{3}\Delta_{\|}\right) & 0 \\ 0 & 0 & 0 & 0 \end{bmatrix},$$

$$H_2 \equiv \begin{bmatrix} 0 & -f_{,+} & 0 & f_{,-} \\ f_{,+} & 0 & 0 & 0 \\ 0 & 0 & 0 & 0 \\ f_{,+} & 0 & 0 & 0 \end{bmatrix}$$

in which E_{g0} is the band gap in the absence of any field, $P_{||}$ and $P_{\perp}$ are the momentum matrix elements parallel and perpendicular to the direction of crystal axis respectively, δ is the crystal field splitting constant, $\Delta_{||}$ and $\Delta_{\perp}$ are the spin-orbit splitting constants parallel and perpendicular to the C-axis respectively, $f_{,\pm} \equiv (P_{\perp}/\sqrt{2})(k_x \pm ik_y)$ and $i = \sqrt{-1}$. Thus, neglecting the contribution of the higher bands and the free electron term, the diagonalization of the above matrix leads to the dispersion relation of the conduction electrons in bulk specimens of nonlinear optical materials as

$$\gamma(E) = f_1(E)k_s^2 + f_2(E)k_z^2 \tag{1.2}$$

where

$$\gamma(E) \equiv E(E + E_{g0})\left[(E + E_{g0})(E + E_{g0} + \Delta_{||}) + \delta\left(E + E_{g0} + \frac{2}{3}\Delta_{||}\right) + \frac{2}{9}(\Delta_{||}^2 - \Delta_{\perp}^2)\right],$$

E is the total energy of the electron as measured from the edge of the conduction band in the vertically upward direction in the absence of any quantization, $k_s^2 = k_x^2 + k_y^2$,

$$f_1(E) \equiv \frac{\hbar^2 E_{g0}(E_{g0} + \Delta_{\perp})}{\left[2m_{\perp}^*\left(E_{g0} + \frac{2}{3}\Delta_{\perp}\right)\right]}\left[\delta\left(E + E_{g0} + \frac{1}{3}\Delta_{||}\right) + (E + E_{g0}) \times \left(E + E_{g0} + \frac{2}{3}\Delta_{||}\right) + \frac{1}{9}(\Delta_{||}^2 - \Delta_{||}^2)\right],$$

$$f_2(E) \equiv \frac{\hbar^2 E_{g0} + (E_{g0} + \Delta_{||})}{\left[2m_{||}^*\left(E_{g0} + \frac{2}{3}\Delta_{||}\right)\right]}\left[(E + E_{g0})\left(E + E_{g0} + \frac{2}{3}\Delta_{||}\right)\right],$$

$\hbar = \mathrm{h}/2\pi$, h is Planck's constant and $m_{||}^*$ and $m_{\perp}^*$ are the longitudinal and transverse effective electron masses at the edge of the conduction band respectively.

Thus the generalized unperturbed electron energy spectrum for the bulk specimens of the nonlinear optical materials in the absence of band tails can be expressed following (1.2) as

$$\frac{\hbar^2 k_z^2}{2m_{||}^*} + \left(\frac{b_{||}}{b_\perp}\frac{c_\perp}{c_{||}}\right)\frac{\hbar^2 k_s^2}{2m_\perp^*}$$
$$= \left\{\frac{E(\alpha E+1)(b_{||}E+1)}{(c_{||}E+1)} + \frac{\alpha b_{||}}{c_{||}}\left[\delta E + \frac{2}{9}(\Delta_{||}^2 - \Delta_\perp^2)\right]\right.$$
$$\left. - \left(\frac{2}{9}\right)\frac{\alpha b_{||}}{c_{||}}\frac{\left(\Delta_{||}^2 - \Delta_\perp^2\right)}{(c_{||}E+1)}\right\} - \left(\frac{\hbar^2 k_s^2}{2m_\perp^*}\right)\left\{\left(\frac{b_{||}}{b_\perp}\frac{c_\perp}{c_{||}}\right)\right.$$
$$\times\left[\left(\frac{\delta}{2} + \frac{\Delta_{||}^2 - \Delta_\perp^2}{6\Delta_{||}}\right)\frac{\alpha_{||}}{\alpha_{||}E+1}\right.$$
$$\left.\left. + \left(\frac{\delta}{2} - \left\{\frac{\Delta_{||}^2 - \Delta_\perp^2}{6\Delta_{||}}\right\}\right)\frac{c_{||}}{c_{||}E+1}\right]\right\} \tag{1.3}$$

where, $b_{||} \equiv 1/(E_g + \Delta_{||})$, $c_\perp \equiv 1/(E_g + \frac{2}{3}\Delta_\perp)$, $b_\perp \equiv 1/(E_g + \Delta_\perp)$, $c_{||} \equiv 1/(E_g + \frac{2}{3}\Delta_{||})$ and $\alpha \equiv 1/E_g$

The Gaussian distribution $F(V)$ of the impurity potential is given by [122]

$$F(V) = (\pi\eta_g^2)^{1/2}\exp(-V^2/\eta_g^2) \tag{1.4}$$

where, η_g is the impurity screening potential. It appears from (1.4) that the variance parameter η_g is not equal to zero, but the mean value is zero. Further, the impurities are assumed to be uncorrelated and the band mixing effect has been neglected in this simplified theoretical formalism.

We have to average the kinetic energy in the order to obtain the EM in nonlinear optical materials in the presence of band tails. Using (1.3) and (1.4), we get

$$\left[\frac{\hbar^2 k_z^2}{2m_{||}^*}\int_{-\infty}^{E} F(V)dV\right] + \left[\left(\frac{b_{||}}{b_\perp}\frac{c_\perp}{c_{||}}\right)\frac{\hbar^2 k_s^2}{2m_\perp^*}\int_{-\infty}^{E} F(V)dV\right]$$

$$= \left\{\int_{-\infty}^{E}\frac{(E-V)\left[\alpha(E-V)+1\right]\left[b_{||}(E-V)+1\right]}{\left[c_{||}(E-V)+1\right]}F(V)dV\right.$$

$$+\frac{\alpha b_{||}}{c_{||}}\left[\delta\int_{-\infty}^{E}(E-V)F(V)dV+\frac{2}{9}\left(\Delta_{||}^2-\Delta_{\perp}^2\right)\int_{-\infty}^{E}F(V)dV\right]$$

$$-\left(\frac{2}{9}\right)\frac{\alpha b_{||}}{c_{||}}\left(\Delta_{||}^2-\Delta_{\perp}^2\right)\int_{-\infty}^{E}\frac{F(V)dV}{[c_{||}(E-V)+1]}\Bigg\}-\left(\frac{\hbar^2k_s^2}{2m_{\perp}^*}\right)$$

$$\times\left\{\left(\frac{b_{||}}{b_{\perp}}\frac{c_{\perp}}{c_{||}}\right)\left[\left(\frac{\delta}{2}+\frac{\Delta_{||}^2-\Delta_{\perp}^2}{6\Delta_{||}}\right)\alpha\int_{-\infty}^{E}\frac{F(V)dV}{[\alpha(E-V)+1]}\right.\right.$$

$$\left.\left.+\left(\frac{\delta}{2}-\frac{\Delta_{||}^2-\Delta_{\perp}^2}{6\Delta_{||}}\right)c_{||}\int_{-\infty}^{E}\frac{F(V)dV}{[c_{||}(E-V)+1]}\right]\right\} \tag{1.5}$$

The (1.5) can be rewritten as

$$\frac{\hbar^2k_z^2}{2m_{||}^*}I(1)+\left(\frac{b_{||}}{b_{\perp}}\frac{c_{\perp}}{c_{||}}\right)\frac{\hbar^2k_s^2}{2m_{\perp}^*}I(1)$$

$$=\left\{I_3\left(c_{||}\right)+\frac{\alpha b_{||}}{c_{||}}\left[\delta I\left(4\right)+\frac{2}{9}(\Delta_{||}^2-\Delta_{\perp}^2)I(1)\right]\right.$$

$$\left.-\left(\frac{2}{9}\right)\frac{\alpha b_{||}}{c_{||}}\left(\Delta_{||}^2-\Delta_{\perp}^2\right)I\left(c_{||}\right)\right\}$$

$$-\left(\frac{\hbar^2k_s^2}{2m_{\perp}^*}\right)\left\{\left(\frac{b_{||}}{b_{\perp}}\frac{c_{\perp}}{c_{||}}\right)\left[\left(\frac{\delta}{2}+\frac{\Delta_{||}^2-\Delta_{\perp}^2}{6\Delta_{||}}\right)\alpha I\left(\alpha\right)\right.\right.$$

$$\left.\left.+\left(\frac{\delta}{2}-\left\{\frac{\Delta_{||}^2-\Delta_{\perp}^2}{6\Delta_{||}}\right\}\right)c_{||}I\left(c_{||}\right)\right]\right\} \tag{1.6}$$

where,

$$I(1)\equiv\int_{-\infty}^{E}F(V)dV \tag{1.7}$$

$$I_3\left(c_{||}\right) \equiv \int_{-\infty}^{E} \frac{(E-V)\left[\alpha(E-V)+1\right]\left[b_{||}(E-V)+1\right]}{\left[c_{||}(E-V)+1\right]} F(V)dV \tag{1.8}$$

$$I(4) \equiv \int_{-\infty}^{E} (E-V)F(V)dV \tag{1.9}$$

$$I\left(\alpha\right) \equiv \int_{-\infty}^{E} \frac{F(V)dV}{\left[\alpha(E-V)+1\right]} \tag{1.10}$$

From (1.10) we can obtain $I(c_{||})$ by replacing α by c_{11} in both sides of (1.10). Substituting $E-V \equiv x$ and $x/\eta_g \equiv t_0$, we get from (1.7)

$$I(1) = (\exp(-E^2/\eta_g^2)/\sqrt{\pi}) \int_{0}^{\infty} \exp[-t_0^2 + (2Et_0/\eta_g)]dt_0$$

Thus,

$$I(1) = \left[\frac{1 + Erf(E/\eta_g)}{2}\right] \tag{1.11}$$

where, $Erf(E/\eta_g)$ is the error function of (E/η_g).

From (1.9), one can write

$$I(4) = \left(1/\eta_g\sqrt{\pi}\right) \int_{-\infty}^{E} (E-V)\exp(-V^2/\eta_g^2)dV$$

$$= \frac{E}{2}[1 + Erf(E/\eta_g)] - \left\{\frac{1}{\sqrt{\pi\eta_g^2}} \int_{-\infty}^{E} V\exp(-V^2/\eta_g^2)dV\right\} \tag{1.12}$$

After computing this simple integration, one obtains

Thus,

$$I(4) = \eta_g \exp(-E^2/\eta_g^2)(2\sqrt{\pi})^{-1} + \frac{E}{2}(1 + Erf(E/\eta_g)) = \gamma_0(E, \eta_g) \tag{1.13}$$

From (1.10), we can write

$$I(\alpha) = \frac{1}{\sqrt{\pi\eta_g^2}} \int_{-\infty}^{E} \frac{\exp(-V^2/\eta_g^2)dV}{[\alpha(E-V)+1]} \tag{1.14}$$

When, $V \to \pm\infty$, $\dfrac{1}{[\alpha(E-V)+1]} \to 0$ and $\exp(-V^2/\eta_g^2) \to 0$;

Thus (1.14) can be expressed as

$$I(\alpha) = (1/\alpha\eta_g\sqrt{\pi})\int_{-\infty}^{\infty} \exp(-t^2)(u-t)^{-1}dt \tag{1.15}$$

where, $\dfrac{V}{\eta_g} \equiv t$ and $u \equiv \left(\dfrac{1+\alpha E}{\alpha\eta}\right)$.

It is well known that [123, 124]

$$W(Z) = (i/\pi)\int_{-\infty}^{\infty} (Z-t)^{-1}\exp(-t^2)dt \tag{1.16}$$

In which $i = \sqrt{-1}$ and Z, in general, is a complex number.

We also know [123, 124],

$$W(Z) = \exp(-Z^2)Erf\,c(-iZ) \tag{1.17}$$

where, $Erf\,c(Z) \equiv 1 - Erf(Z)$.

Thus, $Erf\,c(-iu) = 1 - Erf(-iu)$

Since, $Erf(-iu) = -Erf(iu)$

Therefore, $Erf\,c(-iu) = 1 + Erf(iu)$.

Thus,

$$I(\alpha) = [-i\sqrt{\pi}/\alpha\eta_g]\exp(-u^2)[1 + Erf(iu)] \tag{1.18}$$

We also know that [123, 124]

$$Erf(x+iy) = Erf(x) + \left(\frac{e^{-x^2}}{2\pi x}\right)\left[(1-\cos(2xy)) + i\sin(2xy)\right.$$
$$\left. + \frac{2}{\pi}e^{-x^2}\sum_{p=1}^{\infty}\frac{\exp(-p^2/4)}{(p^2+4x^2)}\right][f_p(x,y) + ig_p(x,y) + \varepsilon(x,y)] \tag{1.19}$$

where,

$$f_p(x,y) \equiv [2x - 2x\cosh(py)\cos(2xy) + p\sinh(py)\sin(2xy)],$$

$$g_p(x,y) \equiv [2x\cosh(py)\sin(2xy) + p\sinh(py)\cos(2xy)], |\varepsilon(x,y)|$$

$$\approx 10^{-16}|Erf\,(x+iy)|$$

Substituting $x = 0$ and $y = u$ in (2.19), one obtains,

$$Erf\,(iu) = \left(\frac{2i}{\pi}\right)\sum_{p=1}^{\infty}\left\{\frac{\exp(-p^2/4)}{p}\sinh(pu)\right\} \tag{1.20}$$

Therefore, one can write

$$I(\alpha) = C_{21}(\alpha, E, \eta_g) - iD_{21}(\alpha, E, \eta_g) \tag{1.21}$$

where,

$$C_{21}(\alpha, E, \eta_g) \equiv \left[\frac{2}{\alpha\eta_g\sqrt{\pi}}\right]\exp\left(-u^2\right)$$

$$\times\left[\sum_{p=1}^{\infty}\left\{\frac{\exp\left(-p^2/4\right)}{p}\sinh\left(pu\right)\right\}\right] \quad \text{and}$$

$$D_{21}\left(\alpha, E, \eta_g\right) \equiv \left[\frac{\sqrt{\pi}}{\alpha\eta_g}\exp\left(-u^2\right)\right].$$

(1.21) consists of both real and imaginary parts and therefore, $I(\alpha)$ is complex, which can also be proved by using the method of analytic continuation of Complex Analysis.

The integral $I_3(c_{||})$ in (1.8) can be written as

$$I_3(c_{||}) = \left(\frac{\alpha b_{||}}{c_{||}}\right)I(5) + \left(\frac{\alpha c_{||} + b_{||}c_{||} - \alpha b_{||}}{c_{||}^2}\right)I(4)$$

$$+\frac{1}{c_{||}}\left(1-\frac{\alpha}{c_{||}}\right)\left(1-\frac{b_{||}}{c_{||}}\right)I(1)$$

$$-\left\{\frac{1}{c_{||}}\left(1-\frac{\alpha}{c_{||}}\right)\left(1-\frac{b_{||}}{c_{||}}\right)I\left(c_{||}\right)\right\} \tag{1.22}$$

where

$$I(5) \equiv \int_{-\infty}^{E} (E-V)^2 F(V) dV \tag{1.23}$$

From (1.23) one can write

$$I(5) = \frac{1}{\sqrt{\pi\eta_g^2}} \left[E^2 \int_{-\infty}^{E} \exp\left(\frac{-V^2}{\eta_g^2}\right) dV - 2E \int_{-\infty}^{E} V \exp\left(\frac{-V^2}{\eta_g^2}\right) dV \right.$$
$$\left. + \int_{-\infty}^{E} V^2 \exp\left(\frac{-V^2}{\eta_g^2}\right) dV \right]$$

The evaluations of the component integrals lead us to write

$$I(5) = \frac{\eta_g E}{2\sqrt{\pi}} \exp\left(\frac{-E^2}{\eta_g^2}\right) + \frac{1}{4}\left(\eta_g^2 + 2E^2\right)\left[1 + Erf\left(\frac{E}{\eta_g}\right)\right]$$
$$= \theta_0(E, \eta_g) \tag{1.24}$$

Thus combining the aforementioned equations, $I_3(c_{||})$ can be expressed as

$$I_3(c_{||}) = A_{21}(E, \eta_g) + iB_{21}(E, \eta_g) \tag{1.25}$$

where,

$$A_{21}(E, \eta) \equiv \left[\frac{\alpha b_{||}}{c_{||}} \left[\frac{\eta_g E}{2\sqrt{\pi}} \exp\left(\frac{-E^2}{\eta_g^2}\right) + \frac{1}{4}(\eta_g^2 + 2E^2) \right.\right.$$
$$\left.\left. \times \left\{ 1 + Erf\left(\frac{E}{\eta_g}\right) \right\} \right] + \left[\frac{\alpha c_{||} + b_{||}c_{||} - \alpha b_{||}}{c_{||}^2} \right]$$
$$\times \left\{ \frac{E}{2}\left[1 + Erf\left(E/\eta\right)\right] + \frac{\eta_g \exp\left(-E^2/\eta_g^2\right)}{2\sqrt{\pi}} \right\}$$
$$+ \frac{1}{c_{||}}\left(1 - \frac{\alpha}{c_{||}}\right)\left(1 - \frac{b_{||}}{c_{||}}\right)\frac{1}{2}\left[1 + Erf\left(E/\eta\right)\right]$$

$$-\left\{\frac{2}{c_{||}^2\eta_g\sqrt{\pi}}\left(1-\frac{\alpha}{c_{||}}\right)\left(1-\frac{b_{||}}{c_{||}}\right)\exp(-u_1^2)\right\}$$

$$\times\left[\sum_{p=1}^{\infty}\left\{\frac{\exp\left(-p^2/4\right)}{p}sinh\left(pu_1\right)\right\}\right],$$

$$u_1 \equiv \left[\frac{1+c_{||}E}{c_{||}\eta_g}\right] \quad \text{and}$$

$$B_{21}(E,\eta_g) \equiv \frac{\sqrt{\pi}}{c_{||}^2\eta_g}\left(1-\frac{\alpha}{c_{||}}\right)\left(1-\frac{b_{||}}{c_{||}}\right)\exp(-u_1^2).$$

Therefore, the combination of all the appropriate integrals together with algebraic manipulations leads to the expression of the dispersion relation of the conduction electrons of HD nonlinear optical materials forming Gaussian band tails as

$$\frac{\hbar^2k_z^2}{2m_{||}^*T_{21}(E,\eta_g)}+\frac{\hbar^2k_s^2}{2m_{\perp}^*T_{22}(E,\eta_g)}=1 \tag{1.26}$$

where, $T_{21}(E,\eta_g)$ and $T_{22}(E,\eta_g)$ have both real and complex parts and are given by

$$T_{21}(E,\eta_g) \equiv \left[T_{27}(E,\eta_g)+iT_{28}(E,\eta_g)\right],$$

$$T_{27}(E,\eta_g) \equiv \left[\frac{T_{23}(E,\eta_g)}{T_5(E,\eta_g)}\right],$$

$$T_{23}(E,\eta_g) \equiv \left[A_{21}(E,\eta_g)+\frac{\alpha b_{||}}{c_{||}}\left[\delta\gamma_0(E,\eta_g)\right.\right.$$

$$\left.+\frac{1}{9}(\Delta_{||}^2-\Delta_{\perp}^2)[1+Erf(E/\eta_g)]\right]$$

$$\left.-\left\{\frac{2}{9}\left(\frac{\alpha b_{||}}{c_{||}}\right)\left(\Delta_{||}^2-\Delta_{\perp}^2\right)G_{21}\left(c_{||},E,\eta_g\right)\right\}\right],$$

$$G_{21}(E,\eta_g) \equiv \frac{2}{c_{||}\eta_g\sqrt{\pi}}\exp\left(-u_1^2\right)\sum_{p=1}^{\infty}\left\{\frac{\exp\left(-p^2/4\right)}{p} sinh(pu_1)\right\},$$

$$T_5(E,\eta_g) \equiv \frac{1}{2}\left[1+Erf(E/\eta_g)\right], \quad T_{28}(E,\eta_g) \equiv \left[\frac{T_{24}(E,\eta_g)}{T_5(E,\eta_g)}\right],$$

$$T_{24}(E,\eta_g) \equiv \left[B_{21}(E,\eta_g) + \frac{2}{9}\frac{\alpha b_{||}}{c_{||}}\left(\Delta_{||}^2-\Delta_{\perp}^2\right)H_{21}\left(\mathrm{c}_{||},E,\eta_g\right)\right],$$

$$H_{21}\left(\mathrm{c}_{||},E,\eta_g\right) \equiv \left[\frac{\sqrt{\pi}}{\eta_g \mathrm{c}_{||}}\exp(-u_1^2)\right],$$

$$T_{22}(E,\eta_g) \equiv [T_{29}(E,\eta_g) + iT_{30}(E,\eta_g)],$$

$$T_{29}(E,\eta_g) \equiv \frac{T_{23}(E,\eta_g)T_{25}(E,\eta_g) - T_{24}(E,\eta_g)T_{26}(E,\eta_g)}{[(T_{25}(E,\eta_g))^2 + (T_{26}(E,\eta_g))^2]},$$

$$\begin{aligned}T_{25}(E,\eta_g) \equiv \Bigg[&\left(\frac{b_{||}}{b_{\perp}}\frac{c_{\perp}}{c_{||}}\right)\frac{1}{2}\left[1+Erf\left(\frac{E}{\eta_g}\right)\right]\\ &+\left(\frac{b_{||}}{b_{\perp}}\frac{c_{\perp}}{c_{||}}\right)\left(\frac{\delta}{2}+\left[\frac{\Delta_{||}^2-\Delta_{\perp}^2}{6\Delta_{||}}\right]\right)\alpha_{||}C_{21}(\alpha_{||},E,\eta_g)\\ &+\left(\frac{b_{||}c_{\perp}}{b_{\perp}}\right)\left(\frac{\delta}{2}-\left[\frac{\Delta_{||}^2-\Delta_{\perp}^2}{6\Delta_{||}}\right]\right)G_{21}\left(\alpha_{||},E,\eta_g\right)\Bigg],\end{aligned}$$

$$C_{21}(\alpha,E,\eta_g) \equiv \left[\frac{2}{\alpha\sqrt{\pi}\eta_g}\exp\left(-u^2\right)\left[\sum_{p=1}^{\infty}\frac{\exp\left(-p^2/4\right)}{p} sinh(pu)\right]\right],$$

$$\begin{aligned}T_{26}(E,\eta_g) \equiv &\left(\frac{b_{||}}{b_{\perp}}\frac{c_{\perp}}{c_{||}}\right)\left(\frac{\delta}{2}-\frac{\Delta_{||}^2-\Delta_{\perp}^2}{6\Delta_{||}}\right)\alpha D_{21}\left(\alpha,E,\eta_g\right)\\ &+\frac{b_{||}c_{\perp}}{b_{\perp}}\left(\frac{\delta}{2}-\frac{\Delta_{||}^2-\Delta_{\perp}^2}{6\Delta_{||}}\right)H_{21}\left(c_{||},E,\eta_g\right)\end{aligned}$$

And

$$T_{30}(E,\eta_g) \equiv \frac{T_{24}(E,\eta_g)T_{25}(E,\eta_g) + T_{23}(E,\eta_g)T_{26}(E,\eta_g)}{\left[(T_{25}(E,\eta_g))^2 + (T_{26}(E,\eta_g))^2\right]}$$

From (1.26), it appears that the energy spectrum in HD nonlinear optical materials is complex. **The complex nature of the electron dispersion law in HD materials occurs from the existence of the essential poles in the corresponding electron energy spectrum in the absence of band tails.** It may be noted that the complex band structures have already been studied for bulk materials and super lattices without heavy dopingand bears no relationship with the complex electron dispersion law as indicated by (1.26). The physical picture behind the formulation of the complex energy spectrum in HDS is the interaction of the impurity atoms in the tails with the splitting constants of the valance bands. More is the interaction; more is the prominence of the complex part than the other case. In the absence of band tails, $\eta_g \to 0$, and there is no interaction of the impurity atoms in the tails with the spin orbit constants. As a result, there exist no complex energy spectrum and (1.26) gets converted into (1.2) when $\eta_g \to 0$. Besides, the complex spectra are not related to same evanescent modes in the band tails and the conduction bands.

The transverse and the longitudinal EMs at the Fermi energy (E_{F_h}) of HD nonlinear optical materials can, respectively, be expressed as

$$m_\perp^*(E_{F_h},\eta_g) = m_\perp^*\{T_{29}(E,\eta_g)\}'\big|_{E=E_{F_h}} \tag{1.27}$$

and

$$m_{||}^*(E_{F_h},\eta_g) = m_{||}^*\{T_{27}(E,\eta_g)\}'\big|_{E=E_{F_h}} \tag{1.28}$$

where E_{F_k} is the Fermi energy of HDS in the presence of band tails as measured from the edge of the conduction band in the vertically upward direction in the absence of band tails and the primes denote the differentiations of the differentiable functions with respect to Fermi energy in the appropriate case.

In the absence of band tails $\eta_g \to 0$ and we get

$$m_{\perp}^{*}(E_F, 0) = \frac{\hbar^2}{2}\left[\frac{\psi_2(E)\{\psi_1(E)\}' - \psi_1(E)\{\psi_2(E)\}'}{\{\psi_2(E)\}^2}\right]\Bigg|_{E=E_F} \tag{1.29}$$

and

$$m_{\|}^{*}(E_F, 0) = \frac{\hbar^2}{2}\left[\frac{\psi_3(E)\{\psi_1(E)\}' - \{\psi_1(E)\}\{\psi_3(E)\}'}{\{\psi_3(E)\}^2}\right]\Bigg|_{E=E_F} \tag{1.30}$$

where E_F is the Fermi energy as measured from the edge of the conduction band in the vertically upward direction in the absence of any perturbation, $\psi_1(E) = \gamma(E)$, $\psi_2(E) = f_1(E)$ and $\psi_3(E) = f_2(E)$.

Comparing the aforementioned equations, one can infer that **the effective masses exist in the forbidden zone, which is impossible without the effect of band tailing. For semiconductors, in the absence of band tails the effective mass in the band gap is infinity.**

The density-of-states (DOS) function is given by

$$N_{HD}(E, \eta_g) = \frac{2g_v m_{\perp}^{*}\sqrt{2m_{\|}^{*}}}{3\pi^2\hbar^3} R_{11}(E, \eta_g)\cos\left[\psi_{11}(E, \eta_g)\right] \tag{1.31a}$$

where,

$$R_{11}(E, \eta_g)$$

$$\equiv \Bigg[\Bigg[\{T_{29}(E, \eta_g)\}'\sqrt{x(E, \eta_g)} + \frac{T_{29}(E, \eta_g)\{x(E, \eta_g)\}'}{2\sqrt{x(E, \eta_g)}}$$

$$- \{T_{30}(E, \eta_g)\}'\sqrt{y(E, \eta_g)} - \frac{T_{30}(E, \eta_g)\{y(E, \eta_g)\}'}{2\sqrt{y(E, \eta_g)}}\Bigg]^2$$

$$+ \Bigg[\{T_{29}(E, \eta_g)\}'\sqrt{y(E, \eta_g)} + \frac{T_{29}(E, \eta_g)\{y(E, \eta_g)\}'}{2\sqrt{y(E, \eta_g)}}$$

$$+\{T_{30}(E,\eta_g)\}'\sqrt{x(E,\eta_g)}-\frac{T_{30}(E,\eta_g)\{x(E,\eta_g)\}'}{2\sqrt{x(E,\eta_g)}}\Bigg]^2\Bigg]^{1/2},$$

$$x(E,\eta_g)\equiv\frac{1}{2}\left[T_{27}(E,\eta_g)+\sqrt{\{T_{27}(E,\eta_g)\}^2+\{T_{28}(E,\eta_g)\}^2}\right],$$

$$y(E,\eta_g)\equiv\frac{1}{2}\left[\sqrt{\{T_{27}(E,\eta_g)\}^2+\{T_{28}(E,\eta_g)\}^2}-T_{27}(E,\eta_g)\right]$$

and

$$\psi_{11}(E,\eta_g)\equiv\tan^{-1}\Bigg[\Bigg[\{T_{29}(E,\eta_g)\}'\sqrt{y(E,\eta_g)}+\frac{T_{29}(E,\eta_g)}{2\sqrt{y(E,\eta_g)}}$$

$$+\{T_{30}(E,\eta_g)\}'\sqrt{x(E,\eta_g)}+\frac{T_{30}\{x(E,\eta_g)\}'}{2\sqrt{x(E,\eta_g)}}\Bigg]$$

$$\times\Bigg[\{T_{29}(E,\eta_g)\}'\sqrt{x(E,\eta_g)}+\frac{T_{29}(E,\eta_g)\{x(E,\eta_g)\}'}{2\sqrt{x(E,\eta_g)}}$$

$$-\{T_{30}(E,\eta_g)\}'\sqrt{y(E,\eta_g)}+\frac{T_{30}\{y(E,\eta_g)\}'}{2\sqrt{y(E,\eta_g)}}\Bigg]^{-1}\Bigg].$$

The oscillatory nature of the DOS for HD nonlinear optical materials is apparent from (1.31a). For, $\psi_{11}(E,\eta_g)\geq\pi$, the cosine function becomes negative leading to the negative values of the DOS. The electrons cannot exist for the negative values of the DOS and therefore, this region is forbidden for electrons, which indicates that in the band tail, **there appears a new forbidden zone in addition to the normal band gap of the semiconductor.**

The use of (1.31a) leads to the expression of the electron concentration as

$$n_0=\frac{2g_v m_\perp^*\sqrt{2m_{\parallel}^*}}{3\pi^2\hbar^3}\left[I_{11}(E_{F_h},\eta_g)+\sum_{r=1}^{s}L(r)[I_{11}(E_{F_h},\eta_g)]\right]\quad(1.31b)$$

where,

$$I_{11}(E_{F_h},\eta_g) \equiv \left[T_{29}(E_{F_h},\eta_g)\sqrt{x(E_{F_h},\eta_g)} - T_{30}(E_{F_h},\eta_g)\sqrt{y(E_{F_h},\eta_g)}\right],$$

$$L(r) = 2(k_B T)^{2r}(1-2^{1-2r})\xi(2r)\frac{\partial^{2r}}{\partial E_{Fs}^{2r}},$$

r is the set of real positive integers whose upper s and $\xi(2r)$ is the Zeta function of order 2r [123, 124].

For dimensional quantization along z- direction, the dispersion relation of the 2D electrons in this case can be written following (1.26) as

$$\frac{\hbar^2(n_z\pi/d_z)^2}{2m_{||}^* T_{21}(E,\eta_g)} + \frac{\hbar^2 k_s^2}{2m_\perp^* T_{22}(E,\eta_g)} = 1 \tag{1.32}$$

where, $n_z(=1,2,3,\ldots)$ and d_z are the size quantum number and the nano-thickness along the z-direction respectively.

The general expression of the total 2D DOS ($N_{2DT}(E)$) can, in general, be expressed as

$$N_{2DT}(E) = \frac{2g_v}{(2\pi)^2}\sum_{n_z=1}^{n_{z\max}}\frac{\partial A(E,n_z)}{\partial E}H(E-E_{n_z}) \tag{1.33}$$

where $A(E,n_z)$ is the area of the constant energy 2D wave vector space and in this case it is for QWs, $H(E-E_{n_z})$ is the Heaviside step function and E_{n_z} is the corresponding sub-band energy. Using (1.32) and (1.33), the expression of the $N_{2DT}(E)$ for QWs of HD nonlinear optical materials can be written as

$$N_{2DT}(E) = \frac{m_\perp^* g_v}{\pi\hbar^2}\sum_{n_z=1}^{n_{z\max}} T'_{1D}(E,\eta_g,n_z)H(E-E_{n_Z D1}) \tag{1.34}$$

where, $T_{1D}(E,\eta_g,n_z) = [1-\frac{h^2(n_z\pi/d_z)^2}{2m_{||}^* T_{21}(E,\eta_g)}]T_{22}(E,\eta_g)$ and the sub-band energies $E_{n_z D1}$ in this case is given by the following equation

$$\frac{h^2(n_z\pi/d_z)^2}{2m_{||}^* T_{21}(E_{n_z D1},\eta_g)} = 1 \tag{1.35}$$

Thus we observe that both the total DOS and sub-band energies of QWs of HD nonlinear optical materials are complex due to the presence of the pole in energy axis of the corresponding materials in the absence of band tails.

The EM in this case is given by

$$m^*(E_{F1HD}, \eta_g, n_z) = m_\perp^*[\text{Real part of } T'_{1D}(E_{F1HD}, \eta_g, n_z)] \quad (1.36a)$$

Although the importance of the EM is already well-known in the literature [125–144], we observe that the EM is the function of size quantum number and the Fermi energy due to the combined influence of the crystal filed splitting constant and the anisotropic spin-orbit splitting constants respectively. Besides it is a function of η_g due to which the EM exists in the band gap, which is otherwise impossible.

Combining (1.34) with the Fermi-Dirac occupation probability factor, integrating between $E_{n_z D1}$ to infinity and applying the generalized Sommerfeld's lemma [143], the 2D carrier statistics in this case assumes the form

$$n_{2D} = \frac{m_\perp^* g_v}{\pi\hbar^2} \sum_{n_z=1}^{n_{z\max}} [\text{Real part of } [T_{1D}(E_{F1HD}, \eta_g, n_z) + T_{2D}(E_{F1HD}, \eta_g, n_z)]] \quad (1.36b)$$

where, $T_{2D}(E_{F1HD}, \eta_g, n_z) = \sum_{r=1}^{s} L(r)[T_{1D}(E_{F1HD}, \eta_g, n_z)]$, E_{F1HD} is the Fermi energy in the presence of size quantization of the QWs of HD non-linear optical materials as measured from the edge of the conduction band in the vertically upward direction in the absence of any perturbation.

In the absence of band-tails, the 2D EM in the x-y plane at the Fermi level, the total 2D DOS, the sub-band energy $E_{n_{z1}}$, of non-linear optical materials and the surface electron concentration can, respectively, be written as

$$\psi_1(E) = \psi_2(E)k_s^2 + \psi_3(E)(n_z\pi/d_z)^2 \quad (1.37)$$

$$m^*(E_{Fs}, n_z) = \left(\frac{\hbar^2}{2}\right)[\psi_2(E_{Fs})]^{-2}$$

$$\times \left[\psi_2(E_{Fs}) \left\{ \{\psi_1(E_{Fs})\}' - \{\psi_3(E_{Fs})\}' \left(\frac{n_z\pi}{d_z}\right)^2 \right\} - \left\{ \psi_1(E_{Fs}) - \psi_3(E_{Fs}) \left(\frac{n_z\pi}{d_z}\right)^2 \right\} \{\psi_2(E_{Fs})\}' \right] \tag{1.38}$$

$$N_{2DT}(E) = \left(\frac{g_v}{2\pi}\right) \sum_{n_z=1}^{n_{z\max}} [\psi_2(E)]^{-2} \times \left[\psi_2(E) \left\{ \{\psi_1(E)\}' - \{\psi_3(E)\}' \left(\frac{n_z\pi}{d_z}\right)^2 \right\} - \left\{ \psi_1(E) - \psi_3(E) \left(\frac{n_z\pi}{d_z}\right)^2 \right\} \{\psi_2(E)\}' \right] \times H(E - E_{n_{z1}}) \tag{1.39}$$

$$\psi_1(E_{n_{z1}}) = \psi_2(E_{n_{z1}})(n_z\pi/d_z)^2 \tag{1.40a}$$

$$n_{2D} = \frac{g_v}{2\pi} \sum_{n_z=1}^{n_{z\max}} [T_{51}(E_{Fs}, n_z) + T_{52}(E_{Fs}, n_z)] \tag{1.40b}$$

where

$$\psi_1(E) = \gamma(E),\ \psi_2(E) = f_1(E),\ \psi_3(E) = f_2(E)$$

E_{Fs} is the Fermi energy in the 2-D sized quantized material in the presence of size quantization and in the absence of band-tails as measured from the edge of the conduction band in the vertically upward direction in the absence of any quantization,

$$T_{51}(E_{Fs}, n_z) \equiv \left[\frac{\psi_1(E_{Fs}) - \psi_3(E_{Fs})\,(n_z\pi/d_z)^2}{\psi_2(E_{Fs})} \right] \quad \text{and}$$

$$T_{52}(E_{Fs}, n_z) \equiv \sum_{r=1}^{s} L(r)[T_{51}(E_{Fs}, n_z)].$$

In the absence of band-tails, the DOS for bulk specimens of non-linear optical materials is given by

$$D_0(E) = g_v(3\pi^2)^{-1}\psi_4(E) \tag{1.41a}$$

$$\psi_4(E) \equiv \left[\frac{3}{2}\frac{\sqrt{\psi_1(E)}\,[\psi_1(E)]'}{\psi_2(E)\sqrt{\psi_3(E)}} - \frac{[\psi_2(E)]'\,[\psi_1(E)]^{3/2}}{[\psi_2(E)]^2\sqrt{\psi_3(E)}} - \frac{1}{2}\frac{[\psi_3(E)]'\,[\psi_1(E)]^{3/2}}{\psi_2(E)\,[\psi_3(E)]^{3/2}}\right],$$

$$[\psi_1(E)]' \equiv \left[(2E+E_g)\psi_1(E)\,[E\,(E+E_g)]^{-1} + E(E+E_g)\left(2E+2E_g+\delta+\Delta_{||}\right)\right],$$

$$[\psi_2(E)]' \equiv \left[2m_\perp^*\left(E_g+\frac{2}{3}\Delta_\perp\right)\right]^{-1}\left[\hbar^2 E_g\,(E_g+\Delta_\perp)\right] \times \left[\delta+2E+2E_g+\frac{2}{3}\Delta_{||}\right]$$

and

$$[\psi_3(E)]' \equiv \left[2m_{||}^*\left(E_g+\frac{2}{3}\Delta_{||}\right)\right]^{-1}\left[\hbar^2 E_g\,(E_g+\Delta_{||})\right] \times \left[2E+2E_g+\frac{2}{3}\Delta_{||}\right]$$

Combining (1.41a) with the Fermi-Dirac occupation probability factor and using the generalized Somm*Erf*eld's lemma, the electron concentration can be written as

$$n_0 = g_v(3\pi^2)^{-1}[M(E_F)+N(E_F)] \tag{1.41b}$$

where, $M(E_F) \equiv \left[\frac{[\psi_1(E_F)]^{\frac{3}{2}}}{\psi_2(E_F)\sqrt{\psi_3(E_F)}}\right]$, E_F is the Fermi energy of the bulk specimen in the absence of band tails as measured from the edge of the conduction band in the vertically upward direction and $N(E_F) \equiv \sum_{r=1}^{s} L(r)M(E_F)$

1.2.2 *The EM in Quantum Wells (QWs) of HD III-V materials*

The EM of the conduction electrons of III-V compounds are described by the models of Kane (both three and two bands) [36,37], Stillman *et al.* [38], Newson [39] and Palik *et al.* [40] respectively. For the purpose of complete and coherent presentation and relative comparison, the EMs in QWs of HD III-V materials have also been investigated.

(a) The Three Band Model of Kane

Under the conditions, $\delta = 0$, $\Delta_{||} = \Delta_{\perp} = \Delta$ (isotropic spin orbit splitting constant) and $m_{||}^* = m_{\perp}^* = m_c$ (isotropic effective electron mass at the edge of the conduction band), (1.2) gets simplified as

$$\frac{\hbar^2 k^2}{2m_c} = I_{11}(E),\ I_{11}(E) \equiv= \frac{E(E+E_{g0})(E+E_{g0}+\Delta)(E_{g0}+\frac{2}{3}\Delta)}{E_{g0}(E_{g0}+\Delta)(E+E_{g0}+\frac{2}{3}\Delta)} \tag{1.42}$$

which is known as the three band model of Kane [37] and is often used to investigate the physical properties of III-V materials.

Under the said conditions, the HD electron dispersion law in this case can be written as

$$\frac{\hbar^2 k^2}{2m_c} = T_{31}(E,\eta_g) + iT_{32}(E,\eta_g) \tag{1.43}$$

where,

$$T_{31}(E,\eta_g) \equiv \left(\frac{2}{1+Erf\,(E/\eta_g)}\right)\left[\frac{\alpha b}{c}\theta_0(E,\eta_g) + \left[\frac{\alpha c + bc - \alpha b}{c^2}\right]\right.$$
$$\times\,\gamma_0(E,\eta_g) + \frac{1}{c}\left(1-\frac{\alpha}{c}\right)\left(1-\frac{b}{c}\right)\frac{1}{2}\left[1+Erf\left(\frac{E}{\eta_g}\right)\right]$$
$$-\frac{1}{c}\left(1-\frac{\alpha}{c}\right)\left(1-\frac{b}{c}\right)\frac{2}{c\eta_g\sqrt{\pi}}\exp\left(-u_2^2\right)$$

$$\times \left[\sum_{p=1}^{\infty} \frac{\exp\left(-p^2/4\right)}{p} sinh\left(pu_2\right) \right] \Bigg],$$

$$b \equiv (E_g + \Delta)^{-1}, \quad c \equiv \left(E_g + \frac{2}{3}\Delta\right)^{-1},$$

$$u_2 \equiv \frac{1+cE}{c\eta_g} \quad \text{and} \quad T_{32}(E, \eta_g)$$

$$\equiv \left(\frac{2}{1 + Erf\,(E/\eta_g)}\right) \frac{1}{c} \left(1 - \frac{\alpha}{c}\right) \left(1 - \frac{b}{c}\right) \frac{\sqrt{\pi}}{c\eta_g} \exp(-u_2^2).$$

Thus, the complex energy spectrum occurs due to the term $T_{32}(E, \eta_g)$ and this imaginary band is quite different from the forbidden energy band.

The EM at the Fermi level is given by

$$m^*(E_{F_h}, \eta_g) = m_c\{T_{31}(E, \eta_g)\}'|_{E=E_{F_h}} \tag{1.44}$$

Thus, the EM in HD III-V, ternary and quaternary materials exists in the band gap, which is the new attribute of the theory of band tailing.

In the absence of band tails, $\eta_g \to 0$ and the EM assumes the form

$$m^*(E_F) = m_c\{I_{11}(E)\}'|_{E=E_F} \tag{1.45}$$

The DOS function in this case can be written as

$$N_{HD}(E, \eta_g) = \frac{g_v}{3\pi^2}\left(\frac{2m_c}{\hbar^2}\right)^{3/2} R_{21}(E, \eta_g)\cos[\vartheta_{21}(E, \eta_g)] \tag{1.46}$$

where,

$$R_{21}(E, \eta_g) \equiv \left[\frac{\left[\{\alpha_{11}(E, \eta_g)\}'\right]^2}{4\alpha_{11}(E, \eta_g)} + \frac{\left[\{\beta_{11}(E, \eta_g)\}'\right]^2}{4\beta_{11}(E, \eta_g)}\right]^{1/2},$$

$$\alpha_{11}(E,\eta_g) \equiv \frac{1}{2}\left[T_{33}(E,\eta_g) + \sqrt{\{T_{33}(E,\eta_g)\}^2 + \{T_{34}(E,\eta_g)\}^2}\right],$$

$$T_{33}(E,\eta_g) \equiv \left[\{T_{31}(E,\eta_g)\}^3 - 3T_{31}(E,\eta_g)\{T_{32}(E,\eta_g)\}^2\right],$$

$$T_{34}(E,\eta_g) \equiv \left[3T_{32}(E,\eta_g)\{T_{31}(E,\eta_g)\}^2 - \{T_{32}(E,\eta_g)\}^3\right],$$

$$\beta_{11}(E,\eta_g) \equiv \frac{1}{2}\left[\sqrt{\{T_{33}(E,\eta_g)\}^2 + \{T_{34}(E,\eta_g)\}^2} - T_{33}(E,\eta_g)\right]$$

and

$$\vartheta_{21}(E,\eta_g) \equiv \tan^{-1}\left[\frac{\{\beta_{11}(E,\eta_g)\}'}{\{\alpha_{11}(E,\eta_g)\}'}\sqrt{\frac{\alpha_{11}(E,\eta_g)}{\beta_{11}(E,\eta_g)}}\right].$$

Thus, the oscillatory DOS function becomes negative for $\vartheta_{21}(E,\eta_g) \geq \pi$ and a new forbidden zone will appear in addition to the normal band gap.

The electron concentration can be expressed as

$$n_0 = \frac{g_v}{3\pi^2}\left(\frac{2m_c}{\hbar^2}\right)\left[\bar{I}_{111e}(E_{F_h},\eta_g) + \sum_{r=1}^{s} L(r)[\bar{I}_{111e}(E_{F_h},\eta_g)]\right] \tag{1.47}$$

where

$$\bar{I}_{111e}(E_{F_h},\eta_g) = \{\gamma_2(E_{F_h},\eta_g)\}^{3/2}$$

For dimensional quantization along z-direction, the dispersion relation of the 2D electrons in this case can be written as

$$\frac{\hbar^2(n_z\pi/d_z)^2}{2m_c} + \frac{\hbar^2(k_s)^2}{2m_c} = T_{31}(E,\eta_g) + iT_{32}(E,\eta_g) \tag{1.48}$$

The expression of the $N_{2DT}(E)$ in this case assumes the form

$$N_{2DT}(E) = \frac{m_c g_v}{\pi\hbar^2}\sum_{n_z=1}^{n_{z\max}} T'_{5D}(E,\eta_g,n_z)H(E - E_{n_Z D5}) \tag{1.49}$$

where, $T_{5D}(E,\eta_g,n_z) = [T_{31}(E,\eta_g) + iT_{32}(E,\eta_g) - \hbar^2(n_z\pi/d_z)^2 (2m_c)^{-1}]$ and the sub-band energies E_{n_zD5} in this case given by

$$\{\hbar^2(n_z/d_z)^2\}(2m_c)^{-1} = T_{31}(E_{n_zD5},\eta_g) \tag{1.50}$$

Thus we observe that both the total DOS in QWs of HD III-V compounds and the sub-band energies are complex due to the presence of the pole in energy axis of the corresponding materials in the absence of band tails.

The EM in this case is given by

$$m^*(E_{F1HD},\eta_g,n_z) = m_c[T'_{31}(E_{F1HD},\eta_g,n_z)] \tag{1.51}$$

The carrier statistics in this case can be expressed as

$$n_{2D} = \frac{m_c g_v}{\pi\hbar^2} \sum_{n_z=1}^{n_{z\max}} [\text{Real part of } [T_{5D}(E_{F1HD},\eta_g,n_z) + T_{6D}(E_{F1HD},\eta_g,n_z)]] \tag{1.52}$$

where, $T_{6D}(E_{F1HD},\eta_g,n_z) = \sum_{r=1}^{s} L(r)[T_{5D}(E_{F1HD},\eta_g,n_z)]$,

In the absence of band tails, the 2D dispersion relation, EM in the x-y plane at the Fermi level, the total 2D DOS, the sub-band energy and the electron concentration for QWs of III-V materials assume the following forms

$$\frac{\hbar^2 k_s^2}{2m_c} + \frac{\hbar^2}{2m_c}(n_z\pi/d_z)^2 = I_{11}(E) \tag{1.53}$$

$$m^*(E_{Fs}) = m_c\{I_{11}(E_{Fs})\}' \tag{1.54}$$

It is worth noting that the EM in this case is a function of Fermi energy alone and is independent of size quantum number.

$$N_{2DT}(E) = \left(\frac{m_c g_v}{\pi\hbar^2}\right) \sum_{n_z=1}^{n_{z\max}} \{[I_{11}(E)]' H(E - E_{n_{z2}})\} \tag{1.55}$$

where, the sub-band energies $E_{n_{z2}}$ can be expressed as

$$I_{11}(E_{n_{z2}}) = \frac{\hbar^2}{2m_c}(n_z\pi/d_z)^2 \tag{1.56}$$

$$n_{2D} = \frac{m_c g_v}{\pi\hbar^2} \sum_{n_z=1}^{n_{z\max}} [T_{53}(E_{Fs},n_z) + T_{54}(E_{Fs},n_z)] \tag{1.57}$$

where

$$T_{53}(E_{Fs}, n_z) \equiv \left[I_{11}(E_{Fs}) - \frac{h^2}{2m_c}\left(\frac{n_z\pi}{d_z}\right)^2 \right],$$

$$T_{54}(E_{Fs}, n_z) \equiv \sum_{r=1}^{s} L(r) T_{53}(E_{Fs}, n_z),$$

In the absence of band tails, the DOS function, the EM and the electron concentration in bulk III-V, ternary and quaternary materials in accordance with the unperturbed three band model of Kane assume the following forms

$$D_0(E) = 4\pi g_v \left(\frac{2m_c}{h^2}\right)^{3/2} \sqrt{I_{11}(E)}[I'_{11}(E)] \tag{1.58a}$$

$$m^*(E_F) = m_c\{I_{11}(E_F)\}' \tag{1.58b}$$

$$n_0 = \frac{g_v}{3\pi^2}\left(\frac{2m_c}{h^2}\right)^{3/2} [M_1(E_F) + N_1(E_F)] \tag{1.59}$$

where,

$$I'_{11}(E) \equiv I_{11}(E) \left[\frac{1}{E} + \frac{1}{E+E_g} + \frac{1}{E+E_g+\Delta} - \frac{1}{E+E_g+\frac{2}{3}\Delta} \right],$$

$$M_1(E_F) \equiv [I_{11}(E_F)]^{3/2}, \quad N_1(E_F) \equiv \sum_{r=1}^{s} L(r) M_1(E_F)$$

Under the inequalities $\Delta \gg E_{g0}$ or $\Delta \ll E_{g0}$, (1.42) can be expressed as

$$E(1+\alpha E) = \frac{\hbar^2 k^2}{2m_c} \tag{1.60}$$

where $\alpha \equiv (E_{g0})^{-1}$ and is known as band non-parabolicity.

It may be noted that (1.60) is the well-known two band model of Kane and is used in the literature to study the physical properties of those III-V and opto-electronic materials whose energy band structures obey the aforementioned inequalities.

The dispersion relation in HD III-V, ternary and quaternary materials whose energy spectrum in the absence of band tails obeys the two band model of Kane as defined by (1.60), can be written as

$$\frac{\hbar^2 k^2}{2m_c} = \gamma_2(E, \eta_g) \tag{1.61}$$

where,

$$\gamma_2(E, \eta_g) \equiv \left[\frac{2}{1 + Erf(E/\eta_g)}\right][\gamma_0(E, \eta_g) + \alpha\theta_0(E, \eta_g)]\,.$$

The EM in this case can be written as

$$m^*(E_{F_h}, \eta_g) = m_c\{\gamma_2(E, \eta_g)\}'|_{E=E_{F_h}} \tag{1.62a}$$

Thus, one again observes that the EM in this case exists in the band gap.

In the absence of band tails, $\eta_g \to 0$ and the EM assumes the well-known form

$$m^*(E_F) = m_c\{1 + 2alphaE\}|_{E=E_F} \tag{1.62b}$$

The DOS function in this case can be written as

$$N_{HD}(E, \eta_g) = \frac{g_v}{2\pi^2}\left(\frac{2m_c}{\hbar^2}\right)^{3/2}\sqrt{\gamma_2(E, \eta_g)\{\gamma_2(E, \eta_g)\}'} \tag{1.62c}$$

Since, the poles of the original two band Kane model are at infinity and and no finite poles with respect to energy, therefore the HD counterpart will be totally real and the complex band vanishes.

The electron concentration is given by

$$n_0 = \frac{g_v}{3\pi^2}\left(\frac{2m_c}{\hbar^2}\right)^{3/2}\left[\bar{I}_{111}(E_{F_h}, \eta_g) + \sum_{r=1}^{s} L(r)[\bar{I}_{111}(E_{F_h}, \eta_g)]\right] \tag{1.63}$$

where,

$$\bar{I}_{111}(E_{F_h}, \eta_g) = \{\gamma_2(E_{F_h}, \eta_g)\}^{3/2}$$

For dimensional quantization along z-direction, the dispersion relation of the 2D electrons in this case can be written following (1.61) as

$$\frac{\hbar^2(n_z\pi/d_Z)^2}{2m_c} + \frac{\hbar^2(k_s)^2}{2m_C} = \gamma_2(E,\eta_g) \tag{1.64}$$

The expression of the $N_{2DT}(E)$ in this case can be written as

$$N_{2DT}(E) = \frac{m_c g_v}{\pi\hbar^2}\sum_{n_z=1}^{n_{z\max}} T'_{7D}(E,\eta_g,n_z)H(E - E_{n_zD7}) \tag{1.65}$$

where, $T_{7D}(E,\eta_g,n_z) = [\gamma_2(E,\eta_g) - \hbar^2(n_z\pi/d_z)^2(2m_c)^{-1}]$,

The sub-band energies E_{n_zD7} in this case given by

$$\{\hbar^2(n_z\pi/d_z)^2\}(2m_c)^{-1} = \gamma^2(E_{n_zD7},\eta_g) \tag{1.66}$$

Thus, we observe that both the total DOS and sub-band energies of QWs of HD III-V compounds in accordance with two band model of Kane are not at all complex since the dispersion relation in accordance with the said model has no pole in the finite complex plane.

The EM in this case is given by

$$m^*(E_{F1HD},\eta_g,n_z) = m_c[\gamma'_2(E_{F1HD},\eta_g,n_z)] \tag{1.67}$$

The electron statistics in this case assumes the form

$$n_{2D} = \frac{m_c g_v}{\pi\hbar^2}\sum_{n_z=1}^{n_{z\max}} [T_{7D}(E_{F1HD},\eta_g,n_z) + T_{8D}(E_{F1HD},\eta_g,n_z)] \tag{1.68}$$

where, $T_{8D}(E_{F1HD},\eta_g,n_z) = \sum_{r=1}^{s} L(r)[T_{7D}(E_{F1HD},\eta_g,n_z)]$,

Under the inequalities $\Delta \gg E_{g0}$ or $\Delta \ll E_{g0}$, (1.60) assumes the form

$$E(1+\alpha E) = \frac{\hbar^2}{2m_c} + \frac{\hbar^2}{2m_c}\left(\frac{n_z\pi}{d_z}\right)^2 \tag{1.69a}$$

The EM can be written from (1.69a) as

$$m^*(E_{F_s}) = m_c(1 + 2\alpha E_{F_s}) \tag{1.69b}$$

The total 2D DOS function assumes the form

$$N_{2DT}(E) = \frac{m_c g_v}{\pi\hbar^2}\sum_{n_z=1}^{n_{z\max}}(1+2\alpha E)H(E-E_{n_{z3}}) \tag{1.70}$$

where, the sub-band energy ($E_{n_{z3}}$) can be expressed as

$$\frac{\hbar^2}{2m_c}(n_z\pi/d_z)^2 = E_{n_{z3}}(1_\alpha E_{n_{z3}}) \tag{1.71}$$

The 2D electron statistics can be written as

$$\begin{aligned} n_{2D} &= \frac{m_c g_v}{\pi\hbar^2}\sum_{n_z=1}^{n_{z\max}}\int_{E_{n_{z3}}}^{\infty}\frac{(1+2\alpha E)dE}{1+\exp\left(\frac{E-E_{F_s}}{k_BT}\right)} \\ &= \frac{m_c k_B T g_v}{\pi\hbar^2}\sum_{n_z=1}^{n_{z\max}}[(1+2\alpha E_{n_{z3}})F_0(\eta_{n1})+2\alpha k_BTF_1(\eta_{n1})] \end{aligned} \tag{1.72}$$

where, $\eta_{n_1} \equiv (E_{Fs}-E_{n_{z3}})/k_BT$ and $F_j(\eta)$ is the one parameter Fermi-Dirac integral of order j which can be written [145] as

$$F_j(\eta) = \left(\frac{1}{\Gamma(j+1)}\right)\int_0^{\infty}\frac{x^j\,dx}{1+\exp(x-\eta)}, \quad j>-1 \tag{1.73}$$

or for all j, analytically continued as a complex contour integral around the negative x-axis

$$F_j(\eta) = \left(\frac{\Gamma(-j)}{2\pi\sqrt{-1}}\right)\int_{-\infty}^{+0}\frac{x^j\,dx}{1+\exp(-x-\eta)} \tag{1.74}$$

where η is the dimensionless quantity and x is independent variable.

The forms of the DOS the EM and the electron statistics for bulk specimens of III-V materials in the absence of band tails whose energy band structures are defined by the two-band model of Kane can, respectively, be written as

$$D_0(E) = 4\pi g_v \left(\frac{2m_c}{h^2}\right)^{3/2} \sqrt{I_{11e}(E)}[I'_{11e}(E)] \tag{1.75a}$$

$$m*(E_F) = m_c[I'_{11e}(E_F)] \tag{1.75b}$$

$$n_0 = \frac{g_v}{3\pi^2}\left(\frac{2m_c}{h^2}\right)^{3/2}[M_2(E_F) + N_2(E_F)] \tag{1.76}$$

where,

$$I_{11e}(E) \equiv E(1+\alpha E), \quad I'_{11e}(E) \equiv (1+2\alpha E),$$
$$M_2(E_F) \equiv [I_{11e}(E_F)]^{3/2}$$

and

$$N_2(E_F) \equiv \sum_{r=1}^{s} L(r) M_2(E_F)$$

(c) Under the constraints $\Delta >> E_g$ or $\Delta << E_g$ together with the inequality $\alpha E_F << 1$, the (1.76) assumes the forms as

$$n_0 = g_v N_c \left[F_{1/2}(\eta) + \left(\frac{15\alpha k_B T}{4}\right) F_{3/2}(\eta)\right] \tag{1.77}$$

where,

$$N_C \equiv 2\left(\frac{2\pi m^* k_B T}{h^2}\right)^{3/2} \quad \text{and} \quad \eta \equiv \frac{E_F}{k_B T}$$

The dispersion relation in HDS whose energy spectrum in the absence of band tails obeys the parabolic energy bands is given by

$$\frac{\hbar^2 k^2}{2m_c} = \gamma_3(E, \eta_g) \tag{1.78}$$

where,

$$\gamma_3(E, \eta_g) \equiv \left[\frac{2}{(1 + Erf(E/\eta_g))}\right] \gamma_0(E, \eta_g).$$

Since the dispersion relation in accordance with the said model is an all zero function with no pole in the finite complex plane, therefore the HD counterpart will be totally real, which is also apparent form the expression (1.78).

The EM in this case can be written as

$$m^*(E_{F_h}, \eta_g) = m_c\{\gamma_3(E_{F_h}, \eta_g)\}' \tag{1.79}$$

In the absence of band tails, $\eta_g \to 0$ and the EM assumes the form

$$m^*(E_F) = m_c \tag{1.80}$$

It is well-known that the EM in unperturbed parabolic energy bands is a constant quantity in general excluding cross-fields configuration. However, the same mass in the corresponding HD bulk counterpart becomes a complicated function of Fermi energy and the impurity potential together with the fact that the EM also exists in the band gap solely due to the presence of finite η_g.

The DOS function in this case can be written as

$$N_{HD}(E, \eta_g) = \frac{g_v}{2\pi^2}\left(\frac{2m_c}{\hbar^2}\right)^{3/2}\sqrt{\gamma_3(E, \eta_g)\{\gamma_3(E, \eta_g)\}'} \tag{1.81}$$

The electron concentration is given by

$$n_0 = \frac{g_v}{3\pi^2}\left(\frac{2m_c}{\hbar^2}\right)^{3/2}\left[\bar{I}_{113}(E_{F_k}, \eta_g) + \sum_{r=1}^{s} L(r)[\bar{I}_{113}(E_{F_k}, \eta_g)]\right] \tag{1.82}$$

where,

$$\bar{I}_{113}(E_{F_k}, \eta_g) = \{\gamma_3(E_{F_k}, \eta_g)\}^{3/2}$$

For dimensional quantization along z-direction, the dispersion relation of the 2D electrons in this case can be written following (1.78) as

$$\frac{\hbar^2\left(\frac{n_z\pi}{d_z}\right)^2}{2m_c} + \frac{\hbar^2(k_s)^2}{2m_c} = \gamma_3(E, \eta_g) \tag{1.83}$$

the expression of the $N_{2DT}(E)$ in this case can be written as

$$N_{2DT}(E)\frac{m_c g_v}{\pi\hbar^2}\sum_{n_z=1}^{n_z\max} T'_{9D}(E, \eta_g, \mathrm{n}_z)H(\mathrm{E} - \mathrm{E}_{n_z D9}) \tag{1.84}$$

where,

$$T_{9D}(E,\eta_g,\mathrm{n}_z) = \left[\gamma_3(E,\eta_g) - \hbar^2\left(\frac{\mathrm{n}_z\pi}{\mathrm{d}_z}\right)^2(2m_c)^{-1}\right].$$

The sub-band energies $\mathrm{E}_{n_z D9}$ in this case given by

$$\left\{\hbar\left(\frac{n_z\pi}{d_z}\right)^2\right\}(2m_c)^{-1} = \gamma_3(E_{n_z D9},\eta_g) \tag{1.85}$$

The EM in this case can be written as

$$m^*(E_{F1HD},\eta_g,n_z) = m_c[\gamma_3'(E_{F1HD},\eta_g)] \tag{1.86}$$

The carrier statistics in this case assumes the form

$$n_{2D} = \frac{m_c g_v}{\pi\hbar^2}\sum_{n_z=1}^{n_{z\max}}[T_{9D}(E_{F1HD},\eta_g,\mathrm{n}_z) + T_{10D}(E_{F1HD},\eta_g,\mathrm{n}_z)] \tag{1.87}$$

where,

$$T_{10D}(E_{F1HD},\eta_g,\mathrm{n}_z) = \sum_{r=1}^{s} L(r)[T_{9D}(E_{F1HD},\eta_g,\mathrm{n}_z)],$$

Under the condition $\alpha \to 0$, the expressions of total 2D DOS, for semiconductors without forming band tails whose bulk electrons are defined by the isotropic parabolic energy bands can, be written as

$$N_{2DT}(E) = \frac{m_c g_v}{\pi\hbar^2}\sum_{n_z=1}^{n_{z\max}} H(E - E_{n_{zp}}) \tag{1.88}$$

The sub-band energy ($E_{n_{zp}}$), the n_{2D} and the EM can, respectively, be expressed as

$$E_{n_{zp}} = \frac{\hbar^2}{2m_c}\left(\frac{n_z\pi}{d_z}\right)^2 \tag{1.89}$$

$$n_{2D} = \frac{m_c k_B T g_v}{\pi\hbar^2}\sum_{n_z=1}^{n_{z\max}} F_0(\eta_{n2}) \tag{1.90a}$$

$$m^*(E_F) = m_c \tag{1.90b}$$

where,

$$\eta_{n_z} \equiv \frac{1}{k_B T}\left[E_{F_s} - \frac{\hbar^2}{2m_c}\left(\frac{n_z \pi}{d_z}\right)^2\right]$$

(b) The Model of Stillman *et al.*

In accordance with the model of Stillman *et al.* [38], the electron dispersion law of III–V materials assumes the form

$$E = \bar{t}_{11}k^2 - \bar{t}_{12}k^4 \tag{1.91}$$

where,

$$\bar{t}_{11} \equiv \frac{\hbar^2}{2m_c};$$

$$\bar{t}_{12} \equiv \left(1 - \frac{m_c}{m_0}\right)^2 \left(\frac{\hbar^2}{2m_c}\right)^2 \times \left[\left(3E_{g_0} + 4\Delta + \frac{2\Delta^2}{E_{g_0}}\right) \cdot \{(E_{g_0} + \Delta)(2\Delta + 3E_{g_0})\}^{-1}\right]$$

and m_0 is the free electron mass

In the presence of band tails, (1.91) gets transformed as

$$\frac{\hbar^2 k^2}{2m_c} = I_{12}(E, \eta_g) \tag{1.92}$$

where,

$$I_{12}(E, \eta_g) = a_{11}\left[1 - (1 - a_{12}\gamma_3(E, \eta_g))^{\frac{1}{2}}\right], \quad a_{11} \equiv \left(\frac{\hbar^2 \bar{t}_{11}}{4m_c \bar{t}_{12}}\right)$$

and

$$a_{12} \equiv \frac{4\bar{t}_{12}}{\bar{t}_{11}^2}$$

The EM can be written as

$$m^*(E_{F_h}, \eta_g) = m_c\{I_{12}(E_{F_h}, \eta_g)\}' \tag{1.93}$$

The DOS function in this case can be written as

$$N_{HD}(E, \eta_g) = \frac{g_v}{2\pi^2}\left(\frac{2m_c}{\hbar^2}\right)^{3/2} \sqrt{I_{12}(E, \eta_g)}\{I_{12}(E, \eta_g)\}' \tag{1.94}$$

The electron concentration is given by

$$n_0 = \frac{g_v}{3\pi^2}\left(\frac{2m_c}{\hbar^2}\right)^{3/2}\left[\bar{I}_{121}(E_{F_k},\eta_g) + \sum_{r=1}^{s} L(r)[\bar{I}_{121}(E_{F_k},\eta_g)]\right] \tag{1.95}$$

where,

$$\bar{I}_{121}(E_{F_k},\eta_g) = \{I_{12}(E_{F_k},\eta_g)\}^{3/2}$$

For dimensional quantization along z-direction, the dispersion relation of the 2D electrons in this case can be written following (1.108) as

$$\frac{\hbar^2\left(\frac{\mathrm{n}_z\pi}{\mathrm{d}_z}\right)^2}{2m_c} + \frac{\hbar^2(k_s)^2}{2m_c} = I_{12}(E,\eta_g) \tag{1.96}$$

the expression of the $N_{2DT}(E)$ in this case can be written as

$$N_{2DT}(E) = \frac{m_c g_v}{\pi\hbar^2}\sum_{n_z=1}^{n_{z\,\max}} T'_{11D}(E,\eta_g,\mathrm{n}_z)H(\mathrm{E}-\mathrm{E}_{n_z D11}) \tag{1.97}$$

where,

$$T_{11D}(E,\eta_g,\mathrm{n}_z) = \left[I_{12}(E,\eta_g) - \hbar^2\left(\frac{\mathrm{n}_z\pi}{\mathrm{d}_z}\right)^2(2m_c)^{-1}\right],$$

The sub-band energies $\mathrm{E}_{n_z D11}$ in this case given by

$$\left\{\hbar^2\left(\frac{\mathrm{n}_z\pi}{\mathrm{d}_z}\right)^2\right\}(2m_c)^{-1} = I_{12}(\mathrm{E}_{n_z D11},\eta_g) \tag{1.98}$$

The EM in this case assumes the form

$$m^*(E_{F1HD},\eta_g,\mathrm{n}_z) = m_c[I'_{12}(E_{F1HD},\eta_g,\mathrm{n}_z)] \tag{1.99}$$

The 2-D electron statistics in this case can be written as

$$n_{2D} = \frac{m_c g_v}{\pi\hbar^2}\sum_{n_z=1}^{n_{z\,\max}}[T_{11D}(E_{F1HD},\eta_g,\mathrm{n}_z) + T_{12D}(E_{F1HD},\eta_g,\mathrm{n}_z)] \tag{1.100}$$

where,

$$T_{12D}(E_{F1HD},\eta_g,\mathrm{n}_z) = \sum_{r=1}^{s} L(r)[T_{11D}(E_{F1HD},\eta_g,\mathrm{n}_z)],$$

For unperturbed material, the 2-D EM can be expressed as

$$m^*(E_{Fs}) = m_c\{I_{12}(E_{Fs})\}' \tag{1.101}$$

where

$$I_{12}(E) \equiv a_{11}[1 - (1 - a_{12}(E))^{\frac{1}{2}}]$$

It appears that the EM in this case is a function of Fermi energy alone and is independent of size quantum number. The total 2D DOS function in the absence of band tails in this case can be written as

$$N_{2DT}(E) = \left(\frac{m_c g_v}{\pi\hbar^2}\right) \sum_{n_z=1}^{n_{z\max}} \{[I_{12}(E)]' H(E - E_{n_{z3}})\} \tag{1.102}$$

where, the sub-band energies $E_{n_{z3}}$ can be expressed as

$$I_{12}(E_{n_{z3}}) = \frac{\hbar^2}{2m_c}\left(\frac{n_z\pi}{d_z}\right)^2 \tag{1.103}$$

The 2D electron concentration assumes the form

$$n_{2D} = \frac{m_c g_v}{\pi\hbar^2} \sum_{n_z=1}^{n_{z\max}} [T_{55}(E_{Fs}, n_z) + T_{56}(E_{Fs}, n_z)] \tag{1.104}$$

where

$$T_{55}(E_{Fs}, n_z) \equiv \left[I_{12}(E_{Fs}) - \frac{\hbar^2}{2m_c}\left(\frac{n_z\pi}{d_z}\right)^2\right]$$

and

$$T_{56}(E_{Fs}, n_z) \equiv \sum_{r=1}^{s} L(r) T_{55}(E_{Fs}, n_z)$$

The expression of electron concentration for bulk specimens of III-V semiconductors (in the absence of band tails) can be written in accordance with the model of Stillman *et al.* as

$$n_0 = \frac{g_v}{3\pi^2}\left(\frac{2m_c}{\hbar^2}\right)^{3/2} [M_{A_{10}}(E_F) + N_{A_{10}}(E_F)] \tag{1.105}$$

where,

$$M_{A_{10}}(E_F) = [I_{12}(E_F)]^{3/2}$$

and

$$N_{A_{10}}(E_F) = \sum_{r=1}^{s} L(r)[M_{A_{10}}(E_F)]$$

(c) Model of Palik *et al.*

The energy spectrum of the conduction electrons in III–V semiconductors up to the fourth order in effective mass theory, taking into account the interactions of heavy hole, light hole and the split-off holes can be expressed in accordance with the model of Palik *et al.* [40] as

$$E = \frac{\hbar^2 k^2}{2m_c} - \bar{B}_{11} k^4 \tag{1.106}$$

where

$$\bar{B}_{11} = \left[\frac{\hbar^4}{4E_{g_0}(m_c)^2}\right]\left[\frac{1+\frac{x_{11}^2}{2}}{1+\frac{x_{11}}{2}}\right](1-y_{11})^2, \quad x_{11} = \left[1+\left(\frac{\Delta}{E_{g0}}\right)\right]^{-1}$$

and

$$y_{11} = \frac{m_c}{m_0}$$

The (1.106) gets simplified as

$$\frac{\hbar^2 k^2}{2m_c} = I_{13}(E) \tag{1.107}$$

where

$$I_{13}(E) = \bar{b}_{12}\left[\bar{a}_{12} - ((\bar{a}_{12})^2 - 4E\bar{B}_{11})^{1/2}\right], \quad \bar{a}_{12}\left(\frac{\hbar^2}{2m_c}\right)$$

and

$$\bar{b}_{12} = \left[\frac{\bar{a}_{12}}{2\bar{B}_{11}}\right]$$

Under the condition of heavy doping forming Gaussian band tails, (1.107) assumes the form

$$\frac{\hbar^2 k^2}{2m_c} = I_{13}(E, \eta_g) \tag{1.108}$$

where,

$$I_{13}(E,\eta_g)=\bar{b}_{12}[\bar{a}_{12}-((\bar{a}_{12})^2-4\bar{B}_{11}\gamma_3(E,\eta_g))^{1/2}]$$

The EM can be written as

$$m^*(E_{F_h},\eta_g)=m_c\{I_{13}(E_{F_k},\eta_g)\}' \tag{1.109}$$

The DOS function in this case can be expressed as

$$N_{HD}(E,\eta_g)=\frac{g_v}{2\pi^2}\left(\frac{2m_c}{\hbar^2}\right)^{3/2}\sqrt{I_{13}(E,\eta_g)\{I_{13}(E,\eta_g)\}'} \tag{1.110}$$

Since, the original band model in this case is a no pole function, in the finite complex plane therefore, the HD counterpart will be totally real and the complex band vanishes.

The electron concentration is given by

$$n_0=\frac{g_v}{3\pi^2}\left(\frac{2m_c}{\hbar^2}\right)^{3/2}\left[\bar{I}_{123}(E_{F_k},\eta_g)+\sum_{r=1}^{s}L(r)[\bar{I}_{123}(E_{F_k},\eta_g)]\right] \tag{1.111}$$

where,

$$\bar{I}_{123}(E_{F_k},\eta_g)=\{I_{13}(E_{F_k},\eta_g)\}^{3/2}$$

For dimensional quantization along z-direction, the dispersion relation of the 2D electrons in this case can be written following (1.107) as

$$\frac{\hbar^2\left(\frac{\mathrm{n}_z\pi}{\mathrm{d}_z}\right)^2}{2m_c}+\frac{\hbar^2(k_s)^2}{2m_c}=I_{13}(E,\eta_g) \tag{1.112}$$

the expression of the $N_{2DT}(E)$ in this case can be written as

$$N_{2DT}(E)=\frac{m_c g_v}{\pi\hbar^2}\sum_{n_z}^{n_{z\max}}T'_{13D}(E,\eta_g,n_z)H(E-E_{n_zD13}) \tag{1.113}$$

where,

$$T13D(E,\eta_g,\mathrm{n}_z)=\left[I_{13}(E,\eta_g)-\hbar^2\left(\frac{\mathrm{n}_z\pi}{\mathrm{d}_z}\right)^2(2m_c)^{-1}\right],$$

The sub-band energies E_{n_zD13} in this case given by

$$\left\{\hbar\left(\frac{n_z\pi}{d_z}\right)^2\right\}(2m_c)^{-1} = I_{13}(E_{n_zD13},\eta_g) \tag{1.114}$$

The EM in this case can be expressed as

$$m^*(E_{F1HD},\eta_g,n_z) = m_c[I'_{13}(E_{F1HD},\eta_g,n_z)] \tag{1.115}$$

The 2-D electron statistics in this case can be written as

$$n_{2D} = \frac{m_c g_v}{\pi\hbar^2}\sum_{n_z=1}^{n_{z\max}}[T_{13D}(E_{F1HD},\eta_g,\mathrm{n}_z) + T_{14D}(E_{F1HD},\eta_g,\mathrm{n}_z)] \tag{1.116}$$

where,

$$T_{14D}(E_{F1HD},\eta_g,\mathrm{n}_z) = \sum_{r=1}^{s} L(r)[T_{13D}(E_{F1HD},\eta_g,\mathrm{n}_z)],$$

The 2D electron dispersion relation in the absence of band tails this case assumes the form

$$\frac{\hbar^2k_s^2}{2m_c} + \frac{\hbar^2}{2m_c}\left(\frac{n_z\pi}{d_z}\right)^2 = I_{13}(E) \tag{1.117a}$$

The EM in this case can be written from (1.117a) as

$$m^*(E_{F_s}) = m_c[I_{13}(E_{F_s})]' \tag{1.117b}$$

The total 2D DOS function can be written as

$$N_{2DT}(E) = \left(\frac{m_c g_v}{\pi\hbar^2}\right)\sum_{n_z=1}^{n_{z\max}}\{[I_{13}(E)]'H(E - E_{n_{z4}})\} \tag{1.118}$$

where, the sub-band energies $E_{n_{z4}}$ can be expressed as

$$I_{13}(E_{n_{z4}}) = \frac{\hbar^2}{2m_c}\left(\frac{n_z\pi}{d_z}\right)^2 \tag{1.119}$$

The 2D electron concentration assumes the form

$$n_{2D} = \frac{m_c g_v}{\pi\hbar^2}\sum_{n_z=1}^{n_{z\max}}[T_{57}(E_{Fs},n_z) + T_{58}(E_{Fs},n_z)] \tag{1.120}$$

where

$$T_{57}(E_{Fs}, n_z) \equiv \left[I_{13}(E_{Fs}) - \frac{\hbar^2}{2m_c} \left(\frac{n_z \pi}{d_z} \right)^2 \right]$$

and

$$T_{58}(E_{Fs}, n_z) \equiv \sum_{r=1}^{s} L(r) T_{57}(E_{Fs}, n_z)$$

The expression of electron concentration for bulk specimens of III–V semiconductors (in the absence of band tails) can be written in accordance with the model of Stillman *et al.* as

$$n_0 = \frac{g_v}{3\pi^2} \left(\frac{2m_c}{\hbar^2} \right)^{3/2} [\bar{M}_{A_{10}}(E_F) + \bar{N}_{A_{10}}(E_F)] \quad (1.121)$$

where,

$$\bar{M}_{A_{10}}(E_F) = [I_{13}(E_F)]^{3/2}$$

and

$$\bar{N}_{A_{10}}(E_F) = \sum_{r=1}^{s} L(r) [\bar{M}_{A_{10}}(E_F)]$$

1.2.3 *The EM in Quantum Wells (QWs) of HD II-VI materials*

The carrier energy spectra in bulk specimens of II-VI compounds in accordance with Hopfield model [57] can be written as

$$E = a_0' k_s^2 + b_0' k_z^2 \pm \bar{\lambda}_0 k_s \quad (1.122)$$

where $a_0' \equiv \hbar^2/2m_\perp^*, b_0' \equiv \hbar^2/2m_\parallel^*$, and $\bar{\lambda}_0$ represents the splitting of the two-spin states by the spin orbit coupling and the crystalline field.

Therefore the dispersion relation of the carriers in HD II–VI materials in the presence of Gaussian band tails can be expressed as

$$\gamma_3(E, \eta_g) = a_0' k_s^2 + b_0' k_z^2 \pm \bar{\lambda}_0 k_s \quad (1.123)$$

Thus, the energy spectrum in this case is real since the corresponding E-k relation in the absence of band tails as given by (1.123) is a no pole function in the finite complex plane.

The transverse and the longitudinal EMs masses are, respectively, given by

$$m_{\perp}^*(E_{F_h},\eta_g) = m_{\perp}^*\{\gamma_3(E,\eta_g)\}'\left[1 \pm \left(\frac{\bar{\lambda}_0}{\sqrt{(\bar{\lambda}_0)^2 + 4a_0'\gamma_3(E,\eta_g)}}\right)\right]\Bigg|_{E=E_F} \tag{1.124}$$

and

$$m_{\parallel}^*(E_{F_h},\eta_g) = m_{\parallel}^*\ \{\gamma_3(E,\eta_g)\}'|_{E=E_{F_h}} \tag{1.125}$$

Thus the transverse EM in HD II–VI semiconductors is a function of electron energy and is double valued due to the presence of $\bar{\lambda}_0$ and due to heavy doping the same mass exists in the band gap.

In the absence of band tails, $\eta_g \to 0$, we get

$$m_{\perp}^*(E_F) = m_{\perp}^*\left[1 \pm \left(\frac{\bar{\lambda}_0}{\sqrt{(\bar{\lambda}_0)^2 + 4a_0'E}}\right)\right]\Bigg|_{E=E_F} \tag{1.126}$$

and

$$m_{\parallel}^*(E_F) = m_{\parallel}^* \tag{1.127}$$

The volume in k-space as enclosed (1.123) can be expressed as

$$V(E,\eta_g) = \frac{4\pi}{3a_0'\sqrt{b_0'}}\left[\{\gamma_3(E,\eta_g)\}^{3/2} + \frac{3}{8}\frac{(\bar{\lambda}_0)^2\sqrt{\gamma_3(E,\eta_g)}}{a_0'} \pm \left(\frac{3}{4}\frac{\bar{\lambda}_0}{\sqrt{a_0'}}\right)\left(\gamma_3(E,\eta_g) + \frac{(\bar{\lambda}_0)^2}{4a_0'}\right)\sin^{-1} \times \left[\frac{\sqrt{\gamma_3(E,\eta_g)}}{\sqrt{\gamma_3(E,\eta_g) + \frac{(\bar{\lambda}_0)^2}{4a_0'}}}\right]\right] \tag{1.128}$$

Therefore, the electron concentration can be written as

$$n_0 = \frac{g_v}{3\pi^2 a_0'\sqrt{b_0'}}\left[\bar{I}_{124}(E_{F_k},\eta_g) + \sum_{r=1}^{s} L(r)[\bar{I}_{124}(E_{F_k},\eta_g)]\right] \tag{1.129}$$

where,

$$\bar{I}_{124}(E_{F_k},\eta_g) = \left[\{\gamma_3(E_{F_k},\eta_g)\}^{3/2} + \frac{3}{8}\frac{(\bar{\lambda}_0)^2\sqrt{\gamma_3(E_{F_k},\eta_g)}}{a_0'}\right]$$

The dispersion relation of the conduction electrons of QWs of HD II–VI materials for dimensional quantization along z-direction can be written following (1.123) as

$$\gamma_3(E,\eta_g) = a_0' k_s^2 + b_0'\left(\frac{\pi n_z}{d_z}\right)^2 \pm \bar{\lambda}_0 k_s \tag{1.130}$$

The EM can be expressed following (1.130) as

$$m^*(E_{F1HD},n_z,\eta_g) = m_\perp^*\left[1 \mp \frac{(\bar{\lambda}_0)}{\left[(\bar{\lambda}_0)^2 - 4a_0'b_0'\left(\frac{n_z\pi}{d_z}\right)^2 + 4a_0'\gamma_3(E_{F1HD},\eta_g)\right]^{1/2}}\right] \times \gamma_3'(E_{F1HD},\eta_g) \tag{1.131}$$

Thus we observe that the doubled valued effective mass in 2-D QWs of HD II-VI materials is a function of Fermi energy, size quantum number and the screening potential respectively together with the fact that the same mass exists in the band gap due to the sole presence of the splitting of the two-spin states by the spin orbit coupling and the crystalline field.

The sub-band energy in this case is given by

$$\gamma_3(E_{n_z D14},\eta_g) = b_0'\left(\frac{\pi n_z}{d_z}\right)^2 \tag{1.132}$$

The surface electron concentration at low temperatures assumes the form

$$n_{2D} = \frac{g_v m_\perp^*}{\pi\hbar^2}\sum_{n_z=1}^{n_{z\max}}(\gamma_3(E_{F1HD},\eta_g) - E_{n_z D14} + (\bar{\lambda}_0)^2 m_\perp^* \hbar^{-2}) \tag{1.133}$$

The dispersion relation of the conduction electrons of QWs of II–VI materials for dimensional quantization along z-direction in the absence of band tails can be written following (1.122) as

$$E = a_0' k_s^2 + b_0' \left(\frac{n_z \pi}{d_z} \right)^2 \pm \bar{\lambda}_0 k_s \tag{1.134}$$

Using (1.134), the EM in this case can be written as

$$m^*(E_{Fs}, n_z) = m_\perp^* \left[1 \mp \frac{(\bar{\lambda}_0)}{\left[(\bar{\lambda}_0)^2 - 4a_0' b_0' \left(\frac{n_z \pi}{d_z} \right)^2 + 4a_0' E_{Fs} \right]^{1/2}} \right] \tag{1.135}$$

The sub-band energy $E_{n_z}5$ assumes the form

$$E_{n_{z5}} = b_0' (n_z / \pi / d_z)^2 \tag{1.136}$$

The area of constant energy 2D quantized surface in this case is given by where

$$A_\pm(E, n_z) = \left[\frac{\pi}{2(a_0')^2} \left[(\bar{\lambda}_0)^2 + 2a_0'(E - E_{n_{z5}}) \pm \bar{\lambda}_0 [(\bar{\lambda}_0)^2 + 4a_0'(E - E_{n_{z5}})]^{1/2} \right] \right]$$

The surface electron concentration can be expressed in this case as

$$n_{2D} = \frac{-2g_v}{2(2\pi)^2} \sum_{n_z=1}^{n_{z\max}} \int_{E_{n_{z5}}}^{\infty} [A_+(E_{Fs}, n_z) + A_-(E_{Fs}, n_z)] \frac{\partial}{\partial E} \{f_0(E)\} dE \tag{1.137}$$

where $f_0(E)$ is the Fermi-Dirac occupation probability factor.

From (1.137) we get

$$n_{2D} = \frac{g_v m_\perp^* k_B T}{\pi \hbar^2} \sum_{n_z=1}^{n_{z\max}} F_0(\eta_{n_{z8}}) \tag{1.138}$$

where

$$\eta_{n_{z8}} = (E_{Fs} - E_{n_{z5}} (\bar{\lambda})^2 m_\perp^* \hbar^{-2})(k_B T)^{-1}$$

1.2.4 The EM in Quantum Wells (QWs) of HD IV–VI materials

(a) The Model of Dimmock

The dispersion relation of the conduction electrons in IV–VI semiconductors can be expressed in accordance with Dimmock [146] as

$$\left[\bar{\varepsilon} - \frac{E_{g0}}{2} - \frac{\hbar^2 k_s^2}{2m_t^-} - \frac{\hbar^2 k_z^2}{2m_l^-}\right]\left[\bar{\varepsilon} + \frac{E_{g0}}{2} + \frac{\hbar^2 k_s^2}{2m_t^-} + \frac{\hbar^2 k_z^2}{2m_l^-}\right] = P_\perp^2 k_s^2 + P_\parallel^2 k_z^2 \tag{1.139}$$

where, $\bar{\varepsilon}$ is the energy as measured from the center of the band gap E_{g0}, $m_t^\pm$ and $m_t^\pm$ represent the contributions to the transverse and longitudinal effective masses of the external L_6^+ and L_6^- bands arising from the $\vec{k}\cdot\vec{p}$ perturbations with the other bands taken to the second order.

Substituting,

$$P_\perp^2 \equiv (\hbar^2 E_g/2m_l^*), \quad P_\parallel^2 \equiv \left(\frac{\hbar^2 E_g}{2m_l^*}\right)$$

and

$$\bar{\varepsilon} \equiv \left[E + \left(\frac{E_g}{2}\right)\right]$$

(where, m_t^* and m_l^* are the transverse and the longitudinal effective masses at k = 0), (1.139) gets transformed as

$$\left[E - \frac{\hbar^2 k_s^2}{2m_t^-} - \frac{\hbar^2 k_z^2}{2m_l^-}\right]\left[1 + \alpha E + \alpha\frac{\hbar^2 k_s^2}{2m_t^+} + \alpha\frac{\hbar^2 k_z^2}{2m_l^+}\right] = \frac{\hbar^2 k_s^2}{2m_t^*} + \frac{\hbar^2 k_z^2}{2m_l^*} \tag{1.140}$$

From (1.140), we can write

$$\frac{\alpha\hbar^4 k_s^4}{4m_t^+ m_t^-} + \hbar^2 k_s^2\left[\left(\frac{1}{2m_t^*} - \frac{1}{2m_t^-}\right) + \alpha E\left(\frac{1}{2m_t^-} - \frac{1}{2m_t^+}\right) + \frac{\alpha\hbar^2 k_z^2}{4m_l^- m_t^+}\right]$$
$$+ \left[\left(\frac{\hbar^2 k_z^2}{2m_l^*} + \frac{\hbar^2 k_z^2}{2m_l^-}\right) + \frac{\alpha E}{2}\hbar^2 k_z^2\left(\frac{1}{m_l^-} - \frac{1}{m_l^+}\right)\right.$$
$$\left. + \frac{\alpha\hbar^4 k_z^4}{4m_l^+ m_t^-} - E(1 + \alpha E)\right] = 0 \tag{1.141}$$

Using (1.141), the dispersion relation of the conduction electrons in HD IV–VI materials can be expressed as

$$\frac{\alpha\hbar^4 k_s^4}{4m_t^+ m_l^-} Z_0(E,\eta_g) + \hbar^2 k_s^2 \left[\lambda_{71}(E,\eta_g)k_z^2 + \lambda_{72}(E,\eta_g)\right] + \left[\lambda_{73}(E,\eta_g)k_z^2 + \lambda_{74}(E,\eta_g)k_z^4 - \lambda_{75}(E,\eta_g)\right] = 0 \quad (1.142)$$

where,

$$Z_0(E,\eta_g) \equiv \frac{1}{2}\left[1 + Erf\left(\frac{E}{\eta_g}\right)\right],$$

$$\lambda_{70}(E,\eta_g) \equiv \frac{\alpha}{4m_t^+ m_l^-} Z_0(E,\eta_g)$$

$$\lambda_{71}(E,\eta_g) \equiv \left[\frac{\alpha\hbar^2}{4m_t^- m_l^+} Z_0(E,\eta_g) + \frac{\alpha\hbar^2}{4m_l^- m_t^+} Z_0(E,\eta_g)\right],$$

$$\lambda_{72}(E,\eta_g) \equiv \left[\left(\frac{1}{2m_t^*} - \frac{1}{2m_t^-}\right) Z_0(E,\eta_g) + \alpha\left(\frac{1}{2m_t^-} - \frac{1}{2m_t^+}\right)\gamma_0(E,\eta_g)\right],$$

$$\lambda_{73}(E,\eta_g) \equiv \left[\left(\frac{\hbar^2}{2m_l^*} + \frac{\hbar^2}{2m_l^-}\right) Z_0(E,\eta_g) + \frac{\alpha\hbar^2}{2}\left(\frac{1}{m_l^-} - \frac{1}{2m_l^+}\right)\gamma_0(E,\eta_g)\right],$$

$$\lambda_{74}(E,\eta_g) \equiv \frac{\alpha\hbar^4 Z_0(E,\eta_g)}{4m_l^+ m_l^-}$$

and

$$\lambda_{75}(E,\eta_g) \equiv \left[\gamma_0(E,\eta_g) + \alpha\theta_0(E,\eta_g)\right]$$

Thus, the energy spectrum in this case is real since the corresponding dispersion relation in the absence of band tails as given by (1.142) is a pole-less function with respect to energy axis in the finite complex plane.

The respective transverse and the longitudinal EMs' in this case can be written as

$$m_{\perp}^{*}(E_{F_h},\eta_g) = \{2Z_0(E,\eta_g)\}^{-2}\left[Z_0(E,\eta_g)\left[-\{\lambda_{72}(E,\eta_g)\}' + \frac{\{\lambda_{78}(E,\eta_g)\}'}{2\sqrt{\lambda_{78}(E,\eta_g)}}\right] - \{Z_0(E,\eta_g)\}'\left[-\lambda_{72}(E,\eta_g) + \sqrt{\lambda_{78}(E,\eta_g)}\right]\right]\Bigg|_{E=E_{F_h}} \quad (1.143)$$

where,

$$\lambda_{78}(E,\eta_g) \equiv [4\lambda_{70}(E,\eta_g)\lambda_{75}(E,\eta_g)]$$

and

$$m_{\parallel}^{*}(E_{F_h},\eta_g) = \frac{\hbar^2}{4}\left[-\{\lambda_{84}(E,\eta_g)\}' + \frac{\{\lambda_{84}(E,\eta_g)\}'\lambda_{84}(E,\eta_g) + 2\{\lambda_{85}(E,\eta_g)\}'}{\sqrt{(\lambda_{84}(E,\eta_g))^2 + 4\lambda_{85}(E,\eta_g)}}\right]\Bigg|_{E=E_{F_h}} \quad (1.144)$$

in which,

$$\lambda_{84}(E,\eta_g) \equiv \frac{\lambda_{73}(E,\eta_g)}{\lambda_{74}(E,\eta_g)}$$

and

$$\lambda_{85}(E,\eta_g) \equiv \frac{\lambda_{75}(E,\eta_g)}{\lambda_{74}(E,\eta_g)}$$

Thus, we can see that the both the EMs' in this case exist in the band gap.

In the absence of band tails, $\eta_g \to 0$, we get

$$m_{\perp}^{*}(E_F) = \frac{\hbar^2}{2}\left[-\{\alpha_{11}(E)\}' + \frac{\alpha_{511}\{T_{311}(E)\}'}{2\sqrt{T_{311}(E)}}\right]\Bigg|_{E=E_F} \quad (1.145)$$

where

$$\alpha_{11}(E) \equiv \frac{2m_t^+ m_t^-}{\alpha\hbar^2}\alpha_{211}(E), \quad \alpha_{211}(E) \equiv \left[\frac{1}{2m_t^*} - \frac{\alpha E}{2m_t^+} + \frac{1+\alpha E}{2m_t^-}\right],$$

$$\alpha_{511} \equiv \frac{2m_t^+ m_t^-}{\alpha\hbar^2}\omega_{11}$$

$$(\omega_{11}) \equiv \left[\frac{\alpha^2}{16}\left[\frac{1}{m_t^- m_l^+} + \frac{1}{m_l^- m_t^+}\right]^2 - \frac{\alpha^2}{4m_t^- m_t^+ m_l^- m_l^+}\right]^{1/2},$$

$$T_{311}(E) \equiv \frac{\omega_{311}(E)}{(\omega_{11})^2},$$

$$\omega_{311}(E) \equiv \left[\frac{\alpha E(1+\alpha E)}{m_t^+ m_t^-} + \left[\frac{1}{2m_t^*} - \left(\frac{\alpha E}{2m_t^+}\right) + \frac{(1+\alpha E)}{2m_t^-}\right]^2\right]$$

and

$$m_{\|}^*(E_F) = \left(\frac{m_t^+ m_l^-}{\alpha}\right)\left[\left(\frac{\alpha}{2m_l^+} - \frac{\alpha}{2m_l^-}\right)\right.$$
$$\left.\left.+\frac{1}{2}\left\{\frac{2\left[\frac{1}{2m_l^*} + \frac{1+\alpha E}{2m_l^-} - \frac{\alpha E}{2m_l^+}\right]\left(\frac{\alpha}{2m_l^-} - \frac{\alpha}{2m_l^+}\right) + \frac{\alpha(1+2\alpha E)}{m_l^- m_l^+}}{\left[\left[\frac{1}{2m_l^*} + \frac{1+\alpha E}{2m_l^-} - \frac{\alpha E}{2m_l^+}\right]^2 + \frac{\alpha E(1+\alpha E)}{m_l^- m_l^+}\right]^{1/2}}\right\}\right]\right|_{E=E_F} \tag{1.146}$$

The volume in k-space as enclosed by (1.142) can be written through the integral as

$$V(E,\eta_g) = 2\pi\int_0^{\lambda_{86}(E,\eta_g)} [-[\lambda_{79}(E,\eta_g)k_z^2 + \lambda_{80}(E,\eta_g)]$$
$$+\sqrt{\lambda_{81}(E,\eta_g)k_z^4 + \lambda_{82}(E,\eta_g)k_z^2 + \lambda_{83}(E,\eta_g)}]dk_z \tag{1.147}$$

where,

$$\lambda_{86}(E,\eta_g) \equiv \left[\frac{\sqrt{[\lambda_{84}(E,\eta_g)]^2 + 4\lambda_{85}(E,\eta_g)} - \lambda_{84}(E,\eta_g)}{2}\right]^{1/2},$$

$$\lambda_{79}(E,\eta_g) \equiv \frac{\lambda_{71}(E,\eta_g)}{2\hbar^2 Z_0(E,\eta_g)}$$

$$\lambda_{81}(E,\eta_g) \equiv \frac{\lambda_{76}(E,\eta_g)}{4\hbar^4[Z_0(E,\eta_g)]^2}, \quad \lambda_{76}(E,\eta_g) \equiv [\lambda_{71}(E,\eta_g)]^2,$$

$$\lambda_{76}(E,\eta_g) \equiv [\lambda_{71}(E,\eta_g)]^2$$

$$\lambda_{77}(E,\eta_g) \equiv [2\lambda_{71}(E,\eta_g)\lambda_{72}(E,\eta_g) - 4\lambda_{70}(E,\eta_g)\lambda_{73}(E,\eta_g) - 4\lambda_{70}(E,\eta_g)\lambda_{74}(E,\eta_g)]$$

$$\lambda_{83}(E,\eta_g) \equiv \frac{\lambda_{78}(E,\eta_g)}{9\hbar^4[Z_0(E,\eta_g)]^2}$$

and

$$\lambda_{78}(E,\eta_g) \equiv [4\lambda_{70}(E,\eta_g)\lambda_{75}(E,\eta_g)]$$

Thus,

$$V(E,\eta_g) = [\lambda_{87}(E,\eta_g)]\int_0^{\lambda_{86}(E,\eta_g)} \times\left[\sqrt{k_z^4 + \lambda_{88}(E,\eta_g)k_z^2 + \lambda_{89}(E,\eta_g)} - \lambda_{90}(E,\eta_g)\right]dk_z \tag{1.148}$$

where,

$$\lambda_{87}(E,\eta_g) \equiv 2\pi\sqrt{\lambda_{81}(E,\eta_g)}, \quad \lambda_{88}(E,\eta_g) \equiv \frac{\lambda_{82}(E,\eta_g)}{\lambda_{81}(E,\eta_g)},$$

$$\lambda_{89}(E,\eta_g) \equiv \frac{\lambda_{83}(E,\eta_g)}{\lambda_{81}(E,\eta_g)}$$

and

$$\lambda_{90}(E,\eta_g) \equiv 2\pi\left[\frac{\lambda_{79}(E,\eta_g)\{\lambda_{86}(E,\eta_g)\}^3}{3} + \lambda_{80}(E,\eta_g)\lambda_{89}(E,\eta_g)\right].$$

The (1.148) can be written as

$$V(E,\eta_g) = [\lambda_{87}(E,\eta_g)\,\lambda_{95}(E,\eta_g) - \lambda_{90}(E,\eta_g)] \tag{1.149}$$

in which,

$$\begin{aligned}
\lambda_{95}(E,\eta_g) \equiv \Bigg[&\frac{\lambda_{91}(E,\eta_g)}{3}[-E_t[\lambda_{93}(E,\eta_g),\lambda_{94}(E,\eta_g)]] \\
&\times[\{\lambda_{91}(E,\eta_g)\}^2+\{\lambda_{92}(E,\eta_g)\}^2 \\
&+2\{\lambda_{92}(E,\eta_g)\}^2F_i[\lambda_{93}(E,\eta_g),\lambda_{94}(E,\eta_g)]] \\
&+\left(\frac{\{\lambda_{86}(E,\eta_g)\}}{3}\right)[\{\lambda_{86}(E,\eta_g)\}^2 \\
&+\{\lambda_{91}(E,\eta_g)\}^2+2\{\lambda_{92}(E,\eta_g)\}^2] \\
&[\{\lambda_{91}(E,\eta_g)\}^2+\{\lambda_{86}(E,\eta_g)\}^2]^{1/2} \\
&\times[\{\lambda_{92}(E,\eta_g)\}^2+\{\lambda_{86}(E,\eta_g)\}^2]^{-1/2}\Bigg],
\end{aligned}$$

$$\begin{aligned}
\{\lambda_{91}(E,\eta_g)\}^2 \equiv \frac{1}{2}\Bigg[&\sqrt{\{\lambda_{88}(E,\eta_g)\}^2-4\lambda_{89}(E,\eta_g)} \\
&+\lambda_{88}(E,\eta_g)]\Bigg], E_i[\lambda_{93}(E,\eta_g),\lambda_{94}(E,\eta_g)],
\end{aligned}$$

is the incomplete elliptic integral of the 2nd kind and is given by [123, 124],

$$\begin{aligned}
&E_i[\lambda_{93}(E,\eta_g),\lambda_{94}(E,\eta_g)] \\
&\quad\equiv \int_0^{\lambda_{93}(E,\eta_g)}[\{1-\{\lambda_{94}(E,\eta_g)\}^2\sin^2\xi\}^{1/2}]d\xi,
\end{aligned}$$

ξ is the variable of integration in this case,

$$\lambda_{93}(E,\eta_g) \equiv \tan^{-1}\left[\frac{\lambda_{86}(E,\eta_g)}{\lambda_{92}(E,\eta_g)}\right],$$

$$\{\lambda_{92}(E,\eta_g)\}^2 \equiv \frac{1}{2}\left[\lambda_{88}(E,\eta_g)-\sqrt{\{\lambda_{88}(E,\eta_g)\}^2-4\lambda_{89}(E,\eta_g)}\right],$$

$$\lambda_{94}(E,\eta_g) \equiv \frac{\sqrt{\{\lambda_{91}(E,\eta_g)\}^2 - \{\lambda_{92}(E,\eta_g)\}^2}}{\lambda_{91}(E,\eta_g)},$$

$$F_i\left[\lambda_{93}(E,\eta_g), \lambda_{94}(E,\eta_g)\right]$$

is the incomplete elliptic integral of the 1st kind and is given by [123, 124],

$$F_i\left[\lambda_{93}(E,\eta_g), \lambda_{94}(E,\eta_g)\right] \equiv \int_0^{\lambda_{93}(E,\eta_g)} \left[\left\{1 - \{\lambda_{94}(E,\eta_g)\}^2 \sin^2\xi\right\}^{-1/2}\right] d\xi.$$

The DOS function in this case is given by

$$N_{HD}(E,\eta_g) = \frac{g_v}{4\pi^3}[\{\lambda_{87}(E,\eta_g)\}'\lambda_{95}(E,\eta_g) + \{\lambda_{95}(E,\eta_g)\}'\lambda_{87}(E,\eta_g) - \{\lambda_{90}(E,\eta_g)\}'] \quad (1.150)$$

Therefore the electron concentration can be expressed as

$$n_0 = \frac{g_v}{4\pi^3}\left[\bar{I}_{125}(E_{F_h},\eta_g) + \sum_{r=1}^{s} L(r)[\bar{I}_{125}(E_{F_h},\eta_g)]\right] \quad (1.151)$$

where,

$$\bar{I}_{125}(E_{F_h},\eta_g) = [\{\lambda_{87}(E_{F_h},\eta_g)\}\lambda_{95}(E_{F_h},\eta_g) - \{\lambda_{90}(E_{F_h},\eta_g)\}]$$

The 2D dispersion relation of the conduction electrons in QWs of IV–VI materials in the absence of band tails for the dimensional quantization along z direction can be expressed as

$$E(1+\alpha E) + \alpha E\left(\frac{\hbar^2 k_x^2}{2x_4} + \frac{\hbar^2 k_y^2}{2x_5}\right) + \alpha E\frac{\hbar^2}{2x_6}\left(\frac{n_z\pi}{d_z}\right)$$
$$- (1+\alpha E)\left(\frac{\hbar^2 k_x^2}{2x_1} + \frac{\hbar^2 k_y^2}{2x_2}\right)$$
$$- \alpha\left(\frac{\hbar^2 k_x^2}{2x_1} + \frac{\hbar^2 k_y^2}{2x_2}\right)\left(\frac{\hbar^2 k_x^2}{2x_4} + \frac{\hbar^2 k_y^2}{2x_5}\right)$$

$$- \alpha \left(\frac{\hbar^2 k_x^2}{2x_1} + \frac{\hbar^2 k_y^2}{2x_2} \right) \frac{\hbar^2}{2x_6} \left(\frac{n_z \pi}{d_z} \right)^2$$

$$- (1 + \alpha E) \frac{\hbar^2}{2x_3} \left(\frac{n_z \pi}{d_z} \right)^2$$

$$- \alpha \frac{\hbar^2}{2x_3} \left(\frac{n_z \pi}{d_z} \right)^2 \left(\frac{\hbar^2 k_x^2}{2x_4} + \frac{\hbar^2 k_y^2}{2x_5} \right)$$

$$- \alpha \frac{\hbar^2}{2x_3} \left(\frac{n_z \pi}{d_z} \right)^2 \frac{\hbar^2}{2x_6} \left(\frac{n_z \pi}{d_z} \right)^2$$

$$= \frac{\hbar^2 k_x^2}{2m_1} + \frac{\hbar^2 k_y^2}{2m_2} + \frac{\hbar^2}{2m_3} \left(\frac{n_z \pi}{d_z} \right)^2 \tag{1.152}$$

where

$$x_4 = m_t^+, \quad x_5 = \frac{m_t^+ + 2m_l^+}{3}, \quad x_6 = \frac{3m_t^+ m_l^+}{2m_l^+ + m_t^+},$$

$$x_1 = m_t^-, x_2 = \frac{m_t^- + 2m_l^-}{3}, \quad x_3 = \frac{3m_t^- m_l^-}{2m_l^- + m_t^-},$$

$$m_1 = m_t^*, m_2 = \frac{m_t^* + 2m_l^*}{3}$$

and

$$m_3 = \frac{3m_l^* m_t^*}{m_t^* + 2m_l^*}.$$

Therefore, the HD 2-D dispersion relation In this case assumes the form

$$\gamma_2 (E, \eta_g) + \alpha \gamma_3 (E, \eta_g) \left(\frac{\hbar^2 k_x^2}{2x_4} + \frac{\hbar^2 k_y^2}{2x_5} \right)$$

$$+ \alpha \gamma_3 (E, \eta_g) \frac{\hbar^2}{2x_6} \left(\frac{n_z \pi}{d_z} \right)^2$$

$$- (1 + \alpha \gamma_3 (E, \eta_g)) \left(\frac{\hbar^2 k_x^2}{2x_1} + \frac{\hbar^2 k_y^2}{2x_2} \right)$$

$$
\begin{aligned}
& -\alpha\left(\frac{\hbar^2 k_x^2}{2x_1}+\frac{\hbar^2 k_y^2}{2x_2}\right)\left(\frac{\hbar^2 k_x^2}{2x_4}+\frac{\hbar^2 k_y^2}{2x_5}\right) \\
& -\alpha\left(\frac{\hbar^2 k_x^2}{2x_1}+\frac{\hbar^2 k_y^2}{2x_2}\right)\frac{\hbar^2}{2x_6}\left(\frac{n_z\pi}{d_z}\right)^2 \\
& -\left(1+\alpha\gamma_3\left(E,\eta_g\right)\right)\frac{\hbar^2}{2x_3}\left(\frac{n_z\pi}{d_z}\right)^2 \\
& -\alpha\frac{\hbar^2}{2x_3}\left(\frac{n_z\pi}{d_z}\right)^2\left(\frac{\hbar^2 k_x^2}{2x_4}+\frac{\hbar^2 k_y^2}{2x_5}\right) \\
& -\alpha\frac{\hbar^2}{2x_3}\left(\frac{n_z\pi}{d_z}\right)^2\frac{\hbar^2}{2x_6}\left(\frac{n_z\pi}{d_z}\right)^2 \\
= & \frac{\hbar^2 k_x^2}{2m_1}+\frac{\hbar^2 k_y^2}{2m_2}+\frac{\hbar^2}{2m_3}\left(\frac{n_z\pi}{d_z}\right)^2 \qquad (1.153)
\end{aligned}
$$

Substituting, $k_x = r\cos\theta$ and $k_y = r\sin\theta$ (where r and θ are 2D polar coordinates in 2D wave vector space) in (1.153), we can write

$$
\begin{aligned}
& r^4\left[\alpha\frac{1}{4}\left(\frac{\hbar^2\cos^2\theta}{x_1}+\frac{\hbar^2\sin^2\theta}{x_2}\right)\left(\frac{\hbar^2\cos^2\theta}{x_4}+\frac{\hbar^2\sin^2\theta}{x_5}\right)\right] \\
& \quad +r^2\frac{1}{2}\left[\left(\frac{\hbar^2\cos^2\theta}{m_1}+\frac{\hbar^2\sin^2\theta}{m_2}\right)\right. \\
& \quad +\alpha\frac{\hbar^2}{2x_3}\left(\frac{n_z\pi}{d_z}\right)^2\left(\frac{\hbar^2\cos^2\theta}{x_4}+\frac{\hbar^2\sin^2\theta}{x_5}\right) \\
& \quad +\alpha\left(\frac{\hbar^2\cos^2\theta}{x_1}+\frac{\hbar^2\sin^2\theta}{x_2}\right)\frac{\hbar^2}{2x_6}\left(\frac{n_z\pi}{d_z}\right)^2 \\
& \quad +\hbar^2\left(1+\alpha\gamma_3\left(E,\eta_g\right)\right)\left(\frac{\cos^2\theta}{x_1}+\frac{\sin^2\theta}{x_2}\right) \\
& \quad -\hbar^2\alpha\gamma_3\left(E,\eta_g\right)\left(\frac{\cos^2\theta}{x_4}+\frac{\sin^2\theta}{x_5}\right)
\end{aligned}
$$

$$
-\left[\gamma_2\left(E,\eta_g\right)+\alpha\gamma_3\left(E,\eta_g\right)\frac{\hbar^2}{2x_6}\left(\frac{n_z\pi}{d_z}\right)^2\right.
$$

$$
\left.-\left(1+\alpha\gamma_3\left(E,\eta_g\right)\right)\frac{\hbar^2}{2x_3}\left(\frac{n_z\pi}{d_z}\right)^2-\alpha\left(\frac{\hbar^4}{4x_3x_6}\left(\frac{n_z\pi}{d_z}\right)^4\right)\right]=0 \tag{1.154}
$$

The area $A(E,n_z)$ of the 2D wave vector space can be expressed as

$$
A(E,n_z)=\bar{J}_1-\bar{J}_2 \tag{1.155}
$$

where

$$
\bar{J}_1\equiv 2\int_0^{\pi/2}\frac{c_1}{b_1}d\theta \tag{1.156}
$$

and

$$
\bar{J}_2\equiv 2\int_0^{\pi/2}\frac{ac_1^2}{b_1^3}d\theta \tag{1.157}
$$

in which

$$
a\equiv\left[\alpha\left(\frac{\hbar^4}{4}\right)\left(\frac{\cos^2\theta}{x_1}+\frac{\sin^2\theta}{x_2}\right)\left(\frac{\cos^2\theta}{x_4}+\frac{\sin^2\theta}{x_5}\right)\right],
$$

$$
b_1\equiv\left(\frac{\hbar^2}{2}\right)\left[\left(\frac{\cos^2\theta}{m_1}+\frac{\sin^2\theta}{m_2}\right)\right.
$$

$$
+\alpha\left(\frac{\hbar^2}{2x_3}\right)\left(\frac{n_z\pi}{d_z}\right)^2\left(\frac{\cos^2\theta}{x_4}+\frac{\sin^2\theta}{x_5}\right)
$$

$$
+\alpha\left(\frac{\hbar^2}{2x_6}\right)\left(\frac{n_z\pi}{d_z}\right)^2\left(\frac{\cos^2\theta}{m_1}+\frac{\sin^2\theta}{m_2}\right)
$$

$$
+\left(1+\alpha\gamma_3\left(E,\eta_g\right)\right)\left(\frac{\cos^2\theta}{x_1}+\frac{\sin^2\theta}{x_2}\right)
$$

$$
\left.-\alpha\gamma_3\left(E,\eta_g\right)\left(\frac{\cos^2\theta}{x_4}+\frac{\sin^2\theta}{x_5}\right)\right]
$$

and

$$c_1 \equiv \left[\gamma_2(E,\eta_g) + \alpha\gamma_3(E,\eta_g)\left(\frac{\hbar^2}{2x_6}\right)\left(\frac{n_z\pi}{d_z}\right)^2 - (1+\alpha\gamma_3(E,\eta_g))\left(\frac{\hbar^2}{2x_3}\right)\left(\frac{n_z\pi}{d_z}\right)^2 - \alpha\left(\frac{\hbar^4}{4x_3x_6}\right)\left(\frac{n_z\pi}{d_z}\right)^4\right]$$

The (1.156) can be expressed as

$$\bar{J}_1 = 2\int_0^{\pi/2} \frac{t_{31}(E,n_z)d\theta}{A_{11}(E,n_z)\cos^2\theta + B_{11}(E,n_z)\sin^2\theta}$$

where,

$$t_{31}(E,n_z) \equiv c_1, \quad A_{11}(E,n_z) \equiv \frac{\hbar^2}{2m_1}t_{11}(E,n_z),$$

$$t_{11}(E,n_z) \equiv \left[1 + m_1\left[\frac{1}{x_4}\frac{\alpha\hbar^2}{2x_3}\left(\frac{n_z\pi}{d_z}\right)^2 + \frac{\alpha\hbar^2}{2x_1x_6}\left(\frac{n_z\pi}{d_z}\right)^2 + \frac{1+\alpha\gamma_2(E,\eta_g)}{x_1} - \frac{\alpha\gamma_3(E,\eta_g)}{x_4}\right]\right]$$

$$B_{11}(E,n_z) \equiv \frac{\hbar^2}{2m_2}t_{21}(E,n_z)$$

and

$$t_{21}(E,n_z) \equiv \left[1 + m_2\left[\frac{\alpha\hbar^2}{2x_3x_5}\left(\frac{n_z\pi}{d_z}\right)^2 + \frac{\alpha\hbar^2}{2x_2x_6}\left(\frac{n_z\pi}{d_z}\right)^2 + \frac{1+\alpha\gamma_3(E,\eta_g)}{x_2} - \frac{\alpha\gamma_3(E,\eta_g)}{x_5}\right]\right].$$

Performing the integration, we get

$$\bar{J}_1 = \pi t_{31}(E,n_z)\left[A_{11}(E,n_z)B_{11}(E,n_z)\right]^{-1/2} \tag{1.158}$$

From (1.158) we can write

$$\bar{J}_2 = \frac{\alpha t_{31}^2(E,n_z)\hbar^4}{2B_{11}^3(E,n_z)}I \tag{1.159}$$

where,

$$I \equiv \int_0^\infty \frac{\left(a_1 + a_2 z^2\right)\left(a_3 + a_4 z^2\right) dz}{\left[(\bar{a})^2 + z^2\right]^3}, \quad (\bar{a})^2 = \left(\frac{A_{11}(E,n_z)}{B_{11}(E,n_z)}\right), \tag{1.160}$$

in which

$$a_1 \equiv \frac{1}{x_1}, \quad a_2 \equiv \frac{1}{x_2}, \quad z = \tan\theta, \theta$$

is a new variable,

$$a_3 \equiv \frac{1}{x_4}, \quad a_4 \equiv \frac{1}{x_5}$$

and

$$(\bar{a})^2 \equiv \left(\frac{A_1(E,n_z)}{B_1(E,n_z)}\right).$$

The use of the Residue theorem leads to the evaluation of the integral in (1.160) as

$$I = \frac{\pi}{4\bar{a}}\left[a_1 a_4 + 3a_2 a_4\right], \tag{1.161}$$

Therefore, the 2D area of the 2D wave vector space can be written as

$$A_{HD}(E,n_z) = \frac{\pi t_{31}(E,n_z)}{\sqrt{A_{11}(E,n_z)B_{11}(E,n_z)}} \times \left[1 - \frac{1}{x_5}\left(\frac{1}{x_1} + \frac{3}{x_2}\right)\frac{\alpha t_{31}(E,n_z)\hbar^4}{8B_{11}^2(E,n_z)}\right] \tag{1.162}$$

The EM for the HD QWs of IV–VI materials can thus be written as

$$m^*(E,n_z) = \frac{\hbar^2}{2}\left[\theta_{5HD}(E,n_z)\right]\Big|_{E=E_{F1HD}} \tag{1.163}$$

where,

$$\theta_{5HD}(E,n_z) \equiv \left[1 - \frac{1}{x_5}\left(\frac{1}{x_1} + \frac{3}{x_2}\right)\frac{\alpha t_{31}(E,n_z)\hbar^4}{8\left[B_{11}(E,n_z)\right]^2}\right] \times \left[A_{11}(E,n_z)B_{11}(E,n_z)\right]^{-1}$$

$$\times\left[\sqrt{A_{11}(E,n_z)B_{11}(E,n_z)}\,\{t_{31}(E,n_z)\}'\right.$$

$$-\,t_{31}(E,n_z)\left\{\frac{1}{2}\{A_{11}(E,n_z)\}'\left[\frac{B_{11}(E,n_z)}{A_{11}(E,n_z)}\right]^{1/2}\right.$$

$$\left.\left.+\frac{1}{2}\{B_{11}(E,n_z)\}'\left[\frac{A_{11}(E,n_z)}{B_{11}(E,n_z)}\right]^{1/2}\right\}\right]$$

$$-\frac{1}{8}\frac{t_{31}(E,n_z)\alpha\hbar^4}{\sqrt{A_{11}(E,n_z)B_{11}(E,n_z)}}\frac{1}{x_5}\left(\frac{1}{x_1}+\frac{3}{x_2}\right)$$

$$\times[B_{11}(E,n_z)]^{-4}[\{B_{11}(E,n_z)\}^2\{t_{31}(E,n_z)\}'$$

$$-\,2B_{11}(E,n_z)\{B_{11}(E,n_z)\}'t_{31}(E,n_z)]$$

Thus, the EM is a function of Fermi energy and the quantum number due to the band non-parabolicity.

The total DOS function can be written as

$$N_{2DT}(E)=\left(\frac{g_v}{2\pi}\right)\sum_{n_z=1}^{n_{z\max}}\theta_{5HD}(E,n_z)H\left(E-E_{n_{z7HD}}\right) \tag{1.164}$$

where the sub-band energy $(E_{n_{z7}})$ in this case can be written as

$$\begin{aligned}&\gamma_2\left(E_{n_{z7HD}},\eta_g\right)+\alpha\gamma_3\left(E_{n_{z7HD}},\eta_g\right)\frac{\hbar^2}{2x_6}\left(\frac{n_z\pi}{d_z}\right)^2\\&\quad-\left(1+\alpha\gamma_3\left(E_{n_{z7HD}},\eta_g\right)\right)\frac{\hbar^2}{2x_3}\left(\frac{n_z\pi}{d_z}\right)^2\\&\quad-\alpha\frac{\hbar^2}{2x_3}\left(\frac{n_z\pi}{d_z}\right)^2\frac{\hbar^2}{2x_6}\left(\frac{n_z\pi}{d_z}\right)^2-\left[\frac{\hbar^2}{2m_3}\left(\frac{n_z\pi}{d_z}\right)^2\right]=0\end{aligned} \tag{1.165}$$

The use (1.164) leads to the expression of 2D electron statistics as

$$n_{2D}=\frac{g_v}{2\pi}\sum_{n_z=1}^{n_{\max}}[T_{55HD}(E_{F1HD},n_z)+T_{56HD}(E_{F1HD},n_z)] \tag{1.166}$$

where

$$T_{55HD}(E_{F1HD}, n_z) \equiv \frac{A_{HD}(E_{F1HD}, n_z)}{\pi}$$

and

$$T_{56HD}(E_{F1HD}, n_z) \equiv \sum_{r=1}^{s} L(r) T_{55HD}(E_{F1HD}, n_z).$$

In the absence of band-tails the EM in QWs of IV–VI materials can be written as

$$m^*(E, n_z) = \frac{\hbar^2}{2} [\theta_5(E, n_z)]\Big|_{E=E_{Fs}} \tag{1.167}$$

where,

$$\begin{aligned}
\theta_5(E, n_z) \equiv & \left[1 - \frac{1}{x_5}\left(\frac{1}{x_1} + \frac{3}{x_2}\right)\frac{\alpha t_{30}(E, n_z)\hbar^4}{8\,[B_{10}(E, n_z)]^2}\right] \\
& \times [A_{10}(E, n_z) B_{10}(E, n_z)]^{-1} \\
& \times \left[\sqrt{A_{10}(E, n_z) B_{10}(E, n_z)}\,\{t_{30}(E, n_z)\}'\right. \\
& - \{t_{30}(E, n_z)\}\left\{\frac{1}{2}\{A_{10}(E, n_z)\}'\left[\frac{B_{10}(E, n_z)}{A_{10}(E, n_z)}\right]^{1/2}\right. \\
& \left.\left. + \frac{1}{2}\{B_{10}(E, n_z)\}'\left[\frac{A_{10}(E, n_z)}{B_{10}(E, n_z)}\right]^{1/2}\right\}\right] \\
& - \frac{1}{8}\frac{t_{30}(E, n_z)\alpha\hbar^4}{\sqrt{A_{10}(E, n_z) B_{10}(E, n_z)}}\frac{1}{x_5}\left(\frac{1}{x_1} + \frac{3}{x_2}\right)[B_{10}(E, n_z)]^{-4} \\
& \times [\{B_{10}(E, n_z)\}^2 \{t_{30}(E, n_z)\}' \\
& - 2B_{10}(E, n_z)\{B_{10}(E, n_z)\}' t_{30}(E, n_z)]
\end{aligned}$$

where

$$t_{30}(E, n_z) = c_0,$$

$$c_0 \equiv \left[E(1+\alpha E) + \alpha E\left(\frac{\hbar^2}{2x_6}\right)\left(\frac{n_z\pi}{d_z}\right)^2 - (1+\alpha E)\left(\frac{\hbar^2}{2x_3}\right)\left(\frac{n_z\pi}{d_z}\right)^2 - \alpha\left(\frac{\hbar^2}{4x_3x_6}\right)\left(\frac{n_z\pi}{d_z}\right)^4\right],$$

$$A_{10}(E,n_z) \equiv \frac{\hbar^2}{2m_1}t_{10}(E,n_z),$$

$$t_{10}(E,n_z) \equiv \left[1 + m_1\left[\frac{1}{x_4}\frac{\alpha\hbar^2}{2x_3}\left(\frac{n_z\pi}{d_z}\right)^2 + \frac{\alpha\hbar^2}{2x_1x_6}\left(\frac{n_z\pi}{d_z}\right)^2 + \frac{1+\alpha E}{x_1} - \frac{\alpha E}{x_4}\right]\right],$$

$$B_{10}(E,n_z) \equiv \frac{\hbar^2}{2m_2}t_{20}(E,n_z)$$

and

$$t_{20}(E,n_z) \equiv \left[1 + m_2\left[\frac{\alpha\hbar^2}{2x_3x_5}\left(\frac{n_z\pi}{d_z}\right)^2 + \frac{\alpha\hbar^2}{2x_2x_6}\left(\frac{n_z\pi}{d_z}\right)^2 + \frac{1+\alpha E}{x_2} - \frac{\alpha E}{x_5}\right]\right]$$

Thus, the EM is a function of Fermi energy and the quantum number due to the band non-parabolicity.

The total DOS function can be written as

$$N_{2DT}(E) = \left(\frac{g_v}{2\pi}\right)\sum_{n_z=1}^{n_{z\max}} \theta_5(E,n_z)H(E-E_{n_{z7}}) \tag{1.168}$$

where the sub-band energy $(E_{n_{z7}})$ in this case can be written as

$$E_{n_{z7}}(1+\alpha E_{n_{z7}}) + \alpha E_{n_{z7}}\frac{\hbar^2}{2x_6}\left(\frac{n_z\pi}{d_z}\right)^2 - (1+\alpha E_{n_{z7}})\frac{\hbar^2}{2x_3}\left(\frac{n_z\pi}{d_z}\right)^2 - \alpha\frac{\hbar^2}{2x_3}\left(\frac{n_z\pi}{d_z}\right)^2\frac{\hbar^2}{2x_6}\left(\frac{n_z\pi}{d_z}\right)^2 - \left[\frac{\hbar^2}{2m_3}\left(\frac{n_z\pi}{d_z}\right)^2\right] = 0 \tag{1.169}$$

In the absence of band-tails, the expression of 2D electron statistics can be written as

$$n_{2D} = \frac{g_v}{2\pi} \sum_{n_z=1}^{n_{z\max}} [T_{550}(E_{Fs}, n_z) + T_{560}(E_{Fs}, n_z)] \tag{1.170}$$

where,

$$T_{550}(E_{Fs}, n_z) \equiv \frac{A_0(E_{Fs}, n_z)}{\pi},$$

$$A_0(E, n_z) = \frac{\pi t_{30}(E, n_z)}{\sqrt{A_{10}(E, n_z)B_{10}(E, n_z)}} \times \left[1 - \frac{1}{x_5}\left(\frac{1}{x_1} + \frac{3}{x_2}\right)\frac{\alpha t_{30}(E, n_z)\hbar^4}{8B_{10}^2(E, n_z)}\right],$$

and

$$T_{560}(E_{Fs}, n_z) = \sum_{r=1}^{s} L(r)[T_{550}(E_{Fs}, n_z)] \tag{1.171}$$

For bulk specimens of IV–VI materials, the expression of electron concentration assumes the forms

$$n_0 = \left(\frac{g_v}{2\pi^2}\right)[M_{A4}(E_{F_b}) + N_{A_4}(E_{F_b})] \tag{1.172}$$

$$\omega_{A_1} = \left[\frac{\alpha^2}{16}\left[\frac{1}{m_t^- m_l^+} + \frac{1}{m_l^- m_t^+}\right]^2 - \frac{\alpha^2}{4m_l^+ m_t^- m_l^- m_t^+}\right],$$

$$J_{A_1}(E_{F_b}) = \frac{A_A(E_{F_b})}{3}[-(A_A^2(E_{F_b}) + B_A^2(E_{F_b}))E(\lambda, q) + 2B_A^2(E_{F_b})F(\lambda, q)] + \frac{\bar{\tau}_{A_1}(E_{F_b})}{3} \times \left[(\bar{\tau}_{A_1}(E_{F_b}))^2 + A_A^2(E_{F_b}) + 2B_A^2(E_{F_b})\right] \times \left[A_A^2(E_{F_b}) + \bar{\tau}_{A_1}^2(E_{F_b})\right]^{1/2}\left[B_A^2(E_{F_b}) + \bar{\tau}_{A_1}^2(E_{F_b})\right]^{-1/2}$$

$$\lambda = \tan^{-1}\frac{\bar{\tau}_A(E_{F_b})}{B_A(E_{F_b})}, \quad q = \left[\frac{\sqrt{A_A^2(E_{F_b}) - B_A^2(E_{F_b})}}{A_A(E_{F_b})}\right],$$

$$A_A(E_{F_b}) = \frac{[\tau_{A_2}(E_{F_b}) + \sqrt{\tau_{A_2}^2(E_{F_b}) - 4\tau_{A_3}(E_{F_b})}]^{1/2}}{\sqrt{2}},$$

$$B_A(E_{F_b}) = \frac{\left[\tau_{A_2}(E_{F_b}) - \sqrt{\tau_{A_2}^2(E_{F_b}) - 4\tau_{A_3}(E_{F_b})}\right]^{1/2}}{\sqrt{2}},$$

$$\tau_{A_2}(E_{F_b}) = \frac{\omega_{A_2}(E_{F_b})}{\omega_{A_1}^2}, \quad \tau_{A_3}(E_{F_b}) = \frac{\omega_{A_3}(E_{F_b})}{\omega_{A_1}^2},$$

$$\omega_{A_2}(E_{F_b}) = \left[\frac{\alpha}{2}\left[\frac{1}{2m_t^*} - \frac{\alpha \cdot E_{F_b}}{2m_t^+} + \frac{1+\alpha \cdot E_{F_b}}{2m_t^-}\right] \cdot \left[\frac{1}{m_t^- m_l^+} + \frac{1}{m_l^- m_t^+}\right]\right.$$
$$\left. - \frac{\alpha}{m_t^+ m_t^-}\left[\frac{1}{2m_l^*} + \frac{\alpha \cdot E_{F_b}}{2m_l^+} + \frac{1+\alpha \cdot E_{F_b}}{2m_l^-}\right]\right]$$

$$\omega_{A_3}(E_{F_b}) = \left[\frac{\alpha \cdot E_{F_b}(1+\alpha \cdot E_{F_b})}{m_t^+ m_t^-} + \left[\frac{1}{2m_t^*} - \frac{\alpha \cdot E_{F_b}}{2m_t^+} + \frac{1+\alpha \cdot E_{F_b}}{2m_t^-}\right]^2\right],$$

$$\alpha_2(E_{F_b}) = \left[\frac{1}{2m_t^*} - \frac{\alpha.E_{F_b}}{2m_t^+} + \frac{1+\alpha.E_{F_b}}{2m_t^-}\right],$$

$$\alpha_3 = \frac{\alpha\hbar^2}{4}\left[\frac{1}{m_t^- m_l^+} + \frac{1}{m_l^- m_t^+}\right],$$

$$\tau_{A_1}(E_{F_b}) = \left[\frac{2m_l^+ m_l^-}{\alpha\hbar^2}\right]^{1/2}\left[-\left[\frac{1}{2m_l^*} + \frac{1+\alpha \cdot E_{F_b}}{m_l^-} - \frac{\alpha \cdot E_{F_b}}{2m_l^+}\right]\right.$$
$$+ \left[\left[\frac{1}{2m_l^*} + \frac{1+\alpha \cdot E_{F_b}}{m_l^-} - \frac{\alpha \cdot E_{F_b}}{2m_l^+}\right]^2\right.$$
$$\left.\left. + \frac{\alpha \cdot E_{F_b}(1+\alpha \cdot E_{F_b})}{m_l^- m_l^+}\right]^{1/2}\right]^{1/2} \quad E(\lambda, q)$$

is the in complete Elliptic integral of second kind, $F(\lambda, q)$ is the incomplete Elliptic integral of first kind,

$$N_{A_4}(E_{Fb}) = \sum_{r=1}^{s} L(r)\,[M_{A_4}(E_{Fb})],$$

$$I_{17}\left(E'\right) = \frac{-I_{15}\left(E'\right) + \sqrt{I_{15}^2\left(E'\right) + 4I_{16}\left(E'\right)}}{2},$$

$$I_{15}\left(E'\right) = \frac{2m_l^+ m_t^-}{\alpha\hbar^2}\left[\frac{1}{m_l^*} + \frac{1}{m_l^-} + \alpha E'\left(\frac{1}{m_l^-} - \frac{1}{m_l^+}\right)\right]$$

and

$$I_{16}\left(E'\right) = \frac{4m_l^+ m_t^-}{\alpha\hbar^4} E'(1 + \alpha E')$$

(b) The Model of Bangert and Kastner

The dispersion relation of the conduction electrons in bulk specimens of IV–VI semiconductors in accordance with the model of Bangert and Kastner [147] is given by

$$\omega_1(E)k_s^2 + \omega_2(E)k_z^2 = 1 \tag{1.173}$$

where

$$\omega_1(\mathrm{E}) = (2E)^{-1}\left[\frac{(\bar{\mathrm{R}})^2}{\mathrm{E_{g_0}}(1+\alpha_1\mathrm{E})} + \frac{(\bar{\mathrm{S}})^2}{\Delta_c'(1+\alpha_2\mathrm{E})} + \frac{(\bar{\mathrm{Q}})^2}{\Delta_c''(1+\alpha_3\mathrm{E})}\right] \text{ and}$$

$$\omega_2(\mathrm{E}) = (2E)^{-1}\left[\frac{(\bar{\mathrm{A}})^2}{\mathrm{E_{g0}}(1+\alpha_1\mathrm{E})} + \frac{(\bar{\mathrm{S}}+\bar{\mathrm{Q}})^2}{\Delta_c''(1+\alpha_3\mathrm{E})}\right],$$

$$(\overline{R})^2 = 2.3\times 10^{-10}(eVm)^2, \quad (\bar{S})^2 = 4.6(\bar{R})^2,$$

$$\alpha_1 = \frac{1}{\mathrm{E_{g0}}}, \quad \alpha_2 = \frac{1}{\Delta_c'}, \quad \alpha_3 = \frac{1}{\Delta_c''},$$

$$\Delta_c'' = 3.28eV, \quad \Delta_c' = 3.07eV, \quad (\bar{Q})^2 - 1.3(\bar{R})^2,$$

$$(\bar{A})^2 = 0.8\times 10^{-4}(eVm)^2$$

The electron energy spectrum in heavily doped IV–VI materials in this case can be expressed as

$$2I(4) = k_s^2\left[\{c_1(\alpha_1, E, E_g) - iD_1(\alpha_1, E, E_g)\}\frac{\left(\bar{R}\right)^2}{E_{g_0}}\right.$$

$$+ \{c_2(\alpha_2, E, E_g) - iD_2(\alpha_2, E, E_g)\}\frac{\left(\bar{S}\right)^2}{\Delta_c'}$$

$$+\{c_3(\alpha_3,E,E_g)-iD_3(\alpha_3,E,E_g)\}\frac{(\bar{Q})^2}{\Delta_c''}\Bigg]$$

$$+k_z^2\left[\frac{2(\bar{A})^2}{E_{g0}}\{c_1(\alpha_1,E,E_g)-iD_1(\alpha_1,E,E_g)\}\right.$$

$$\left.+\frac{(\bar{S}+\bar{Q})^2}{\Delta_c''}\{c_3(\alpha_3,E,E_g)-iD_3(\alpha_3,E,E_g)\}\right] \quad (1.174)$$

where

$$\alpha_1=\frac{1}{E_g},\quad \alpha_2=\frac{1}{\Delta_c'},\quad \alpha_3=\frac{1}{\Delta_c''},\quad G_i=\frac{1+\alpha_i E}{\eta_g\alpha_i},$$

$$c_i(\alpha_i,E,\eta_g)=\left[\frac{2}{\alpha_i\eta_g\sqrt{\pi}}\right]\exp\left(-u_i^2\right)$$

$$\times\left[\sum_{p=1}^{\infty}\left\{\exp\left(\frac{-p^2}{4}\right)(\sinh(pu_i))\right\}p^{-1}\right]$$

$$i=1,2,3\quad D_i(\alpha_i,E,\eta_g)=\left[\frac{\sqrt{\pi}}{\alpha_i\eta_g}\right]\exp(-u_i^2),$$

Therefore (1.174) can be written as,

$$F_1(E,\eta_g)k_s^2+F_2(E,\eta_g)k_z^2=1 \quad (1.175)$$

where,

$$F_1(E,\eta_g)=[2\gamma_0(E,\eta_g)]^{-1}\left[\frac{(\bar{R})^2}{E_g}\{C_1(\alpha_1,E,E_g)-iD_1(\alpha_1,E,E_g)\}\right.$$

$$+\frac{(\bar{S})^2}{\Delta_c'}\{C_2(\alpha_2,E,E_g)-iD_2(\alpha_2,E,E_g)\}$$

$$\left.+\frac{(\bar{Q})^2}{\Delta_c''}\{C_3(\alpha_3,E,E_g)-iD_3(\alpha_3,E,E_g)\}\right] \quad \text{and}$$

$$F_2(E,\eta_g) = [2\gamma_0(E,\eta_g)]^{-1}\left[\frac{2(\bar{A})^2}{E_g}\{C_1(\alpha_1,E,\eta_g) - iD_1(\alpha_1,E,\eta_g)\}\right.$$
$$\left. + \frac{(\bar{S}+\bar{Q})^2}{\Delta_c''}\{C_3(\alpha_3,E,\eta_g) - iD_3(\alpha_3,E,\eta_g)\}\right]$$

Since $F_1(E,\eta_g)$ and $F_2(E,\eta_g)$ are complex, the energy spectrum is also complex in the presence of Gaussian band tails.

The EMs can be written as

$$m_{\perp}^{*}(E_{F_h},\eta_g) = \left(\frac{\hbar^2}{2}\right) \text{ Real part of } \left(\frac{F_1'(E_{F_h},\eta_g)}{F_1^2(E_{F_h},\eta_g)}\right) \tag{1.176}$$

$$m_{11}^{*}(E_{F_h},\eta_g) = \left(\frac{\hbar^2}{2}\right) \text{ Real part of } \left(\frac{F_2'(E_{F_h},\eta_g)}{F_2^2(E_{F_h},\eta_g)}\right) \tag{1.177}$$

It appears then that, the evolution of the masses needs an expression of the carrier concentration, which in turn is determined by the DOS function.

The DOS function in this case can be expressed as

$$N_{HD}(\mathrm{E},\eta_g)\frac{g_v}{3\pi^2}F_3'(\mathrm{E},\eta_g), F_3(\mathrm{E},\eta_g) = [F_1(\mathrm{E},\eta_g)\sqrt{F_2}(\mathrm{E},\eta_g)]^{-1} \tag{1.178}$$

The electron concentration is given by

$$n_0 = \frac{g_v}{3\pi^2} \text{ Real part of } \left[F_3(\mathrm{E}_{F_h},\eta_g) + \sum_{r=1}^{s} L(r)[F_3(\mathrm{E}_{F_h},\eta_g)]\right] \tag{1.179}$$

The 2D dispersion relation in this case assumes the form

$$k_s^2 = F_6(E,\eta_g,n_z) \tag{1.180}$$

where,

$$F_6(E,\eta_g,n_z) = \left[\frac{[1 - F_2(E,\eta_g)(n_z\pi/d_z)^2]}{F_1(E,\eta_2)}\right]$$

The EM in this case is given by

$$m^*(E_{F1HD},\eta_g,n_z)=\frac{\hbar^2}{2}\text{ Real part of }[F_6'(E_{F1HD},\eta_g,n_z)] \quad (1.181)$$

The total DOS function can be written as

$$N_{2DT}(\mathrm{E})\frac{g_v}{2\pi}\sum_{n_z=1}^{n_{z\max}}F_6'(E,\eta_g,n_z)H(E-E_{n_z71HD}) \quad (1.182)$$

where E_{n_z71HD} the quantized energy in this case and is given by

$$1=F_2(E_{nz71HD},\eta_g)\left(\frac{\pi n_z}{d_z}\right)^2 \quad (1.183)$$

The surface electron concentration can be expressed as

$$n_0=\frac{g_v}{2\pi}\text{ Real part of }\left[\sum_{n_z=1}^{n_{z\max}}\left[F_6(\mathrm{E}_{F1HD},\eta_g,\mathrm{n}_z)\right.\right.$$
$$\left.\left.+\sum_{r=1}^{s}L(r)[F_6(\mathrm{E}_{F1HD},\eta_g,\mathrm{n}_z)]\right]\right] \quad (1.184)$$

In the absence of band-tails the EMs can be written as

$$m_\perp^*(E_F)=\left(\frac{\hbar^2}{2}\right)\left(\frac{F_{11}'(E_F)}{F_{11}^2(E_F)}\right) \quad (1.185)$$

$$m_{11}^*(E_F)=\left(\frac{\hbar^2}{2}\right)\left(\frac{F_{12}'(E_F)}{F_{12}^2(E_F)}\right) \quad (1.186)$$

where

$$F_{11}(E)=\left[\frac{(\bar{\mathrm{R}})^2}{E_{g0}(1+\alpha_1E)}+\frac{(\bar{S})^2}{\Delta_c'(1+\alpha_2E)}+\frac{(\bar{Q})^2}{\Delta_{g0}''(1+\alpha_3E)}\right][2E^{-1}]$$

and

$$F_{12}(E)=\left[\frac{(\bar{A})^2}{E_{g0}(1+\alpha_1E)}+\frac{(\bar{S}+\bar{Q})^2}{\Delta_{g0}''(1+\alpha_3E)}\right][2E^{-1}]$$

It appears then that, the evolution of the masses needs an expression of the carrier concentration, which in turn is determined by the DOS function.

The DOS function in this case can be expressed as

$$N(\mathrm{E}) = \frac{g_v}{3\pi^2} F'_{13}(\mathrm{E}), \quad F_{13}(\mathrm{E}) = [F_{11}(\mathrm{E})\sqrt{F_{12}(\mathrm{E})}]^{-1} \tag{1.187}$$

The electron concentration is given by

$$n_0 = \frac{g_v}{3\pi^2}\left[F_{13}(\mathrm{E}_F) + \sum_{r=1}^{s} L(\mathrm{r})[F_{13}(\mathrm{E}_F)]\right] \tag{1.188}$$

In the absence of band-tails, the 2D dispersion relation in this case assumes the form

$$k_s^2 = F_{16}(E, n_z) \tag{1.189}$$

where,

$$F_{16}(E, n_z) = \left[\frac{[1 - F_{12}(E)(n_z\pi/d_z)^2]}{F_{11}(E)}\right]$$

The EM in this case is given by

$$m^*(E_{F_s}, n_z) = \frac{\hbar^2}{2}[F'_{16}(E_{f_s}, n_z)] \tag{1.190}$$

The total DOS function can be written as

$$N_{2DT}(\mathrm{E}) = \frac{g_v}{2\pi}\sum_{n_z=1}^{n_{z\max}} F'_{16}(E, n_z)H(E - E_{n_z711}) \tag{1.191}$$

where E_{n_z711} is the quantized energy in this case and is given by

$$1 = F_{12}(E_{nz711}, \eta_g)(\pi n_z/d_z)^2 \tag{1.192}$$

The surface electron concentration can be expressed as

$$n_0 = \frac{g_v}{2\pi}\left[\sum_{n_z=1}^{n_{z\max}} [F_{16}(\mathrm{E}_{F_s}, \mathrm{n}_z) + \sum_{r=1}^{s} L(r)[F_{16}(\mathrm{E}_{F_s}, \mathrm{n}_z)]]\right] \tag{1.193}$$

1.2.5 *The EM in Quantum Wells (QWs) of HD stressed Kane type materials*

The electron energy spectrum in stressed Kane type semiconductors can be written [148] as

$$\left(\frac{k_x}{\bar{a}_0(E)}\right)^2 + \left(\frac{k_y}{\bar{b}_0(E)}\right)^2 + \left(\frac{k_z}{\bar{c}_0(E)}\right)^2 = 1 \tag{1.194}$$

where,

$$[\bar{a}_0(E)]^2 \equiv \frac{\bar{K}_0(E)}{\bar{A}_0(E) + \frac{1}{2}\bar{D}_0(E)},$$

$$\bar{K}_0(E) \equiv \left[E - C_1\varepsilon - \frac{2C_2^2\varepsilon_{xy}^2}{3E_g'}\right]\left(\frac{3E_g'}{2B_2^2}\right),$$

C_1 is the conduction band deformation potential, ε is the trace of the strain tensor $\hat{\varepsilon}$ which can be written as

$$\hat{\varepsilon} = \begin{bmatrix} \varepsilon_{xx} & \varepsilon_{xy} & 0 \\ \varepsilon_{xy} & \varepsilon_{yy} & 0 \\ 0 & 0 & \varepsilon_{zz} \end{bmatrix},$$

C_2 is a constant which describes the strain interaction between the conduction and valance bands, $E_g' \equiv E_g + E - C_1\varepsilon, B_2$ is the momentum matrix element,

$$\bar{A}_0(E) \equiv \left[1 - \frac{(\bar{a}_0 + C_1)}{E_g'} + \frac{3\bar{b}_0\varepsilon_{xx}}{2E_g'} - \frac{\bar{b}_0\varepsilon}{2E_g'}\right], \quad \bar{a}_0 \equiv -\frac{1}{3}(\bar{b}_0 + 2\bar{m}),$$

$$\bar{b}_0 \equiv \frac{1}{3}(\bar{l} - \bar{m}), \quad \bar{d}_0 \equiv \frac{2\bar{n}}{\sqrt{3}},$$

$\bar{l}, \bar{m}, \bar{n}$ are the matrix elements of the strain perturbation operator, $\bar{D}_0(E) \equiv (\bar{d}_0\sqrt{3})\frac{\varepsilon_{xy}}{E_g'}$,

$$\left[\bar{b}_0(E)\right]^2 \equiv \frac{\bar{K}_0(E)}{\bar{A}_0(E) - \frac{1}{2}\bar{D}_0(E)}, \quad [\bar{c}_0(E)]^2 \equiv \frac{\bar{K}_0(E)}{\bar{L}_0(E)} \quad \text{and}$$

$$\bar{L}_0(E) \equiv \left[1 - \frac{(\bar{a}_0 + C_1)}{E_g'} + \frac{3\bar{b}_0\varepsilon_{zz}}{E_g'} - \frac{\bar{b}_0\varepsilon}{2E_g'}\right]$$

The use of (1.194) can be written as

$$(E-\alpha_1)k_x^2+(E-\alpha_2)k_y^2+(E-\alpha_3)k_z^2=t_1E^3-t_2E^2+t_3E+t_4 \tag{1.195a}$$

Where

$$\alpha_1 \equiv \left[E_g - C_1\varepsilon - (\bar{a}_0+C_1)\varepsilon + \frac{3}{2}\bar{b}_0\varepsilon_{xx} - \frac{\bar{b}_0}{2}\varepsilon + \left(\frac{\sqrt{3}}{2}\right)\varepsilon_{xy}\bar{d}_0\right],$$

$$\alpha_2 \equiv \left[E_g - C_1\varepsilon - (\bar{a}_0+C_1)\varepsilon + \frac{3}{2}\bar{b}_0\varepsilon_{xx} - \frac{\bar{b}_0}{2}\varepsilon - \left(\frac{\sqrt{3}}{2}\right)\varepsilon_{xy}\bar{d}_0\right],$$

$$\alpha_3 \equiv \left[E_g - C_1\varepsilon - (\bar{a}_0+C_1)\varepsilon + \frac{3}{2}\bar{b}_0\varepsilon_{zz} - \frac{\bar{b}_0}{2}\varepsilon\right],\quad t_1 \equiv \left(\frac{3}{2}B_2^2\right),$$

$$t_2 \equiv \left(\frac{1}{2}B_2^2\right)[6(E_g - C_1\varepsilon) + 3C_1\varepsilon],$$

$$t_3 \equiv \left(\frac{1}{2}B_2^2\right)\left[3(E_g - C_1\varepsilon)^2 + 6C_1\varepsilon(E_g - C_1\varepsilon) - 2C_2^2\varepsilon_{xy}^2\right] \quad \text{and}$$

$$t_4 \equiv \left(\frac{1}{2}B_2^2\right)\left[-3C_1\varepsilon(E_g - C_1\varepsilon)^2 + 2C_2^2\varepsilon_{xy}^2\right].$$

(1.195a) can be written as

$$Ek^2 - T_{17}k_x^2 - T_{27}k_y^2 - T_{37}k_z^2 = [q_{67}E^3 - R_{67}E^2 + V_{67}E + \rho_{67}] \tag{1.195b}$$

where,

$$\mathrm{T}_{17} = \alpha_1, \quad \mathrm{T}_{27} = \alpha_2, \quad \mathrm{T}_{37} = \alpha_3, \quad \mathrm{t}_1 = \mathrm{q}_{67},$$

$$\mathrm{t}_2 = \mathrm{R}_{67}, \quad \mathrm{t}_3 = \mathrm{V}_{67} \quad \text{and} \quad \mathrm{t}_4 = \rho_{67}$$

Under the condition of heavy doping, (1.186b) can be written as

$$I(4)k^2 - T_{17}I(1)k_x^2 - T_{27}I(1)k_y^2 - T_{37}k_z^2I(1) = [q_{67}I(6) - R_{67}I(5) + V_{67}I(4) + \rho_{67}I(1)] \tag{1.195c}$$

where,

$$I(6) = \int_{-\infty}^{E}(E-V)^3F(V)dV \tag{1.196}$$

(1.196) can be written as

$$I(6) = E^3 I(1) - 3E^2 I(7) + 3EI(8) - I(9) \tag{1.197}$$

In which,

$$I(7) = \int_{-\infty}^{E} VF(V)dV \tag{1.198}$$

$$I(8) = \int_{-\infty}^{E} V^2 F(V)dV \tag{1.199}$$

$$I(9) = \int_{-\infty}^{E} V^3 F(V)dV \tag{1.200}$$

Using (1.4), together with simple algebraic manipulations, one obtains

$$I(7) = \frac{-\eta_g}{2\sqrt{\pi}} \exp\left(\frac{-E^2}{\eta_g^2}\right) \tag{1.201}$$

$$I(8) = \frac{\eta_g^2}{4}\left[1 + Erf\left(\frac{E}{\eta_g}\right)\right] \tag{1.202}$$

and

$$I(9) = \frac{-\eta_g^3}{2\sqrt{\pi}} \exp\left(\frac{-E^2}{\eta_g^2}\right)\left[1 + \frac{E^2}{\eta_g^2}\right] \tag{1.203}$$

Thus (1.197) can be written as

$$I(6) = \left[\frac{E}{2}\left[1 + Erf\left(\frac{E}{\eta_g}\right)\right]\left[E^2 + \frac{3}{2}\eta_g^2\right] + \frac{\eta_g}{2\sqrt{\pi}} \exp\left(\frac{-E^2}{\eta_g^2}\right)\left[4E^2 + \eta_g^2\right]\right] \tag{1.204}$$

Thus, combining the appropriate equations, the dispersion relations of the conduction electrons in HD stressed materials can be expressed as

$$P_{11}(E,\eta_g)k_x^2 + Q_{11}(E,\eta_g)k_y^2 + S_{11}(E,\eta_g)k_z^2 = 1 \tag{1.205}$$

where,

$$P_{11}(E,\eta_g) \equiv \left[\frac{\gamma_0(E,\eta_g) - \left(\frac{T_{17}}{2}\right)\left[1 + Erf\left(\frac{E}{\eta_g}\right)\right]}{\Delta_{14}(E,\eta_g)}\right],$$

$$\Delta_{14}(E,\eta_g) \equiv \left[q_{67}\left\{\frac{E}{2}\left[1 + Erf\left(\frac{E}{\eta_g}\right)\right]\left[E^2 + \frac{3}{2}\eta_g^2\right] + \frac{\eta_g}{2\sqrt{\pi}}\exp\left(\frac{-E^2}{\eta_g^2}\right)[4E^2 + \eta_g^2]\right\} - R_{67}\theta_0(E,\eta_g) + V_{67}\gamma_0(E,\eta_g) + \frac{\rho_{67}}{2}\left[1 + Erf\left(\frac{E}{\eta_g}\right)\right]\right],$$

$$Q_{11}(E,\eta_g) \equiv \left[\frac{\gamma_0(E,\eta_g) - \left(\frac{T_{27}}{2}\right)\left[1 + Erf\left(\frac{E}{\eta_g}\right)\right]}{\Delta_{14}(E,\eta_g)}\right]$$

and

$$S_{11}(E,\eta_g) \equiv \left[\frac{\gamma_0(E,\eta_g) - \left(\frac{T_{37}}{2}\right)\left[1 + Erf\left(\frac{E}{\eta_g}\right)\right]}{\Delta_{14}(E,\eta_g)}\right].$$

Thus, the energy spectrum in this case is real since the dispersion relation of the corresponding materials in the absence of band tails as given by (1.194) has no poles in the finite complex plane.

The EMs along x, y and z directions in this case can be written as

$$m_{xx}^*(E_{F_h},\eta_g) = \frac{\hbar^2}{2}\left[[\gamma_0(E_{F_h},\eta_g) - (T_{17}/2)[1 + Erf(E_{F_h}/\eta_g)]]^{-2} \times [\{\Delta_{14}(E_{F_h},\eta_g)\}'[\gamma_0(E_{F_h},\eta_g) - (T_{17}/2)[1 + Erf(E_{F_h}/\eta_g)]]]\right.$$

$$
- \Delta_{14}(E_{F_h}, \eta_g)\left[\frac{1}{2}\left[1 + Erf\left(\frac{E_{F_h}}{\eta_g}\right)\right]\right.
$$

$$
\left.\left.- \left\{\frac{T_{17}}{\eta_g\sqrt{\pi}}\exp\left(\frac{-E_{F_h}^2}{\eta_g^2}\right)\right\}\right]\right] \tag{1.206}
$$

$$
m_{yy}^*(E_{F_h}, \eta_g) = \frac{\hbar^2}{2}\left[[\gamma_0(E_{F_h}, \eta_g) - (T_{27}/2)[1 + Erf(E_{F_h}/\eta_g)]]^{-2}\right.
$$

$$
\times [\{\Delta_{14}(E_{F_h}, \eta_g)\}'[\gamma_0(E_{F_h}, \eta_g)
$$

$$
- (T_{27}/2)[1 + Erf(E_{F_h}/\eta_g)]]]
$$

$$
- \Delta_{14}(E_{F_h}, \eta_g)\left[\frac{1}{2}\left[1 + Erf\left(\frac{E_{F_h}}{\eta_g}\right)\right]\right.
$$

$$
\left.\left.- \left\{\frac{T_{27}}{\eta_g\sqrt{\pi}}\exp\left(\frac{-E_{F_h}^2}{\eta_g^2}\right)\right\}\right]\right] \tag{1.207}
$$

and

$$
m_{zz}^*(E_{F_h}, \eta_g) = \frac{\hbar^2}{2}\left[[\gamma_0(E_{F_h}, \eta_g) - (T_{37}/2)[1 + Erf(E_{F_h}/\eta_g)]]^{-2}\right.
$$

$$
\times [\{\Delta_{14}(E_{F_h}, \eta_g)\}'[\gamma_0(E_{F_h}, \eta_g)
$$

$$
- (T_{37}/2)[1 + Erf(E_{F_h}/\eta_g)]]]
$$

$$
- \Delta_{14}(E_{F_h}, \eta_g)\left[\frac{1}{2}\left[1 + Erf\left(\frac{E_{F_h}}{\eta_g}\right)\right]\right.
$$

$$
\left.\left.- \left\{\frac{T_{37}}{\eta_g\sqrt{\pi}}\exp\left(\frac{-E_{F_h}^2}{\eta_g^2}\right)\right\}\right]\right] \tag{1.208}
$$

Thus, we can see that the EMs in this case exist within the band gap.

In the absence of band tails, $\eta_g \to 0$ we get

$$
m_{xx}^*(E_F) = \hbar^2\bar{a}_0(E_F)\{\bar{a}_0(E_F)\}' \tag{1.209}
$$

$$
m_{xx}^*(E_F) = \hbar^2\bar{b}_0(E_F)\{\bar{b}_0(E_F)\}' \tag{1.210}
$$

and

$$m^*_{xx}(E_F) = \hbar^2 \bar{c}_0(E_F)\{\bar{c}_0(E_F)\}' \tag{1.211}$$

The DOS function in this case can be written as

$$N_{HD}(E, \eta_g) = \frac{g_v}{3\pi^2} \Delta_{100}(E, \eta_g) \tag{1.212}$$

where

$$\Delta_{100}(E, \eta_g)\{\Delta_{15}(E, \eta_g)\}^{-2} \left[\frac{3}{2} \{\Delta_{15}(E, \eta_g)\} \sqrt{\Delta_{14}(E, \eta_g)} \{\Delta_{14}(E, \eta_g)\}' - \{\Delta_{14}(E, \eta_g)\}^{3/2} \{\Delta_{15}(E, \eta_g)\}' \right]$$

and

$$\begin{aligned} \Delta_{15}(E, \eta_g) \equiv [[\gamma_0(E, \eta_g) - (T_{17}/2)[1 + Erf(E/\eta_g)]] \\ \times [\gamma_0(E, \eta_g) - (T_{27}/2)[1 + Erf(E/\eta_g)]] \\ \times [\gamma_0(E, \eta_g) - (T_{37}/2)[1 + Erf(E/\eta_g)]]]^{1/2}. \end{aligned}$$

Using (1.212), the electron concentration at can be written as

$$n_0 = \frac{g_v}{3\pi^2} \left[\bar{I}_{126}(E_{F_k}, \eta_g) + \sum_{r=1}^{s} L(r)[\bar{I}_{126}(E_{F_h}, \eta_g)] \right] \tag{1.213}$$

where,

$$\bar{I}_{126}(E_{F_k}, \eta_g) = \left[\frac{\{\Delta_{14}(E_{F_k}, \eta_g)\}^{3/2}}{\Delta_{15}(E_{F_k}, \eta_g)} \right]$$

The dispersion relation of the conduction electrons in HD QWs of Kane type semiconductors can be written as

$$P_{11}(E, \eta_g)k_x^2 + Q_{11}(E, \eta_g)k_y^2 + S_{11}(E, \eta_g)(\pi n_z/d_z)^2 = 1 \tag{1.214}$$

The EM can be expressed as

$$m^*(E_{F1HD}, \eta_g, n_z) = \frac{\hbar^2}{2} A'_{56}(E_{F1HD}, \eta_g, n_z)$$

where,

$$A_{56}(E, \eta_g, n_z) = \frac{\pi[1 - S_{11}(E, \eta_g)(n_z\pi/d_z)^2]}{\sqrt{P_{11}(E, \eta_g), Q_{11}(E, \eta_g)}} \tag{1.215}$$

From (1.215), it appears that the EM is a function of Fermi energy, and size quantum number and the same mass exists in the band gap.

Thus, the total 2D DOS function can be expressed as

$$N_{2DT}(E) = \left(\frac{g_v}{2\pi}\right) \sum_{n_z=1}^{n_{z\max}} A'_{56}(E_{F1HD}, \eta_g, n_z) \tag{1.216}$$

The sub-band energies $(E_{n_{z8}HD})$ are given by

$$S_{11}(E_{n_{z8}HD}, \eta_g)(\pi n_z/d_z)^2 = 1 \tag{1.217}$$

The 2D surface electron concentration per unit area for QWs of stressed HD Kane type compounds can be written as

$$n_{2D} = \frac{g_v}{2\pi} \sum_{n_z=1}^{n_{z\max}} [T_{57HD}(E_{Fs1HD}, \eta_g, n_z) + T_{58HD}(E_{Fs1HD}, \eta_g, n_z)] \tag{1.218}$$

where,

$$T_{57}(E_{F1HD}, \eta_g, n_z) \equiv A_{56}(E_{F1HD}, \eta_g, n_z)$$

and

$$T_{58HD}(E_{F1HD}, \eta_g, n_z) \equiv \sum_{r=1}^{s} L(r) T_{57HD}(E_{F1HD}, \eta_g, n_z).$$

In the absence of band tails, the 2D electron energy spectrum in QWs of stressed materials assumes the form

$$\frac{k_x^2}{[\bar{a}_0(E)]^2} + \frac{k_y^2}{[\bar{b}_0(E)]^2} + \frac{1}{[\bar{c}_0(E)]^2}(n_z\pi/d_z)^2 = 1 \tag{1.219}$$

The area of 2D wave vector space enclosed by (1.219) can be written as

$$A(E, n_z) = \pi P^2(E, n_z)\bar{a}_0(E)\bar{b}_0(E)$$

where

$$P^2(E, n_z) = [1 - [n_z\pi/d_z\bar{c}_0(E)]^2].$$

From (1.219), the EM can be written as

$$m^*(E_{F_s}, n_z) = \frac{\hbar^2}{2}[P^2(E_{F_s}, n_z)\bar{a}_0(E_{F_s})\bar{b}_0(E_{F_s})]' \tag{1.220}$$

Thus, the total 2D DOS function can be expressed as

$$N_{2DT}(E) = \left(\frac{g_v}{2\pi}\right) \sum_{n_z=1}^{n_{z\max}} \theta_6(E, n_z) H(E - E_{n211}) \tag{1.221}$$

in which,

$$\begin{aligned}\theta_6(E, n_z) = {} & [2P(E, n_z)\{P(E, n_z)\}' \bar{a}_0(E)\bar{b}_0(E) \\ & + \{P(E, n_z)\}^2 \{\bar{a}_0(E)\}' \bar{b}_0(E) \\ & + \{P(E, n_z)\}^2 \{\bar{b}_0(E)\}' \bar{a}_0(E)]\end{aligned}$$

The sub-band energies ($E_{nz_{11}}$) are given by

$$\bar{c}_0(E_{n_{z11}}) = n_z\pi/d_z \tag{1.222}$$

The 2D surface electron concentration per unit area for QWs of stressed Kane type compounds can be written as

$$n_{2D} = \frac{g_v}{2\pi} \sum_{n_z=1}^{n_{z\max}} [T_{61}(E_{Fs}, n_z) + T_{62}(E_{Fs}, n_z)] \tag{1.223}$$

where

$$T_{61}(E_{Fs}, n_z) \equiv [P^2(E_{Fs}, n_z)\bar{a}_0(E_{Fs})\bar{b}_0(E_{Fs})]$$

and

$$T_{62}(E_{Fs}, n_z) \equiv \sum_{r=1}^{s} L(r) T_{61}(E_{Fs}, n_z).$$

The DOS function for bulk specimens of stressed Kane type semiconductors in the absence of band tail can be written as

$$D_0(E) = g_v(3\pi^2)^{-1}\bar{T}_0(E) \tag{1.224}$$

where

$$\begin{aligned}\bar{T}_0(E) = {} & [\bar{a}_0(E)\bar{b}_0(E)[\bar{c}_0(E)]' + \bar{a}_0(E)[\bar{b}_0(E)]'\bar{c}_0(E) \\ & + [\bar{a}_0(E)]'\bar{b}_0(E)\bar{c}_0(E)]\end{aligned}$$

Combining (1.224) *with the Fermi-Dirac occupation probability factor and using the generalized Sommerfeld lemma the electron concentration in this case can be expressed as*

$$n_0 = g_v(3\pi^2)^{-1}[M_4(E_F) + N_4(E_F)] \tag{1.225}$$

where,

$$M_4(E_F) \equiv [\bar{a}_0(E_F)\bar{b}_0(E_F)\bar{c}_0(E_F)]$$

and

$$N_4(E_F) \equiv \sum_{r=1}^{s} L(r)M_4(E_F).$$

1.2.6 *The EM in Quantum Wells (QWs) of HD Te*

The dispersion relation of the conduction electrons in Te can be expressed as [149]

$$E = \psi_1 k_z^2 + \psi_2 k_s^2 \pm [\psi_3^2 k_z^2 + \psi_4^2 k_s^2]^{1/2} \tag{1.226}$$

where, the values of the system constants are given in Table 1.1.

The carrier energy spectrum in HD Te can be written as

$$\gamma_3(E, \eta_g) = \psi_1 k_z^2 + \psi_2 k_s^2 \pm [\psi_3^2 k_z^2 + \psi_4^2 k_s^2]^{1/2} \tag{1.227}$$

The EMs along k_z and k_s directions assume the forms

$$m_z^*(E_{F_h}, \eta_g) = \frac{\hbar^2}{2\psi_1}\left[1 - \frac{\psi_3}{\sqrt{\psi_3^2 + 4\psi_1\gamma_3(E_{F_k}, \eta_g)}}\right]\gamma_3'(E_{F_h}, \eta_g) \tag{1.228}$$

and

$$m_s^*(E_{F_h}, \eta_g) = \frac{\hbar^2}{2\psi_2}\left[1 - \frac{\psi_4}{\sqrt{\psi_4^2 + 4\psi_2\gamma_3(E_{F_h}, \eta_g)}}\right]\gamma_3'(E_{F_h}, \eta_g) \tag{1.229}$$

The investigations of EMs require the expression of electron concentration, which can be written from (1.227) as

$$n_0 = \frac{g_v}{3\pi^2}[t_{1HD}(E_{F_h}, \eta_g) + t_{2HD}(E_{F_h}, \eta_g)] \tag{1.230}$$

where,

$$t_{1HD}(E_{F_h}, \eta_g) = [3\psi_{5HD}(E_{F_h}, \eta_g)\Gamma_{3HD}(E_{F_h}, \eta_g) - \psi_6\Gamma_{3HD}^3(E_{F_h}, \eta_g)],$$

$$\psi_{5HD}(E_{F_h},\eta_g) = \left[\frac{\gamma_3(E_{F_h},\eta_g)}{\psi_2} + \frac{\psi_4^2}{2\psi_2^2}\right],$$

$$\Gamma_{3HD}(E_{F_h},\eta_g) = \frac{\sqrt{\psi_3^2 + 4\psi_1\gamma_3(E_{F_h},\eta_g)}}{2\psi_1}, \quad \psi_6 = \frac{\psi_1}{\psi_2}$$

and

$$t_{2HD}(E_{F_h},\eta_g) = \sum_{r=1}^{s} L(r)t_{1HD}(E_{F_h},\eta_g)$$

The 2D electron energy spectrum in HD QW of Te can be written using (1.227) as

$$k_s^2 = \psi_{5HD}(E,\eta_g) - \psi_6\left(\frac{\pi n_z}{d_z}\right)^2 \pm \psi_7\left[\psi_{8HD}^2(E,\eta_g) - \left(\frac{\pi n_z}{d_z}\right)^2\right]^{1/2} \tag{1.231}$$

where,

$$\psi_7 = \frac{\psi_4\sqrt{\psi_1}}{\psi_2^{3/2}}$$

and

$$\psi_{8HD}^2(E,\eta_g) = \left[\frac{\psi_4^4 + 4\gamma_3(E,\eta_g)\psi_2\psi_4^2 + 4\psi_2^2\psi_3^2}{4\psi_1\psi_2\psi_4^2}\right]$$

The EM in this case is given by

$$m^*\left(E_{F1HD},\eta_g,n_z\right) = \frac{\hbar^2}{2}\left[\psi'_{5HD}\left(E_{F1HD},\eta_g\right) \pm \frac{\psi_{8HD}\left(E_{F1HD},\eta_g\right)\psi'_{8HD}\left(E_{F1HD},\eta_g\right)}{\sqrt{\psi_{8HD}^2\left(E_{F1HD},\eta_g\right) - (\pi n_z/d_z)^2}}\right] \tag{1.232}$$

The total DOS function in this case can be expressed as

$$N_{2DT}(E) = \frac{g_v}{\pi}\sum_{n_z=1}^{n_{z\max}} \psi'_{5HD}\left(E,\eta_g\right)H\left(E - E_{n_{z59HD}}\right) \tag{1.233}$$

where $E_{n_{z59HD}}$ is the lowest positive root of the equation

$$\psi_{5HD}(E_{n_{z59HD}}, \eta_g) - \psi_6 \left(\frac{\pi n_z}{d_z}\right)^2 \pm \psi_7 \left[\psi_{8HD}^2(E_{n_{z59HD}}, \eta_g) - \left(\frac{\pi n_z}{d_z}\right)^2\right]^{1/2} = 0 \quad (1.234)$$

The surface electron concentration is given by

$$n_{2D} = \frac{g_v}{\pi} \sum_{n_z=1}^{n_{z\max}} [t_{1HDTe}(E_{F1HD}, \eta_g, n_z) + t_{2HDTe}(E_{F1HD}, \eta_g, n_z)] \quad (1.235)$$

where,

$$t_{1HDTe}(E_{F1HD}, \eta_g, n_z) = \left[\psi_{5HD}(E_{F1HD}, \eta_g, n_z) - \psi_6 \left(\frac{\pi n_z}{d_z}\right)^2\right]$$

and

$$t_{2HDTe}(E_{F1HD}, \eta_g, n_z) = \sum_{r=1}^{s} L(r)[t_{1HDTe}(E_{F1HD}, \eta_g, n_z)]$$

The 2D electron energy spectrum in QWs of Te in the absence of band tails assumes the form

$$k_s^2 = \psi_5(E) - \psi_6 \left(\frac{\pi n_z}{d_z}\right)^2 \pm \psi_7 \left[\psi_8^2(E) - \left(\frac{\pi n_z}{d_z}\right)^2\right]^{1/2} \quad (1.236)$$

where,

$$\psi_5(E) = \left[\frac{E}{\psi_2} + \frac{\psi_4^2}{2\psi_2^2}\right]$$

and

$$\psi_8^2(E) = \frac{\psi_4^4 + 4E\psi_2\psi_4^2 + 4\psi_2^2\psi_3^2}{4\psi_1\psi_2\psi_4^2}$$

Thus, the total 2D DOS function can be expressed as

$$N_{2DT}(E) = \left(\frac{g_v}{\pi}\right) \sum_{n_z=1}^{n_{z\max}} t'_{40}(E, n_z) H\left(E - E_{n_{z12}}\right) \quad (1.237a)$$

where,

$$t_{40}(E, n_z) = \left[\psi_5(E) - \psi_6\left(\frac{\pi n_z}{d_z}\right)^2 \pm \psi_7\left[\psi_8^2(E) - \left(\frac{\pi n_z}{d_z}\right)^2\right]^{1/2}\right]^{1/2}$$

The sub-band energies ($E_{n_{z12}}$) are given by

$$E_{n_{z12}} = \psi_1(n_z\pi/d_z)^2 \pm \psi_3(n_z\pi/d_z) \quad (1.237b)$$

Using (1.236) the EM can be expressed as

$$m^*(E_{F_s}, n_z) = \frac{\hbar^2}{2} t'_{40}(E_{F_s}, n_z) \quad (1.237c)$$

The 2D surface electron concentration per unit area for QWs of Te can be written as

$$n_{2D} = \frac{g_v}{\pi} \sum_{n_z=1}^{n_{z\max}} [t_{40}(E_{Fs}, n_z) + t_{41}(E_{Fs}, n_z)] \quad (1.237d)$$

where

$$t_{41}(E_{Fs}, n_z) \equiv \sum_{r=1}^{s} L(r) t_{40}(E_{Fs}, n_z).$$

1.2.7 *The EM in Quantum Wells (QWs) of HD Gallium Phosphide*

The energy spectrum of the conduction electrons in n-GaP can be written as [150]

$$E = \frac{\hbar^2 k_s^2}{2m_\perp^*} + \frac{\hbar^2}{2m_\parallel^*}\left[\bar{A}' k_s^2 + k_z^2\right] - \left[\frac{\hbar^4 k_0^2}{m_\parallel^{*2}}\left(k_s^2 + k_z^2\right) + |V_G|^2\right]^{1/2} + |V_G| \quad (1.238)$$

where, k_0 and $|V_G|$ are constants of the energy spectrum and $\bar{A}' = 1$.

The dispersion relation of the conduction electrons in HD n-GaP can be expressed as

$$\gamma_3(E,\eta_g) = \frac{\hbar^2 k_s^2}{2m_\perp^*} + \frac{\hbar^2}{2m_\parallel^*}\left[\bar{A}' k_s^2 + k_z^2\right] - \left[\frac{\hbar^4 k_0^2}{m_\parallel^{*2}}\left(k_s^2 + k_z^2\right) + |V_G|^2\right]^{1/2} - |V_G| \quad (1.239)$$

The EMs assume the forms as

$$m_z^*(E_{F_h},\eta_g) = \frac{\hbar^2 \gamma_3'(E_{F_h},\eta_g)}{b}[1 \pm (C + bD) \times [C^2 + 4bD^2 + 4bC\gamma_3(E_{F_h},\eta_g) - 4bCD + 4b^2\gamma_3(E_{F_h},\eta_g)D]^{-1/2}] \quad (1.240)$$

and

$$m_s^*(E_{F_h},\eta_g) = \frac{\hbar^2}{2}[t_{11}\gamma_3'(E_{F_h},\eta_g) - t_{41}t_5'(E_{F_h},\eta_g)] \quad (1.241)$$

where,

$$b = \frac{\hbar^2}{2m_\parallel^*}, \quad C = \left(\frac{\hbar^2 k_0}{m_\parallel^*}\right)^2, \quad D = |V_G|, t_{11} = \frac{1}{a},$$

$$a = \frac{\hbar^2}{2m_\perp^*} + \bar{A}' b, \quad t_{41} = \frac{\sqrt{g_3}}{2a^2},$$

$$g_3 = (4abc + 4a^2 c), \quad t_5^2(E_{F_h},\eta_g) = [g_2 - 4aC\gamma_3(E_{F_h},\eta_g)](g_3)^{-1},$$

$$g_2 = \left(4a^2 b^2 + C^2 + 4aCD\right)$$

The electron concentration can be expressed as

$$n_0 = \frac{g_v}{4\pi^2}\left[\bar{I}_{127}(E_{F_h},\eta_g) + \sum_{r=1}^{s} L(r)[\bar{I}_{127}(E_{F_h},\eta_g)]\right] \quad (1.242)$$

where,

$$\bar{I}_{127}(E_{F_h},\eta_g) = [M_{1HD}(E_{F_h},\eta_g)],$$

$$M_{1HD}(E_{F_h},\eta_g) = \Bigg[2(t_{11}\gamma_3(E_{F_h},\eta_g)+t_{21})\sqrt{t_{81}+t_{91}\gamma_3(E_{F_h},\eta_g)}$$
$$+\left(\frac{t_{31}}{3}\right)\theta^3_{,-}(E_{F_h},\eta_g)$$
$$+\left(\frac{t_{41}}{2}\right)[\theta_{,-}(E_{F_h},\eta_g)\sqrt{\theta^2_{-}(E_{F_h},\eta_g)+t_5(E_{F_h},\eta_g)}$$
$$-\sqrt{t_5(E_{F_h},\eta_g)}]+\left(\frac{t_{41}t_5(E_{F_h},\eta_g)}{2}\right)\ln$$
$$\times\left|\frac{\theta_{,-}(E_{F_h},\eta_g)+\sqrt{\theta^2_{-}(E_{F_h},\eta_g)+t_5(E_{F_h},\eta_g)}}{\sqrt{t_5(E_{F_h},\eta_g)}}\right|\Bigg],$$

$$t_{21}=\frac{g_1}{2a^2},\quad g_1=-(C+2aD),$$

$$t_{81}=[t^4_{41}+4t^2_{41}t_{21}t_{31}+(4t^2_{31}t^2_{41}g_2)(g_3)^{-1}],\quad t_{31}=\frac{b}{a},$$

$$t_{91}=[4t_{11}t_{31}t^2_{41}+8t_{11}t_{21}t^2_{31}-(16t^2_{31}t^2_{41}aC)(g_3)^{-1}],$$

$$\theta_{-}(E_{F_h},\eta_g)=(t_{31}\sqrt{2})^{-1}[t_{61}+t_{71}\gamma_3(E_{F_h},\eta_g)$$
$$-\sqrt{t_{81}+t_{91}\gamma_3(E_{F_h},\eta_g)}],$$

$$t_{61}=(t^2_{41}+2t_{21}t_{31})$$

and

$$t_{71}=(2t_{11}t_{31})$$

The 2D dispersion relation in QW of *HD* GaP can be expressed following (1.239) as

$$k_s^2=t_{11}\gamma_3(E,\eta_g)+t_{21}-t_{31}\left(\frac{\pi n_z}{d_z}\right)^2-t_{41}\left[\left(\frac{\pi n_z}{d_z}\right)^2+t_5^2(E,\eta_g)\right]^{1/2} \tag{1.243}$$

The EM in this case can be written following (1.243) as

$$m^*(E_{F1HD},\eta_g,n_z) = \frac{\hbar^2}{2}\left[t_{11}\gamma_3'(E_{F1HD},\eta_g) - t_{41}t_5(E_{F1HD},\eta_g)t_5'(E_{F1HD},\eta_g) \times \left[\left(\frac{\pi n_z}{d_z}\right)^2 + t_5^2(E_{F1HD},\eta_g)\right]^{-1/2}\right] \quad (1.244)$$

The total DOS function assumes the form

$$N_{2DT}(E,\eta_g) = \frac{g_v}{2\pi}\sum_{n_z=1}^{n_{z\max}}\left[t_{11}\gamma_3'(E_{F1HD},\eta_g) - t_{41}t_5(E_{F1HD},\eta_g)t_5'(E_{F1HD},\eta_g) \times \left[\left(\frac{\pi n_z}{d_z}\right) + t_5^2(E_{F1HD},\eta_g)\right]^{1/2}\right] H(E - E_{nz8THD}) \quad (1.245)$$

where, $E_{n_{z8THD}}$ is given by the equation

$$t_{11}\gamma_3(E_{n_{z8THD}}) + t_{21} - t_{31}\left(\frac{\pi n_z}{d_z}\right)^2 - t_{41}\left[\left(\frac{\pi n_z}{d_z}\right)^2 + t_5^2(E_{n_{z8THD}},\eta_g)\right]^{1/2} = 0 \quad (1.246)$$

The surface electron concentration in QW of HD n-GaP can be written as

$$n_s = \frac{g_v}{\pi}\sum_{n_z=1}^{n_{z\max}}[t_{3HDGaP}(E_{F1HD},\eta_g,n_z) + t_{4HDGaP}(E_{F1HD},\eta_g,n_z)] \quad (1.247)$$

where,

$$t_{1HDGap}(E_{F1HD},\eta_g,n_z) = t_{11}\gamma_3(E_{F1HD},\eta_g,n_z) + t_{21}\left(\frac{\pi n_z}{d_z}\right)^2$$

$$- t_{41}\left[\left(\frac{\pi n_z}{d_z}\right)^2 + t_5^2(E_{F1HD},\eta_g,n_z)\right]^{1/2}$$

$$t_{4HDGap}(E_{F1HD},\eta_g,n_z) = \sum_{r=1}^{s} L(r)[t_{3HDGap}(E_{F1HD},\eta_g,n_z)]$$

The 2D electron dispersion relation in size-quantized n-GaP in the absence of band tails assumes the form

$$E = ak_s^2 + c(n_z\pi/d_z)^2 + |V_G| - [Dk_s^2 + |V_G|^2 + D(n_z\pi/d_z)^2]^{1/2} \quad (1.248)$$

The sub-band energy ($E_{n_{z13}}$) are given by

$$E_{n_{z13}} = c(\pi n_z/d_z)^2 + |V_G| - [|V_G|^2 + D(\pi n_z/d_z)^2]^{1/2} \quad (1.249)$$

(1.248) can be expressed as

$$k_s^2 = t_{42}(E,n_z) \quad (1.250)$$

in which,

$$t_{42}(E,n_z) \equiv \Big[\{2a(E-t_1)+D\} - \{[2a(E-t_1)+D]^2 - 4a^2[(E-t_1)^2 - t_2]\}^{1/2}\Big],$$

$$t_1 \equiv |V_G| + C(\pi n_z/d_z)^2$$

and

$$t_2 \equiv |V_G|^2 + C(\pi n_z/d_z)^2.$$

The total DOS function is given by

$$N_{2DT}(E) = \frac{g_v}{4\pi a^2}\sum_{n_z=1}^{n_{z\max}}[t'_{42}(E,n_z)]H(E - E_{n_z13}) \quad (1.251a)$$

Using (1.251a) the EM can be expressed as

$$m^*(E_{F_s},n_z) = \frac{\hbar^2}{2}t'_{42}(E_{F_s},n_z) \quad (1.251b)$$

The electron statistics in QWs in n-GaP assumes the form

$$n_{2D} = \left[\left(\frac{g_v}{4\pi a^2}\right)\sum_{n_z=1}^{n_{z\max}}[t_{42}(E_{Fs},n_z) + t_{43}(E_{Fs},n_z)]\right] \quad (1.252)$$

where

$$t_{43}(E_{Fs}, n_z) = \sum_{r=1}^{s} L(r)[t_{42}(E_{Fs}, n_z)].$$

The EMs in bulk specimens of n-GaP in the absence of band tails can be written as

$$m_s^*(E_F) = \frac{\hbar^2}{2}[t_{11} - t_{41}t_5'(E_F)] \quad (1.253)$$

and

$$m_z^*(E_F) = \frac{\hbar^2}{b}[1 - C[4bCE_F + 4b^2D^2 + C^2 - 4bCD]^{-1/2}] \quad (1.254)$$

where

$$t_5(E_F) = \left[\frac{g_2 - 4aCE_F}{g_3}\right]^{1/2}$$

The electron concentration in this case assumes the form

$$n_0 = \frac{g_v}{4\pi^2}[M_1(E_F) + N_1(E_F)] \quad (1.255)$$

where

$$M_1(E_F) = \left[2(t_{11}E_F + t_{21})\sqrt{t_{91}E_F + t_{81}} + \frac{t_{31}}{3}\phi^3(E_F)\right.$$
$$+ \frac{t_{41}}{2}\left[\phi(E_F)\sqrt{\phi^2(E_F) + t_5(E_F)}\right.$$
$$\left.\left. + \frac{t_{41}t_5(E_F)}{2}\left[\ln\left|\frac{\phi(E_F) + \sqrt{\phi^2(E_F) + t_5(E_F)}}{\sqrt{t_5(E_F)}}\right|\right]\right]\right],$$

$$\phi(E_F) = \left(t_{31}\sqrt{2}\right)^{-1}\left[t_{61} + E_F t_{71} - [t_{81} + t_{91}E_F]^{1/2}\right]$$

$$N_1(E_F) = \sum_{r=1}^{s_0}[L(r)M_1(E_F)]$$

1.2.8 *The EM in Quantum Wells (QWs) of HD Platinum Antimonide*

The dispersion relation for the n-type $PtSb_2$ can be written as [151]

$$\left[E+\lambda_0\frac{a^2}{4}k^2-lk_s^2\frac{a^2}{4}\right]\left[E+\delta_0-v\frac{a^2}{4}k^2-n'k_s^2\frac{a^2}{4}\right]=I\left(\frac{a^4}{16}\right)k^4 \tag{1.256}$$

(1.256) assumes the form

$$[E+\omega_1k_s^2+\omega_2k_z^2][E+\delta_0+\omega_3k_s^2-\omega_4k_z^2]=I_1(k_s^2+k_z^2)^2 \tag{1.257}$$

where

$$\omega_1=\left[\lambda_0\frac{a^2}{4}+l\frac{a^2}{4}\right],\quad \omega_2=\lambda_0\frac{a^2}{4},\omega_3=\left[n'\frac{a^2}{4}-\nu\frac{a^2}{4}\right],$$
$$\omega_4=\nu\frac{a^2}{4},I_1=I\left(\frac{a^2}{4}\right)^2,$$

$\lambda_0, l, \delta_0, v, n'$ and a are the band constants.

The carrier dispersion law in HD $PtSb_2$ can be written as

$$T_{11}k_s^4-k_s^2[T_{21}(E,\eta_g)-T_{31}k_z^2]+[T_{41}k_z^4-T_{51}(E,\eta_g)k_z^2-T_{61}(E,\eta_g)] = 0 \tag{1.258}$$

where,

$$T_{11}=(I_1-\omega_2\omega_3),\quad T_{21}(E,\eta_g)=[\omega_1\delta_0+\omega_1\gamma_3(E,\eta_g)+\omega_3\gamma_3(E,\eta_g)],$$
$$T_{31}=[2I_1+\omega_2\omega_4-\omega_2\omega_3],\quad T_{41}=[2I_1+\omega_2\omega_4],$$
$$T_{51}(E,\eta_g)=[\omega_2\gamma_0-\omega_4\gamma_3(E,\eta_g)+\omega_2\gamma_3(E,\eta_g)],$$
$$T_{61}(E,\eta_g)=[\gamma_8(E,\eta_g)+\gamma_0\gamma_3(E,\eta_g)]$$

and

$$\gamma_8(E,\eta_g)=2\theta_0(E,\eta_g)[1+Erf(E/\eta_g)]^{-1}$$

The EMs are given by

$$m_s^*(E_{F_h},\eta_g) = \frac{\hbar^2}{2T_{11}}\left[T_{21}'(E_{F_h},\eta_g) + \frac{(T_{21}(E_{F_h},\eta_g)T_{21}'(E_{F_h},\eta_g) + 2T_{11}T_{61}'(E_{F_h},\eta_g)}{\sqrt{T_{21}^2(E_{F_h},\eta_g) + 4T_{11}T_{61}'(E_{F_h},\eta_g)}}\right] \tag{1.259}$$

and

$$m_z^*(E_{F_h},\eta_g) = \left(\frac{\hbar^2}{2T_{41}}\right)[T_{51}'(E_{F_h},\eta_g) + [T_{51}'(E_{F_h},\eta_g)T_{51}'(E_{F_h},\eta_g) + 2T_{41}T_{61}'(E_{F_h},\eta_g)] \times [T_{51}^2(E_{F_h},\eta_g) + 4T_{41}T_{61}(E_{F_h},\eta_g)]^{-1/2}] \tag{1.260}$$

The electron concentration assumes the form

$$n_0 = \frac{g_v}{3\pi^2}\left[\bar{I}_{128}(E_{F_h},\eta_g) + \sum_{r=1}^{s} L(r)[\bar{I}_{128}(E_{F_h},\eta_g)]\right] \tag{1.261}$$

where,

$$\bar{I}_{128}(E_{F_h},\eta_g) = [M_{6HD}(E_{F_h},\eta_g)],$$

$$M_{6HD}(E_{F_h},\eta_g) = \left[T_{91HD}(E_{F_h},\eta_g)\rho_{2HD}(E_{F_h},\eta_g) - T_{101}\frac{\rho_{2HD}^3(E_{F_h},\eta_g)}{3} - T_{11}J_3(E_{F_h},\eta_g)\right]$$

$$T_{91HD}(E_{F_h},\eta_g) = \frac{T_{21}(E_{F_h},\eta_g)}{2T_{11}},$$

$$\rho_{2HD}(E_{F_h},\eta_g = [(2T_{41})^{-1}[T_{51}(E_{F_h},\eta_g) + \sqrt{T_{51}^2(E_{F_h},\eta_g) + 4T_{41}T_{61}(E_{F_h},\eta_g)}]]^{1/2}$$

$$T_{101} = [T_{31}/2T_{11}]$$

$$J_3(E_{F_h}, \eta_g) = \frac{\rho_{2HD}(E_{F_h}, \eta_g)}{3}[[A_{3HD}^2(E_{F_h}, \eta_g) + B_{3HD}^2(E_{F_h}, \eta_g)]E_0(\eta(E_{F_h}, \eta_g), t(E_{F_h}, \eta_g)) - [A_{3HD}^2(E_{F_h}, \eta_g) - B_{3HD}^2(E_{F_h}, \eta_g)]F_0(\eta(E_{F_h}, \eta_g), t(E_{F_h}, \eta_g))] + \frac{\rho_{2HD}(E_{F_h}, \eta_g)}{3}[(A_{3HD}^2(E_{F_h}, \eta_g) - \rho_{2HD}^2(E_{F_h}, \eta_g))(B_{3HD}^2(E_{F_h}, \eta_g) - \rho_{2HD}^2(E_{F_h}, \eta_g))]^{1/2}, \quad E_0(\eta(E_{F_h}, \eta_g), \quad t(E_{F_h}, \eta_g))$$

and

$$F_0(\eta(E_{F_h}, \eta_g), \quad t(E_{F_h}, \eta_g))$$

are the incomplete elliptic integrals of second and first respectively.

$$A_{3HD}^2(E_{F_h}, \eta_g) = \frac{1}{2}\left[T_{12}(E_{F_h}, \eta_g) + \sqrt{T_{12}^2(E_{F_h}, \eta_g) - 4T_{13}(E_{F_h}, \eta_g)}\right],$$

$$T_{12}(E_{F_h}, \eta_g) = \left[\frac{T_7(E_{F_h}, \eta_g)}{\bar{T}_{61}}\right]$$

$$\bar{T}_{61} = [T_{31}^2 - 4T_{11}T_{41}],$$

$$T_7(E_{F_h}, \eta_g) = [2T_{31}T_{21}(E_{F_h}, \eta_g) - 4T_{11}T_{51}(E_{F_h}, \eta_g)],$$

$$T_{13}(E_{F_h}, \eta_g) = \left(\frac{T_8(E_{F_h}, \eta_g)}{\bar{T}_8}\right),$$

$$T_8(E_{F_h}, \eta_g) = [T_{21}^2(E_{F_h}, \eta_g) + 4T_{11}T_{61}(E_{F_h}, \eta_g)],$$

$$B_{3HD}^2(E_{F_h}, \eta_g) = \frac{1}{2}\left[T_{12}(E_{F_h}, \eta_g) - \sqrt{T_{12}^2(E_{F_h}, \eta_g) - 4T_{13}(E_{F_h}, \eta_g)}\right],$$

$$\bar{T}_{11} = \left[\frac{\sqrt{\bar{T}_{61}}}{2T_{11}}\right]$$

$$t(E_{F_h}, \eta_g) = \left[\frac{B_3(E_{F_h}, \eta_g)}{A_3(E_{F_h}, \eta_g)}\right], \quad \eta(E_{F_h}, \eta_g) = \sin^{-1}\left[\frac{\rho_2(E_{F_h}, \eta_g)}{B_3(E_{F_h}, \eta_g)}\right]$$

From (1.258) the dispersion relation in QWs of HD $PtSb_2$ can be expressed as

$$T_{11}k_s^4 - P_{1HD}(E,\eta_g,n_z)k_s^2 + P_{2HD}(E,\eta_g,n_z) = 0 \qquad (1.262)$$

where,

$$P_{1HD}(E,\eta_g,n_z) = \left[T_{21}(E_{F_h},\eta_g) - T_{31}\left(\frac{\pi n_z}{d_z}\right)\right]$$

$$P_{2HD}(E,\eta_g,n_z) = \left[T_{41}\left(\frac{\pi n_z}{d_z}\right)^4 - T_{51}(E_{F_h},\eta_g)\left(\frac{\pi n_z}{d_z}\right)^2 - T_{61}(E_{F_h},\eta_g)\right]$$

(1.262) can be written as

$$k_s^2 = A_{60}(E,\eta_g,n_z) \qquad (1.263)$$

where,

$$A_{60}(E,\eta_g,n_z) = \Big[P_{1HD}(E,\eta_g,n_z) - \sqrt{P_{1HD}^2(E,\eta_g,n_z) - 4T_{11}P_{2HD}(E,\eta_g,n_z)}\Big]$$

The EM assumes the form

$$m^*(E_{F1HD},\eta_g,n_z) = \frac{\hbar^2}{2}A'_{60}(E_{F1HD},\eta_g,n_z) \qquad (1.264)$$

The surface electron concentration is given by

$$n_0 = \frac{g_v}{2\pi}\sum_{n_z=1}^{n_{z\max}}[A_{60}(E_{F1HD},\eta_g,n_z) + B_{60}(E_{F1HD},\eta_g,n_z)] \qquad (1.265)$$

where,

$$B_{60}(E_{F1HD},\eta_g,n_z) = \sum_{r=1}^{s_0} L(r)[A_{60}(E_{F1HD},\eta_g,n_z)]$$

From (1.257), we can write the expression of the 2D dispersion law in QWs of n-$PtSb_2$ in the absence of band tails as

$$k_s^2 = t_{44}(E,\eta_z) \qquad (1.266)$$

where,

$$t_{44}(E,n_z) = [2A_9]^{-1}\left[-A_{10}(E,n_z) + \sqrt{A_{10}^2(E,n_z) + 4A_9A_{11}(E,n_z)}\right],$$

$$A_9 \equiv [I_1 + \omega_1\omega_3],$$

$$A_{10}(E,n_z) \equiv \left[\omega_3 E + \omega_1\left\{E + \delta_0 - \omega_4\left(\frac{\pi n_z}{d_z}\right)^2\right\} + \omega_2\omega_3\left(\frac{\pi n_z}{d_z}\right)^2 + 2I_1\left(\frac{\pi n_z}{d_z}\right)^2\right]$$

and

$$A_{11}(E,n_z) \equiv \left[E\left[E + \delta_0 - \omega_4\left(\frac{\pi n_z}{d_z}\right)^2\right] + \omega_2\left(\frac{\pi n_z}{d_z}\right)^2\left[E + \delta_0 - \omega_4\left(\frac{\pi n_z}{d_z}\right)^2\right] - I_1\left(\frac{\pi n_z}{d_z}\right)^4\right]$$

The area of k_s space can be expressed as

$$A(E,n_z) = \pi t_{44}(E,n_z) \tag{1.267}$$

The total DOS function assumes the form

$$N_{2DT}(E) = \frac{g_v}{2\pi}\sum_{n_z=1}^{n_{z\max}} [t'_{44}(E,n_z)]H(E - E_{n_{z14}}) \tag{1.268}$$

where the quantized levels $E_{n_{z14}}$ can be expressed through the equation

$$E_{n_{z14}} = (2)^{-1}\left[-\left[\omega_2\left(\frac{\pi n_z}{d_z}\right)^2 + \delta_0 - \omega_4\left(\frac{\pi n_z}{d_z}\right)^2\right] + \left\{\left[\omega_2\left(\frac{\pi n_z}{d_z}\right)^2 + \delta_0 - \omega_4\left(\frac{\pi n_z}{d_z}\right)^2\right]^2 + 4\left[I_1\left(\frac{\pi n_z}{d_z}\right)^4 + \omega_2\omega_4\left(\frac{\pi n_z}{d_z}\right)^4 - \omega_2\delta_0\left(\frac{\pi n_z}{d_z}\right)^2\right]\right\}^{1/2}\right] \tag{1.269}$$

Using (1.269), the EM in this case can be written as

$$m^*(E_{F_s}, n_z) = \frac{\hbar^2}{2} t'_{44}(E_{F_s}, n_z) \tag{1.270}$$

The electron statistics can be written as

$$n_{2D} = \frac{g_v}{2\pi} \sum_{n_z=1}^{n_{z\max}} [t_{44}(E_{Fs}, n_z) + t_{45}(E_{Fs}, n_z)] \tag{1.271}$$

where

$$t_{45}(E_{F2D}, n_z) \equiv \sum_{r=1}^{s} L(r)[t_{44}(E_{F2D}, n_z)]$$

1.2.9 *The EM in Quantum Wells (QWs) of HD Bismuth Telluride*

The dispersion relation of the conduction electrons in Bi_2Te_3 can be written as [152]

$$E(1 + \alpha E) = \bar{\omega}_1 k_x^2 + \bar{\omega}_2 k_y^2 + \bar{\omega}_3 k_z^2 + 2\bar{\omega}_4 k_z k_y \tag{1.272}$$

where

$$\bar{\omega}_1 = \frac{\hbar^2}{2m_0}\bar{\alpha}_{11}, \quad \bar{\omega}_2 = \frac{\hbar^2}{2m_0}\bar{\alpha}_{22}, \quad \bar{\omega}_3 = \frac{\hbar^2}{2m_0}\bar{\alpha}_{33}, \quad \bar{\omega}_4 = \frac{\hbar^2}{2m_0}\bar{\alpha}_{23},$$

in which $\bar{\alpha}_{11}, \bar{\alpha}_{22}, \bar{\alpha}_{33}$ and $\bar{\alpha}_{23}$ are system constants.

The dispersion relation in HD Bi_2Te_3 assumes the form

$$\gamma_2(E, \eta_g) = \bar{\omega}_1 k_x^2 + \bar{\omega}_2 k_y^2 + \bar{\omega}_3 k_z^2 + 2\bar{\omega}_4 k_z k_y \tag{1.273}$$

The EMs can, respectively, be expressed as

$$m_x^*(E_{F_h}, \eta_g) = \frac{\hbar^2}{2\bar{w}_1} \gamma'_2(E_{F_h}, \eta_g) \tag{1.274}$$

$$m_y^*(E_{F_h}, \eta_g) = \frac{\hbar^2}{2\bar{w}_2} \gamma'_2(E_{F_h}, \eta_g) \tag{1.275}$$

$$m_z^*(E_{F_h}, \eta_g) = \frac{\hbar^2}{2\bar{w}_3} \gamma'_2(E_{F_h}, \eta_g) \tag{1.276}$$

The DOS function in this case is given by

$$N(E) = 4\pi g_v \left(\frac{2m_0}{h^2}\right)^{3/2} \frac{\sqrt{\gamma_2(E,\eta_g)}\gamma_2'(E,\eta_g)}{\sqrt{\alpha_{11}\alpha_{22}\alpha_{33} - 4\alpha_{11}\alpha_{23}^2}} \tag{1.277}$$

Thus combining (1.277) with the Fermi Dirac occupation probability factor, the electron concentration can be written as

$$n_0 = \frac{g_v}{3\pi^2}\left(\frac{2m_0}{\hbar^2}\right)^{3/2} (\alpha_{11}\alpha_{22}\alpha_{33} - 4\alpha_{11}\alpha_{23}^2)^{-1/2} \times [U_{1HD}(E_{F_h},\eta_g) + U_{2HD}(E_{F_h},\eta_g)] \tag{1.278}$$

where,

$$U_{1HD}(E_{F_h},\eta_g) = [\gamma_2(E_{F_h},\eta_g)]^{3/2},$$

$$U_{2HD}(E_{F_h},\eta_g) = \sum_{r=1}^{s} L(r)[U_{1HD}(E_{F_h},\eta_g)]$$

The dispersion relation in QWs of HD Bi_2Te_3 can be expressed as

$$\gamma_2(E,\eta_g) = \bar{\omega}_1\left(\frac{\pi n_x}{d_x}\right)^2 + \bar{\omega}_2 k_y^2 + \bar{\omega}_3 k_z^2 + 2\bar{\omega}_4 k_z k_y \tag{1.279}$$

The EM can be expressed as

$$m^*(E_{F1HD},\eta_g) = \frac{m_0}{\sqrt{\alpha_{11}\alpha_{33} - 4\alpha_{23}^2}} \gamma_2'(E_{F1HD},\eta_g) \tag{1.280}$$

The surface electron concentration can be written as

$$n_{2D} = \frac{g_v}{2\pi}\left[\sum_{n_x=1}^{n_{x\max}} R_{60}(\mathrm{E}_{F1HD},\eta_g,\mathrm{n}_x) + R_{61}(\mathrm{E}_{F1HD},\eta_g,\mathrm{n}_x)\right] \tag{1.281}$$

$$R_{60}(E_{F1HD},\eta_g,n_z) = \frac{1}{\sqrt{a_{11}\alpha_{33} - 4\alpha_{23}^2}} \times \left[\frac{2m_0\gamma_2'(E_{F1HD},\eta_g)}{\hbar^2} - \frac{2m_0}{\hbar^2}\left(\frac{\pi n_x}{d_x}\right)^2 \bar{\alpha}_{11}\right]$$

and

$$R_{61}(E_{F1HD},\eta_g,n_z) = \sum_{r=1}^{s} L(r)[R_{60}(E_{F1HD},\eta_g,n_z)]$$

The 2D electron dispersion law in QWs of Bi_2Te_3 in the absence of band tails assumes the form

$$E(1+\alpha E) = \bar{\omega}_1 \left(\frac{n_x \pi}{d_x}\right)^2 + \bar{\omega}_2 k_y^2 + \bar{\omega}_3 k_z^2 + 2\bar{\omega}_4 k_z k_y \tag{1.282}$$

The area of the ellipse is given by

$$A_n(E, n_x) = \frac{\pi}{\sqrt{\bar{\alpha}_{22}\bar{\alpha}_{23} - 4\bar{\alpha}_{23}^2}} \left[\frac{2m_0 E(1+\alpha E)}{\hbar^2} - \bar{\omega}_1 \left(\frac{n_x \pi}{d_x}\right)^2\right] \tag{1.283}$$

The total DOS function assumes the form

$$N_{2DT}(E) = \frac{g_v m_0}{\pi \hbar^2 \sqrt{\bar{\alpha}_{22}\bar{\alpha}_{33} - 4\bar{\alpha}_{23}^2}} \sum_{n_x=1}^{n_{x\max}} (1+2\alpha E) H(E - E_{n_{z15}}) \tag{1.284}$$

where, $(E_{n_{z15}})$ can be expressed through the equation

$$E_{n_{z15}}(1+\alpha E_{n_{z15}}) = \bar{\omega}_1 \left(\frac{n_x \pi}{d_x}\right)^2 \tag{1.285a}$$

The EM in this case assumes the form as

$$m^*(E_{F_s}) = \frac{m_0(1+2\alpha(E_{F_s}))}{\sqrt{\bar{\alpha}_{22}\bar{\alpha}_{33} - 4\bar{\alpha}_{23}^2}} \tag{1.285b}$$

The electron concentration can be written as

$$n_{2D} = \frac{k_B T g_v}{\pi \hbar^2} \left(\frac{m_0}{\sqrt{\bar{\alpha}_{22}\bar{\alpha}_{33} - 4\bar{\alpha}_{23}^2}}\right) \times \sum_{n_z=1}^{n_{z\max}} [(1+2\alpha E_{n_{\eta 5}}) F_0(\eta_{n15}) + 2\alpha k_B T F_1(\eta_{n15})] \tag{1.286}$$

where

$$\eta_{n15} = \frac{E_{F_s} - E_{n_z 15}}{K_B T}.$$

1.2.10 *The EM in Quantum Wells (QWs) of HD Germanium*

It is well known that the conduction electrons of n-Ge obey two different types of dispersion laws since band non-parabolicity has been included in two different ways as given in the literature [153, 154].

(a) The model Cardona *et al.*

The energy spectrum of the conduction electrons in bulk specimens of n-Ge can be expressed in accordance with Cardona *et al.* [153] as

$$E = -\frac{E_{g0}}{2} + \frac{\hbar^2 k_z^2}{2m_{\parallel}^*} + \left[\frac{E_{g0}^2}{4} + E_{g0} k_s^2 \left(\frac{\hbar^2}{2m_{\perp}^*}\right)\right]^{1/2} \quad (1.287)$$

where in this case $m_{\parallel}^*$ and $m_{\perp}^*$ are the longitudinal and transverse effective masses along <111> direction at the edge of the conduction band respectively

(1.287) can be written as

$$\frac{\hbar^2 k_s^2}{2m_{\perp}^*} = E\left(1+\alpha E\right) + \alpha\left(\frac{\hbar^2 k_z^2}{2m_{\parallel}^*}\right)^2 - \left(1+2\alpha E\right)\left(\frac{\hbar^2 k_z^2}{2m_{\parallel}^*}\right) \quad (1.288)$$

The dispersion relation under the condition of heavy doping can be expressed from (1.288) as

$$\frac{\hbar^2 k_s^2}{2m_{\perp}^*} = \gamma_2(E,\eta_g) + \alpha\left(\frac{\hbar^2 k_z^2}{2m_{\parallel}^*}\right)^2 - (1+2\alpha\gamma_3(E,\eta_g))\left(\frac{\hbar^2 k_z^2}{2m_{\parallel}^*}\right) \quad (1.289)$$

The EMs can be written as

$$m_s^*(E_{Fh},\eta_g) = m_{\perp}^* \gamma_2'(E_{Fh},\eta_g) \quad (1.290)$$

and

$$m_z^*(E_{Fh},\eta_g) = m_{\parallel}^* \left[\gamma_3'(E_{Fh},\eta_g) - \frac{\gamma_3'(E_{Fh},\eta_g)[1+2\alpha\gamma_3(E_{Fh},\eta_g)] - \gamma_2'(E_{Fh},\eta_g)}{\sqrt{[1+2\alpha\gamma_3(E_{Fh},\eta_g)]^2 - 4\alpha\gamma_2(E_{Fh},\eta_g)}}\right] \quad (1.291)$$

The electron concentration can be written as

$$n_0 = \frac{8\pi g_v m_{\perp}^* \sqrt{2m_{\parallel}^*}}{h^3}\left[\bar{I}_{129}(E_{Fh},\eta_g) + \sum_{r=1}^{s} L(r)[\bar{I}_{129}(E_{Fh},\eta_g)]\right] \quad (1.292)$$

where,

$$[\bar{I}_{129}(E_{F_h},\eta_g)] = [M_{8HD}(E_{F_h},\eta_g)],$$

$$[M_{8HD}(E_{F_h},\eta_g)] = [\gamma_3(E_{F_h},\eta_g)]^{1/2}\left[\gamma_2(E_{F_h},\eta_g) + \frac{\alpha}{5}\gamma_3^2(E_{F_h},\eta_g)\right] - \frac{\gamma_3(E_{F_h},\eta_g)}{3}[1 + 2\alpha\gamma_3(E_{F_h},\eta_g)]$$

In the presence of size quantization, the dispersion law in QW of HD Ge can be written as

$$\frac{\hbar^2 k_s^2}{2m_\perp^*} = \gamma_2(E,\eta_g) + \alpha\left(\frac{\hbar^2\left(\frac{n_z\pi}{d_z}\right)^2}{2m_\parallel^*}\right)^2 - (1 + 2\alpha\gamma_3(E,\eta_g))\frac{\hbar^2\left(\frac{n_z\pi}{d_z}\right)^2}{2m_\parallel^*} \tag{1.293a}$$

The EM assumes the from

$$m_s^*(E_{F1HD},\eta_g,n_z) = m_\perp^*\left[\gamma_\perp'(E_{F1HD},\eta_g) - \frac{\alpha\hbar^2}{m_\parallel^*}\left(\frac{n_z\pi}{d_z}\right)^2\gamma_3'(E_{F_h},\eta_g)\right] \tag{1.293b}$$

The surface electron concentration per unit area is given by

$$n_{2D} = \frac{g_v m_\perp^*}{\pi\hbar^2}\sum_{n_z=1}^{n_{z\max}}[R_1(E_{F1HD},\eta_g,n_z) + S_1(E_{F1HD},\eta_g,n_z)] \tag{1.294}$$

where,

$$R_1(E_{F1HD},\eta_g,n_z) = \left[\gamma_2(E_{F1HD},\eta_g) + \alpha\left(\frac{\hbar^2\left(\frac{n_z\pi}{d_z}\right)^2}{2m_\parallel^*}\right)^2 - (1 + 2\alpha\gamma_3(E_{F1HD},\eta_g))\frac{\hbar^2\left(\frac{n_z\pi}{d_z}\right)^2}{2m_\parallel^*}\right]$$

and

$$S_1(E_{F1HD},\eta_g,n_z) = \sum_{r=1}^{s} L(r)[R_1(E_{F1HD},\eta_g,n_z)]$$

In the presence of size quantization along k_z direction, the 2D dispersion relation of the conduction relations in QWs of n-Ge in the absence of band tails can be written by extending the method as given in [153] as

$$\frac{\hbar^2 k_x^2}{2m_1^*} + \frac{\hbar^2 k_y^2}{2m_2^*} = \gamma\left(E, n_z\right) \tag{1.295}$$

where,

$$m_1^* \equiv m_\perp^*, \quad m_2^* = \frac{m_\perp^* + 2m_\parallel^*}{3},$$

$$\gamma\left(E, n_z\right) = \left[E(1+\alpha E) - (1+2\alpha E)\frac{\hbar^2}{2m_3^*}\left(\frac{n_z\pi}{d_z}\right)^2 + \alpha\left[\frac{\hbar^2}{2m_3^*}\left(\frac{n_z\pi}{d_z}\right)^2\right]^2\right]$$

and

$$m_3^* \equiv \frac{3m_\parallel^* m_\perp^*}{2m_\parallel^* + m_\perp^*}.$$

The area of ellipse of the 2D surface as given by (1.295) can be written as

$$A(E, n_z) = \frac{2\pi\sqrt{m_1^* m_2^*}}{\hbar^2}\gamma(E, n_z) \tag{1.296a}$$

The EM in this case can be written as

$$m^*(E_{F_s}, n_z) = (\sqrt{m_1^* m_2^*})[\gamma(E_{F_s}, n_z)]' \tag{1.296b}$$

The DOS function per sub-band can be expressed as

$$N_{2D}(E) = \frac{4\sqrt{m_1^* m_2^*}}{\pi\hbar^2}\left[1 + 2\alpha E - 2\alpha\left(\frac{\hbar^2}{2m_3^*}\left(\frac{\pi n_z}{d_z}\right)^2\right)\right] \tag{1.297a}$$

The total DOS function is given by

$$N_{2DT}(E) = \frac{4}{\pi\hbar^2}\sqrt{m_1^* m_2^*}\sum_{n_z=1}^{n_{z\max}}\left[1 + 2\alpha E - 2\alpha\left(\frac{\hbar^2}{2m_3^*}\left(\frac{\pi n_z}{d_z}\right)^2\right)\right] \times H\left(E - E_{n_{z16}}\right) \tag{1.297b}$$

where, $E_{n_{z16}}$ is the positive root of the following equation

$$E_{n_{z16}}(1+\alpha E_{n_{z16}}) - (1+2\alpha E_{n_{z16}})\left(\frac{\hbar^2}{2m_3^*}\left(\frac{\pi n_z}{d_z}\right)^2\right) + \alpha\left(\frac{\hbar^2}{2m_3^*}\left(\frac{\pi n_z}{d_z}\right)^2\right)^2 = 0 \quad (1.298)$$

Thus combining (1.297a) with the Fermi Dirac occupation probability factor, the 2D electron statistics in this case can be written as

$$n_{2D} = \frac{4\sqrt{m_1^* m_2^*} k_B T}{\pi\hbar^2} \times \sum_{n_z=1}^{n_{z\max}} [(\mathrm{A}_1(\mathrm{n}_z) + 2\alpha\eta_{n_{z16}})\mathrm{F}_0(\eta_{n16}) + 2\alpha k_B T \mathrm{F}_1(\eta_{n_{z16}})] \quad (1.299)$$

where

$$A_1(n_z) \equiv \left[1 + 2\alpha\left(\frac{\hbar^2}{2m_3^*}\right)\left(\frac{\pi n_z}{d_z}\right)^2\right]$$

and

$$\eta_{n_{z16}} \equiv \frac{1}{k_B T}[E_{F2D} - E_{n_{z16}}].$$

The expressions of EMs' in bulk specimens of Ge in the absence of band tails can be written as

$$m_z^*(E_F) = m_\parallel^* \quad (1.300)$$

$$m_s^*(E_F) = m_\perp^*(1 + 2\alpha E_F) \quad (1.301)$$

The DOS function for bulk specimens of Ge in the absence of band tails can be written as

$$N(E) = 4\pi g_v \left(\frac{2m_D^*}{h^2}\right)^{\frac{3}{2}} \left[E^{\frac{1}{2}} - \frac{5}{6}\alpha E^{\frac{3}{2}} + \frac{18\alpha}{5}\left(\frac{m_{11}^*}{\hbar^2}\right)^2 E^{\frac{7}{2}}\right];$$

$$m_D = \left(m_\perp^{*2} \cdot m_\parallel^*\right)^{\frac{1}{3}} \quad (1.302)$$

Using (1.302), the electron concentration in bulk specimens of Ge can be written as

$$n_0 = N_{\text{cl}} \left[F_{\frac{1}{2}}(\eta) - \frac{5}{4}\alpha k_B T F_{\frac{3}{2}}(\eta) + \frac{189}{4}\alpha k_B T \left(\frac{m_{11}^* k_B T}{\hbar^2} \right)^2 F_{\frac{7}{2}}(\eta) \right];$$

$$N_{\text{cl}} = 2g_v \left(\frac{2\pi m_D^* k_B T}{\hbar^2} \right)^{\frac{3}{2}} \tag{1.303}$$

(b) The Model of Wang and Ressler

The dispersion relation of the conduction electron in bulk specimens of n-Ge can be expressed in accordance with the model of Wang and Ressler [154] can be written as

$$E = \frac{\hbar^2 k_z^2}{2m_{\|}^*} + \frac{\hbar^2 k_s^2}{2m_{\perp}^*} - \bar{\alpha}_4 \left(\frac{\hbar^2 k_s^2}{2m_{\perp}^*} \right)^2 - \bar{\alpha}_5 \left(\frac{\hbar^2 k_s^2}{2m_{\perp}^*} \right) \left(\frac{\hbar^2 k_z^2}{2m_{\|}^*} \right) - \bar{\alpha}_6 \left(\frac{\hbar^2 k_z^2}{2m_{\|}^*} \right)^2 \tag{1.304}$$

where,

$$\bar{\alpha}_4 = \beta_4 \left(\frac{2m_{\perp}^*}{\hbar^2} \right), \quad \beta_4 = 1.4\beta_5,$$

$$\beta_5 = \frac{\alpha \hbar^4}{4} \left[(m_{\perp}^*)^{-1} - (m_0)^{-1} \right]^2, \quad \bar{\alpha}_5 = \bar{\alpha}_7 \left(\frac{4m_{\perp}^* m_{\|}^*}{\hbar^4} \right), \quad \bar{\alpha}_7 = 0.8\beta_5$$

and

$$\bar{\alpha}_6 = (0.005\beta_5) \left(\frac{2m_{\|}^*}{\hbar^2} \right)^2$$

The energy spectrum under the condition of heavy doping can be written as

$$\gamma_3(E, \eta_g) = \frac{\hbar^2 k_z^2}{2m_{\|}^*} + \frac{\hbar^2 k_s^2}{2m_{\perp}^*} - \bar{\alpha}_4 \left(\frac{\hbar^2 k_s^2}{2m_{\perp}^*} \right)^2 - \bar{\alpha}_5 \left(\frac{\hbar^2 k_s^2}{2m_{\perp}^*} \right) \left(\frac{\hbar^2 k_z^2}{2m_{\|}^*} \right) - \bar{\alpha}_6 \left(\frac{\hbar^2 k_z^2}{2m_{\|}^*} \right)^2 \tag{1.305}$$

(1.305) can be expressed as

$$\left[\bar{\alpha}_8 - \bar{\alpha}_9\left(\frac{n_z\pi}{d_z}\right)^2 - \bar{\alpha}_{10}\left[\left(\frac{n_z\pi}{d_z}\right)^4 + \bar{\alpha}_{11}\left(\frac{n_z\pi}{d_z}\right)^2 + \bar{\alpha}_{11}(E_{n_{zHD24}},\eta_g)\right]^{1/2}\right] = 0$$

$$\frac{\hbar^2 k_s^2}{2m_\perp^*} = \bar{\alpha}_8 - \bar{\alpha}_9 k_z^2 - \bar{\alpha}_{10}\left[k_z^4 + \bar{\alpha}_{11}k_z^2 + \bar{\alpha}_{12}(E,\eta_g)\right]^{1/2} \tag{1.306}$$

where,

$$\overline{\alpha_8} = \frac{1}{2\bar{\alpha}_4}, \quad \bar{\alpha}_9 = \frac{\bar{\alpha}_5}{2\bar{\alpha}_4}\left(\frac{\hbar^2}{2m_\parallel^*}\right), \quad \bar{\alpha}_{10} = \frac{1}{2\bar{\alpha}_4}\left(\frac{\hbar^2}{2m_\parallel^*}\right)\sqrt{\bar{\alpha}_5^2 - 4\bar{\alpha}_4\bar{\alpha}_6},$$

$$\bar{\alpha}_{11} = \frac{2m_\parallel^*}{\hbar^2}\left[\frac{4\bar{\alpha}_4 - 2\bar{\alpha}_5}{\bar{\alpha}_5^2 - 4\bar{\alpha}_4\bar{\alpha}_6}\right]$$

and

$$\bar{\alpha}_{12}(E,\eta_g) = \left(\frac{2m_\parallel^*}{\hbar^2}\right)^2\left[\frac{(1 - 4\bar{\alpha}_4\gamma_3(E,\eta_g))}{\bar{\alpha}_5^2 - 4\bar{\alpha}_4\bar{\alpha}_6}\right]$$

The EMs can be written as

$$m_z^*(E_{Fh},\eta_g) = \left[\frac{m_\parallel^*\gamma_3'(E_{Fh},\eta_g)}{\sqrt{1 - 4\bar{\alpha}_6\gamma_3(E_{Fh},\eta_g)}}\right] \tag{1.307}$$

$$m_\perp^*(E_{Fh},\eta_g) = \left[\frac{m_\perp^*\gamma_3'(E_{Fh},\eta_g)}{\sqrt{1 - 4\bar{\alpha}_4\gamma_3(E_{Fh},\eta_g)}}\right] \tag{1.308}$$

The electron concentration in HD Ge in accordance with the model of Wang and Ressler can be expressed as

$$n_0 = \frac{m_\perp^* g_v}{\pi^2\hbar^2}[I_3(E_{F_h},\eta_s) + I_4(E_{F_h},\eta_s)] \tag{1.309}$$

where

$$I_3(E_{F_h},\eta_s) = \left[\bar{\alpha}_8\rho_{10}(E_{F_h},\eta_s) - \frac{\bar{\alpha}_9}{3}\rho_{10}^3(E_{F_h},\eta_s)\bar{\alpha}_{10}J_{10}(E_{F_h},\eta_s)\right],$$

$$\rho_{10}(E_{F_h},\eta_g) = \frac{1}{\hbar}\left[\frac{m_\parallel^*}{\bar{\alpha}_6}\right]^{1/2}\left[1-\sqrt{1-4\bar{\alpha}_6\gamma_3(E_{F_h},\eta_g)}\right]^{1/2}$$

$$J_{10}(E_{F_h},\eta_g) = \frac{\bar{A}_1(E_{F_h},\eta_g)}{3}[-E_0(\lambda(E_{F_h},\eta_g),q(E_{F_h},\eta_g)).$$
$$\times[\bar{A}_1^2(E_{F_h},\eta_g)+\bar{B}_1^2(E_{F_h},\eta_g)]$$
$$+2\bar{B}_1^2(E_{F_h},\eta_g)F_0(\lambda(E_{F_h},\eta_g),q(E_{F_h},\eta_g))]$$
$$+\frac{\bar{A}_1(E_{F_h},\eta_g)}{3}[\rho_{10}^2(E_{F_h},\eta_g)+\bar{A}_1^2(E_{F_h},\eta_g)$$
$$+2\bar{B}_1^2(E_{F_h},\eta_g)]\left[\frac{\bar{A}_1^2(E_{F_h},\eta_g)+\rho_{10}^2(E_{F_h},\eta_g)}{\bar{B}_1^2(E_{F_h},\eta_g)+\rho_{10}^2(E_{F_h},\eta_g)}\right]^{1/2}$$

$$\bar{A}_1^2(E_{F_h},\eta_g) = \frac{1}{2}[\bar{\alpha}_{11}+\sqrt{\bar{\alpha}_{11}^2-4\alpha_{12}^{-2}(E_{F_h},\eta_g)}],$$

$$\bar{B}_1^2(E_{F_h},\eta_g) = \frac{1}{2}[\bar{\alpha}_{11}-\sqrt{\bar{\alpha}_{11}^2-4\bar{\alpha}_{12}^2(E_{F_h},\eta_g)}],$$

$$\lambda(E_{F_h},\eta_g) = \tan^{-1}\left[\frac{\rho_{10}(E_{F_h},\eta_g)}{\bar{B}_1(E_{F_h},\eta_g)}\right],$$

$$q(E_{F_h},\eta_g) = \left[\frac{\bar{A}_1^2(E_{F_h},\eta_g)-\bar{B}_1^2(E_{F_h},\eta_g)}{\bar{A}_1^2(E_{F_h},\eta_g)}\right]$$

and

$$I_4(E_{F_h},\eta_g) = \sum_{r=1}^{s} L(r)[I_3(E_{F_h},\eta_g)]$$

The dispersion relation in QW of HD Ge can be written as

$$\frac{\hbar^2k_s^2}{2m_\perp^*} = \bar{a}_8-\bar{a}_9\left(\frac{n_z\pi}{d_z}\right)^2-\bar{a}_{10}\left[\left(\frac{n_z\pi}{d_z}\right)^4+\bar{a}_{11}\left(\frac{n_z\pi}{d_z}\right)^2+\bar{a}_{12}(E,\eta_g)\right]^{1/2}, \tag{1.310}$$

(1.310) can be expressed as

$$\frac{\hbar^2 k_s^2}{2m_\perp^*} = A_{75}(E, \eta_g, n_z) \tag{1.311}$$

where,

$$A_{75}(E, \eta_g, n_z) = \left[\bar{\alpha}_8 - \bar{\alpha}_9\left(\frac{\pi n_z}{d_z}\right)^2 - \bar{\alpha}_{10}\left[\left(\frac{\pi n_z}{d_z}\right)^4 + \bar{\alpha}_{11}\left(\frac{\pi n_z}{d_z}\right)^2 + \bar{\alpha}_{12}(E, \eta_g)\right]^{1/2}\right]$$

The EM is given by

$$m_s^*(E_{F1HD}, \eta_g, n_z) = m_\perp^* A'_{75}(E_{F1HD}, \eta_g, n_z) \tag{1.312}$$

The electron concentration per unit area assumes the form

$$N_{2D} = \frac{m_\perp^* g_v}{\pi\hbar^2}\sum_{n_z=1}^{n_{z\max}}[A_{75}(E_{F1HD}, \eta_g, n_z) + A_{76}(E_{F1HD}, \eta_g, n_z)] \tag{1.313}$$

where,

$$A_{76}(E_{F1HD}, \eta_g, n_z) = \sum_{r=1}^{s} L(r)[A_{75}(E_{F1HD}, \eta_g, n_z)]$$

The 2D dispersion law in the absence of band tails can be expressed as

$$E = A_5(n_z) + A_6(n_z)\beta - \bar{\alpha}_4\beta^2 \tag{1.314}$$

where,

$$A_5(n_z) = \frac{\hbar^2}{2m_3^*}\left(\frac{\pi n_z}{d_z}\right)^2\left[1 - \bar{\alpha}_6\left(\frac{\hbar^2}{2m_3^*}\right)\left(\frac{\pi n_z}{d_z}\right)^2\right],$$

$$A_6(n_z) = \left[1 - \bar{\alpha}_5\left(\frac{\hbar^2}{2m_3^*}\right)\left(\frac{\pi n_z}{d_z}\right)^2\right]$$

and

$$\beta \equiv \frac{\hbar^2 k_x^2}{2m_1^*} + \frac{\hbar^2 k_y^2}{2m_2^*}.$$

(1.314) can be written as

$$\frac{\hbar^2 k_x^2}{2m_1^*} + \frac{\hbar^2 k_y^2}{2m_2^*} = I_1(E, n_z) \tag{1.315}$$

where,

$$I_1(E, n_z) = (2\bar{\alpha}_4)^{-1}[A_6(n_z) - [A_6^2(n_z) - 4\bar{\alpha}_4 E + 4\bar{\alpha}_4 A_5(n_z)]^{1/2}]$$

From (1.315), the area of the 2D k_s-space is given by

$$A(E, n_z) = \frac{2\pi\sqrt{m_1^* m_2^*}}{\hbar^2} I_1(E, n_z) \tag{1.316}$$

Using (1.316) in this case can be expressed as

$$m^*(E_{F_s}, n_z) = (\sqrt{m_1^* m_2^*})\,[I_1(E_{F_z}, n_z)]' \tag{1.317}$$

The DOS function per sub-band can be written as

$$N_{2D}(E) = \frac{4}{\pi}\frac{\sqrt{m_1^* m_2^*}}{\hbar^2}\{I_1(E, n_z)\}' \tag{1.318}$$

where

$$\{I_1(E, n_z)\}' \equiv \frac{\partial}{\partial E}[I_1(E, n_z)]$$

The total DOS function assumes the form

$$N_{2DT}(E) = \frac{4\sqrt{m_1^* m_2^*}}{\pi\hbar^2}\sum_{n_z=1}^{n_{z\max}}\{I_1(E, n_z)\}' H(E - E_{n_{z17}}) \tag{1.319}$$

where, the sub-band energy ($E_{n_{z17}}$) are given by

$$E_{n_{z17}} = \left(\frac{\hbar^2}{2m_3^*}\right)\left(\frac{\pi n_z}{d_z}\right)^2\left[1 - \bar{\alpha}_6\left(\frac{\hbar^2}{2m_3^*}\right)\left(\frac{\pi n_z}{d_z}\right)^2\right] \tag{1.320}$$

The electron statistics can be written as

$$n_{2D} = \frac{4\sqrt{m_1^* m_2^*}}{\pi\hbar^2}\sum_{n_z=1}^{n_{z\max}}[t_{46}(E_{Fs}, \eta_z) + t_{47}(E_{Fs}, \eta_z)] \tag{1.321}$$

where

$$t_{46}(E_{F_s}, n_z) \equiv I_1(E_{F_s}, n_z), \quad t_{47}(E_{F_s}, n_z) \equiv \sum_{r=1}^{S} L(r)(t_{46}(E_{F_s}, n_z))$$

1.2.11 *The EM in Quantum Wells (QWs) of HD Gallium Antimonide*

The dispersion relation of the conduction electrons in n-GaSb can be written as [155]

$$E = \frac{\hbar^2 k^2}{2m_0} - \frac{\bar{E}'_{g0}}{2} + \frac{\bar{E}'_{g0}}{2}\left[1 + \frac{2\hbar^2 k^2}{\bar{E}'_{g0}}\left(\frac{1}{m_c} - \frac{1}{m_0}\right)\right]^{1/2} \tag{1.322}$$

where

$$\bar{E}'_{g0} = \left[E_{g0} + \frac{5.10^{-5}T^2}{2(112+T)}\right] eV$$

(1.322) can be expressed as

$$\frac{\hbar^2 k^2}{2m_c} = I_{36}(E) \tag{1.323}$$

where

$$I_{36}(E) = \left[E + \bar{E}'_{g0} - \left(\frac{m_c}{m_0}\right)\left(\frac{\bar{E}'_{g0}}{2}\right) - \left[\left(\frac{\bar{E}'_{g0}}{2}\right)^2 + \left[\left(\frac{(\bar{E}'_{g0})^2}{2}\right)\left(1 - \left(\frac{m_c}{m_0}\right)\right)\right] + \left[\left(\frac{\bar{E}'_{g0}}{2}\right)\left(1 - \left(\frac{m_c}{m_0}\right)\right)\right]^2 + E\bar{E}'_{g0}\left(1 - \left(\frac{m_c}{m_0}\right)\right)\right]^{1/2}\right]$$

Under the condition of heavy doping (1.323) assumes the form

$$\frac{\hbar^2 k^2}{2m_c} = I_{36}(E, \eta_g) \tag{1.324}$$

where,

$$I_{36}(E, \eta_g) = \left[\gamma_3(E, \eta_g) + E'_g - \frac{m_c}{m_0} \cdot \frac{E'_g}{2} - \right.$$

$$\times\left[\left(\frac{E_g'}{2}\right)^2+\left[\frac{E_g'}{2}\left(1-\frac{m_c}{m_0}\right)\right]^2\left(\frac{E_g'}{2}\right)^2\left(1-\frac{m_c}{m_0}\right)\right.$$
$$\left.+\gamma_3\left(E,\eta_g\right)E_g'\left(1-\frac{m_c}{m_0}\right)\right]^{1/2}\Bigg]$$

The EM can be written as

$$m^*(E_{F_h},\eta_g)=m_c\{I_{36}(E_{F_h},\eta_g)\}' \tag{1.325}$$

The DOS function in this case can be written as

$$N_{HD}(E,\eta_g)=\frac{g_v}{2\pi^2}\left(\frac{2m_c}{\hbar^2}\right)^{3/2}\sqrt{I_{36}(E,\eta_g)}\{I_{36}(E,\eta_g)\}' \tag{1.326}$$

Since, the original band model in this case is a no pole function, therefore, the HD counterpart will be totally real, and the complex band vanishes.

The electron concentration is given by

$$n_0=\frac{g_v}{3\pi^2}\left(\frac{2m_c}{\hbar^2}\right)^{3/2}\left[\{I_{36}(E_{F_h},\eta_g)\}^{3/2}+\sum_{r=1}^{s}L(r)[\{I_{36}(E_{F_h},\eta_g)\}^{3/2}]\right] \tag{1.327}$$

For dimensional quantization along z-direction, the dispersion relation of the 2D electrons in QWs of HD GaSb can be written following (1.324) as

$$\frac{\hbar^2\left(\frac{\mathrm{n}_z\pi}{\mathrm{d}_z}\right)^2}{2m_c}+\frac{\hbar^2(k_s)^2}{2m_c}=I_{36}\left(E,\eta_g\right) \tag{1.328}$$

The expression of the $N_{2DT}(E)$ in this case can be written as

$$N_{2DT}(E)=\frac{m_c g_v}{\pi\hbar^2}\sum_{n_z=1}^{n_{z\max}}T'_{119D}(E,\eta_g,\mathrm{n}_z)H(\mathrm{E}-\mathrm{E}_{n_zD119}) \tag{1.329}$$

where,

$$T_{119D}(E,\eta_g,\mathrm{n}_z)=\left[I_{36}(E,\eta_g)-\hbar^2\left(\frac{\mathrm{n}_z\pi}{\mathrm{d}_z}\right)^2(2m_c)^{-1}\right],$$

The sub-band energies E_{n_zD119} in this case given by

$$\left\{\hbar^2\left(\frac{n_z\pi}{d_z}\right)^2\right\}(2m_c)^{-1} = I_{36}(\mathrm{E}_{n_zD119},\eta_g) \tag{1.330}$$

The EM in this case assumes the form

$$m^*(E_{F1HD},\eta_g,\mathrm{n}_z) = m_c[I'_{36}(E_{F1HD},\eta_g,\mathrm{n}_z)] \tag{1.331}$$

The 2-D electron statistics in this case can be written as

$$\begin{aligned} n_{2T} = &\left(\frac{m_c g_v}{\pi\hbar^2}\right) \\ &\times \sum_{n_z=1}^{n_{z\max}} [T_{119D}(E_{F1HD},\eta_g,\mathrm{n}_z) + T_{129D}(E_{F1HD},\eta_g,\mathrm{n}_z)] \end{aligned} \tag{1.332}$$

where,

$$T_{129D}(E_{F1HD},\eta_g,\mathrm{n}_z) = \sum_{r=1}^{s} L(r)[T_{119D}(E_{F1HD},\eta_g,\mathrm{n}_z)],$$

The total 2D DOS function in the absence of band tails in this case can be written as

$$N_{2DT}(E) = \left(\frac{m_c g_v}{\pi\hbar^2}\right)\sum_{n_z=1}^{n_{\max}}\{[I_{36}(E)]'H(E-E_{n_{z44}})\} \tag{1.333}$$

where, the sub-band energies $E_{n_{z44}}$ can be expressed as

$$I_{36}(E_{n_{z44}}) = \frac{\hbar^2}{2m_c}\left(\frac{\pi n_z}{d_z}\right)^2 \tag{1.334}$$

The EM in this case can be written as

$$m^*(E_{F_s}) = (m_c)[I_{36}(E_{F_s})]' \tag{1.335}$$

The 2D carrier concentration assumes the form

$$N_{2D} = \left(\frac{m_c g_v}{\pi\hbar^2}\right)\sum_{n_z=1}^{n_{\max}}[\bar{T}_{55}(E_{Fs},n_z) + \bar{T}_{56}(E_{Fs},n_z)] \tag{1.336}$$

where

$$\bar{T}_{55}(E_{Fs},n_z) = \left[I_{36}(E_{Fs}) - \frac{\hbar^2}{2m_c}\left(\frac{\pi n_z}{d_z}\right)^2\right]$$

and

$$\bar{T}_{56}(E_{Fs}, n_z) = \sum_{r=1}^{s} L(r)[\bar{T}_{55}(E_{Fs}, n_z)]$$

The expression of electron concentration for bulk specimens of GaSb (in the absence of band tails) can be expressed as

$$n_0 = \frac{g_v}{3\pi^2}\left(\frac{2m_c}{\hbar^2}\right)^{3/2}[\bar{M}_{A_{10}}(E_F) + \bar{N}_{A_{10}}(E_F)] \tag{1.337}$$

where,

$$\bar{M}_{A_{10}}(E_F) = [I_{36}(E_F)]^{3/2}$$

and

$$\bar{N}_{A_{10}}(E_F) = \sum_{r=1}^{s} L(r)[\bar{M}_{A_{10}}(E_F)]$$

1.2.12 *The EM in Quantum Wells (QWs) of HD II–V materials*

The dispersion relation of the holes are given by [156]

$$E = \theta_1 k_x^2 + \theta_2 k_y^2 + \theta_3 k_z^2 + \delta_4 k_x \mp [\{\theta_5 k_x^2 + \theta_6 k_y^2 + \theta_7 k_z^2 + \delta_5 k_3\}^2 + G_3^2 k_y^2 + \Delta_3^2]^{\frac{1}{2}} \pm \Delta_3 \tag{1.338a}$$

where, k_x, k_y and k_z are expressed in the units of 10^{10} m^{-1},

$$\theta_1 = \frac{1}{2}(a_1 + b_1), \quad \theta_2 = \frac{1}{2}(a_2 + b_2), \quad \theta_3 = \frac{1}{2}(a_3 + b_3),$$

$$\delta_4 = \frac{1}{2}(A + B),$$

$$\theta_5 = \frac{1}{2}(a_1 - b_1), \quad \theta_6 = \frac{1}{2}(a_2 - b_2), \quad \theta_7 = \frac{1}{2}(a_3 - b_3),$$

$$\delta_5 = \frac{1}{2}(A - B),$$

$a_i (i = 1, 2, 3, 4), b_i, A, B, G_3$ *and* Δ_3 *are system constants*

The EM of the holes in II–V compounds in accordance with Yamada [129] can be expressed as

$$E = A_{10}k_x^2 + A_{11}k_y^2 + A_{12}k_z^2 + A_{13}k_x \pm [(A_{14}k_x^2 + A_{15}k_y^2 + A_{16}k_z^2 + A_{17}k_x)^2 + A_{18}k_y^2 + A_{19}^2]^{1/2} \tag{1.338b}$$

where,

$$\theta_1 = A_{10}, \quad \theta_2 = A_{11}, \quad \theta_3 = A_{12}, \quad \delta_4 = A_{13}, \quad \theta_5 = A_{14},$$
$$\theta_6 = A_{15}, \quad \theta_7 = A_{16}, \quad \delta_5 = A_{17}, \quad G_3^2 = A_{18}, \quad \Delta_3^2 = A_{19}^2$$

and E in Eq (1.338b) is E in Eq (1.338a) $\mp\Delta_3$

The EM under the condition of formation of band tails can be written in this case as

$$\gamma_3(E,\eta_g) = A_{10}k_x^2 + A_{11}k_y^2 + A_{12}k_z^2 + A_{13}k_x \pm [(A_{14}k_x^2 + A_{15}k_y^2 + A_{16}k_z^2 + A_{17}k_x)^2 + A_{18}k_y^2 + A_{19}^2]^{1/2} \tag{1.339}$$

The hole energy spectrum in this case assumes the form

$$\gamma_3(E,\eta_g) = A_{10}\left(\frac{n_x\pi}{d_x}\right)^2 + A_{11}k_y^2 + A_{12}k_z^2 + A_{13}\left(\frac{n_x\pi}{d_x}\right) \pm \left[\left(A_{14}\left(\frac{n_x\pi}{d_x}\right)^2 + A_{15}k_y^2 + A_{16}k_z^2 + A_{17}\left(\frac{n_x\pi}{d_x}\right)^2\right) + A_{18}k_y^2 + A_{19}^2\right]^{1/2} \tag{1.340}$$

The sub-band energy ($E_{n_z HD401}$) is the lowest positive root of the following equation

$$\gamma_3(E_{n_z HD401},\eta_g) = A_{10}\left(\frac{n_x\pi}{d_x}\right)^2 + A_{13}\left(\frac{n_x\pi}{d_x}\right) \pm \left[\left(A_{14}\left(\frac{n_x\pi}{d_x}\right)^2 + A_{17}\left(\frac{n_x\pi}{d_x}\right)\right)^2 + A_{19}^2\right]^{1/2} \tag{1.341}$$

The EM, DOS function and the electron concentration should be calculated numerically.

1.2.13 *The EM in Quantum Wells (QWs) of HD Lead Germanium Telluride*

The dispersion relation of the carriers in n-type $Pb_{1-x}Ga_xTe$ with x = 0.01 can be written following Vassilev [157] as

$$[E - 0.606k_s^2 - 0.0722k_z^2][E + \bar{E}_{g_0} + 0.411k_s^2 + 0.0377k_z^2] = 0.23k_s^2 + 0.02k_z^2 \pm [0.06\bar{E}_g + 0.061k_s^2 + 0.0066k_z^2]k_s \quad (1.342)$$

where, $\bar{E}_{g_0}(= 0.21\text{eV})$, is the energy gap for the transition point, the zero of the energy E is at the edge of the conduction band of the Γ point of the Brillouin zone and is measured positively upwards, k_s, k_y and k_z are in the units of $10^9 m^-1$.

The electron energy spectrum in n-type $Pb_{1-x}Ge_xTe$ under the condition of formation of band tails can be written as

$$\left[\frac{2}{1 + Erf\left(\frac{E}{\eta_g}\right)}\right]\theta_0(E, \eta_g) + \gamma_3(E_{nzHD406}, \eta_g) \times \left[\bar{E}_{g_0} - 0.195k_s^2 - 0.345k_z^2\right] = \left[0.23k_s^2 + 0.02k_z^2 \pm \left[0.06\bar{E}_{g_0} + 0.061k_s^2 + 0.0066k_z^2\right]k_s + \left[\bar{E}_{g_0} + 0.411k_s^2 + 0.377k_z^2\right]\left[0.606k_s^2 + 0.722k_z^2\right]\right] \quad (1.343)$$

The $E - k_s$ relation in HD QWs of n-type HD $Pb_{1-x}Ge_xTe$ assumes the form

$$\left[\frac{2}{1 + Erf\left(\frac{E}{\eta_g}\right)}\right]\theta_0(E, \eta_g) + \gamma_3(E, \eta_g) \times \left[\bar{E}_{g_0} - 0.195k_s^2 - 0.345\left(\frac{n_z\pi}{d_z}\right)^2\right]$$

$$= \Bigg[0.23k_s^2 + 0.02\left(\frac{\mathrm{n}_z\pi}{d_z}\right)^2 \pm \left[0.06\bar{E}_{g_0} + 0.061k_s^2 + 0.0066\left(\frac{\mathrm{n}_z\pi}{d_z}\right)^2\right]k_s + \left[\bar{E}_{g_0} + 0.411k_s^2 + 0.377\left(\frac{\mathrm{n}_z\pi}{d_z}\right)^2\right] \times \left[0.606k_s^2 + 0.722\left(\frac{\mathrm{n}_z\pi}{d_z}\right)^2\right]\Bigg] \tag{1.344}$$

The sub-band energy ($E_{n_{zHD400}}$) is the lowest positive root of the following equation

$$\left[\frac{2}{1+Erf\left(\frac{E_{n_{zHD400}}}{\eta_g}\right)}\right]\theta_0\left(E_{n_{zHD400}},\eta_g\right)+\gamma_3\left(E_{n_{zHD400}},\eta_g\right) \times \left[\bar{E}_{g_0} - 0.345\left(\frac{\mathrm{n}_z\pi}{d_z}\right)^2\right] = \Bigg[0.02\left(\frac{\mathrm{n}_z\pi}{d_z}\right)^2 \pm \left[\bar{E}_{g_0} + 0.377\left(\frac{\mathrm{n}_z\pi}{d_z}\right)^2\right]\left[0.722\left(\frac{\mathrm{n}_z\pi}{d_z}\right)^2\right]\Bigg] \tag{1.345}$$

The EM, DOS function and the electron concentration should be calculated numerically.

1.2.14 *The EM in Quantum Wells (QWs) of HD Zinc and Cadmium Diphosphides*

The dispersion relation of the carriers of cadmium and zinc diphosphides are given by [158]

$$E = \left[\beta_1 + \frac{\beta_2\beta_3(k)}{8\beta_4}\right]k^2 \pm \left\{\left[\beta_4\beta_3(k)x\left(\beta_5 - \frac{\beta_2\beta_3(k)}{8\beta_4}\right)k^2\right] + 8\beta_4^2\left(1-\frac{\beta_3^2(k)}{4}\right) - \beta_2\left(1-\frac{\beta_3^2(k)}{4}\right)k^2\right\}^{1/2} \tag{1.346}$$

where $\beta_1, \beta_2, \beta_4$ and β_5 are system constants and

$$\beta_3(k) = \frac{k_x^2 + k_y^2 - 2k_z^2}{k^2}$$

Under the condition of formation of band tail, the above equation assumes the form

$$\gamma_3(E, \eta_g) = \left[\beta_1 + \frac{\beta_2\beta_3(k)}{8\beta_4}\right]k^2 \pm \left\{\left[\beta_4\beta_3(k)\left(\beta_5 - \frac{\beta_2\beta_3(k)}{8\beta_4}\right)k^2\right] + 8\beta_4^2\left(1 - \frac{\beta_3^2(k)}{4}\right) - \beta_2\left(1 - \frac{\beta_3^2(k)}{4}\right)k^2\right\}^{1/2} \quad (1.347)$$

The EM in HD QWs of Zinc and Cadmium diphosphides can be written as

$$\begin{aligned}\gamma_3(E, \eta_g) = {} & \left[\beta_1 + \frac{\beta_2\beta_{31}(k)}{8\beta_4}\right]\left[k_x^2 + k_y^2 + \left(\frac{n_z\pi}{d_z}\right)^2\right] \\ & \pm \left\{\left[\beta_4\beta_{31}(k)\left(\beta_5 - \frac{\beta_2\beta_{31}(k)}{8\beta_4}\right)\left[k_x^2 + k_y^2 + \left(\frac{n_z\pi}{d_z}\right)^2\right]\right]\right. \\ & + 8\beta_4^2\left(1 - \frac{\beta_{31}^2(k)}{4}\right) - \beta_2\left(1 - \frac{\beta_{31}^2(k)}{4}\right) \\ & \left.\times\left[k_x^2 + k_y^2 + \left(\frac{n_z\pi}{d_z}\right)^2\right]\right\}^{1/2} \end{aligned} \quad (1.348)$$

where

$$\beta_{31}(k) = \left[\frac{k_x^2 + k_y^2 - 2\left(\frac{n_z\pi}{d_z}\right)^2}{\left[k_x^2 + k_y^2 + \left(\frac{n_z\pi}{d_z}\right)^2\right]}\right]$$

The sub-band energy $(E_{n_{zHD401}})$ is the lowest positive root of the following equation

$$\gamma_3\left(E_{n_{zHD402}}, \eta_g\right) = \left[\beta_1 - \frac{\beta_2}{4\beta_4}\right]\left[\left(\frac{n_z\pi}{d_z}\right)^2\right] \pm \left\{\left[-2\beta_4\left(\beta_5 - \frac{\beta_2}{4\beta_4}\right)\left[\left(\frac{n_z\pi}{d_z}\right)^2\right]\right]\right\}^{1/2}. \tag{1.349}$$

The EM, DOS function and the electron concentration should be calculated numerically.

Thus, we can summarize the whole mathematical background in the following way.

In this chapter, we have investigated the 3D and 2D EMs from HD bulk and QWs of non-linear optical materials on the basis of a newly formulated electron dispersion law considering the anisotropies of the effective electron masses, the spin orbit splitting constants and the influence of crystal field splitting within the framework of **k.p** formalism. The results for 3D and 2D EMs from HD bulk and QWs of III–V, ternary and quaternary compounds in accordance with the three and two band models of Kane form a special case of our generalized analysis. We have also studied the EM in accordance with the models of Stillman *et al.* and Palik *et al.* respectively since these models find use to describe the electron energy spectrum of the aforesaid materials. The 3D and 2D EMs has also been derived for HD bulk and QWs of II–VI, IV–VI, stressed materials, Te, n-GaP, p-$PtSb_2$, Bi_2Te_3, n-Ge and n-GaSb compounds by using the models of Hopfield, Dimmock, Bangert and Kastner, Seiler, Bouat and Thuillier, Rees, Emtage, Kohler, Cardona, Wang *et al.* and Mathur *et al.* respectively and transforming each and on the basis of the appropriate carrier energy spectra. The well-known expressions of the EMs in the absence of band tails for wide gap materials have been obtained as special cases of our generalized analysis under certain limiting conditions. This indirect test not only exhibits the mathematical compatibility of our formulation but also shows the fact that our simple analysis is a more generalized one, since one can obtain the corresponding results for relatively wide gap materials having parabolic energy bands under certain limiting conditions from our present derivation.

1.3 Results and Discussion

Using the appropriate equations and taking the energy band constants as given in the Appendix A, we have Since Cd3As 2 crystals plotted the EM in QWs of HD Cd_3As_2 as a function of film thickness and have been shown in Fig. 1.1. For comparison, we have also plotted the EM in the absence of the crystal field splitting for the three and the two band model of Kane. Figure 1.1 exhibits the effect of 1D quantization on the EM in general, and bears a good amount of discussion. It appears that the effect of van Hove singularity makes the EM to suffer severe discontinuities. Assuming a carrier degeneracy of 10^{18} m^{-2}, Fig. 1.1a shows that the EM can reach upto about 10% of its free mass at a film thickness of 5 nm, which is quite high from its bulk value and may degrade the carrier mobility to a great extent. In the same figure, we have also demonstrated the effect of assuming only the lowest level subband. It appears that with this

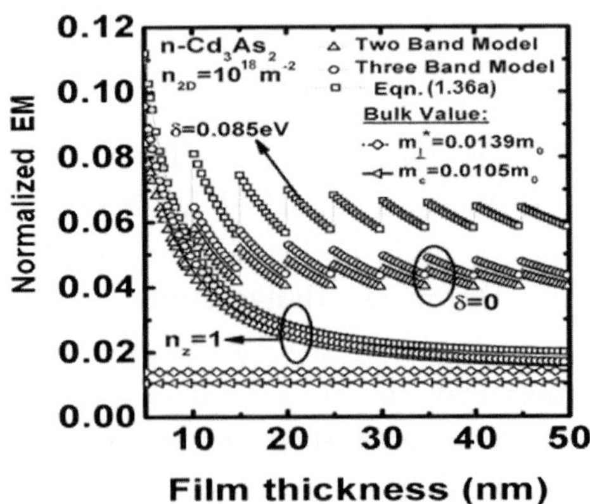

Fig. 1.1a. Plot of the normalized EM as function of film thickness for QWs of HD n-Cd_3As_2 considering (1.36a). The plots for three and two band models of Kane have also been exhibited in which, $m^*_\perp = 0.0139m_0$ and $m_c = \frac{1}{2}(m^*_{11} + m^*_\perp)$ $m_0 = 0.0105m_0$ are the corresponding bulk values.

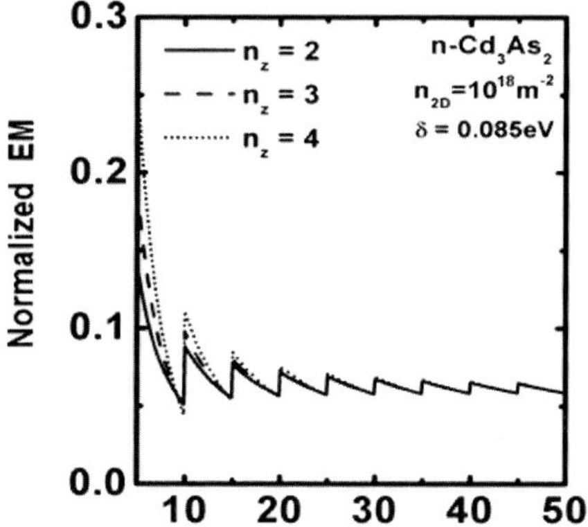

Fig. 1.1b. Plot of the normalized EM as function of film thickness for QWs of HD n-Cd_3As_2 for all cases of Fig. 1.1a at different three sub-band levels.

approximation, the EM approaches to the bulk value $m^*_\perp = 0.0139m_0$ more quickly than that of by considering the sub-bands. With this, it is now more obvious to note that the assumption of a single subband occupancy throughout leads to the practical approach to the determination of EM. All the models of the single subband occupied curves tend to merge with the bulk value near 50 nm thickness.

The increase in the EM with the reduction of film thickness is due to the increased Fermi energy of the material. It must be noted that with such a highly doped system, the Fermi energy is determined by the carrier statistics equation. It is this Fermi energy which should be used in the determination of the EM. This is not with the case in an intrinsic material. In such a case, the Fermi energy coincides with the intrinsic energy level, which is very near to the energy band gap of the material.

Thus all the curves below such thickness are expected to suffer deviation with our existing theoretical model, if plotted. In all the subsequent geometry dependent curves, we have restricted ourselves above sub-4 nm regime. Since Cd_3As_2 crystals are usually grown as degenerate n-type specimens, the Fermi level mass will be the effective mass of consideration for transport in Cd_3As_2. Hence in the quantum limit, the effective mass at the Fermi level corresponding to the lowest electric sub-band will be the effective conductivity mass for electron transport in Cd_3As_2. It appears from these figures that the Fermi level mass is significantly influenced by the effects of 1D quantization particularly in tetragonal semiconductors like n-Cd_3As_2 having crystalline field-effects and energydependent anisotropy of the effective mass. It has been found that the effective mass at the Fermi level depends on the size quantum number due to the combined influence of crystal-field splitting and the anisotropic spin orbit splitting constant, resulting in different effective masses at the Fermi level corresponding to different electric sub-bands (the different effective masses being the same in the absence of field splitting as can be seen from the Figs. 1.1a and 1.1b). It has further been observed that the different effective masses corresponding to different electric sub-bands closely approach each other, for a given film thickness, with increasing electron concentration and, for a given electron concentration, with increasing film thickness. These are in conformity with expectations since both with increasing electron concentration at a given film thickness and with increasing film thickness for a given electron concentration, the effects of size quantization gradually become less and less significant. As in bulk specimens, the Fermi level mass increases with increasing carrier concentration at a given value of the film thickness. Besides, for particular values of the film thickness and electron concentration, the combined effect of $\delta \neq 0$ and $\Delta_{11} \neq \Delta_{\perp}$ effect of crystal-field splitting is to reduce the effective mass corresponding to any particular sub-band. It may further be noted that if the direction normal to the film is taken as one of the transverse directions of the single ellipsoid at the zone centre and not as the longitudinal direction as assumed in the present chapter, the effective mass at the Fermi level corresponding to any given sub-band would be somewhat different. Nevertheless, since the mass anisotropy in Cd_3As_2 is indeed small as can be seen from the values of P_{11} and $P_{\perp}$ which are very close to each other, the arbitrary choice of the direction normal to the film with respect to the major axis of the ellipsoid would not result in a significant change in the effective mass at the Fermi level corresponding to a particular sub-band. The Fermi level mass should gradually become closer to that of bulk specimens with increasing film thickness since, for such

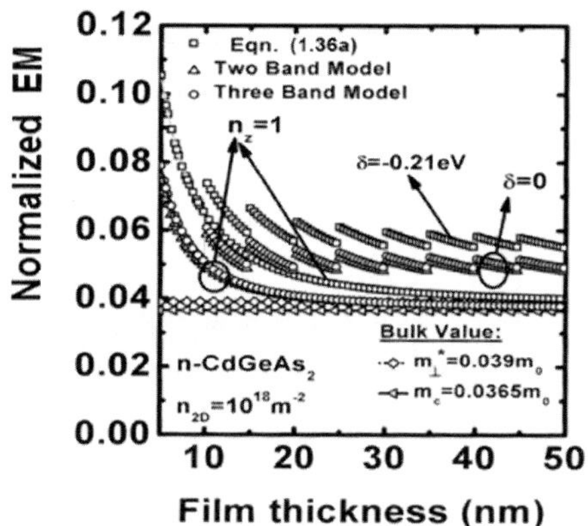

Fig. 1.2. Plot of the normalized EM as function of film thickness for QWs of HD n-$CdGeAs_2$ for all the cases of Fig. 1.1 in which, $m_{\perp}^{*} = 0.039m_0$ and $m_c = \frac{1}{2}(m_{11}^{*} + m_{\perp}^{*})m_0 = 0.0105m_0$ are the corresponding bulk values.

thicknesses, the effects of size quantization are greatly diminished. This has also been confirmed in our present work. Furthermore, the general features of the effects of size quantization on the effective mass as discussed here would also be valid with the only exception that the effective mass at the Fermi level will be independent of the size quantum number in the absence of crystal-field splitting and anisotropic spin orbit splitting constant for the III–V small-gap semiconductors since these semiconductors have non-parabolic energy bands obeying Kane's dispersion relation and the present chapter is based on the generalized Kane's model.

Figure 1.2 exhibits the plot EM in QWs of HD n-$CdGeAs_2$ as a function of film thickness in accordance with the HD three and two band models of Kane together with the incorporation of the crystal field parameter. It appears that the effect of the crystal field splitting increases the EM sharply below sub-20 nm. The EM also increases about 7% at 5 nm and converges to its bulk value beyond 20 nm at the same value of electron degeneracy. The effect of film thickness on the EM of HD III–V semiconductors, most important with respect to extremely low field high mobility of which are HD n-InAs, n-InSb and n-GaAs has been exhibited in Figs. 1.3, 1.4 and 1.5 respectively.

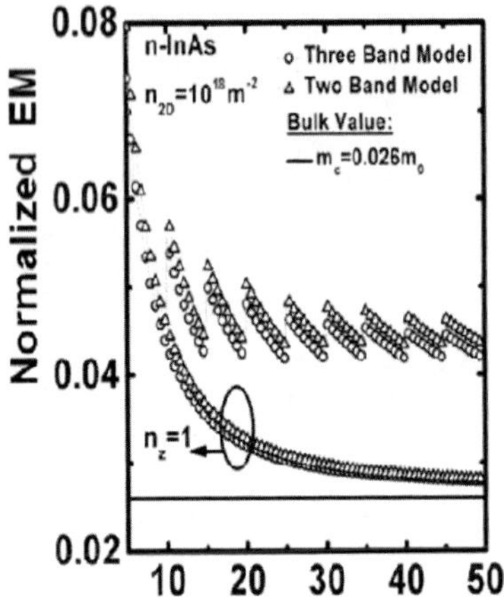

Fig. 1.3. Plot of the normalized EM as function of film thickness for QWs of HD n-InAs in accordance with the three and the two band model of Kane.

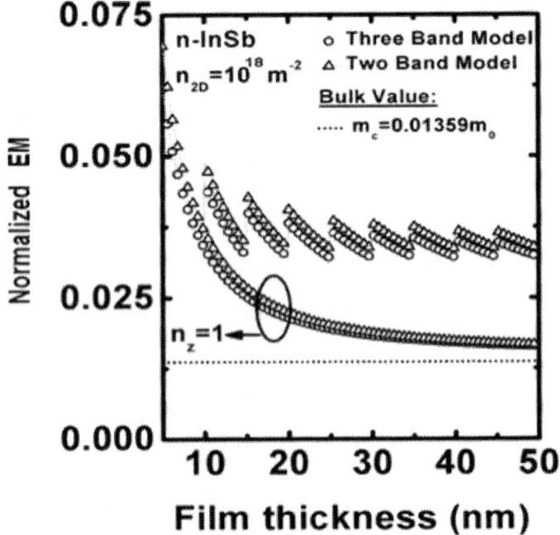

Fig. 1.4. Plot of normalized EM as function of film thickness for QWs of HD n-InSb for all the cases of Fig. 1.3.

The effect of non-linearity of the energy band structure on the respective EMs has been clearly indicated. It appears that in the determination of the EM, it is sufficient to take the HD two band model of Kane to explain the variation of the EM over a wide range of thinckess. The deviation from the HD three band model of Kane is much less; indicating that the complexity in the energy

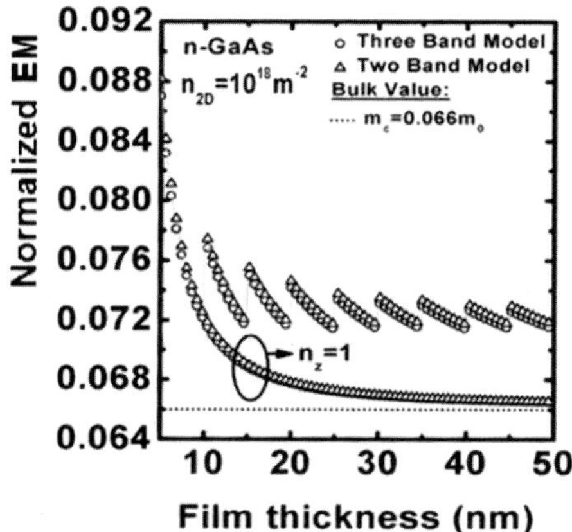

Fig. 1.5. Plot of the normalized EM as function of film thickness for QWs of HD n-GaAs for all the cases of Fig. 1.3.

band model can be reduces to large extent by considering only HD two band model of Kane. This is extremely important with respect to the numerical computation in device analyses preformance where sufficient longer computation time affects the efficiency in characterizing the compact model with respect to the said materials. In all the Figs. 1.3–1.5, we have demonstrated the effect of two widely known models *viz.* the three and the two band models. Figures 1.6 and 1.7 exhibit the variation of the EM with respect to the film thickness for the HD ternary and quaternary materials at same carrier degeneracy level. It appears that at an alloy composition $x = 0.3$, the EM in both the cases tends to about 0.1 times the rest mass at film thickness of 5 nm. Th effect of variation of EM on the alloy compostion for these two materials has been exhibited in Fig. 1.8.

The effect of increasing the alloy composition increases the EM for the said two materials. For the purpose of comparison, we have also plotted the the variation of the bulk effective mass with the alloy compostion. For the quaternary material, the difference between the two energy band model is not much, as also can be seen from Figs. 1.6 and 1.7. The increment in EM is rather linear in case of InGaAsP than that of HgCdTe. This also exhibits the variation of the electron

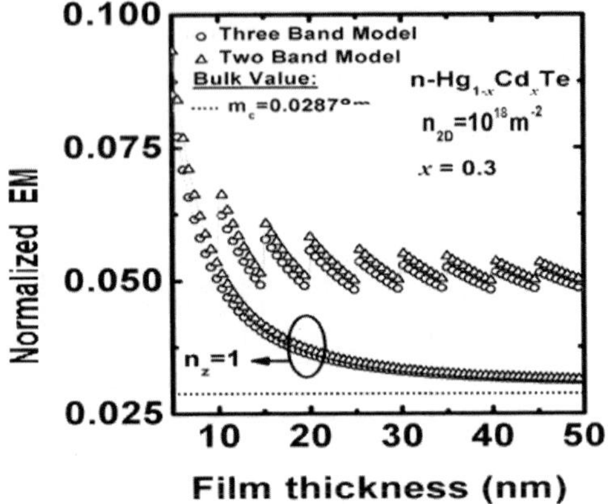

Fig. 1.6. Plot of the normalized EM as function of film thickness for QWs of HD n-$Hg_{0.3}Cd_{0.7}Te$ for all the cases of Fig. 1.3.

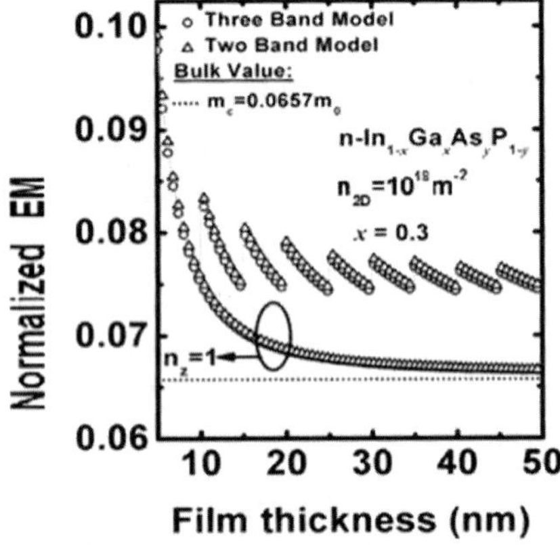

Fig. 1.7. Plot of the normalized EM as function of film thickness for QWs of HD n-$In_{1-x}Ga_xAs_yP_{1-y}$ for all the cases of Fig. 1.3.

mobility in these systems as the alloy composition changes. It appears that with increase in x, the mobility falls down assuming a constant relaxation rate.

The effect of carrier degeneracy on the EM in nonlinear optical, III–V, ternary and quaternary materials have been exhibited in the Figs. 1.9–1.15. It appears that the EM for all the aforementioned materials at 10 nm film thickness are almost invariant below sub

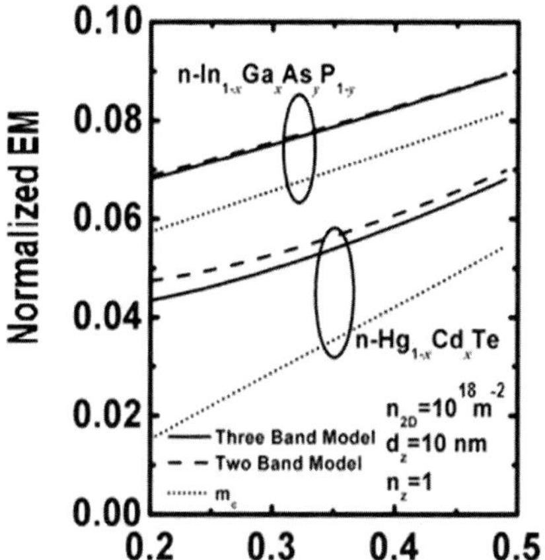

Fig. 1.8. Plot of the normalized EM at the lowest subband as function of alloy compostion in QWs of HD n-$Hg_{1-x}Cd_xTe$ and HD n-$In_{1-x}Ga_xAs_yP_{1-y}$ for the three and the two band models of Kane respectively.

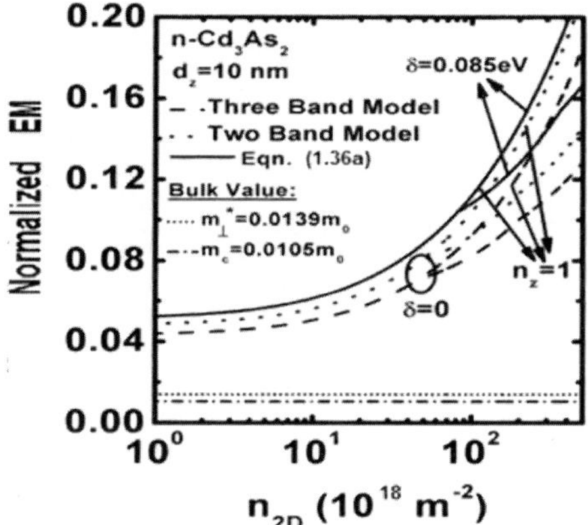

Fig. 1.9. Plot of the normalized EM as function of surface electron concentration in QWs of HD n-Cd_3As_2. The plots for three and two band models of Kane have also been exhibited in which, $m_{\perp}^{*} = 0.0139m_0$ and $m_c = \frac{1}{2}(m_{11}^{*} + m_{\perp}^{*})m_0 = 0.0105m_0$ are the corresponding bulk values.

10^{15} m^{-2}. The effect of inclusion of both the higher order subbands and lowest subband has been exhibited. From all the curves, it appears that the EM bears almost exponential relation with the carrier degeneracy.

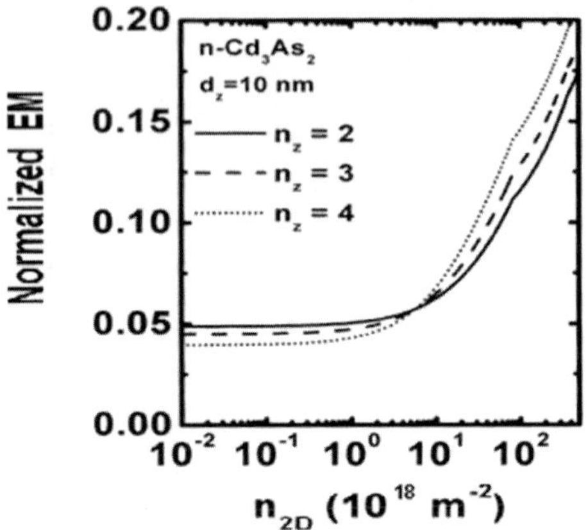

Fig. 1.9b. Plot of the normalized EM as function of surface electron concentration in QWs of HD n-Cd_3As_2 at different subband levels for all cases of Fig. 9a.

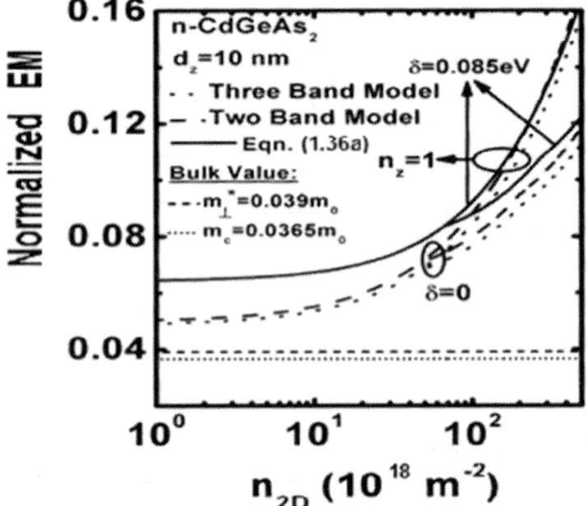

Fig. 1.10. Plot of the normalized EM as function of surface electron concentration in QWs of HD n-$CdGeAs_2$ for all the cases of Fig. 1.9.

The variation of the EM in II–VI materials like HD p-CdS has been exhibited in Figs. 1.16 and 1.17 as functions of film thickness and Fermi energy respectively. In these two figures, instead of obtaining the Fermi energy from the corresponsing carrier statistics, we have followed the opposite route, i.e., what values of the Fermi energy makes the EM to be very low or very high. A corresponding concentration of that order can then be evaluated. A decision of this

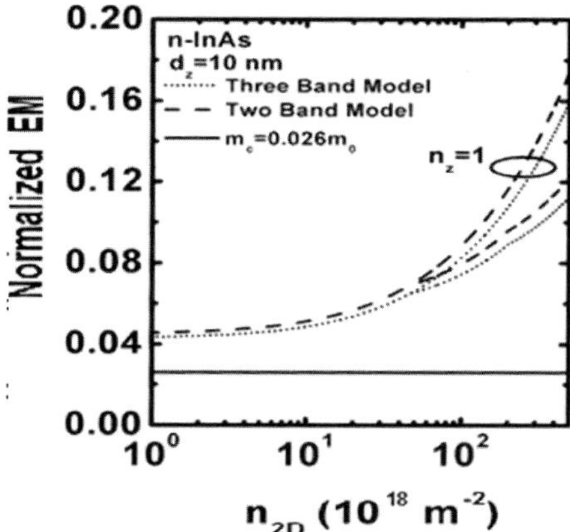

Fig. 1.11. Plot of the normalized EM as function of surface electron concentration in QWs of HD n-InAs.

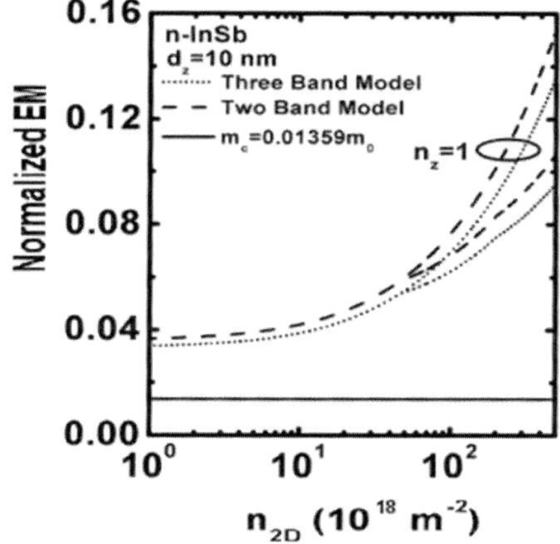

Fig. 1.12. Plot of the normalized EM as function of surface electron concentration in QWs of HD n-InSb.

kind aids a good amount of estimation in the optimization. Using this approach, we estimate that the EM can soar up to 0.77 times rest mass in the higher valley, while for the lower valley, it may plunge upto about 0.55 times rest mass. The effect of valley degeneracy as we see from this two curves expresses much in the understanding the electron transport direction.

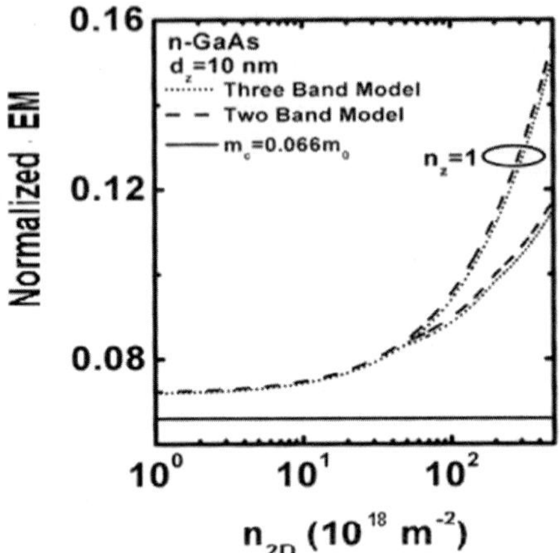

Fig. 1.13. Plot of the normalized EM as function of surface electron concentration in QWs of HD n-GaAs.

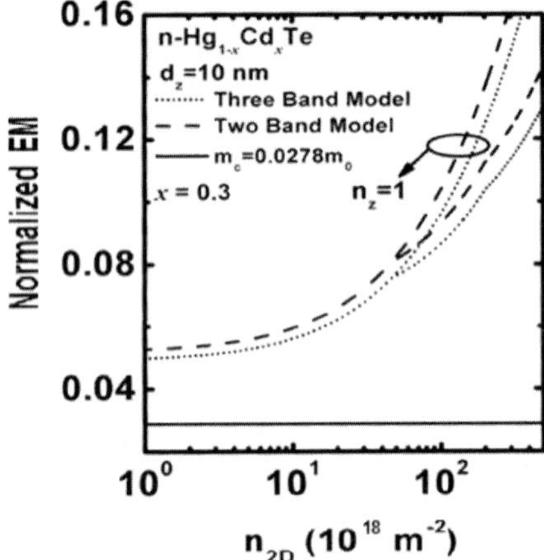

Fig. 1.14. Plot of the normalized EM as function of surface electron concentration in QWs of HD n-HgCdTe.

It appears from the two curves that the channel oriented along the lower valley direction will most probably results an increased value of current due to the low EM. It would have been of much interest to figure out how the energy band gap at the two valleys changes with respect to the thickness and is left as an exercise to the reader.

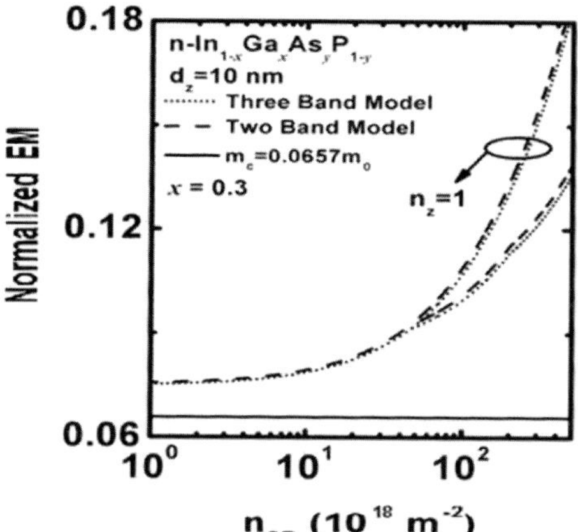

Fig. 1.15. Plot of the normalized EM as function of surface electron concentration in QWs of HD n-InGaAsP.

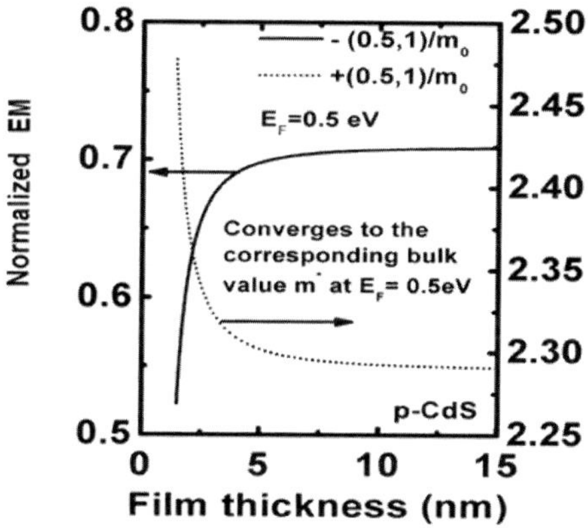

Fig. 1.16. Plot of the normalized EM as function of film thicknesss in QWs of HD p-CdS in two different conduction band valleys.

Figures 1.18 and 1.19 exhibit the effect of film thickness and the carrier concentration on the EM of QWs of PbSe and PbS respectively. The effect of increasing the carrier degeneracy has also been exhibited in Fig. 1.18. It appears that the EM increases from its corresponding bulk value sharply at the 5 nm film thickness implying a tremendous decrease in the carrier mobility.

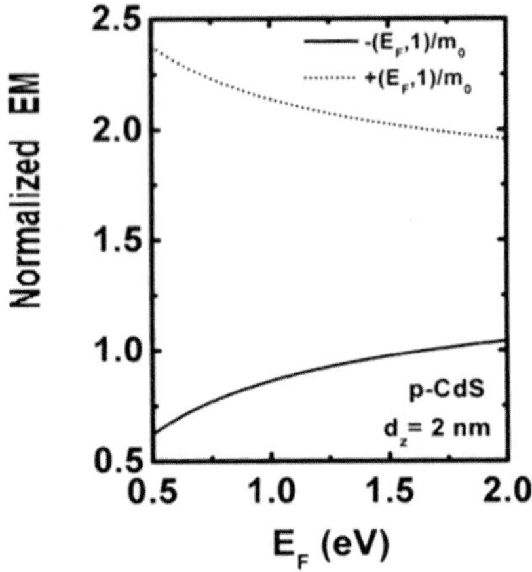

Fig. 1.17. Plot of the normalized EM as function of Fermi energy for all the cases of Fig. 1.16.

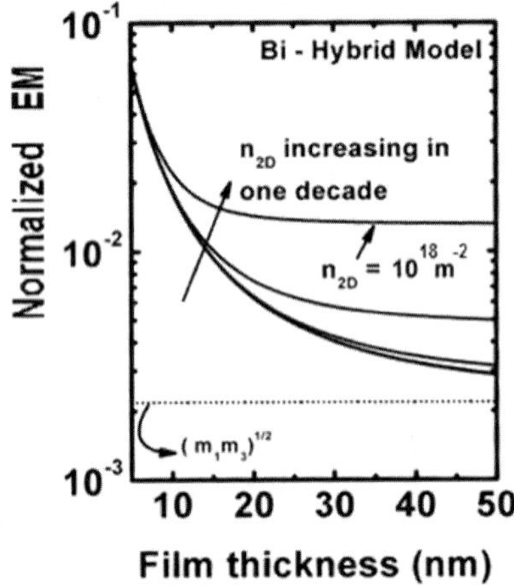

Fig. 1.18. Plot of the normalized EM as function of film thicknesss in QWs of HD PbSe for different carrier concentration values using the Dimmock model.

Figure 1.19 exhibits the effect of different temperatures on EM in QWs of HD PbS for a varying surface electron concentration. Figures 1.20, 1.21 and 1.22 exhibits the variation of the EM at the lowest subband level for QWs of PbTe strained InSb and Ge. The effective mass in IV–VI materials exhibits strong variation for PbTe,

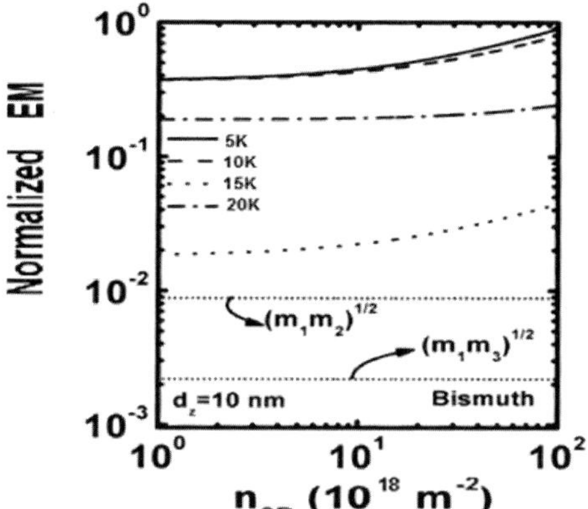

Fig. 1.19. Plot of the normalized EM at the lowest subband as function of surface electron coencentration in QWs of HD PbS at four different values of temperature.

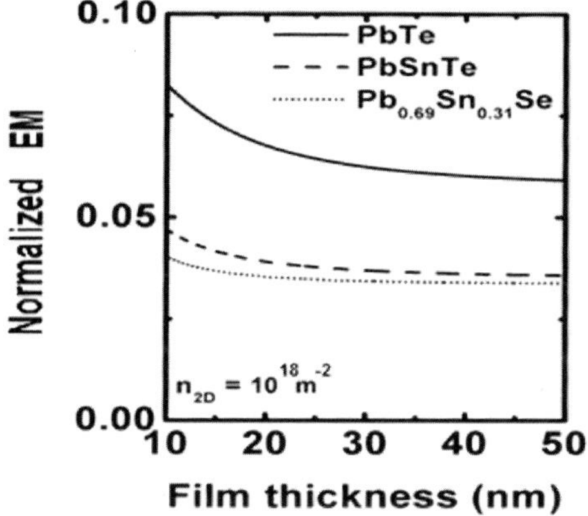

Fig. 1.20. Plot of the normalized EM at the lowest subband as function of film thickness in QWs of HD IV–VI materials.

an excellent thermoelectric material, whereas least for PbSnSe. It also appears that the EM of PbTe is higher than that of PbSnSe and PbSnTe. With the advent of strained quantum effect devices, the analysis in EM in strained quantum wells becomes very much important. It appears that the compressive and tensile strain doesnot tend to modify the respective magnitude of the EM in strained quantum wells of HD InSb. It should be noted that the EM in Fig. 1.21 has been evaluated by considering the momentum matrix

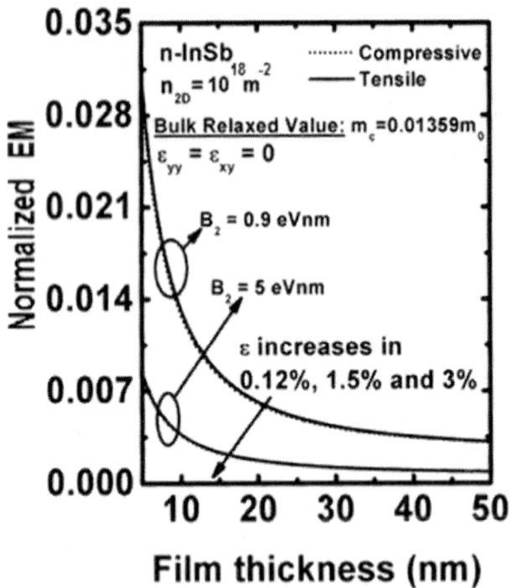

Fig. 1.21. Plot of the normalized EM at the lowest subband as function of film thickness in QWs of uniaxial strained HD InSb.

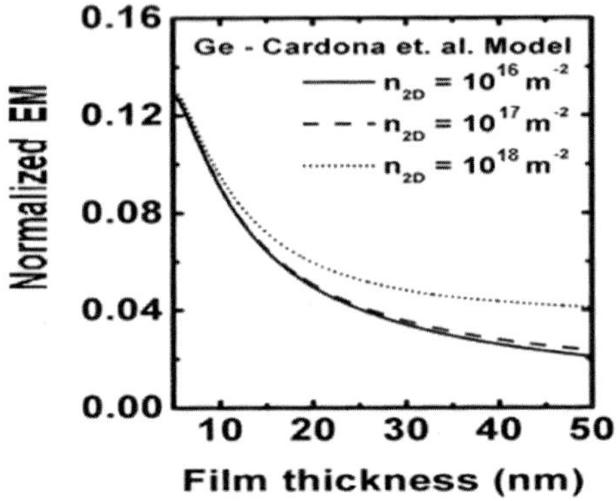

Fig. 1.22. Plot of the normalized EM at the lowest subband as function of film thickness in QWs of HD Ge.

element $B_2 = 0.9$ eVnm. This is a bulk value. However, an arbitrary increase in this geometry dependent parameter sufficiently reduces the EM and finds thus extensive use in strained film transitors. The variation of the EM in Qws of HD Ge has been exhibited in Fig. 1.22 as function of film thickness for the model of Cardona *et al.* The general trend of increase in the EM has also been exhibited here at least 6 times the bulk value $\sqrt{m_1^* m_2^*}$ for three different carrier concentration levels.

We observe that considering the various subband levels, the EM exhibits a step-functional decreasing dependence with increase in film thickness for QWs of all the single valley materials. The combined influence of the anisotropies of the energy band constants and the crystal field splitting is to enhance the EM as compared with the corresponding which is based on two band model of Kane in the whole range of thicknesses as considered in Fig. 1.1. The periodicity with respect to the film thickness is same in both the cases and is invariant of the energy band constants.

The influence of quantum confinement is immediately apparent from Figs. 1.1–1.7 and 1.9–1.15 since the EM depends strongly on the thickness of the quantum-confined materials in contrast with the corresponding bulk specimens. The EM changes with increasing carrier concentration suffering discontinuities with different numerical magnitudes. It appears from the aforementioned figures that the EM exhibits spikes for particular values of film thickness which, in turn, depends on the particular band structure of the specific semiconductor. Moreover, the EM from QWs of different compounds can be smaller than of bulk specimens of the same materials having multi valley conduction band like in case of p-CdS, which is also a direct signature of quantum confinement. This effect of the discontinuity on the EM will be less and less prominent with increasing film thickness. For bulk specimens of the same material, the EM will be found to increase continuously with increasing electron degeneracy in a non-oscillatory manner. The appearance of the discrete jumps in the respective figures is due to the redistribution of the electrons among the quantized energy levels when the size quantum number corresponding to the highest occupied level changes from one fixed value to the others.

With varying electron degeneracy, a change is reflected in the EM through the redistribution of the electrons among the size-quantized levels. It may be noted that at the transition zone from one sub band to another, the height of the peaks between any two sub-bands decreases with the increasing in the degree of quantum confinement and is clearly shown in the respective figures. It should be noted that although, the EM changes in various manners with all the variables as evident from all the figures, the rates of variations are totally band-structure dependent.

It is imperative to state that the present investigation excludes the many-body, hot electron, broadening and the allied effects in the simplified theoretical formalism due to the absence of proper analytical techniques for including them for generalized systems as considered here. We have also approximated the variation of value of the work function from its bulk value in the present system. Our simplified approach will be appropriate for the purpose of comparisons when the methods of tackling the formidable problems after inclusion of the said effects for the generalized systems emerge. The results of this simplified approach get transformed to the well-known formulation of the EM for wide gap materials having parabolic energy bands. This indirect test not only exhibits the mathematical compatibility of the formulation but also shows the fact that this simple analysis is more generalized one, since one can obtain the corresponding results for materials having parabolic energy bands under certain limiting conditions from the present derivation. For the purpose of computer simulations for obtaining the plots of EM versus various external variables, we have taken very low temperatures since the quantization effects are basically low temperature phenomena together with the fact that the temperature dependence of all the energy band constants of all the semiconductors and their nanostructures as considered in this chapter are not available in the literature. Our results as formulated in this chapter are valid for finite temperatures and are useful in comparing the results for temperature variations of EM after the availibility of the temperature dependences of such constants of various dispersion relations in this context. It is worth noting that the nature of the curves of EM with various physical variables based on our simplified formulations as presented here would be useful to analyze the experimental results when they materialize. The inclusion of the said effects would certainly increase the accuracy of the results although the qualitative features of EM would not change in the presence of the aforementioned effects.

It can be noted that on the basis of the dispersion relations of the various quantized structures as discussed above, the low field carrier mobility, drive currents in field effect transistors, Fowler Nordhiem field current, the Debye screening length, the plasma frequency,the activity coeficient, the carrier contribution to the elastic constants,the diffussion coefficient of minority carrers, the third-order non-linear optical susceptibility, the heat capacity, the dia and paramagnetic susceptibilities and the various important dc/ac transport coefficients can be probed for all types of HDQWs as considered here. Thus, our theoretical formulation comprises the dispersion relation dependent properties of various technologically important quantum-confined semiconductors having different band-structures. We have not considered other types of compounds in order to keep the presentation concise and succinct. With different sets of energy band parameters, one gets different numerical values of the EM. The nature of variations of the EM as shown here would be similar for the other types of materials and the simplified analysis of this chapter exhibits the basic qualitative features of the EM. The reader can also explore the EM for the left over 2D materials to enjoy the introcate computer programming and the 2D physics in this context. It may be noted that the basic aim of this chapter is not solely to demonstrate the influence of quantum confinement on the EM for a wide class of quantized materials but also to formulate the appropriate carrier statistics in the most generalized form, since the transport and other phenomena in modern nano-structured devices having different band structures and the derivation of the expressions of many important carrier properties are based on the temperature-dependent carrier statistics in such systems.

1.4 Open Research Problems

The problems under these sections of this monograph are by far the most important part and few open research problems from this chapter till end are being presented. The numerical values of the energy band constants for various

semiconductors are given in Appendix A for the related computer simulations.

(R.1.1) Investigate the EM for the bulk HD semiconductors whose respective dispersion relations of the carriers in the absence of doping are given below:

(a) The electron dispersion law in n-GaP can be written as [159]

$$E = \frac{\hbar^2 k_s^2}{2m_{\parallel}^*} + \frac{\hbar^2 k_s^2}{2m_{\perp}^*} \mp \frac{\bar{\Delta}}{2} \pm \left[\left(\frac{\bar{\Delta}}{2}\right)^2 + P_1 k_z^2 + D_1 k_x^2 k_y^2\right]^{1/2} \tag{R1.1}$$

where, $\bar{\Delta} = 335\, meV, P_1 = 2 \times 10^{-10}\, eVm, D_1 = P_1 a_1$ and $a_1 = 5.4 \times 10^{-10} m$.

(b) The carrier energy spectrum of IV–VI semiconductors in accordance with Foley *et al.* [160] can be written as

$$E + \frac{E_g}{2} = E_-(k) + \left[\left[E_+(k) + \frac{E_g}{2}\right]^2 + P_\perp^2 k_s^2 + P_\parallel^2 k_z^2\right]^{1/2} \tag{R1.2}$$

where,

$$E_+(k) = \frac{\hbar^2 k_s^2}{2m_\perp^+} + \frac{\hbar^2 k_z^2}{2m_\parallel^+}, \quad E_-(k) = \frac{\hbar^2 k_s^2}{2m_\perp^-} + \frac{\hbar^2 k_z^2}{2m_\parallel^-}$$

represents the contribution from the interaction of the conduction and the valance band edge states with the more distant bands and the free electron term,

$$\frac{1}{m_\perp^\pm} = \frac{1}{2}\left[\frac{1}{m_{tc}} \pm \frac{1}{m_{tv}}\right], \quad \frac{1}{m_\parallel^\pm} = \frac{1}{2}\left[\frac{1}{m_{1c}} \pm \frac{1}{m_{tv}}\right],$$

For n-PbTe, $P_\perp = 4.61 \times 10^{-10} eVm, P_\parallel = 1.48 \times 10^{-10}$ $eVm, \frac{m_0}{m_{tv}} = 10.36, \frac{m_0}{m_{1v}} = 0.75, \frac{m_0}{m_{tc}} = 11.36, \frac{m_0}{m_{1c}} = 1.20$ and $g_v = 4$.

(c) The conduction electrons of n-GaSb obey the following two dispersion relations:

(i) In accordance with the model of Seiler *et al.* [161]

$$E = \left[-\frac{E_g}{2} + \frac{E_g}{2} \left[1 + \alpha_4 k^2\right]^{1/2} + \frac{\bar{\zeta}_0 \hbar^2 k^2}{2m_0} + \frac{\bar{v}_0 f_1(k)\hbar^2}{2m_0} \pm \frac{\bar{\omega}_0 f_2(k)\hbar^2}{2m_0} \right] \quad \text{(R1.3)}$$

where

$$\alpha_4 \equiv 4P^2 \left(E_g + \frac{2}{3}\Delta \right) \left[E_g^2 \left(E_g + \Delta \right) \right]^{-1},$$

P is the isotropic momentum matrix element, $f_1(k) \equiv k^{-2}[k_x^2 k_y^2 + k_y^2 k_z^2 + k_z^2 k_x^2$ represents the warping of the Fermi surface, $f_2(k) \equiv [\{k^2(k_x^2 k_y^2 + k_y^2 k_z^2 + k_z^2 k_x^2) - 9k_x^2 k_y^2 k_z^2\}^{1/2} k^{-1}]$ represents the inversion asymmetry splitting of the conduction band and $\bar{\varsigma}_0, \bar{v}_0$ and $\hat{\omega}$ represent the constants of the electron spectrum in this case.

(ii) In accordance with the model of Zhang *et al.* [162]

$$E = \left[E_2^{(1)} + E_2^{(2)} K_{4,1} \right] k^2 + \left[E_4^{(1)} + E_4^{(2)} K_{4,1} \right] k^4 + k^6 \left[E_6^{(1)} + E_6^{(2)} K_{4,1} + E_6^{(3)} K_{6,1} \right] \quad \text{(R1.4)}$$

where

$$K_{4,1} \equiv \frac{5}{4}\sqrt{21} \left[\frac{k_x^4 + k_y^4 + k_z^4}{k^4} - \frac{3}{5} \right],$$

$$K_{6,1} \equiv \sqrt{\frac{639639}{32}} \left[\frac{k_x^2 k_y^4 k_z^2}{k^6} + \frac{1}{22} \left(\frac{k_x^4 + k_y^4 + k_z^4}{k^4} - \frac{3}{5} \right) - \frac{1}{105} \right],$$

the coefficients are in eV, the values of k are $10(\frac{a}{2\pi})$ times those of k in atomic units (a is the lattice constant), $E_2^{(1)} = 1.0239620$, $E_2^{(2)} = 0$, $E_4^{(1)} = -1.1320772$, $E_4^{(2)} = 0.05658$, $E_6^{(1)} = 1.1072073$, $E_6^{(2)} = -0.1134024$ and $E_6^{(3)} = -0.0072275$.

(d) In addition to the well known band models as discussed in this monograph, the conduction electrons of

III–V semiconductors obey the following three dispersion relations:

(i) In accordance with the model of Rossler [163]

$$E = \frac{\hbar^2 k^2}{2m*} + \bar{\alpha}_{10} k^4 + \bar{\beta}_{10}\left[k_x^2 k_y^2 + k_y^2 k_z^2 + k_z^2 k_x^2\right] \pm \bar{\gamma}_{10}[k^2(k_x^2 k_y^2 + k_y^2 k_z^2 + k_z^2 k_x^2) - 9k_x^2 k_y^2 k_z^2]^{1/2} \quad \text{(R1.5)}$$

where $\bar{\alpha}_{10} = \bar{\alpha}_{11} + \bar{\alpha}_{12} k, \bar{\beta}_{10} = \bar{\beta}_{11} + \bar{\beta}_{12} k$, and $\bar{\gamma}_{10} = \bar{\gamma}_{11} + \bar{\gamma}_{12}$ in which, $\alpha_{11} = -2132 \times 10^{-40} eVm^4$, $\alpha_{12} = 9030 \times 10^{-50}$ $eVm^5, \bar{\beta}_{11} = -2493 \times 10^{-40} eVm^4$, $\bar{\beta}_{12} = 12594 \times 10^{-50} eVm^5$, $\bar{\gamma}_{11} = 30 \times 10^{-30} eVm^3$ and $\bar{\gamma}_{12} = -154 \times 10^{-42} eVm^4$.

(ii) In accordance with Johnson and Dickey [164], the electron energy spectrum assumes the form

$$E = -\frac{E_g}{2} + \frac{\hbar^2 k^2}{2}\left[\frac{1}{m_0} + \frac{1}{m_{\gamma b}}\right] + \frac{E_g}{2}\left[1 + 4\frac{\hbar^2 k^2}{2m_c'}\frac{\bar{f}_1(E)}{E_g}\right]^{1/2} \quad \text{(R1.6)}$$

where,

$$\frac{m_0}{m_c'} \equiv P^2\left[\frac{\left(E_g + \frac{2\Delta}{3}\right)}{E_g\left(E_g + \Delta\right)}\right],$$

$$\bar{f}_1(E) \equiv \frac{\left(E_g + \Delta\right)\left(E + E_g + \frac{2\Delta}{3}\right)}{\left(E_g + \frac{2\Delta}{3}\right)\left(E + E_g + \Delta\right)}, \quad m_c' = 0.139\, m_0$$

and

$$m_{\gamma b} = \left[\frac{1}{m_c'} - \frac{2}{m_0}\right]^{-1}.$$

(iii) In accordance with Agafonov *et al.* [165], the electron energy spectrum can be written as

$$E = \frac{\bar{\eta} - E_g}{2}\left[1 - \frac{\hbar k^2}{2\bar{\eta} m^*}\left\{\frac{D\sqrt{3} - 3\bar{B}}{2\left(\frac{\hbar^2}{2m^*}\right)}\right\}\left[\frac{k_x^4 + k_y^4 + k_z^4}{k^4}\right]\right] \quad \text{(R1.7)}$$

where,

$$\bar{\eta} \equiv \left(E_g^2 + \frac{8}{3}P^2k^2\right)^{1/2}$$

and

$$\bar{B} \equiv -21\frac{\hbar^2}{2m_0}$$

and

$$D \equiv -40\left(\frac{\hbar^2}{2m_0}\right).$$

(e) The energy spectrum of the carriers in the two higher valance bands and the single lower valance band of Te can, respectively, be expressed as [166]

$$\bar{E} = A_{10}k_z^2 + B_{10}k_s^2 \pm [\Delta_{10}^2 + (\beta_{10}k_z)^2]^{1/2} \quad \text{and}$$
$$\bar{E} = \Delta_{\|} + A_{10}k_z^2 + B_{10}k_s^2 \pm \beta_{10}k_z \tag{R1.8}$$

where, $\bar{E}$ is the energy of the hole as measured from the top of the valance and within it, $A_{10} = 3.77 \times 10^{-19}eVm^2$, $B_{10} = 3.57 \times 10^{-19}eVm^2$, $\Delta_{10} = 0.628eV$, $(\beta_{10})^2 = 6 \times 10^{-20}(eVm)^2$ and $\Delta_{\|} = 1004 \times 10^{-5}eV$ are the spectrum constants.

(f) The dispersion relation of the holes in p-InSb can be written in accordance with Cunningham [167] as

$$\bar{E} = c_4(1+\gamma_4 f_4)k^2 \pm \frac{1}{3}[2\sqrt{2}\sqrt{c_4}\sqrt{16+5\gamma_4}\sqrt{E_4}g_4k] \tag{R1.9}$$

where,

$$c_4 = \frac{\hbar^2}{2m_0} + \theta_4, \quad \theta_4 \equiv 4.7\frac{\hbar^2}{2m_0}, \quad \gamma_4 \equiv \frac{b_4}{c_4},$$
$$b_4 \equiv \frac{3}{2}b_5 + 2\theta_4, \quad b_5 \equiv 2.4\frac{\hbar^2}{2m_0},$$

$f_4 \equiv \frac{1}{4}[\sin^2 2\theta + \sin^4\theta \sin^2 2\phi], \theta$ is measured from the positive z-axis,

$$g_4 \equiv \sin\theta\left[\cos^2\theta + \frac{1}{4}\sin^4\theta\sin^2 2\phi\right] \quad \text{and} \quad E_4 = 5\times10^{-4}eV.$$

(g) The energy spectrum of the valance bands of CuCl in accordance with Yekimov *et al.* [168] can be written as

$$E_h = (\gamma_6 - 2\gamma_7)\frac{\hbar^2 k^2}{2m_0} \tag{R1.10}$$

and

$$E_{l,s} = (\gamma_6 + \gamma_7)\frac{\hbar^2 k^2}{2m_0} - \frac{\Delta_1}{2} \pm \left[\frac{\Delta_1^2}{4} + \gamma_7\Delta_1\frac{\hbar^2 k^2}{2m_0} + 9\left(\frac{\gamma_7\hbar^2 k^2}{2m_0}\right)^2\right]^{1/2} \tag{R1.11}$$

where, $\gamma_6 = 0.53, \gamma_7 = 0.07, \Delta_1 = 70meV$.
(h) In the presence of stress, χ_6 along <001> and <111> directions, the energy spectra of the holes in semiconductors having diamond structure valance bands can be respectively expressed following Roman [169] *et al.* as

$$E = A_6k^2 \pm \left[\bar{B}_7^2k^4 + \delta_6^2 + B_7\delta_6(2k_z^2 - k_s^2)\right]^{1/2} \tag{R1.12}$$

and

$$E = A_6k^2 \pm \left[\bar{B}_7^2k^4 + \delta_7^2 + \frac{D_6}{\sqrt{3}}\delta_7(2k_z^2 - k_s^2)\right]^{1/2} \tag{R1.13}$$

where, A_6, B_7, D_6 and C_6 are inverse mass band parameters in which $\delta_6 \equiv l_7(\bar{S}_{11} - \bar{S}_{12})\chi_6, \bar{S}_{ij}$ are the usual elastic compliance constants, $\bar{B}_7^2 \equiv (B_7^2 + \frac{c_6^2}{5})$ and $\delta_7 \equiv (\frac{d_g S_{44}}{2\sqrt{3}})\chi_6$. For gray tin, $d_8 = -4.1eV$, $l_7 = -2.3eV$, $A_6 = 19.2\frac{\hbar^2}{2m_0}$, $B_7 = 26.3\frac{\hbar^2}{2m_0}$, $D_6 = 31\frac{\hbar^2}{2m_0}$ and $c_6^2 = -1112\frac{\hbar^2}{2m_0}$.

(R.1.2) Investigate the EM for HDQWs of all the semiconductors as considered in R.1.1.

(R.1.3) Investigate the EM for bulk specimens of the heavily–doped semiconductors in the presences of exponential, Kane, Halperian, Lax and Bonch-Burevich types of Band tails [57] for all systems whose unperturbed carrier energy spectra are defined in R.1.1 respectively.

(R.1.4) Investigate the EM as defined in (R.1.3) for QWs of HD of all the heavily doped semiconductors as considered in R.1.3.

(R.1.5) Investigate the EM same set of masses for HD bulk specimens of the negative refractive index, organic, magnetic and other advanced optical materials in the presence of an arbitrarily oriented alternating electric field.

(R.1.6) Investigate the EM for HDQWs of the negative refractive index, organic, magnetic and other advanced optical materials in the presence of an arbitrarily oriented alternating electric field.

(R.1.7) Investigate the EM for multiple QWs of semiconductors whose unperturbed carrier energy spectra are defined in R1.1 and heavily–doped semiconductors in the presences of exponential, Kane, Halperian, Lax and Bonch-Burevich types of Band tails [37] for all systems whose unperturbed carrier energy spectra are defined in the same problems respectively.

(R.1.8) Investigate the EM for all the appropriate HD low dimensional systems of this chapter in the presence of finite potential wells.

(R.1.9) Investigate the EM for all the appropriate low dimensional HD systems of this chapter in the presence of parabolic potential wells.

(R.1.10) Investigate the EM for all the appropriate low dimensional HD systems of this chapter forming quantum rings.

(R.1.11) Investigate the EM for all the appropriate low dimensional HD systems for all the above appropriate problems in the presence of elliptical Hill and quantum square rings.

(R.1.12) Investigate all the appropriate problems of this chapter by removing all the mathematical approximations and establishing the respective appropriate uniqueness conditions.

References

1. P. M. Petroff, A. C. Gossard, W. Wiegmann, Appl. Phys. Lett. **45**, 620 (1984);
2. J. M. Gaines, P. M. Petroff, H. Kroemar, R. J. Simes, R. S. Geels, J. H. English, J. Vac. Sci. Technol. B **6**, 1378 (1988)

3. J. Cilbert, P. M. Petroff, G. J. Dolan, S. J. Pearton, A. C. Gossard, J. H. English, Appl. Phys. Lett. **49**, 1275 (1986)
4. T. Fujui, H. Saito, Appl. Phys. Lett. **50**, 824 (1987)
5. H. Sasaki, Jpn. J. Appl. Phys. **19**, 94 (1980)
6. P. M. Petroff, A. C. Gossard, R. A. Logan, W. Weigmann, Appl. Phys. Lett. **41**, 635 (1982)
7. H. Temkin, G. J. Dolan, M. B. Panish, S. N. G. Chu, Appl. Phys. Lett. **50**, 413 (1988)
8. I. Miller, A. Miller, A. Shahar, U. Koren, P. J. Corvini, Appl. Phys. Lett. **54**, 188 (1989)
9. L. L. Chang, H. Esaki, C. A. Chang, L. Esaki, Phys. Rev. Lett. **38**, 1489 (1977)
10. K. Hess, M. S. Shur, J. J. Drunnond, H. Morkoc, IEEE Trans. Electron. Devices **ED-30**, 07 (1983)
11. G. Bastard, *Wave Mechanics Applied to Semiconductor Hetero-structures*, Les Editions de Physique(Halsted, Les Ulis, New York, 1988)
12. M. J. Kelly, *Low dimensional semiconductors: materials, physics, technology, devices* (Oxford University Press, Oxford, 1995)
13. C. Weisbuch, B. Vinter, *Quantum Semiconductor Structures* (Boston Academic Press, Boston, 1991)
14. N. T. Linch, Festkorperprobleme **23**, 27 (1985)
15. D. R. Sciferes, C. Lindstrom, R. D. Burnham, W. Streifer, T. L. Paoli, Electron. Lett. **19**, 170 (1983)
16. P. M. Solomon, Proc. IEEE **70**, 489 (1982)
17. T. E. Schlesinger, T. Kuech, Appl. Phys. Lett. **49**, 519 (1986)
18. D. Kasemet, C. S. Hong, N. B. Patel, P. D. Dapkus, Appl. Phys. Lett. **41**, 912 (1982)
19. K. Woodbridge, P. Blood, E. D. Pletcher, P. J. Hulyer, Appl. Phys. Lett. **45**, 16 (1984)
20. S. Tarucha, H. O. Okamoto, Appl. Phys. Lett. **45**, 16 (1984)
21. H. Heiblum, D. C. Thomas, C. M. Knoedler, M. I. Nathan, Appl. Phys. Lett. **47**, 1105 (1985)
22. O. Aina, M. Mattingly, F. Y. Juan, P. K. Bhattacharyya, Appl. Phys. Lett. **50**, 43 (1987)
23. I. Suemune, L. A. Coldren, IEEE J. Quant. Electron. **24**, 1178 (1988)
24. D. A. B. Miller, D. S. Chemla, T. C. Damen, J. H. Wood, A. C. Burrus, A. C. Gossard, W. Weigmann, IEEE J. Quant. Electron. **21**, 1462 (1985)
25. J. W. Rowe, J. L. Shay, Phys. Rev. B **3**, 451 (1973)
26. H. Kildal, Phys. Rev. B **10**, 5082 (1974)
27. J. Bodnar, in *Proceedings of the International Conference on Physics of Narrow-gap Semiconductors* (Polish Science Publishers, Warsaw, 1978)
28. G. P. Chuiko, N. N. Chuiko, Sov. Phys. Semicond. **15**, 739 (1981)
29. K. P. Ghatak, S. N. Biswas, Proc. SPIE **1484**, 149 (1991)
30. A. Rogalski, J. Alloys Comp. **371**, 53 (2004)
31. A. Baumgartner, A. Chaggar, A. Patanè, L. Eaves, M. Henini, Appl. Phys. Lett. **92**, 091121 (2008)

32. J. Devenson, R. Teissier, O. Cathabard, A. N. Baranov, Proc. SPIE **6909**, 69090U (2008)
33. B. S. Passmore, J. Wu, M. O. Manasreh, G. J. Salamo, Appl. Phys. Lett. **91**, 233508 (2007)
34. M. Mikhailova, N. Stoyanov, I. Andreev, B. Zhurtanov, S. Kizhaev, E.Kunitsyna, K. Salikhov, Y. Yakovlev, Proc. SPIE **6585**, 658526 (2007)
35. W. Kruppa, J. B. Boos, B. R. Bennett, N. A. Papanicolaou, D. Park, R. Bass, Electron. Lett. **42**, 688 (2006)
36. E. O. Kane, in *Semiconductors and Semimetals*, vol. 1, ed. by R. K. Willardson, A. C. Beer, Academic Press, New York, **75** (1966)
37. B. R. Nag, *Electron Transport in Compound Semiconductors* (Springer, Heidelberg, 1980); A. N. Chakravarti, K. P. Ghatak, A. Dhar, K. K. Ghosh, S. Ghosh, Acta Phys. Polon A **60**, 151 (1981)
38. G. E. Stillman, C. M. Wolfe, J. O. Dimmock, in *Semiconductors and Semimetals*, **12**, ed. by R. K. Willardon, A. C. Beer Academic Press, New York, **169** (1977)
39. D. J. Newson, A. Karobe, Semicond. Sci. Tech. **3**, 786 (1988)
40. E. D. Palik, G. S. Picus, S. Teither, R. E. Wallis, Phys. Rev., 475 (1961)
41. P. Y. Lu, C. H. Wung, C. M. Williams, S. N. G. Chu, C. M. Stiles, Appl. Phys. Lett. **49**, 1372 (1986)
42. N. R. Taskar, I. B. Bhat, K. K. Prat, D. Terry, H. Ehasani, S. K. Ghandhi, J. Vac. Sci. Tech. **7A**, 281 (1989)
43. F. Koch, *Springer Series in Solid States Sciences* (Springer, Germany, 1984)
44. L. R. Tomasetta, H. D. Law, R. C. Eden, I. Reyhimy, K. Nakano, IEEE J. Quant. Electron. **14**, 800 (1978)
45. T. Yamato, K. Sakai, S. Akiba, Y. Suematsu, IEEE J. Quantum Electron. **14**, 95 (1978)
46. T. P. Pearsall, B. I. Miller, R. J. Capik, Appl. Phys. Lett. **28**, 499 (1976)
47. M. A. Washington, R. E. Nahory, M. A. Pollack, E. D. Beeke, Appl. Phys. Lett. **33**, 854 (1978)
48. M. I. Timmons, S. M. Bedair, R. J. Markunas, J. A. Hutchby, in *Proceedings of the 16th IEEE Photovoltaic Specialist Conference* (IEEE, San Diego, California 666, 1982)
49. J. A. Zapien, Y. K. Liu, Y. Y. Shan, H. Tang, C. S. Lee, S. T. Lee, Appl. Phys. Lett. **90**, 213114 (2007)
50. M. Park, Proc. SPIE **2524**, 142 (1995)
51. S.-G. Hur, E. T. Kim, J. H. Lee, G. H. Kim, S. G. Yoon, Electrochem. Solid-State Lett. **11**, H176 (2008)
52. H. Kroemer, Rev. Mod. Phys. **73**, 783 (2001)
53. T. Nguyen Duy, J. Meslage, G. Pichard, J. Crys. Growth **72**, 490 (1985)
54. T. Aramoto, F. Adurodija, Y. Nishiyama, T. Arita, A. Hanafusa, K. Omura, A. Morita, Solar Energy Mater. Solar Cells **75**, 211 (2003)
55. H. B. Barber, J. Electron. Mater. **25**, 1232 (1996)
56. S. Taniguchi, T. Hino, S. Itoh, K. Nakano, N. Nakayama, A. Ishibashi, M. Ikeda, Electron. Lett. **32**, 552 (1996)
57. J. J. Hopfield, J. Appl. Phys. **32**, 2277 (1961)

58. G. P. Agrawal, N. K. Dutta, *Semicond. Lasers* (Van Nostrand Reinhold, New York, 1993)
59. S. Chatterjee, U. Pal, Opt. Eng. (Bellingham), **32**, 2923 (1993)
60. T. K. Chaudhuri, Int. J. Energy Res. **16**, 481 (1992)
61. J. H. Dughaish, Phys. B **322**, 205 (2002)
62. C. Wood, Rep. Prog. Phys. **51**, 459 (1988)
63. K. F. Hsu, S. Loo, F. Guo, W. Chen, J. S. Dyck, C. Uher, T. Hogan, E. K. Polychroniadis, M. G. Kanatzidis, Science **303**, 818 (2004)
64. J. Androulakis, K. F. Hsu, R. Pcionek, H. Kong, C. Uher, J. J. D'Angelo, A. Downey, T. Hogan, M. G. Kanatzidis, Adv. Mater. **18**, 1170 (2006)
65. P. F. P. Poudeu, J. D'Angelo, A. D. Downey, J. L. Short, T. P. Hogan, M. G. Kanatzidis, Angew. Chem. Int. Ed. **45**, 3835 (2006)
66. P. F. Poudeu, J. D'Angelo, H. Kong, A. Downey, J. L. Short, R. Pcionek, T. P. Hogan, C. Uher, M. G. Kanatzidis, J. Am. Chem. Soc. **128**, 14347 (2006)
67. J. R. Sootsman, R. J. Pcionek, H. Kong, C. Uher, M. G. Kanatzidis, Chem. Mater. **18**, 4993 (2006)
68. A. J. Mountvala, G. Abowitz, J. Am. Ceram. Soc. **48**, 651 (1965)
69. E. I. Rogacheva, I. M. Krivulkin, O. N. Nashchekina, AYu. Sipatov, V. A. Volobuev, M. S. Dresselhaus, Appl. Phys. Lett. **78**, 3238 (2001)
70. H. S. Lee, B. Cheong, T. S. Lee, K. S. Lee, W. M. Kim, J. W. Lee, S. H. Cho, J. Y. Huh, Appl. Phys. Lett. **85**, 2782 (2004)
71. K. Kishimoto, M. Tsukamoto, T. Koyanagi, J. Appl. Phys. **92**, 5331 (2002)
72. E. I. Rogacheva, O. N. Nashchekina, S. N. Grigorov, M. A. Us, M. S. Dresselhaus, S. B. Cronin, Nanotechnology **14**, 53 (2003)
73. E. I. Rogacheva, O. N. Nashchekina, A. V. Meriuts, S. G. Lyubchenko, M. S. Dresselhaus, G. Dresselhaus, Appl. Phys. Lett. **86**, 063103 (2005)
74. E. I. Rogacheva, S. N. Grigorov, O. N. Nashchekina, T. V. Tavrina, S. G. Lyubchenko, AYu. Sipatov, V. V. Volobuev, A. G. Fedorov, M. S. Dresselhaus, Thin Solid Films **493**, 41 (2005)
75. X. Qiu, Y. Lou, A. C. S. Samia, A. Devadoss, J. D. Burgess, S. Dayal, C. Burda, Angew. Chem. Int. Ed. **44**, 5855 (2005)
76. C. Wang, G. Zhang, S. Fan, Y. Li, J. Phys. Chem. Solids **62**, 1957 (2001)
77. B. Poudel, W. Z. Wang, D. Z. Wang, J. Y. Huang, Z. F. Ren, J. Nanosci. Nanotechnol. **6**, 1050 (2006)
78. B. Zhang, J. He, T. M. Tritt, Appl. Phys. Lett. **88**, 043119 (2006)
79. W. Heiss, H. Groiss, E. KaQWmann, G. Hesser, M. Böberl, G. Springholz, F. Schäffler, K. Koike, H. Harada, M. Yano, Appl. Phys. Lett. **88**, 192109 (2006)
80. B. A. Akimov, V. A. Bogoyavlenskiy, L. I. Ryabova, V. N. Vasil'kov, Phys. Rev. B **61**, 16045 (2000)
81. Ya. A. Ugai, A. M. Samoilov, M. K. Sharov, O. B. Yatsenko, B. A. Akimov, Inorg. Mater. **38**, 12 (2002)
82. Ya. A. Ugai, A. M. Samoilov, S. A. Buchnev, Yu. V. Synorov, M. K. Sharov, Inorg. Mater. **38**, 450 (2002)

83. A. M. Samoilov, S. A. Buchnev, Yu. V Synorov, B. L. Agapov, A. M. Khoviv. Inorg. Mater. **39**, 1132 (2003)
84. A. M. Samoilov, S. A. Buchnev, E. A. Dolgopolova, Yu. V Synorov, A. M. Khoviv, Inorg. Mater. **40**, 349 (2004)
85. H. Murakami, W. Hattori, R. Aoki, Phys. C **269**, 83 (1996)
86. H. Murakami, W. Hattori, Y. Mizomata, R. Aoki, Phys. C **273**, 41 (1996)
87. H. Murakami, R. Aoki, K. Sakai, Thin Solid Films **27**, 343 (1999)
88. B. A. Volkov, L. I. Ryabova, D. R. Khokhlov, Phys. Usp. **45**, 819 (2002), and references therein
89. F. Hüe, M. Hÿtch, H. Bender, F. Houdellier, A. Claverie, Phys. Rev. Lett. **100**, 156602 (2008)
90. S. Banerjee, K. A. Shore, C. J. Mitchell, J. L. Sly, M. Missous, IEE Proc. Circuits Devices Syst. **152**, 497 (2005)
91. M. Razeghi, A. Evans, S. Slivken, J. S. Yu, J. G. Zheng, V. P. Dravid, Proc. SPIE **5840**, 54 (2005)
92. R. A. Stradling, Semicond. Sci. Technol. **6**, C52 (1991)
93. P. K. Weimer, Proc. IEEE **52**, 608 (1964)
94. G. Ribakovs, A. A. Gundjian, IEEE J. Quant. Electron. **QE-14**, 42 (1978)
95. S. K. Dey, J. Vac. Sci. Technol. **10**, 227 (1973)
96. S. J. Lynch, Thin Solid Films **102**, 47 (1983)
97. V. V. Kudzin, V. S. Kulakov, D. R. Pape', S. V. Kulakov, V. V. Molotok, IEEE. Ultrason. Symp. **1**, 749 (1997)
98. F. Hatami, V. Lordi, J. S. Harris, H. Kostial, W. T. Masselink, J. Appl. Phys. **97**, 096106 (2005)
99. B. W. Wessels, J. Electrochem. Soc. **722**, 402 (1975)
100. D. W. L. Tolfree, J. Sci. Instrum. **41**, 788 (1964)
101. P. B. Hart, Proc. IEEE **61**, 880 (1973)
102. M. A. Hines, G. D. Scholes, Adv. Mater. **15**, 1844 (2003)
103. C. A. Wang, R. K. Huang, D. A. Shiau, M. K. Connors, P. G. Murphy, P. W. O'Brien, A. C. Anderson, D. M. DePoy, G. Nichols, M. N. Palmisiano, Appl. Phys. Lett. **83**, 1286 (2003)
104. C. W. Hitchcock, R. J. Gutmann, J. M. Borrego, I. B. Bhat, G. W. Charache, IEEE Trans. Electron. Devices **46**, 2154 (1999)
105. H. J. Goldsmid, R. W. Douglas, Br. J. Appl. Phys. **5**, 386 (1954)
106. F. D. Rosi, B. Abeles, R. V. Jensen, J. Phys. Chem. Sol. **10**, 191 (1959)
107. T. M. Tritt (ed.), *Semiconductors and Semimetals*, vol. **69**, **70** and **71**: Recent Trends in Thermoelectric Materials Research I, II and III (Academic Press, New York, 2000)
108. D. M. Rowe (ed.), *CRC Handbook of Thermoelectrics* (CRC Press, Boca Raton, 1995)
109. D. M. Rowe, C. M. Bhandari, *Modern Thermoelectrics* (Reston Publishing Company,Virginia, 1983)
110. D. M. Rowe (ed.), *Thermoelectrics Handbook: Macro to Nano* (CRC Press, Boca Raton, 2006)
111. H. Choi, M. Chang, M. Jo, S. J. Jung, H. Hwang, Electrochem. Solid-State Lett. **11**, H154 (2008)

112. S. Cova, M. Ghioni, A. Lacaita, C. Samori, F. Zappa, Appl. Opt. **35**, 1956 (1996)
113. H. W. H. Lee, B. R. Taylor, S. M. Kauzlarich, *Nonlinear Optics: Materials, Fundamentals, and Applications* **12**, (Technical Digest, 2000)
114. E. Brundermann, U. Heugen, A. Bergner, R. Schiwon, G. W. Schwaab, S. Ebbinghaus, D. R. Chamberlin, E. E. Haller, M. Havenith, IN *29th International Conference on Infrared and MillimeterWaves and 12th International Conference on Terahertz, Electronics*, vol **283** (2004)
115. A. N. Baranov, T. I. Voronina, N. S. Zimogorova, L. M. Kauskaya, Y. P. Yakoviev, Sov. Phys. Semicond. **19**, 1676 (1985)
116. M. Yano, Y. Suzuki, T. Ishii, Y. Matsushima, M. Kimata, Jpn. J. Appl. Phys. **17**, 2091 (1978)
117. F. S. Yuang, Y. K. Su, N. Y. Li, Jpn. J. Appl. Phys. **30**, 207 (1991)
118. F. S. Yuang, Y. K. Su, N. Y. Li, K. J. Gan, J. Appl. Phys. **68**, 6383 (1990)
119. Y. K. Su, S. M. Chen, J. Appl. Phys. **73**, 8349 (1993)
120. S. K. Haywood, A. B. Henriques, N. J. Mason, R. J. Nicholas, P. J. Walker, Semicond. Sci. Technol. **3**, 315 (1988)
121. A. N. Chakravarti, K. P. Ghatak, K. K. Ghosh, S. Ghosh, H. M. Mukherjee, Phys. Stat. Sol. b **108**, 609 (1981); K. P. Ghatak, M. Mondal, Z. fur Physik B, **64**, 223 (1986)
122. P. K. Chakraborty, A. Sinha, S. Bhattacharya and K. P. Ghatak, Physica B: Condensed Matter **390**, 325 (2007); P. K. Chakraborty and K. P. Ghatak, J. Phys. Chem. Solids **62**, 1061 (2001); Phys. Letts. A, **288**, 335 (2001); Phys. D, Appl. Phys. **32**, 2438 (1999)
123. M. Abramowitz and I. A. Stegun, Handbook of Mathematical Functions with Formulas, Graphs and Mathematical Tables (Wiley, New York, 1964)
124. I. S. Gradshteyn and I. M. Ryzhik, Tables of Integrals, Series and Products (Academic Press, New York, 1965)
125. V. Heine, Proc. Phys. Soc. **81**, 300 (1963); J. N. Schulman and Y. C. Chang, Phys. Rev. B, **24**, 4445 (1981)
126. S. Adachi, J. Appl. Phys. **58**, R11 (1985)
127. K. P. Ghatak, J. P. Banerjee, D. Bhattacharya, Nanotechnology **7**, 110 (1996); S. Bhattacharya, S. Choudhury, K. P. Ghatak, Superlatt. and Microstruct. **48**, 257 (2010); K. P. Ghatak, S. Bhattacharya, S. Pahari, D. De, S. Ghosh, M. Mitra, Ann. Phys., **17**, 195 (2008); S. Pahari, S. Bhattacharya, K. P. Ghatak, J. Comput. Theor. Nanosci. **(Invited Paper)**, **6**, 2088 (2009)
128. M. Krieehbaum, P. Kocevar, H. Pascher, G. Bauer, IEEE QE **24**, 1727 1988); M. S. Lundstrom, J. Guo, *Nanoscale Transistors, Device Physics, Modeling and Simulation* (Springer, USA, 2006); R. Saito, G. Dresselhaus, M. S. Dresselhaus, *Physical Properties of Carbon Nanotubes* (Imperial College Press, London, **1998**); X. Yang, J. Ni, Phys. Rev. B **72**, 195426 (2005); W. Mintmire, C. T. White, Phys. Rev. Letts. **81**, 2506 (1998)
129. G. L. Bir, G. E. Pikus, *Symmetry and Strain–Induced effects in Semiconductors* (Nauka, Russia, 1972); M. Mondal, K. P. Ghatak, Phys. Stat. Sol.

(*b*) **135**, K21 (1986); C. C. Wu, C. J. Lin, J. Low. Temp. Phys. **57**, 469 (1984); Y. Yamada, J. Phys. Soc. Jpn. **35**, 1600 (1973)

130. A. N. Chakravarti, K. P. Ghatak, K. K. Ghosh, S. Ghosh, H. M. Mukherjee, Phys. Stat. Sol. b **118**, 843 (1983); S. Bhattacharya,, D. De, S. M. Adhikari, K. P. Ghatak, Superlatt. and Microstruc. **51**, 203 (2012); K. P. Ghatak, M. Mondal, Z. F. Naturforschung **41A**, 821 (1986)
131. A. N. Chakravarti, A. K Choudhury, K. P. Ghatak, S. Ghosh, A. Dhar, Appl. Phys. **25**, 105 (1981); P. K. Chakraborty, G. C. Datta , K. P. Ghatak, Phys. Scrip. **68**, 368 (2003)
132. A. N. Chakravarti, K. P. Ghatak, A. Dhar, K. K. Ghosh, S. Ghosh, Appl. Phys. **A26**, 165 (1981); K. P. Ghatak, S. Bhattacharya, S. K. Biswas, A. Dey and A. K. Dasgupta, Phys. Scrip. **75**, 820 (2007)
133. K. P. Ghatak, M. Mondal, Z. F. Physik B. **B69**, 471 (1988); A. N. Chakravarti, K. P. Ghatak, K. K. Ghosh, S. Ghosh, A. Dhar, Z. F, Physik B. **47**, 149 (1982)
134. H. A. Lyden, Phys. Rev. **135**, A514 (1964); E. D Palik, G. B. Wright, in *Semiconductors and Semimetals*, edited by R. K. Willardson and A. C. Beer **3**, (Academic Press, New York, USA,1967), p-421
135. M. Mondal, K. P. Ghatak, Phys. Letts. **131 A**, 529 (1988); K. P. Ghatak, B. Mitra, Int. J. Electron. **72**, 541 (1992); B. Mitra, A. Ghoshal, K. P. Ghatak, Nouvo Cimento D **12D**, 891 (1990), K. P. Ghatak, S. N. Biswas, Nonlinear Optics and Quantum Optics **4**, 347 (1993)
136. K. P. Ghatak, A. Ghoshal. B. Mitra., Nouvo Cimento **14D**, 903 (1992); K. P Ghatak, A. Ghoshal, B. Mitra, Nouvo Cimento. **13D**, 867 (1991); B. Mitra, K. P. Ghatak, Solid State Electron. **32**, 177 (1989); M. Mondal, N. Chattapadhyay, K. P. Ghatak, J. Low Temp. Phys. **66** 131 (1987)
137. P. N. Hai, W. M. Chen, I. A. Buyanova, H. P. Xin , C. W. Tu; Appl. Phys. Lett. **77**, 1843 (2000); D. P. DiVincenzo, E. J Mele; Phys. Rev. B, **29**, 1685 (1984); P. Perlin, E. Litwin-Staszewska, B. Suchanek, W. Knap, J. Camassel, T. Suski, R. Piotrzkowski, I. Grzegory, S. Porowski, E. Kaminska, J. C. Chervin; Appl. Phys. Lett. **68**, 1114 (1996);
138. V. K. Arora, H. Jeafarian, Phys. Rev. B. **13**, 4457 (1976); S. E. Ostapov, V. V. Zhikharevich, V. G. Deibuk; Semicond. Phys. Quan. Electron. and Optoelectron **9**, 29 (2006)
139. M. J. Aubin, L. G. Caron, J. P. Jay Gerin; Phys. Rev. B, **15**, 3872 (1977); S. L. Sewall, R. R. Cooney, P. Kambhampati; Appl. Phys. Lett, **94**, 243116 (2009)
140. K. Tanaka, N. Kotera; in 20th Internat. *Conf. on Indium Phosphide and Related Materials*, (Versailles, France, 2008), p. 1–4.
141. M. Singh, P. R. Wallace, S. D. Jog, J. Erushanov, J. Phys. Chem. Solids. **45**, 409 (1984)
142. W. Zawadzki, Adv. Phys. **23**, 435, (1974)
143. G. E. Smith; Phys. Rev. Lett. **9**, 487 (1962)
144. D. Schneider, D. Rurup, A. Plichta, H.-U. Grubert, A. Schlachetzki, K. Hansen; Z. Phys. B, **95**, 281 (1994);
145. K. P. Ghatak, B. Mitra, D. K. Basu, B. Nag, Nolinear Optics **17**, 171 (1997)

146. J. O. Dimmock, *in The Physics of Semimetals and Narrowgap Semiconductors*, ed. by D. L. Carter, R. T. Bates (Pergamon Press, Oxford, 1971)
147. E. Bangert, P. Kastner, Phys. Stat. Sol. (b) **61**, 503 (1974);
148. D. G. Seiler, B. D. Bajaj and A. E. Stephens, Phys. Rev. **B 16**, 2822 (1977); A. V. Germaneko and G. M. Minkov, Phys. Stat. Sol. (b) **184**, 9 (1994); G. L. Bir and G. E. Pikus, *Symmetry and Strain–Induced effects in Semiconductors* (Nauka, Russia, 1972)
149. J. Bouat, J. C. Thuillier, Surf. Sci. **73**, 528(1978)
150. G. J. Rees, *Physics of Compounds, in Proceedings of the 13th International Conference* ed. By F. G. Fumi (North Holland Company, 1976), p. 1166
151. P. R. Emtage, Phys. Rev. **138,** A246 (1965)
152. M. Stordeur, W. kuhnberger, Phys. Stat. Sol. **69**, 377 (1975); D. R. Lovett, *Semimetals and Narrow-Bandgap Semiconductor* (Pion Limited, UK, 1977); H. Kohler, Phys. Stat. Sol. **74**, 591 (1976)
153. M. Cardona, W. Paul, H. Brooks Helv, Acta Phys. **33**, 329 (1960); A. F. Gibson in *Proceeding of International School of Physics*, ENRICO FEPMI, course XIII, ed. By R. A. Smith (Academic Press, New York, 1963), p. 171
154. C. C. Wang, N. W. Ressler, Phys. Rev. **2**, 1827 (1970)
155. P. C. Mathur, S. Jain, Phys. Rev. **19**, 1359 (1979)
156. E. I. Rogacheva, S. N. Grigorov, O. N. Nashchekina, T. V. Tavrina, S. G. Lyubchenko, AYu. Sipatov, V. V. Volobuev, A. G. Fedorov, M. S. Dresselhaus, Thin Solid Films **493**, 41 (2005)
157. L. A. Vassilev, Phys. State Sol. (b), **121**, 203 (1984)
158. G. P. Chuiko, Sov. Phys. Semiconduct. **19**, 1381 (1985)
159. E. L. Ivchenko, G. E. Pikus Sov, Phys. Semicond. **13**, 579 (1979);
160. G. M. T. Foley, P. N. Langenberg, Phys. Rev. B **15B**, 4850 (1977)
161. D. G. Seiler, W. M. Beeker, L. M. Roth, Phys. Rev. **1**, 764 (1970)
162. H. I. Zhang, Phys. Rev. B **1**, 3450 (1970)
163. U. Rossler, Solid State Commun. **49**, 943 (1984)
164. J. Johnson, D. H. Dickey, Phys. Rev. **1**, 2676 (1970)
165. V. G. Agafonov, P. M. Valov, B. S. Ryvkin, I. D. Yarashetskin, Sov. Phys. Semiconduct. **12**, 1182 (1978)
166. N. S. Averkiev, V. M. Asnin, A. A. Bakun, A. M. Danishevskii, E. L. Ivchenko, G. E. Pikus, A. A. Rogachev, Sov. Phys. Semicond. **18**, 379 (1984)
167. R. W. Cunningham, Phys. Rev. **167**, 761 (1968)
168. A. I. Yekimov, A. A. Onushchenko, A. G. Plyukhin, AI. L. Efros, J. Expt. Theor. Phys. **88**, 1490 (1985)
169. B. J. Roman, A. W. Ewald, Phys. Rev. **B5**, 3914 (1972)

Chapter 2

The EM in Doping Superlattices of HD Non-Parabolic Semiconductors

2.1 Introduction

The technological importance of super-lattices in general, and specifically doping super-lattices [1–20] has already been stated in the preface and also in the references [1–20] of this chapter. In Section 2.2.1, under theoretical background, the EM in doping superlattices of HD non-linear optical semiconductors has been investigated. Section 2.2.2 contains the results for doping superlattices of HD III-V, ternary and quaternary semiconductors in accordance with the three and the two band models of Kane together with parabolic energy bands and they form the special cases of Section 2.2.1. Sections 2.2.3, 2.2.4 and 2.2.5 contain the study of the EM for doping superlattices of HD II-VI, IV-VI and stressed Kane type semiconductors respectively. Sections 2.3 and 2.4 contain the results and discissions and 11 open research problems for this chapter.

2.2 Theoretical Background

2.2.1 *The EM in doping superlattices of HD nonlinear optical semiconductors*

The DR of the conduction electrons in doping superlattices of HD nonlinear optical materials can be expressed by using (1.26) and following the method as given in [19, 20] as

$$\frac{(n_i+\frac{1}{2})}{\hbar T_{21}(E,\eta_g)}\omega_{8HD}(E,\eta_g)+\frac{\hbar^2k_s^2}{2m_\perp^* T_{22}(E,\eta_g)}=1 \tag{2.1}$$

where

$$\omega_{8HD}(E,\eta_g)\equiv \text{Real part of }\left(\frac{n_{2D}|e|^2}{d_0\varepsilon_{sc}[m_\parallel^* T'_{21}(E,\eta_g)]}\right)^{1/2},$$

$n_i(=0,1,2\ldots)$ is the mini-band index for nipi structures and d_0 is the superlattice period.

The EM in this case assumes the form

$$m^*(E_{FnHD},\mathrm{n}_i\eta_g)=\text{ Real part of }\left(\frac{\hbar^2}{2}\right)\mathrm{G}'_{21HD}(E_{FnHD},\mathrm{n}_i\eta_g) \tag{2.2}$$

where,

$$G_{21HD}(E,\eta_g,\mathrm{n}_i)=\frac{2m_\perp^* T_{22}(E,\eta_g)}{\hbar^2}\left[1-\frac{(n_i+\frac{1}{2})}{\hbar T_{21}(E,\eta_g)}\omega_{8HD}(E,\eta_g)\right]$$

and $\bar{E}_{FnHD}$ is the Fermi energy in the present case as measured from the edge of the conduction band in vertically upward direction in the absence of any quantization.

From (2.2), we observe that the EM is a function of the Fermi energy, nipi subband index, scattering potential and the other material constants which is the characteristic feature of doping superlattices of HD non-linear optical materials.

The subband energy $(E_{1_{n_i}HD})$ can be written as

$$\frac{(n_i+\frac{1}{2})}{\hbar T_{21}(E_{1_{n_i}HD},\eta_g)}\omega_{8HD}(E_{1_{n_i}HD},\eta_g)=1 \tag{2.3}$$

The DOS function for doping superlattices of HD non-linear optical materials can be expressed as

$$N_{nipiHD}(E,\eta_g)=\frac{g_v}{2\pi}\sum_{n_i=0}^{n_i\max}G'_{21HD}(E,\eta_g,n_i)H(E-E_{1n_iHD}) \tag{2.4a}$$

The 2D electron statistics can be written as

$$n_{2D} = \frac{g_V}{2\pi} \text{ Real Part of } \left[\sum_{n_i=0}^{n_{i\max}} \left[G_{21HD}(E_{FnHD}, n_i, \eta_g) + \sum_{r=1}^{s} L(r)[G_{21HD}(E_{FnHD}, n_i \eta_g)]\right]\right] \tag{2.4b}$$

The dispersion relation of the conduction electrons in nipi structures of nonlinear optical materials can be expressed by using (1.2) and following the method as given in [19, 20] as

$$\psi_1(E) = \psi_2(E)k_s^2 + \psi_3(E)\left(n_i + \frac{1}{2}\right)\frac{2m_\parallel^*}{\hbar}\omega_8(E) \tag{2.4c}$$

where

$$\omega_8(E) \equiv \left(\frac{n_{2D}|e|^2}{\varepsilon_{sc}[\theta_1(E)]}\right)^{1/2}$$

and

$$\theta_1(E) \equiv \frac{\hbar^2}{2}\left\{\frac{\psi_3(E)[\psi_1(E)]' - \psi_1(E)[\psi_3(E)]'}{[\psi_3(E)]^2}\right\}$$

The EM in this case assumes the form

$$m^*(E_{Fn}, n_i) = \left(\frac{\hbar^2}{2}\right) R_{81}(E, n_i)\Big|_{E=\bar{E}_{Fn}} \tag{2.5}$$

where,

$$R_{81}(E, n_i) \equiv [\psi_2(E)]^{-2}\left[\psi_2(E)\left\{[\psi_1(E)]' - \left(\frac{2m_\parallel^*}{\hbar}\right)[\psi_3(E)]'\left(n_i + \frac{1}{2}\right)[\omega_8(E)] - \left(\frac{2m_\parallel^*}{\hbar}\right)[\psi_3(E)]\left(n_i + \frac{1}{2}\right)[\omega_8(E)]'\right\}\right.$$

$$- \left\{ [\psi_1(E)] - \left(\frac{2m_{\|}^*}{\hbar}\right) [\psi_3(E)] \left(n_i + \frac{1}{2}\right) [\omega_8(E)] \right\}$$

$$\times [\psi_2(E)]' \Bigg]$$

and $\bar{E}_{Fn}$ is the Fermi energy in the present case as measured from the edge of the conduction band in vertically upward direction in the absence of any quantization.

The sub-band energy (E_{1ni}) can be written as

$$\psi_1(E_{1ni}) = \psi_3(E_{1ni}) \left(n_i + \frac{1}{2}\right) \frac{2m_{\|}^*}{\hbar} \omega_8(E_{1ni}) \tag{2.6}$$

The density-of-states function for nipi structures of non-linear optical materials can be expressed as

$$N_{nipi}(E) = \frac{g_v}{2\pi} \sum_{n_i=0}^{n_i \max} R_{81}(E, n_i) H(E - E_{1ni}) \tag{2.7}$$

The electron concentration, can be written as

$$n_0 = \frac{g_v}{2\pi} \sum_{n_i=0}^{n_i \max} \left[T_{81}(\bar{E}_{Fn}, n_i) + T_{82}(\bar{E}_{Fn}, n_i)\right] \tag{2.8}$$

where,

$$T_{81}(\bar{E}_{Fn}, n_i) \equiv \left[\psi_1(\bar{E}_{Fn}) - \psi_3(\bar{E}_{Fn}) \left(n_i + \frac{1}{2}\right) \frac{2m_{\|}^*}{\hbar} \omega_8(\bar{E}_{Fn})\right] [\psi_2(\bar{E}_{Fn})]^{-1}$$

and

$$T_{82}(\bar{E}_{Fn}, n_i) \equiv \sum_{r=1}^{s} L(r) T_{81}(\bar{E}_{Fn}, n_i).$$

2.2.2 *The EM in doping superlattices of HD III-V, ternary and quaternary semiconductors*

(a) $\Delta_{\|} = \Delta_{\perp} = \Delta, \delta = 0$ and $m_{\|}^* = m_{\perp}^* = m_c$

The electron energy spectrum in doping superlattices of HD III-V, ternary and quaternary materials can be expressed from (2.1) under the conditions $\Delta_{\parallel} = \Delta_{\perp} = \Delta, \delta = 0$, and $m_{\parallel}^* = m_{\perp}^* = m_c$, as

$$\frac{\hbar^2 k_s^2}{2m_c} = \left[T_{31}(E,\eta_g) + \mathrm{i}T_{32}(E,\eta_g) - \left(n_i + \frac{1}{2}\right)\hbar\omega_{9HD}(E,\eta_g)\right] \quad (2.9)$$

where

$$\omega_{9HD}(E,\eta_g) \equiv \left(\frac{n_{2D}|e|^2}{d_0\varepsilon_{sc}T'_{31}(E,\eta_g)m_c}\right)^{1/2}$$

The EM in this case assumes the form

$$m^*(E_{FnHD}, n_i, \eta_g) = \text{ Real Part of } \left(\frac{\hbar^2}{2}\right) G'_{23HD}(E_{FnHD}, \eta_g, n_i) \quad (2.10)$$

where

$$G'_{23HD}(E_{FnHD}, \eta_g, n_i) = \frac{2m_c}{\hbar^2}\left[T_{31}(E_{FnHD}, \eta_g) + \mathrm{i}T_{32}(E_{FnHD}, \eta_g) - \left(n_i + \frac{1}{2}\right)\hbar\omega_{9HD}(E_{FnHD}, \eta_g)\right]$$

The sub-band energy E_{2n_iHD} can be written as

$$\left[T_{31}(E_{2n_iHD}, \eta_g) + \mathrm{i}T_{32}(E_{2n_iHD}, \eta_g) - \left(n_i + \frac{1}{2}\right)\hbar\omega_{9HD}(E_{2n_iHD}, \eta_g)\right] = 0 \quad (2.11)$$

The DOS function for doping superlattics of HD III-V, ternary and quaternary materials can be expressed as

$$N_{nipiHD}(E, \eta_g) = \frac{g_v m_c}{\pi\hbar^2}\sum_{n_i=0}^{n_i\max} G'_{23HD}(E, \eta_g, n_i)H(E - E_{2n_iHD}) \quad (2.12a)$$

The electron concentration can be written as

$$n_{2D} = \frac{g_V}{2\pi} \text{ Real Part of } \left[\sum_{n_i=0}^{n_{i\max}}\left[G_{23HD}(E_{FnHD}, n_i, \eta_g) + \sum_{r=1}^{s} L(r)[G_{23HD}(E_{FnHD}, n_i, \eta_g)]\right]\right] \quad (2.12b)$$

In the absence of band tails, the electron energy spectrum in nipi structures of III-V, ternary and quaternary materials can be expressed under the conditions $\Delta_{\|} = \Delta_{\perp} = \Delta$, $\delta = 0$ and $m_{\|}^* = m_{\perp}^* = m_c$, as

$$I_{11}(E) = \left(n_i + \frac{1}{2}\right)\hbar\omega_9(E) + \frac{\hbar^2 k_s^2}{2m_c} \tag{2.13}$$

where

$$\omega_9(E) \equiv \left(\frac{n_{2D}|e|^2}{\varepsilon_{sc} I'(E) m_c}\right)^{1/2}.$$

The EM in this case can be written as

$$m^*(E_{Fn}, n_i) = m_c R_{82}(E, n_i)|_{E=E_{Fn}} \tag{2.14}$$

in which,

$$R_{82}(E, n_i) \equiv \left\{[I_{11}(E)]' - \left(n_i + \frac{1}{2}\right)\hbar[\omega_9(E)]'\right\}.$$

From (2.14) we observe that the EM in this case is a function of the Fermi energy, nipi subband index and the other material constants which is the characteristic feature of nipi structures of III-V, ternary and quaternary compounds whose bulk dispersion relations is defined by the three band model of Kane.

The sub-band energies (E_{2ni}) can be written as

$$I_{11}(E_{2ni}) = \left(n_i + \frac{1}{2}\right)\hbar\omega_9(E_{2ni}) \tag{2.15}$$

The density-of-states function in this case can be expressed as

$$N_{nipi}(E) = \frac{m_c g_v}{\pi\hbar^2}\sum_{n_i=0}^{n_i \max} R_{82}(E, n_i) H(E - E_{2ni}) \tag{2.16a}$$

The use of (2.16a) leads to the expression of the electron concentration as

$$n_0 = \frac{m_c g_v}{\pi\hbar^2}\sum_{n_i=0}^{n_i \max}[T_{83}(\bar{E}_{Fn}, n_i) + T_{84}(\bar{E}_{Fn}, n_i)] \tag{2.16b}$$

where

$$T_{83}(\bar{E}_{Fn}, n_i) \equiv \left[I_{11}(\bar{E}_{Fn}) - \left(n_i + \frac{1}{2}\right)\hbar\omega_9(\bar{E}_{Fn})\right]$$

and

$$T_{84}(\bar{E}_{Fn}, n_i) \equiv \sum_{r=1}^{s} L(r) T_{83}(\bar{E}_{Fn}, n_i).$$

(b) The Two Band Model of Kane

The electron energy spectrum in doping superlattices of HD III-V, ternary and quaternary materials whose energy band structures in the absence of band tails are described by the two band model of Kane can be expressed under the conditions $\Delta >> E_g$ or $\Delta << E_g$, as

$$\frac{\hbar^2 k_s^2}{2m_c} = \left[\gamma_2(E, \eta_g) - \left(n_i + \frac{1}{2}\right) \hbar\omega_{10HD}(E, \eta_g)\right] \tag{2.17}$$

where

$$\omega_{10HD}(E) \equiv \left(\frac{n_{2D}|e|^2}{d_0 \varepsilon_{sc} \gamma_2'(E, \eta_g) m_c}\right)^{\frac{1}{2}}$$

The EM in this case assumes the form

$$m^*(\mathrm{E}_{FnHD}, n_i, \eta_g) = \left(\frac{\hbar^2}{2}\right) G'_{25HD}(E_{FnHD}, \eta_g, n_i) \tag{2.18}$$

where

$$G_{25HD}(E_{FnHD}, \eta_g, n_i) = \frac{2m_c}{\hbar^2} \left[\gamma_2(\mathrm{E}_{FnHD}, \eta_g) - \left(n_i + \frac{1}{2}\right) \hbar\omega_{10HD}(E_{FnHD}, \eta_g)\right]$$

The subband energy E_{3n_iHD} can be written as

$$\left[\gamma_2(E_{3n_iHD}, \eta_g) - \left(n_i + \frac{1}{2}\right) \hbar\omega_{9HD}(E_{3n_iHD}, \eta_g)\right] = 0 \tag{2.19}$$

The DOS function in this case is given by

$$N_{nipiHD}(E, \eta_g) = \frac{g_v m_c}{\pi\hbar^2} \sum_{n_i=0}^{n_i\,\max} [G_{25HD}(E, \eta_g, n_i) H(E - E_{3n_iHD}) \tag{2.20a}$$

The electron concentration can be written as

$$n_{2D} = \frac{g_V}{2\pi}\left[\sum_{n_i=0}^{n_{i\max}}\left[G_{25HD}(E_{FnHD}, n_i, \eta_g) + \sum_{r=1}^{s} L(r)[G_{25HD}(E_{FnHD}, n_i, \eta_g)]\right]\right] \quad (2.20b)$$

In the absence of band tails, the EM in this case assumes the form

$$E(1+\alpha E) = \left(n_i + \frac{1}{2}\right)\hbar\omega_{20}(E) + \frac{\hbar^2 k_s^2}{2m_c} \quad (2.21)$$

where

$$\omega_{20}(E) \equiv \left(\frac{n_{2D}|e|^2}{d_0\varepsilon_{sc}(1+2\alpha E)m_c}\right)^{\frac{1}{2}}$$

The EM in this case can be written as

$$m^*(E_{Fn}, n_i) = m_c R_{182}(E, n_i)|_{E=E_{fn}} \quad (2.22)$$

in which,

$$R_{182}(E, n_i) \equiv \left\{[1+2\alpha E] - \left(n_i + \frac{1}{2}\right)\hbar[\omega_{19}(E)]'\right\}.$$

From (2.22), we observe that the EM in this case is a function of the Fermi energy, nipi subband index and the other material constants which is the characteristic feature of doping superlattices of III-V, ternary and quaternary compounds whose bulk EMs is defined by the three band model of Kane.

The sub-band energies (E_{3ni}) can be written as

$$E_{3ni}(1+\alpha E_{3ni}) = \left(n_i + \frac{1}{2}\right)\hbar\omega_{20}(E_{3ni}) \quad (2.23)$$

The DOS function in this case can be expressed as

$$N_{nipi}(E) = \frac{m_c g_v}{\pi\hbar^2}\sum_{n_i=0}^{n_i\max} R_{182}(E, n_i)H(E - E_{3ni}) \quad (2.24a)$$

The use of (2.24a) leads to the expression of the electron concentration as

$$n_0 = \frac{m_c g_v}{\pi \hbar^2} \sum_{n_i=0}^{n_i \max} [T_{831}(\bar{E}_{Fn}, n_i) + T_{841}(\bar{E}_{Fn}, n_i)] \tag{2.24b}$$

where

$$T_{831}(\bar{E}_{Fn}, n_i) \equiv \left[(1 + \alpha \bar{E}_{Fn})(\bar{E}_{Fn}) - \left(n_i + \frac{1}{2}\right) \hbar \omega_{91}(\bar{E}_{Fn})\right],$$

$$\omega_{91}(\bar{E}_{Fn}) = \left[\frac{n_{2D} e^2}{\varepsilon_{sc} m_c (1 + 2\alpha \bar{E}_{Fn})}\right]$$

and

$$T_{841}(\bar{E}_{Fn}, n_i) \equiv \sum_{r=1}^{s} L(r) T_{831}(\bar{E}_{Fn}, n_i).$$

(c) Parabolic Energy Bands

The electron energy spectrum in nipi structures of HD III-V, ternary and quaternary materials whose energy band structures in the absence of band tails are described by the parabolic energy bands can be expressed as

$$\frac{\hbar^2 k_s^2}{2m_c} = \left[\gamma_3(E, \eta_g) - \left(n_i + \frac{1}{2}\right) \hbar \omega_{11HD}(E, \eta_g)\right] \tag{2.25}$$

where

$$\omega_{11HD}(E) \equiv \left(\frac{n_{2D}|e|^2}{d_0 \varepsilon_{sc} \gamma_3'(E, \eta_g) m_c}\right)^{\frac{1}{2}}$$

The EM in this case assumes the form

$$m^*(E_{FnHD}, n_i, \eta_g) = \left(\frac{\hbar^2}{2}\right) G'_{27HD}(E_{FnHD}, \eta_g, n_i) \tag{2.26}$$

where

$$G_{27HD}(E_{FnHD}, \eta_g, n_i)$$
$$= \frac{2m_c}{\hbar^2} \left[\gamma_3(E_{FnHD}, \eta_g) - \left(n_i + \frac{1}{2}\right) \hbar \omega_{11HD}(E_{FnHD}, \eta_g)\right]$$

The sub-band energy E_{4n_iHD} can be expressed as

$$\left[\gamma_3(E_{4n_iHD},\eta_g)-\left(n_i+\frac{1}{2}\right)\hbar\omega_{11HD}(E_{4n_iHD},\eta_g)\right]=0 \qquad (2.27)$$

The DOS function in this case is given by

$$N_{nipiHD}(E,\eta_g)=\frac{g_v m_c}{\pi\hbar^2}\sum_{n_i=0}^{n_i\max}G'_{27HD}(E,\eta_g,n_i)H(E-E_{4n_iHD}) \qquad (2.28a)$$

The electron concentration can be written as

$$n_{2D}=\frac{g_V}{2\pi}\left[\sum_{n_i=0}^{n_i\max}\left[G_{27HD}(E_{FnHD},n_i,\eta_g)+\sum_{r=1}^{s}L(r)[G_{27HD}(E_{FnHD},n_i,\eta_g)]\right]\right] \qquad (2.28b)$$

In the absence of band tails, the EM in this case assumes the form

$$E=\left(n_i+\frac{1}{2}\right)\hbar\omega_{21}+\frac{\hbar^2k_s^2}{2m_c} \qquad (2.29)$$

where

$$\omega_{21}\equiv\left(\frac{n_{2D}|e|^2}{d_0\varepsilon_{sc}m_c}\right)^{\frac{1}{2}}$$

The EM in this case can be written as

$$m^*(E_{Fn},n_i)=m_c \qquad (2.30)$$

Thus the EM in this case is a constant quantity.

The sub-band energies (E_{4ni}) can be written as

$$E_{4ni}=\left(n_i+\frac{1}{2}\right)\hbar\omega_{21} \qquad (2.31)$$

The DOS function in this case can be expressed as

$$N_{nipi}(E)=\frac{m_cg_v}{\pi\hbar^2}\sum_{n_i=0}^{n_i\max}H(E-E_{4ni}) \qquad (2.32a)$$

The electron concentration can be written as

$$n_{2D} = \frac{m_c g_v}{\pi \hbar^2} \sum_{n_i=0}^{n_{i\max}} F_0 \left(\frac{E_{Fn} - E_{4ni}}{k_B T} \right) \tag{2.32b}$$

2.2.3 *The EM in doping superlattices of HD II-VI semiconductors*

The 2D electron dispersion law in doping superlattices of HD II-VI semiconductors can be expressed as

$$\gamma_3(\mathrm{E}, \eta_g) = a_0' k_s^2 + \left(n_i + \frac{1}{2}\right) \hbar\omega_{30}(\mathrm{E}, \eta_g) \pm \bar{\lambda}_0 k_s,$$

$$\omega_{20}(\mathrm{E}, \eta_g) \equiv \left(\frac{n_{2D}|e|^2}{d_0 \gamma_3(E, \eta_g) \varepsilon_{sc} m_{\|}^*} \right)^{\frac{1}{2}} \tag{2.33}$$

The EM in this case assumes the form as

$$m^*(\mathrm{E}_{FnHD}, n_i, \eta_g) = m_\perp^* \left\{ 1 - \bar{\lambda}_0 \left[(\bar{\lambda}_0)^2 + 4a_0' \gamma_3(\mathrm{E}_{FnHD}, \eta_g) \right.\right.$$
$$\left.\left. - 4d_0 \left(n_i + \frac{1}{2}\right) \hbar\omega_{30}(\mathrm{E}_{FnHD}, \eta_g) \right]^{-1/2} \right\} \gamma_3'(\mathrm{E}_{FnHD}, \eta_g) \tag{2.34}$$

The sub-band energy can be written as

$$\gamma_3(E_{6n_iHD}, \eta_g) = \left(n_i + \frac{1}{2}\right) \hbar\omega_{30}(E_{6n_iHD}, \eta_g) \tag{2.35}$$

The DOS function in this case is given by

$$N_{nipiHD}(E) = \frac{g_\nu}{4\pi (a_0')^2} \sum_{n_i=0}^{n_{i\max}} [G_{30HD}(E, \eta_g, n_i)]' H(E - E_{6n_iHD}) \tag{2.36a}$$

where

$$G_{30HD}(E, \eta_g, n_i) = \left[(\bar{\lambda}_0) - 2d_0 \left\{ \left(n_i + \frac{1}{2} \right) \hbar\omega_{30}(E, \eta_g, n_i) - \gamma_3(E, \eta_g, n_i) \right\} \right]$$

The electron concentration can be written as

$$n_{2D} = \frac{g_\nu}{4\pi(a_0')^2} \sum_{n_i=0}^{n_{i\max}} \int_{E_{6n_iHD}}^{\infty} [G_{30HD}(E, \eta_g, n_i)]' f(E) dE \quad (2.36\text{b})$$

In the absence of band-tails, the carrier dispersion law in doping superlattices of II-VI compounds can be expressed as

$$E = a_0' k_s^2 + \left(n_i + \frac{1}{2} \right) \hbar\bar{\omega}_{10} \pm \bar{\lambda}_0 k_s, \quad \bar{\omega}_{10} = \left(\frac{n_{2D}|e|^2}{d_0 \varepsilon_{sc} m_\parallel^*} \right)^{\frac{1}{2}} \quad (2.37)$$

Using (2.37), the EM in this case can be written as

$$m^*(E_{Fn}, n_i) = m_\perp^* \left\{ 1 - \bar{\lambda}_0 \left[(\bar{\lambda}_0)^2 + 4a_0' E_{Fn} - 4a_0' \left(n_i + \frac{1}{2} \right) \hbar\bar{\omega}_{10} \right]^{-1/2} \right\} \quad (2.38)$$

Thus, the EM in this case is a function of the Fermi energy, the nipi subband index number and the energy spectrum constants due to the only presence of $\bar{\lambda}_0$.

The sub-band energies (E_{8ni}) assume the form as

$$E_{8ni} = \left(n_i + \frac{1}{2} \right) \hbar\bar{\omega}_{10} \quad (2.39)$$

The DOS function in this case can be expressed as

$$N_{nipi}(E) = \frac{m_\perp^* g_v}{\pi\hbar^2} \sum_{n_i=0}^{n_i \max} \left[1 - \frac{a_{81}}{\sqrt{E + b_{81}(n_i)}} \right] H(E - E_{8ni}) \quad (2.40\text{a})$$

in which,

$$a_{81} \equiv \frac{\bar{\lambda}_0}{2\sqrt{a_0'}}$$

and

$$b_{81}(n_i) \equiv \left[\frac{1}{4a_0'}\left[(\bar{\lambda}_0)^2 - 4a_0'\left(n_i + \frac{1}{2}\right)\hbar\bar{\omega}_{10}\right]\right].$$

The use of the (2.40a) leads to the electron concentration under the condition of extreme degeneracy as

$$n_{2D} = \frac{g_v m_\perp^*}{\pi\hbar^2}\sum_{ni=0}^{ni_{\max}}\left(E_{Fn} - E_{3ni} + (\bar{\lambda_0})^2 m_\perp^* \hbar^{-2}\right) \tag{2.40b}$$

2.2.4 *The EM in doping superlattices of HD IV-VI semiconductors*

The 2D electron dispersion law in this case is given by

$$k_s^2 = \delta_{15}(E, \eta_g, n_i) \tag{2.41}$$

where

$$\delta_{15}(E, \eta_g, n_i) = [2\delta_{12}(E, \eta_g)]^{-1}[-\delta_{13}(E, \eta_g, n_i) + \sqrt{\delta_{13}^2(E, \eta_g, n_i) - 4\delta_{12}(E, \eta_g)\delta_{14}(E, \eta_g, n_i)}],$$

$$\delta_{12}(E, \eta_g) = \frac{\alpha\hbar^4 Z_0(E, \eta_g)}{4m_t^+ m_l^-},$$

$$\delta_{13}(E, \eta_g, n_i) = \hbar^2[\lambda_{71}(E, \eta_g)\delta_{11}(E, \eta_g, n_i) + \lambda_{12}(E, \eta_g)],$$

$$\delta_{14}(E, \eta_g, n_i) = [\lambda_{73}(E, \eta_g)\delta_{11}^2(E, \eta_g, n_i) + \lambda_{74}(E, \eta_g)\delta_{11}^4(E, \eta_g, n_i) - \lambda_{74}(E, \eta_g)],$$

$$\delta_{11}(E, \eta_g, n_i) \equiv \frac{2}{\hbar}m_{HD}^*(0, \eta_g)\left(n_i + \frac{1}{2}\right)\left[\frac{e^2 n_{2D}}{d_0\varepsilon_{sc}m_{HD}^*(E, \eta_g)}\right]^{\frac{1}{2}}$$

and

$$m^*_{HD}(E,\eta_g) = \frac{\hbar^2}{4\lambda_{76}^2(E,\eta_g)}\left[2\lambda_{74}(E,\eta_g)\left\{-\lambda'_{73}(E,\eta_g) + \frac{\lambda_{73}(E,\eta_g)\lambda'_{73}(E,\eta_g) + 2\lambda'_{74}(E,\eta_g)\lambda_{75}(E,\eta_g) + 2\lambda_{74}(E,\eta_g)\lambda'_{75}(E,\eta_g)}{\sqrt{\lambda_{73}^2(E,\eta_g) + 4\lambda_{74}(E,\eta_g)\lambda_{75}(E,\eta_g)}}\right\} - 2\lambda'_{74}(E,\eta_g)\{-\lambda_{73}(E,\eta_g) + \sqrt{\lambda_{73}^2(E,\eta_g) + 4\lambda_{74}(E,\eta_g)\lambda_{75}(E,\eta_g)}\}\right]$$

The EM in this case assumes the form

$$m^*(\mathrm{E}_{FnHD}, n_i, \eta_g) = \left(\frac{\hbar^2}{2}\right)\delta'_{15}(E_{FnHD}, \eta_g, n_i) \tag{2.42}$$

The sub-band energy E_{9n_iHD} can be expressed as in this case as

$$0 = \delta_{15}(E_{9n_iHD}, \eta_g, n_i) \tag{2.43}$$

The DOS function in this case is given by

$$N_{nipiHD}(E) = \frac{g_\nu}{2\pi}\sum_{n_i=0}^{n_{i\max}}[\delta_{15}(E,\eta_g,n_i)]'H(E - E_{9n_iHD}) \tag{2.44a}$$

The electron concentration can be written as

$$n_{2D} = \frac{g_\nu}{2\pi}\sum_{n_i=0}^{n_{i\max}}\int_{E_{9n_iHD}}^{\infty}[\delta_{15}(E,\eta_g,n_i)]'f(E)dE \tag{2.44b}$$

The carrier energy spectrum in doping superlattices of IV-VI compounds in the absence of band tails can be written as

$$k_s^2 = (\hbar^2 S_{19})^{-1}\left[-S_{20}(E,n_i) + \sqrt{S_{20}^2(E,n_i) + 4S_{19}S_{21}(E,n_i)}\right] \tag{2.45}$$

In which,

$$S_{19} \equiv \left(\frac{\alpha}{m_t^+ m_t^-}\right),$$

$$S_{20}(E,n_i) \equiv \left\{\frac{1}{m_t^*} - \left(\frac{\alpha E}{m_t^+}\right) + \frac{1+\alpha E}{m_t^-} + \frac{\alpha\hbar^2}{2m_l^+ m_t^-}\left(n_i + \frac{1}{2}\right)T(E)\right.$$
$$\left. + \frac{\alpha\hbar^2}{2m_l^- m_t^+}\left(n_i + \frac{1}{2}\right)T(E)\right\}$$

$$T(E) \equiv \frac{2m^*(0)}{\hbar}\omega_{11}(E), \quad m^*(0) \equiv \left(\frac{m_l^* m_l^-}{m_l^* + m_l^-}\right),$$

$$\omega_{11}(E) \equiv \left(\frac{n_{2D}|e|^2}{d_0 \varepsilon_{sc} m^*(E)}\right)^{\frac{1}{2}},$$

$$m^*(E) \equiv \frac{1}{4t_1}\left[-(t_2(E))' + \frac{t_2(E)(t_2(E))' + 2t_1(1+2\alpha E)}{\sqrt{t_2^2(E) + 4Et_1(1+\alpha E)}}\right],$$

$$t_1 \equiv \left(\frac{\alpha}{4m_l^+ m_l^-}\right),$$

$$t_2(E) \equiv \frac{1}{2}\left[\left(\frac{1}{m_l^*}\right) - \left(\frac{\alpha E}{m_l^+}\right) + \left(\frac{1+\alpha E}{m_l^-}\right)\right],$$

$$(t_2(E))' \equiv \frac{\alpha}{2}\left(\frac{1}{m_l^-} - \left(\frac{1}{m_l^+}\right)\right)$$

and

$$S_{21}(E,n_i) \equiv \left[E(1+\alpha E) + \frac{\alpha E\hbar^2}{2m_l^+}\left(n_i+\frac{1}{2}\right)T(E)\right.$$
$$+\frac{\hbar^2}{2m_l^-}\left(n_i+\frac{1}{2}\right)T(E)(1+\alpha E)$$
$$\left.+\frac{\hbar^4}{4m_l^- m_l^+}\left(n_i+\frac{1}{2}\right)T(E) - \left(\frac{\hbar^2}{2m_l^*}\right)T(E)\left(n_i+\frac{1}{2}\right)\right].$$

Using (2.45) the EM in this case can be written as

$$m^*(E_{Fn},n_i) = R_{84}(E,n_i)|_{E=E_{Fn}} \tag{2.46}$$

where,

$$R_{84}(E,n_i) \equiv (2S_{19})^{-1}\left[-(S_{20}(E,n_i))'\right.$$
$$\left.+\frac{S_{20}(E,n_i)[S_{20}(E,n_i)]' + 2S_{19}[S_{21}(E,n_i)]'}{[\{[S_{20}(E,n_i)]'\}^2 + 4S_{19}S_{21}(E,n_i)]^{1/2}}\right].$$

Thus, one can observe that the EM in this case is a function of both the Fermi energy and the nipi sub-band index number together with the spectrum constants of the system due to the presence of band non-parabolicity.

The sub-band energies (E_{10ni}) can be written as

$$\left[E_{10ni} - \frac{\hbar^2}{2m_l^-}T(E_{10ni})\left(n_i+\frac{1}{2}\right)\right]$$
$$\times\left[1+\alpha E_{10ni} + +\alpha\frac{\hbar^2}{2m_l^+}T(E_{10ni})\left(n_i+\frac{1}{2}\right)\right]$$
$$= \left[\frac{\hbar^2}{2m_l^*}T(E_{10ni})\left(n_i+\frac{1}{2}\right)\right] \tag{2.47}$$

The DOS function in this case assumes the form as

$$N_{nipi}(E) = \frac{g_v}{\pi\hbar^2}\sum_{n_i=0}^{n_i\,\max} R_{84}(E,n_i)H(E-E_{10ni}) \tag{2.48a}$$

The electron concentration can be written as

$$n_{2D} = \frac{g_v}{\pi\hbar^2} \sum_{n_i=0}^{n_{i\max}} \int_{E_{10ni}}^{\infty} R_{84}(E, n_i) f(E) dE \tag{2.48b}$$

2.2.5 *The EM in doping superlattices of HD stressed Kane type semiconductors*

The 2D DR in this case is given by

$$P_{11}(E, \eta_g)\mathrm{k}_x^2 + Q_{11}(E, \eta_g)\mathrm{k}_y^2 + S_{11}(E, \eta_g)\delta_{19}(E, \eta_g, n_i) = 1 \tag{2.49}$$

where

$$\delta_{19}(E, \eta_g, n_i) \equiv \frac{2}{\hbar} m_{zz}^*(0, \eta_g)\left(n_i + \frac{1}{2}\right)\left[\frac{n_{2D}e^2}{d_0 \varepsilon_{sc} m_{zz}(E, \eta_g)}\right]^{\frac{1}{2}}$$

The EM in this case assumes the form

$$m^*(\mathrm{E}_{FnHD}, n_i, \eta_g) = \left(\frac{\hbar^2}{2}\right)\delta_{20}'(E_{FnHD}, \eta_g, n_i) \tag{2.50}$$

where

$$\delta_{20}(E_{FnHD}, \eta_g, n_i) = \frac{[1 - S_{11}(E_{FnHD}, \eta_g)\delta_{19}(E_{FnHD}, \eta_g, n_i)]}{\sqrt{P_{11}(E_{FnHD}, \eta_g) Q_{11}(E_{FnHD}, \eta_g)}}.$$

The sub-band energy E_{15n_iHD} can be expressed as in this case as

$$S_{11}(E_{15n_iHD}, \eta_g)\delta_{19}(E_{15n_iHD}, \eta_g, n_i) = 1 \tag{2.51}$$

The DOS function in this case is given by

$$N_{nipiHD}(E) = \frac{g_\nu}{2\pi}\sum_{n_i=0}^{n_{i\max}} [\delta_{20}(E, \eta_g, n_i)]' H(E - E_{15n_iHD}). \tag{2.52a}$$

The electron concentration can be written as

$$n_{2D} = \frac{g_\nu}{2\pi}\sum_{n_i=0}^{n_{i\max}} \int_{E_{15n_iHD}}^{\infty} [\delta_{20}(E, \eta_g, n_i)]' f(E) dE. \tag{2.52b}$$

The electron dispersion law in the doping superlattices of stressed Kane type semiconductors in the absence of band tails can be written as

$$\frac{k_x^2}{[\bar{a}_0(E)]^2} + \frac{k_y^2}{[\bar{b}_0(E)]^2} + \frac{1}{[\bar{c}_0(E)]^2}\frac{2m_z^*(0)}{\hbar}\left(n_i + \frac{1}{2}\right)\omega_{12}(E) = 1 \quad (2.53)$$

where

$$\omega_{12}(E) \equiv \left(\frac{n_{2D}|e|^2}{d_0\varepsilon_{sc}m_z^*(E)}\right)^{\frac{1}{2}}$$

and

$$m_z^*(E) \equiv \hbar^2\bar{c}_0(E)\frac{\partial}{\partial E}\left[\bar{c}_0(E)\right].$$

The use of (2.53) leads to the expression of the EM as

$$m^*(E_{Fn}, n_i) = \left(\frac{\hbar^2}{2}\right) R_{85}(E, n_i)\bigg|_{E=E_{Fn}} \quad (2.54)$$

where,

$$\begin{aligned} R_{85}(E, n_i) \equiv \Bigg[& \left[(\bar{a}_0(E))'\bar{b}_0(E) + (\bar{b}_0(E))'\bar{a}_0(E)\right] \\ & \times \left[1 - \frac{1}{[\bar{c}_0(E)]^2}\frac{2m_z^*(0)}{\hbar}\left(n_i + \frac{1}{2}\right)\omega_{12}(E)\right] \\ & - \left[\frac{\bar{a}_0(E)\,\bar{b}_0(E)}{[\bar{c}_0(E)]^2}\frac{2m_z^*(0)}{\hbar}\left(n_i + \frac{1}{2}\right)[\omega_{12}(E)]'\right] \\ & + \left[\frac{\bar{a}_0(E)\,\bar{b}_0(E)\,[\bar{c}_0(E)]'}{[\bar{c}_0(E)]^3}\frac{4m_z^*(0)}{\hbar}\left(n_i + \frac{1}{2}\right)[\omega_{12}(E)]\right]\Bigg] \end{aligned}$$

Thus, the EM is a function of the Fermi energy and the nipi subband index due to the presence of stress and band non-parabolicity only.

The sub-band energies (E_{25ni}) can be written as

$$\frac{1}{[\bar{c}_0(E_{25ni})]^2}\frac{2m_z^*(0)}{\hbar}\left(n_i + \frac{1}{2}\right)\omega_{12}(E_{25ni}) = 1 \quad (2.55)$$

The DOS function can be written as

$$N_{nipi}(E) = \frac{g_v}{\pi\hbar^2}\sum_{n_i=0}^{n_i\max} R_{85}(E, n_i)H(E - E_{25ni}) \quad (2.56)$$

The electron concentration can be written as

$$n_{2D} = \frac{g_v}{\pi \hbar^2} \sum_{n_i=0}^{n_{i\max}} \int_{E_{25ni}}^{\infty} R_{85}(E, n_i) f(E) dE \tag{2.57}$$

2.3 Result and Discussions

The effect of nipi superlattice period on the EM has been exhibited in Figs. 2.1 to 2.9 for different materials. Using (2.2) and (2.5) together with the energy band constants as given in Appendix A, we have plotted the EM in nipi structures of non-linear optical materials taking HD Cd_3As_2 and HD $CdGeAs_2$ as examples in Figs. 2.2 and 2.3. From both the Figs. 2.1 and 2.2, it appears that the effect of increment of the superlattice period increases the EM in the presence of extreme carrier degeneracy of the order of 10^{18} m^{-2}. For comparison with the bulk anisotropic effective masses, we have also exhibited the same in the said figures. It appears that the EM can be much less than that of the corresponding bulk values below 10 nm period for HD Cd_3As_2. Thus in such condition, one can expect an increase in the carrier mobility to a great extent, in fact almost double. The effect of crystal field splitting has also been exhibited in the same Figs. 2.1 and 2.2. It appears that the effect of δ on

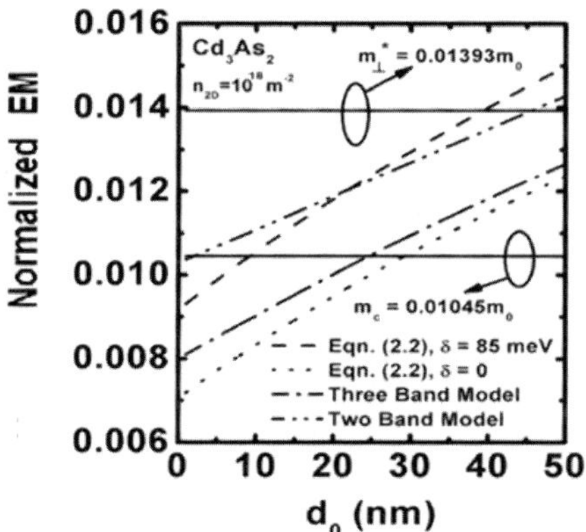

Fig. 2.1. Plot of the normalized EM as function of superlattice period for HD n-Cd_3As_2 considering Eqn. (2.2) The plots for three and two band models of Kane have also been exhibited with their corresponding bulk values.

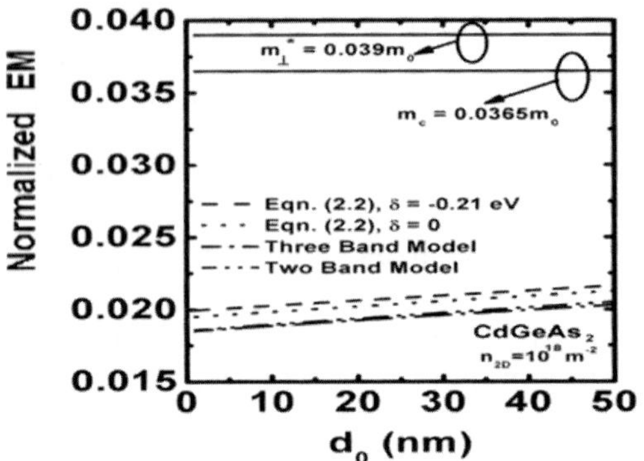

Fig. 2.2. Plot of the normalized EM as function of superlattice period for HD n-$CdGeAs_2$ for all cases of Fig. 2.1.

the EM is largest in case of HD Cd_3As_2. The approximation in the energy band structure also makes a significant deviation of the EM in case of HD Cd_3As_2. However for HD $CdGeAs_2$, the EM exhibits a slow variation over the superlattice period as compare with Fig. 2.1. At this point it should be noted that with the increase with the superlattice period, the EM in HD Cd_3As_2 by considering the energy dispersion relation with the absence of the crystal field splitting and the three band model of Kane actually tends to the anisotropic bulk value $0.01393m_0$.

This is not with the case when the effect of crystal field splitting and the two band equivalent model is considered. In these two cases, the EM is overestimated against the bulk value. This is not with the case of Fig. 2.2 of HD $CdGeAs_2$, where the EM converges to the bulk anisotropic value at larger superlattice period. The effect of superlattice period on the EM in the ground state subband in III-V materials has been evaluated using the three and the two band model of Kane in Figs. 2.3–2.5 for HD InAs, InSb and GaAs respectively.

It appears from these figures that the effect of the variation of the energy dispersion relation model on EM is almost insignificant for HD GaAs nipi structures, whereas for HD InSb, the EM exhibits a significant deviation. In almost all the cases of about 1 nm period, the EM approximately becomes half of the respective isotropic effective bulk masses indicating the mobility rise of upto 200%.

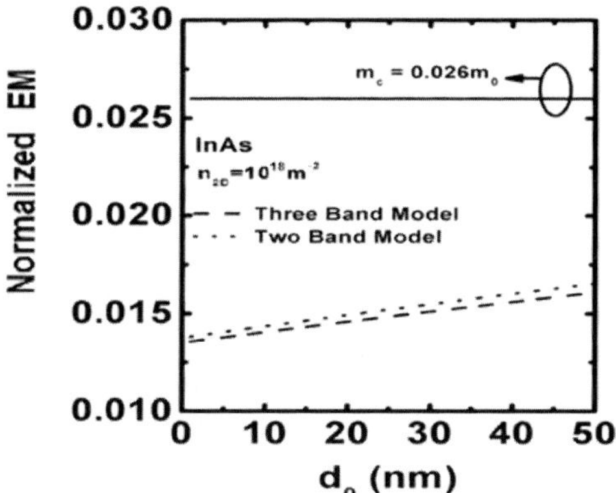

Fig. 2.3. Plot of the normalized EM as function of superlattice period for HD n-InAs considering the three and two band model of Kane with the corresponding bulk value.

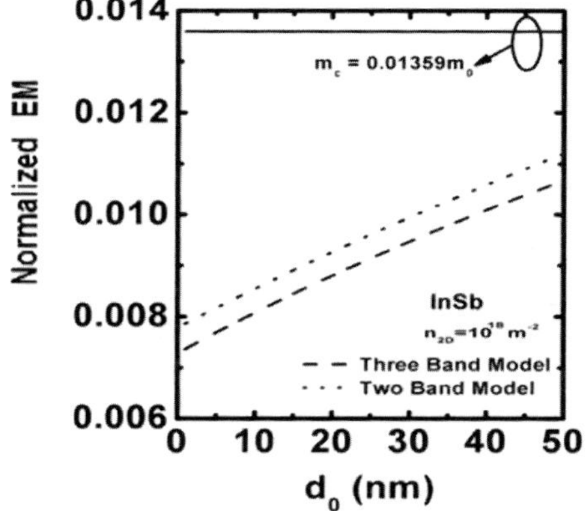

Fig. 2.4. Plot of the normalized EM as function of superlattice period for HD n-InSb considering the three and two band model of Kane with the corresponding bulk value.

In Figs. 2.6 and 2.7, the EM as function of the periods has been further evaluated for the ternary and quaternary materials like HD $Hg_{1-x}Cd_xTe$ and HD $In_{1-x}Ga_xAs_{1-y}P_y$, where the energy band gap in these materials can be modulated by changing the alloy fraction x. We see that the EM in case of HD $In_{1-x}Ga_xAs_{1-y}P_y$ almost exhibits no significant variation an approaches quickly its bulk normalized value 0.0287 at x = 0.3 as compared to HD $Hg_{1-x}Cd_xTe$. Figs. 2.8 and 2.9 exhibit the EM at the lowest subband in II-VI and IV-VI

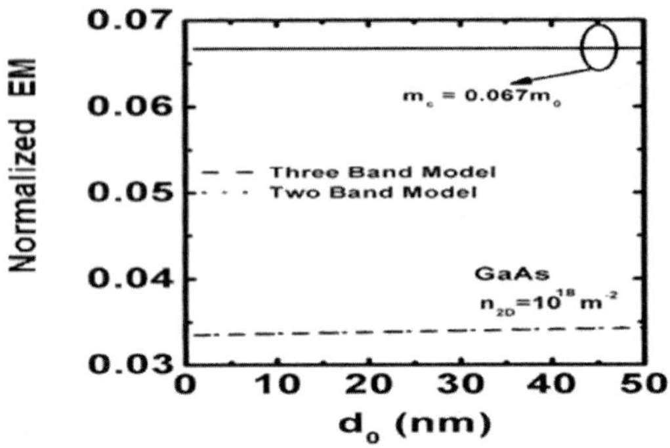

Fig. 2.5. Plot of the normalized EM as function of superlattice period for HD n-GaAs considering the three and two band model of Kane with the corresponding bulk value.

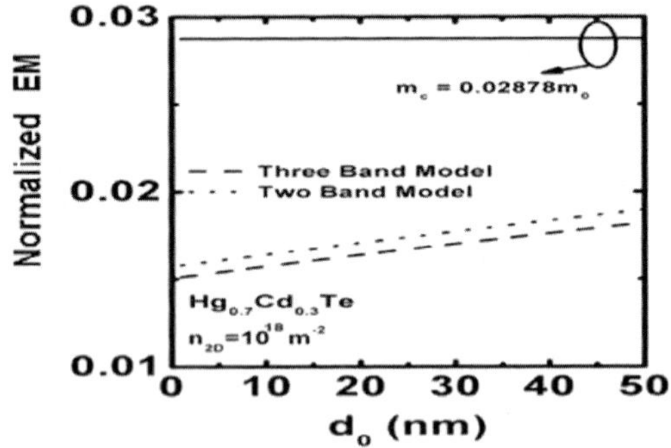

Fig. 2.6. Plot of the normalized EM as function of superlattice period for HD n-$Hg_{1-x}Cd_xTe$ considering the three and two band model of Kane with the corresponding bulk value at $x = 0.3$.

nipi structures of HD CdS and HD PbTe respectively. The effect of increasing the doping concentration from 4×10^{25} to $6 \times 10^{25}m^{-3}$ on EM in HD CdS has been also exhibited in Fig. 2.8 for a period bandwidth of $50 - 100\mu m$. It appears that with the increase in the doping concentration, the EM in HD CdS increases and approaches the bulk longitudinal normalized value 1.5. However in case of HD PbTe, we see that the EM saturates above superlattice period of about $20\mu m$.

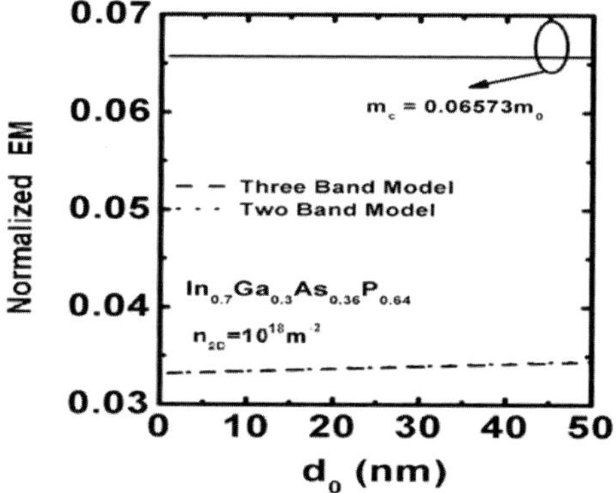

Fig. 2.7. Plot of the normalized EM as function of superlattice period for HD n-$In_{1-x}Ga_xAs_{1-y}P_y$ considering the three and two band model of Kane with the corresponding bulk value at $x = 0.3$.

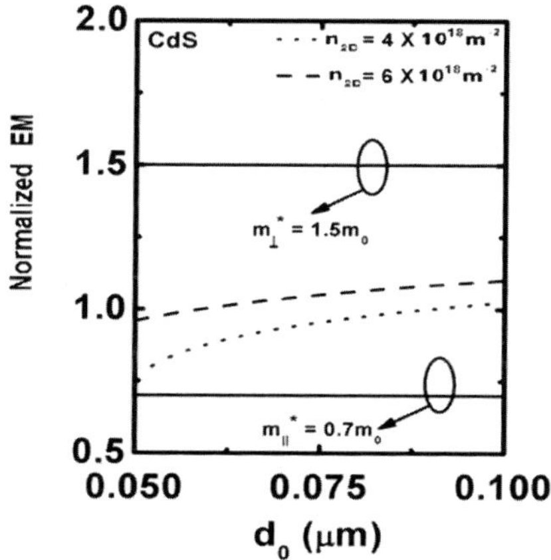

Fig. 2.8. Plot of the normalized EM as function of superlattice period for HD p-CdS.

The effect of doping concentration on the EM in the lowest subband level in all the aforementioned materials has been exhibited in Figs. 2.10–2.17. It appears that the EM increases with the increases in carrier degeneracy for all the cases and may becomes even large than that of their corresponding bulk value along the proper transport direction. From Fig. 2.11 we see that the EM is almost constant below the degeneracy of about 10^{18} m^{-2}. The effect

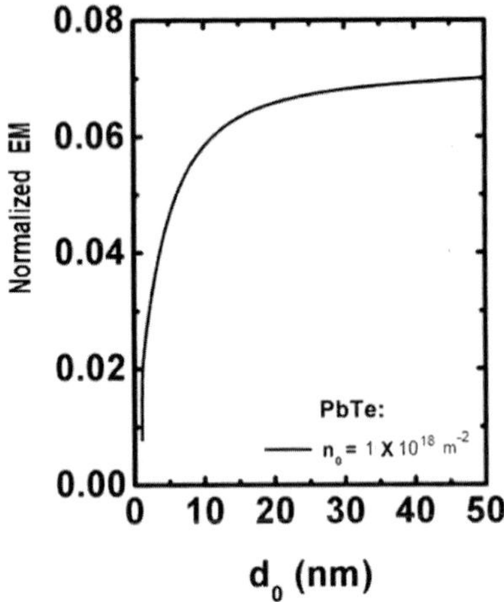

Fig. 2.9. Plot of the normalized EM as function of superlattice period for PbTe.

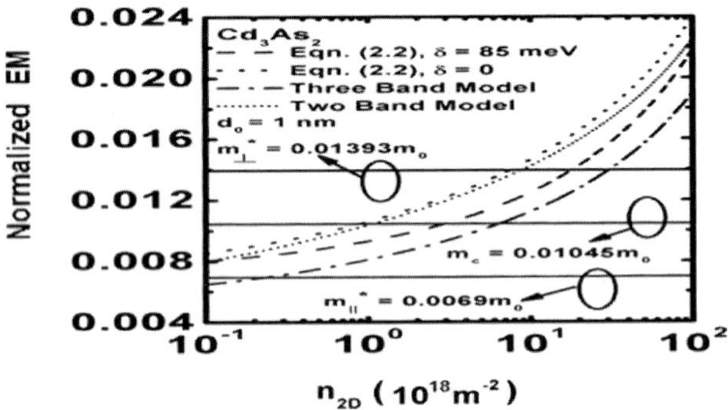

Fig. 2.10. Plot of the normalized EM as function of doping concentration for HD n-Cd_3As_2 considering all cases of Fig. 2.1.

of different models of energy band structures has been exhibited to present the dependency of the EM on the same. It appears from Figs. 2.14 and 2.16 that the second and third order Kane model almost exhibits no differences of the EM from the two. The respective saturation of the EM with the decrease in the degeneracy is different for the materials as this depends on the Fermi energy which is a function of the energy band parameters.

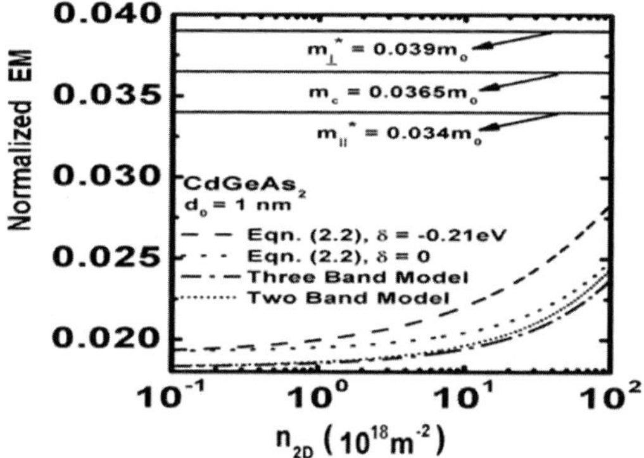

Fig. 2.11. Plot of the normalized EM as function of doping concentration for HD n-$CdGeAs_2$ considering all cases of Fig. 2.2.

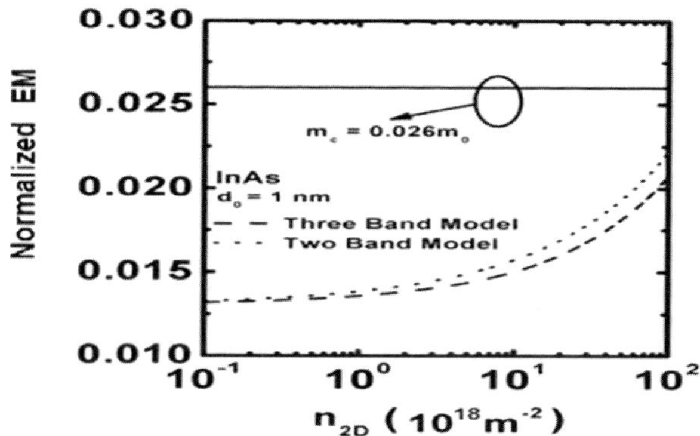

Fig. 2.12. Plot of the normalized EM as function of doping concentration for HD n-InAs considering all cases of Fig. 2.3.

Figure 2.18 exhibits the variation of the EM with increasing degeneracy for HD PbTe nipi. Anomalous behavior in the variation of the EM has been exhibited as one increases the degeneracy. It appears that above 2×10^{18} m^{-2}, the EM decreases. This should not be in general confused with other plots since an increase in the degeneracy increases the Fermi energy which increases the EM. However, in this case, the effect of the different spectrum constants defines the variation of the EM.

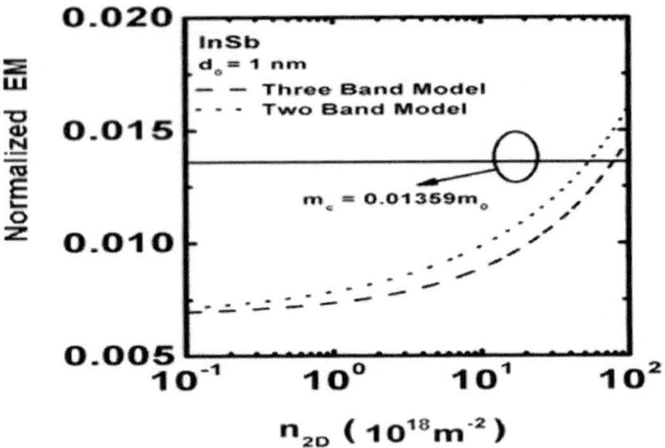

Fig. 2.13. Plot of the normalized EM as function of doping concentration for HD n-InSb considering all cases of Fig. 2.4.

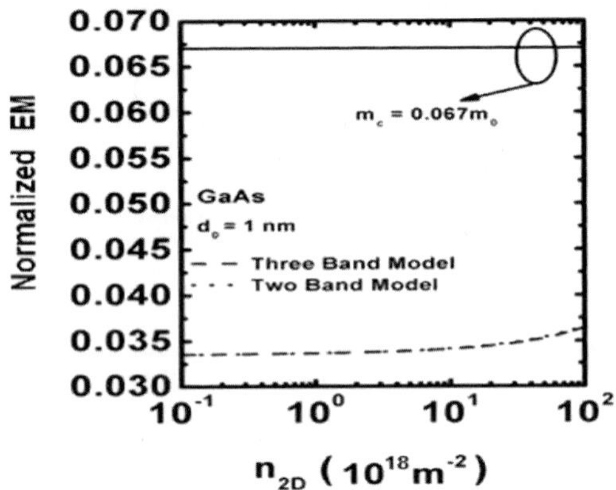

Fig. 2.14. Plot of the normalized EM as function of doping concentration for HD n-GaAs considering all cases of Fig. 2.5.

The variation of the EM with alloy composition for the ternary and quaternary materials has been exhibited in Fig. 2.18 for the three and the two energy band model of Kane. Almost no difference in the two energy band model in this case is exhibited. The variation of the EM for the quaternary is rather slow than that of the ternary

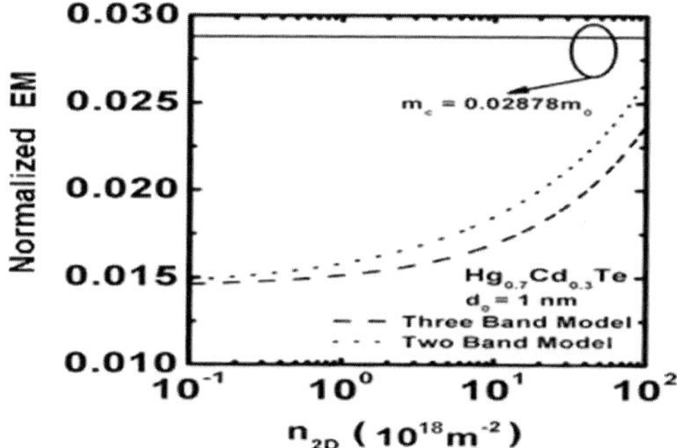

Fig. 2.15. Plot of the normalized EM as function of doping concentration for HD n-$Hg_{1-x}Cd_xTe$ considering all cases of Fig. 2.6.

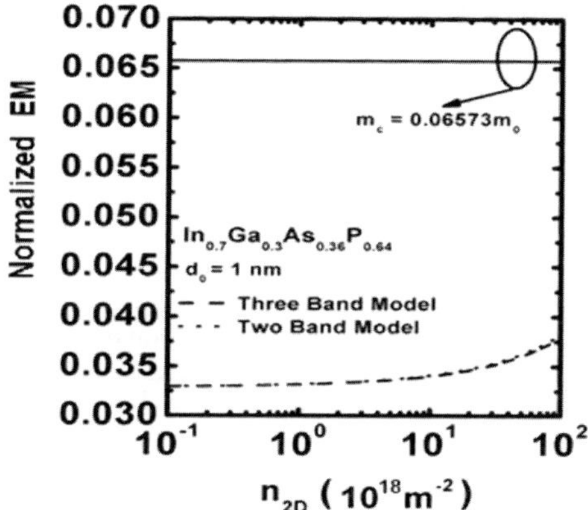

Fig. 2.16. Plot of the normalized EM as function of doping concentration for HD n-$In_{1-x}Ga_xAs_{1-y}P_y$.

which is due to the variation of the energy band gap through the alloy composition. The joy of executing the intricate computer programming for the variation of EM in case of stressed HD InSb nipi structure whose energy band parameters have been given in Appendix A has been left to the reader.

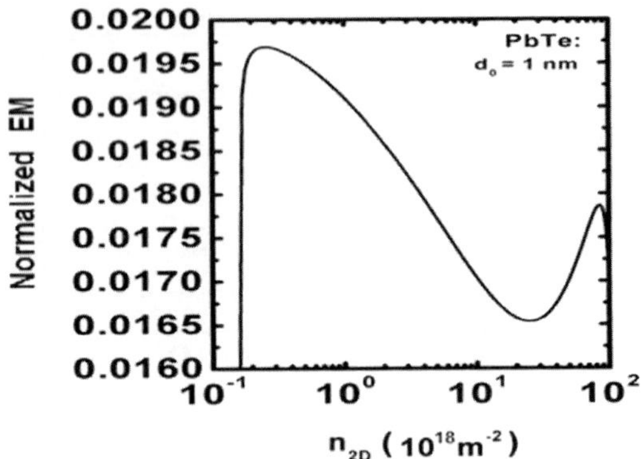

Fig. 2.17. Plot of the normalized EM as function of doping concentration for HD PbTe.

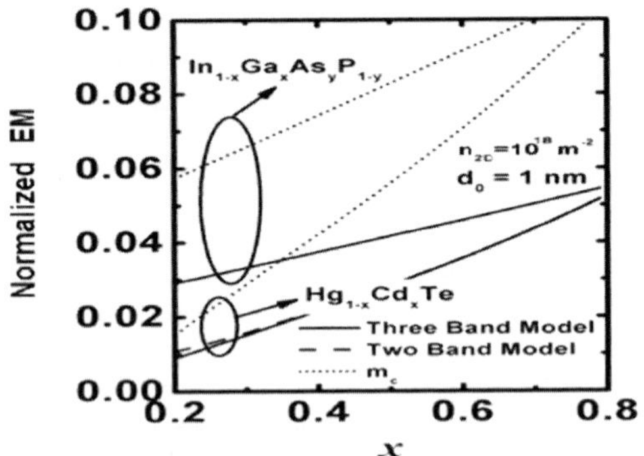

Fig. 2.18. Plot of the normalized EM as function of alloy composition for HD $Hg_{1-x}Cd_xTe$ and HD n-$In_{1-x}Ga_xAs_{1-y}P_y$ considering the three and two band model of Kane with the corresponding bulk value at different x.

2.4 Open Research Problems

(R.2.1) Investigate the EM in the presence of an arbitrarily oriented non-quantizing magnetic field for nipi structures of HD non-linear optical semiconductors by including the electron spin. Study all the special cases for HD III-V, ternary and quaternary materials in this context.

(R.2.2) Investigate the EM in nipi structures of HD IV-VI, II-VI and stressed Kane type compounds in the presence of an arbitrarily oriented non-quantizing magnetic field by including the electron spin.

(R.2.3) Investigate the EM for HD nipi structures of all the materials as stated in R.1.1 of chapter 1.

(R.2.4) Investigate the HD for all the problems from R.2.1 to R.2.3 in the presence of an additional arbitrarily oriented electric field.

(R.2.5) Investigate the EM as defined in (R.2.1) for all the problems from R.2.1 to R.2.3 in the presence of arbitrarily oriented crossed electric and magnetic fields.

(R.2.6) Investigate the EM as defined in (R.2.1) for nipi structures of the heavily–doped semiconductors in the presences of Gaussian, exponential, Kane, Halperian, Lax and Bonch-Burevich types of Band tails for all systems whose unperturbed carrier energy spectra are defined in R.1.1 and R.1.2 respectively.

(R.2.7) Investigate the EM as defined in (R.2.1) for HD nipi structures of the negative refractive index, organic, magnetic and other advanced optical materials in the presence of an arbitrarily oriented alternating electric field.

(R.2.8) Investigate the EM for all the HD nipi systems of this chapter in the presence of finite potential wells.

(R.2.9) Investigate the EM for all the HD nipi systems of this chapter in the presence of parabolic potential wells.

(R.2.10) Investigate all the appropriate problems of this chapter by including the many body, image force, broadening and hot carrier effects respectively.

(R.2.11) Investigate all the appropriate problems of this chapter by removing all the mathematical approximations and establishing the respective appropriate uniqueness conditions.

References

1. N. G. Anderson, W. D. Laidig, R. M. Kolbas and Y. C. Lo, J. Appl. Phys. **60**, 2361 (1986).
2. F. Capasso, Semiconductors and Semimetals **22**, 2 (1985).

3. F. Capasso, K. Mohammed, A. Y. Cho, R. Hull and A. L. Hutchinson, Appl. Phys. Letts., **47**, 420 (1985).
4. F. Capasso and R. A. Kiehl, J. Appl. Phys. **58**, 1366 (1985).
5. K. Ploog and G. H. Doheler, Adv. Phys. **32**, 285 (1983).
6. F. Capasso, K. Mohammed and A. Y. Cho, Appl. Phys. Lett. 478 (1986).
7. R. Grill, C. Metzner, and G. H. Döhler, Phys. Rev. B, **63**, 235316, (2001); J. Z. Wang, Z. G. Wang, Z. M. Wang, S. L. Feng, and Z. Yang, Phys. Rev. B, **62**, 6956 (2000); Phys. Rev. B, **61**, 15614 (2000).
8. A. R. Kost, M. H. Jupina, T. C. Hasenberg and E. M. Garmire, J. Appl. Phys., **99**, 023501, (2006).
9. A. G. Smirnov, D. V. Ushakov and V. K. Kononenko, Proc. SPIE, **4706**, 70, (2002).
10. D. V. Ushakov, V. K. Kononenko and I. S. Manak, Proc. SPIE, **4358**, 171, (2001).
11. J. Z. Wang, Z. G. Wang, Z. M. Wang, S. L. Feng and Z. Yang, Phys. Rev. B, **62**, 6956, (2000).
12. A. R. Kost, L. West, T. C. Hasenberg, J. O. White, M. Matloubian and G. C. Valley, Appl. Phys. Lett., **63**, 3494, (1993).
13. S. Bastola, S. J. Chua and S. J. Xu, J. Appl. Phys., **83**, 1476, (1998).
14. Z. J. Yang, E. M. Garmire and D. Doctor, J. Appl. Phys., **82**, 3874, (1997).
15. G. H. Avetisyan, V. B. Kulikov, I. D. Zalevsky and P. V. Bulaev, Proc. SPIE, **2694**, 216, (1996).
16. U. Pfeiffer, M. Kneissl, B. Knüpfer, N. Müller, P. Kiesel, G. H. Döhler, and J. S. Smith, Appl. Phys. Lett., **68**, 1838, (1996).
17. H. L. Vaghjiani, E. A. Johnson, M. J. Kane, R. Grey, and C. C. Phillips, J. Appl. Phys., **76**, 4407, (1994).
18. P. Kiesel, K. H. Gulden, A. Hoefler, M. Kneissl, B. Knuepfer, S. U. Dankowski, P. Riel, X. X. Wu, J. S. Smith and G. H. Doehler, Proc. SPIE, **1985**, 278, (1993)
19. G.H. Doheler, Phys. Script. **24**, 430 (1981).
20. S. Mukherjee, S. N. Mitra, P. K. Bose, A. R. Ghatak, A. Neoigi, J. P. Banerjee, A. Sinha, M. Pal, S. Bhattacharya and K. P. Ghatak, J. Compu. Theor. Nanosc., **4**, 550, (2007).

Chapter 3

The EM in Accumulation and Inversion Layers of Non-Parabolic Semiconductors

3.1 Introduction

It is well known that the electrons in bulk semiconductors in general, have three dimensional freedom of motion. When, these electrons are confined in a one dimensional potential well whose width is of the order of the carrier wavelength, the motion in that particular direction gets quantized while that along the other two directions remains as free. Thus, the energy spectrum appears in the shape of discrete levels for the one dimensional quantization, each of which has a continuum for the two dimensional free motion. The transport phenomena of such one dimensional confined carriers have recently studied [1–20] with great interest. For the metal-oxide-semiconductor (MOS) structures, the work functions of the metal and the semiconductor substrate are different and the application of an external voltage at the metal-gate causes the change in the charge density at the oxide semiconductor interface leading to a bending of the energy bands of the semiconductor near the surface. As a result, a one dimensional potential well is formed at the semiconductor interface. The spatial variation of the potential profile is so sharp that for considerable large values of the electric field, the width of the potential well becomes of the order of the de Broglie wavelength of the carriers. The Fermi energy, which is near the edge of the conduction band in the bulk, becomes nearer to the edge of the valance band at the surface creating accumulation layers. The energy levels of the carriers bound within the potential well get quantized and form electric subbands. Each of the subband corresponds to a quantized level in a plane perpendicular to the surface leading to a quasi two dimensional electron gas. Thus, the extreme band bending at low temperature allows us to observe the quantum effects at the surface. Though considerable work has already been done, nevertheless it appears from the literature that the EM in accumulation layers of non-parabolic semiconductors has yet to be investigated in details. For the purpose of comparison we shall also study the EM for inversion layers of non-parabolic compounds.

In what follows in Section 3.2.1, of the theoretical background, the EM in accumulation and Inversion layers of nonlinear optical semiconductors has been studied under weak electric field limit. Section 3.2.2 contains the results for accumulation and Inversion layers of III–V, ternary and quaternary semiconductors for the weak electric field limit whose bulk electrons obey the three and the two band models of Kane together with parabolic energy bands and they form the special cases of Section 3.2.1. Section 3.2.3 contains the study of the EM for accumulation and Inversion layers of II–VI semiconductors, which is valid for all values of electric field. Sections 3.2.4 and 3.2.5 contain the study of the EM in accumulation and Inversion layers of IV–VI and stressed semiconductors respectively. Section 3.2.6 contains the study of the EM in accumulation and Inversion layers of Ge. Section 3.3 contains the summary and conclusion in this context. Section 3.4 contains a dozen of open research problems.

3.2 Theoretical Background

3.2.1 *The EM in accumulation and inversion layers of non-linear optical semiconductors*

In the presence of a surface electric field F_s along z direction and perpendicular to the surface, (1.26) assumes the form

$$\frac{\hbar^2 k_z^2}{2m_\parallel^*} + \frac{\hbar^2 k_s^2}{2m_\perp^*} \frac{T_{21}(E - |e|F_s z, \eta_g)}{T_{22}(E - |e|F_s z, \eta_g)} = T_{21}(E - |e|F_s z, \eta_g) \quad (3.1)$$

where, for this section, E represents the electron energy as measured from the edge of the conduction band at the surface in the vertically upward direction.

The quantization rule for 2D carriers in this case, is given by [229]

$$\int_0^{z_t} k_z dz = \frac{2}{3}(S_i)^{3/2} \tag{3.2}$$

where, z_t is the classical turning point and S_i is the zeros of the Airy function ($Ai(-S_i) = 0$).

Using (3.1) and (3.2) leads to the DR of the 2D electrons in accumulation layers of HD non-linear optical materials under the condition of weak electric field limit as

$$\frac{\hbar^2 k_s^2}{2m_{\parallel}^*} = L_6(E, i, \eta_g) \tag{3.3}$$

where

$$L_6(E, i, \eta_g) = \frac{T_{21}(E, \eta_g) - L_3(E, i, \eta_g)}{L_4(E, i, \eta_g)},$$

$$L_3(E, i, \eta_g) = S_i[T'_{21}(E, \eta_g)]^{2/3}\left[\frac{\hbar|e|F_s}{\sqrt{2m_{\parallel}^*}}\right]^{2/3}$$

and

$$L_4(E, i, \eta_g)\left[\frac{T_{21}(E, \eta_g)}{T_{22}(E, \eta_g)} + L_3(E, i, \eta_g)\frac{T_{21}(E, \eta_g)}{T'_{21}(E, \eta_g)T_{22}(E, \eta_g)}\right.$$
$$\left.\frac{2}{3}\left\{\frac{T'_{21}(E, \eta_g)}{T_{21}(E, \eta_g)} - \frac{T'_{22}(E, \eta_g)}{T_{22}(E, \eta_g)}\right\}\right]$$

The EM in this case can be written as

$$m^*(E'_f, i, \eta_g) = m_{\parallel}^* \text{ Real part of } [L'_6(E'_f, i, \eta_g)] \tag{3.4}$$

where

$$E'_f = eV_g - \frac{e^2 n_s d_{ex}}{\varepsilon_{ox}} + E_{FB},$$

V_g is the gate voltage, n_s is the surface electron concentration, d_{ox} is the thickness of the oxide layer, ε_{ox} is the permittivity of the oxide layer, $F_s = \frac{en_s}{\varepsilon_{sc}}$, ε_{sc} is the semiconductor permittivity and E_{FB} should be determined from the equation

$$n_B = \frac{2g_v}{(2\pi)^3} \times \frac{2m_\perp^* \sqrt{2m_\parallel^*}}{\hbar^3} \text{ Real Part of } \left[T_{22}(E_{FB}, \eta_g)\sqrt{T_{21}(E_{FB}, \eta_g)}\right] \tag{3.5}$$

and n_B is the bulk electron concentration.

The sub-band energy E_i can be determined from the equation

$$0 = \text{ Real part of } L_6(E_i, i, \eta_g) \tag{3.6}$$

The surface electron concentration in the regime of very low temperatures where the quantum effects become prominent can be written as

$$n_s = 2g_v \text{ Real part of } \sum_{i=0}^{i_{\max}} \left[\left[\frac{m_\perp^*}{2\pi\hbar^2} L_6(E_f', i, \eta_g)\right] + \frac{1}{(2\pi)^3} \frac{2m_\perp^* \sqrt{2m_\parallel^*}}{\hbar^3} t_i \left[T_{22}(E_{FB}, \eta_g)\sqrt{T_{21}(E_{FB}, \eta_g)}\right]\right] \tag{3.7}$$

where $t_i = \frac{E_{i\max}}{eF_s(1+i_{\max})}$, $E_{i\max}$, is the root of the Real part of the equation

$$T_{21}(E_{i\max}, \eta_g) - L_3(E_{i\max}, i_{\max}, \eta_g) = 0 \tag{3.8a}$$

In what follows, we shall discuss the EM in inversion layers of non-linear optical materials for the purpose of relative comparison. In the presence of a surface electric field F_s along z direction and perpendicular to the surface, (1.2) assumes the form

$$\psi_1(E - |e|F_s z) = \psi_2(E - |e|F_s z).k_s^2 + \psi_3(E - |e|F_s z)k_z^2 \tag{3.8b}$$

where

$$\psi_1(E) = \gamma(E), \psi_2(E) = f_2(E) \quad \text{and} \quad \psi_3(E) = f_2(E)$$

Using (3.2) and (3.8b), under the weak electric field limit, one can write,

$$\int_0^{z_t} \sqrt{A_7(E) - |e|F_s z D_7(E)}dz = \frac{2}{3}(S_i)^{3/2} \tag{3.9}$$

in which,

$$A_7(E) \equiv \left[\frac{\psi_1(E) - \psi_2(E)k_s^2}{\psi_3(E)}\right], \quad D_7(E) \equiv [B_7(E) - A_7(E)C_7(E)],$$

$$B_7(E) \equiv \left[\frac{(\psi_1(E))' - (\psi_2(E))'k_s^2}{\psi_3(E)}\right] \quad \text{and} \quad C_7(E) \equiv \left[\frac{(\psi_3(E))'}{\psi_3(E)}\right].$$

Thus, the 2D electron dispersion law in inversion layers of nonlinear optical materials under the weak electric field limit can approximately be written as

$$\psi_1(E) = P_7(E,i)k_s^2 + Q_7(E,i) \tag{3.10}$$

where

$$t_2(E) \equiv \left[\frac{[\psi_2(E)]'}{\psi_3(E)} - \left(\frac{\psi_2(E)[\psi_3(E)]'}{[\psi_3(E)]^2}\right)\right],$$

$$t_1(E) = \left[\frac{[\psi_1(E)]'}{\psi_3(E)} - \left(\frac{\psi_3(E)[\psi_3(E)]'}{[\psi_3(E)]^2}\right)\right],$$

$$P_7(E,i) \equiv \left[\psi_2(E) - \left(\frac{2t_2(E)}{3[t_1(E)]^{1/3}}\right)\psi_3(E)S_i(|e|F_2)^{2/3}\right] \quad \text{and}$$

$$Q_7(E,i) \equiv S_i\psi_3(E)[|e|F_s t_1(E)]^{2/3}.$$

The EM in the x-y plane can be expressed as

$$m^*(E_{Fiw}, i) = \left(\frac{\hbar^2}{2}\right) G_7(E,i)\bigg|_{E=E_{Fw}} \tag{3.11}$$

where, $G_7(E,i) \equiv [P_7(E,i)]^{-2}[P_7(E,i)\{(\psi_1(E))' - (Q_7(E,i))'\} - \{\psi_1(E) - (Q_7(E,i))\}(P_7(E,i))']$ and E_{Fiw} is the Fermi energy under the weak electric field limit as measured from the edge of the conduction band at the surface in the vertically upward direction. Thus, we observe that the EM is the function of subband index, the Fermi energy and other band constants due to the combined influence of the crystal filed splitting constant and the anisotropic spin-orbit splitting constants respectively.

The sub-band energy ($E_{n_{iw1}}$) in this case can be obtained from (3.10) as

$$\psi_1(E_{n_{iw1}}) = Q_7(E_{n_{iw1}}, i) \tag{3.12}$$

The general expression of the 2D total DOS function in this case can be written as

$$N_{2D_i}(E) = \frac{2g_v}{(2\pi)^2} \sum_{i=0}^{i_{\max}} \frac{\partial}{\partial E}[A(E,i)H(E - E_{n_i})] \tag{3.13}$$

where, $A(E, i)$ is the area of the constant energy 2D wave vector space for inversion layers and E_{n_i} is the corresponding sub-band energy.

Using (3.10) and (3.13), the total 2D DOS function under the weak electric field limit can be expressed as

$$N_{2D_i}(E) = \frac{g_v}{(2\pi)} \sum_{i=0}^{i_{\max}} [G_7(E,i)H(E - E_{n_{iw1}})] \tag{3.14a}$$

The surface electron concentration per unit area is given by

$$n_{2D} = \frac{g_v}{(2\pi)} \sum_{i=0}^{i_{\max}} [G_{70}(E_{Fiw}, i) + \sum_{r=1}^{s} L(r)[G_{70}(E_{Fiw}, i)]] \tag{3.14b}$$

where

$$G_{70}(E_{Fiw}, i) = [[\psi_1(E_{Fiw}) - Q_7(E_{Fiw}, i)](P_7(E_{Fiw}, i))^{-1}]$$

3.2.2 *The EM in accumulation and inversion layers of III–V, ternary and quaternary semiconductors*

$$\textbf{(a)}\ \delta = 0, \quad \Delta_{\|} = \Delta_{\perp} = \Delta \quad \text{and} \quad m_{\|}^* = m_{\perp}^* = m_c,$$

Using the substitutions $\delta = 0$, $\Delta_{\|} = \Delta_{\perp} = \Delta$ and $m_{\|}^* = m_{\perp}^* = m_c$, (3.3) under the condition of weak electric field limit, assumes the form

$$T_{90}(E, \eta_g) = \frac{\hbar^2 k_s^2}{2m_c} + S_i \left[\frac{\hbar|e|F_s[T_{90}(E, \eta_g)]'}{\sqrt{2m_c}}\right]^{2/3} \tag{3.15}$$

where, $T_{90}(E, \eta_g) = T_{31}(E, \eta_g) + iT_{32}(E, \eta_g)$ (3.15) represents the EM of the 2D electrons in accumulation layers of HD III–V, ternary and quaternary materials under the weak electric field limit whose bulk electrons obey the HD three band model of Kane. Since the electron energy spectrum in accordance with the HD three-band model of Kane is complex in nature, (3.15) will also be complex. The both complexities occur due to the presence of poles in the finite complex plane of the dispersion relation of the materials in the absence of band tails.

The EM can be expressed as

$$m^*(E'_f, i, \eta_g) = m_c \text{ Real part of } P'_{3HD}(E'_f, i, \eta_g) \tag{3.16}$$

where,

$$P_{3HD}(E'_f, i, \eta_g) = \left[T_{90}(E'_f, i, \eta_g) - S_i\left[\frac{\hbar|e|F_s[T_{90}(E'_f, \eta_g)]'}{\sqrt{2m_c}}\right]^{2/3}\right]$$

Thus, one can observe that the EM is a function of the sub-band index, surface electric field, the Fermi energy and the other spectrum constants due to the combined influence of E_g and Δ.

The sub-band energy E_{i1} is given by

$$\begin{aligned} 0 = {} & \text{Real part of } [T_{90}(E_{i1}, \eta_g) \\ & - S_i[\hbar|e|F_s[T_{90}(E_{i1}, \eta_g)]'.(2m_c)^{-1/2}]^{2/3}] \end{aligned} \tag{3.17}$$

The DOS function can be written as

$$N_{2D_i}(E) = \frac{m_c g_v}{\pi\hbar^2}\sum_{i=0}^{i_{\max}}[P_{3HD}(E, i, \eta_g)H(E - E_{i1})] \tag{3.18}$$

Thus the DOS function is complex in nature.

The surface electron concentration is given by

$$\begin{aligned} n_s = g_v \text{Real part of } & \sum_{i=0}^{i_{\max}}\left[\left[\frac{m_c}{\pi\hbar^2}P_{3HD}(E'_f, i, \eta_g)\right]\right. \\ & \left. + \frac{1}{3\pi^2}\left(\frac{2m_c}{\hbar^2}\right)^{3/2} t_i[T_{90}(E_{FB}, \eta_g)]^{3/2}\right] \end{aligned} \tag{3.19}$$

where E_{FB} should be determined from the following equation

$$n_B = \frac{g_v}{3\pi^2}\left(\frac{2m_c}{\hbar^2}\right)^{3/2} \text{Real part of } [T_{90}(E_{FB}, \eta_g)]^{3/2} \tag{3.20}$$

Using the substitutions $\delta = 0, \Delta_{\|} = \Delta_{\perp} = \Delta$ and $m_{\|}^* = m_{\perp}^* = m_c$, (3.10) under the condition of weak electric field limit, assumes the form

$$I_{11}(E) = \frac{\hbar^2 k_s^2}{2m_c} + S_i \left[\frac{\hbar|e|F_s[I_{11}(E)]'}{\sqrt{2m_c}}\right]^{2/3} \quad (3.21)$$

(3.21) represents the dispersion relation of the 2D electrons in inversion layers of III–V, ternary and quaternary materials under the weak electric field limit whose bulk electrons obey the three band model of Kane.

The EM can be expressed as

$$m^*(E_{Fiw}, i) = m_c[P_3(E,i)]|_{E=E_{Fiw}} \quad (3.22)$$

where,

$$P_3(E,i) = \left\{[I_{11}(E)]' - \left\{\frac{2}{3}S_i\left[\frac{\hbar|e|F_s}{\sqrt{2m_c}}\right]^{2/3}\{[I_{11}(E))]'\}^{-1/3}[I_{11}(E))]''\right\}\right\}.$$

Thus, one can observe that the EM is a function of the sub-band index, surface electric field, the Fermi energy and the other spectrum constants due to the combined influence of E_g and Δ.

The sub-band energy $(E_{n_{iw2}})$ in this case can be obtained from (3.21) as

$$I_{11}(E_{n_{iw2}}) = S_i \left[\frac{\hbar|e|F_s[I_{11}(E_{n_{iw2}})]'}{\sqrt{2m_c}}\right]^{2/3} \quad (3.23)$$

Thus the 2D total DOS function in weak electric field limit can be expressed as

$$N_{2D_i}(E) = \frac{m_c g_v}{\pi\hbar^2}\sum_{i=0}^{i_{\max}}[P_3(E,i)H(E - E_{n_{iw2}})] \quad (3.24a)$$

The surface electron concentration per unit area is given by

$$n_{2D} = \frac{m_c g_v}{\pi\hbar^2}\sum_{i=0}^{i_{\max}}[P_{30}(E_{Fiw}, i) + \sum_{r=1}^{s} L(r)[P_{30}(E_{Fiw}, i)]] \quad (3.24b)$$

where

$$P_{30}(E_{Fiw}, i) = \left[I_{11}(E_{Fiw}) - S_i\left[\frac{\hbar|e|F_s[I_{11}(E_{Fiw})]'}{\sqrt{2m_c}}\right]^{2/3}\left(\frac{\hbar^2}{2m_c}\right)^{-1}\right]$$

(b) $\Delta \gg E_g$ or $\Delta \ll E_g$

Using the constraints $\Delta \gg E_g$ or $\Delta \ll E_g$, then (3.21) under the low electric field limit assumes the form

$$\gamma_2(E, \eta_g) = \frac{\hbar^2 k_s^2}{2m_c} + S_i\left[\frac{\hbar|e|F_s[\gamma_2(E, \eta_g)]'}{\sqrt{2m_c}}\right]^{2/3} \tag{3.25}$$

(3.25) represents the dispersion relation of the 2D electrons in accumulation layers of HD III–V, ternary and quaternary materials under the weak electric field limit whose bulk electrons obey the HD two band model of Kane.

The EM can be expressed as

$$m^*(E'_f, i, \eta_g) = m_c P'_{3HD1}(E'_f, i, \eta_g) \tag{3.26}$$

where

$$P_{3HD1}(E'_f, i, \eta_g) = \left[\gamma_2(E'_f, \eta_g) - S_i\left[\frac{\hbar|e|F_s[\gamma_2(E'_f, \eta_g)]'}{\sqrt{2m_c}}\right]^{2/3}\right]$$

Thus, one can observe that the EM is a function of the sub-band index, surface electric field, the Fermi energy and the other spectrum constants due to the combined influence of E_g and Δ.

The sub-band energy E_{i1} is given by

$$0 = [\gamma_2(E_{i2}, \eta_g) - S_i[\hbar|e|F_s[\gamma_2(E_{i2}, \eta_g)]'.(2m_c)^{-1/2}]^{2/3}] \tag{3.27}$$

The DOS function can be written as

$$N_{2D_i}(E) = \frac{m_c g_v}{\pi\hbar^2}\sum_{i=0}^{i_{\max}}[P_{3HD1}(E, i, \eta_g)H(E - E_{i2})] \tag{3.28}$$

Thus the DOS function is complex in nature.

The surface electron concentration is given by

$$n_S = g_v \sum_{i=0}^{i_{\max}} \left[\left[\frac{m_c}{\pi\hbar^2} P_{3HD1}(E'_f, i, \eta_g) \right] + \frac{1}{3\pi^2} \left(\frac{2m_c}{\hbar^2} \right)^{3/2} t_i [\gamma_2(E_{FB}, \eta_g)]^{3/2} \right] \quad (3.29)$$

where E_{FB} should be determined from the following equation

$$n_B = \frac{g_v}{3\pi^2} \left(\frac{2m_c}{\hbar^2} \right)^{3/2} [\gamma_2(E_{FB}, \eta_g)]^{3/2} \quad (3.30)$$

Using the constraints $\Delta \gg E_g$ or $\Delta \ll E_g$, (3.21) under the low electric field limit assumes the form

$$E(1 + \alpha E) = \frac{\hbar^2 k_s^2}{2m_c} + S_i \left[\frac{\hbar|e|F_s(1 + 2\alpha E)}{\sqrt{2m_c}} \right]^{2/3} \quad (3.31)$$

For large values of i, $S_i \to [\frac{3\pi}{2}(i + \frac{3}{4})]^{2/3}$ [5], (3.31) gets simplified as

$$E(1 + \alpha E) = \frac{\hbar^2 k_s^2}{2m_c} + \left[\frac{3\pi\hbar|e|F_s}{2} \left(i + \frac{3}{4} \right) \frac{(1 + 2\alpha E)}{\sqrt{2m_c}} \right]^{2/3} \quad (3.32)$$

(3.32) was derived ***for the first time by Antcliffe et al.*** [3].

The EM in this case is given by

$$m^*(E_{Fiw}, i) = m_c [P_6(E, i)]|_{E=E_{Fiw}} \quad (3.33)$$

where,

$$P_6(E, i) \equiv \left\{ 1 + 2\alpha E - \frac{4\alpha}{3} S_i \left[\frac{\hbar|e|F_s}{\sqrt{2m_c}} \right]^{2/3} \{1 + 2\alpha E\}^{-1/3} \right\}.$$

Thus, one can observe that the EM is a function of the subband index, surface electric field and the Fermi energy due to the presence of band non-parabolicity only.

The sub-band energies ($E_{n_{iw3}}$) are given by

$$(E_{n_{iw3}})(1+\alpha E_{n_{iw3}}) = S_i \left[\frac{\hbar|e|F_s(1+2\alpha E_{n_{iw3}})}{\sqrt{2m_c}}\right]^{2/3} \tag{3.34}$$

The total 2D DOS function can be written as

$$N_{2D}(E) = \frac{m_c g_v}{\pi\hbar^2}\sum_{i=0}^{i_{\max}}\left\{\left[1+2\alpha E - \frac{4\alpha}{3}S_i\left[\frac{\hbar|e|F_s}{\sqrt{2m_c}}\right]^{2/3}(1+2\alpha E)^{-1/3}\right]\right.$$
$$\left.\times H(E-E_{n_{iw3}})\right\} \tag{3.35a}$$

Under the condition $\alpha E \ll 1$, the use of (3.35a) and the Fermi-Dirac integral lead to the expression of n_{2Dw} as

$$n_{2Dw} = \left(\frac{g_v m_c k_B T}{\pi\hbar^2}\right)\sum_{i=0}^{i_{\max}}$$
$$\times\{[1+D_i+2\alpha E_{n_{iw3}}].F_0(\eta_{iw})+2\alpha k_B T F_1(\eta_{iw})\} \tag{3.35b}$$

where,

$$D_i \equiv \frac{4\alpha S_i}{3}\left(\frac{\hbar|e|F_s}{\sqrt{2m_c}}\right)^{2/3} \quad \text{and} \quad \eta_{iw} \equiv \left[\frac{E_{Fiw}-E_{n_{iw3}}}{k_B T}\right].$$

For all values of αE_{Fiw}, the n_{2Dw} can be written by using (3.35a) as

$$n_{2Dw} = \left(\frac{g_v m_c}{\pi\hbar^2}\right)\sum_{i=0}^{i_{\max}}[P_{5w}(E_{Fiw},i)+Q_{5w}(E_{Fiw},i)] \tag{3.35c}$$

where,

$$P_{5w}(E_{Fiw},i) \equiv \left[E_{Fiw}(1+\alpha E_{Fiw}) - S_i\left[\frac{\hbar\,|e|\,F_s}{\sqrt{2m_c}}(1+2\alpha E_{Fiw})\right]^{2/3}\right]$$

and

$$Q_{5w}(E_{Fiw},i) \equiv \sum_{r=1}^{s} L(r)P_{5w}(E_{Fiw},i).$$

The electron concentration under the weak electric field quantum limit assumes the form

$$\bar{n}_{2Dw} = \frac{g_v m_c}{\pi\hbar^2}\left\{\bar{E}_{Fw}(1+\alpha\bar{E}_{Fw}) - S_0\left(\frac{\hbar|e|F_s}{\sqrt{2m_c}}\right)^{2/3}(1+2\alpha\bar{E}_{Fw})^{2/3}\right\}. \tag{3.35d}$$

(c) $\alpha \to 0$

Using the constraints $\alpha \to 0$, (3.25) under the low electric field limit assumes the form

$$\gamma_3(E,\eta_g) = \frac{\hbar^2 k_s^2}{2m_c} + S_i\left[\frac{\hbar|e|F_s[\gamma_3(E,\eta_g)]'}{\sqrt{2m_c}}\right]^{2/3}. \tag{3.36}$$

The (3.36) represents the dispersion relation of the 2D electrons in accumulation layers of HD III–V, ternary and quaternary materials under the weak electric field limit whose bulk electrons obey the HD parabolic band model.

The EM can be expressed as

$$m^*(E'_f,i,\eta_g) = m_c P'_{3HD2}(E'_f,i,\eta_g) \tag{3.37}$$

where,

$$P_{3HD2}(E'_f,i,\eta_g) = \left[\gamma_3(E'_f,\eta_g) - S_i\left[\frac{\hbar|e|F_s\left[\gamma_3(E'_f,\eta_g)\right]'}{\sqrt{2m_c}}\right]^{2/3}\right]$$

Thus, one can observe that the EM is a function of the sub-band index, surface electric field, the Fermi energy and the other spectrum constants due to the combined influence of E_g and Δ.

The sub-band energy E_{i1} is given by

$$0 = [\gamma_3(E_{i2},\eta_g) - S_i[\hbar|e|F_s[\gamma_3(E_{i2},\eta_g)]'(2m_c)^{-1/2}]^{2/3}]. \tag{3.38}$$

The DOS function can be written as

$$N_{2D_i}(E) = \frac{m_c g_v}{\pi\hbar^2}\sum_{i=0}^{i_{\max}}[P_{3HD2}(E,i,\eta_g)H(E-E_{i3})]. \tag{3.39}$$

The surface electron concentration is given by

$$n_S = g_v \sum_{i=0}^{i_{\max}} \left[\left[\frac{m_c}{\pi\hbar^2} P_{3HD2}(E'_f, i, \eta_g) \right] + \frac{1}{3\pi^2} \left(\frac{2m_c}{\hbar^2} \right)^{3/2} t_i [\gamma_3(E_{FB}, \eta_g)]^{3/2} \right] \tag{3.40}$$

where E_{FB} should be determined from the following equation

$$n_B = \frac{g_v}{3\pi^2} \left(\frac{2m_c}{\hbar^2} \right)^{3/2} [\gamma_3(E_{FB}, \eta_g)]^{3/2} \tag{3.41}$$

For $\alpha \to 0$, as for inversion layers, whose bulk electrons are defined by the parabolic energy bands, from (3.32), we can write,

$$E = \frac{\hbar^2 k_s^2}{2m_c} + S_i \left[\frac{\hbar |e| F_s}{\sqrt{2m_c}} \right]^{2/3} \tag{3.42}$$

(3.42) is valid for all values of the surface electric field [3].

The electric sub-band energy ($E_{n_{i4}}$) assumes the form, from (3.42) as

$$E_{n_{i4}} = S_i \left[\frac{\hbar |e| F_s}{\sqrt{2m_c}} \right]^{2/3} \tag{3.43}$$

The total DOS function can be written using (3.42) as

$$N_{2D}(E) = \frac{m_c g_v}{\pi\hbar^2} \sum_{i=0}^{i_{\max}} H(E - E_{n_{i4}}). \tag{3.44a}$$

The use of equation (3.44a) leads to the expression of n_{2Di} as

$$n_{2Di} = \frac{g_v m_c k_B T}{\pi\hbar^2} \sum_{i=0}^{i_{\max}} F_0(\eta_i) \tag{3.44b}$$

where, $\eta_i \equiv (k_B T)^{-1}[E_{Fi} - S_i[\frac{\hbar|e|F_s}{\sqrt{2m_c}}]^{2/3}]$, E_{Fi} is the Fermi energy as measured from the edge of the conduction band at the surface.

3.2.3 *The EM in accumulation and inversion layers of II–VI semiconductors*

The use of (1.105a) and (3.2) leads to the expression of the quantization integral as

$$\frac{\sqrt{2m_{\|}^*}}{\hbar}\int_0^{z_t}[\gamma_3(E,\eta_g)-|e|F_s z\gamma_3'(E,\eta_g)-a_0'k_s^2\mp(\bar{\lambda}_0)k_s]^{1/2} \tag{3.45}$$

where,

$$z_t\equiv(|e|F_s\gamma_3'(E,\eta_g))^{-1}[\gamma_3(E,\eta_g)-a_0'k_s^2\mp(\bar{\lambda}_0)k_s].$$

Therefore, the 2D electron dispersion law for accumulation layers of HD II–VI semiconductors can be expressed as

$$\gamma_2(E,\eta_g)=a_0'k_s^2\pm(\bar{\lambda}_0)k_s+S_i\left[\frac{\hbar|e|F_s\gamma_3'(E,\eta_g)}{\sqrt{2m_{\|}^*}}\right]^{2/3} \tag{3.46}$$

The area of the 2D surface as enclosed by (3.46) can be expressed as

$$A(E,\eta_g,i)=\frac{\pi}{a_0'^2}\Delta_{10}(E,\eta_g,i) \tag{3.47}$$

where

$$\Delta_{10}(E,\eta_g,i)=\left[(\bar{\lambda}_0)^2-2a_0'\left[-\gamma_3(E,\eta_g)+S_i\left(\frac{\hbar|e|F_s\gamma_3'(E,\eta_g)}{\sqrt{2m_{\|}^*}}\right)^{2/3}\right]\right]$$

The EM in this case assumes the form

$$m^*(E_f',\eta_g,i)=m_\perp^*\Delta_{10}(E_f',\eta_g,i) \tag{3.48}$$

The sub-band energy E_{i2} can be written as

$$\gamma_3(E_{i2},\eta_g)=S_i\left(\frac{\hbar|e|F_s\gamma_3'(E_{i2},\eta_g)}{\sqrt{2m_{\|}^*}}\right)^{2/3} \tag{3.49}$$

The surface electron concentration can be written as

$$n_S = g_v \sum_{i=0}^{i_{\max}} \left[\left[\left(\frac{m_\perp^*}{\pi\hbar^2}\right)[\Delta_{10}(E'_f, i, \eta_g) + \Delta_{11}(E'_f, i, \eta_g)]\right]\right.$$

$$+ \frac{t_i}{2}\left(\frac{k_0 T}{\pi b'_0}\right)^{3/2}\left(\frac{b'_0}{a'_0}\right)$$

$$\left.\times\left[F_{1/2}\left(\frac{E_{FB}}{k_B T}\right) + \frac{(\bar{\lambda}_0)^2}{2a'_0 k_B T} F_{-1/2}\left(\frac{E_{FB}}{k_B T}\right)\right]\right] \quad (3.50)$$

The E_{FB} can be determined from the following equation

$$n_B = \frac{g_v}{2}\left(\frac{k_0 T}{\pi b'_0}\right)^{3/2}\left(\frac{b'_0}{a'_0}\right)\left[F_{1/2}\left(\frac{E_{FB}}{k_B T}\right) + \frac{(\bar{\lambda}_0)^2}{2a'_0 k_B T} F_{-1/2}\left(\frac{E_{FB}}{k_B T}\right)\right] \quad (3.51)$$

The use of (1.132) and (3.2) leads to the expression of the quantization integral in this case as

$$\frac{\sqrt{2m_\parallel^*}}{\hbar}\int_0^{z_t}[E - |e|F_s z - a'_0 k_s^2 \mp (\bar{\lambda}_0)k_s]^{1/2} dz = \frac{2}{3}(S_i)^{3/2} \quad (3.52)$$

where, $z_t \equiv (|e|F_s)^{-1}[E - a'_0 k_s^2 \mp (\bar{\lambda})k_s]$.

Therefore, the 2D electron dispersion law for inversion layers of II–VI semiconductors can be expressed for all values of F_s as

$$E = a'_0 k_s^2 \pm (\bar{\lambda}_0)k_s + S_i\left(\frac{\hbar|e|F_s}{\sqrt{2m_\parallel^*}}\right)^{2/3} \quad (3.53)$$

The area of the 2D surface as enclosed by (3.53) can be expressed as

$$A(E,i) = \frac{\pi(m_\perp^*)^2}{\hbar^4}\left[\left\{2(\bar{\lambda}_0)^2 - \frac{2\hbar^2}{m_\perp^*}S_i\left(\frac{\hbar|e|F_s}{\sqrt{2m_\parallel^*}}\right)^{2/3} + \frac{2\hbar^2 E}{m_\perp^*}\right\}\right.$$

$$\left. - 2(\bar{\lambda}_0)\left[(\bar{\lambda}_0)^2 - \frac{2\hbar^2}{m_\perp^*}S_i\left(\frac{\hbar|e|F_s}{\sqrt{2m_\parallel^*}}\right)^{2/3} + \frac{2\hbar^2 E}{m_\perp^*}\right]^{1/2}\right] \quad (3.54)$$

The EM is given by

$$m^*(E_{Fi}, i) = m_\perp^* \left[1 - \frac{\rho_{71}}{\sqrt{E_{Fi} + \rho_{72}}}\right] \tag{3.55}$$

where, E_{Fi} is the Fermi energy in this case,

$$\rho_{71} \equiv \frac{\bar{\lambda}_0}{2\sqrt{a_0'}} \quad \text{and} \quad \rho_{72} \equiv \left[(\rho_{71})^2 - \left(\frac{\hbar|e|F_s}{\sqrt{2m_\parallel^*}}\right)^{2/3}\right].$$

Thus, the EM depends on both the Fermi energy and the sub-band index due to the presence of the term $\bar{\lambda}_0$.

The sub-band energy $(E_{n_{i6}})$ can be written as

$$E_{n_{i6}} = S_i \left(\frac{\hbar|e|F_s}{\sqrt{2m_\parallel^*}}\right)^{2/3} \tag{3.56a}$$

The total 2D density-of-states function can be written as

$$N_{2D_i}(E) = \frac{m_\perp^* g_v}{\pi\hbar^2} \sum_{i=0}^{i_{\max}} \left\{\left[1 - \frac{\rho_{71}}{\sqrt{E + \rho_{72}}}\right] H(E - E_{n_{i6}})\right\} \tag{3.56b}$$

The surface electron concentration assumes the form

$$n_{2Di} = \frac{g_v m_\perp^* k_B T}{\pi\hbar^2} \left\{\sum_{i=0}^{i_{\max}} \left[F_0(\eta_i) - \left\{\frac{\bar{\lambda}_0 f_7(E_{Fi}, i)}{2\sqrt{a_0'} k_B T}\right\}\right]\right\} \tag{3.56c}$$

where,

$$\eta_i \equiv \left[\frac{E_{Fi} - E_{n_{i6}}}{k_B T}\right],$$

$$f_7(E_{Fi}, i) \equiv \left[2[\sqrt{\eta_i + \delta_{72}} - \sqrt{\delta_{72}}] + \sum_{r=1}^{s} \left\{2(1 - 2^{1-2r})\zeta(2r) \frac{(-1)^{2r-1}(2r-1)!}{(\eta_i + \delta_{72})^{2r}}\right\}\right] \quad \text{and}$$

$$\delta_{72} \equiv \frac{(\bar{\lambda}_0)^2}{4a_0' k_B T}.$$

The electron concentration in the electric field quantum limit can be obtained by

$$\bar{n}_{2Di} = \frac{g_v}{2a_0'}\left[\bar{E}_{F0} + \frac{(\bar{\lambda}_0)^2}{4a_0'} - \bar{E}_0 - \left\{\frac{\bar{\lambda}_0}{2\sqrt{a_0'}}\left[\bar{E}_{F0} + \frac{(\bar{\lambda}_0)^2}{4a_0'} - \bar{E}_0\right]^{1/2}\right\}\right] \tag{3.56d}$$

$\bar{E}_0$ is the sub-band energy in this case and is given by equation $\bar{E}_0 \equiv S_0(\frac{\hbar|e|F_s}{\sqrt{2m_\parallel^*}})^{2/3}$.

3.2.4 *The EM in accumulation and inversion layers of IV–VI semiconductors*

The 2D electron dispersion relation in accumulation layers of IV-VI semiconductors can be written as

$$\theta_1(E,i,\eta_g)k_x^2 + \theta_2(E,i,\eta_g)k_y^2 = \theta_3(E,i,\eta_g) \tag{3.57}$$

where,

$$\theta_1(E,i,\eta_g) = [F_1(E,\eta_g) + S_i(eF_s a_1(E,\eta_g))^{2/3}F_2(E,\eta_g)]$$

$$a_1(E,\eta_g) = \frac{1}{F_2(E,\eta_g)}\left[\frac{F_2'(E,\eta_g)}{F_2(E,\eta_g)}F_1(E,\eta_g) - F_1'(E,\eta_g)\right]$$

$$\theta_2(E,i,\eta_g) = [F_1(E,\eta_g) + \frac{2a_2(E,\eta_g)}{3a_1(E,\eta_g)}(eF_s a_1(E,\eta_g))^{2/3}S_i F_1(E,\eta_g)]$$

$$a_2(E,\eta_g) = \frac{1}{F_2(E,\eta_g)}\left[\frac{F_2'(E,\eta_g)}{F_2(E,\eta_g)}F_1(E,\eta_g) - F_1'(E,\eta_g)\right]$$

$$\theta_3(E,i,\eta_g) = \left[1 + \frac{2C(E,\eta_g)}{3a_1(E,\eta_g)}S_i(eF_s a_1(E,\eta_g))^{2/3}F_2(E,\eta_g)\right] \quad \text{and}$$

$$C(E,\eta_g) = \left[\frac{F_2'(E,\eta_g)}{F_2^2(E,\eta_g)}\right]$$

The EM can be expressed as

$$m^*(E,i,\eta_g) = \text{ Real Part of } \frac{\hbar^2}{2}\theta_4'(E_f',i,\eta_g) \tag{3.58}$$

where

$$\theta_4(E'_f, i, \eta_g) = \frac{\theta_3(E'_f, i, \eta_g)}{\sqrt{\theta_1(E'_f, i, \eta_g)\theta_2(E'_f, i, \eta_g)}}$$

The sub-band energy E_{i3} is given by

$$\theta_3(E_{i3}, i, \eta_g) = 0 \tag{3.59}$$

The 2D electron concentration in accumulation layer of IV–VI materials under the condition of extreme degeneracy and low electric field limit can be written as

$$n_S = g_v \text{ Real part of } \sum_{i=0}^{i_{\max}} \left[\theta_4(E'_f, i, \eta_g) + \frac{t_i}{3\pi^2}\left[F_1(E_{FB}, \eta_g)\sqrt{F_2(E_{FB}, \eta_g)}\right]^{-1}\right] \tag{3.60}$$

where E_{FB} can be determined from the equation

$$n_B = g_v \text{ Real part of } \left[\frac{1}{3\pi^2}\left[F_1(E_{FB}, \eta_g)\sqrt{F_2(E_{FB}, \eta_g)}\right]^{-1}\right] \tag{3.61}$$

The 2D electron dispersion relation of the inversion layers of IV–VI semiconductors in the low electric field limit can be written as

$$k_s^2 = \beta_3(E, i) \tag{3.62}$$

where

$$\beta_3(E, i) = \frac{\beta_1(E, i)}{\beta_2(E, i)},$$

$$\beta_1(E, i) = 1 - \left[\frac{eF_s V'_2(E)}{V_2^2(E)}\right]^{2/3} S_i V_2(E),$$

$$V_2(E) = \left[\frac{2(\bar{A})^2}{E_{g0}(1 + \alpha_1 E)} + \frac{(\bar{S} + \bar{Q})^2}{\Delta''_c(1 + \alpha_3 E)}\right](2E)^{-1},$$

$$\beta_2(E,i) = \left[V_1(E) + \left[\frac{eF_sV_2'(E)}{V_2^2(E)}\right]^{2/3} S_iV_2(E)\frac{2}{3}\frac{V_2^2(E)}{V_2'(E)}\right.$$
$$\left.\left[\frac{V_1(E)V_2'(E)}{V_2^2(E)} - \frac{V_1'(E)}{V_2(E)}\right]\right] \quad \text{and}$$
$$V_1(E) = \left[\frac{(\bar{R})^2}{E_{g0}(1+\alpha_1E)} + \frac{(\bar{S})^2}{\Delta_c'(1+\alpha_2E)} + \frac{(\bar{Q})^2}{\Delta_c''(1+\alpha_3E)}\right](2E)^{-1}.$$

The EM can be expressed as

$$m^*(E_{Fi},i) = \frac{\hbar^2}{2}\beta_3'(E_{Fi},i) \tag{3.63}$$

The sub-band energy (E_{i4}) can be written as

$$0 = \beta_3(E_{i4},i) \tag{3.64a}$$

The surface electron concentration per unit area

$$n_{2D} = \frac{g_v}{2\pi}\sum_{i=0}^{i_{\max}}\left[\beta_3(E_{Fi},i) + \sum_{r=1}^{s}L(r)[\beta_3(E_{Fi},i)]\right]. \tag{3.64b}$$

3.2.5 *The EM in accumulation and inversion layers of stressed Kane type semiconductors*

The 2D electron EM in accumulation layers of stressed III–V semiconductors can be written as

$$\theta_{13}(E,i,\eta_g)k_x^2 + \theta_{23}(E,i,\eta_g)k_y^2 = \theta_{33}(E,i,\eta_g) \tag{3.65}$$

where

$$\theta_{13}(E,i,\eta_g) = [f_1(E,\eta_g) + S_i(eF_sa_{13}(E,\eta_g))^{2/3}f_3(E,\eta_g)]$$
$$a_{13}(E,\eta_g) = \frac{1}{f_3(E,\eta_g)}\left[\frac{f_3'(E,\eta_g)}{f_3(E,\eta_g)}f_1(E,\eta_g) - f_1'(E,\eta_g)\right]$$
$$\theta_{23}(E,i,\eta_g) = \left[f_2(E,\eta_g) + \frac{2a_{23}(E,\eta_g)}{3a_{13}(E,\eta_g)}(eF_sa_{13}(E,\eta_g))^{2/3}S_if_2(E,\eta_g)\right]$$
$$a_{23}(E,\eta_g) = \frac{1}{f_3(E,\eta_g)}\left[\frac{f_3'(E,\eta_g)}{f_3(E,\eta_g)}f_2(E,\eta_g) - f_2'(E,\eta_g)\right]$$

$$\theta_{33}(E,i,\eta_g) = \left[1 + \frac{2C_3(E,\eta_g)}{3a_{13}(E,\eta_g)} S_i(eF_s a_{13}(E,\eta_g))^{2/3} f_3(E,\eta_g)\right] \quad \text{and}$$

$$C_3(E,\eta_g) = \left[\frac{f_3'(E,\eta_g)}{f_3^2(E,\eta_g)}\right].$$

The EM can be expressed as

$$m^*(E_f',i,\eta_g) = \frac{\hbar^2}{2}\theta_{43}'(E_f',i,\eta_g) \tag{3.66}$$

where

$$\theta_{43}(E_f',i,\eta_g) = \frac{\theta_{33}(E_f',i,\eta_g)}{\sqrt{\theta_{13}(E_f',i,\eta_g)\theta_{23}(E_f',i,\eta_g)}}.$$

The sub-band energy E_{i33} is given by

$$\theta_{33}(E_{i33},i,\eta_g) = 0 \tag{3.67}$$

The 2D electron concentration in accumulation layers of stressed III–V materials under the condition of extreme degeneracy and low electric field limit can be written as

$$n_S = g_v \sum_{i=0}^{i_{\max}} \left[\theta_{43}(E_f',i,\eta_g) + \frac{t_i}{3\pi^2}[f_1(E_{FB},\eta_g)f_2(E_{FB},\eta_g)f_3(E_{FB},\eta_g)]^{-1/2}\right] \tag{3.68}$$

The E_{FB} can be determined from the following equation

$$n_B = g_v[f_1(E_{FB},\eta_g)f_2(E_{FB},\eta_g)f_3(E_{FB},\eta_g)]^{-1/2} \tag{3.69}$$

The expression of the EM of the 2D electrons in inversion layers of stressed III–V materials under the low electric field limit as

$$[T_{57}(E,i)]k_x^2 + [T_{67}(E,i)]k_y^2 = T_{77}(E,i) \tag{3.70}$$

where,

$$T_{57}(E,i) = \left[E - \alpha_1 + \frac{2}{3}S_i\left(\frac{|e|^2}{\varepsilon_{sc}}\right)^{2/3}(n_{2Dw})^{2/3}L_{17}(E)\right],$$

$$[\bar{T}_{47}(E)] = \left[\{\rho_5(E))\}' - \left(\frac{\rho(E)}{E-\alpha_3}\right)\right],$$

$$T_{67}(E,i) = \left[E - T_2 + \frac{2}{3}S_i\left(\frac{|e|^2}{\varepsilon_{sc}}\right)^{2/3}(n_{2Dw})^{2/3}L_{27}(E)\right],$$

$$L_{27}(E) = \left[\frac{(E-\alpha_2)}{(E-\alpha_3)^{2/3}[\bar{T}_{47}(E)]^{1/3}} - \left(\frac{(E-\alpha_3)^{1/3}}{[\bar{T}_{47}(E)]^{1/3}}\right)\right],$$

$$T_{77}(E,i) = \left[\rho_5(E) - S_i\left(\frac{|e|^2}{\varepsilon_{sc}}\right)^{2/3}(n_{2Dw})^{2/3}L_{37}(E)\right],$$

$$L_{37}(E) \equiv (E-\alpha_3)^{1/3}[\bar{T}_{47}(E)]^{2/3} \quad \text{and}$$

$$\rho_5(E) \equiv [t_1E^3 - t_2E^2 + t_3E + t_4].$$

The area of the 2D surface under the weak electric field limit can be written as

$$A(E,i) = \frac{\pi T_{77}(E,i)}{\sqrt{T_{57}(E,i)T_{67}(E,i)}}. \tag{3.71}$$

The sub-band energies ($E_{n_{iw8}}$) in this case are defined by

$$T_{47}(E_{n_{iw8}}) = S_i\left(\frac{|e|^2}{\varepsilon_{sc}}\right)^{2/3}(n_{2Dw})^{2/3}L_{37}(E_{n_{iw8}}). \tag{3.72}$$

The expression of the EM in this case can be written as

$$m^*(E_{Fiw},i) = \frac{\hbar^2}{2}L_{47}(E,i)\bigg|_{E=E_{Fiw}} \tag{3.73}$$

where,

$$L_{47}(E,i) \equiv \left[\frac{1}{T_{57}(E,i)T_{67}(E,i)}\right]$$
$$\times\left[\{T_{77}(E,i)\}'[T_{57}(E,i)T_{67}(E,i)]^{1/2} - \left(\frac{T_{77}(E,i)}{2}\right)\right.$$
$$\left\{\{T_{57}(E,i)\}'\left[\frac{T_{67}(E,i)}{T_{57}(E,i)}\right]^{1/2}\right.$$
$$\left.\left.+\{T_{67}(E,i)\}'\left[\frac{T_{57}(E,i)}{T_{67}(E,i)}\right]^{1/2}\right\}\right].$$

The total 2D DOS function can be expressed as

$$N_{2D}(E) = \frac{g_v}{2\pi}\sum_{i=0}^{i_{\max}}\{L_{47}(E,i)H(E - E_{n_{iw8}})\} \tag{3.74a}$$

The surface electron concentration under the weak electric field quantum limit assumes the form

$$n_{2Dw} = \frac{g_v}{(2\pi)}\left\{\sum_{i=0}^{i_{\max}}[P_{8w}(E_{Fwi},i) + Q_{8w}(E_{Fwi},i)]\right\} \tag{3.74b}$$

where,

$$P_{8w}(E_{Fwi},i) \equiv \frac{T_{77}(E_{Fwi},i)}{\sqrt{T_{57}(E_{Fwi},i)T_{67}(E_{Fwi},i)}}$$

and

$$Q_{8w}(E_{Fiw},i) \equiv \sum_{r=1}^{s} L(r)P_{8w}(E_{Fiw},i).$$

Using (3.74b), the expression of the 2D surface electron concentration under the weak electric field quantum limit can be written as

$$\bar{n}_{2Dw} = \frac{g_v}{2\pi}\frac{T_{77}(\bar{E}_{Fw},0)}{\sqrt{T_{57}(\bar{E}_{Fw},0)T_{67}(\bar{E}_{Fw},0)}}. \tag{3.74c}$$

3.2.6 *The EM in accumulation and inversion layers of germanium*

The 2D EM in accumulation layers of Ge can be written as

$$\frac{\hbar^2 k_x^2}{2m_1^*} + \frac{\hbar^2 k_y^2}{2m_2^*} = \gamma_{10}(E, i, \eta_g) \tag{3.75}$$

where

$$\gamma_{10}(E, i, \eta_g) = \left[\gamma_3(E, \eta_g)[1 + \alpha\gamma_3(E, \eta_g)] - S_i\left[\frac{\hbar^2 eFs\gamma_3'(E, \eta_g)}{\sqrt{2m_3^*}}\right]^{2/3}\right.$$
$$\left.\times [1 + 2\alpha\gamma_3(E, \eta_g)] + \alpha\left[S_i\left[\frac{\hbar eFs\gamma_3'(E, \eta_g)}{\sqrt{2m_3^*}}\right]^{2/3}\right]^2\right]$$

The EM can be expressed as

$$m^*(E_f', i, \eta_g) = \sqrt{m_1^* m_2^*}[\gamma_{10}'(E, i, \eta_g)] \tag{3.76}$$

The band non-parabolicity and heavy doping makes the mass quantum number dependent.

The sub-band energy E_{i14} can be written as

$$\gamma_{10}(E_{i14}, i, \eta_g) = 0. \tag{3.77}$$

The surface electron concentration in accumulation layers can be written as

$$n_s = 2g_v \sum_{i=0}^{i_{\max}} \left[\left[\frac{\sqrt{m_1^* m_2^*}}{\pi\hbar^2}[\gamma_{10}(E_f', i, \eta_g)]\right.\right.$$
$$\left.+ t_i \frac{8\pi m_\perp^* \sqrt{2m_\parallel^*}}{\hbar^3}[2\gamma_3(E_{FB}, \eta_g)]^{3/2}\left[1 + \frac{4\alpha}{5}\gamma_3(E_{FB}, \eta_g)\right]\right] \tag{3.78}$$

where E_{FB} can be determined from the following equation

$$n_B = g_v\left[\frac{8\pi m_\perp^* \sqrt{2m_\parallel^*}}{\hbar^3}[2\gamma_3(E_{FB}, \eta_g)]^{3/2}[1 + \frac{4\alpha}{5}\gamma_3(E_{FB}, \eta_g)]\right]. \tag{3.79}$$

The 2D electron dispersion law in inversion layers of Ge at low electric field limit can be expressed as

$$\frac{\hbar^2 k_x^2}{2m_1^*} + \frac{\hbar^2 k_y^2}{2m_2^*} = [E(1+\alpha E) + \alpha E_{i20}^2 - E_{i20}(1+2\alpha E)] \quad (3.80)$$

where,

$$E_{i20} = S_i \left(\frac{\hbar e F_s}{\sqrt{2m_3}}\right)^{2/3}.$$

The area of 2D space is

$$A = \frac{2\pi\sqrt{m_1 m_2}}{\hbar^2}[E(1+\alpha E) + \alpha E_{i20}^2 - E_{i20}(1+2\alpha E)] \quad (3.81)$$

The EM assumes the form

$$m^*(E_{Fiw}, i) = \sqrt{m_1 m_2}[1 + 2\alpha E_{Fiw} - E_{i20} 2\alpha] \quad (3.82)$$

Thus the EM is the function of both Fermi energy and quantum number due to band non-parabolicity.

The DOS function is given by

$$N_{2D}(E) = \frac{2g_v}{(2\pi)^2} \cdot \frac{2\pi\sqrt{m_1 m_2}}{\hbar^2} \sum_{i=0}^{i_{\max}} [1 + 2\alpha E - 2\alpha E_{i20}] H(E - E_{i20}) \quad (3.83a)$$

The surface electron concentration per unit area is given by

$$n_{2D} = \frac{2g_v}{(2\pi)^2} \cdot \frac{2\pi\sqrt{m_1 m_2}}{\hbar^2} k_B T \sum_{i=0}^{i_{\max}} [F_0(\eta_x) + 2\alpha k_B T F_1(\eta_x)] \quad (3.83a')$$

where

$$\eta_x = \frac{E_{Fiw} - E_{i20}}{k_B T}.$$

3.3 Results and Discussion

The effect of surface electric field on the EM at the quantum limit in accumulation layers of Cd_3As_2 and $CdGeAs_2$ has been exhibited in Figs. 3.1–3.4. In Figs. 3.1 and 3.2, we have demonstrated the variation

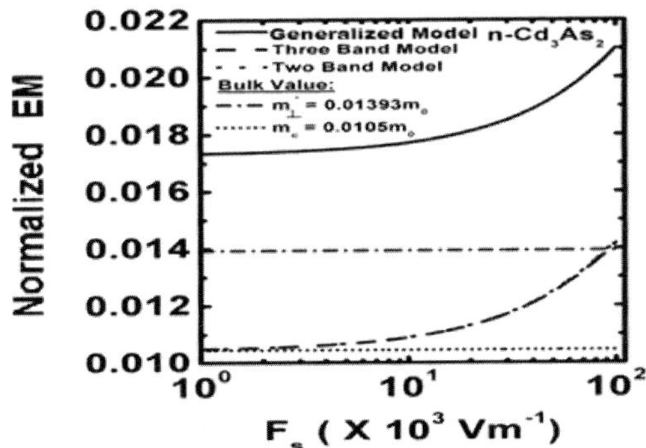

Fig. 3.1. Plot of the normalized EM under weak electric field limit as the function of surface electric field for accumulation layers of Cd_3As_2 in accordance with the generalized theory. The simplified results for three and two band models of Kane have also been exhibited in which, $m_{\perp}^{*} = 0.0139m_0$ and $m_c = \frac{1}{2}(m_{\perp}^{*} + m_{\parallel}^{*}) = 0.0105m_0$ are the corresponding bulk values.

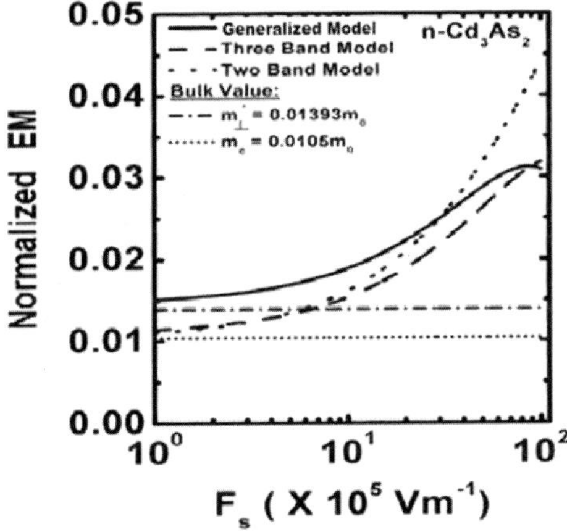

Fig. 3.2. Plot of the normalized EM at strong electric field limit as function of surface electric field for accumulation layers of Cd_3As_2 for all cases of Fig. 3.1.

of the EM with electric field in the weak electric field regime which was extended up to $10^5 \mathrm{Vm}^{-1}$. It appears that with the increase in the electric field, the EM in Fig. 3.1 increases considering the generalized energy band model, the three and two band models of Kane which are the special cases of our generalized analysis. It appears that in the weak field regime, the deviation between the three and two

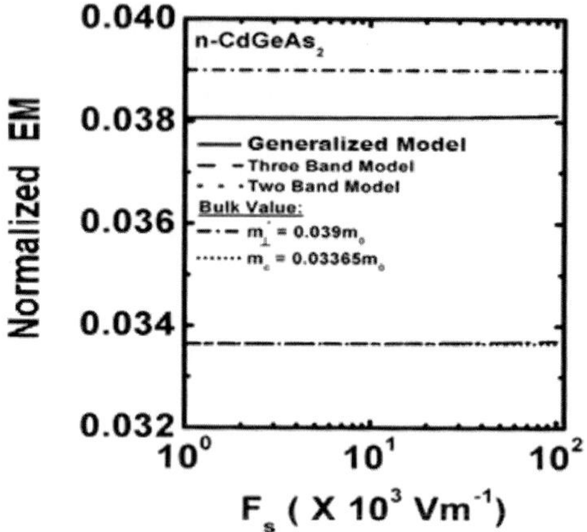

Fig. 3.3. Plot of the normalized EM at weak electric field limit as function of surface electric field for accumulation layers of $CdGeAs_2$ in accordance with the generalized theory. The simplified results for three and two band models of Kane have also been exhibited in which, $m_\perp^* = 0.039m_0$ and $m_c = \frac{1}{2}(m_\perp^* + m_\parallel^*) = 0.03365m_0$ are the corresponding bulk values.

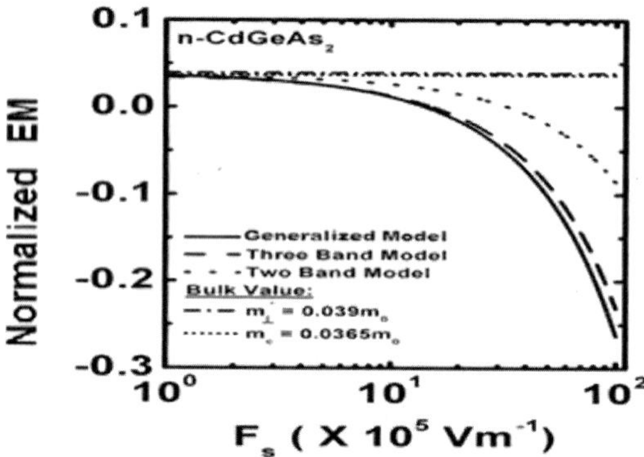

Fig. 3.4. Plot of the normalized EM at strong electric filed limit as function of surface electric field for accumulation layers of $CdGeAs_2$ for all cases of Fig. 3.3.

band models of Kane is less however significant difference is with the consideration of the crystal field. It should be noted that it is these two models which tends to the isotropic bulk effective mass value $0.0105m_0$, rather than the generalized model.

In the high field regime, Fig. 3.2, the difference in the three and two band models of Kane appears which marks a significant variation in the value of the EM. The effect of crystal field splitting tends to decrease the EM considering the generalized energy band model.

A closer look at the two figures reveal more interesting features of the continuity of the weak accumulation energy band model in the strong field and strong accumulation energy band model in the weak field. It is due to this non-convergence there is a slight mismatch of the EM at the boundary of 10^5 Vm^{-1} in both the Figs. 3.1 and 3.2 and needs more attention towards the development of the generalized theory valid for all values of electric field is still a formidable problem in this case. Figures 3.3 and 3.4 exhibit the EM in n-CdGeAs$_2$ for all the cases of Figs. 3.1 and 3.2 respectively. It appears that in the weak electric field regime the EM is almost invariant of the electric field; however with the increase in the field, the EM sharply decreases and tends to take negative values which challenge the applicability of the quantization condition (3.2) at strong electric field for accumulation layers of CdGeAs$_2$. The effect of electric field on the EM of accumulation layers of InAs has been exhibited in Figs. 3.5 and 3.6. Almost no variation of the EM in weak field appears for accumulation layers of n-InAs while for higher fields, the EM tends to decrease. The effect of surface electric field on accumulation layers of GaAs and InSb at weak and strong electric field has been exhibited in Fig. 3.7. Same trend as InAs in weak

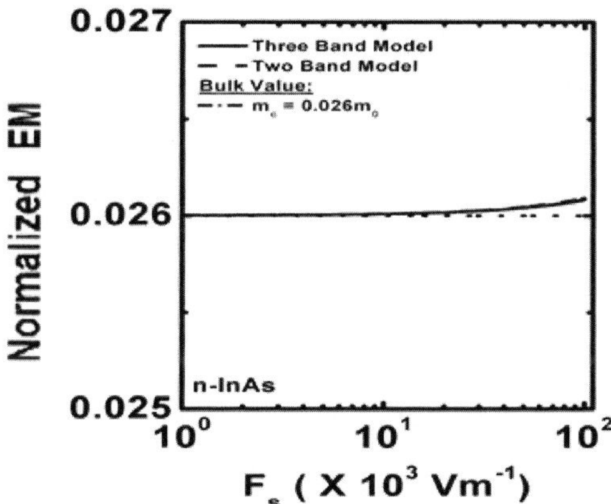

Fig. 3.5. Plot of the normalized EM at low electric field limit as function of surface electric field for accumulation layers of n-InAs.

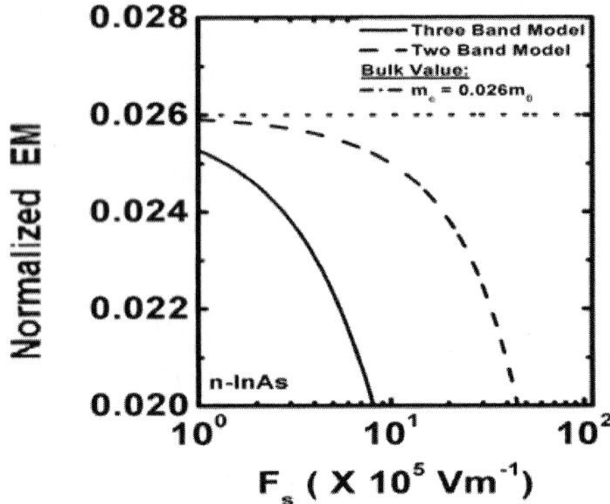

Fig. 3.6. Plot of the normalized EM at high electric field limit as function of surface electric field for accumulation layers of n-InAs.

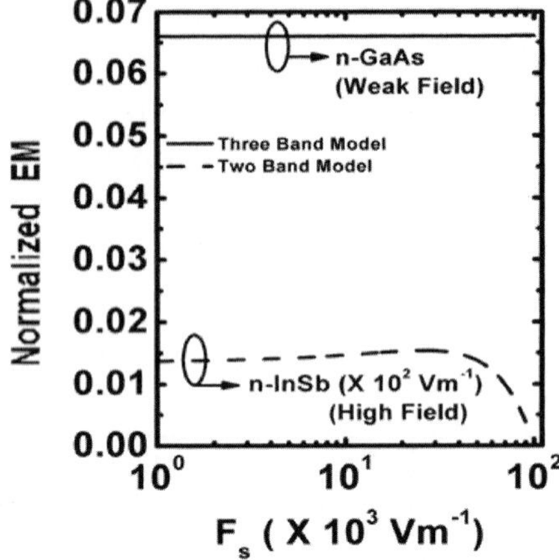

Fig. 3.7. Plot of the normalized EM at low and high electric field limits as function of surface electric field for accumulation layers of GaAs and InSb respectively.

field again follows for GaAs, where the difference in the energy band model in determining the EM is vanishing small. With the increase in the electric field at high value, the EM in accumulation layers of InSb tends to fall down. This is not with the case of CdS in Fig. 3.8 where the effect of increasing the electric field increases the EM monotonically presenting a significant change.

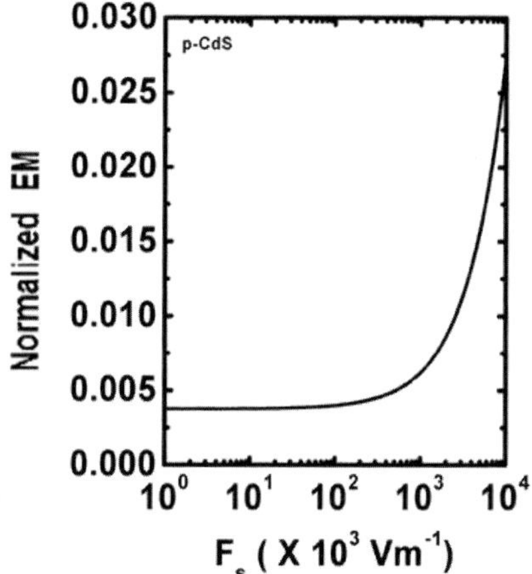

Fig. 3.8. Plot of the normalized EM as function of surface electric field for accumulation layers of CdS.

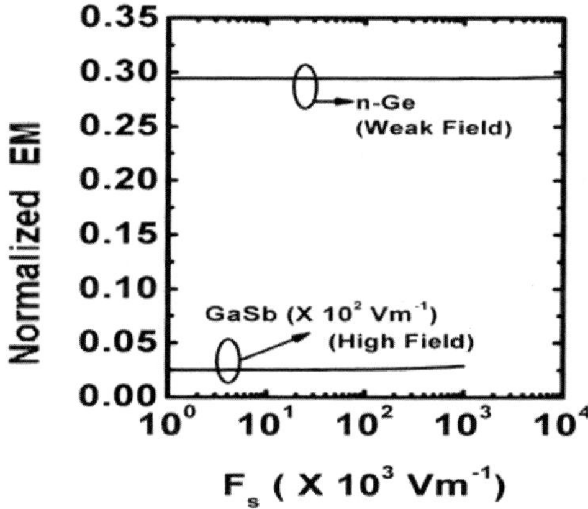

Fig. 3.9. Plot of the normalized EM at weak and strong electric field limits as function of surface electric field for accumulation layers of Ge and GaSb respectively.

Finally, in Fig. 3.9, we present the variation of the EM in accumulation layers of Ge following Cardona model under weak inversion regime and accumulation layers of GaSb under strong inversion regime. It appears that the EM in case of GaSb increases within the regime of 10^8 Vm^{-1}, while the EM in Ge stays almost constant.

3.4 Open Research Problems

(R.3.1) Investigate the EM in the presence of an arbitrarily oriented electric quantization for accumulation layers of tetragonal semiconductors. Study all the special cases for III–V, ternary and quaternary materials in this context.

(R.3.2) Investigate the EM in accumulation layers of IV–VI, II–VI and stressed Kane type compounds in the presence of an arbitrarily oriented quantizing electric field.

(R.3.3) Investigate the EM in accumulation layers of all the materials as stated in R.1.1 of chapter 1 in the presence of an arbitrarily oriented quantizing electric field.

(R.3.4) Investigate the EM in the presence of an arbitrarily oriented non-quantizing magnetic field in accumulation layers of tetragonal semiconductors by including the electron spin. Study all the special cases for III–V, ternary and quaternary materials in this context.

(R.3.5) Investigate the EM in accumulation layers of IV–VI, II–VI and stressed Kane type compounds in the presence of an arbitrarily oriented non-quantizing magnetic field by including the electron spin.

(R.3.6) Investigate the EM in accumulation layers of all the materials as stated in R.1.1 of chapter 1 in the presence of an arbitrarily oriented non-quantizing magnetic field by including electron spin.

(R.3.7) Investigate the EM in accumulation layers for all the problems from R.3.1 to R.3.6 in the presence of an additional arbitrarily oriented electric field.

(R.3.8) Investigate the EM in accumulation layers for all the problems from R.3.1 to R.3.3 in the presence of arbitrarily oriented crossed electric and magnetic fields.

(R.3.9) Investigate the EM in accumulation layers for all the problems from R.3.1 to R.3.8 in the presence of surface states.

(R.3.10) Investigate the EM in accumulation layers for all the problems from R.3.1 to R.3.8 in the presence of hot electron effects.

(R.3.11) Investigate the EM in accumulation layers for all the problems from R.3.1 to R.3.6 by including the occupancy of the electrons in various electric subbands.

(R.3.12) Investigate the problems from R.3.1 to R.3.11 for the appropriate p-channel accumulation layers.

References

1. T. Ando, H. Fowler and F. Stern, *Rev. Mod. Phys.*, **54**, 437 (1982)
2. J.J. Quinn and P.J. Styles, (ed.) *Electronic Properties of Quasi Two Dimensional Systems, North Holland, Amsterdam,* (1976)
3. G.A. Antcliffe, R.T. Bate and R.A. *Reynolds, Proceedings of the International Conference, Physics of Semi-metals and Narrow-Gap semiconductors* (ed.) D. L. Carter and R. T. Bate, Pergamon Press, *Oxford*, (499) (1971)
4. Z. A. Weinberg, *Sol. Stat. Electron.*, **20**, 11 (1977)
5. G. Paasch, T. Fiedler, M. Kolar and I. Bartos, *Phys. Stat. Sol.* (b) **118**, 641 (1983)
6. S. Lamari, *Phys. Rev.* B, **64**, 245340 (2001)
7. T. Matsuyama, R. Kürsten, C. *Meißner, and U. Merkt, Phys. Rev.* B, **61**, 15588 (2000)
8. P.V. Santos and M. Cardona, *Phys. Rev. Lett.*, **72**, 432 (1994)
9. L. Bu, Y. Zhang, B.A. Mason, R.E. Doezema, and J.A. Slinkman *Phys. Rev.* B, **45**, 11336 (1992)
10. P.D. Dresselhaus, C.M. Papavassiliou, R.G. Wheeler, and R.N. Sacks, *Phys. Rev. Lett.*, **68**, 106 (1992)
11. U. Kunze, *Phys. Rev. B,* **41**, 1707 (1990)
12. E. Yamaguchi, *Phys. Rev. B*, **32**, 5280 (1985)
13. Th. Lindner and G. Paasch, *J. Appl. Phys.*, **102**, 054514 (2007)
14. S. Lamari, *J. Appl. Phys.* **91**, 1698 (2002)
15. K.P. Ghatak and M. Mondal, *J. Appl. Phys.*, **70**, 299 (1991).
16. K.P. Ghatak and S.N. Biswas, *J. Vac. Sc. and Tech.*, **7B**, 104 (1989)
17. B. Mitra and K.P. Ghatak, i., **32**, 177 (1989)
18. K.P. Ghatak and M. Mondal, *J. Appl. Phys.*, **62**, 922 (1987)
19. M. Mondal and K.P. Ghatak, *J. Magnet. Magnetic Mat.*, **62**, 115 (1986); M. Mondal and K.P. Ghatak**,** *Phys. Script.*, **31**, 613 (1985)
20. K.P. Ghatak and M. Mondal, *Z. fur Physik B,* **64**, 223 (1986); K.P. Ghatak and S.N. Biswas, *Sol. State Electron.*, **37**, 1437 (1994). D.R. Choudhury, A.K. Chowdhury, K.P. Ghatak, A.N. Chakravarti, *Phys. Stat.* Sol. (b) **98**, (K141) (1980); A.N. Chakravarti, A.K. Chowdhury, K.P. Ghatak, *Phys. Stat. Sol.* (a) **63**, K97 (1981); M. Mondal, K.P. Ghatak, *Acta Phys. Polon.* A **67**, 983 (1985); M. Mondal, K.P. Ghatak, *Phys. Stat. Sol.* (b) **128**, K21 (1985); M. Mondal, K.P. Ghatak, *Phys. Stat. Sol.* (a) **93**, 377 (1986); K.P. Ghatak, M. Mondal, *Phys. Stat. Sol.* (b) **135**, 819 (1986); M. Mondal, K.P. Ghatak,

Phys. Stat. Sol. (b) **139**, 185 (1987); K.P. Ghatak, N. Chattopadhyay, S.N. Biswas, OE/Fibers' **87**, 203 (1987); K.P. Ghatak, N. Chatterjee, M. Mondal, *Phys. Stat. Sol.* (b) **139**, K25 (1987); K.P. Ghatak, M. Mondal, *Phys. Stat. Sol.* **(b) 148**, 645 (1988); K.P. Ghatak, A. Ghosal, *Phys. Stat. Sol.* (b) **151**, K135 (1989); K.P. Ghatak, N. Chattopadhyay, M. Mondal, *Appl. Phys.* **A 48**, 365 (1989)

Chapter 4

The EM in Nano-Wires (NWs) of Heavily Doped (HD) Non-Parabolic Materials

4.1 Introduction

It is well-known that in quantum wires (QWs), the restriction of the motion of the carriers along two directions may be viewed as carrier confinement by two infinitely deep 1D rectangular potential wells, along any two orthogonal directions leading to quantization of the wave vectors along the said directions, allowing 1D carrier transport [1]. With the help of modern experimental techniques, such one dimensional quantized structures have been experimentally realized and enjoy an enormous range of important applications in the realm of nano-science in the quantum regime. They have generated much interest in the analysis of nano-structured devices for investigating their electronic, optical and allied properties in general [2–4]. Examples of such new applications are based on the different transport properties of ballistic charge carriers which include quantum resistors [5, 9, 10], resonant tunneling diodes and band filters [11, 12], quantum switches [13], quantum sensors [14–16], quantum logic gates [17–18], quantum transistors and sub tuners [19–21], heterojunction FETs [22], high-speed digital networks [23], high-frequency microwave circuits [24], optical modulators [25], optical switching systems [26, 27], and other devices.

In this chapter in Sections 4.2.1–4.2.14, we have investigated the EMs in NWs of HD non-linear optical, III-V, II-VI, stressed Kane type, Te, GaP, $PtSb_2$, Bi_2Te_3, Ge, GaAs, II-V, lead germanium telluride and zinc and cadmium phosphides respectively. Section 4.3 contains the summary and conclusion pertaining to this chapter. Section 4.4 presents 19 open research problems.

4.2 Theoretical Background

4.2.1 *The EM in NWs of HD nonlinear optical materials*

The DR of the 1D electrons in this case can be written following (1.32) as

$$\frac{\hbar^2(n_z\pi/d_z)^2}{2m_\parallel^* T_{21}(E,\eta_g)} + \frac{\hbar^2(n_y\pi/d_y)^2}{2m_\parallel^* T_{22}(E,\eta_g)} + \frac{\hbar^2 k_x^2}{2m_\parallel^* T_{21}(E,\eta_g)} = 1 \tag{4.1}$$

where, $n_z(=1,2,3,\ldots), d_z$ are the size quantum number and the nano-thickness along the z-direction respectively, $n_y(=1,2,3,\ldots)$ and d_y are the size quantum number and the nano-thickness along the y-direction respectively

The 1D DOS function per sub-band is given by

$$N_{1D}(E) = \frac{2g_v}{\pi}\frac{\partial k_x}{\partial E}. \tag{4.2}$$

Thus by using (4.1) and (4.2) the total DOS function in this case can be written as

$$N_{1DHD\Gamma}(E,\eta_g) = \frac{2g_v}{\pi}\sum_{n_y=1}^{n_{y\max}}\sum_{n_z=1}^{n_{z\max}} T_1'(E,n_y,n_z,\eta_g)H(E-E'_{1HDNW}) \tag{4.3}$$

where, E'_{1HDNW} is the complex sub-band energy which can be expressed in this case as

$$\frac{\hbar^2(n_z\pi/d_z)^2}{2m_\parallel^* T_{21}(E'_{1HDNW},\eta_g)} + \frac{\hbar^2(n_y\pi/d_y)^2}{2m_\parallel^* T_{22}(E'_{1HDNW},\eta_g)} = 1 \tag{4.4}$$

and

$$T_{1HDNW}(E,n_y,n_z,\eta_g)$$

$$= \left[\left[1 - \frac{\hbar^2(n_z\pi/d_z)^2}{2m_\parallel^* T_{21}(E,\eta_g)} + \frac{\hbar^2(n_y\pi/d_y)^2}{2m_\parallel^* T_{22}(E,\eta_g)}\right]\frac{2m_\parallel^* T_{21}(E,\eta_g)}{\hbar^2}\right]^{1/2}.$$

The EM in this case in given by

$$m^*(E_{F1HDNW}, n_y, n_z, \eta_g) = \frac{\hbar^2}{2}\left[\text{Real part of } \frac{\partial}{\partial(E_{F1HDNW})}[T_{1HDNW}(E, n_y, n_z, \eta_g)]^2\right] \tag{4.5a}$$

where E_{F1HDNW} in the Fermi energy in this case

Thus, we observe that the EM is the function of size quantum numbers in both the directions and the Fermi energy due to the combined influence of the crystal filed splitting constant and the anisotropic spin-orbit splitting constants respectively. Besides it is a function of η_g due to which the EM exists in the band gap, which is otherwise impossible.

The electron concentration per unit length in this case can be written as

$$n_{01D} = \frac{2g_v}{\pi} \text{ Real part of } \sum_{n_y=1}^{n_{y\max}} \sum_{n_z=1}^{n_{z\max}} \left[T_1(E_{F1HDNW}, n_y, n_z, \eta_g) + \sum_{r=1}^{s} L(r)[T_1(E_{F1HDNW}, n_y, n_z, \eta_g)]\right] \tag{4.5b}$$

In the absence of band-tails, for electron motion along x-direction only, the 1D electron dispersion law in this case can be written following (2.2) as

$$\gamma(E) = f_1(E)k_x^2 + f_1(E)(\pi n_y/d_y)^2 + f_2(E)(\pi n_z/d_z)^2 \tag{4.6}$$

Using (4.2) and (4.6) the total DOS function in this case can be expressed as.

$$N_{1D\Gamma}(E) = \frac{2g_v}{\pi} \sum_{n_y=1}^{n_{y\max}} \sum_{n_z=1}^{n_{z\max}} f'_{10}(E, n_y, n_z) H(E - E'_1) \tag{4.7}$$

where

$$f_{10}(E,n_y,n_z) = [f_1(E)]^{-1/2}\left[\gamma(E) - f_1(E)\left(\frac{n_y\pi}{d_y}\right)^2 - f_2(E)\left(\frac{n_z\pi}{d_z}\right)^2\right]^{1/2}$$

In (4.7), the sub-band energy (E'_1) are given by the equation

$$\gamma(E'_1) = f_1(E'_1)(\pi n_y/d_y)^2 + f_2(E'_1)(\pi n_z/d_z)^2 \tag{4.8a}$$

The EM in this case assumes the form

$$m^*(E_{F1D},n_y,n_z) = \frac{\hbar^2}{2}f'_{10}(E_{F1D},n_y,n_z) \tag{4.8b}$$

where E_{F1D} is the Fermi energy in this case.

The electron concentration per unit length can be written as

$$n_{01D} = \frac{2g_v}{\pi}\sum_{n_y=1}^{n_{y\max}}\sum_{n_z=1}^{n_{z\max}} \times \left[f_{10}(E_{F1D},n_y,n_z) + \sum_{r=1}^{s} L(r)[f_{10}(E_{F1D},n_y,n_z)]\right]. \tag{4.8c}$$

4.2.2 *The EM in NWs of HD III-V materials*

(i) Three band model of Kane

The dispersion relation of the 1D electrons in this case can be written following (1.48) as

$$\frac{\hbar^2(n_z\pi/d_z)^2}{2m_c} + \frac{\hbar^2(n_y\pi/d_y)^2}{2m_c} + \frac{\hbar^2 k_x^2}{2m_c} = T_{31}(E,\eta_g) + iT_{32}(E,\eta_g). \tag{4.9}$$

The DOS function in this case assumes the form

$$N_{1DHD\Gamma}(E,\eta_g) = \frac{2g_v}{\pi}\text{ Real part of }\sum_{n_y=1}^{n_{y\max}}\sum_{n_z=1}^{n_{z\max}} \times T'_2(E,n_y,n_z,\eta_g)H(E - E'_{2HDNW}) \tag{4.10}$$

where

$$T_2(E,n_y,n_z,\eta_g) = \left[\left[T_{31}(E,\eta_g) + iT_{32}(E,\eta_g) - \left[\frac{\hbar^2(n_z\pi/d_z)^2}{2m_c} + \frac{\hbar^2(n_y\pi/d_y)^2}{2m_c}\right]\right]\frac{2m_c}{\hbar^2}\right]^{1/2}.$$

In (4.10), E'_{2HDNW} is the sub-band energy in this case which can be expressed as

$$\frac{\hbar^2(n_z\pi/d_z)^2}{2m_c} + \frac{\hbar^2(n_y\pi/d_y)^2}{2m_c} = T_{31}(E'_{2HDNW},\eta_g) + iT_{32}(E'_{2HDNW},\eta_g). \tag{4.11}$$

The EM in this case is given by

$$m^*(E'_{2HDNW},\eta_g) = m_c[T'_{31}(E'_{2HDNW},\eta_g)]. \tag{4.12a}$$

The electron concentration per unit length can be written as

$$n_{01D} = \frac{2g_v}{\pi}\text{ Real part of }\sum_{n_y=1}^{n_{y\max}}\sum_{n_z=1}^{n_{z\max}}\left[T_2(E'_{2HDNW},n_y,n_z,\eta_g) + \sum_{r=1}^{s} L(r)[T_2(E'_{2HDNW},n_y,n_z,\eta_g)]\right]. \tag{4.12b}$$

In the absence of band tails the DR in this case assumes the form

$$I_{11}(E) = \frac{\hbar^2 k_x^2}{2m_c} + G_2(n_y,n_z) \tag{4.13}$$

where,

$$G_2(n_y,n_z) = \frac{\hbar^2\pi^2}{2m_c}\left[\left(\frac{n_y\pi}{d_y}\right)^2 + \left(\frac{n_z\pi}{d_z}\right)^2\right].$$

The DOS function can be expressed as

$$N_{1D\Gamma}(E) = \frac{2g_v}{\pi}\sum_{n_y=1}^{n_{y\max}}\sum_{n_z=1}^{n_{z\max}} f'_{11}(E,n_y,n_z)H(E-E'_2) \tag{4.14}$$

where,

$$f_{11}(E,n_y,n_z) = \left[\frac{2m_c}{\hbar^2}\left[I_{11}(E) - G_2(n_y,n_z)\right]\right]^{1/2}$$

In (4.14), the sub-band energy E'_2 can be written as

$$G_2(n_y, n_z) = I_{11}(E'_2). \tag{4.15}$$

The EM in this case assumes the form

$$m^*(E_{F1D}) = m_c I'_{11}(E_{F1D}). \tag{4.16a}$$

The electron concentration per unit length can be written as

$$n_{01D} = \frac{2g_v}{\pi} \sum_{n_y=1}^{n_{y\max}} \sum_{n_z=1}^{n_{z\max}} \times \left[f_{11}(E_{F1D}, n_y, n_z) + \sum_{r=1}^{s} L(r)[f_{11}(E_{F1D}, n_y, n_z)] \right]. \tag{4.16b}$$

(ii) Two band model of Kane

The DR of the 1D electrons in this case can be written as

$$\frac{\hbar^2(n_z\pi/d_z)^2}{2m_c} + \frac{\hbar^2(n_y\pi/d_y)^2}{2m_c} + \frac{\hbar^2 k_x^2}{2m_c} = \gamma_2(E, \eta_g). \tag{4.17}$$

The DOS function in this case assumes the form

$$N_{1DHD\Gamma}(E, \eta_g) = \frac{2g_v}{\pi} \sum_{n_y=1}^{n_{y\max}} \sum_{n_z=1}^{n_{z\max}} T'_3(E, n_y, n_z, \eta_g) H(E - E'_{3HDNW}) \tag{4.18}$$

where,

$$T_3(E, n_y, n_z, \eta_g) = \left[\left[\gamma_2(E, \eta_g) - \left[\frac{\hbar^2(n_z\pi/d_z)^2}{2m_c} + \frac{\hbar^2(n_y\pi/d_y)^2}{2m_c} \right] \right] \frac{2m_c}{\hbar^2} \right]^{1/2}.$$

In (4.18), E'_{3HDNW} is the sub-band energy in this case which can be expressed as

$$\frac{\hbar^2(n_z\pi/d_z)^2}{2m_c} + \frac{\hbar^2(n_y\pi/d_y)^2}{2m_c} = \gamma_2(E'_{3HDNW}, \eta_g). \tag{4.19}$$

The EM in this case is given by

$$m^*(E_{F1HDNW}, \eta_g) = m_c[\gamma_2(E_{F1HDNW}, \eta_g)]. \tag{4.20a}$$

The electron concentration per unit length can be written as

$$n_{01D} = \frac{2g_v}{\pi} \sum_{n_y=1}^{n_{y\max}} \sum_{n_z=1}^{n_{z\max}} \left[T_3(E_{F1HDNW}, n_y, n_z, \eta_g) + \sum_{r=1}^{s} L(r)[T_3(E_{F1HDNW}, n_y, n_z, \eta_g)] \right]. \quad (4.20b)$$

The expression of 1D DR, for NWs of III-V materials whose energy band structures are defined by the two-band model of Kane in the absence of band tailing assumes the form

$$E(1 + \alpha E) = \frac{\hbar^2 k_x^2}{2m_c} + G_2(n_y, n_z). \quad (4.21)$$

The DOS function can be expressed as

$$N_{1D\Gamma}(E) = \frac{2g_v}{\pi} \sum_{n_y=1}^{n_{y\max}} \sum_{n_z=1}^{n_{z\max}} f'_{12}(E, n_y, n_z) H(E - E'_3) \quad (4.22)$$

where,

$$f_{12}(E, n_y, n_z) = \left[\frac{2m_c}{\hbar^2} [E(1 + \alpha E) - G_2(n_y, n_z)] \right]^{1/2}.$$

In (4.22), the sub-band energy E'_3 can be written as

$$E'_3 = (2\alpha)^{-1}[-1 + \sqrt{1 + 4\alpha G_2(n_y n_z)}] \quad (4.23)$$

The EM in this case assumes the form

$$m^*(E_{F1D}) = m_c(1 + 2\alpha E_{F1D}) \quad (4.24a)$$

The electron concentration per unit length can be expressed as

$$n_{01D} = \frac{2g_v}{\pi} \sum_{n_y=1}^{n_{y\max}} \sum_{n_z=1}^{n_{z\max}} \left[f_{12}(E_{F1D}, n_y, n_z) + \sum_{r=1}^{s} L(r)[f_{12}(E_{F1D}, n_y, n_z)] \right] \quad (4.24b)$$

(iii) Parabolic energy bands

The DR of the 1D electrons in this case can be written as

$$\frac{\hbar^2(n_z\pi/d_z)^2}{2m_c}+\frac{\hbar^2(n_y\pi/d_y)^2}{2m_c}+\frac{\hbar^2k_x^2}{2m_c}=\gamma_3(E,\eta_g). \tag{4.25}$$

The DOS function in this case assumes the form

$$N_{1DHD\Gamma}(E,\eta_g)=\frac{2g_v}{\pi}\sum_{n_y=1}^{n_{y\max}}\sum_{n_z=1}^{n_{z\max}}T_4'(E,n_y,n_z,\eta_g)H(E-E'_{4HDNW}) \tag{4.26}$$

where

$$T_4(E,n_y,n_z,\eta_g)$$
$$=\left[\left[\gamma_3(E,\eta_g)-\left[\frac{\hbar^2(n_z\pi/d_z)^2}{2m_c}+\frac{\hbar^2(n_y\pi/d_y)^2}{2m_c}\right]\right]\frac{2m_c}{\hbar^2}\right]^{1/2}.$$

In (4.26), E'_{3HDNW} is the sub-band energy in this case which can be expressed as

$$\frac{\hbar^2(n_z\pi/d_z)^2}{2m_c}+\frac{\hbar^2(n_y\pi/d_y)^2}{2m_c}=\gamma_3(E'_{4HDNW},\eta_g). \tag{4.27}$$

The EM in this case is given by

$$m^*(E_{F1HDNW},\eta_g)=m_c[\gamma_3'(E_{F1HDNW},\eta_g)] \tag{4.28a}$$

The electron concentration per unit length can be written as

$$n_{01D}=\frac{2g_v}{\pi}\sum_{n_y=1}^{n_{y\max}}\sum_{n_z=1}^{n_{z\max}}\Bigg[T_4(E_{F1HDNW},n_y,n_z,\eta_g)$$
$$+\sum_{r=1}^{s}L(r)[T_4(E_{F1HDNW},n_y,n_z,\eta_g)]\Bigg] \tag{4.28b}$$

The expression of 1D DR, for NWs of III-V materials whose energy band structures are defined by the parabolic energy bandsin the absence of band tailing assumes the form

$$E = \frac{\hbar^2 k_x^2}{2m_c} + G_2(n_y, n_z) \tag{4.29}$$

The DOS function can be expressed as

$$N_{1D\Gamma}(E) = \frac{2g_v}{\pi} \sum_{n_y=1}^{n_{y\max}} \sum_{n_z=1}^{n_{z\max}} f'_{13}(E, n_y, n_z) H(E - E'_4) \tag{4.30}$$

where

$$f_{13}(E, n_y, n_z) = \left[\frac{2m_c}{\hbar^2}\left[E - G_2(n_y, n_z)\right]\right]^{1/2}$$

In (4.30), the sub-band energy E'_4 can be written as

$$E'_4 = G_2(n_y, n_z) \tag{4.31}$$

The EM in this case assumes the form

$$m^*(E_{F1D}) = m_c \tag{4.32a}$$

The electron concentration per unit length can be written as

$$n_{01D} = \frac{2g_v}{\pi} \sum_{n_y=1}^{n_{y\max}} \sum_{n_z=1}^{n_{z\max}} \left[f_{13}(E_{F1D}, n_y, n_z) + \sum_{r=1}^{s} L(r)[f_{13}(E_{F1D}, n_y, n_z)] \right] \tag{4.32b}$$

(iv) The Model of Stillman *et al.*

The DR of the 1D electrons in this case can be written as

$$\frac{\hbar^2 (n_z \pi / d_z)^2}{2m_c} + \frac{\hbar^2 (n_y \pi / d_y)^2}{2m_c} + \frac{\hbar^2 k_x^2}{2m_c} = \theta_4(E, \eta_g) \tag{4.33}$$

where

$$\theta_4(E, \eta_g) = I_{12}(E, \eta_g).$$

The DOS function in this case assumes the form

$$N_{1DHD\Gamma}(E, \eta_g) = \frac{2g_v}{\pi} \sum_{n_y=1}^{n_{y\max}} \sum_{n_z=1}^{n_{z\max}} T_5'(E, n_y, n_z, \eta_g) H(E - E_{5HDNW}') \tag{4.34}$$

where

$$T_5(E, n_y, n_z, \eta_g) = \left[[\theta_4(E, \eta_g) - [G_2(n_y, n_z)]] \frac{2m_c}{\hbar^2} \right]^{1/2}.$$

In (4.34), E_{5HDNW}' is the sub-band energy in this case which can be expressed as

$$G_2(n_y, n_z) = \theta_4(E_{5HDNW}', \eta_g). \tag{4.35}$$

The EM in this case is given by

$$m^*(E_{F1HDNW}, \eta_g) = m_c[\theta_4'(E_{F1HDNW}, \eta_g)]. \tag{4.36a}$$

The electron concentration per unit length can be written as

$$n_{01D} = \frac{2g_v}{\pi} \sum_{n_y=1}^{n_{y\max}} \sum_{n_z=1}^{n_{z\max}} \left[T_5(E_{F1HDNW}, n_y, n_z, \eta_g) + \sum_{r=1}^{s} L(r)[T_5(E_{F1HDNW}, n_y, n_z, \eta_g)] \right]. \tag{4.36b}$$

The expression of 1D DR, for NWs of III-V materials whose energy band structures are defined by the model of Stillman *et al.* in the absence of band tailing assumes the form

$$I_{12}(E) = \frac{\hbar^2 k_x^2}{2m_c} + G_2(n_y, n_z). \tag{4.37}$$

In this case, the quantized energy E_9' is given by

$$I_{12}(E_9') = G_2(n_y, n_z). \tag{4.38}$$

The DOS function can be expressed as

$$N_{1D\Gamma}(E) = \frac{2g_v}{\pi} \sum_{n_y=1}^{n_{y\max}} \sum_{n_z=1}^{n_{z\max}} f_{14}'(E, n_y, n_z) H(E - E_9') \tag{4.39}$$

where

$$f_{14}(E, n_y, n_z) = \left[\frac{2m_c}{\hbar^2}[I_{12}(E) - (G_2(n_y, n_z))]\right]^{1/2}.$$

The EM in this case assumes the form

$$m^*(E_{F1D}) = m_c I_{12}'(E_{F1D}). \tag{4.40a}$$

The electron concentration per unit length can be written as

$$n_{01D} = \frac{2g_v}{\pi} \sum_{n_y=1}^{n_{y\max}} \sum_{n_z=1}^{n_{z\max}} \left[f_{14}(E_{F1D}, n_y, n_z) + \sum_{r=1}^{s} L(r)[f_{14}(E_{F1D}, n_y, n_z)]\right]. \tag{4.40b}$$

(v) The Model of Palik *et al.*

The DR of the 1D electrons in this case can be written as

$$\frac{\hbar^2(n_z\pi/d_z)^2}{2m_c} + \frac{\hbar^2(n_y\pi/d_y)^2}{2m_c} + \frac{\hbar^2 k_x^2}{2m_c} = \theta_5(E, \eta_g) \tag{4.41}$$

where

$$\theta_5(E, \eta_g) = I_{13}(E, \eta_g).$$

The DOS function in this case assumes the form

$$N_{1DHD\Gamma}(E, \eta_g) = \frac{2g_v}{\pi} \sum_{n_y=1}^{n_{y\max}} \sum_{n_z=1}^{n_{z\max}} T_6'(E, n_y, n_z, \eta_g) H(E - E_{6HDNW}') \tag{4.42}$$

where

$$T_6(E, n_y, n_z, \eta_g) = \left[[\theta_5(E, \eta_g) - (G_2(n_y, n_z))]\frac{2m_c}{\hbar^2}\right]^{1/2}.$$

In (4.42), E'_{6HDNW} is the sub-band energy in this case which can be expressed as

$$G_2(n_y, n_z) = \theta_5(E'_{6HDNW}, \eta_g). \tag{4.43}$$

The EM in this case is given by

$$m^*(E_{F1HDNW}, \eta_g) = m_c[\theta'_5(E_{F1HDNW}, \eta_g)]. \tag{4.44a}$$

The electron concentration per unit length can be written as

$$n_{01D} = \frac{2g_v}{\pi} \sum_{n_y=1}^{n_{y\max}} \sum_{n_z=1}^{n_{z\max}} \left[T_6(E_{F1HDNW}, n_y, n_z, \eta_g) + \sum_{r=1}^{s} L(r)[T_6(E_{F1HDNW}, n_y, n_z, \eta_g)] \right]. \tag{4.44b}$$

The expression of 1D DR, for NWs of III-V materials whose energy band structures are defined by the model of Palik *et al.* in the absence of band tailing assumes the form

$$I_{13}(E) = \frac{\hbar^2 k_x^2}{2m_c} + G_2(n_y, n_z) \tag{4.45}$$

In this case, the quantized energy E'_{10} is given by

$$I_{13}(E'_{10}) = G_2(n_y, n_z). \tag{4.46}$$

The DOS function can be expressed as

$$N_{1D\Gamma}(E) = \frac{2g_v}{\pi} \sum_{n_y=1}^{n_{y\max}} \sum_{n_z=1}^{n_{z\max}} f'_{15}(E, n_y, n_z) H(E - E'_{10}) \tag{4.47}$$

where

$$f_{15}(E, n_y, n_z) = \left[\frac{2m_c}{\hbar^2} [I_{13}(E) - (G_2(n_y, n_z))] \right]^{1/2}.$$

The EM in this case assumes the form

$$m^*(E_{F1D}) = m_c I'_{13}(E_{F1D}). \tag{4.48a}$$

The electron concentration per unit length can be written as

$$n_{01D} = \frac{2g_v}{\pi} \sum_{n_y=1}^{n_{y\max}} \sum_{n_z=1}^{n_{z\max}} \left[f_{15}(E_{F1D}, n_y, n_z) + \sum_{r=1}^{s} L(r)[f_{15}(E_{F1D}, n_y, n_z)] \right]. \quad (4.48b)$$

4.2.3 *The EM in NWs of HD II-VI materials*

The 1D DR in NW of HD II-VI materials can be written as

$$\gamma_3(E, \eta_g) = a_0' \left[\left(\frac{n_x \pi}{d_x} \right)^2 + \left(\frac{n_y \pi}{d_y} \right)^2 \right] \pm \bar{\lambda}_0 \left[\left(\frac{n_x \pi}{d_x} \right)^2 + \left(\frac{n_y \pi}{d_y} \right)^2 \right]^{1/2} + \frac{\hbar^2 k_z^2}{2m_{\parallel}^*}. \quad (4.49)$$

The DOS function in this case assumes the form

$$N_{1DHD\Gamma}(E, \eta_g) = \frac{g_v}{\pi} \sum_{n_x=1}^{n_{x\max}} \sum_{n_y=1}^{n_{y\max}} T_7'(E, n_x, n_y, \eta_g) H(E - E_{13HDNW}') \quad (4.50)$$

where

$$T_7(E, n_y, n_z, \eta_g) = \left[[\gamma_3(E, \eta_g) - [G_{3,\pm}(n_x, n_y)]] \frac{2m_{\parallel}^*}{\hbar^2} \right]^{1/2}$$

and

$$G_{3,\pm}(n_x, n_y) = a_0' \left[\left(\frac{n_x \pi}{d_x} \right)^2 + \left(\frac{n_y \pi}{d_y} \right)^2 \right] \pm \bar{\lambda}_0 \left[\left(\frac{n_x \pi}{d_x} \right)^2 + \left(\frac{n_y \pi}{d_y} \right)^2 \right]^{1/2}$$

and E_{13HDNW}' is the sub-band energy in this case which can be expressed as

$$\gamma_3(E_{13HDNW}', \eta_g) = a_0' \left[\left(\frac{n_x \pi}{d_x} \right)^2 + \left(\frac{n_y \pi}{d_y} \right)^2 \right] \pm \bar{\lambda}_0 \left[\left(\frac{n_x \pi}{d_x} \right)^2 + \left(\frac{n_y \pi}{d_y} \right)^2 \right]^{1/2}. \quad (4.51)$$

The EM in this case is given by

$$m^*(E_{F1HDNW}, \eta_g) = m_{\|}^* \gamma_3'(E_{F1HDNW}, \eta_g). \tag{4.52a}$$

The electron concentration per unit length can be written as

$$n_{01D} = \frac{g_v}{\pi} \sum_{n_x=1}^{n_{x\max}} \sum_{n_y=1}^{n_{y\max}} \left[T_7(E_{F1HDNW}, n_x, n_y, \eta_g) + \sum_{r=1}^{s} L(r)[T_7(E_{F1HDNW}, n_x, n_y, \eta_g)] \right]. \tag{4.52b}$$

The 1D DR for NWs of II-VI materials in the absence of band-tails can be written as

$$E = b_0' k_z^2 + G_{3,\pm}(n_x, n_y). \tag{4.53}$$

The DOS function can be expressed as

$$N_{1D\Gamma}(E) = \frac{g_v}{\pi} \sum_{n_x=1}^{n_{x\max}} \sum_{n_y=1}^{n_{y\max}} f_{20}'(E, n_x, n_y) H(E - E_{20}') \tag{4.54}$$

where

$$f_{20}(E, n_x, n_y) = \left[\frac{2m_{\|}^*}{\hbar^2} [E - G_{3,\pm}(n_x, n_y)] \right]^{1/2}.$$

In (4.54), the sub-band energy E_{20}' can be written as

$$E_{20}' = G_{3,\pm}(n_x, n_y). \tag{4.55}$$

The EM in this case assumes the form

$$m^*(E_{F1D}) = m_{\|}^*. \tag{4.56a}$$

The electron concentration per unit length can be written as

$$n_{01D} = \frac{g_v}{\pi} \sum_{n_x=1}^{n_{x\max}} \sum_{n_y=1}^{n_{y\max}} \left[f_{20}(E_{F1D}, n_x, n_y) + \sum_{r=1}^{s} L(r)[f_{20}(E_{F1D}, n_x, n_y)] \right]. \tag{4.56b}$$

4.2.4 *The EM in NWs of HD IV-VI materials*

(i) Dimmock Model

The 1D electron dispersion law in NW of HD IV-VI materials in accordance with the Dimmock model can be expressed as

$$
\begin{aligned}
&\gamma_2(E,\eta_g)+\alpha\gamma_3(E,\eta_g)\left(\frac{\hbar^2}{2x_4}\left(\frac{n_x\pi}{d_x}\right)^2+\frac{\hbar^2}{2x_5}\left(\frac{n_y\pi}{d_y}\right)^2\right)\\
&\quad+\alpha\gamma_3(E,\eta_g)\frac{\hbar^2}{2x_6}k_z^2-(1+\alpha\gamma_3(E,\eta_g))\\
&\quad\times\left(\frac{\hbar^2}{2x_1}\left(\frac{n_x\pi}{d_x}\right)^2+\frac{\hbar^2}{2x_2}\left(\frac{n_y\pi}{d_y}\right)^2\right)\\
&\quad-\alpha\left(\frac{\hbar^2}{2x_1}\left(\frac{n_x\pi}{d_x}\right)^2+\frac{\hbar^2}{2x_2}\left(\frac{n_y\pi}{d_y}\right)^2\right)\\
&\quad\times\left(\frac{\hbar^2}{2x_4}\left(\frac{n_x\pi}{d_x}\right)^2+\frac{\hbar^2}{2x_5}\left(\frac{n_y\pi}{d_y}\right)^2\right)\\
&\quad-\alpha\left(\frac{\hbar^2}{2x_1}\left(\frac{n_x\pi}{d_x}\right)^2+\frac{\hbar^2}{2x_2}\left(\frac{n_y\pi}{d_y}\right)^2\right)\frac{\hbar^2}{2x_6}k_z^2\\
&\quad-(1+\alpha\gamma_3(E,\eta_g))\frac{\hbar^2}{2x_3}k_z^2\\
&\quad-\alpha\frac{\hbar^2}{2x_3}k_z^2\left(\frac{\hbar^2}{2x_4}\left(\frac{n_x\pi}{d_x}\right)^2+\frac{\hbar^2}{2x_5}\left(\frac{n_y\pi}{d_y}\right)^2\right)-\alpha\frac{\hbar^4k_z^4}{4x_3x_6}\\
&=\frac{\hbar^2}{2m_1}\left(\frac{n_x,\pi}{d_x}\right)^2+\frac{\hbar^2}{2m_2}\left(\frac{n_y,\pi}{d_y}\right)^2\frac{\hbar^2}{2m_3}k_z^2.
\end{aligned}
\tag{4.57}
$$

Equation (4.57) can be written as

$$k_z = T_{36}(E,\eta_g,n_x,n_y) \tag{4.58}$$

where,

$$T_{36}(E,\eta_g,n_x,n_y) = \left[(2C_{22})^{-1}\left[-B_{HD}(E,\eta_g,n_x,n_y) + \sqrt{B_{HD}^2(E,\eta_g,n_x,n_y)+4C_{22}A_{HD}(E,\eta_g,n_x,n_y)}\right]\right]^{1/2}$$

$$C_{22} = \left(\alpha\frac{\hbar^4}{4x_3x_6}\right), \quad B_{HD}(E,\eta_g,n_x,n_y)$$

$$= \left[\alpha\left(\frac{\hbar^2}{2x_1}\left(\frac{n_x\pi}{d_x}\right)^2+\frac{\hbar^2}{2x_2}\left(\frac{n_y\pi}{d_y}\right)^2\right)\frac{\hbar^2}{2x_6} + (1+\alpha\gamma_3(E,\eta_g))\frac{\hbar^2}{2x_3} - \alpha\gamma_3(E,\eta_g)\frac{\hbar^2}{2x_6}+\frac{\hbar^2}{2m_3} + \alpha\frac{\hbar^2}{2x_3}\left(\frac{\hbar^2}{2x_4}\left(\frac{n_x\pi}{d_x}\right)^2+\frac{\hbar^2}{2x_5}\left(\frac{n_y\pi}{d_y}\right)^2\right)\right]$$

and

$$A_{HD}(E,\eta_g,n_x,n_y) = \left[-\left[\frac{\hbar^2}{2m_1}\left(\frac{n_x\pi}{d_x}\right)^2+\frac{\hbar^2}{2m_2}\left(\frac{n_x\pi}{d_2}\right)^2\right]\gamma_2(E,\eta_g)+\alpha\gamma_3(E,\eta_g)\right] \times\left(\frac{\hbar^2}{2x_4}\left(\frac{n_x\pi}{d_x}\right)^2+\frac{\hbar^2}{2x_5}\left(\frac{n_y\pi}{d_y}\right)^2\right) - \alpha\left(\frac{\hbar^2}{2x_1}\left(\frac{n_x\pi}{d_x}\right)^2+\frac{\hbar^2}{2x_2}\left(\frac{n_y\pi}{d_y}\right)^2\right) \times\left(\frac{\hbar^2}{2x_4}\left(\frac{n_x\pi}{d_x}\right)^2+\frac{\hbar^2}{2x_5}\left(\frac{n_y\pi}{d_y}\right)^2\right) - (1+\alpha\gamma_3(E,\eta_g))\left(\frac{\hbar^2}{2x_1}\left(\frac{n_x\pi}{d_x}\right)^2+\frac{\hbar^2}{2x_2}\left(\frac{n_y\pi}{d_y}\right)^2\right).$$

The DOS function in this case can be written as

$$N_{1DHD\Gamma}(E,\eta_g)=\frac{2g_v}{\pi}\sum_{n_x=1}^{n_{x\max}}\sum_{n_y=1}^{n_{y\max}}T'_{36}(E,n_x,n_y,\eta_g)H(E-E'_{14HDNW}) \tag{4.59}$$

In (4.59), E'_{14HDNW} is the sub-band energy in this case which can be expressed as

$$0=T_{36}(E'_{14HDNW},\eta_g,n_x,n_y). \tag{4.60}$$

The EM in this case is given by

$$m^*(E_{F1HDNW},n_g,n_x,\eta_y)=\frac{\hbar^2}{2}\frac{\partial}{\partial E}\left[T^2_{36}(E_{F1HDNW},n_g,n_x,\eta_y)\right]. \tag{4.61a}$$

The electron concentration per unit length can be written as

$$n_{01D}=\frac{2g_v}{\pi}\sum_{n_x=1}^{n_{x\max}}\sum_{n_y=1}^{n_{y\max}}\left[T_{36}(E_{F1HDNW},n_x,n_y,\eta_g)\right.$$
$$\left.+\sum_{r=1}^{s}L(r)[T_{36}(E_{F1HDNW},n_x,n_y,\eta_g)]\right]. \tag{4.61b}$$

The 1D DR in NW of IV-VI materials in the absence of band tails can be expressed as

$$E(1+\alpha E)+\alpha E\left(\frac{\hbar^2}{2x_4}\left(\frac{n_x\pi}{d_x}\right)^2+\frac{\hbar^2}{2x_5}\left(\frac{n_y\pi}{d_y}\right)^2\right)$$
$$+\alpha E\frac{\hbar^2}{2x_6}k_z^2-(1+\alpha E)\left(\frac{\hbar^2}{2x_1}\left(\frac{n_x\pi}{d_x}\right)^2+\frac{\hbar^2}{2x_2}\left(\frac{n_y\pi}{d_y}\right)^2\right)$$
$$-\alpha\left(\frac{\hbar^2}{2x_1}\left(\frac{n_x\pi}{d_x}\right)^2+\frac{\hbar^2}{2x_2}\left(\frac{n_y\pi}{d_y}\right)^2\right)$$
$$\times\left(\frac{\hbar^2}{2x_4}\left(\frac{n_x\pi}{d_x}\right)^2+\frac{\hbar^2}{2x_5}\left(\frac{n_y\pi}{d_y}\right)^2\right)$$

$$-\alpha\left(\frac{\hbar^2}{2x_1}\left(\frac{n_x\pi}{d_x}\right)^2+\frac{\hbar^2}{2x_2}\left(\frac{n_y\pi}{d_y}\right)^2\right)\frac{\hbar^2}{2x_6}k_z^2$$
$$-(1+\alpha E)\frac{\hbar^2}{2x_3}k_z^2-\alpha\frac{\hbar^2}{2x_3}k_z^2\left(\frac{\hbar^2}{2x_4}\left(\frac{n_x\pi}{d_x}\right)^2+\frac{\hbar^2}{2x_5}\left(\frac{n_y\pi}{d_y}\right)^2\right)$$
$$-\alpha\frac{\hbar^4k_z^4}{4x_3x_6}=\frac{\hbar^2}{2m_1}\left(\frac{n_x\pi}{d_x}\right)^2+\frac{\hbar^2}{2m_2}\left(\frac{n_y\pi}{d_y}\right)^2+\frac{\hbar^2}{2m_3}k_z^2 \quad (4.62)$$

Equation (4.62) can be written as

$$k_z=T_{40}(E,n_x,n_y) \quad (4.63)$$

where,

$$T_{40}(E,n_x,n_y)=\left[(2C_{22})^{-1}\left[-B_0(E,n_x,n_y)+\sqrt{B_0^2(E,n_x,n_y)+4C_{22}A_0(E,n_x,n_y)}\right]\right]^{1/2}$$

where,

$$B_0(E,n_x,n_y)=\left[\alpha\left(\frac{\hbar^2}{2x_1}\left(\frac{n_x\pi}{d_x}\right)^2+\frac{\hbar^2}{2x_2}\left(\frac{n_y\pi}{d_y}\right)^2\right)\frac{\hbar^2}{2x_6}+(1+\alpha E)\frac{\hbar^2}{2x_3}-\alpha E\frac{\hbar^2}{2x_6}+\frac{\hbar^2}{2m_3}+\alpha\frac{\hbar^2}{2x_3}\left(\frac{\hbar^2}{2x_4}\left(\frac{n_x\pi}{d_x}\right)^2+\frac{\hbar^2}{2x_5}\left(\frac{n_y\pi}{d_y}\right)^2\right)\right]$$

and

$$A_0(E,n_x,n_y)=\left[-\left[\left(\frac{\hbar^2}{2m_1}\left(\frac{n_x\pi}{d_x}\right)^2+\frac{\hbar^2}{2m_2}\left(\frac{n_y\pi}{d_y}\right)^2\right)\right]E(1+\alpha E)+\alpha E\left(\frac{\hbar^2}{2x_4}\left(\frac{n_x\pi}{d_x}\right)^2+\frac{\hbar^2}{2x_5}\left(\frac{n_y\pi}{d_y}\right)^2\right)\right.$$

$$-\alpha\left(\frac{\hbar^2}{2x_1}\left(\frac{n_x\pi}{d_x}\right)^2+\frac{\hbar^2}{2x_2}\left(\frac{n_y\pi}{d_y}\right)^2\right)$$

$$\times\left(\frac{\hbar^2}{2x_4}\left(\frac{n_x\pi}{d_x}\right)^2+\frac{\hbar^2}{2x_5}\left(\frac{n_y\pi}{d_y}\right)^2\right)$$

$$\left.-(1+\alpha E)\left(\frac{\hbar^2}{2x_1}\left(\frac{n_x\pi}{d_x}\right)^2+\frac{\hbar^2}{2x_2}\left(\frac{n_y\pi}{d_y}\right)^2\right)\right]$$

The DOS function can be expressed as

$$N_{1D\Gamma}(E)=\frac{g_v}{\pi}\sum_{n_x=1}^{n_{x\max}}\sum_{n_y=1}^{n_{y\max}}T'_{40}(E,n_x,n_y)H(E-E'_{22}) \tag{4.64}$$

In (4.64), E'_{22} is the sub-band energy in this case which can be expressed as

$$0=T_{40}(E'_{22},n_x,n_y) \tag{4.65}$$

The EM in this case is given by

$$m^*(E_{F1D},n_x,\eta_y)=\frac{\hbar^2}{2}\frac{\partial}{\partial E}\left[T^2_{40}(E_{F1D},n_x,\eta_y)\right] \tag{4.66a}$$

The electron concentration per unit length can be written as

$$n_{01D}=\frac{g_v}{\pi}\sum_{n_x=1}^{n_{x\max}}\sum_{n_y=1}^{n_{y\max}}\left[T_{40}(E_{F1D},n_x,n_y)+\sum_{r=1}^{s}[T_{40}(E_{F1D},n_x,n_y)]\right] \tag{4.66b}$$

(ii) Bangert and Kastner Model

The 1D DR in NW of IV-VI materials in accordance with the present model can be written as

$$F_1(E,\eta_g)\left[\left(\frac{n_x\pi}{d_x}\right)^2+\left(\frac{n_y\pi}{d_y}\right)^2\right]+F_2(E,\eta_g)k_z^2=1 \tag{4.67}$$

The (4.67) can be written as

$$k_z=T_{60}(E,\eta_g,n_x,n_y) \tag{4.68}$$

where

$$T_{60}(E,\eta_g,n_x,n_y)$$
$$=\left[\left[1-F_1(E,\eta_g)\left[\left(\frac{n_x\pi}{d_x}\right)^2+\left(\frac{n_y\pi}{d_y}\right)\right]\right][F_2(E,\eta_g)]^{-1}\right]^{1/2}$$

The DOS function in this case can be written as

$$N_{1DHD\Gamma}(E,\eta_g)=\frac{2g_v}{\pi}\sum_{n_x=1}^{n_{x\max}}\sum_{n_y=1}^{n_{y\max}}T_{40}'(E,n_x,n_y,\eta_g)H(E-E_{15HDNW}') \tag{4.69}$$

In (4.69), E'_{15HDNW} is the sub-band energy in this case which can be expressed as

$$0=T_{60}(E'_{15HDNW},\eta_g,n_x,n_y) \tag{4.70}$$

The EM in this case is given by

$$m^*(E_{F1HDNW},\eta_g,n_x,\eta_y)=\frac{\hbar^2}{2}\frac{\partial}{\partial E}\left[T_{40}^2(E_{F1HDNW},\eta_g,n_x,\eta_y)\right] \tag{4.71a}$$

The electron concentration per unit length can be written as

$$n_{01D}=\frac{2g_v}{\pi}\sum_{n_x=1}^{n_{x\max}}\sum_{n_y=1}^{n_{y\max}}\left[T_{40}(E_{F1HDNW},n_x,n_y,\eta_g)\right.$$
$$\left.+\sum_{r=1}^{s}L(r)[T_{40}(E_{F1HDNW},n_x,n_y,\eta_g)]\right] \tag{4.71b}$$

The 1D DR in the absence of band tailing can be written in this case as

$$\omega_1(E)\left[\left(\frac{\pi n_x}{d_x}\right)^2+\left(\frac{n\pi_y}{d_y}\right)^2\right]+\omega_2(E)\,k_z^2=1 \tag{4.72}$$

The (4.72) can be written as

$$k_z=T_{61}(E,n_x,n_y) \tag{4.73}$$

where

$$T_{61}(E,n_x,n_y)=\left[\left[1-\omega_1(E)\left[\left(\frac{n_x\pi}{d_x}\right)^2\right]\right][\omega_2(E)]^{-1}\right]^{1/2}.$$

The DOS function can be expressed as

$$N_{1D\Gamma}(E) = \frac{2g_v}{\pi} \sum_{n_x=1}^{n_{x\max}} \sum_{n_y=1}^{n_{y\max}} T'_{61}(E, n_x, n_y) H(E - E'_{24}). \tag{4.74}$$

In (4.74), E'_{24} is the sub-band energy ,which in this case can be expressed as

$$0 = T_{40}\left(E'_{24}, n_x, n_y\right) \tag{4.75}$$

The EM in this case is given by

$$m^*(E_{F1D}, n_x, \eta_y) = \frac{\hbar^2}{2} \frac{\partial}{\partial E} \left[T_{61}^2(E_{F1D}, n_x, n_y)\right] \tag{4.76a}$$

The electron concentration per unit length can be written as

$$n_{01D} = \frac{2g_v}{\pi} \sum_{n_x=1}^{n_{x\max}} \sum_{n_y=1}^{n_{y\max}} \left[T_{61}(E_{F1D}, n_x, n_y) + \sum_{r=1}^{s} L(r)[T_{61}(E_{F1D}, n_x, n_y)]\right] \tag{4.76b}$$

4.2.5 *The EM in NWs of HD stressed Kane type materials*

The 1D DR in this case can be written as

$$P_{11}(E, \eta_g) \left(\frac{\pi n_x}{d_x}\right)^2 + Q_{11}(E, \eta_g) \left(\frac{\pi n_y}{d_y}\right)^{2'} + S_{11}(E, \eta_g) k_z^2 = 1. \tag{4.77}$$

(4.77) can be written as

$$k_z = T_{70}(E, \eta_g, n_x, n_y) \tag{4.78}$$

where

$$T_{70}(E, \eta_g, n_x, n_y) = \left[\left[1 - P_{11}(E, \eta_g) \left(\frac{\pi n_x}{d_x}\right)^2 + Q_{11}(E, \eta_g) \left(\frac{\pi n_y}{d_y}\right)^2\right] [S_{11}(E, \eta_g)]^{-1}\right]^{-1/2}$$

The DOS function in this case can be written as

$$N_{1DHD\Gamma}(E,\eta_g) = \frac{2g_v}{\pi} \sum_{n_x=1}^{n_{x\max}} \sum_{n_y=1}^{n_{y\max}} T'_{70}(E,n_x,n_y,\eta_g)H(E-E'_{30HDNW}) \tag{4.79}$$

In (4.79), E'_{30HDNW} is the sub-band energy in this case which can be expressed as

$$0 = T_{70}(E'_{30HDNW},\eta_g,n_x,n_y) \tag{4.80}$$

The EM in this case is given by

$$m^*(E_{F1HDNW},\eta_g,n_x,\eta_y) = \frac{\hbar^2}{2}\frac{\partial}{\partial E}\left[T^2_{70}(E_{F1HDNW},\eta_g,n_x,\eta_y)\right] \tag{4.81a}$$

The electron concentration per unit length can be written as

$$n_{01D} = \frac{2g_v}{\pi} \sum_{n_x=1}^{n_{x\max}} \sum_{n_y=1}^{n_{y\max}} \left[T_{70}(E_{F1HDNW},n_x,n_y,\eta_g)\right.$$

$$\left. + \sum_{r=1}^{s} L(r)[T_{70}(E_{F1HDNW},n_x,n_y,\eta_g)]\right] \tag{4.81b}$$

In the absence of band tailing the 1D DR in this case assumes the form

$$k_z = t_{701}(E,n_x,n_y) \tag{4.82}$$

where

$$t_{701}\left(E,n_x,n_y\right) = \left[\left[\bar{c}_0(E)\left[1-\left(\frac{\pi n_x}{d_x\bar{a}_0(E)}\right)^2-\left(\frac{\pi n_y}{d_y\bar{b}_0(E)}\right)^2\right]\right]\right]^{1/2}.$$

The total 1D DOS function can be written as

$$N_{1D\Gamma}(E) = \frac{2g_v}{\pi} \sum_{n_x=1}^{n_{x\max}} \sum_{n_y=1}^{n_{y\max}} T'_{701}(E,n_x,n_y)H(E-E'_{26}). \tag{4.83}$$

In (4.83), E'_{26} is the sub-band energy,which in this case can be expressed as

$$0 = t_{701}(E'_{26},n_x,n_y) \tag{4.84}$$

The EM in this case is given by

$$m^*(E_{F1D}, n_x, \eta_y) = \frac{\hbar^2}{2}\frac{\partial}{\partial E}\left[t_{701}^2(E_{F1D}, n_x, n_y)\right]. \tag{4.85a}$$

The electron concentration per unit length can be written as

$$n_{01D} = \frac{2g_v}{\pi}\sum_{n_x=1}^{n_{x\max}}\sum_{n_y=1}^{n_{y\max}}\Bigg[T_{701}(E_{F1D}, n_x, n_y) + \sum_{r=1}^{s} L(r)[T_{701}(E_{F1D}, n_x, n_y)]\Bigg]. \tag{4.85b}$$

4.2.6 *The EM in NWs of HD Te*

The 1D DR may be written in this case as

$$k_x = t_{72}(E, n_y, n_z, \eta_g) \tag{4.86}$$

where

$$t_{72}(E, \eta_g, n_x, n_y) = \left[-\left(\frac{n_y\pi}{d_y}\right)^2 + \psi_{5HD}(E, \eta_g) - \psi_6\left(\frac{\pi n_z}{d_z}\right)^2 \pm \psi_7\left[\psi_{8HD}^2(E, \eta_g) - \left(\frac{\pi n_z}{d_z}\right)^2\right]^{1/2}\right]^{1/2}$$

The DOS function in this case can be written as

$$N_{1DHD\Gamma}(E, \eta_g) = \frac{2g_v}{\pi}\sum_{n_x=1}^{n_{x\max}}\sum_{n_y=1}^{n_{y\max}} t'_{72}(E, n_x, n_y, \eta_g) H(E - E'_{31HDNW}). \tag{4.87}$$

In (4.87), E'_{31HDNW} is the sub-band energy in this case which can be expressed as

$$0 = t_{72}(E'_{31HDNW}, \eta_g, n_x, n_y). \tag{4.88}$$

The EM in this case is given by

$$m^*(E_{F1HDNW}, \eta_g, n_x, \eta_y) = \frac{\hbar^2}{2}\frac{\partial}{\partial E}\left[t_{72}^2(E_{F1HDNW}, \eta_g, n_x, \eta_y)\right]. \tag{4.89a}$$

The electron concentration per unit length can be written as

$$n_{01D} = \frac{2g_v}{\pi} \sum_{n_x=1}^{n_{x\max}} \sum_{n_y=1}^{n_{y\max}} \left[t_{72}(E_{F1HDNW}, n_x, n_y, \eta_g) + \sum_{r=1}^{s} L(r)[t_{72}(E_{F1HDNW}, n_x, n_y, \eta_g)] \right]. \tag{4.89b}$$

In the absence of band tailing the 1D DR in this case assumes the form

$$k_x = H_{70}(E, n_y, n_z) \tag{4.90}$$

where

$$H_{70}(E, n_x, n_y) = \left[-\left(\frac{n_y \pi}{d_y}\right)^2 + \psi_5(E) - \psi_6 \left(\frac{\pi n_z}{d_z}\right)^2 \pm \psi_7 \left[\psi_8^2(E) - \left(\frac{\pi n_z}{d_z}\right)^2 \right]^{1/2} \right]^{1/2}$$

$$N_{1D\Gamma}(E) = \frac{2g_v}{\pi} \sum_{n_x=1}^{n_{x\max}} \sum_{n_y=1}^{n_{y\max}} H'_{70}(E, n_y, n_z) H(E - E'_{44}) \tag{4.91}$$

In (4.91), E'_{44} is the sub-band energy in this case which can be expressed as

$$0 = H_{70}(E'_{44}, n_x, n_y). \tag{4.92}$$

The EM in this case is given by

$$m^*(E_{F1D}, n_y, \eta_z) = \frac{\hbar^2}{2} \frac{\partial}{\partial E} [H_{70}^2(E_{F1D}, n_y, n_z)] \tag{4.93a}$$

The electron concentration per unit length can be written as

$$n_{01D} = \frac{2g_v}{\pi} \sum_{n_x=1}^{n_{x\max}} \sum_{n_y=1}^{n_{y\max}} \left[H_{70}(E_{F1D}, n_y, n_z) + \sum_{r=1}^{s} L(r)[H_{70}(E_{F1D}, n_y, n_z)] \right] \tag{4.93b}$$

4.2.7 *The EM in NWs of HD gallium phosphide*

The 1D DR may be written in this case as

$$k_x = u_{70}(E, n_y, n_z, \eta_g) \tag{4.94}$$

where,

$$u_{70}(E, \eta_g, n_y, n_z) = \left[-\left(\frac{n_y \pi}{d_y}\right)^2 + t_{11}\gamma_3(E, \eta_g) + t_{21} - t_{31}\left(\frac{\pi n_z}{d_z}\right)^2 - t_{41}\left[\left(\frac{n_z \pi}{d_z}\right)^2 + t_5^2(E, \eta_g)\right]^{1/2}\right]^{1/2}$$

The DOS function in this case can be written as

$$N_{1DHD\Gamma}(E, \eta_g) = \frac{2g_v}{\pi} \sum_{n_y=1}^{n_{y\max}} \sum_{n_z=1}^{n_{z\max}} u'_{70}(E, n_y, n_z, \eta_g) H(E - E'_{32HDNW}) \tag{4.95}$$

In (4.95), E'_{32HDNW} is the sub-band energy in this case which can be expressed as

$$0 = u_{70}(E'_{32HDNW}, \eta_g, n_y, n_z). \tag{4.96}$$

The EM in this case is given by

$$m^*(E_{F1HDNW}, \eta_g, n_y, \eta_z) = \frac{\hbar^2}{2} \frac{\partial}{\partial E} \left[u_{70}^2(E_{F1HDNW}, \eta_g, n_y, \eta_z)\right]. \tag{4.97a}$$

The electron concentration per unit length can be written as

$$n_{01D} = \frac{2g_v}{\pi} \sum_{n_y=1}^{n_{y\max}} \sum_{n_z=1}^{n_{z\max}} \left[u_{70}(E_{F1HDNW}, n_y, n_z, \eta_g) + \sum_{r=1}^{s} L(r)[u_{70}(E_{F1HDNW}, n_y, n_z, \eta_g)] \right]. \tag{4.97b}$$

In the absence of band tailing the 1D EM in this case can be written as

$$k_x = X_{71}(E, n_y, n_z) \tag{4.98}$$

where

$$X_{71}(E, n_y, n_z) = \left[-\left(\frac{n_y\pi}{d_y}\right)^2 + t_{42}(E, n_z)\right]^{1/2}.$$

The DOS function is given by

$$N_{1D\Gamma}(E) = \frac{2g_v}{\pi}\sum_{n_y=1}^{n_{y\max}}\sum_{n_z=1}^{n_{z\max}} X'_{71}(E, n_y, n_z)H(E - E'_{46}) \tag{4.99}$$

In (4.99), E'_{46} is the sub-band energy in this case which can be expressed as

$$0 = X_{71}(E'_{46}, n_y n_z). \tag{4.100}$$

The EM in this case is given by

$$m^*(E_{F1D}, n_y, \eta_z) = \frac{\hbar^2}{2}\frac{\partial}{\partial E}[X_{71}^2(E_{F1D}, n_y, n_z)]. \tag{4.101a}$$

The electron concentration per unit length can be written as

$$n_{01D} = \frac{2g_v}{\pi}\sum_{n_y=1}^{n_{y\max}}\sum_{n_z=1}^{n_{z\max}}\left[X_{71}(E_{F1D}, n_y, n_z) + \sum_{r=1}^{s} L(r)[X_{71}(E_{F1D}, n_y, n_z)]\right]. \tag{4.101b}$$

4.2.8 *The EM in NWs of HD platinum antimonide*

The 1D DR may be written in this case as

$$k_x = V_0(E, n_y, n_z, \eta_g) \tag{4.102}$$

where

$$V_{70}(E, n_y, n_z, \eta_g) = \left[-\left(\frac{n_y\pi}{d_y}\right)^2 + A_{60}(E, \eta_g, n_y)\right]^{1/2}.$$

The DOS function in this case can be written as

$$N_{1DHD\Gamma}(E, \eta_g) = \frac{2g_v}{\pi}\sum_{n_y=1}^{n_{y\max}}\sum_{n_z=1}^{n_{z\max}} V'_{70}(E, n_y, n_z, \eta_g)H(E - E'_{34HDNW}). \tag{4.103}$$

In (4.103), E'_{34HDNW} is the sub-band energy in this case which can be expressed as

$$0 = V_{70}(E'_{34HDNW}, \eta_g, n_y, n_z) \tag{4.104}$$

The EM in this case is given by

$$m^*(E_{F1HDNW}, \eta_g, n_y, \eta_z) = \frac{\hbar^2}{2}\frac{\partial}{\partial E}\left[V_{70}^2(E_{F1HDNW}, \eta_g, n_y, \eta_z)\right] \tag{4.105a}$$

The electron concentration per unit length can be written as

$$n_{01D} = \frac{2g_v}{\pi}\sum_{n_y=1}^{n_{y\max}}\sum_{n_z=1}^{n_{z\max}}\left[V_{70}(E_{F1HDNW}, n_y, n_z, \eta_g) + \sum_{r=1}^{s} L(r)[V_{70}(E_{F1HDNW}, n_y, n_z, \eta_g)]\right] \tag{4.105b}$$

In the absence of band tailing the 1D DR in this case can be written as

$$k_x = D_{71}(E, n_y, n_z) \tag{4.106}$$

where

$$D_{71}(E, n_y, n_z) = \left[-\left(\frac{n_y\pi}{d_y}\right)^2 + t_{44}(E, n_z)\right]^{1/2}.$$

The DOS function is given by

$$N_{1D\Gamma}(E) = \frac{2g_v}{\pi}\sum_{n_y=1}^{n_{y\max}}\sum_{n_z=1}^{n_{z\max}} D'_{71}(E, n_y, n_z)H(E - E'_{48}). \tag{4.107}$$

In (4.107), E'_{48} is the sub-band energy in this case which can be expressed as

$$0 = D_{71}(E'_{48}, n_y, n_z) \tag{4.108}$$

The EM in this case is given by

$$m^*(E_{F1D}, n_y, \eta_z) = \frac{\hbar^2}{2}\frac{\partial}{\partial E}\left[D_{71}^2(E_{F1D}, n_y, n_z)\right]. \tag{4.109a}$$

The electron concentration per unit length can be written as

$$n_{01D} = \frac{2g_v}{\pi} \sum_{n_y=1}^{n_{y\max}} \sum_{n_z=1}^{n_{z\max}} \left[D_{71}(E_{F1D}, n_y, n_z) + \sum_{r=1}^{s} L(r)[D_{71}(E_{F1D}, n_y, n_z)] \right]. \tag{4.109b}$$

4.2.9 *The EM in NWs of HD bismuth telluride*

The DR in this case can be written as

$$k_x = J_{70}(E, n_y, n_z, \eta_g) \tag{4.110}$$

where

$$J_{70}(E, n_y, n_z, \eta_g) = \left[\left[\gamma_2(E, \eta_g) - \bar{\omega}_2 \left(\frac{n_y \pi}{d_y} \right)^2 - \bar{\omega}_3 \left(\frac{n_z \pi}{d_z} \right)^2 - 2\bar{\omega}_4 \left(\frac{n_y n_z \pi}{d_y d_z} \right)^2 \right] (\bar{\omega}_1)^{-1} \right]^{1/2}.$$

The DOS function in this case can be written as

$$N_{1DHD\Gamma}(E, \eta_g) = \frac{2g_v}{\pi} \sum_{n_y=1}^{n_{y\max}} \sum_{n_z=1}^{n_{z\max}} J'_{70}(E, n_y, n_z, \eta_g) H(E - E'_{50HDNW}) \tag{4.111}$$

and E'_{50HDNW} is the sub-band energy in this case which can be expressed as

$$0 = J_{70}(E'_{50HDNW}, \eta_g, n_y, n_z). \tag{4.112}$$

The EM in this case is given by

$$m^*(E_{F1HDNW}, \eta_g, n_y, \eta_z) = \frac{\hbar^2}{2} \frac{\partial}{\partial E} \left[J_{70}^2(E_{F1HDNW}, \eta_g, n_y, \eta_z) \right]. \tag{4.113a}$$

The electron concentration per unit length can be written as

$$n_{01D} = \frac{2g_v}{\pi} \sum_{n_y=1}^{n_{y\max}} \sum_{n_z=1}^{n_{z\max}} \left[J_{70}(E_{F1HDNW}, n_y, n_z, \eta_g) \right.$$
$$\left. + \sum_{r=1}^{s} L(r)[J_{70}(E_{F1HDNW}, n_y, n_z, \eta_g)] \right]. \quad (4.113\text{b})$$

In the absence of band tailing, the 1D DR in this case can be written as

$$k_x = B_{71}(E, n_y, n_z) \quad (4.114)$$

where

$$B_{71}(E, n_y, n_z) = \left[\left[E(1 + \alpha E) - \bar{\omega}_2 \left(\frac{n_y \pi}{d_y} \right)^2 - \bar{\omega}_3 \left(\frac{n_z \pi}{d_z} \right)^2 \right.\right.$$
$$\left.\left. - 2\bar{\omega}_4 \left(\frac{n_y n_z \pi}{d_y d_z} \right)^2 \right] (\bar{\omega}_1)^{-1} \right]^{1/2}.$$

The DOS function is given by

$$N_{1D\Gamma}(E) = \frac{2g_v}{\pi} \sum_{n_y=1}^{n_{y\max}} \sum_{n_z=1}^{n_{z\max}} B'_{71}(E, n_y, n_z) H(E - E'_{50}) \quad (4.115)$$

In (4.115), E'_{50} is the sub-band energy in this case which can be expressed as

$$0 = B_{71}(E'_{50}, n_y, n_z) \quad (4.116)$$

The EM in this case is given by

$$m^*(E_{F1D}, n_y, \eta_z) = \frac{\hbar^2}{2} \frac{\partial}{\partial E} \left[B_{71}^2(E_{F1D}, n_y, n_z) \right] \quad (4.117\text{a})$$

The electron concentration per unit length can be written as

$$n_{01D} = \frac{2g_v}{\pi} \sum_{n_y=1}^{n_{y\max}} \sum_{n_z=1}^{n_{z\max}} \left[B_{71}(E_{F1D}, n_y, n_z) \right.$$
$$\left. + \sum_{r=1}^{s} L(r)[B_{71}(E_{F1D}, n_y, n_z)] \right] \quad (4.117\text{b})$$

4.2.10 *The EM in NWs of HD germanium*

(a) Model of Cardona *et al.*

The DR in accordance with this model in the present case can be written as

$$k_x = L_{70}(E, n_y, n_z, \eta_g) \tag{4.118}$$

where

$$L_{70}(E, n_y, n_z, \eta_g) = \left[\left[\gamma_2(E, \eta_g) + \alpha\left[\frac{\hbar^2}{2m_{||}^*}\left(\frac{n_z\pi}{d_z}\right)^2\right]^2\right.\right.$$
$$\left.\left. - (1 + 2\alpha\gamma_3(E, \eta_g))\frac{\hbar^2}{2m_{||}^*}\left(\frac{n_z\pi}{d_z}\right)^2\right]\left(\frac{2m_{||}^*}{\hbar^2}\right)\right]^{1/2}$$

The DOS function in this case can be written as

$$N_{1DHD\Gamma}(E, \eta_g) = \frac{2g_v}{\pi}\sum_{n_y=1}^{n_{y\max}}\sum_{n_z=1}^{n_{z\max}} L'_{70}(E, n_y, n_z, \eta_g)H(E - E'_{52HDNW}) \tag{4.119}$$

In (4.119), E'_{52HDNW} is the sub-band energy in this case which can be expressed as

$$0 = L_{70}(E'_{52HDNW}, \eta_g, n_{y,} n_z) \tag{4.120}$$

The EM in this case is given by

$$m^*(E_{F1HDNW}, \eta_g, n_y, \eta_z) = \frac{\hbar^2}{2}\frac{\partial}{\partial E}\left[L_{70}^2(E_{F1HDNW}, \eta_g, n_y, \eta_z)\right] \tag{4.121a}$$

The electron concentration per unit length can be written as

$$n_{01D} = \frac{2g_v}{\pi}\sum_{n_y=1}^{n_{y\max}}\sum_{n_z=1}^{n_{z\max}}\left[L_{70}(E_{F1HDNW}, n_y, n_z, \eta_g)\right.$$
$$\left. + \sum_{r=1}^{s} L(r)[L_{70}(E_{F1HDNW}, n_y, n_z, \eta_g)]\right] \tag{4.121b}$$

In the absence of band tailing, the 1D DR in this case can be written as

$$k_x = B_{77}(E, n_y, n_z) \tag{4.122}$$

where,

$$B_{77}(E,n_y,n_z)=\left[\left[E(1+\alpha E)+\alpha\left[\frac{\hbar^2}{2m_{||}^*}\left(\frac{n_z\pi}{d_z}\right)^2\right]^2\right.\right.$$
$$\left.\left.-(1+2\alpha E)\frac{\hbar^2}{2m_{||}^*}\left(\frac{n_z\pi}{d_z}\right)^2\right]\left(\frac{2m_{||}^*}{\hbar^2}\right)\right]^{1/2}$$

The DOS function is given by

$$N_{1D\Gamma}(E)=\frac{2g_v}{\pi}\sum_{n_y=1}^{n_{y\max}}\sum_{n_z=1}^{n_{z\max}}B'_{77}(E,n_y,n_z)H(E-E'_{60}). \quad (4.123)$$

In (4.123), E'_{60} is the sub-band energy in this case which can be expressed as

$$0=B_{77}(E'_{60},n_y,n_z). \quad (4.124)$$

The EM in this case is given by

$$m^*(E_{F1D},n_y,\eta_z)=\frac{\hbar^2}{2}\frac{\partial}{\partial E}\left[B_{77}^2(E_{F1D},n_y,n_z)\right]. \quad (4.125a)$$

The electron concentration per unit length can be written as

$$n_{01D}=\frac{2g_v}{\pi}\sum_{n_y=1}^{n_{y\max}}\sum_{n_z=1}^{n_{z\max}}\left[B_{77}(E_{F1D},n_y,n_z)\right.$$
$$\left.+\sum_{r=1}^{s}L(r)[B_{77}(E_{F1D},n_y,n_z)]\right] \quad (4.125b)$$

(b) Model of Wang *et al*

The 1D DR in accordance with this model in the present case can be written as

$$k_x=\beta_{70}(E,n_y,n_z,\eta_g) \quad (4.126)$$

where

$$\beta_{70}(E,n_y,n_z,\eta_g) = \left[-\left(\frac{n_y\pi}{d_y}\right)^2 + \frac{2m_\perp^*}{\hbar^2}\left[\bar{\alpha}_8 - \bar{\alpha}_9\left(\frac{\pi n_z}{d_z}\right)^2 - \bar{\alpha}_{10}\left(\frac{\pi n_z}{d_z}\right)^2 + \bar{\alpha}_{11}\left(\frac{\pi n_z}{d_z}\right)^2 + \bar{\alpha}_{12}(E,\eta_g)\right]^{1/2}\right]^{1/2}$$

The DOS function in this case can be written as

$$N_{1DHD\Gamma}(E,\eta_g) = \frac{2g_v}{\pi}\sum_{n_y=1}^{n_{y\max}}\sum_{n_z=1}^{n_{z\max}}\beta'_{70}(E,n_y,n_z,\eta_g)H(E-E'_{54HDNW}). \tag{4.127}$$

In (4.127), E'_{54HDNW} is the sub-band energy in this case which can be expressed as

$$0 = \beta_{70}(E'_{54HDNW},\eta_g,n_y,n_z). \tag{4.128}$$

The EM in this case is given by

$$m^*(E_{F1HDNW},\eta_g,n_y,\eta_z) = \frac{\hbar^2}{2}\frac{\partial}{\partial E}\left[\beta_{70}^2(E_{F1HDNW},\eta_g,n_y,\eta_z)\right]. \tag{4.129a}$$

The electron concentration per unit length can be written as

$$n_{01D} = \frac{2g_v}{\pi}\sum_{n_y=1}^{n_{y\max}}\sum_{n_z=1}^{n_{z\max}}\left[\beta_{70}(E_{F1HDNW},n_y,n_z,\eta_g) + \sum_{r=1}^{s}L(r)[\beta_{70}(E_{F1HDNW},n_y,n_z,\eta_g)]\right] \tag{4.129b}$$

In the absence of band tailing the 1D DR in this case can be written as

$$k_x = P_{77}(E,n_y,n_z) \tag{4.130}$$

where

$$P_{77}(E,n_y,n_z) = \left[\left[I_1(E,n_z) - \frac{\hbar^2}{2m_2^*}\left(\frac{n_y\pi}{d_y}\right)^2\right]\left(\frac{2m_1^*}{\hbar^2}\right)\right]^{1/2}.$$

The DOS function is given by

$$N_{1D\Gamma}(E) = \frac{2g_v}{\pi} \sum_{n_y=1}^{n_{y\max}} \sum_{n_z=1}^{n_{z\max}} P'_{77}(E, n_y, n_z) H(E - E'_{80}) \qquad (4.131)$$

and E'_{80} is the sub-band energy in this case which can be expressed as

$$0 = P_{77}(E'_{80}, n_y, n_z). \qquad (4.132)$$

The EM in this case is given by

$$m^*(E_{F1D}, n_y, \eta_z) = \frac{\hbar^2}{2} \frac{\partial}{\partial E} \left[P^2_{77}(E_{F1D}, n_y, n_z)\right] \qquad (4.133a)$$

The electron concentration per unit length can be written as

$$n_{01D} = \frac{2g_v}{\pi} \sum_{n_y=1}^{n_{y\max}} \sum_{n_z=1}^{n_{z\max}} \left[P_{77}(E_{F1D}, n_y, n_z) + \sum_{r=1}^{s} L(r)[P_{77}(E_{F1D}, n_y, n_z)]\right] \qquad (4.133b)$$

4.2.11 *The EM in NWs of HD gallium antimonide*

The DR of the 1D electrons in this case can be written as

$$\frac{\hbar^2(n_z\pi/d_z)^2}{2m_c} + \frac{\hbar^2(n_y\pi/d_y)^2}{2m_c} + \frac{\hbar^2 k_x^2}{2m_c} = I_{36}(E, \eta_g). \qquad (4.134)$$

The DOS function in this case can be written as

$$N_{1DHD\Gamma}(E, \eta_g) = \frac{2g_v}{\pi} \sum_{n_y=1}^{n_{y\max}} \sum_{n_z=1}^{n_{z\max}} N'_{100}(E, n_y, n_z, \eta_g) H(E - E'_{100HDNW}) \qquad (4.135)$$

where

$$N_{100}(E, n_y, n_z, \eta_g) = \left[[I_{36}(E, \eta_g) - G_2(n_y, n_z)]\left(\frac{2m_c}{\hbar^2}\right)\right]^{1/2}$$

and $E'_{100HDNW}$ is the sub-band energy in this case which can be expressed as

$$\frac{\hbar^2(n_z\pi/d_z)^2}{2m_c} + \frac{\hbar^2(n_y\pi/d_y)^2}{2m_c} = I_{36}(E'_{100HDNW}, \eta_g) \qquad (4.136)$$

The EM in this case is given by

$$m^*(E_{F1HDNW}, \eta_g) = m_c\left[I'_{36}(E_{F1HDNW}, \eta_g)\right] \quad (4.137a)$$

The electron concentration per unit length can be written as

$$n_{01D} = \frac{2g_v}{\pi}\sum_{n_y=1}^{n_{y\max}}\sum_{n_z=1}^{n_{z\max}}\left[N_{100}(E_{F1HDNW}, n_y, n_z, \eta_g) + \sum_{r=1}^{s} L(r)[N_{100}(E_{F1HDNW}, n_y, n_z, \eta_g)]\right] \quad (4.137b)$$

The expression of 1D DR, for NWs of GaSb whose energy band structures in the absence of band tailing assumes the form

$$I_{36}(E) = \frac{\hbar^2 k_x^2}{2m_c} + G_2(n_y, n_z). \quad (4.138)$$

The DOS function can be expressed as

$$N_{1D\Gamma}(E) = \frac{2g_v}{\pi}\sum_{n_y=1}^{n_{y\max}}\sum_{n_z=1}^{n_{z\max}} f'_{151}(E, n_y, n_z)H(E - E'_{101}) \quad (4.139)$$

where

$$f_{151}(E, n_y, n_z) = \left[\frac{2m_c}{\hbar^2}\left[I_{36}(E) - G_2(n_y, n_z)\right]\right]^{1/2}.$$

In this case, the quantized energy E'_{101} is given by

$$I_{36}(E'_{101}) = G_2(n_y, n_z) \quad (4.140a)$$

The EM can be written as

$$m^*(E_{F1D}) = m_c\left[I'_{36}(E_{F1D})\right] \quad (4.140b)$$

The electron concentration per unit length can be written as

$$n_{01D} = \frac{2g_v}{\pi}\sum_{n_y=1}^{n_{y\max}}\sum_{n_z=1}^{n_{z\max}}\left[f_{151}(E_{F1D}, n_y, n_z) + \sum_{r=1}^{s} L(r)[f_{151}(E_{F1D}, n_y, n_z)]\right] \quad (4.140c)$$

4.2.12 *The EM in NWs of HD II-V materials*

The DR of the 1D holes in II-V compounds can be expressed as

$$\gamma_3(E,\eta_g) = A_{10}\left(\frac{n_x\pi}{d_x}\right)^2 + A_{11}\left(\frac{n_y\pi}{d_y}\right)^2 + A_{12}k_z^2 + A_{13}\left(\frac{n_x\pi}{d_x}\right)$$
$$\pm\left[\left(A_{14}\left(\frac{n_x\pi}{d_x}\right)^2 + A_{15}\left(\frac{n_y\pi}{d_y}\right)^2\right.\right.$$
$$\left.\left. + A_{16}k_z^2 + A_{17}\left(\frac{n_x\pi}{d_x}\right)\right)^2 + A_{18}\left(\frac{n_y\pi}{d_y}\right)^2 + A_{19}^2\right]^{1/2} \tag{4.141}$$

where the numerical values of the energy band constants are given in appendix A. The sub-band energy (E_{n_zHD401}) is the lowest positive root of the following equation

$$\gamma_3(E_{n_zHD401},\eta_g) = A_{10}\left(\frac{n_x\pi}{d_x}\right)^2 + A_{13}\left(\frac{n_x\pi}{d_x}\right)$$
$$\pm\left[\left(A_{14}\left(\frac{n_x\pi}{d_x}\right)^2 + A_{17}\left(\frac{n_x\pi}{d_x}\right)\right)^2 + A_{19}^2\right]^{1/2}. \tag{4.142}$$

(4.142) can be expressed as

$$k_z = \Delta_{27}(E,\eta_g,n_x,n_y) \tag{4.143}$$

where

$$\Delta_{27}(E,\eta_g,n_x,n_y) = [(A_{12}^2 - A_{16}^2)^{-1}$$
$$\times [[\Delta_{21}(E,\eta_g,n_x,n_y)A_{12} + A_{16}\Delta_{22}(n_x,n_y)]$$
$$+ [[\Delta_{21}(E,\eta_g,n_x,n_y)A_{12} + A_{16}\Delta_{22}(n_x,n_y)]^2$$
$$- (A_{12}^2 - A_{16}^2)\Delta_{25}(E,\eta_g,n_x,n_y)]^2]],$$

$$\Delta_{21}(E,\eta_g,n_x,n_y) = \left[\gamma_3(E,\eta_g) - A_{10}\left(\frac{n_x\pi}{d_x}\right)^2 - A_{11}\left(\frac{n_y\pi}{d_y}\right)^2 - A_{13}\left(\frac{n_x\pi}{d_x}\right)^2\right],$$

$$\Delta_{22}(n_x,n_y) = \left[A_{14}\left(\frac{n_x\pi}{d_x}\right)^2 + A_{15}\left(\frac{n_y\pi}{d_y}\right)^2 + A_{17}\left(\frac{n_x\pi}{d_x}\right)^2\right],$$

$$\Delta_{25}(n_x,n_y,E,\eta_g) = [\Delta_{21}^2(E,\eta_g,n_x,n_y) - \Delta_{24}(n_x,n_y)],$$

$$\Delta_{24}(n_x,n_y) = [\Delta_{22}^2(n_x,n_y) + \Delta_{23}(n_y)]$$

and

$$\Delta_{23}(n_y) = \left[A_{18}\left(\frac{n_y\pi}{d_y}\right)^2 + A_{19}^2\right].$$

The DOS function in this case can be written as

$$N_{1DHD\Gamma}(E,\eta_g) = \frac{g_v}{\pi}\sum_{n_x=1}^{n_{x\max}}\sum_{n_y=1}^{n_{y\max}} \Delta'_{27}(E,n_y,n_z,\eta_g)H(E-E'_{200HDNW}) \tag{4.144}$$

In (4.144), $E'_{200HDNW}$ is the sub-band energy in this case which can be expressed as

$$0 = \Delta_{27}(E'_{200HDNW},n_x,n_y,\eta_g) \tag{4.145}$$

The EM in this case is given by

$$m^*(E_{F1HDNW},\eta_g,n_y,\eta_z) = \frac{\hbar^2}{2}\frac{\partial}{\partial E}[\Delta_{27}^2(E_{F1HDNW},\eta_g,n_y,\eta_z)] \tag{4.146a}$$

The electron concentration per unit length can be written as

$$n_{01D} = \frac{g_v}{\pi}\sum_{n_x=1}^{n_{x\max}}\sum_{n_y=1}^{n_{y\max}}\left[\Delta_{27}(E_{F1HDNW},n_y,n_z,\eta_g) + \sum_{r=1}^{s} L(r)[\Delta_{27}(E_{F1HDNW},n_y,n_z,\eta_g)]\right] \tag{4.146b}$$

In the absence of band-tailing, the 1D hole energy spectrum in this case assumes the form

$$E = A_{10}\left(\frac{n_x\pi}{d_x}\right)^2 + A_{11}\left(\frac{n_y\pi}{d_y}\right)^2 + A_{12}k_z^2 + A_{13}\left(\frac{n_x\pi}{d_x}\right)$$
$$\pm\left[\left(A_{14}\left(\frac{n_x\pi}{d_x}\right)^2 + A_{15}\left(\frac{n_y\pi}{d_y}\right)^2 + A_{16}k_z^2 + A_{17}\left(\frac{n_x\pi}{d_x}\right)^2\right)\right.$$
$$\left. + A_{18}\left(\frac{n_y\pi}{d_y}\right)^2 + A_{19}^2\right]^{1/2} \quad (4.147)$$

The sub-band energy (E'_{300}) is the lowest positive root of the following equation

$$E'_{300} = A_{10}\left(\frac{n_x\pi}{d_x}\right)^2 + A_{13}\left(\frac{n_x\pi}{d_x}\right)^2$$
$$\pm\left[\left(A_{14}\left(\frac{n_x\pi}{d_x}\right)^2 + A_{17}\left(\frac{n_x\pi}{d_x}\right)^2\right) + A_{19}^2\right]^{1/2} \quad (4.148)$$

(4.147) can be expressed as

$$k_z = \Delta_{271}(E, n_x, n_y) \quad (4.149)$$

where

$$\Delta_{271}(E, n_x, n_y) = [(A_{12}^2 - A_{16}^2)^{-1}$$
$$\times [[\Delta_{211}(E, n_x, n_y)A_{12} + A_{16}\Delta_{22}(n_x, n_y)]$$
$$+ [[\Delta_{211}(E, \eta_g, n_x, n_y)A_{12} + A_{16}\Delta_{22}(n_x, n_y)]^2$$
$$- (A_{12}^2 - A_{16}^2)\Delta_{251}(E, n_x, n_y)]^{1/2}]],$$

$$\Delta_{211}(E, n_x, n_y) = \left[E - A_{10}\left(\frac{n_x\pi}{d_x}\right)^2 - A_{11}\left(\frac{n_y\pi}{d_y}\right)^2 - A_{13}\left(\frac{n_x\pi}{d_x}\right)^2\right],$$

$$\Delta_{22}(n_x, n_y) = \left[A_{14}\left(\frac{n_x\pi}{d_x}\right)^2 + A_{15}\left(\frac{n_y\pi}{d_y}\right)^2 + A_{17}\left(\frac{n_x\pi}{d_x}\right)^2\right],$$

$$\Delta_{251}(n_x, n_y, E) = \left[\Delta_{211}^2(E, n_x, n_y) - \Delta_{24}(n_x, n_y)\right],$$
$$\Delta_{24}(n_x, n_y) = \left[\Delta_{22}^2(n_x, n_y) + \Delta_{23}(n_y)\right]$$

and

$$\Delta_{23}(n_y) = \left[A_{18}\left(\frac{n_y\pi}{d_y}\right)^2 + A_{19}^2\right]$$

The DOS function in this case can be written as

$$N_{1D\Gamma}(E) = \frac{g_v}{\pi}\sum_{n_x=1}^{n_{x\max}}\sum_{n_y=1}^{n_{y\max}} \Delta'_{271}(E, n_x, n_y)H(E - E'_{300}) \quad (4.150)$$

In (4.150), E'_{300} is the sub-band energy in this case which can be expressed as

$$0 = \Delta_{271}(E'_{300}, n_x, n_y) \quad (4.151)$$

The EM in this case is given by

$$m^*(E_{F1D}, n_x, n_y) = \frac{\hbar^2}{2}\frac{\partial}{\partial E}\left[\Delta_{271}^2(E_{F1D}, n_x, n_y)\right] \quad (4.152a)$$

The electron concentration per unit length can be written as

$$n_{01D} = \frac{g_v}{\pi}\sum_{n_x=1}^{n_{x\max}}\sum_{n_y=1}^{n_{y\max}}\left[\Delta_{271}(E_{F1D}, n_x, n_y) + \sum_{r=1}^{s} L(r)[\Delta_{271}(E_{F1D}, n_x, n_y)]\right] \quad (4.152b)$$

4.2.13 *The EM in NWs of HD lead germanium telluride*

The 1D electron energy spectrum in n-type $Pb_{1-x}Ge_xTe$ under the condition of formation of band tails can be written as

$$\left[\frac{2}{1 + Erf\left(\frac{E}{\eta_g}\right)}\right]\theta_0(E, \eta_g) + \gamma_3(E, \eta_g)$$
$$\times\left[\bar{E}_{g0} - 0.195\left[\left(\frac{n_x\pi}{d_x}\right)^2 + \left(\frac{n_y\pi}{d_y}\right)^2\right] - 0.345k_z^2\right]$$

$$= \left[0.23\left[\left(\frac{n_x\pi}{d_x}\right)^2+\left(\frac{n_y\pi}{d_y}\right)^2\right]+0.02k_z^2\right]$$
$$\pm\left[0.06\bar{E}_{g0}+0.061\left[\left(\frac{n_x\pi}{d_x}\right)^2+\left(\frac{n_y\pi}{d_y}\right)^2\right]+0.0066k_z^2\right]$$
$$\times\left[\left(\frac{n_x\pi}{d_x}\right)^2+\left(\frac{n_y\pi}{d_y}\right)^2\right]^{1/2}$$
$$+\left[\bar{E}_{g0}+0.411\left[\left(\frac{n_x\pi}{d_x}\right)^2+\left(\frac{n_y\pi}{d_y}\right)^2\right]+0.377k_z^2\right]$$
$$\times\left[0.606\left[\left(\frac{n_x\pi}{d_x}\right)^2+\left(\frac{n_y\pi}{d_y}\right)^2\right]+0.722k_z^2\right]\Bigg] \quad (4.153)$$

The sub-band energy ($E_{n_{zHD500}}$) is the lowest positive root of the following equation

$$\left[\frac{2}{1+Erf\left(\frac{E_{n_{zHD500}}}{\eta_g}\right)}\right]\theta_0\left(E_{n_{zHD500}},\eta_g\right)+\gamma_3\left(E_{n_{zHD500}},\eta_g\right)$$
$$\times\left[\bar{E}_{g0}-0.195\left[\left(\frac{n_x\pi}{d_x}\right)^2+\left(\frac{n_y\pi}{d_y}\right)^2\right]\right]$$
$$=\left[0.23\left[\left(\frac{n_x\pi}{d_x}\right)^2+\left(\frac{n_y\pi}{d_y}\right)^2\right]\right.$$
$$\pm\left[0.06\bar{E}_{g0}+0.061\left[\left(\frac{n_x\pi}{d_x}\right)^2+\left(\frac{n_y\pi}{d_y}\right)^2\right]\right.$$
$$\times\left[\left(\frac{n_x\pi}{d_x}\right)^2+\left(\frac{n_y\pi}{d_y}\right)^2\right]^{1/2}$$
$$+\left[\bar{E}_{g0}+0.411\left[\left(\frac{n_x\pi}{d_x}\right)^2+\left(\frac{n_y\pi}{d_y}\right)^2\right]\right]$$
$$\times\left[0.606\left(\frac{n_x\pi}{d_x}\right)^2+\left(\frac{n_y\pi}{d_y}\right)^2\right]\Bigg]\Bigg]. \quad (4.154)$$

The EM and the 1D concentration per unit length for both the cases should be calculated numerically.

The 1D dispersion law of n-type $Pb_{1-x}Ge_xTe$ with $x = 0.01$ in the absence of band-tails can be expressed as

$$\left[E - 0.606\left[\left(\frac{n_x\pi}{d_x}\right)^2 + \left(\frac{n_y\pi}{d_y}\right)^2\right] - 0.722k_z^2\right]$$
$$\times \left[E + \overline{E}_{g_0} + 0.411\left[\left(\frac{n_x\pi}{d_x}\right)^2 + \left(\frac{n_y\pi}{d_y}\right)^2\right] + 0.377k_z^2\right]$$
$$= 0.23\left[\left(\frac{n_x\pi}{d_x}\right)^2 + \left(\frac{n_y\pi}{d_y}\right)^2\right] + 0.02k_z^2$$
$$\pm \left[0.06\overline{E}_{g_0} + 0.061\left[\left(\frac{n_x\pi}{d_x}\right)^2 + \left(\frac{n_y\pi}{d_y}\right)^2\right] + 0.0066k_z^2\right]$$
$$\times \left[\left(\frac{n_x\pi}{d_x}\right)^2 + \left(\frac{n_y\pi}{d_y}\right)^2\right]^{1/2}. \tag{4.155}$$

The sub-band energy $\bar{E}_{500}$ in this can be written as

$$\left[\bar{E}_{500} - 0.606\left[\left(\frac{n_x\pi}{d_x}\right)^2 + \left(\frac{n_y\pi}{d_y}\right)^2\right]\right]$$
$$\times \left[\bar{E}_{500} + \overline{E}_{g_0} + 0.411\left[\left(\frac{n_x\pi}{d_x}\right)^2 + \left(\frac{n_y\pi}{d_y}\right)^2\right]\right]$$
$$= 0.23\left[\left(\frac{n_x\pi}{d_x}\right)^2 + \left(\frac{n_y\pi}{d_y}\right)^2\right]$$
$$\pm \left[0.06\overline{E}_{g_0} + 0.061\left[\left(\frac{n_x\pi}{d_x}\right)^2 + \left(\frac{n_y\pi}{d_y}\right)^2\right]\right]$$
$$\times \left[\left(\frac{n_x\pi}{d_x}\right)^2 + \left(\frac{n_y\pi}{d_y}\right)^2\right]^{1/2}. \tag{4.156}$$

The EM and 1D electron statistics can be numerically calculated in this case.

4.2.14 *The EM in NWs of HD zinc and cadmium diphosphides*

The DR in HD NWs of Zinc and Cadmium diphosphides can be written as

$$\begin{aligned}\gamma_3(E,\eta_g) = &\left[\beta_1 + \frac{\beta_2\beta_{31}(k_x,n_y,n_z)}{8\beta_4}\right]\left[k_x^2 + \left(\frac{n_y\pi}{d_y}\right)^2 + \left(\frac{n_z\pi}{d_z}\right)^2\right] \\ &\pm\left\{\left[\beta_4\beta_{31}(k_x,n_y,n_z)\left(\beta_5 - \frac{\beta_2\beta_{31}(k_x,n_y,n_z)}{8\beta_4}\right)\right.\right. \\ &\times\left.\left[k_x^2 + \left(\frac{n_y\pi}{d_y}\right)^2 + \left(\frac{n_z\pi}{d_z}\right)^2\right]\right] \\ &+8\beta_4^2\left(1 - \frac{\beta_{31}^2(k_x,n_y,n_z)}{4}\right) - \beta_2\left(1 - \frac{\beta_{31}^2(k_x,n_y,n_z)}{4}\right) \\ &\times\left.\left[k_x^2 + \left(\frac{n_y\pi}{d_y}\right)^2 + \left(\frac{n_z\pi}{d_z}\right)^2\right]\right\}^{1/2} \end{aligned} \tag{4.157}$$

where

$$\beta_{31}(k_x,n_y,n_z) = \left[\frac{k_x^2 + \left(\frac{n_y\pi}{d_y}\right)^2 - 2\left(\frac{n_z\pi}{d_z}\right)^2}{\left[k_x^2 + \left(\frac{n_y\pi}{d_y}\right)^2 + \left(\frac{n_z\pi}{d_z}\right)^2\right]}\right].$$

The sub-band energy $E_{n_{zHD600}}$ in this case assumes the form

$$\begin{aligned}\gamma_3(E_{n_{zHD600}},\eta_g) = &\left[\beta_1 + \frac{\beta_2\beta_{31}(0,n_y,n_z)}{8\beta_4}\right]\left[\left(\frac{n_y\pi}{d_y}\right)^2 + \left(\frac{n_z\pi}{d_z}\right)^2\right] \\ &\pm\left\{\left[\beta_4\beta_{31}(0,n_y,n_z)\left(\beta_5 - \frac{\beta_2\beta_{31}(0,n_y,n_z)}{8\beta_4}\right)\right.\right.\end{aligned}$$

$$\times\left[\left(\frac{n_y\pi}{d_y}\right)^2+\left(\frac{n_z\pi}{d_z}\right)^2\right]\right]$$

$$+8\beta_4^2\left(1-\frac{\beta_{31}^2\left(0,n_y,n_z\right)}{4}\right)$$

$$-\beta_2\left(1-\frac{\beta_{31}^2\left(0,n_y,n_z\right)}{4}\right)$$

$$\times\left[\left(\frac{n_y\pi}{d_y}\right)^2+\left(\frac{n_z\pi}{d_z}\right)^2\right]\right]\Bigg\}^{1/2} \tag{4.158}$$

where

$$\beta_{31}\left(0,n_y,n_z\right)=\left[\frac{\left(\frac{n_y\pi}{d_y}\right)^2-2\left(\frac{n_z\pi}{d_z}\right)^2}{\left[\left(\frac{n_y\pi}{d_y}\right)^2+\left(\frac{n_z\pi}{d_z}\right)^2\right]}\right]$$

The EM and the 1D concentration per unit length should be obtained numerically.

The 1D DR in NWs of Zinc and Cadmium diphosphides in the absence of band-tails can be written as

$$E=\left[\beta_1+\frac{\beta_2\beta_{311}\left(k_x,n_y,n_z\right)}{8\beta_4}\right]\left[k_x^2+\left(\frac{n_y\pi}{d_y}\right)^2+\left(\frac{n_z\pi}{d_z}\right)^2\right]$$

$$\pm\Bigg\{\left[\beta_4\beta_{311}\left(k_x,n_y,n_z\right)\left(\beta_5-\frac{\beta_2\beta_{311}\left(k_x,n_y,n_z\right)}{8\beta_4}\right)\right.$$

$$\times\left[k_x^2+\left(\frac{n_y\pi}{d_y}\right)^2+\left(\frac{n_z\pi}{d_z}\right)^2\right]\right]$$

$$+8\beta_4^2\left(1-\frac{\beta_{311}^2\left(k_x,n_y,n_z\right)}{4}\right)-\beta_2\left(1-\frac{\beta_{311}^2\left(k_x,n_y,n_z\right)}{4}\right)$$

$$\times\left[k_x^2+\left(\frac{n_y\pi}{d_y}\right)^2+\left(\frac{n_z\pi}{d_z}\right)^2\right]\right]\Bigg\}^{1/2} \tag{4.159}$$

The sub-band energy (E_{700}) is the lowest positive root of the following equation

$$E_{700} = \left[\beta_1 + \frac{\beta_2\beta_{311}(0,n_y,n_z)}{8\beta_4}\right]\left[\left(\frac{n_y\pi}{d_y}\right)^2 + \left(\frac{n_z\pi}{d_z}\right)^2\right]$$
$$\pm \left\{ \left[\beta_4\beta_{311}(0,n_y,n_z)\left(\beta_5 - \frac{\beta_2\beta_{311}(0,n_y,n_z)}{8\beta_4}\right)\right.\right.$$
$$\times \left.\left[\left(\frac{n_y\pi}{d_y}\right)^2 + \left(\frac{n_z\pi}{d_z}\right)^2\right]\right]$$
$$+ 8\beta_4^2\left(1 - \frac{\beta_{311}^2(0,n_y,n_z)}{4}\right) - \beta_2\left(1 - \frac{\beta_{311}^2(0,n_y,n_z)}{4}\right)$$
$$\times \left.\left[\left(\frac{n_y\pi}{d_y}\right)^2 + \left(\frac{n_z\pi}{d_z}\right)^2\right]\right\}^{1/2} \quad (4.160)$$

The EM and the 1D concentration per unit length for both the cases should be calculated numerically.

4.3 Results and Discussion

The variation of the EM in HD NWs of different materials along the transport direction has been exhibited in the figures below at various conditions. Throughout our formalism, we have assumed the HD NWs to be of rectangular cross-sectional dimensions so that the usual "particle-in-a-box" concept can be applied along the quantized directions. The HD NWs are assumed to be degenerately doped with a carrier density starting from 10^8 m^{-1}. While deriving the closed form analytical solutions of EM in all the materials, we have also assumed that the constants of the energy band structures of the materials are independent of thickness in the range beyond 5nm. Keeping this trend in view, we have assumed the invariant property of the material energy spectrum constants and evaluated the EM in HD NWs of Cd_3As_2 and $CdGeAs_2$ as function of wire thickness along their respective transport directions in Figs. 4.1 and 4.2. It appears from Figs. 4.1 and 4.2 that the EM at the lowest subband in both

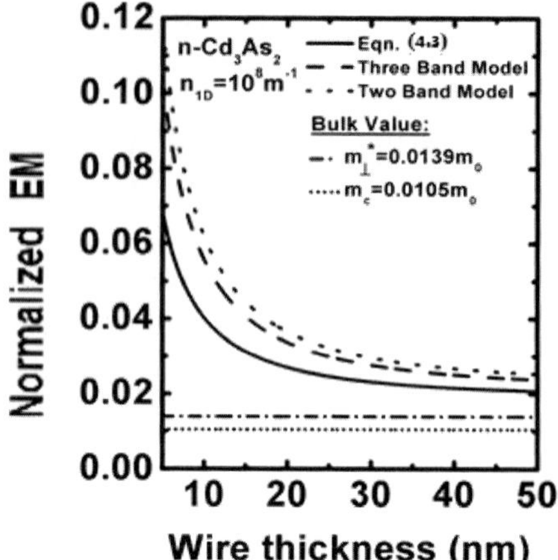

Fig. 4.1. Plot of the normalized EM as function of wire thickness for NWs of HD Cd_3As_2 considering (4.3). The plots for three and two band models of Kane have also been exhibited with their corresponding anisotropic bulk values as presented in Fig. 1.1.

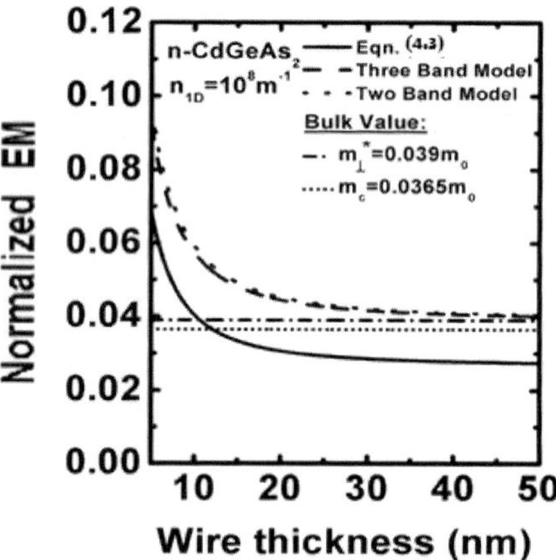

Fig. 4.2. Plot of the normalized EM as a function of wire thickness for NWs of HD $CdGeAs_2$ for all the cases of Fig. 4.1.

the cases are strong cross-sectional functions of the dimensions which converge to their corresponding bulk values at larger dimensions. The effect of crystal field splitting in case of dispersion relation of Cd_3As_2 lets the asymptotic fall to be closer to the bulk value. It should be noted that all the curves have been evaluated at T = 4 K where the average thermal energy i.e., $E_F + k_BT$ is very less than that of the difference of the adjacent subband energies. This leads the carrier to reside in the lowest sub-bands only.

Figures 3.4–3.5 exhibit the variation of the EM with the wire thickness for III-V materials namely InAs, InSb and GaAs in accordance with the well-known standard non-parabolic dispersion

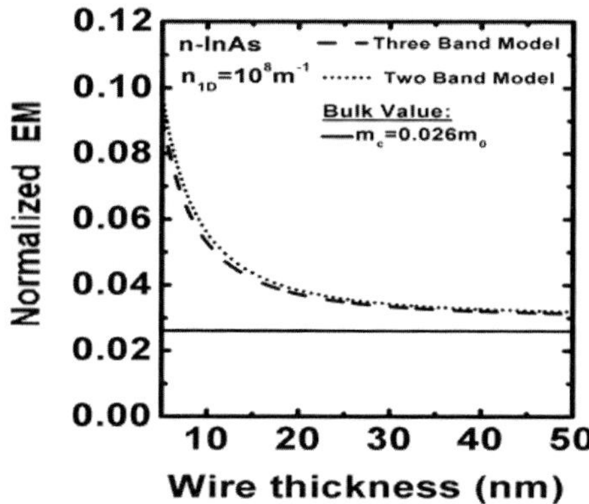

Fig. 4.3. Plot of the normalized EM as function of wire thickness for NWs of HD InAs considering the three and two band models of Kane with the corresponding isotropic bulk value.

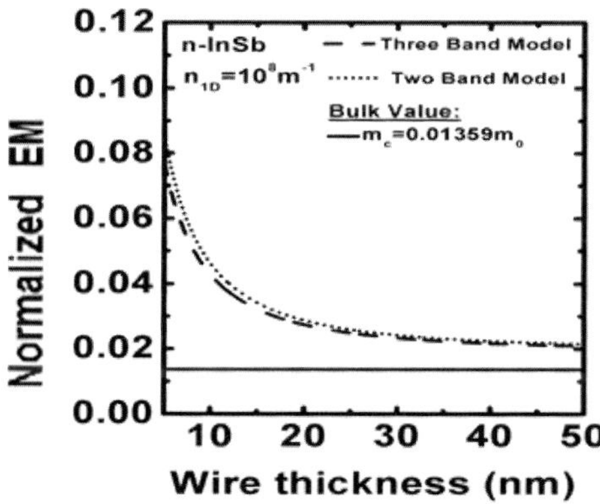

Fig. 4.4. Plot of the normalized EM as function of wire thickness for NWs of HD InSb considering the three and two band models of Kane with the corresponding isotropic bulk value.

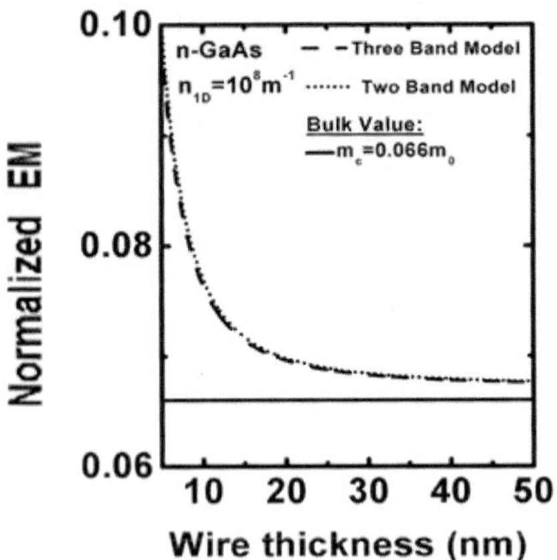

Fig. 4.5. Plot of the normalized EM as function of wire thickness for NWs of HD GaAs considering the three and two band models of Kane with the corresponding isotropic bulk value.

relation of Kane. The reader is expected to evaluate the EM for other models as derived in this chapter for III-V materials.

It appears from these figures that the difference in the energy band models in predicting the EM is almost insignificant. Hence for all practical purposes for determination of EM, the second order model of Kane can fit well. It should be noted that the EM for these materials as presented here can be compared with that of the EM in 2D systems as given in Chapter 1.

A quick view can lead us to interpret that the EM for both the case are same. However, it should also be kept in mind about the difference in the wire thickness and carrier concentration. All the curves in this chapter have been evaluated at those concentrations for which the EM stand close to that of their corresponding 2D systems. Further in deriving the results, we have assumed that the conduction band valley does not splits along the channel transport direction, which is a usual case with Silicon nanowire along [110] and [111] valleys (Chapter 8). Figures 4.6 and 4.7 exhibit the variation of the EM for $Hg_{1-x}Cd_xTe$ and $In_{1-x}Ga_xAs_{1-y}P_y$ considering all the aforementioned cases at $x = 0.3$. In Figs. 4.8–4.14, we have exhibited the variation of the EM as function of carrier degeneracy.

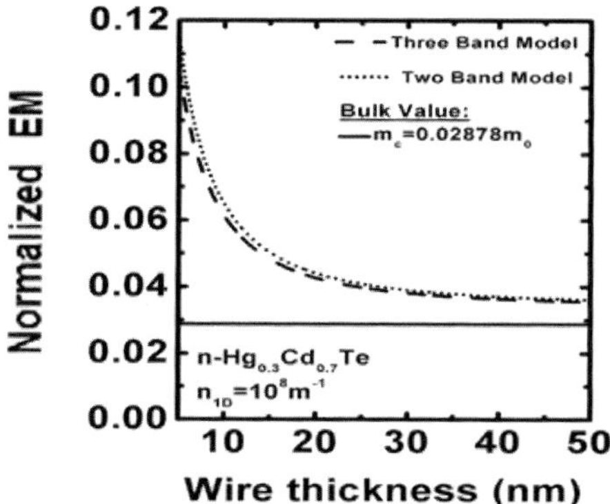

Fig. 4.6. Plot of the normalized EM as function of wire thickness for NWs of HD $Hg_{1-x}Cd_xTe$ considering the three and two band models of Kane with the corresponding isotropic bulk value at x = 0.3.

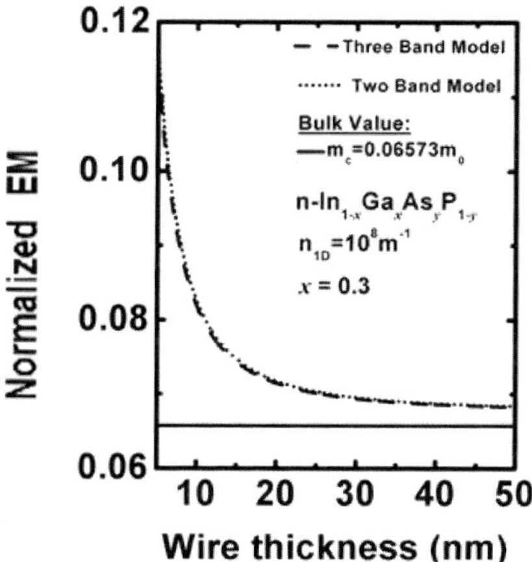

Fig. 4.7. Plot of the normalized EM as function of wire thickness for NWs of HD $In_{1-x}Ga_xAs_{1-y}P_y$ considering the three and two band models of Kane with the corresponding isotropic bulk value at x = 0.3.

It appears from the said figures that the EM increases with the increase in the degeneracy. The EM rises sharply above 10^7 m^{-1} for all the materials in an exponential way due to the presence of the Fermi-Dirac probability factor in the respective carrier concentration equation. Figure 4.15 exhibits the variation of the EM as function of

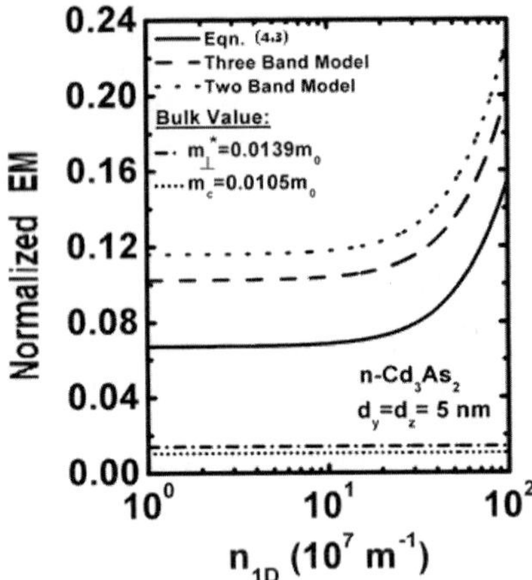

Fig. 4.8. Plot of the normalized EM as function of carrier degeneracy for HD n-Cd_3As_2 NW for all cases of Fig. 4.1.

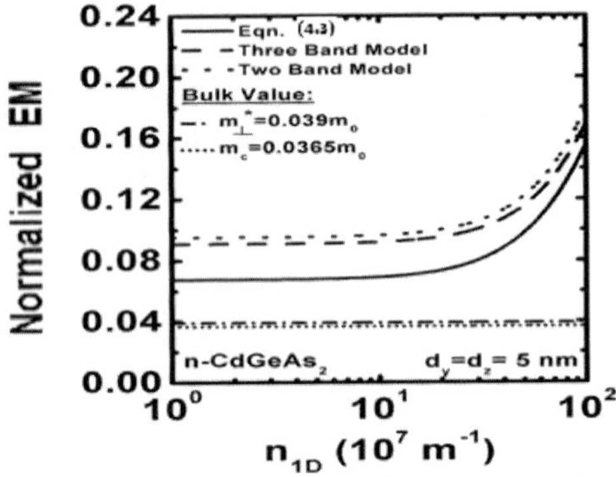

Fig. 4.9. Plot of the normalized EM as function of carrier degeneracy for NWs of HD $CdGeAs_2$ for all cases of Fig. 4.2.

the alloy composition in ternary and quaternary systems. It appears that as in the previous cases of quantum confinements, the EM in this case also exhibits an increasing variation with x. We have also plotted the variation of the bulk effective mass as x varies to present a comparative view. The influence of band non-parabolicity on the EM in these two materials can easily be seen. In both the cases, we

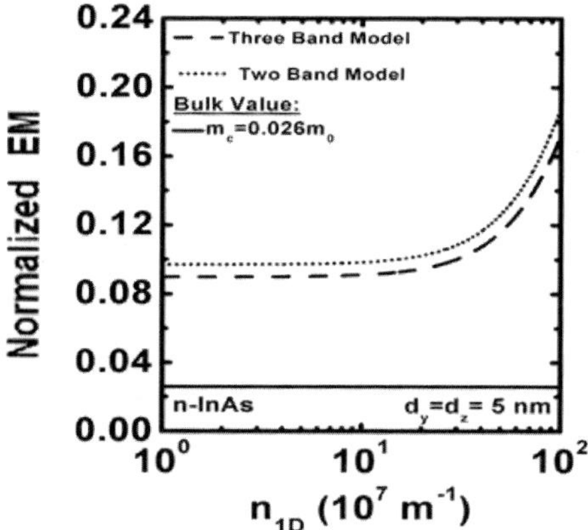

Fig. 4.10. Plot of the normalized EM as function of carrier degeneracy for n-InAs NW for all cases of Fig. 4.3.

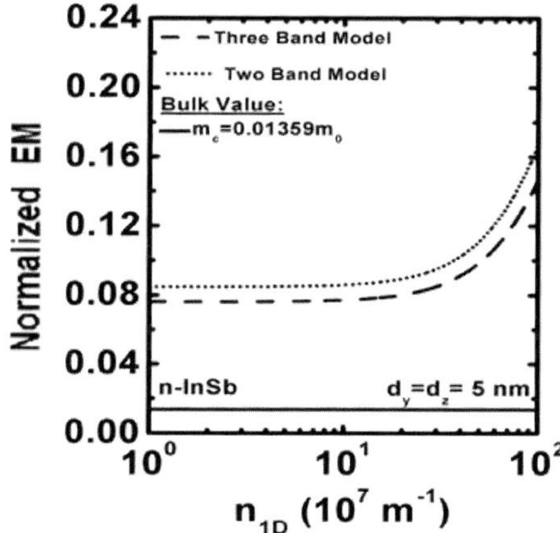

Fig. 4.11. Plot of the normalized EM as function of carrier degeneracy for HD n-InSb NW for all cases of Fig. 4.4.

see that the EM is a slow variation function of x due to the change in band gap.

The effect of dimensionality on nanowire of HD PbS has been exhibited in Figs. 4.16 and 4.17 for the energy band models of Dimmock, Bangert and Kastner, Cohen and Lax. It appears that as the dimension reduces in Fig. 4.16, the EM increases which is

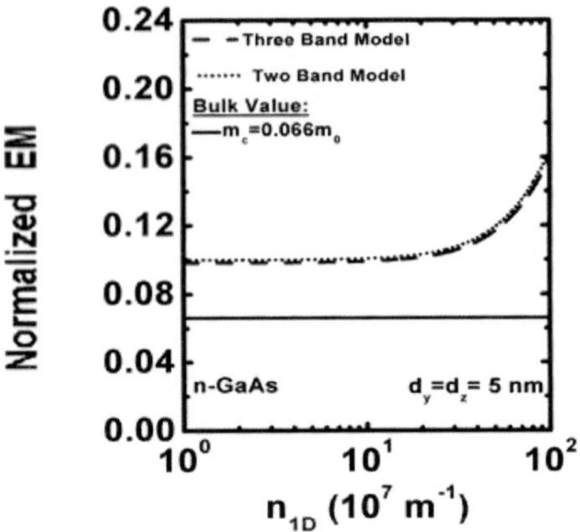

Fig. 4.12. Plot of the normalized EM as function of carrier degeneracy for HD n-GaAs NW for all cases of Fig. 4.5.

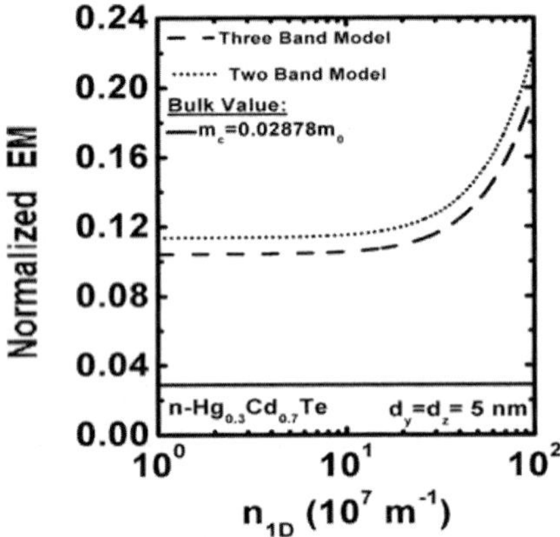

Fig. 4.13. Plot of the normalized EM as function of carrier degeneracy for HD n-$Hg_{1-x}Cd_xTe$ NW for all cases of Fig. 4.6.

generally accepted. However, using Dimmock model the EM tends to decrease in the sub-5 nm regime there by unfolding the validity of the model in this zone. However at large thicknesses, the entire model tends to their corresponding bulk value $0.00194m_0$. The variation of the EM as function of carrier degeneracy has further been plotted using the aforementioned band structure models. It appears from the two figures that the effect of different band structure models

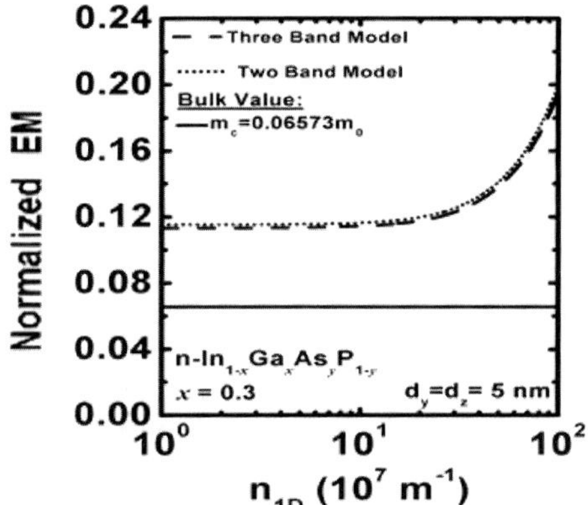

Fig. 4.14. Plot of the normalized EM as function of carrier degeneracy for HD n-$In_{1-x}Ga_xAs_{1-y}P_y$ NW for all cases of Fig. 4.8.

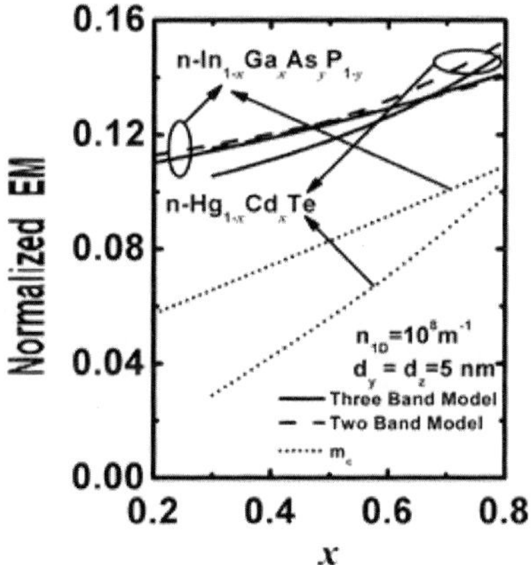

Fig. 4.15. Plot of the normalized EM as function of alloy composition for HD n-$Hg_{1-x}Cd_xTe$ and n-$In_{1-x}Ga_xAs_{1-y}P_y$ NWs.

has significantly less deviation from one another except for model of Dimmock.

The EM is found to increase almost linear with degeneracy $2 \times 10^8\,\mathrm{m}^{-1}$ and beyond using the all the models. In case of PbTe, the EM rises sharply with decreasing wire thickness below sub-15 nm from the bulk value $0.098m_0$ at carrier degeneracy of 10^9 m^{-1}, which can affect the carrier mobility strongly.

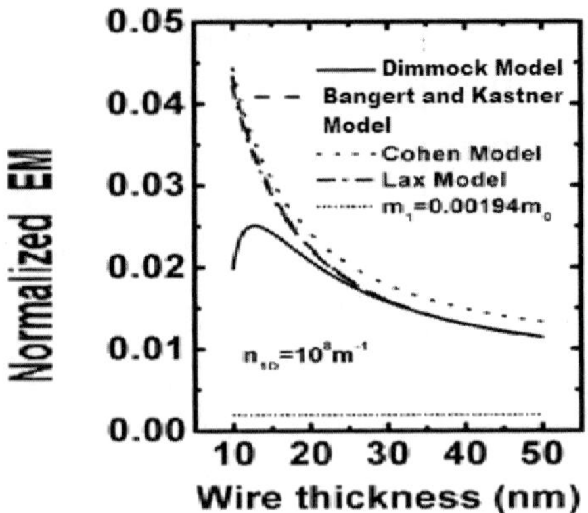

Fig. 4.16. Plot of the normalized EM as function of wire thickness for HD Bismuth NW.

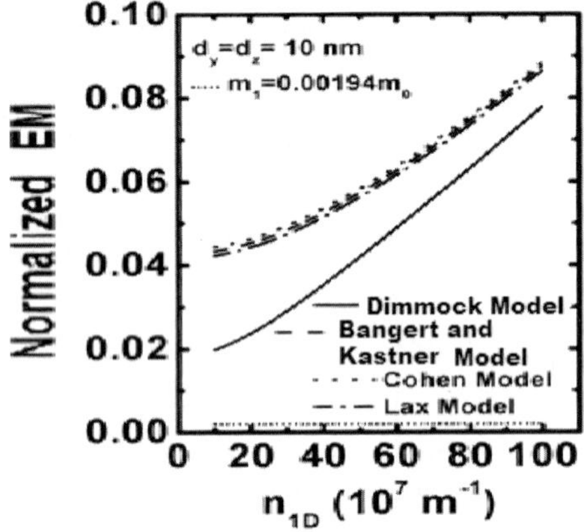

Fig. 4.17. Plot of the normalized EM as function of carrier degeneracy for HD Bismuth NW.

The effect of strain on stressed InSb nanowires has been exhibited in Fig. 4.19 for two different momentum matrix elements to signify its importance as dimension reduces. A compressive strain of 3% has been applied along x and z directions to predict the variation of the EM along the transport direction. At this point, we could not provide the influence of strain on the energy band structure of InSb nanowire due to the lack in both experimental and simulation investigations.

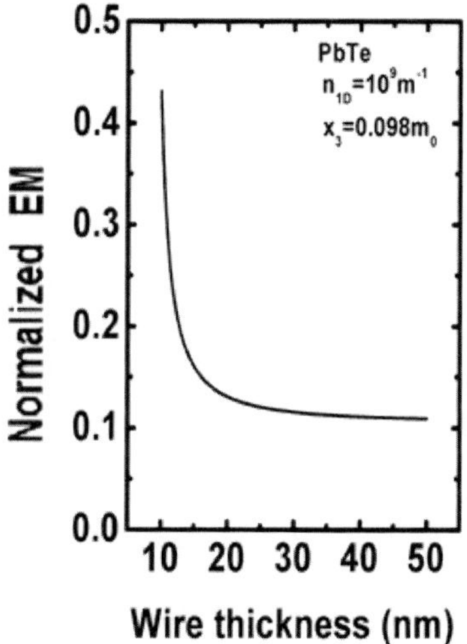

Fig. 4.18. Plot of the normalized EM as function of carrier degeneracy for HD PbTe NW.

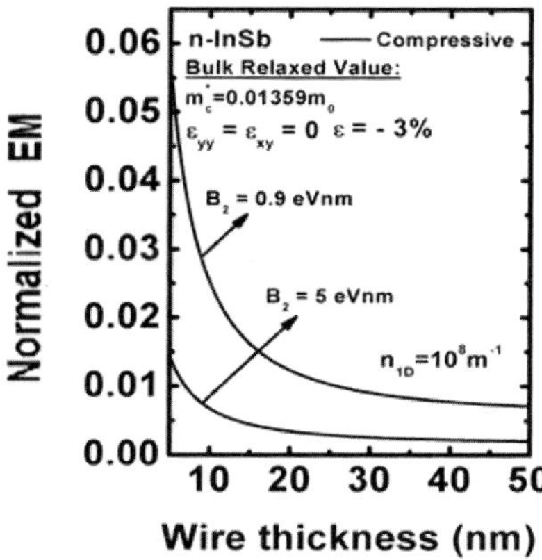

Fig. 4.19. Plot of the normalized EM as function of wire thickness for stressed HD InSb NW.

Figure 4.20 exhibits the effect of cross-sectional dimension on the EM in n-Ge, n-GaSb and topological insulators like n-Bi_2Te_3. We have used the Cardona model to evaluate the EM in Ge nanowires. The reader is left with the corresponding evaluation of the EM for Wang *et al.* model of the same. It appears that the EM in n-Ge and

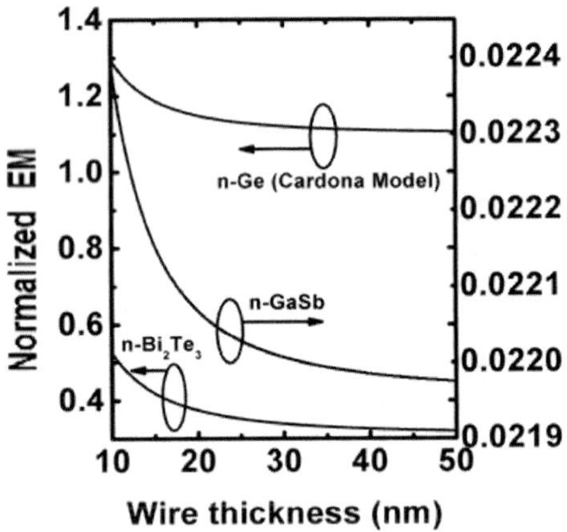

Fig. 4.20. Plot of the normalized EM as function of wire thickness for HD Ge, GaSb and Bi_2Te_3 NWs.

Bi_2Te_3 evolve from its corresponding bulk value significantly from below 15 nm. However in case of GaSb, the EM is almost invariant.

4.4 Open Research Problems

(R.4.1) Investigate the EM for NWs of all of the HDmaterials in the presences of Gaussian, exponential, Kane, Halperian, Lax and Bonch-Burevich types of band tails for all systems whose unperturbed carrier energy spectra are defined in R.1.1.

(R.4.2) Investigate the EM in the presence of strain for NWs of all of the HDmaterials of the negative refractive index, organic, magnetic and other advanced optical materials.

(R.4.3) Investigate the EM for the NWs of HD negative refractive index, organic, magnetic and other advanced optical materials in the presence of an arbitrarily oriented alternating electric field.

(R.4.4) Investigate the EM for the multiple NWs of HD materials whose unperturbed carrier energy spectra are defined in R.1.1.

(R.4.5) Investigate the EM for all the appropriate HD one dimensional systems of this chapter in the presence of finite potential wells.

(R.4.6) Investigate the EM for all the appropriate HD one dimensional systems of this chapter in the presence of parabolic potential wells.

(R.4.7) Investigate the EM for HD one dimensional systems of the negative refractive index and other advanced optical materials in the presence of an arbitrarily oriented alternating electric field and non-uniform light waves and in the presence of strain.

(R.4.8) Investigate the EM for triangular HD one dimensional systems of the negative refractive index, organic, magnetic and other advanced optical materials in the presence of an arbitrarily oriented alternating electric field in the presence of strain.

(R.4.9) Investigate the EM for all the problems of (R.1.1) in the presence of arbitrarily oriented magnetic field.

(R.4.10) Investigate the EM for all the problems of (R.1.1) in the presence of alternating electric field.

(R.4.11) Investigate the EM for all the problems of (R.1.1) in the presence of alternating magnetic field.

(R.4.12) Investigate the EM for all the problems of (R.1.1) in the presence of crossed electric field and quantizing magnetic fields.

(R.4.13) Investigate the EM for all the problems of (R.1.1) in the presence of crossed alternating electric field and alternating quantizing magnetic fields.

(R.4.14) Investigate the EM for HD NWs of the negative refractive index, organic and magnetic materials.

(R.4.15) Investigate the EM for HD NWs of the negative refractive index, organic and magnetic materials in the presence of alternating time dependent magnetic field.

(R.4.16) Investigate the EM for HD NWs of the negative refractive index, organic and magnetic materials in the presence of in the presence of crossed alternating electric field and alternating quantizing magnetic fields.

(R.4.17) (a) Investigate the EM for HD NWs of the negative refractive index, organic, magnetic and other advanced optical materials in the presence of an arbitrarily oriented alternating electric field considering many body effects.

(b) Investigate all the appropriate problems of this chapter for a Dirac electron.

(R.4.18) Investigate all the appropriate problems of this chapter by including the many body, image force, broadening and hot carrier effects respectively.

(R.4.19) Investigate all the appropriate problems of this chapter by removing all the mathematical approximations and establishing the respective appropriate uniqueness conditions.

References

1. P. Harrison, Quantum Wells, Wires and Dots, John Wiley and Sons, Ltd, (2002); B. K. Ridley, Electrons and Phonons in semiconductors multilayers, Cambridge University Press, Cambridge (1997); G. Bastard, Wave mechanics applied to semiconductor heterostructures, Halsted; Les Ulis, Les Editions de Physique, New York (1988); V. V. Martin, A. A. Kochelap and M. A. Stroscio, Quantum Heterostructures, Cambridge University Press, Cambridge (1999).
2. C. S. Lent and D. J. Kirkner, *J. Appl. Phys.* **67**, 6353 (1990); F. Sols, M. Macucci, U. Ravaioli and K. Hess, *Appl. Phys. Lett.* **54**, 350 (1980).
3. C. S. Kim, A. M. Satanin, Y. S. Joe and R. M. Cosby, *Phys. Rev. B*, **60**, 10962 (1999).
4. S. Midgley and J. B. Wang, *Phys. Rev. B* **64**, 153304 (2001).
5. T. Sugaya, J. P. Bird, M. Ogura, Y. Sugiyama, D. K. Ferry and K. Y. Jang, *Appl. Phys. Lett.* **80**, 434 (2002).
6. B. Kane, G. Facer, A. Dzurak, N. Lumpkin, R. Clark, L. PfeiKer and K. West, *Appl. Phys. Lett.* **72**, 3506 (1998).
7. C. Dekker, *Physics Today*, **52**, 22 (1999).
8. A. Yacoby, H. L. Stormer, N. S. Wingreen, L. N. Pfeiffer, K. W. Baldwin and K. W. West, *Phys. Rev. Lett.* **77**, 4612 (1996).
9. Y. Hayamizu, M. Yoshita, S. Watanabe, H. A. L. PfeiKer and K. West, *Appl. Phys. Lett.* **81**, 4937 (2002).
10. S. Frank, P. Poncharal, Z. L. Wang and W. A. Heer, *Science* **280**, 1744 (1998).

11. I. Kamiya, I. I. Tanaka, K. Tanaka, F. Yamada, Y. Shinozuka and H. Sakaki, *Physica E* **13**, 131 (2002).
12. A. K. Geim, P. C. Main, N. LaScala, L. Eaves, T. J. Foster, P. H. Beton, J. W. Sakai, F. W. Sheard, M. Henini, G. Hill et al., *Phys. Rev. Lett.* **72**, 2061 (1994).
13. A. S. Melinkov and V. M. Vinokur, *Nature* **415**, 60(2002).
14. K. Schwab, E. A. henriksen, J. M. Worlock and M. L. Roukes, *Nature* **404**, 974 (2000).
15. L. Kouwenhoven, *Nature* **403**, 374 (2000).
16. S. Komiyama, O. Astafiev, V. Antonov and H. Hirai, *Nature* **403**, 405 (2000).
17. E. Paspalakis, Z. Kis, E. Voutsinas and A. F. Terziz, *Phys. Rev. B* **69**, 155316 (2004).
18. J. H. Jefferson, M. Fearn, D. L. J. Tipton and T. P. Spiller, *Phys. Rev. A* **66**, 042328 (2002).
19. J. Appenzeller, C. Schroer, T. Schapers, A. Hart, A. Froster, B. Lengeler and H. Luth, *Phys. Rev. B* **53**, 9959 (1996).
20. J. Appenzeller and C. Schroer, *J. Appl. Phys.* **87**, 31659 (2002).
21. P. Debray, O.E. Raichev, M. Rahman, R. Akis and W. C. Mitchel, *Appl. Phys. Lett.* **74**, 768 (1999).
22. P. M. Solomon, Proc. IEEE, 70, 489 (1982; T. E. Schlesinger and T. Kuech, *Appl. Phys. Lett.* **49**, 519 (1986).
23. D. Kasemet, C. S. Hong, N. B. Patel and P. D. Dapkus, *Appl. Phys. Letts.* **41**, 912 (1982); K. Woodbridge, P. Blood, E. D. Pletcher and P. J. Hulyer, *Appl. Phys. Lett.* **45**, 16 (1984); S. Tarucha and H. O. Okamoto, *Appl. Phys. Letts.* **45**, 16 (1984); H. Heiblum, D. C. Thomas, C. M. Knoedler, and M. I Nathan, *Appl. Phys. Letts.* **47**, 1105 (1985).
24. O. Aina, M. Mattingly, F. Y. Juan and P. K. Bhattacharyya, *Appl. Phys. Letts.* **50**, 43 (1987).
25. I. Suemune and L. A. Coldren, IEEE, *J. Quant. Electronic.* **24**, 1178 (1988).
26. D. A. B. Miller, D. S. Chemla, T. C. Damen, J. H. Wood, A. C. Burrus, A. C. Gossard and W. Weigmann, IEEE, *J. Quant. Electron.* **21**, 1462 (1985).
27. J. S. Blakemore, Semiconductor Statistics, Dover, New York (1987); K P Ghatak, S Bhattacharya, S K Biswas, A Deyand A K Dasgupta, Phys. Scr., **75**, 820, (2007).

Chapter 5

The EM in Heavily Doped (HD) Non-Parabolic Semiconductors Under Magnetic Quantization

5.1 Introduction

It is well known that the band structure of semiconductors can be drastically changed by applying the external fields. The effects of the quantizing magnetic field on the band structure of compound semiconductors are more striking and can be observed easily in experiments [1–3]. Under magnetic quantization, the motion of the electron parallel to the magnetic field remains unaltered while the area of the wave vector space perpendicular to the direction of the magnetic field gets quantized in accordance with the Landau's rule of area quantization in the wave-vector space [3]. The energy levels of the carriers in a magnetic field (with the component of the wave-vector parallel to the direction of magnetic field be equated with zero) are termed as the Landau levels and the quantized energies are known as the Landau sub-bands. It is important to note that the same conclusion may be arrived either by solving the single-particle time independent Schrödinger differential equation in the presence of a quantizing magnetic field or by using the operator method. The quantizing magnetic field tends to remove the degeneracy and increases the band gap. A semiconductor, placed in a magnetic field B, can absorb radiative energy with the frequency $(\omega_0 = (|e|B/m_c))$. This phenomenon is known as cyclotron or diamagnetic resonance. The effect of energy quantization is experimentally noticeable when the separation between any two consecutive Landau levels is greater than k_BT. A number of interesting transport phenomena originate from the change in the basic band structure of the semiconductor in the presence of quantizing magnetic field. These have been widely been investigated and also served as diagnostic tools for characterizing the different materials having various band structures [4–7]. The discreteness in the Landau levels leads to a whole crop of magneto-oscillatory phenomena, important among which are (i) Shubnikov-de Haas oscillations in magneto-resistance; (ii) de Haas-van Alphen oscillations in magnetic susceptibility; (iii) magneto-phonon oscillations in thermoelectric power, etc.

In this chapter in Section 5.2.1, of the theoretical background, the DR has been investigated in HD non linear optical semiconductors in the presence of a quantizing magnetic field. Section 5.2.2 contains the results for HD III–V, ternary and quaternary compounds in accordance with the three and the two band models of Kane. In the same section the DR in accordance with the models of Stillman *et al.* and Palik *et al.* have also been studied for the purpose of relative comparison. Section 5.2.3 contains the study of the DR for HD II–VI semiconductors under magnetic quantization. In Section 5.2.4, the DR in HD IV–VI materials has been discussed in accordance with the models of Cohen, Lax, Dimmock, Bangert and Kastner and Foley and Landenberg respectively. In Section 5.2.5, the magneto-DR for the stressed HD Kane type semiconductors has been investigated. In Section 5.2.6, the DR in HD Te has been studied under magnetic quantization. In Section 5.2.7, the magneto-DR in n-GaP has been studied. In Section 5.2.8, the DR in HD $PtSb_2$ has been explored under magnetic quantization. In Section 5.2.9, the magneto-DR in HD Bi_2Te_3 has been studied. In Section 5.2.10, the DR in HD Ge has been studied under magnetic quantization in accordance with the models of Cardona *et al.* and Wang and Ressler respectively. In Sections 5.2.11 and 5.2.12, the magneto-DR in HD n-GaSb and II–V compounds has respectively been studied. In Section 5.2.13 the magneto DR in HD $Pb_{1-x}Ge_x$Te has been discussed. Section 5.3 explores the discussion and the last Section 5.4 contains 15 open research problems.

5.2 Theoretical Background

5.2.1 *The EM in HD nonlinear optical semiconductors under magnetic quantization*

The DR under magnetic quantization in non-linear optical materials can be written as

$$\gamma(E) = \frac{\hbar^2 k_s^2}{2m_\perp^*} f_3(E) + \frac{\hbar^2 k_z^2}{2m_\parallel^*} f_4(E) \pm \frac{eB\hbar E_g}{6}\left[\frac{(E_{g_0}+\Delta_\perp)}{(E_{g_0}+\frac{2}{3}\Delta_\perp)}\right]\left[E + E_{g_0} + \delta + \frac{\Delta_\parallel^2 - \Delta_\perp^2}{3\Delta_\parallel}\right] \tag{5.1}$$

where

$$f_3(E) = \frac{E_g(E_g+\Delta_\perp)}{(E_g+\frac{2}{3}\Delta_\perp)}\left[(E+E_g)\left(E+E_g+\frac{2}{3}\Delta_\parallel\right) + \delta\left(E+E_g+\frac{1}{3}\Delta_\parallel\right) + \frac{1}{9}\left(\Delta_\parallel^2-\Delta_\perp^2\right)\right]$$

and

$$f_4(E) = \frac{E_g(E_g+\Delta_\parallel)}{(E_g+\frac{2}{3}\Delta_\parallel)}\left[(E+E_g)\left(E+E_g+\frac{2}{3}\Delta_\parallel\right)\right]$$

(5.1) can be expressed as

$$\frac{\hbar^2 k_z^2}{2m_\parallel^*} + \left(\frac{b_\parallel c_\perp}{b_\perp c_\parallel}\right)\left(\frac{\hbar^2 k_s^2}{2m_\perp^*}\right) = \left\{\left[\frac{ab_\parallel}{c_\parallel}E^2 + \frac{(ac_\parallel + b_\parallel - ab_\parallel)}{c_\parallel^2}E + \frac{1}{c_\parallel}\left(1-\frac{a}{c_\parallel}\right)\left(1-\frac{b_\parallel}{c_\parallel}\right) - \frac{1}{c_\parallel}\left(1-\frac{a}{c_\parallel}\right)\left(1-\frac{b_\parallel}{c_\parallel}\right)\frac{1}{(c_\parallel E+1)}\right] + \frac{ab_\parallel}{c_\parallel}\left[\delta E + \frac{2}{9}\left(\Delta_\parallel^2-\Delta_\perp^2\right)\right] - \frac{2}{9}\frac{ab_\parallel}{c_\parallel}\frac{\left(\Delta_\parallel^2-\Delta_\perp^2\right)}{(c_\parallel E+1)}\right\}$$

$$-\left(\frac{\hbar^2 k_s^2}{2m_\perp^*}\right)\left\{\left(\frac{b_\parallel c_\perp}{b_\perp c_\parallel}\right)\left[\left(\frac{\delta}{2}+\frac{\Delta_\parallel^2-\Delta_\perp^2}{6\Delta_\parallel}\right)\frac{a}{(aE+1)}\right.\right.$$

$$\left.\left.+\left(\frac{\delta}{2}-\frac{\Delta_\parallel^2-\Delta_\perp^2}{6\Delta_\parallel}\right)\frac{c_\parallel}{(c_\parallel E+1)}\right]\right\}$$

$$\pm e_1\left[\frac{\rho_1}{E+E_g}+\frac{\rho_2}{E+E_g+\frac{2}{3}\Delta_\parallel}\right] \tag{5.2}$$

where

$$e_1=\left[\frac{eB\hbar\left(E_g+\frac{2}{3}\Delta_\parallel\right)}{6\left(E_g+\Delta_\parallel\right)}\frac{\left(E_g+\Delta_\perp\right)}{\left(E_g+\frac{2}{3}\Delta_\perp\right)}\right],\quad \rho_1=\frac{\left(-E_g+G_1\right)}{\left(\frac{2}{3}\Delta_\parallel\right)},$$

$$G_1=\left[E_{g0}+\delta+\frac{\Delta_\parallel^2-\Delta_\perp^2}{3\Delta_\parallel}\right]$$

and

$$\rho_2=\frac{3}{2\Delta_\parallel}\left[\frac{\Delta_\parallel}{3}+\frac{\Delta_\perp^2}{3\Delta_\parallel}-\delta\right]$$

Therefore, the DR of the conduction electrons in heavily doped non-linear optical semiconductors in the presence of a quantizing magnetic field B can be written following the methods as developed in Chapter 1 as

$$\frac{\hbar^2 k_z^2}{2m_\parallel^*}=U_{1,\pm}\left(E,n,\eta_g\right)+iU_{2,\pm}\left(E,n,\eta_g\right) \tag{5.3}$$

where

$$U_{1,\pm}\left(E,n,\eta_g\right)=\frac{-eB}{m_\perp^*}\left(n+\frac{1}{2}\right)\left(\frac{b_\parallel c_\perp}{c_\parallel b_\perp}\right)+\left[\frac{1+Erf\left(E/\eta_g\right)}{2}\right]^{-1}$$

$$\cdot\left[\left\{\left[\frac{ab_\parallel}{c_\parallel}\theta_0\left(E,\eta_g\right)+\frac{ac_\parallel+b_\parallel c_\parallel-ab_\parallel}{c_\parallel^2}\gamma_0\left(E,\eta_g\right)\right.\right.\right.$$

$$+\frac{1}{c_{\|}}\left(1-\frac{a}{c_{\|}}\right)\left(1-\frac{b_{\|}}{c_{\|}}\right)\left[\frac{1+Erf\left(E/\eta_g\right)}{2}\right]$$

$$-\frac{1}{c_{\|}}\left(1-\frac{a}{c_{\|}}\right)\left(1-\frac{b_{\|}}{c_{\|}}\right)c\left(\beta_1,E,\eta_g\right)$$

$$+\frac{ab_{\|}}{c_{\|}}\left[\delta\gamma_0\left(E,\eta_g\right)+\frac{2}{9}\left(\Delta_{\|}^2-\Delta_{\perp}^2\right)\right.$$

$$\left.\cdot\left[\frac{1+Erf\left(E/\eta_g\right)}{2}\right]\right]$$

$$\left.-\frac{2}{9}\frac{ab_{\|}}{c_{\|}}\left(\Delta_{\|}^2-\Delta_{\perp}^2\right)c\left(\beta_1,E,\eta_g\right)\right\}-\frac{\hbar eB}{m_{\perp}^*}\left(n+\frac{1}{2}\right)$$

$$\cdot\left\{\left(\frac{b_{\|}c_{\perp}}{c_{\|}b_{\perp}}\right)\left[\left(\frac{\delta}{2}+\frac{\Delta_{\|}^2-\Delta_{\perp}^2}{6\Delta_{\|}}\right)ac\left(\beta_2,E,\eta_g\right)\right.\right.$$

$$\left.\left.+\left(\frac{\delta}{2}+\frac{\Delta_{\|}^2-\Delta_{\perp}^2}{6\Delta_{\|}}\right)c_{\|}c\left(\beta_1,E,\eta_g\right)\right]\right\}$$

$$\pm e_1\left[\frac{\rho_1}{E_g}\left(\beta_2,E,\eta_g\right)+\frac{\rho_2}{\left(E_g+\frac{2}{3}\Delta_{\|}\right)}c\left(\beta_3,E,\eta_g\right)\right],$$

$$U_{2,\pm}\left(E,n,\eta_g\right)=\left[\frac{1+Erf\left(E/\eta_g\right)}{2}\right]^{-1}\left[\frac{1}{c_{\|}}\left(1-\frac{a}{c_{\|}}\right)\left(1-\frac{b_{\|}}{c_{\|}}\right)\right.$$

$$\cdot D\left(\beta_1,E,\eta_g\right)+\frac{2}{9}\frac{ab_{\|}}{c_{\|}}\left(\Delta_{\|}^2-\Delta_{\perp}^2\right)D\left(\beta_1,E,\eta_g\right)$$

$$+\frac{\hbar eB}{m_{\perp}^*}\left(n+\frac{1}{2}\right)\left(\frac{b_{\|}c_{\perp}}{c_{\|}b_{\perp}}\right)\left[\left(\frac{\delta}{2}+\frac{\Delta_{\|}^2-\Delta_{\perp}^2}{6\Delta_{\|}}\right)\right.$$

$$\left.\cdot aD\left(\beta_2,E,\eta_g\right)+\left(\frac{\delta}{2}-\frac{\Delta_{\|}^2-\Delta_{\perp}^2}{6\Delta_{\|}}\right)c_{\|}D\left(\beta_1,E,\eta_g\right)\right]$$

$$\left.\mp e_1\left[\frac{\rho_1}{E_g}D\left(\beta_2,E,\eta_g\right)+\frac{\rho_2}{\left(E_g+\frac{2}{3}\Delta_{\|}\right)}D\left(\beta_3,E,\eta_g\right)\right]\right],$$

$$C(\beta_i, E, \eta_g) = \left[\frac{2}{\beta_i \eta_g \sqrt{\pi}}\right] \exp\left(-u_i^2\right) \times \left[\sum_{p=1}^{\infty} \left\{\frac{\exp\left(\frac{-p^2}{4}\right)}{p}\right\} \sinh(pu_i)\right],$$

$$u_i = \frac{1 + \beta_i E}{\beta_i \eta_g}$$

and

$$D_i(\beta_i, E, \eta_g) = \left[\frac{\sqrt{\pi}}{\beta_i \eta_g} \exp\left(-u_i^2\right)\right]$$

The EEM at the Fermi Level can be written from (6.3) as

$$m_{\pm}^*(E_{FBHD}, n, \eta_g) = m_{\|}^* U'_{1,\pm}(E_{FBHD}, n, \eta_g) \tag{5.4}$$

where (E_{FBHD}) the Fermi energy in this case.

Therefore the double valued EEM in this case is a function of Fermi energy, magnetic field, quantum number and the scattering potential together with the fact that the EEM exists in the band gap which is the general characterstics of HD materials.

The complex density of states function under magnetic quantization is given by

$$N_B(E) = N_{BR}(E) + iN_{BI}(E) = \frac{eB}{2\pi^2\hbar^2}\sqrt{2m_{\|}^*}\sum_{n=0}^{n_{\max}}\left[\frac{x'}{2\sqrt{x}} + \frac{iy'}{2\sqrt{y}}\right] \tag{5.5}$$

where

$$x = \frac{\sqrt{(U_{1,\pm}(E,n,\eta_g))^2 + (U_{2,\pm}(E,n,\eta_g))^2} + (U_{1,\pm}(E,n,\eta_g))}{2}$$

$$y = \frac{\sqrt{(U_{1,\pm}(E,n,\eta_g))^2 + (U_{2,\pm}(E,n,\eta_g))^2} - (U_{1,\pm}(E,n,\eta_g))}{2}$$

and x' and y' are the differentiations of x and y with respect to energy E.

Therefore, from (7.5) we can write

$$N_{BR}(E) = \frac{eB}{4\pi^2\hbar^2}\sqrt{2m_{\|}^*}\sum_{n=0}^{n_{\max}}\frac{x'}{\sqrt{x}} \tag{5.6}$$

and

$$N_{BI}(E) = \frac{eB}{4\pi^2\hbar^2}\sqrt{2m_{\parallel}^*}\sum_{n=0}^{n_{\max}}\frac{y'}{\sqrt{y}} \tag{5.7}$$

where E_{nHD} is the complex Landau sub-band energy which can be obtained from (6.3) by substituting $k_z = 0$ and $E = E_{nHD}$.

5.2.2 *The DR in HD III–V semiconductors under magnetic quantization*

a) The electron energy spectrum in III–V semiconductors under magnetic quantization is given by

$$\frac{E(E+E_g)(E+E_g+\Delta)\left(E_g+\frac{2}{3}\Delta\right)}{E_g(E_g+\Delta)\left(E+E_g+\frac{2}{3}\Delta\right)} = \left(n+\frac{1}{2}\right)\hbar\omega_0 + \frac{\hbar^2k_z^2}{2m_c} \pm \frac{eB\hbar\Delta}{6m_c\left(E+E_g+\frac{2}{3}\Delta\right)} \tag{5.8}$$

(5.8) can be written as

$$\left[\frac{ab}{c}E^2 + \left(\frac{ac+bc-ab}{c^2}\right)E + \frac{1}{c}\left(1-\frac{a}{c}\right)\left(1-\frac{b}{c}\right) - \frac{1}{c}\left(1-\frac{a}{c}\right)\left(1-\frac{b}{c}\right)\frac{1}{(1+cE)}\right] = \left(n+\frac{1}{2}\right)\hbar\omega_0 + \frac{\hbar^2k_z^2}{2m_c} + \frac{eB\hbar\Delta}{6m_c(1+cE)\left(E_g+\frac{2}{3}\Delta\right)}$$

where

$$a = \frac{1}{E_g}, b = \frac{1}{E_g+\Delta} \quad \text{and} \quad c = \frac{1}{E_g+\frac{2}{3}\Delta}$$

Therefore

$$\frac{ab}{c}I(5) + \left(\frac{ab+bc-ab}{c^2}\right)I(4) + \left[\frac{1}{c}\left(1-\frac{a}{c}\right)\left(1-\frac{b}{c}\right) - \left(n+\frac{1}{2}\right)\hbar\omega_0\right]$$

$$I(1) - g_{\pm}\left[G(C,E,\eta_g) - iH(C,E,\eta_g)\right] = \frac{\hbar^2k_z^2}{2m_c}I(1) \tag{5.9}$$

where

$$g_{\pm} = \left[\frac{1}{c}\left(1-\frac{a}{c}\right)\left(1-\frac{b}{c}\right) \pm \frac{eB\hbar\Delta}{6m_c\left(E_g+\frac{2}{3}\Delta\right)}\right],$$

$$G\left(C,E,\eta_g\right) = \left[\frac{2}{C\eta_g\sqrt{\pi}}\right]\exp\left(-u^2\right)$$
$$\times\left[\sum_{p=1}^{\infty}\left\{\frac{\exp\left(\frac{-p^2}{4}\right)}{p}\right\}\sinh(pu)\right],$$

$$u = \frac{1+cE}{C\eta_g},$$

$$H\left(C,E,\eta_g\right) = \left[\frac{\sqrt{\pi}}{C\eta_g}\exp\left(-u^2\right)\right]$$

Therefore

$$\begin{aligned}\frac{\hbar^2k_z^2}{2m_c} = &\left[\frac{ab}{c}\right]\theta_0\left(E,\eta_g\right)\left[\frac{1+Erf\left(E/\eta_g\right)}{2}\right]^{-1} \\ &+\left(\frac{ac+bc-ab}{c^2}\right)\gamma_0\left(E,\eta_g\right)\left[\frac{1+Erf\left(E/\eta_g\right)}{2}\right]^{-1} \\ &+\left[\frac{1}{c}\left(1-\frac{a}{c}\right)\left(1-\frac{b}{c}\right)-\left(n+\frac{1}{2}\right)\hbar\omega_0\right] \\ &-g_{\pm}\left[\frac{1+Erf\left(E/\eta_g\right)}{2}\right]^{-1}\left[G(C,E,\eta_g)-iH(C,E,\eta_g)\right]\end{aligned} \tag{5.10}$$

Therefore the DR is given by

$$\frac{\hbar^2k_z^2}{2m_c} = U_{3,\pm}\left(E,n,\eta_g\right)+iU_{4,\pm}\left(E,\eta_g\right) \tag{5.11}$$

where

$$\begin{aligned}U_{3,\pm}\left(E,n,\eta_g\right) = &\left[\frac{ab}{c}\theta_0\left(E,\eta_g\right)\times\left[\frac{1+Erf\left(E/\eta_g\right)}{2}\right]^{-1}\right. \\ &+\left(\frac{ac+bc-ab}{c^2}\right)\gamma_0\left(E,\eta_g\right)\times\left[\frac{1+Erf\left(E/\eta_g\right)}{2}\right]^{-1}\end{aligned}$$

$$+\frac{1}{c}\left(1-\frac{a}{c}\right)\left(1-\frac{b}{c}\right)-\left(n+\frac{1}{2}\right)\hbar\omega_0$$

$$\left.+g_{\pm}\left[\frac{1+Erf\left(E/\eta_g\right)}{2}\right]^{-1}G\left(C,E,\eta_g\right)\right]$$

and

$$U_{4,\pm}\left(E,\eta_g\right)=g_{\pm}\left[\frac{1+Erf\left(E/\eta_g\right)}{2}\right]^{-1}H\left(C,E,\eta_g\right)$$

The complex Landau energy E_{nHD1} in this case can be obtained by substituting $k_z=0$ and $E=E_{nHD}$ in (6.11). The EEM at the Fermi Level can be written from (7.11) as

$$m_{\pm}^{*}\left(E_{FBHD},n,\eta_g\right)=m_{\|}^{*}U_{3,\pm}'\left(E_{FBHD},n,\eta_g\right) \tag{5.12}$$

Thus the EEM is a function of Fermi energy,Landau quantum number and scattering potential together with the fact it is double valued due to spin.

The complex density of states function under magnetic quantization is given by

$$N_B\left(E\right)=N_{BR1}\left(E\right)+iN_{BI1}\left(E\right)=\frac{eB}{2\pi^2\hbar^2}\sqrt{2m_c}\sum_{n=0}^{n_{\max}}\left[\frac{x'}{2\sqrt{x}}+\frac{iy'}{2\sqrt{y}}\right] \tag{5.13}$$

where

$$x=\frac{\sqrt{\left(U_{3,\pm}\left(E,n,\eta_g\right)\right)^2+\left(U_{4,\pm}\left(E,n,\eta_g\right)\right)^2}+\left(U_{3,\pm}\left(E,n,\eta_g\right)\right)}{2}$$

$$y=\frac{\sqrt{\left(U_{3,\pm}\left(E,n,\eta_g\right)\right)^2+\left(U_{4,\pm}\left(E,n,\eta_g\right)\right)^2}-\left(U_{3,\pm}\left(E,n,\eta_g\right)\right)}{2}$$

and x_1' and y_1' are the differentiations of x and y with respect to energy E.

From (7.13) we can write

$$N_{BR1}(E) = \frac{eB}{4\pi^2\hbar^2}\sqrt{2m_c}\sum_{n=0}^{n_{\max}}\frac{x_1'}{\sqrt{x_1}} \tag{5.14}$$

and

$$N_{BI1}(E) = \frac{eB}{4\pi^2\hbar^2}\sqrt{2m_c}\sum_{n=0}^{n_{\max}}\frac{y_1'}{\sqrt{y_1}} \tag{5.15}$$

(b) Two band model of Kane

The magneto-dispersion law in this case is given by

$$\frac{\hbar^2 k_z^2}{2m_c} = \gamma_2(E,\eta_g) - \left(n+\frac{1}{2}\right)\hbar\omega_0 \mp \frac{1}{2}g^*\mu_0 B \tag{5.16}$$

where g^* is the magnitude of the effective g factor at the edge of the conduction band and μ_0 is the Bohr magnetron.

The EEM at the Fermi Level can be written from (7.16) as

$$m^*(E_{FBHD},\eta_g) = m_c\gamma_2'(E_{FBHD},\eta_g) \tag{5.17}$$

Thus EEM is independent of quantum number.

(c) Parabolic Energy Bands

The magneto-dispersion law in this case is given by

$$\frac{\hbar^2 k_z^2}{2m_c} = \gamma_3(E,\eta_g) - \left(n+\frac{1}{2}\right)\hbar\omega_0 \mp \frac{1}{2}g^*\mu_0 B \tag{5.18}$$

The EEM at the Fermi Level can be written from (7.18) as

$$m^*(E_{FBHD},\eta_g) = m_c\gamma_3'(E_{FBHD},\eta_g) \tag{5.19}$$

Thus the EEM in heavily doped parabolic energy bands is a function of Fermi energy and scattering potential whereas in the absence of heavy doping the same mass is a constant quantity invariant of any variables.

(d) The model of Stillmanet *et al.*

The (2.79) under the condition of band tailing assumes the form

$$k^2 = \left[\frac{\left[\bar{t}_{11} - \sqrt{(\bar{t}_{11})^2 - \bar{t}_{12}\gamma_3(E,\eta_g)}\right]}{2\bar{t}_{12}}\right] \tag{5.20}$$

Therefore the magneto DR is given by

$$k_z^2 = U_7(E, n, \eta_g) \tag{5.21}$$

where

$$U_7(E, n, \eta_g) = \left[\frac{\left[\bar{t}_{11} - \sqrt{(\bar{t}_{11})^2 - 4\bar{t}_{12}\gamma_3(E, \eta_g)}\right]}{2\bar{t}_{12}} - \frac{2eB}{\hbar}\left(n + \frac{1}{2}\right) \right]$$

The EEM at the Fermi Level can be written from (7.21) as

$$m^*(E_{FBHD}, \eta_g) = \frac{\hbar^2}{2} U_7'(E_{FBHD}, n, \eta_g) \tag{5.22}$$

(e) The model of Palik *et al.*

To the fourth order in effective mass theory and taking into account the interactions of the conduction, light hole, heavy-hole and split-off hole bands, the electron energy spectrum in III–V semiconductors in the presence of a quantizing magnetic field $\vec{B}$ can be written as

$$\begin{aligned} E = {} & J_{31} + \left(n + \frac{1}{2}\right)\hbar\omega_0 + \frac{\hbar^2 k_z^2}{2m_c} \pm \frac{1}{4}\left(\frac{m_c}{m_0}\right)\hbar\omega_0 g_0^* \\ & \pm k_{30}\alpha\left(n + \frac{1}{2}\right)(\hbar\omega_0)^2 \pm k_{31}\alpha\hbar\omega_0\left(\frac{\hbar^2 k_z^2}{2m_c}\right) \\ & + k_{32}\alpha\left[\hbar\omega_0\left(n + \frac{1}{2}\right) + \frac{\hbar^2 k_z^2}{2m_c}\right]^2 \end{aligned} \tag{5.23}$$

where

$$J_{31} = -\frac{1}{2}\alpha\hbar\omega_0\left[(1 - y_{11})/(2 + x_{11})^2\right], \quad J_{32},$$

$$\begin{aligned} J_{32} = {} & \left\{\left[\frac{1}{3}(1 - x_{11})^2 - \left(2 + x_{11}^2\right)\right](2 + x_{11}).y_{11} \right. \\ & \left. + \frac{1}{2}\left(1 - x_{11}^2\right)(1 + x_{11})(1 + y_{11})\right\}, \end{aligned}$$

$$g_0^* = 2\left\{1 - \left[\frac{(1 - x_{11})}{(2 + x_{11})}\right]\left[\frac{(1 - y_{11})}{y_{11}}\right]\right\},$$

$$k_{30} = (1 - y_{11})(1 - x_{11})\left\{\left[\left(2 + \frac{3}{2}x_{11} + x_{11}^2\right) \cdot \frac{(1 - y_{11})}{(2 + x_{11})^2}\right] - \frac{2}{3}y_{11}\right\}$$

$$k_{31} = (1 - y_{11})\left[\frac{(1 - x_{11})}{(2 + x_{11})}\right] \cdot \left\{\left[\left(2 + \frac{3}{2}x_{11} + x_{11}^2\right) \cdot \frac{(1 - y_{11})}{(2 + x_{11})^2}\right]\right.$$

$$\left. - \frac{2}{3}(1 - x_{11})\, y_{11}\right\},$$

$$k_{32} = -\left[\left(1 + \frac{1}{2}x_{11}^2\right) \Big/ \left(1 + \frac{1}{2}x_{11}\right)\right](1 - y_{11})^2 x_{11}$$

$$= \left[1 + \left(\frac{\Delta}{E_g}\right)\right]^{-1} \quad \text{and} \quad y_{11} = \frac{m_c}{m_0}$$

Under the condition of heavy doping, (7.23) assumes the form

$$J_{34}k_z^4 + J_{35,\pm}(n)\, k_z^2 + J_{36,\pm}(n) - \gamma_3(E, \eta_g) = 0 \tag{5.24}$$

where

$$J_{34} = \alpha k_{32}\left(\frac{\hbar^2}{2m_c}\right)^2,$$

$$J_{35,\pm}(n) = \left[\frac{\hbar^2}{2m_c} \pm \alpha k_{31}\hbar\omega_0 . \frac{\hbar^2}{2m_c} + \alpha k_{32}\hbar\omega_0 . \frac{\hbar^2}{2m_c}\left(n + \frac{1}{2}\right)\right],$$

$$J_{36,\pm}(n) = \left[J_{31} \pm \frac{1}{4}\left(\frac{m_c}{m_0}\right)\hbar\omega_0 g_0^* \pm k_{30}\alpha(\hbar\omega_0)^2\left(n + \frac{1}{2}\right)\right.$$

$$\left. + k_{32}\alpha\left[(\hbar\omega_0)\left(n + \frac{1}{2}\right)\right]^2\right]$$

The (7.24) can be written as

$$k_z^2 = A_{35,\pm}(E, n, \eta_g) \tag{5.25}$$

where

$$A_{35,\pm}(E, n, \eta_g) = (2J_{34})^{-1}\left[- J_{35,\pm}(n)\right.$$

$$\left. + \sqrt{(J_{35,\pm}(n))^2 - 4J_{34}\left[J_{36,\pm}(n) - \gamma_3(E, \eta_g)\right]}\right]$$

The EEM at the Fermi Level can be written from (7. 25) as

$$m_{\pm}^{*}\left(E_{FBHD},n,\eta_g\right)=\frac{\hbar^2}{2}A'_{35,\pm}\left(E_{FBHD},n,\eta_g\right) \tag{5.26}$$

Thus, the EEM is a function of Fermi energy, Landau quantum number and the scattering potential.

5.2.3 *The DR in HD II–VI semiconductors under magnetic quantization*

The magneto dispersion relation of the carriers in heavily doped II–VI semiconductors are given by

$$\gamma_3\left(E,\eta_g\right)=a'_0\frac{2eB}{\hbar}\left(n+\frac{1}{2}\right)+b'_0k_z^2\pm\lambda'_0\left[\frac{2eB}{\hbar}\left(n+\frac{1}{2}\right)\right]^{1/2} \tag{5.27}$$

(5.27) can be written as

$$k_z^2=U_{8,\pm}\left(E,n,\eta_g\right) \tag{5.28}$$

where

$$U_{8,\pm}\left(E,n,\eta_g\right)=\left(b'_0\right)^{-1}\left[\gamma_3\left(E,\eta_g\right)-\frac{2eBa'_0}{\hbar}\left(n+\frac{1}{2}\right)\right.$$
$$\left.\mp\bar{\lambda}_0\left[\frac{2eB}{\hbar}\left(n+\frac{1}{2}\right)\right]^{1/2}\right]$$

The EEM at the Fermi Level can be written from (7.28) as

$$m^{*}\left(E_{FBHD},\eta_g\right)=\frac{\hbar^2}{2}U'_{8,\pm}\left(E_{FBHD},n,\eta_g\right) \tag{5.29}$$

5.2.4 *The DR in HD IV–VI semiconductors under magnetic quantization*

The electron energy spectrum in IV–VI semiconductors are defined by the models of Cohen, Lax, Dimmock and Bangert and Kastner respectively. The magneto DR in HD IV–VI semiconductors is discussed in accordance with the said model for the purpose of relative comparison.

(a) Cohen Model

In accordance with the Cohen model, the dispersion law of the carriers in IV–VI semiconductors is given by

$$E\left(1+\alpha E\right)=\frac{p_x^2}{2m_1}+\frac{p_z^2}{2m_3}-\frac{\alpha E p_y^2}{2m_2'}+\frac{p_y^2\left(1+\alpha E\right)}{2m_2}+\frac{\alpha p_y^4}{4m_2m_2'} \tag{5.30}$$

where, $p_i \equiv \hbar k_i, i = x, y, z, m_1, m_2$, and m_3 are the effective carrier masses at the band-edge along x, y and z directions respectively and m_2' is the effective-mass tensor component at the top of the valence band (for electrons) or at the bottom of the conduction band (for holes).

The magneto electron energy spectrum in IV–VI semiconductors in the presence of quantizing magnetic field B along z- direction can be written as

$$E\left(1+\alpha E\right)=\left(n+\frac{1}{2}\right)\hbar\omega\left(E\right)\pm\frac{1}{2}g^*\mu_0 B + \frac{3}{8}\alpha\left(n^2+n+\frac{1}{2}\right)\hbar^2\omega^2\left(E\right)+\frac{\hbar^2k_z^2}{2m_3} \tag{5.31}$$

where,

$$\omega\left(E\right)\equiv\frac{|e|B}{\sqrt{m_1m_2}}\left[1+\alpha E\left(1-\frac{m_2}{m_2'}\right)\right]^{1/2}.$$

Therefore the magneto dispersion law in heavily doped IV–VI materials can be expressed as

$$\frac{\hbar^2k_z^2}{2m_3}=U_{16,\pm}\left(E,n,\eta_g\right) \tag{5.32}$$

where

$$U_{16,\pm}\left(E,n,\eta_g\right)=\left[\gamma_2\left(E,\eta_g\right)-\left(n+\frac{1}{2}\right)\frac{\hbar eB}{\sqrt{m_1m_2}} \mp\frac{1}{2}g^*\mu_0B-\frac{3\alpha}{8}\left(n^2+n+\frac{1}{2}\right)\left(\frac{\hbar eB}{\sqrt{m_1m_2}}\right)^2 -\gamma_3\left(E,\eta_g\right)\left[\frac{\alpha}{2}\left(n+\frac{1}{2}\right)\frac{\hbar eB}{\sqrt{m_1m_2}}\left(1-\frac{m_2}{m_2'}\right) +\frac{3\alpha^2}{8}\left(n^2+n+\frac{1}{2}\right)\left(\frac{\hbar eB}{\sqrt{m_1m_2}}\right)^2\left(1-\frac{m_2}{m_2'}\right)\right]\right]$$

The EEM at the Fermi Level can be written from (7.32) as

$$m^*_{\pm}\left(E_{FBHD}, n, \eta_g\right) = m_3 U'_{16,\pm}\left(E_{FBHD}, n, \eta_g\right) \tag{5.33}$$

Thus, the EEM is a function of Fermi energy, Landau quantum number and the scattering potential.

(b) Lax Model

In accordance with this model, the magneto dispersion relation assumes the form

$$E\left(1+\alpha E\right) = \left(n+\frac{1}{2}\right)\hbar\omega_{03}\left(E\right) + \frac{\hbar^2 k_z^2}{2m_3} \pm \frac{1}{2}\mu_0 g^* B \tag{5.34}$$

where,

$$\omega_{03}\left(E\right) \equiv \frac{eB}{\sqrt{m_1 m_2}}$$

The magneto dispersion relation in heavily doped IV–VI materials, can be written following (7.34) as

$$\gamma_2\left(E, \eta_g\right) = \left(n+\frac{1}{2}\right) + \hbar\omega_{03}\left(E\right) + \frac{\hbar^2 k_z^2}{2m_3} \pm \frac{1}{2} g^* \mu_0 B \tag{5.35}$$

(5.35) can be written as

$$\frac{\hbar^2 k_z^2}{2m_3} = U_{17,\pm}\left(E, n, \eta_g\right) \tag{5.36a}$$

where

$$U_{17,\pm}\left(E, n, \eta_g\right) = \gamma_2\left(E, \eta_g\right) - \left(n+\frac{1}{2}\right)\hbar\omega_{03}\left(E\right) \pm \frac{1}{2} g^* \mu_0 B$$

The EEM at the Fermi Level can be written from (7.36a) as

$$m^*\left(E_{FBHD}, \eta_g\right) = m_3 U'_{17,\pm}\left(E_{FBHD}, n, \eta_g\right) \tag{5.36b}$$

(c)Dimmock Model

The dispersion relation under magnetic quantization in HD IV–VI semiconductors can be expressed in accordance with Dimmock model as

$$\gamma_2\left(E, \eta_g\right) + \alpha\gamma_3\left(E, \eta_g\right)\frac{2eB}{\hbar}\left(n+\frac{1}{2}\right)\frac{\hbar^2}{2}\left(\frac{1}{m_t^+} - \frac{1}{m_t^-}\right)$$

$$+\alpha\gamma_3\left(E, \eta_g\right) x\frac{\hbar^2}{2}\left(\frac{1}{m_l^+} - \frac{1}{m_l^-}\right)$$

$$= \frac{\hbar^2 k_s^2}{2m_t^*} + \frac{\hbar^2 k_z^2}{2m_l^*} + \frac{\hbar^2 k_s^2}{2m_t^-} + \frac{\hbar^2 k_z^2}{2m_l^-}$$
$$+ \alpha \left[\frac{\hbar^4 k_s^4}{4m_t^- m_t^+} + \frac{\hbar^2 k_s^2 k_z^2}{4m_l^- m_l^+} + \frac{\hbar^2 k_z^2 k_s^2}{4m_t^+ m_t^-} + \frac{\hbar^4 k_z^4}{4m_l^- m_l^+} \right]$$
$$= \frac{2eB}{\hbar}\left(n + \frac{1}{2}\right)\frac{\hbar^2}{2}\left(\frac{1}{m_t^*} - \frac{1}{m_t^-}\right) + x\frac{\hbar^2}{2}\left(\frac{1}{m_l^*} - \frac{1}{m_l^-}\right)$$
$$+ \alpha \left[\frac{\hbar^4}{4m_t^- m_t^+}\left(\frac{2eB}{\hbar}\left(n + \frac{1}{2}\right)\right)^2 \right.$$
$$\left. + \, x \left[\frac{\hbar^4 eB}{2m_l^+ m_t^- \hbar} + \frac{\hbar^4 eB}{2m_t^+ m_l^- \hbar} \right]\left(n + \frac{1}{2}\right) + \frac{\hbar^4}{4m_l^- m_l^+} x^2 \right] \quad (5.37)$$

where $x = k_z^2$

Therefore the magneto dispersion relation in heavily doped IV–VI materials, whose unperturbed carriers obey the DimmockModel can be expressed as

$$k_z^2 = U_{170}(E, n, \eta_g) \quad (5.38)$$

where

$$U_{170}(E, n, \eta_g) = [2p_9]^{-1}[-q_9(E, n, \eta_g) + [q_9^2(E, n, \eta_g) + 4p_9 R_9(E, n, \eta_g)]^{1/2}],$$
$$p_9 = \frac{\alpha\hbar^4}{4m_l^- m_l^+},$$
$$q_9(E, n, \eta_g) = \left[\frac{\hbar^2}{2}\left(\frac{1}{m_l^*} + \frac{1}{m_l^-}\right) + \frac{\alpha\hbar^3 eB}{2}\left(n + \frac{1}{2}\right) \cdot \left(\frac{1}{m_t^- m_l^+} + \frac{1}{m_l^- m_t^+}\right) - \alpha\gamma_3(E, \eta_g)\left(\frac{1}{m_l^+} + \frac{1}{m_l^-}\right)\right]$$

and

$$R_9(E, n, \eta_g) = \gamma_3(E, \eta_g) + \alpha eB\gamma_3(E, \eta_g)\left(n + \frac{1}{2}\right)\hbar\left(\frac{1}{m_t^+} - \frac{1}{m_t^-}\right) - \frac{\alpha\hbar^2}{m_t^- m_t^+}\left[eB\left(n + \frac{1}{2}\right)^2\right]$$

The EEM at the Fermi Level can be written from (7.38) as

$$m^*\left(E_{FBHD}, n, \eta_g\right) = \frac{\hbar^2}{2} U'_{170}\left(E_{FBHD}, n, \eta_g\right) \tag{5.39}$$

Thus,the EEM is a function of Fermi energy, Landau quantum number and the scattering potential.

(d) Model of Bangert and Kastner

In accordence with this model [8], the carrier energy spectrum in HD IV–VI semiconductors can be written following (2.143) as

$$\frac{k_s^2}{\rho_{11}^2\left(E, \eta_g\right)} + \frac{k_y^2}{\rho_{12}^2\left(E, \eta_g\right)} = 1 \tag{5.40}$$

where

$$\rho_{11}\left(E, \eta_g\right) = \frac{1}{\sqrt{S_1\left(E, \eta_g\right)}}, \quad \rho_{12}\left(E, \eta_g\right) = \frac{1}{\sqrt{S_2\left(E, \eta_g\right)}},$$

$$S_1\left(E, \eta_g\right) = \left[2\gamma_0\left(E, \eta_g\right)\right]^{=1}\left[\frac{\left(\bar{R}\right)^2}{E_g}\left\{c_1\left(\alpha_1 E, E_g\right) - iD_1\left(\alpha_1 E, E_g\right)\right\}\right.$$

$$\cdot\frac{\left(\bar{S}\right)^2}{\Delta_c'}\left\{c_2\left(\alpha_2 E, E_g\right) - iD_2\left(\alpha_2 E, E_g\right)\right\}$$

$$\left.+\frac{\left(\bar{Q}\right)^2}{\Delta_c''}\left\{c_3\left(\alpha_3 E, E_g\right) - iD_3\left(\alpha_3 E, E_g\right)\right\}\right]$$

and

$$S_2\left(E, \eta_g\right) + \left[2\gamma_0\left(E, \eta_g\right)\right]^{-1}\left[\frac{\left(2\bar{A}\right)^2}{E_g}\left\{c_1\left(\alpha_1 E, E_g\right) - iD_1\left(\alpha_1 E, E_g\right)\right\}\right.$$

$$\left.+\frac{\left(\bar{S} + \bar{Q}\right)^2}{\Delta_c''}\left\{c_3\left(\alpha_3 E, E_g\right) - iD_3\left(\alpha_3 E, E_g\right)\right\}\right]$$

Since $S_1(E, \eta_g)$ and $S_2(E, \eta_g)$ are complex, the energy spectrum is also complex in the presence of Gaussian band tails.

Therefore the magneto dispersion law in the presence of a quantizing magnetic field B which makes an angle θ with k_z axis can be written as

$$k_z^2 = U_{18}(E, n, \eta_g) \tag{5.41}$$

where

$$\begin{aligned} U_{18}(E, n, \eta_g) = {} & \left[\rho_{11}^2(E, \eta_g)\sin^2\theta + \rho_{12}^2(E, \eta_g)\cos^2\theta\right] \\ & - \left[\frac{2eB}{\hbar}\left(n + \frac{1}{2}\right)\left[\left(\rho_{11}^2(E, \eta_g)\,\rho_{12}^2(E, \eta_g)\right)^{-1}\right.\right. \\ & \left.\left. \cdot \left\{\rho_{11}^2(E, \eta_g)\sin^2\theta + \rho_{12}^2(E, \eta_g)\cos^2\theta\right\}^{3/2}\right]\right] \end{aligned}$$

The EEM at the Fermi Level can be written from (7.41) as

$$m^*(E_{FBHD}, n, \eta_g) = \frac{\hbar^2}{2} \text{ Real part of } \left[U_{18}(E_{FBHD}, n, \eta_g)\right]' \tag{5.42}$$

Thus, the EEM is a function of Fermi energy, Landau quantum number and the scattering potential and the orientation of the applied quantizing magnetic field.

(e) Model of Foley and Langenberg

The dispersion relation of the conduction electrons of IV–VI semiconductors in accordance with Foley *et al.* can be written as [9]

$$E + \frac{E_g}{2} = E_0(k) + \left[\left[E_+(k) + \frac{E_g}{2}\right]^2 + P_\perp^2 k_s^2 + P_\parallel^2 k_z^2\right]^{1/2} \tag{5.43}$$

where

$$E_+(k) = \frac{\hbar^2 k_s^2}{2m_\perp^+} + \frac{\hbar^2 k_z^2}{2m_\parallel^+}, \quad E_-(k) = \frac{\hbar^2 k_s^2}{2m_\perp^-} + \frac{\hbar^2 k_z^2}{2m_\parallel^-}$$

represents the contribution from the interaction of the conduction and the valence band edge states with the more distant bands and the free electrons term,

$$\frac{1}{m_\perp^\pm} = \frac{1}{2}\left[\frac{1}{m_{tc}} \pm \frac{1}{m_{tv}}\right], \quad \frac{1}{m_\parallel^\pm} = \frac{1}{2}\left[\frac{1}{m_{lc}} \pm \frac{1}{m_{lv}}\right]$$

Following the methods as given in Chapter 1, the dispersion relation in heavily doped IV–VI materials in the present case is given by

$$\left[\left[\gamma_3(E,\eta_g)+\frac{E_{g0}}{2}\right]-\left[\frac{\hbar^2k_s^2}{2m_\perp^-}+\frac{\hbar^2k_z^2}{2m_\parallel^-}\right]\right]^2$$
$$=\left[\frac{\hbar^2k_s^2}{2m_\perp^-}+\frac{\hbar^2k_z^2}{2m_\parallel^-}\right]+\frac{E_{g0}^2}{4}+E_{g0}\left[\frac{\hbar^2k_s^2}{2m_\perp^-}+\frac{\hbar^2k_z^2}{2m_\parallel^-}\right]+P_\parallel^2k_z^2+P_\perp^2k_s^2 \tag{5.44}$$

Therefore the magneto-dispersion relation in heavily doped IV–VI materials can be written as

$$\gamma_3^2(E,\eta_g)+\frac{E_g^2}{4}+E_g\gamma_3(E,\eta_g)+\left[\frac{\hbar eB}{m_\perp^-}\left(n+\frac{1}{2}\right)+\frac{\hbar^2x}{2m_\parallel^-}\right]^2$$
$$-2\left[\gamma_3(E,\eta_g)+\frac{E_g}{2}\right]\left[\frac{\hbar eB\left(n+\frac{1}{2}\right)}{m_\perp^-}+\frac{\hbar^2x}{2m_\parallel^-}\right]$$
$$=\left[\frac{\hbar eB\left(n+\frac{1}{2}\right)}{m_\perp^+}+\frac{\hbar^2x}{2m_\parallel^+}\right]+\left[\frac{\hbar eB}{m_\perp^+}\left(n+\frac{1}{2}\right)+\frac{\hbar^2x}{2m_\parallel^+}\right]^2$$
$$+P_\parallel^2x+P_\perp^2\frac{2eB}{\hbar}\left(n+\frac{1}{2}\right) \tag{5.45}$$

where $k_z^2=x$

Therefore the magneto dispersion relation in IV–VI heavily doped materials, where unperturbed carriers follow the model of Foley *et al.* can be expressed as

$$k_z^2=U_{19}(E,n,\eta_g) \tag{5.46}$$

where

$$U_{19}(E,n,\eta_g)=[2p_{91}]^{-1}[-q_{91}(E,n,\eta_g)+\{q_{91}^2(E,n,\eta_g)$$
$$+4p_{91}R_{91}(E,n,\eta_g)\}^{1/2}]$$

$$p_{91}=\frac{\hbar^4}{4}\left[\frac{1}{(m_\parallel^2)^2}-\frac{1}{(m_\parallel^-)^2}\right],$$

$$q_{91}(E,n,\eta_g) = \left[\frac{\hbar^2 eB}{m_\perp^- m_\parallel^+}\left(n+\frac{1}{2}\right) + P_\parallel^2 + \frac{\hbar^2 E_g}{2m_\parallel^+} - \frac{\hbar^2 eB\left(n+\frac{1}{2}\right)}{m_\perp^- m_\parallel^-}\right.$$
$$\left. + \frac{\hbar^2}{m_\parallel^-}\left(\gamma_3(E,\eta_g) + \frac{E_g}{2}\right)\right]$$
$$R_{91}(E,\eta_g,n) = \left[\gamma_3^2(E,\eta_g) + E_g\gamma_3(E,\eta_g)\right.$$
$$- \frac{2\hbar eB}{m_\perp^-}\left(\gamma_3(E,\eta_g) + \frac{E_g}{2}\right)\left(n+\frac{1}{2}\right)$$
$$\left. - E_g\frac{\hbar eB}{m_\perp^+}\left(n+\frac{1}{2}\right) - P_\perp^2 . \frac{2eB}{\hbar}\left(n+\frac{1}{2}\right)\right]$$

The EEM at the Fermi Level can be written from (7.46) as

$$m^*(E_{FBHD},n,\eta_g) = \frac{\hbar^2}{2}U_{19}'(E_{FBHD},n,\eta_g) \tag{5.47}$$

Thus, as noted already in this case also the EEM is a function of Fermi energy, Landau quantum number and the scattering potential.

5.2.5 *The DR in HD stressed Kane type semiconductors under magnetic quantization*

The DR of the conduction electrons in heavily doped Kane type semiconductors can be written following (2.175) of Chapter 2 as

$$\frac{k_x^2}{a_\parallel^2(E,\eta_g)} + \frac{k_y^2}{b_\parallel^2(E,\eta_g)} + \frac{k_z^2}{c_\parallel^2(E,\eta_g)} = 1$$

where

$$a_\parallel(E,\eta_g) = \frac{1}{\sqrt{P_\parallel(E,\eta_g)}},$$

$$b_\parallel(E,\eta_g) = \frac{1}{\sqrt{Q_\parallel(E,\eta_g)}} \quad \text{and} \quad c_\parallel(E,\eta_g) = \frac{1}{\sqrt{S_\parallel(E,\eta_g)}} \tag{5.48a}$$

The electron energy spectrum in heavily doped Kane type semiconductors in the presence of an arbitrarily oriented quantizing magnetic field B which makes an angle $\bar{\alpha}_1, \bar{\beta}_1$ and $\bar{\gamma}_1$ with k_x, k_y and k_z axes respectively, can be written as

$$(k_z')^2 = U_{41}(E, n, \eta_g) \tag{5.48b}$$

where

$$U_{41}(E, n, \eta_g) = I_2(E, \eta_g)\left[1 - I_3(E, n, \eta_g)\right]$$

$$I_2(E, \eta_g) = \left[[a_{11}(E, \eta_g)]^2 \cos^2 \bar{\alpha}_1 + [b_{11}(E, \eta_g)]^2 \cos^2 \bar{\beta}_1 + [c_{11}(E, \eta_g)]^2 \cos^2 \bar{\gamma}_1\right] \quad \text{and}$$

$$I_3(E, n, \eta_g) = \frac{2eB}{\hbar}\left(n + \frac{1}{2}\right)\left[a_{11}(E, \eta_g)\, b_{11}(E, \eta_g)\, c_{11}(E, \eta_g)\right]^{-1} \cdot \left[I_2(E, \eta_g)\right]^{1/2}$$

The EEM at the Fermi Level can be written from (7.48b) as

$$m^*(E_{FBHD}, n, \eta_g) = \frac{\hbar^2}{2} U_{41}'(E_{FBHD}, n, \eta_g) \tag{5.48c}$$

From (7.48c) we observe that the EEM is a function of Fermi energy, Landau quantum number, the scattering potential and the orientation of the applied quantizing magnetic field.

5.2.6 *The DR in HD Te under magnetic quantization*

The magneto dispersion relation of the conduction electrons in HD Te can be expressed following (2.190) of Chapter 2 as

$$k_z^2 = U_{42,\pm}(E, n, \eta_g) \tag{5.49a}$$

where

$$U_{42,\pm}(E, n, \eta_g) = (2\psi_1^2)^{-1}\left[\left\{2\gamma_3(E, \eta_g)\psi_1 + \psi_3^2 - 4\psi_1\psi_2\frac{eB}{\hbar}\left(n + \frac{1}{2}\right)\right\} - \left\{\psi_3^4 + 4\psi_1\psi_3^2\gamma_3(E, \eta_g)\right\} + \frac{8eB}{\hbar}\left(n + \frac{1}{2}\right)\left(\psi_1^2\psi_4^2 - \psi_1\psi_2\psi_3\right)\right\}^{-1/2}\right]$$

The EEM at the Fermi Level can be written from (7.49a) as

$$m_{\pm}^{*}\left(E_{FBHD}, n, \eta_g\right) = \frac{\hbar^2}{2}\ U'_{42,\pm}\left(E_{FBHD}, n, \eta_g\right) \tag{5.49b}$$

Thus from (5.49b) we note that the EEM is a function of three variables namely Fermi energy, Landau quantum number and the scattering potential.

5.2.7 *The DR in HD Gallium Phosphide under magnetic quantization*

The magneto dispersion relation in HD GaP can be written following (2.202) of Chapter 2 as

$$k_z^2 = U_{43}\left(E, n, \eta_g\right) \tag{5.50a}$$

where

$$\begin{aligned}U_{43}\left(E, n, \eta_g\right) = \left(2b^2\right)^{-1}\Bigg[&\left\{2\gamma_3\left(E, \eta_g\right)b + c - 2Db - 4ab\frac{eB}{\hbar}\left(n + \frac{1}{2}\right)\right\}\\ &+\Bigg\{\left[c^2 + 4bc\gamma_3\left(E, \eta_g\right) + 4D^2b^2 - 4cDb\right]\\ &-\frac{8eB}{\hbar}\left(n + \frac{1}{2}\right)\left(2ab^2D + 4\gamma_3\left(E, \eta_g\right)b^2a + abc\right.\\ &\left.- 2b^2a\gamma_3\left(E, \eta_g\right) - b^2c\right)\Bigg\}^{-1/2}\Bigg]\end{aligned}$$

The EEM at the Fermi Level can be expressed from (7.50a) as

$$m^{*}\left(E_{FBHD}, n, \eta_g\right) = \frac{\hbar^2}{2}U'_{43}\left(E_{FBHD}, n, \eta_g\right) \tag{5.50b}$$

Thus, from (7.50b) it appears that the EEM is the function of Fermi energy, Landau quantum number and the scattering potential.

5.2.8 *The DR in HD Platinum Antimonideunder magnetic quantization*

The magneto dispersion relation in HD $PtSb_2$ can be written following (2.218) as

$$k_z^2 = U_{44}(E, n, \eta_g) \tag{5.51a}$$

where

$$U_{44}(E, n, \eta_g) = \frac{1}{2T_{41}}\left[T_{71}(E, n, \eta_g) + \sqrt{T_{71}^2(E, n, \eta_g) + 4T_{41}T_{71}(E, n, \eta_g)}\right],$$

$$T_{71}(E, n, \eta_g) = \left[T_{51}(E, \eta_g) - T_{31}\frac{2eB}{\hbar}\left(n + \frac{1}{2}\right)\right] \quad \text{and}$$

$$T_{72}(E, n, \eta_g) = \left[T_{61}(E, \eta_g) + \frac{2eB}{\hbar}\left(n + \frac{1}{2}\right)T_{21}(E, \eta_g)\right]$$

The EEM at the Fermi Level can be written from (5.51a) as

$$m^*(E_{FBHD}, n, \eta_g) = \frac{\hbar^2}{2}U'_{44}(E_{FBHD}, n, \eta_g) \tag{5.51b}$$

Thus, from the above equation we infer that the EEM is a function of Landau quantum number, the Fermi energy and the scattering potential.

5.2.9 *The DR in HD Bismuth Telluride under magnetic quantization*

The magneto dispersion relation in HD Bi_2Te_3 can be written following (2.231) as

$$k_x^2 = U_{45}(E, \eta_g, n) \tag{5.52a}$$

where

$$U_{45}(E,\eta_g,n) = \frac{\gamma_2(E,\eta_g) - \left(n+\frac{1}{2}\right)\frac{e\hbar B}{M_{31}}}{\bar{\omega}_1} \quad \text{and}$$

$$M_{31} = \frac{m_0}{\left(\bar{\alpha}_{22}\bar{\alpha}_{33} - \frac{(\bar{\alpha}_{23})^2}{4}\right)^{1/2}}$$

The EEM at the Fermi Level can be written from (7.52a) as

$$m^*(E_{FBHD},\eta_g) = \frac{\hbar^2}{2}U'_{45}(E_{FBHD},n,\eta_g) \tag{5.52b}$$

5.2.10 *The DR in HD Germanium under magnetic quantization*

a. Model of Cardona *et al.*

The magneto dispersion relation in HD Gecan be written following (2.243) as

$$k_x^2 = U_{46}(E,\eta_g,n) \tag{5.53a}$$

where

$$U_{46}(E,n,\eta_g) = \frac{2m_{\parallel}^*}{\hbar^2}\left[\gamma_3(E,\eta_g) + \frac{E_{g0}}{2} - \left[\frac{E_{g0}^2}{4} + \frac{E_{g0}\hbar^2}{m_\perp^*}\frac{2eB}{\hbar}\left(n+\frac{1}{2}\right)\right]\right]^{1/2}$$

The EEM at the Fermi Level can be written from (7.53a) as

$$m^*(E_{FBHD},n,\eta_g) = \frac{\hbar^2}{2}U'_{46}(E_{FBHD},n,\eta_g) \tag{5.53b}$$

From (5.53a) it appears that the EEM is a function of Fermi energy and Landau quantum number due to band non-parabolicity.

b. Model of Wang and Ressler

The magneto dispersion relation in HD Gecan be written following (2.258) as

$$k_x^2 = U_{47}(E,\eta_g,n) \tag{5.54a}$$

where

$$U_{47}(E,n,\eta_g) = \left(\frac{m_{\|}^*}{\hbar^2\bar{\alpha}_6}\right)\left[1-\bar{\alpha}_5\left(n+\frac{1}{2}\right)\hbar\omega_{\perp} - \{\theta_7(n) - 4\bar{\alpha}_6\gamma_3(E,\eta_g)\}^{1/2}\right], \quad \omega_{\perp} = \frac{eB}{m_{\perp}^*}$$

and

$$\theta_7(n) = \left[1+(\bar{\alpha}_5)^2\left\{\left(n+\frac{1}{2}\right)\hbar\omega_{\perp}\right\}^2 - 2\bar{\alpha}_5\left(n+\frac{1}{2}\right)\hbar\omega_{\perp} + 4\bar{\alpha}_6\left(n+\frac{1}{2}\right)\hbar\omega_{\perp} - 4\bar{\alpha}_6\bar{\alpha}_4\left\{\left(n+\frac{1}{2}\right)\hbar\omega_{\perp}\right\}^2\right]$$

The EEM at the Fermi Level can be written from (7.54a) as

$$m^*(E_{FBHD},n,\eta_g) = \frac{\hbar^2}{2}U'_{47}(E_{FBHD},n,\eta_g) \quad (5.54b)$$

From (5.54b) we note that the mass is a function of Fermi energy and quantum number due to band non-parabolicity.

5.2.11 *The DR in HD Gallium Antimonide under magnetic quantization*

The magneto dispersion relation in HD GaSb can be written following (2.274) as

$$k_z^2 = U_{48}(E,\eta_g,n) \quad (5.55a)$$

where

$$U_{47}(E,n,\eta_g) = \left[-\frac{2eB}{\hbar}\left(n+\frac{1}{2}\right) + (2\alpha_9^2)^{-1}\left[\left\{2\alpha_9\gamma_3^2(E,\eta_g) + \alpha_9\bar{E}'_{go} + \frac{\alpha_{10}\left(\bar{E}'_{go}\right)^2}{4}\right\}\right.\right.$$

$$-\left\{\alpha_9^2\left(E'_{g0}\right)^2+\frac{\alpha_{10}^2\left(\bar{E}'_{g0}\right)^4}{16}+\alpha_9\alpha_{10}\gamma_3\left(E,\eta_g\right)\left(\bar{E}'_{g0}\right)^2\right.$$

$$\left.\left.\left.+\frac{\alpha_9\alpha_{10}\left(\bar{E}'_{g0}\right)^3}{2}\right\}^{1/2}\right]\right]$$

$$\alpha_9=\frac{\hbar^2}{2m_0}\quad\text{and}\quad\alpha_{10}=\frac{2\hbar^2}{\bar{E}'_{g0}}\left(\frac{1}{m_c}-\frac{1}{m_0}\right)$$

The EEM at the Fermi Level can be written from (7.55a) as

$$m^*(E_{FBHD},\eta_g)=\frac{\hbar^2}{2}U'_{48}(E_{FBHD},n,\eta_g)\tag{5.55b}$$

5.2.12 *The DR in HD II–V materials under magnetic quantization*

The dispersion relation of the holes are given by [10]

$$\mathrm{E}=\theta_1\mathrm{k}_\mathrm{x}^2+\theta_2\mathrm{k}_\mathrm{y}^2+\theta_3\mathrm{k}_\mathrm{z}^2+\delta_4\mathrm{k}_\mathrm{x}\mathrm{m}[\{\theta_5\mathrm{k}_\mathrm{x}^2+\theta_6\mathrm{k}_\mathrm{y}^2+\theta_7\mathrm{k}_\mathrm{z}^2+\delta_5\mathrm{k}_\mathrm{x}\}^2$$
$$+\mathrm{G}_3^2\mathrm{k}_\mathrm{y}^2+\Delta_3^2]^{1/2}\pm\Delta_3\tag{5.56}$$

where, $\mathrm{k_x}, \mathrm{k_y}$ and $\mathrm{k_z}$ are expressed in the units of $10^{10}\ \mathrm{m}^{-1}$,

$$\theta_1=\frac{1}{2}(\mathrm{a}_1+\mathrm{b}_1),\quad\theta_2=\frac{1}{2}(\mathrm{a}_2+\mathrm{b}_2),\quad\theta_3=\frac{1}{2}(\mathrm{a}_3+\mathrm{b}_3),$$
$$\delta_4=\frac{1}{2}(\mathrm{A}+\mathrm{B}),$$
$$\theta_5=\frac{1}{2}(\mathrm{a}_1-\mathrm{b}_1),\quad\theta_6=\frac{1}{2}(\mathrm{a}_2-\mathrm{b}_2),\quad\theta_7=\frac{1}{2}(\mathrm{a}_3-\mathrm{b}_3),$$
$$\delta_5=\frac{1}{2}(\mathrm{A}-\mathrm{B}),$$

$a_i(i=1,2,3,4), b_i,\ A, B, G_3$ and Δ_3 are system constants

The hole energy spectrum in HD II–V semiconductors can be expressed following the method of Chapter 1 as

$$y_3(E, n_g) = \theta_1 k_x^2 + \theta_2 k_y^2 + \theta_3 k_z^2 + \delta_4 k_x \pm [\{\theta_5 k_x^2 + \theta_6 k_y^2 + \theta_7 k_z^2 + \delta_5 k_x\}^2 + G_3^2 k_y^2 + \Delta_3^2]^{\frac{1}{2}} \pm \Delta_3 \tag{5.57}$$

the magneto dispersion law in HD II–V semiconductors assumes the form

$$k_y^2 = U_{49,\pm}(E, \eta_g, n) \tag{5.58}$$

where,

$$U_{49,\pm}(E, n, n_g) = -[I_{35}\gamma_3(E, \eta_g) + I_{36,\pm}(n) \pm [\gamma_3^2(E, \eta_g) + \gamma_3(E, \eta_g) I_{38,\pm}(n) + I_{39,\pm}(n)]^{1/2}]$$

$$I_{35} = \frac{\theta_2}{(\theta_2^2 - \theta_5^2)}, \quad I_{36,\pm}(n) = \frac{I_{33,\pm}(n)}{2(\theta_2^2 - \theta_5^2)},$$

$$I_{38,\pm}(n) = (4\theta_5^2)^{-1}[4\theta_2 I_{33,\pm}(n) + 8\theta_2^2 I_{31,\pm}(n) - \theta_5^2 I_{31,\pm}(n)],$$

$$I_{39,\pm} = (4\theta_5^2)^{-1}[I_{33,\pm}^2(n) + 4\theta_2^3 I_{34,\pm}(n) - 4\theta_5^3 I_{34,\pm}(n)],$$

$$I_{33,\pm}(n) = [G_3^2 + 2\theta_5 I_{32}(n) - 2\theta_2 I_{31,\pm}(n)],$$

$$I_{34,\pm}(n) = [I_{32,\pm}^2(n) + \Delta_3^2 - I_{31,\pm}(n)],$$

$$I_{31,\pm}(n) = \left[\left(n + \frac{1}{2}\right)\hbar\omega_{31} - \frac{\delta_4^2}{4\theta_1} \pm \Delta_3\right],$$

$$I_{32}(n) = \left[\left(n + \frac{1}{2}\right)\hbar\omega_{32} - \frac{\delta_5^2}{4\theta_5}\right],$$

$$\omega_{31} = \frac{eB}{\sqrt{M_{31}M_{32}}}, \quad \omega_{32} = \frac{eB}{\sqrt{M_{33}M_{34}}}, \quad M_{31} = \frac{\hbar^2}{2\theta_1},$$

$$M_{32} = \frac{\hbar^2}{2\theta_3}, \quad M_{33} = \frac{\hbar^2}{2\theta_5} \quad \text{and} \quad M_{34} = \frac{\hbar^2}{2\theta_7}.$$

The EEM at the Fermi Level can be written from (7.58) as

$$m_\pm^*(E_{FBHD}, n, \eta_g) = \frac{\hbar^2}{2} U'_{49,\pm}(E_{FBHD}, n, \eta_g) \tag{5.59}$$

From (7.59) we note that the EEM is a function of Fermi energy, Landau quantum number and the scattering potential.

5.2.13 *The DR in HD Lead Germanium Telluride under magnetic quantization*

The DR of the carriers in n-type $\mathrm{Pb_{1-x}Ga_xTe}$ with $\mathrm{x} = 0.01$ can be written following Vassilev [11] as

$$\begin{aligned}&\left[E - 0.606k_s^2 - 0.0722k_z^2\right]\left[E + \bar{E}_g + 0.411k_s^2 + 0.0377k_z^2\right]\\&\quad = 0.23k_s^2 + 0.02k_z^2\\&\qquad \pm \left[0.06\bar{E}_g + 0.061k_s^2 + 0.0066k_z^2\right]k_s\end{aligned} \tag{5.60}$$

where, $\bar{E}_g (= 0.21eV)$ is the energy gap for the transition point, the zero of the energy E is at the edge of the conduction band of the Γ point of the Brillouin zone and is measured positively upwards, k_x, k_y and k_z are in the units of $10^9 m^{-1}$.

The magneto dispersion law in HD $\mathrm{Pb_{1-x}Ge_xTe}$ can be expressed following the methods as given in Chapter 1 as

$$\begin{aligned}&\left[\frac{2\theta_0(E,\eta_g)}{1+Erf(E/\eta_g)}\right] + \gamma_3(E,\eta_g)\left[\overline{E_{g0}} - 0.345x - 0.390\frac{eB}{\hbar}\left(n+\frac{1}{2}\right)\right]\\&= \frac{0.46eB}{\hbar}\left(n+\frac{1}{2}\right) + 0.02x\\&\pm\left[0.06\overline{E_{g0}} + 0.122\frac{eB}{\hbar}\left(n+\frac{1}{2}\right) + 0.0066x\right]\\&\cdot\left(\frac{2eB}{\hbar}\left(n+\frac{1}{2}\right)\right)^{1/2} + \left[\overline{E_{g0}} + \frac{0.822eB}{\hbar}\left(n+\frac{1}{2}\right) + 0.377x\right]\\&\cdot\left[\frac{1.212eB}{\hbar}\left(n+\frac{1}{2}\right) + 0.722x\right]\end{aligned} \tag{5.61}$$

The (5.61) assumes the form

$$k_z^2 = U_{50,\pm}(E,\eta_g,n) \tag{5.62}$$

where

$$U_{50,\mp}(E,n,\eta_g) = (2p_{10})^{-1}[q_{10}(E,n,\eta_g) - [q_{10}^2(E,n,\eta_g) + 4p_{10}R_{10,\mp}(E,n,\eta_g)]^{1/2}]$$

$$p_{10}(0.377 \times 0.722), q_{10}(E,n,\eta_g) = [0.02 + 0.345\gamma_3(E,\eta_g) \pm 0.0066\left(\frac{2B}{\hbar}\left(n+\frac{1}{2}\right)\right)^{1/2} + 0.377 \times \frac{1.212eB}{\hbar}\left(n+\frac{1}{2}\right) + 0.722\left[\bar{E}_{go} + 0.822\frac{eB}{\hbar}\left(n+\frac{1}{2}\right)\right]] \quad \text{and}$$

$$R_{10,\mp}(E,n,\eta_g) = \left[\frac{2\theta_0(E,\eta_g)}{1+Eof(E/\eta_g)} + \gamma_3(E,\eta_g)\left[\bar{E}_{go} - 0.390\frac{eB}{\hbar}\left(n+\frac{1}{2}\right)\right] \mp \left(0.06\bar{E}_{go} + 0.122\frac{eB}{\hbar}\left(n+\frac{1}{2}\right)\right) \cdot \left(\frac{2eB}{\hbar}\left(n+\frac{1}{2}\right)\right)^{1/2} - \left(\bar{E}_{go} + 0.822\frac{eB}{\hbar}\left(n+\frac{1}{2}\right)\right) \cdot \frac{1.212eB}{\hbar}\left(n+\frac{1}{2}\right) - \frac{0.46eB}{\hbar}\left(n+\frac{1}{2}\right)\right].$$

The EEM at the Fermi Level can be written from (7.62) as

$$m_{\mp}^*(E_{FBHD},\eta_g) = \frac{\hbar^2}{2}U'_{50,\mp}(E_{FBHD},n,\eta_g) \quad (5.63)$$

Thus from (7.63) we note that the EEM is a function of the Fermi energy, Landau quantum number and the scattering potential.

5.3 Results and Discussion

Using (5.4) and (5.7b) together with the energy band constants as given in Appendix A, we have plotted the EM in HD n-Cd_3As_2 and HD $CdGeAs_2$ as functions of inverse magnetic field for the first two magnetic sub-bands in Fig. 5.1. For the purpose of self assessment, in the same figures, we have also plotted the effect of absence of the crystal field splitting together with the simplified three and two band models of Kane.

From these figures, it appears that the EM is an oscillatory function of the inverse quantizing magnetic field. The oscillatory dependence is due to the crossing over of the Fermi level by the Landau sub-bands in steps resulting in successive reduction the number of occupied Landau levels as the magnetic field is increased. For each coincidence of a Landau level, with the Fermi level, there would be a discontinuity in the density-of-states function resulting in a peak of oscillation.

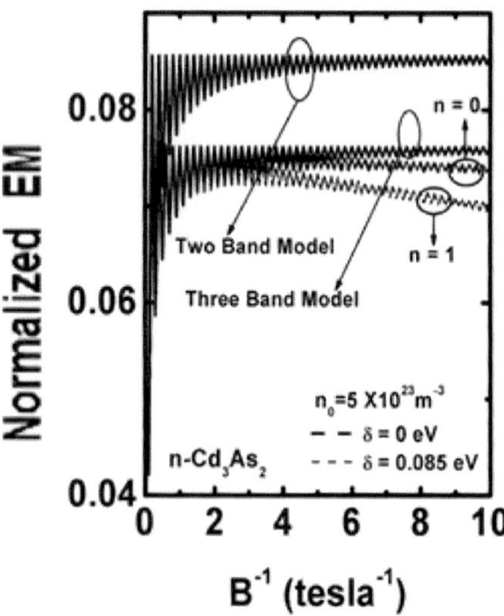

Fig. 5.1. Plot of the magnetic quantum number dependent normalized EM as function of inverse magnetic field for HD n-Cd_3As_2 considering (5.4) both in the presence and absence of crystal field-splitting constant. The graphs for the three and two band models of Kane have also been exhibited.

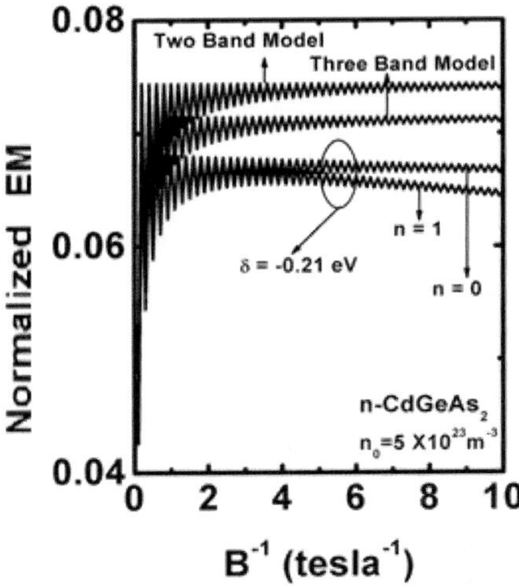

Fig. 5.2. Plot of the magnetic quantum number dependent normalized EM as function of inverse magnetic field for HD n-CdGeAs$_2$ considering (5.4). The plots for the three and two band models of Kane have also been exhibited.

Thus the peaks should occur whenever the Fermi energy is a multiple of energy separation between the two consecutive Landau levels and it may be noted that the origin of oscillations in the EM is the same as that of the Subhnikov-de Hass oscillations. With increase in magnetic field, the amplitude of the oscillation increases and, ultimately, at very large values of the magnetic field, the conditions for the quantum limit is reached when the EM is found to decrease monotonically with increase in magnetic field. Further, in this case we see that the EM is a strong function of the subband quantum number n.

For this reason, we has also plotted the EM for the next higher subband n = 1. It thus appears that the increasing the index decreases the EM for lower values of the field. However at higher field, the difference between them diminishes and all the respective band models tend to coincide with each other which stand out to be a remarkable mathematical simplicity in deriving the analytical expressions of the EM. The presence of the isotropic spin orbit splitting constant in the three band model of Kane changes the value of the EM as compared with the corresponding two band model.

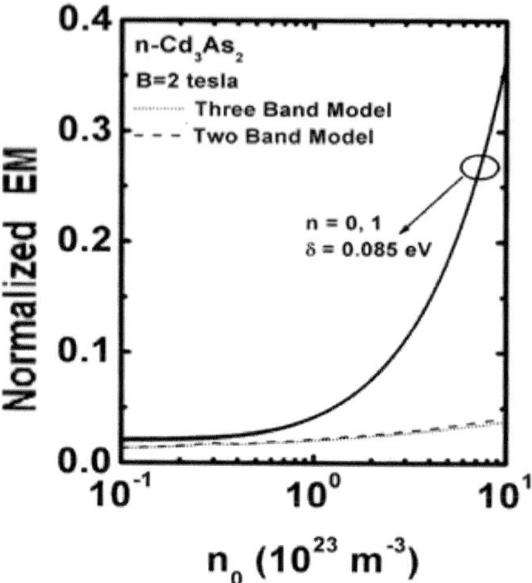

Fig. 5.3. Plot of the magnetic quantum number dependent normalized EM as function of carrier degeneracy for HD n-Cd_3As_2. The plots for the three and two band models of Kane have also been exhibited.

Figures 5.3 and 5.4 exhibit the variation of the EM on the carrier degeneracy in both the aforementioned materials. Oscillatory dependences is exhibited in the case of the equivalent three and the two band model of Kane, the deviation among which for both the materials are almost zero. Further, we also see that there is almost no significant change in the variation of the subband index from $n = 1$ to $n = 2$ in both the cases. An exponential rise in the EM can be observed beyond $10^{23}\,m^{-3}$ for both the materials. In case of HD Cd_3As_2, we see that decreasing the carrier degeneracy converge the EM from all the band models to a unique value. Incidentally, this is not the case of HD $CdGeAs_2$. There is a crossing over of the EM near to the concentration zone of $10^{23}\,m^{-3}$ which overestimates the numerical result. IN addition, the EM exhibits different numerical values for both the materials, the rate of variations of which are different due to the influence of the energy band constants in accordance with all the types of the band models and follow the same trend as shown in Figs. 5.3 and 5.4.

Figures 5.5–5.9 exhibits the variation of the EM on the quantizing magnetic field for HD n-InAs, HD n-GaAs, HD n-InSb, HD

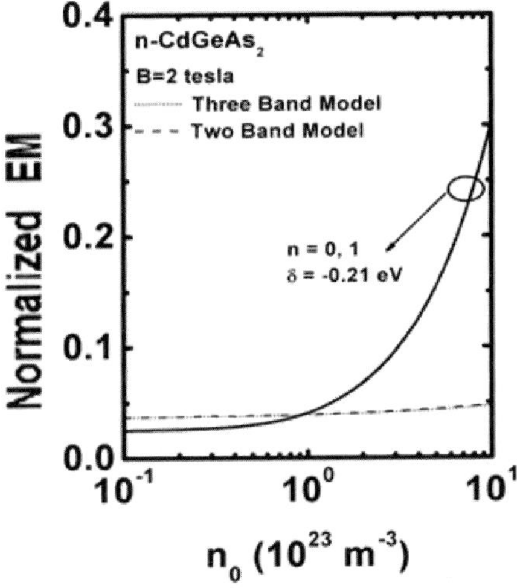

Fig. 5.4. Plot of the magnetic quantum number dependent normalized EM as function of carrier degeneracy for HD n-$CdGeAs_2$. The plots for three and two band models of Kane have also been exhibited.

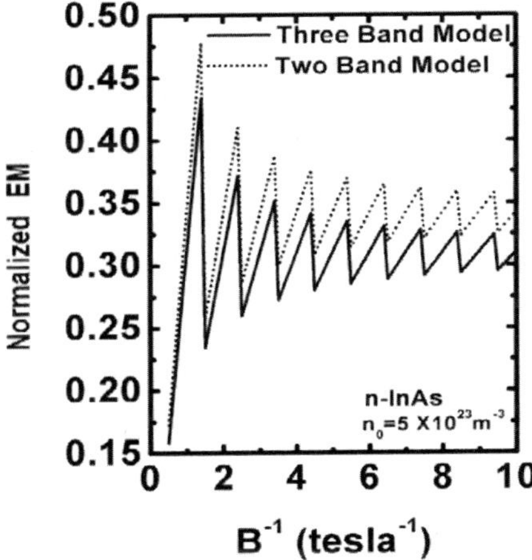

Fig. 5.5. Plot of the normalized EM as function of inverse magnetic field for HD n-InAs considering the three and two band models of Kane.

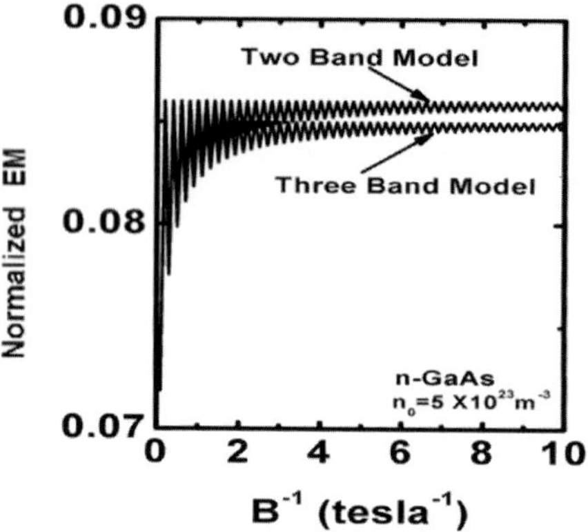

Fig. 5.6. Plot of the normalized EM as function of inverse magnetic field for HD n-GaAs considering the three and two band models of Kane.

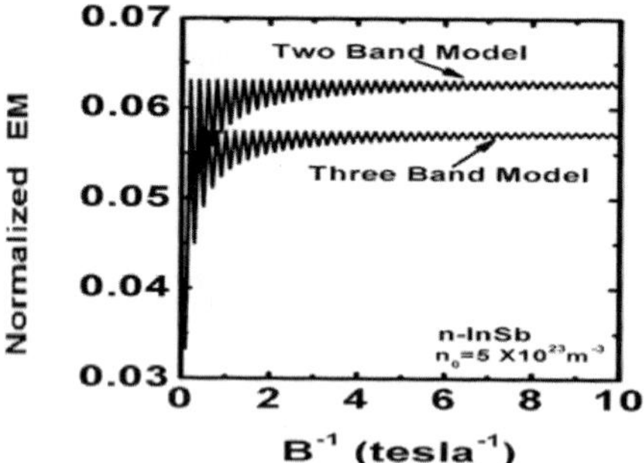

Fig. 5.7. Plot of the normalized EM as function of inverse magnetic field for HD n-InSb considering the three and two band models of Kane.

n-$Hg_{1-x}Cd_xTe$ and HD $In_{1-x}Ga_xAs_yP_{1-y}$ lattice matched to InP in accordance with the three and two band models of Kane respectively. The variations of the EM are periodic and independent of the subband index number with the quantizing magnetic field and the influence of the energy band constants on the EM in accordance with all the band models is apparent from the said figures.

Figures 5.10–5.14 exhibit the concentration dependence of the periodic EM for all the respective aforementioned materials.

It appears from Figs 5.10–5.14 that the periodic oscillatory numerical values of the EM is greatest for the quaternary materials

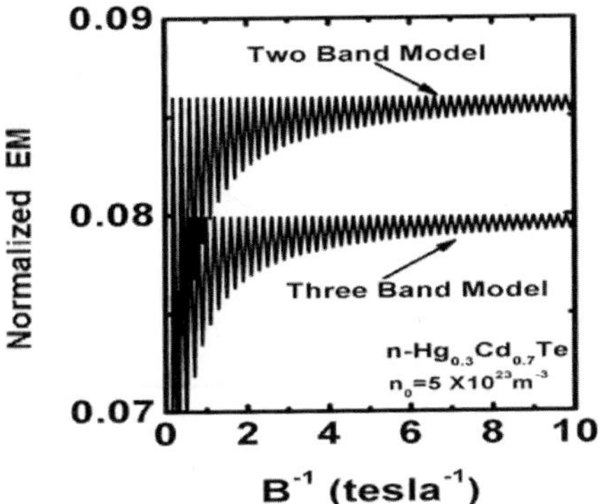

Fig. 5.8. Plot of the normalized EM as function of inverse magnetic field for HD n-$Hg_{0.3}Cd_{0.7}Te$ considering the three and two band models of Kane.

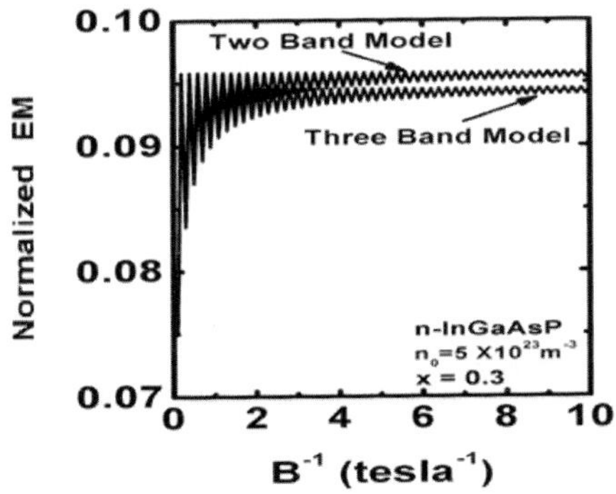

Fig. 5.9. Plot of the normalized EM as function of inverse magnetic field for HD n-$In_{1-x}Ga_xAs_yP_{1-y}$ lattice matched to HD InP considering the three and two band models of Kane.

while the least for InSb for all types of variables in accordance with all types of band models of III–V, ternary and quaternary materials. In Fig. 5.15, we have plotted the variation of the EM as function of alloy composition in HD HgCdTe and HD InGaAsP lattice matched to InP. It appears that the EM increases with the alloy fraction in almost linear way. The result of the EM arising due to the difference in the band structure is also appears to be extremely less.

The numerical computations for the models according to Stillman and Palik have been left as an exercise for the reader.

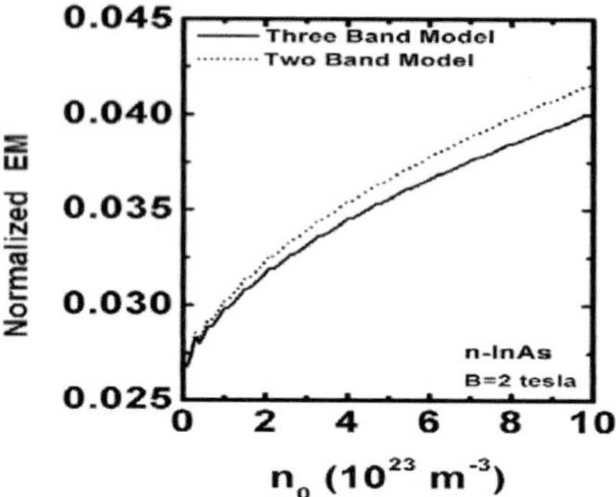

Fig. 5.10. Plot of the normalized EM as function of carrier degeneracy for HD n-InAs considering the three and two band models of Kane.

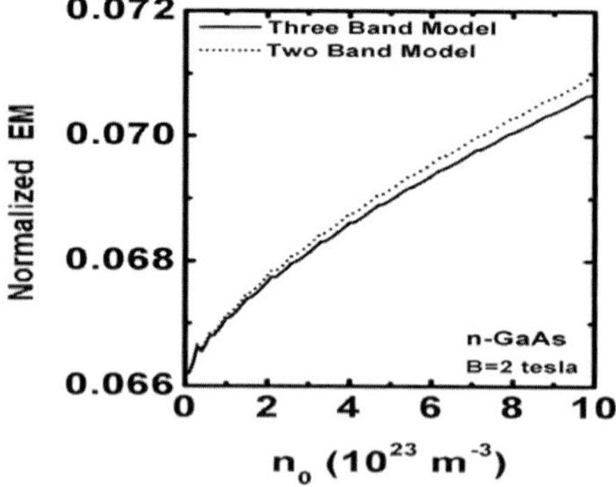

Fig. 5.11. Plot of the normalized EM as function of carrier degeneracy for HD n-GaAs considering the three and two band models of Kane.

Using Eqs. (5.39a) and (5.39b) for Dimmock model, (5.42a) and (5.42b) for the model of Bangert and Kasner, (5.47a) and (5.47b) for the model of Foley and Langenberg, we have plotted the EM for PbS as functions of inverse quantizing magnetic field and carrier degeneracy as shown in Figs 5.16 and 5.17 respectively considering the first two magnetic sub-bands for models of Dimmock.

From Fig. 5.16, it appears that the effect of the energy band structure namely due to the models of Bangert and Kastner and Foley and Langenberg on the EM almost coincides with each other. However, the nonlinear energy dispersion relation of Dimmock tends

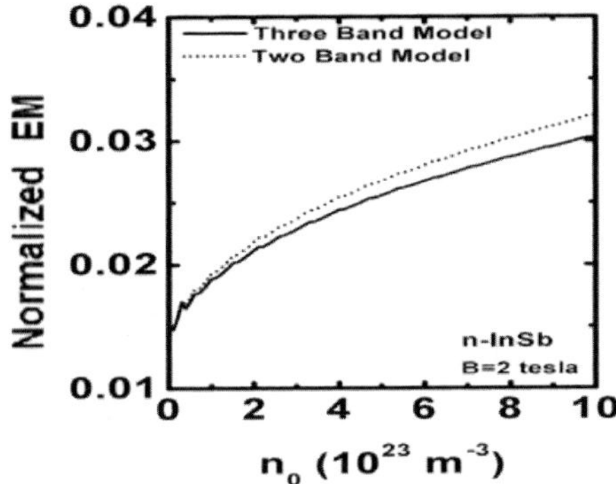

Fig. 5.12. Plot of the normalized EM as function of carrier degeneracy for HD n-InSb considering the three and two band models of Kane.

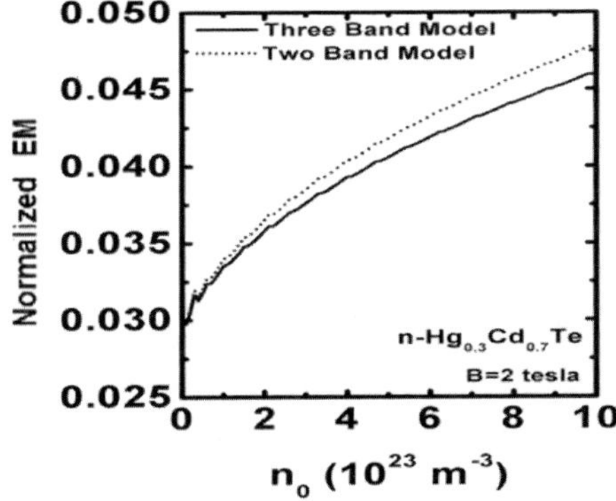

Fig. 5.13. Plot of the normalized EM as function of carrier degeneracy for HD n-$Hg_{0.3}Cd_{0.7}Te$ considering the three and two band models of Kane.

to increase the EM. It appears that the increase in the magnetic subband index increases the EM in this case as compared with that of Figs. 5.1 and 5.2 for nonlinear tetragonal materials. This increase of the EM for the present case results due to the presence of the respective dominant energy spectrum parameters. As the magnetic field increases, we see that with increase in the subband index the EM exhibits a sharp discontinuity and can become a negative quantity, thus questioning the validity of the Dimmock model in the beyond-10 tesla zone.

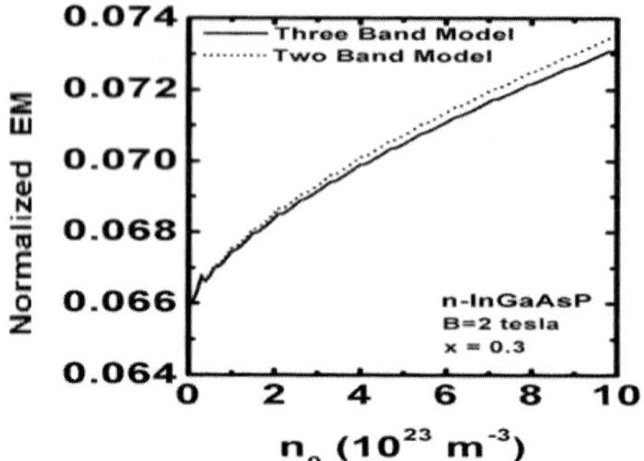

Fig. 5.14. Plot of the normalized EM as function of carrier degeneracy for HD n-$In_{1-x}Ga_xAs_yP_{1-y}$ lattice matched to HD InP considering the three and two band models of Kane.

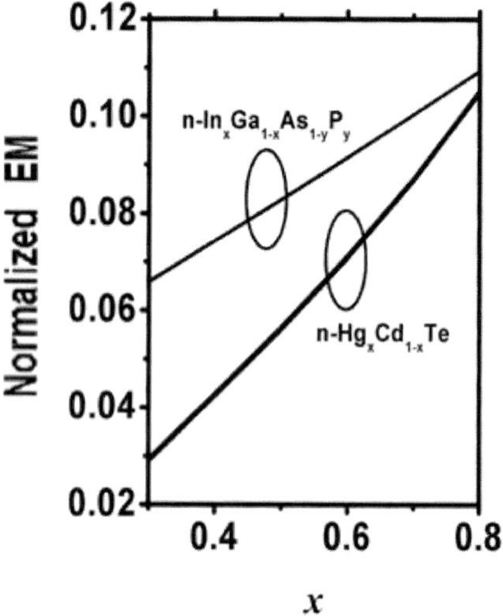

Fig. 5.15. Plot of the normalized EM as function of alloy composition for HD n-$Hg_xCd_{1-x}Te$ and HD n-$In_{1-x}Ga_xAs_yP_{1-y}$ lattice matched to HD InP considering the three and two band models of Kane at a quantizing magnetic field of 2 Tesla and carrier degeneracy of $5 \times 10^{23}\ m^{-3}$.

The variation of the EM on the carrier degeneracy for Bi in Fig. 5.17 is rather slow over $0.1 - 0.5 \times 10^{23}\ m^{-3}$ zone.

Figure 5.18 exhibit the variation of the EM against the quantizing magnetic field for HD PbTe as an example using the dispersion

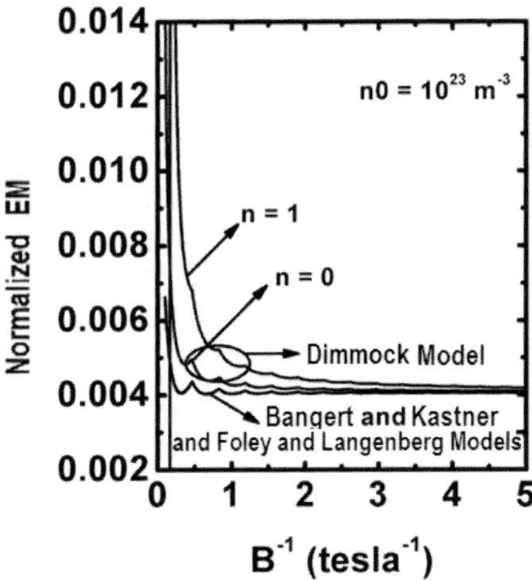

Fig. 5.16. Plot of the normalized EM as function of inverse magnetic field for HD PbS considering the energy band models of Dimmock, Bangert and Kastner and Foley and Langenberg respectively.

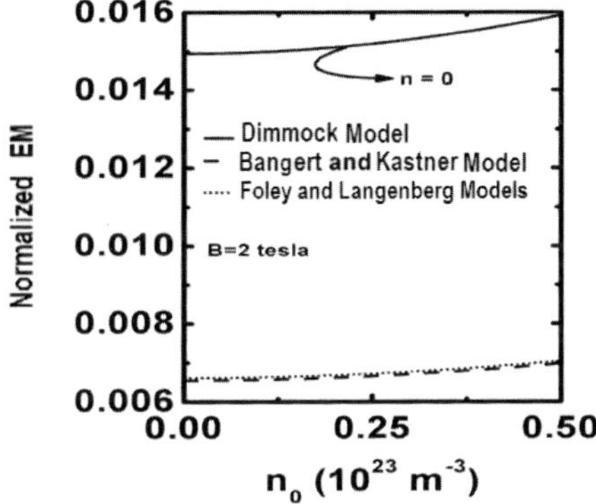

Fig. 5.17. Plot of the normalized EM as function of carrier degeneracy for HD PbS considering the energy band models as considered in Figure 5.16.

relation provided by the Dimmock model at the lowest quantizing subband. In the same figure we have demonstrated the variation of the EM for stressed HD InSb for the first two lowest subbands. Large oscillations are exhibited for HD PbTe case as compared with that of the stressed case, where we have considered the stress to be composed of all the diagonal and off-diagonal strain components as

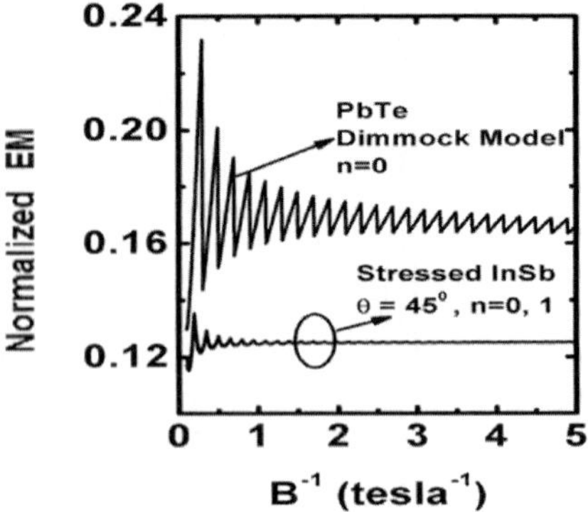

Fig. 5.18. Plot of the normalized EM as function of quantizing magnetic field for HD PbTe and stressed HD InSb at the two lowest magnetic subbands.

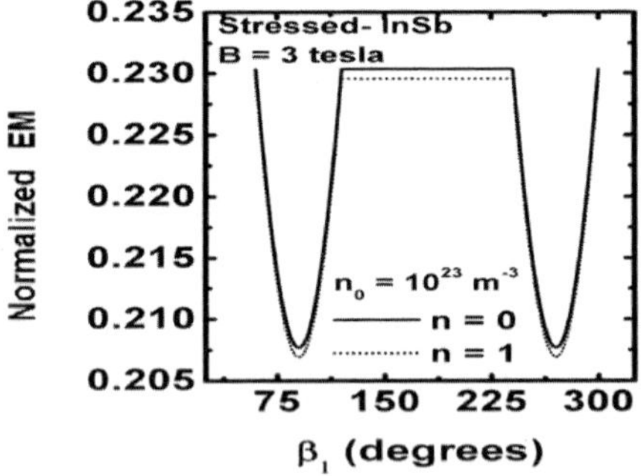

Fig. 5.19. Plot of the normalized EM as function of angular orientation of the magnetic field for stressed HD InSb at the two lowest magnetic subbands.

given in Appindix A. It appears the deviation of the EM from its ground state value is almost zero when the angular dependency is 45°. To exhibit this difference, we have further plotted the EM at the lowest two subbands as function of the angle of orientation of the field in Fig. 5.18. It appears that the EM exhibits periodical variation over the entire angular range as shown in the same figure with the deviation between the subband values at the two minima and the mid angular zone.

Figure 5.20 exhibits the EM in HD Te, HD GaP, HD $PtSb_2$, HD Bi_2Te_3, HD GaSb and as function of quantizing magnetic field at

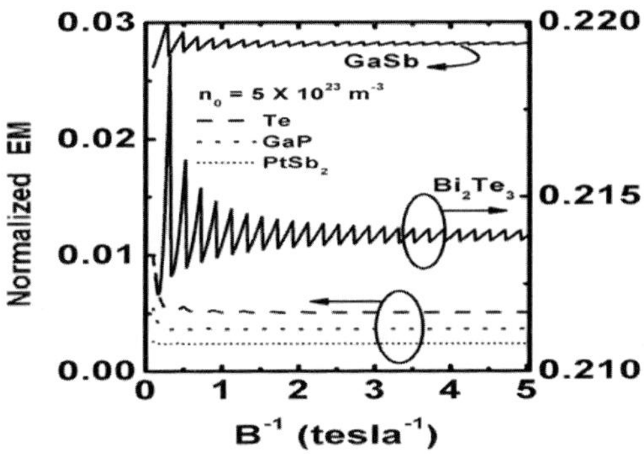

Fig. 5.20. Plot of the normalized EM as function quantizing magnetic field for HD Bi_2Te_3, HD GaSb, HD Te, HD GaP and HD $PtSb_2$ at the lowest magnetic subband.

the lowest subband level. The usual periodical oscillatory nature is exhibited for all the said materials with HD Bi_2Te_3 to exhibit the highest EM numerical values. In case of HD Ge, we observe that the Cardona *et al.* model registers a constant EM, however the Wang *et al.* registers a subband index dependent model. We leave the reader to carry out investigation of the EM using the Wang *et al.* model for HD Ge, other allied models for IV–VI together with that for II–VI materials.

We wish to note that the effect of electron spin has not been considered in obtaining the oscillatory plots. The peaks in all the figures would increase in number with decrease in amplitude if spin splitting term is included in the respective numerical computations. Though, the effects of collisions are usually small at low temperatures, the sharpness of the amplitude of the oscillatory plots would somewhat be reduced by collision broadening. Nevertheless, the present analysis would remain valid since the effects of collision broadening can usually be taken into account by an effective increase in temperature. Although in a more rigorous statement the many body effects should be considered along with the self-consistent procedure, the simplified analysis as presented exhibits the basic qualitative features of the EM in this under the magnetic quantization with reasonable accuracy.

5.4 Open Research Problems

(R.5.1) (a) Investigate the EM in all the bulk semiconductors as considered in this chapter in the absence of any field.
(b) Investigate the same set of masses as defined in (R.5.1) in the presence of an arbitrarily oriented quantizing magnetic field including broadening and the electron spin (applicable under magnetic quantization) for all the bulk semiconductors whose unperturbed carrier energy spectra are defined in Chapter 1.

(R.5.2) Investigate the same set of masses as defined in (R.5.1) in the presence of quantizing magnetic field under an arbitrarily oriented (a) non-uniform electric field and (b) alternating electric field respectively for all the semiconductors whose unperturbed carrier energy spectra are defined in Chapter 1 by including spin and broadening respectively.

(R.5.3) Investigate the same set of masses as defined in (R.5.1) under an arbitrarily oriented alternating quantizing magnetic field by including broadening and the electron spin for all the semiconductors whose unperturbed carrier energy spectra as defined in Chapter 1.

(R.5.4) Investigate the same set of masses as defined in (R.5.1) under an arbitrarily oriented alternating quantizing magnetic field and crossed alternating electric field by including broadening and the electron spin for all the semiconductors whose unperturbed carrier energy spectra as defined in Chapter 1.

(R.5.5) Investigate the same set of masses as defined in (R.5.1) under an arbitrarily oriented alternating quantizing magnetic field and crossed alternating non-uniform electric field by including broadening and the electron spin whose for all the semiconductors unperturbed carrier energy spectra as defined in Chapter 1.

(R.5.6) Investigate the same set of masses as defined in (R.5.1) in the presence and absence of an arbitrarily oriented quantizing magnetic field under exponential, Kane, Halperin, Lax and Bonch-Bruevich band tails [5] for all the semiconductors whose unperturbed carrier energy spectra as defined in Chapter 1 by including spin and broadening (applicable under magnetic quantization).

(R.5.7) Investigate the same set of masses as defined in (R.5.1) in the presence of an arbitrarily oriented quantizing magnetic field for all the semiconductors as defined in (R.5.6) under an arbitrarily oriented (a) non-uniform electric field and (b) alternating electric field respectively whose unperturbed carrier energy spectra as defined in Chapter 1.

(R.5.8) Investigate the same set of masses as defined in (R.5.1) under an arbitrarily oriented alternating quantizing magnetic field by including broadening and the electron spin for all semiconductors whose unperturbed carrier energy spectra as defined in Chapter 1.

(R.5.9) Investigate the same set of masses as defined in (R.5.1) under an arbitrarily oriented alternating quantizing magnetic field and crossed alternating electric field by including broadening and the electron spin for all the semiconductors whose unperturbed carrier energy spectra as defined in Chapter 1.

(R.5.10) Investigate all the appropriate problems of this chapter after proper modifications introducing new theoretical formalisms for functional, negative refractive index, macro molecular, organic and magnetic materials.

(R.5.11) Investigate all the appropriate problems of this chapter for HD p-InSb, HD p-CuCl and stressed semiconductors having diamond structure valence bands whose dispersion relations of the carriers in bulk semiconductors are given by Cunningham [11], Yekimov *et al.* [12] and Roman *et al.* [13] respectively.

(R.5.12) Investigate all the problems of this chapter by removing all the mathematical approximations and establishing the respective appropriate uniqueness conditions.

References

1. N. Miura, *Physics of Semiconductors in High Magnetic Fields*, Series on Semiconductor Science and Technology (Oxford University Press, USA, 2007); K. H. J Buschow, F. R. de Boer, *Physics of Magnetism and Magnetic Materials* (Springer, New York, 2003); D. Sellmyer (Ed.), R. Skomski (Ed.), *Advanced Magnetic Nanostructures* (Springer, New York, 2005); J. A. C. Bland (Ed.), B. Heinrich (Ed.), *Ultrathin Magnetic Structures III: Fundamentals of Nanomagnetism (Pt. 3)* (Springer-Verlag, Germany, 2005); B. K. Ridley, *Quantum Processes in semiconductors*, Fourth Edition (Oxford publications, Oxford, 1999); J. H. Davies, *Physics of low dimensional semiconductors* (Cambridge University Press, UK, 1998); S. Blundell, *Magnetism in Condensed Matter*, Oxford Master Series in Condensed Matter Physics (Oxford University Press, USA, 2001); C. Weisbuch, B. Vinter, *Quantum Semiconductor Structures: Fundamentals and Applications* (Academic Publishers, USA, 1991); D. Ferry, *Semiconductor Transport* (CRC, USA, 2000); M. Reed (Ed.), *Semiconductors and Semimetals: Nanostructured Systems* (Academic Press, USA, 1992); T. Dittrich, *Quantum Transport and Dissipation* (Wiley-VCH Verlag GmbH, Germany, 1998); A. Y. Shik, *Quantum Wells: Physics & Electronics of Two dimensional Systems* (World Scientific, USA, 1997).
2. K. P. Ghatak, M. Mondal, *Zietschrift fur Naturforschung A* **41a**, 881 (1986); K. P. Ghatak, M. Mondal, *J. Appl. Phys.* **62**, 922 (1987); K. P. Ghatak, S. N. Biswas, *Phys. Stat. Sol. (b)* **140**, K107 (1987); K. P. Ghatak, M. Mondal, *Jour. of Mag. and Mag. Mat.* **74**, 203 (1988); K. P. Ghatak, M. Mondal, *Phys. Stat. Sol. (b)* **139**, 195 (1987); K. P. Ghatak, M. Mondal, *Phys. Stat. Sol. (b)* **148**, 645 (1988); K. P. Ghatak, B. Mitra, A. Ghoshal, *Phys. Stat. Sol. (b)* **154**, K121 (1989); K. P. Ghatak, S. N. Biswas, *Jour. of Low Temp. Phys.* **78**, 219 (1990); K. P. Ghatak, M. Mondal, *Phys. Stat. Sol. (b)* **160**, 673 (1990); K. P. Ghatak, B. Mitra, *Phys. Letts. A* **156**, 233 (1991); K. P. Ghatak, A. Ghoshal, B. Mitra, *Nouvo Cimento D* **13D**, 867 (1991); K. P. Ghatak, M. Mondal, *Phys. Stat. Sol. (b)* **148**, 645 (1989); K. P. Ghatak, B. Mitra, *Internat. Jour. of Elect.* **70**, 345 (1991); K. P. Ghatak, S. N. Biswas, *J. Appl. Phys.* **70**, 299 (1991); K. P. Ghatak, A. Ghoshal, *Phys. Stat. Sol. (b)* **170**, K27 (1992); K. P. Ghatak, *Nouvo Cimento D* **13D**, 1321 (1992); K. P. Ghatak, B. Mitra, *Internat. Jour. of Elect.* **72**, 541 (1992); K. P. Ghatak, S. N. Biswas, *Nonlinear Optics* **4**, 347 (1993); K. P. Ghatak, M. Mondal, *Phys. Stat. Sol. (b)* **175**, 113 (1993); K. P. Ghatak, S. N. Biswas, *Nonlinear Optics* **4**, 39 (1993); K. P. Ghatak, B. Mitra, *Nouvo Cimento* **15D**, 97 (1993); K. P. Ghatak, S. N. Biswas, *Nanostructured Materials* **2**, 91 (1993); K. P. Ghatak, M. Mondal, *Phys. Stat. Sol. (b)* **185**, K5 (1994); K. P. Ghatak, B. Goswami, M. Mitra, B. Nag, *Nonlinear Optics* **16**, 9 (1996); K. P. Ghatak, M. Mitra, B. Goswami, B. Nag, *Nonlinear Optics* **16**, 167 (1996); K. P. Ghatak, D. K. Basu, B. Nag, *J. Phys. and Chem. of Sol.* **58**, 133 (1997); K. P. Ghatak, B. Nag, *Nanostructured Materials* **10**, 923 (1998);

3. D. Roy Choudhury, A. K. Choudhury, K. P. Ghatak, A. N. Chakravarti, *Phys. Stat. Sol. (b)* **98**, K141 (1980); A. N. Chakravarti, K. P. Ghatak, A. Dhar and S. Ghosh, *Phys. Stat. Sol. (b)* **105**, K55 (1981); A. N. Chakravarti, A. K. Choudhury, K. P. Ghatak, *Phys. Stat. Sol. (a)* **63**, K97 (1981); A. N. Chakravarti, A. K. Choudhury, K. P. Ghatak, S. Ghosh, A. Dhar, *Appl. Phys.* **25**, 105 (1981); A. N. Chakravarti, K. P. Ghatak, G. B. Rao, K. K. Ghosh, *Phys. Stat. Sol. (b)* **112**, 75 (1982); A. N. Chakravarti, K. P. Ghatak, K. K. Ghosh, H. M. Mukherjee, *Phys. Stat. Sol. (b)* **116**, 17 (1983); M. Mondal, K. P. Ghatak, *Phys. Stat. Sol. (b)* **133**, K143 (1984); M. Mondal, K. P. Ghatak, *Phys. Stat. Sol. (b)* **126**, K47 (1984); M. Mondal, K. P. Ghatak, *Phys. Stat. Sol. (b)* **126**, K41 (1984); M. Mondal, K. P. Ghatak, *Phys. Stat. Sol. (b)* **129**, K745 (1985); M. Mondal, K. P. Ghatak, *Phys. Scr.* **31**, 615 (1985); M. Mondal, K. P. Ghatak, *Phys. Stat. Sol. (b)* **135**, 239 (1986); M. Mondal, K. P. Ghatak, *Phys. Stat. Sol. (b)* **93**, 377 (1986); M. Mondal, K. P. Ghatak, *Phys. Stat. Sol. (b)* **135**, K21 (1986); M. Mondal, S. Bhattacharyya, K. P. Ghatak, *Appl. Phys. A* **42A**, 331 (1987); S. N. Biswas, N. Chattopadhyay,K. P. Ghatak,*Phys. Stat. Sol. (b)* **141**, K47 (1987); B. Mitra, K. P. Ghatak, *Phys. Stat. Sol. (b)* **149**, K117 (1988); B. Mitra, A. Ghoshal, K. P. Ghatak, *Phys. Stat. Sol. (b)* **150**, K67 (1988); M. Mondal, K. P. Ghatak, *Phys. Stat. Sol. (b)* **147**, K179 (1988); M. Mondal, K. P. Ghatak, *Phys. Stat. Sol. (b)* **146**, K97 (1988); B. Mitra, A. Ghoshal, K. P. Ghatak, *Phys. Stat. Sol. (b)* **153**, K209 (1989); B. Mitra, K. P. Ghatak, *Phys. Letts.* **142A**, 401 (1989); B. Mitra, A. Ghoshal,K. P. Ghatak, *Phys. Stat. Sol. (b)* **154**, K147 (1989); B. Mitra, K. P. Ghatak, *Sol. State Elect.* **32**, 515 (1989); B. Mitra, A. Ghoshal, K. P. Ghatak, *Phys. Stat. Sol. (b)* **155**, K23 (1989); B. Mitra, K. P. Ghatak, *Phys. Letts.* **135A**, 397 (1989); B. Mitra, K. P. Ghatak, *Phys. Letts. A* **146A**, 357 (1990); B. Mitra, K. P. Ghatak, *Phys. Stat. Sol. (b)* **164**, K13 (1991); S. N. Biswas, K. P. Ghatak, *Internat. Jour. of Elect.* **70**, 125 (1991);
4. P. R. Wallace, Phys. Stat. Sol. (b), **92**, 49 (1979).
5. B. R. Nag, *Electron Transport in Compound Semiconductors*, Springer Series in Solid-State Sciences, **Vol. 11** (Springer-Verlag, Germany, 1980).
6. K. P. Ghatak, S. Bhattacharya, D. De, *Einstein Relation in Compound Semiconductors and Their Nanostructures*, Springer Series in Materials Science, **Vol. 116** (Springer-Verlag, Germany, 2009).
7. C. C. Wu and C. J. Lin, J. Low Temp. Phys. **57**, 469 (1984); M. H. Chen, C. C. Wu, C. J. Lin, *J. Low Temp. Phys.* **55**, 127 (1984).
8. E. Bangert, P. Kastner, Phys. Stat. Sol (b) **61**, 503 (1974).
9. G. M. T. Foley, P. N. Langenberg, Phys. Rev. B, **15B**, 4850 (1977).
10. M. Singh, P. R. Wallace, S. D. Jog and E. Arushanov, J. Phys. Chem. Solids, **45**, 409, (1984); Y. Yamada, Phys. Soc. Japan, **35**, 1600 (1973), **37**, 606 (1974).
11. R. W. Cunningham, *Phys. Rev.* **167**, 761 (1968).
12. A. I. Yekimov, A. A. Onushchenko, A. G. Plyukhin, Al, L. Efros, *J. Expt. Theor. Phys.* **88**, 1490 (1985).
13. B. J. Roman, A. W. Ewald, *Phys. Rev. B* **5**, 3914 (1972).

Chapter 6

The EM Under Photo Excitation in HD Kane Type Semiconductors

6.1 Introduction

With the advent of nano-photonics, there has been a considerable interest in studying the optical processes in semiconductors and their nanostructures [1]. It appears from the literature, that the investigations have been carried out on the assumption that the carrier energy spectra are invariant quantities in the presence of intense light waves, which is not fundamentally true. The physical properties of semiconductors in the presence of light waves which change the basic dispersion relation have relatively less investigated in the literature [2–4]. In this chapter we shall study the EM in HD III–V, ternary and quaternary semiconductors on the basis of newly formulated electron dispersion law under external photo excitation.

In Section 6.2.1 of the theoretical background 6.2, we have investigated the EM of the conduction electrons of HD III–V, ternary and quaternary materials in the presence of light waves whose unperturbed electron energy spectrum is described by the three-band model of Kane in the absence of band tailing. In Section 6.2.2, we have studied the opto EM in the said HD materials under magnetic quantization. In Section 6.2.3, we have studied the opto EM in the presence of crossed electric and quantizing magnetic fields. In Section 6.2.4, we have studied the EM in QWs in HD Kane type semiconductors. In Section 6.2.5, we have investigated the EM in doping superlattices of HD Kane type semiconductors in the presence of light waves. In Section 6.2.6, we have studied the EM in accumulation and inversion layers of Kane type semiconductors in the presence of light waves. In Section 6.2.7, we have studied the EM in NWs of HD of Kane type semiconductors in the presence of light waves. In Section 6.2.8, we have investigated the EM in QWHD effective mass superlattices of Kane type semiconductors. In Section 6.2.9, we have studied the EM in NWHD effective mass superlattices of Kane type semiconductors. In Section 6.2.10, we have studied the magneto EM in HD effective mass superlattices of Kane type semiconductors in the presence of light waves. In Section 6.2.11, we have studied the EM in QWHD superlattices of Kane type semiconductors with graded interfaces in the presence of light waves. In Section 6.2.12, we have studied the EM in NWHD superlattices of Kane type semiconductors with graded interfaces in the presence of light waves. In Section 6.2.13, we have studied the magneto EM in HD superlattices of Kane type semiconductors with graded interfaces in the presence of light waves. Lastly Section 6.3 presents 6 open research problems.

6.2 Theoretical Background

6.2.1 *The formulation of EM in the presence of light waves in HD III–V, ternary and quaternary semiconductors*

The Hamiltonian ($\hat{H}$) of an electron in the presence of light wave characterized by the vector potential $\vec{A}$ can be written following [3] as

$$\hat{H} = [|(\hat{p} + |e|\vec{A})|^2/2] + V(\vec{r}) \tag{6.1}$$

in which, $\hat{p}$ is the momentum operator, $V(\bar{r})$ is the crystal potential and m is the free electron mass. (6.1) can be expressed as

$$\hat{H} + \hat{H}_0 + \hat{H}' \tag{6.2}$$

where,

$$\hat{H}_0 = \frac{\hat{p}^2}{2m} + V(\vec{r})$$

and

$$\hat{H}' = \frac{|e|}{2m}\vec{A}\cdot\hat{p} \tag{6.3}$$

The perturbed Hamiltonian $\hat{H}'$ can be written as

$$\hat{H}' = \left(\frac{-\hbar|e|}{2m}\right)\left(\vec{A}.\nabla\right). \tag{6.4}$$

where, $i = \sqrt{-1}$ and $\hat{p} = -i\hbar\nabla$.

The vector potential $(\vec{A})$ of the monochromatic light of plane wave can be expressed as

$$\vec{A} = A_0\vec{\varepsilon}_s\cos(\vec{s}_0\cdot\vec{r} - \omega t) \tag{6.5}$$

where A_0 is the amplitude of the light wave, $\vec{\varepsilon}_s$ is the polarization vector, $\vec{s}_0$ is the momentum vector of the incident photon, $\vec{r}$ is the position vector, ω is the angular frequency of light wave and t is the time scale. The matrix element of $\hat{H}'_{nl}$ between initial state, $\psi_1(\vec{q},\vec{r})$ and final state $\psi_n(\vec{k},\vec{r})$ in different bands can be written as

$$\hat{H}'_{nl} = \frac{|e|}{2m}\langle n\vec{k}|\vec{A}\cdot\hat{p}|l\vec{q}\rangle. \tag{6.6}$$

Using (6.4) and (6.5), we can re-write (6.6) as

$$\hat{H}'_{nl} = \left(\frac{-i\hbar|e|A_0}{4m}\right)\vec{\varepsilon}_s\cdot[\{\langle n\vec{k}|e^{(i\vec{s}_0\cdot\vec{r})}\nabla|l\vec{q}\rangle e^{-i\omega t}\} + \{\langle n\vec{k}|e^{(i\vec{s}_0\cdot\vec{r})}\nabla|l\vec{q}\rangle e^{i\omega t}\}] \tag{6.7}$$

The first matrix element of (18.7) can be written as

$$\langle n\vec{k}|e^{(i\vec{s}_0\cdot\vec{r})}\nabla|l\vec{q}\rangle = \int e^{(i[\vec{q}+\vec{s}_0-\vec{k}]\cdot\vec{r})}i\vec{q}u_n^*(\vec{k}\cdot\vec{r})u_l(\vec{q},\vec{r})d^3r + \int e^{(i[\vec{q}+\vec{s}_0-\vec{k}]\cdot\vec{r})}u_n^*(\vec{k},\vec{r})\nabla u_l(\vec{q},\vec{r})d^3r \tag{6.8}$$

The functions $u_n^*u_l$ and $u_n^*\nabla u_l$ are periodic. The integral over all space can be separated into a sum over unit cells times an integral over a single unit cell. It is assumed that the wave length of the electromagnetic wave is sufficiently large so that if $\vec{k}$ and $\vec{q}$ are within the Brillouin zone, $(\vec{q} + \vec{s}_0 - \vec{k})$ is not a reciprocal lattice vector.

Therefore, we can write (6.8) as

$$\langle n\vec{k}|e^{(i\vec{s}_0-\vec{r})}\nabla|l\vec{q}\rangle$$

$$= \left[\frac{(2\pi)^3}{\Omega}\right]\left\{i\vec{q}\delta(\vec{q}+\vec{s}_0-\vec{k})\delta_{nl}\right.$$

$$\left.+\ \delta(\vec{q}+\vec{s}_0-\vec{k})u_n^*\int_{cell}(\vec{k},\vec{r})\nabla u_1(\vec{q},\vec{r})d^3r\right\}$$

$$= \left[\frac{(2\pi)^3}{\Omega}\right]\left\{\delta(\vec{q}+\vec{s}_0-\vec{k})\int_{cell}u_n^*(\vec{k},\vec{r})\nabla u_1(\vec{q},\vec{r})d^3r\right\} \tag{6.9}$$

where, Ω is the volume of the unit cell and $\int u_n^*(\vec{k},\vec{r})u_l(\vec{q},\vec{r})d^3r = \delta(\vec{q}-\vec{k})\delta_{nl} = 0$, since $n \neq l$. The delta function expresses the conservation of wave vector in the absorption of light wave and $\vec{s}_0$ is small compared to the dimension of a typical Brillouin zone and we set $\vec{q}=\vec{k}$.

From (18.8) and (18.9), we can write,

$$\hat{H}'_{nl} = \frac{|e|A_0}{2m}\vec{\varepsilon}_s \cdot \hat{p}_{nl}(\vec{k})\delta(\vec{q}-\vec{k})\cos(\omega t) \tag{6.10}$$

where,

$$\hat{p}_{nl}(\vec{k}) = -i\hbar\int u_n^*\nabla u_1 d^3t = \int u_n^*(\vec{k},\vec{r})\hat{p}u_l(\vec{k},\vec{r})d^3r$$

Therefore, we can write

$$\hat{H}'_{nl} = \frac{|e|A_0}{2m}\vec{\varepsilon}_s \cdot \hat{p}_{nl}(\vec{k}) \tag{6.11}$$

where, $\vec{\varepsilon} = \vec{\varepsilon}_s \cos\omega t$.

When a photon interacts with a semiconductor, the carriers (i.e., electrons) are generated in the bands which are followed by the inter-band transitions. For example, when the carriers are generated in the valence band, the carriers then make inter-band transition to the conduction band. The transition of the electrons within the same band i.e., $\hat{H}'_{nn} = \langle n\vec{k}|\hat{H}'|n\vec{k}\rangle$ is neglected. Because, in such a case, i.e., when the carriers are generated within the same bands by photons, are lost by recombination within the aforementioned band resulting zero carriers.

Therefore,

$$\langle n\vec{k}|\hat{H}'|n\vec{k}\rangle = 0 \tag{6.12}$$

With $n = c$ stands for conduction band and $l = v$ stand for valance band, the energy equation for the conduction electron can approximately be written as

$$I_{11}(E) = \left(\frac{\hbar^2 k^2}{2m_c}\right) + \frac{\left(\frac{|e|A_0}{2m}\right)^2 \langle|\vec{\varepsilon}\cdot\hat{p}_{cv}(\vec{k})|^2\rangle_{av}}{E_c(\vec{k}) - E_v(\vec{k})} \tag{6.13}$$

where, $I_{11}(E) \equiv E(aE+1)(bE+1)/(cE+1), a \equiv 1/E_{g0}$, $a \equiv 1/E_{g0}$, E_{g0} is the un-perturbed band-gap, $b \equiv 1/(E_{g0}+\Delta)$, $c \equiv 1/(E_{g0}+2\Delta/3)$, and $\langle|\vec{\varepsilon}\cdot\hat{p}_{cv}(\vec{k})|^2\rangle_{av}$ represents the average of the square of the optical matrix element(OME).

For the three-band model of Kane, we can write,

$$\xi_{1k} = E_c(\vec{k}) - E_v(\vec{k}) = (E_{g0}^2 + E_{g0}\hbar^2 k^2/m_r)^{1/2} \tag{6.14}$$

where, m_r is the reduced mass and is given by $m_r^{-1} = (m_c)^{-1} + m_v^{-1}$, and m_v is the effective mass of the heavy hole at the top of the valance band in the absence of any field.

The doubly degenerate wave functions $u_1(\vec{k},\vec{r})$ and $u_2(\vec{k},\vec{r})$ can be expressed as [16]

$$u_1(\vec{k},\vec{r}) = a_{k+}[(is)\downarrow'] + b_{k+}\left[\frac{X'-iY'}{\sqrt{2}}\uparrow'\right] + c_{k+}[Z'\downarrow'] \tag{6.15}$$

and

$$u_2(\vec{k},\vec{r}) = a_{k-}[(is)\uparrow'] - b_{k-}\left[\frac{X'+iY'}{\sqrt{2}}\downarrow'\right] + c_{k-}[Z'\uparrow']. \tag{6.16}$$

s is the s-type atomic orbital in both unprimed and primed coordinates, $\downarrow'$ indicates the spin down function in the primed coordinates,

$$a_{k\pm} \equiv \beta[E_{g0} - (\gamma_{0k\pm})^2(E_{g0}-\delta')]^{1/2}(E_{g0}+\delta')^{-1/2},$$

$$\beta \equiv [(6(E_{g0}+2\Delta/3)(E_{g0}+\Delta))/\chi]^{1/2},$$

$$\chi \equiv (6E_{g0}^2 + E_{g0}\Delta + 4\Delta^2), \quad \gamma_{0k\pm} \equiv \left[\frac{(\xi_{1k} \mp E_{g0})}{2(\xi_{1k}+\delta')}\right]^{1/2},$$

$$\xi_{1k} = E_c(\vec{k}) - E_v(\vec{k}) = E_{g0}\left[1 + 2\left(1 + \frac{m_c}{m_v}\right)\frac{I_{11}(E)}{E_{g0}}\right]^{1/2},$$

$$\delta' \equiv (E_{g0}^2\Delta)(\chi)^{-1}, \quad X', Y', \quad \text{and} \quad Z'$$

are the p-type atomic orbitals in the primed coordinates, $\uparrow'$ indicates the spin-up function in the primed coordinates, $b_{k\pm} \equiv \rho\gamma_{0k\pm}, \rho \equiv (4\Delta^2/3\chi)^{1/2}, c_{k\pm} \equiv t\gamma_{0k\pm}$ and $t \equiv [6(E_{g0} + 2\Delta/3)^2/\chi]^{1/2}$.

We can, therefore, write the expression for the optical matrix element (OME) as

$$OME = \hat{p}_{cv}(\vec{k}) = \langle u_1(\vec{k}, \vec{r})|\hat{p}|u_2(\vec{k}, r')\rangle. \tag{6.17}$$

Since the photon vector has no interaction in the same band for the study of inter-band optical transition, we can therefore write

$$\langle S|\hat{p}|S\rangle = \langle X|\hat{p}|X\rangle = \langle Y|\hat{p}|Y\rangle = \langle Z|\hat{p}|Z\rangle = 0$$

and $\langle X|\hat{p}|Y\rangle = \langle Y|\hat{p}|Z\rangle = \langle Z|\hat{p}|X\rangle = 0$

There are finite interactions between the conduction band (CB) and the valance band (VB) and we can obtain

$$\langle S|\hat{p}|X\rangle = \hat{i}\cdot\hat{p} = \hat{i}\cdot\hat{p}_x$$

$$\langle S|\hat{p}|Y\rangle = \hat{j}\cdot\hat{p} = \hat{j}\cdot\hat{p}_x$$

$$\langle S|\hat{p}|Z\rangle = \hat{k}\cdot\hat{p} = \hat{k}\cdot\hat{p}_x$$

where, $\hat{i}, \hat{j}$ and $\hat{k}$ are the unit vectors along x, y and z axes respectively.

It is well known [49] that

$$\begin{bmatrix}\uparrow'\\ \downarrow'\end{bmatrix} = \begin{bmatrix} e^{-i\phi/2}\cos(\theta/2) & e^{i\phi/2}\sin(\theta/2)\\ -e^{-i\phi/2}\sin(\theta/2) & e^{i\phi/2}\cos(\theta/2)\end{bmatrix}\begin{bmatrix}\uparrow\\ \downarrow\end{bmatrix}$$

and

$$\begin{bmatrix}X'\\ Y'\\ Z'\end{bmatrix} = \begin{bmatrix}\cos\theta\cos\phi & \cos\theta\sin\phi & -\sin\theta\\ -\sin\phi & \cos\phi & 0\\ \sin\theta\cos\phi & \sin\theta\sin\phi & \cos\theta\end{bmatrix}\begin{bmatrix}X\\ Y\\ Z\end{bmatrix}$$

Besides, the spin vector can be written as

$$\vec{S} = \frac{\hbar}{2}\vec{\sigma},$$

where,

$$\sigma_x = \begin{bmatrix} 0 & 1 \\ 1 & 0 \end{bmatrix}, \quad \sigma_y = \begin{bmatrix} 0 & -i \\ i & 0 \end{bmatrix}$$

and

$$\sigma_z = \begin{bmatrix} 1 & 0 \\ 0 & -1 \end{bmatrix}.$$

From above, we can write

$$\begin{aligned}
\hat{P}_{CV}(\vec{k}) &= \langle u_1(\vec{k},\vec{r})|\hat{p}|u_2(\vec{k},r')\rangle \\
&= \left\langle \left\{ a_{k_+}[(iS)\downarrow'] + b_{k_+}\left[\left(\frac{X'-iY'}{\sqrt{2}}\right)\uparrow'\right] + c_{k_+}[Z'\downarrow']\right\}\right. \\
&\quad \left. \cdot|\hat{P}|\left\{a_{k_-}(iS)\uparrow' - b_{k_-}\left[\left(\frac{X'+iY'}{\sqrt{2}}\right)\downarrow' + c_{k_-}[Z'\uparrow']\right]\right\}\right\rangle
\end{aligned}$$

Using above relations, we get

$$\begin{aligned}
\hat{p}_{CV}(\vec{k}) &= \left\langle u_1(\vec{k},\vec{r})|\hat{P}|u_2(\vec{k},r')\right\rangle \\
&= \frac{b_{k_+}a_{k_-}}{\sqrt{2}}\{\langle(X'-iY')|\hat{P}|iS\rangle\langle\uparrow'|\uparrow'\rangle\} \\
&\quad + c_{k_+}a_{k_-}\{\langle Z'|\hat{P}|iS\rangle\langle\downarrow'|\uparrow'\rangle\} \\
&\quad - \frac{a_{k_+}b_{k_-}}{\sqrt{2}}\{\langle iS|\hat{P}|(X'+iY')\rangle\langle\downarrow'|\downarrow'\rangle\} \\
&\quad + a_{k_+}c_{k_-}\{\langle iS|\hat{P}|Z'\rangle\langle\downarrow'|\uparrow'\rangle\}
\end{aligned} \tag{6.18}$$

We can also write

$$\begin{aligned}
\langle(X'-iY')|\hat{P}|iS\rangle &= \langle(X')|\hat{P}|iS\rangle - \langle(iY')|\hat{P}|iS\rangle \\
&= i\int u^*_{X+}\hat{P}S - \int -iu^*_{Y+}\hat{P}iu_x \\
&= i\langle X'|\hat{P}|iS\rangle - \langle Y'|\hat{P}|iS\rangle
\end{aligned}$$

From the above relations, for X', Y' and Z', we get

$$|X'\rangle = \cos\theta\cos\phi|X\rangle + \cos\theta\sin\phi|Y\rangle - \sin\theta|Z\rangle$$

Thus,

$$\langle X'|\hat{P}|S\rangle = \cos\theta\cos\phi\langle X'|\hat{P}|S\rangle + \cos\theta\sin\phi\langle Y|\hat{P}|S\rangle - \sin\theta\langle Z|\hat{P}|S\rangle = \hat{P}\hat{r}_1$$

where,

$$\hat{r}_1 = \hat{\imath}\cos\theta\cos\phi + \hat{\jmath}\cos\theta\sin\phi - \hat{k}\sin\theta$$

$$|Y'\rangle = -\sin\phi|\,X\rangle + \cos\phi|\,Y\rangle + 0|\,Z\rangle$$

Thus,

$$\langle Y'|\hat{P}|S\rangle = -\sin\phi\langle X|\hat{P}|S\rangle + \cos\phi\langle Y|\hat{P}|S\rangle + 0\langle Z|\hat{P}|S\rangle = \hat{P}\hat{r}_2$$

where

$$\hat{r}_2 = -\hat{\imath}\sin\phi + \hat{\jmath}\cos\phi$$

so that

$$\langle(X' - iY')|\hat{P}|S\rangle = \hat{P}(i\hat{r}_1 - \hat{r}_2)$$

Thus,

$$\frac{a_{k_-}b_{k_+}}{\sqrt{2}}\langle(X' - iY')|\hat{P}|S\rangle\langle\uparrow'|\uparrow'\rangle = \frac{a_{k_-}b_{k_+}}{\sqrt{2}}(i\hat{r}_1 - \hat{r}_2)\langle\uparrow'|\uparrow'\rangle \tag{6.19}$$

Now since,

$$\langle iS|\hat{P}|(X' + iY')\rangle = i\langle S|\hat{P}|X'\rangle - \langle S|\hat{P}|Y'\rangle = \hat{P}(i\hat{r}_1 - \hat{r}_2)$$

We can write,

$$-\left[\frac{a_{k_+}b_{k_-}}{\sqrt{2}}\{\langle iS|\hat{P}|(X' + iY')\rangle\langle\downarrow'|\downarrow'\rangle\}\right] = \left[\frac{a_{k_+}b_{k_-}}{\sqrt{2}}\hat{P}(i\hat{r}_1 - \hat{r}_2)\langle\downarrow'|\downarrow'\rangle\right] \tag{6.20}$$

Similarly, we get,

$$|Z'\rangle = \sin\theta\cos\phi|X\rangle + \sin\theta\sin\phi|Y\rangle + \cos\theta|Z\rangle$$

So that,

$$\langle Z'|\hat{P}|S\rangle = i\langle Z'|\hat{P}|S\rangle - i\hat{P}\{\sin\theta\cos\phi\hat{\imath} + \sin\theta\sin\phi\hat{\jmath} + \cos\theta\hat{k}\} = i\hat{P}\hat{r}_3$$

where,

$$\hat{r}_3 = \hat{\imath}\sin\theta\cos\phi + \hat{\jmath}\sin\theta\sin\phi + \hat{k}\cos\theta$$

Thus,

$$c_{k_+}a_{k_-}\langle Z'|\hat{P}|S\rangle\langle\downarrow'|\uparrow'\rangle = c_{k_+}a_{k_-}i\hat{P}\hat{r}_3\langle\downarrow'|\uparrow'\rangle \tag{6.21}$$

Similarly, we can write,

$$c_{k_+}a_{k_-}\langle iS|\hat{P}|Z'\rangle\langle\downarrow'|\uparrow'\rangle = c_{k_-}a_{k_+}i\hat{P}\hat{r}_3\langle\downarrow'|\uparrow'\rangle \tag{6.22}$$

Therefore, we obtain

$$\begin{aligned}&\frac{a_{k_-}b_{k_+}}{\sqrt{2}}\{\langle(X'+iY')|\hat{P}|S\rangle\langle\uparrow'|\uparrow'\rangle\} - \frac{a_{k_-}b_{k_+}}{\sqrt{2}}\{\langle iS|\hat{P}|(X'+iY')\rangle\langle\downarrow'|\downarrow'\rangle\}\\ &= \frac{\hat{P}}{\sqrt{2}}(-a_{k_+}b_{k_-}\langle\downarrow'|\downarrow'\rangle + a_{k_-}b_{k_+}\langle\uparrow'|\uparrow'\rangle)(i\hat{r}_2 - \hat{r}_2)\end{aligned} \tag{6.23}$$

Also, we can write,

$$\begin{aligned}&c_{k_+}a_{k_-}\langle Z'|\hat{P}|iS\rangle\langle\downarrow'|\uparrow'\rangle + c_{k_-}a_{k_+}\langle iS|\hat{P}|Z'\rangle\langle\downarrow'|\uparrow'\rangle\\ &\quad = i\hat{P}(c_{k_+}a_{k_-} + c_{k_-}a_{k_+})\hat{r}_3\langle\downarrow'|\downarrow'\rangle\end{aligned} \tag{6.24}$$

Combining (6.23) and (6.24), we find

$$\begin{aligned}\hat{p}_{CV}(\vec{k}) = {}&\frac{\hat{p}}{\sqrt{2}}(i\hat{r}_1 - \hat{r}_2)\{(b_{k_+}a_{k_-})\langle\uparrow'|\uparrow'\rangle - (b_{k_-}a_{k_+})\langle\downarrow'|\downarrow'\rangle\}\\ &+ i\hat{P}\hat{r}_3(c_{k_+}a_{k_-} - c_{k_-}c_{k_+})\langle\downarrow'|\uparrow'\rangle\end{aligned} \tag{6.25}$$

From the above relations, we obtain,

$$\left.\begin{aligned}\uparrow' &= e^{-i\phi/2}\cos(\theta/2)\uparrow + e^{i\phi/2}\sin(\theta/2)\downarrow\\ \downarrow' &= e^{-i\phi/2}\sin(\theta/2)\uparrow + e^{i\phi/2}\cos(\theta/2)\downarrow\end{aligned}\right\} \tag{6.26}$$

Therefore,

$$\begin{aligned}\langle\downarrow'|\uparrow'\rangle_x = {}&-\sin(\theta/2)\cos(\theta/2)\langle\uparrow\mid\uparrow\rangle_x + e^{-i\theta}\cos^2(\theta/2)\langle\downarrow\mid\uparrow\rangle_x\\ &- e^{i\phi}\sin^2(\theta/2)\langle\uparrow\mid\downarrow\rangle_x + \sin(\theta/2)\cos(\theta/2)\langle\downarrow\mid\downarrow\rangle_x\end{aligned} \tag{6.27}$$

But we know from above that

$$\langle\uparrow\mid\uparrow\rangle_x = 0,\quad \langle\downarrow\mid\uparrow\rangle = \frac{1}{2},\quad \langle\downarrow\mid\uparrow\rangle_x = \frac{1}{2}\quad \text{and}\quad \langle\downarrow\mid\downarrow\rangle_x = 0$$

Thus, we get,

$$\begin{aligned}\langle\downarrow'|\uparrow'\rangle_x &= \frac{1}{2}[e^{-i\phi}\cos^2(\theta/2) - e^{i\phi}\sin^2(\theta/2)]\\ &= \frac{1}{2}[(\cos\phi - i\sin\phi)\cos^2(\theta/2) - (\cos\phi + i\sin\phi)\sin^2(\theta/2)]\\ &= \frac{1}{2}[\cos\phi\cos\theta - i\sin\phi] \end{aligned} \quad (6.28)$$

Similarly, we obtain

$$\langle\downarrow'|\uparrow\rangle_y = \frac{1}{2}[i\cos\phi + \sin\phi\cos\theta] \quad \text{and} \quad \langle\downarrow'|\uparrow\rangle_z = \frac{1}{2}[-\sin\theta]$$

Therefore,

$$\begin{aligned}\langle\downarrow'|\uparrow'\rangle &= \hat{\imath}\langle\downarrow'|\uparrow'\rangle_x + \hat{\jmath}\langle\downarrow'|\uparrow'\rangle_y + \hat{k}\langle\downarrow'|\uparrow'\rangle\\ &\cdot\frac{1}{2}\{(\cos\theta\cos\phi - i\sin\phi)\hat{\imath} + (i\cos\phi + \sin\phi\cos\theta)\hat{\jmath} - \sin\theta\hat{k}\}\\ &= \frac{1}{2}[\{(\cos\theta\cos\phi)\hat{\imath} + (\sin\phi\cos\theta)\hat{\jmath}\\ &\quad - \sin\theta\hat{k}\} + i\{-\hat{\imath}\sin\phi + \hat{\jmath}\cos\phi\}]\\ &= \frac{1}{2}[\hat{r}_1 + i\hat{r}_2] = -\frac{1}{2}i[i\hat{r}_1 - \hat{r}_2]\end{aligned}$$

Similarly, we can write

$$\begin{aligned}\langle\uparrow'|\uparrow'\rangle &= \frac{1}{2}[\hat{\imath}\sin\theta\cos\phi + \hat{\jmath}\sin\theta\sin\phi + \hat{k}\cos\theta]\\ &= \frac{1}{2}\hat{r}_3 \quad \text{and} \quad \langle\downarrow'|\downarrow'\rangle = -\frac{1}{2}\hat{r}_3\end{aligned}$$

Using the above results, we can write

$$\begin{aligned}\hat{p}_{CV}(\vec{k}) &= \frac{\hat{p}}{\sqrt{2}}(i\hat{r}_2 - \hat{r}_2)\{(a_{k_-}b_{k_+})\langle\uparrow'|\uparrow'\rangle - (b_{k_-}a_{k_+})\langle\downarrow'|\downarrow'\rangle\}\\ &\quad + i\hat{P}\hat{r}_3\{(c_{k_+}a_{k_-} - c_{k_-}a_{k_+})\langle\downarrow'|\uparrow'\rangle\}\\ &= \frac{\hat{P}}{2}\hat{r}_3\,(i\hat{r}_1 - \hat{r}_2)\left\{\left(\frac{a_{k_-}b_{k_+}}{\sqrt{2}} + \frac{b_{k_-}a_{k_+}}{\sqrt{2}}\right)\right\}\\ &\quad + \frac{\hat{P}}{2}\hat{r}_3(i\hat{r}_2 - \hat{r}_2)\{(c_{k_+}a_{k_-} + c_{k_-}c_{k_+})\}\end{aligned}$$

Thus,

$$\hat{p}_{CV}(\vec{k}) = \frac{\hat{p}}{2}\hat{r}_3(i\hat{r}_2 - \hat{r}_2)\left\{a_{k_+}\left(\frac{b_{k_-}}{\sqrt{2}} + c_{k_-}\right) + a_{k_-}\left(\frac{b_{k_+}}{\sqrt{2}} + c_{k_+}\right)\right\} \tag{6.29}$$

We can write that,

$$|\hat{r}_1| = |\hat{r}_2| = |\hat{r}_2| = 1,$$

also,

$$\hat{P}\hat{r}_3 = \hat{P}_x \sin\theta\cos\phi\hat{i} + \hat{P}_y \sin\theta\sin\phi\hat{j} + \hat{P}_z \cos\theta\hat{k}$$

where,

$$\hat{P} = \langle S|P|X\rangle = \langle S|P|Y\rangle = \langle S|P|Z\rangle,$$

$$\langle S|P'|X\rangle = \int u_C^*(0,\vec{r})\hat{P}u_{VX}(0,\vec{r})d^3r = \hat{P}_{CVX}(0)$$

and

$$\langle S|\hat{P}|Z\rangle = \hat{P}_{CVZ}(0)$$

Thus,

$$\hat{P} = \hat{P}_{CVX}(0) = \hat{P}_{CVY}(0) = \hat{P}_{CVZ}(0) = \hat{P}_{CV}(0)$$

where,

$$\hat{P}_{CV}(0) = \int u_C^*(0,\vec{r})\hat{P}_{u_V}(0,\vec{r})\mathrm{d}^3r \equiv \hat{P}$$

For a plane polarized light wave, we have the polarization vector $\vec{\varepsilon}_s = \hat{k}$, when the light wave vector is traveling along the z-axis. Therefore, for a plane polarized light-wave, we have considered $\vec{\varepsilon}_s = \hat{k}$.

Then, from (6.29) we get

$$(\vec{\varepsilon}\cdot\hat{p}_{CV}(\vec{k})) = \vec{k}\cdot\frac{\hat{P}}{2}\hat{r}_3(i\hat{r}_1 - \hat{r}_2)[A(\vec{k}) + B(\vec{k})]\cos\omega t \tag{6.30}$$

and

$$\left.\begin{aligned} A\left(\vec{k}\right) &= a_{k_-}\left(\tfrac{b_{k_\pm}}{\sqrt{2}} + c_{k_+}\right) \\ B\left(\vec{k}\right) &= a_{k_+}\left(\tfrac{b_{k_-}}{\sqrt{2}} + c_{k_-}\right) \end{aligned}\right\} \tag{6.31}$$

Thus,

$$\begin{aligned} |\vec{\varepsilon}\cdot\hat{p}_{CV}(\vec{k})|^2 &= \left|\vec{k}\cdot\frac{\hat{P}}{2}\hat{r}_3\right|^2 |i\hat{r}_1 - \hat{r}_2|^2[A(\vec{k}) + B(\vec{k})]^2\cos\omega t \\ &= \frac{1}{4}|\hat{P}_z\cos\theta|^2[A(\vec{k}) + B(\vec{k})]^2\cos^2\omega t. \end{aligned} \tag{6.32}$$

So, the average value of $|\vec{\varepsilon}\cdot\hat{p}_{CV}(\vec{k})|^2$ for a plane polarized light-wave is given by

$$\begin{aligned} \langle|\vec{\varepsilon}\cdot\hat{p}_{CV}(\vec{k})|^2\rangle_{av} &= \frac{2}{4}|\hat{P}_z|^2[A(\vec{k}) + B(\vec{k})]^2 \\ &\quad\cdot\left(\int_0^{2\pi} d\phi\int_0^{\pi}\cos^2\theta\sin\theta d\theta\right)\left(\frac{1}{2}\right) \\ &= \frac{2\pi}{3}|\hat{P}_z|^2[A(\vec{k}) + B(\vec{k})]^2 \end{aligned} \tag{6.33}$$

where,

$$|\hat{P}_z|^2 = \left(\frac{1}{2}\right)|\vec{k}\cdot\hat{p}_{cv}(0)|^2$$

and

$$|\vec{k}\cdot\hat{p}_{cv}(0)|^2 = \frac{m^2}{4m_r}\frac{E_{g0}\left(E_{g0} + \Delta\right)}{\left(E_{g0} + \frac{2}{3}\Delta\right)}. \tag{6.34}$$

We shall express $A(\vec{k})$ and $B(\vec{k})$ in terms of constants of the energy spectra in the following way:

Substituting $a_{k_\pm}, b_{k_\pm}, c_{k_\pm}$ and $\gamma_{0k_\pm}$ in $A(\vec{k})$ and $B(\vec{k})$ in (18.31) we get

$$A(\vec{k}) = \beta\left(t + \frac{\rho}{\sqrt{2}}\right)\left\{\left(\frac{E_{g0}}{E_{g0} + \delta'}\right)\gamma_{0k_+}^2 - \gamma_{0k_+}^2\gamma_{0k_-}^2\left(\frac{E_{g0} - \delta'}{E_{g0} + \delta'}\right)\right\}^{1/2} \tag{6.35}$$

$$B(\vec{k}) = \beta\left(t + \frac{\rho}{\sqrt{2}}\right)\left\{\left(\frac{E_{g_0}}{E_{g_0}+\delta'}\right)\gamma^2_{0k_-} - \gamma^2_{0k_+}\gamma^2_{0k_-}\left(\frac{E_{g_0}-\delta'}{E_{g_0}+\delta'}\right)\right\}^{1/2} \tag{6.36}$$

in which,

$$\gamma^2_{0k_+} = \frac{\xi_{1k} - E_{g_0}}{2(\xi_{1k}+\delta')} \equiv \frac{1}{2}\left[1 - \left(\frac{E_{g_0}+\delta'}{\xi_{1k}+\delta'}\right)\right]$$

and

$$\gamma^2_{0k_-} = \frac{\xi_{1k} + E_{g_0}}{2(\xi_{1k}+\delta')} \equiv \frac{1}{2}\left[1 - \left(\frac{E_{g_0}-\delta'}{\xi_{1k}+\delta'}\right)\right]$$

Substituting $x \equiv \xi_{1k} + \delta'$ in $\gamma^2_{0k_\pm}$, we can write,

$$A(\vec{k}) = \beta\left(t + \frac{\rho}{\sqrt{2}}\right)\left\{\left(\frac{E_{g_0}}{E_{g_0}+\delta'}\right)\frac{1}{2}\left(1 - \frac{E_{g_0}+\delta'}{x}\right)\right.$$
$$\left. - \frac{1}{4}\left(\frac{E_{g_0}-\delta'}{E_{g_0}+\delta'}\right)\left(1 - \frac{E_{g_0}+\delta'}{x}\right)\left(1 - \frac{E_{g_0}-\delta'}{x}\right)\right\}^{1/2}$$

Thus,

$$A(\vec{k}) = \frac{\beta}{2}\left(t + \frac{\rho}{\sqrt{2}}\right)\left\{1 - \frac{2a_0}{x} + \frac{a_1}{x^2}\right\}^{1/2}$$

where,

$$a_0 \equiv (E^2_{g_0} + \delta'^2)(E_{g_0}+\delta')^{-1} \quad \text{and} \quad a_1 \equiv (E_{g_0}-\delta')^2.$$

After tedious algebra, one can show that

$$A(\vec{k}) = \frac{\beta}{2}\left(t + \frac{\rho}{\sqrt{2}}\right)(E_{g_0}-\delta')\left[\frac{1}{\xi_{1k}+\delta'} - \frac{1}{E_{g_0}+\delta'}\right]^{1/2}$$
$$\cdot\left[\frac{1}{\xi_{1k}+\delta'}\frac{(E_{g_0}+\delta')}{(E_{g_0}-\delta')^2}\right]^{1/2}. \tag{6.37}$$

Similarly, from (6.36), we can write,

$$B(\vec{k}) = \beta\left(t + \frac{\rho}{\sqrt{2}}\right)\left\{\left(\frac{E_{g_0}}{E_{g_0}+\delta'}\right)\frac{1}{2}\left(1 + \frac{E_{g_0}-\delta'}{x}\right)\right.$$
$$\left. - \frac{1}{4}\left(\frac{E_{g_0}-\delta'}{E_{g_0}+\delta'}\right)\left(1 - \frac{E_{g_0}+\delta'}{x}\right)\left(1 + \frac{E_{g_0}-\delta'}{x}\right)\right\}^{1/2}$$

So that, finally we get,

$$B(\vec{k}) = \frac{\beta}{2}\left(t + \frac{\rho}{\sqrt{2}}\right)\left(1 + \frac{E_{g0} - \delta'}{\xi_{1k} + \delta'}\right). \tag{6.38}$$

Using (6.33), (6.34), (6.37) and (6.38), we can write

$$\left(\frac{|e|A_0}{2m}\right)^2 \frac{\langle|\vec{\varepsilon}\cdot\hat{p}_{CV}(\vec{k})|^2\rangle_{av}}{E_c(\vec{k}) - E_v(\vec{k})}$$

$$= \left(\frac{|e|A_0}{2m}\right)^2 \frac{2\pi}{3}|\vec{k}\cdot\hat{p}_{CV}(0)|^2 \frac{\beta^2}{4}\left(t + \frac{\rho}{\sqrt{2}}\right)^2$$

$$\cdot\frac{1}{\xi_{1k}}\left\{\left(1 + \frac{E_{g0} - \delta'}{\xi_{1k} + \delta'}\right) + (E_{g0} - \delta')\left[\frac{1}{\xi_{1k} + \delta'} - \frac{1}{E_{g0} + \delta'}\right]^{1/2}\right.$$

$$\left.\cdot\left[\frac{1}{\xi_{1k} + \delta'} - \frac{E_{g0} + \delta'}{(E_{g0} - \delta')^2}\right]^{1/2}\right\}^2 \tag{6.39}$$

Following Nag [4], it can be shown that

$$A_0^2 = \frac{I\lambda^2}{2\pi^2 c^3\sqrt{\varepsilon_{sc}\varepsilon_0}} \tag{6.40}$$

where, I is the light intensity of wavelength λ, ε_0 is the permittivity of free space and c is the velocity of light. Thus, the simplified electron energy spectrum in III–V, ternary and quaternary materials in the presence of light waves can approximately be written as

$$\frac{\hbar^2 k^2}{2m_c} = \beta_0(E, \lambda) \tag{6.41}$$

where,

$$\beta_0(E,\lambda) \equiv [I_{11}(E) - \theta_0(E,\lambda)],$$

$$\theta_0(E,\lambda) \equiv \frac{|e|^2}{96 m_r \pi c^3}\frac{I\lambda^2}{\sqrt{\varepsilon_{sc}\varepsilon_0}}\frac{E_{g0}(E_{g0} + \Delta)}{(E_{g0} + \frac{2}{3}\Delta)}\frac{\beta^2}{4}\left(t + \frac{\rho}{\sqrt{2}}\right)^2\frac{1}{\phi_0(E)}$$

$$\cdot\left\{\left(1 + \frac{E_{g0} - \delta'}{\phi_0(E) + \delta'}\right) + (E_{g0} - \delta')\right.$$

$$\cdot\left[\frac{1}{\phi_0(E)+\delta'}-\frac{1}{E_{g_0}+\delta'}\right]^{1/2}$$

$$\left.\cdot\left[\frac{1}{\phi_0(E)+\delta'}-\frac{E_{g_0}+\delta'}{(E_{g_0}-\delta')^2}\right]^{1/2}\right\}^2$$

and

$$\phi_0(E)\equiv E_{g_0}\left(1+2\left(1+\frac{m_c}{m_v}\right)\frac{I_{11}(E)}{E_{g_0}}\right)^{1/2}$$

Thus, under the limiting condition $k'\to 0$, from (18.41), we observe that $E\neq 0$ and is positive. Therefore, in the presence of external light waves, the energy of the electron does not tend to zero when $k'\to 0$, where as for the un-perturbed three band model of Kane, $I_{11}(E)=[\hbar^2k^2/(2m_c)]$ in which $E\to 0$ for $\vec{k}\to 0$. As the conduction band is taken as the reference level of energy, therefore the lowest positive value of E for $k'\to 0$ provides the increased band gap (ΔE_g) of the semiconductor due to photon excitation. The values of the increased band gap can be obtained by computer iteration processes for various values of I and λ respectively.

Special Cases:

(1) For the two-band model of Kane, we have $\Delta\to 0$. Under this condition, $I_{11}(E)\to E(1+aE)=\frac{\hbar^2k^2}{2m_c}$. Since, $\beta\to 1, t\to 1, \rho\to 0, \delta'\to 0$ for $\Delta\to 0$, from (6.41), we can write the energy spectrum of III–V, ternary and quaternary materials in the presence of external photo-excitation whose unperturbed conduction electrons obey the two band model of Kane as

$$\frac{\hbar^2k^2}{2m_c}=\tau_0(E,\lambda) \tag{6.42}$$

where, $\tau_0(E,\lambda)\equiv E(1+aE)-\beta_0(E,\lambda)$,

$$\beta_0(E,\lambda)\equiv\frac{|e|^2}{384\pi c^3 m_r}\frac{I\lambda^2E_{g_0}}{\sqrt{\varepsilon_{sc}\varepsilon_0}}\frac{1}{\phi_1(E)}$$

$$\cdot\left\{\left(1+\frac{E_{g_0}}{\phi_0(E)}\right)+E_{g_0}\left[\frac{1}{\phi_1(E)}-\frac{1}{E_{g_0}}\right]\right\}^2,$$

$$\phi_1(E) \equiv E_{g_0}\left\{1+\frac{2m_c}{m_r}aE(1+aE)\right\}^{1/2}.$$

(2) For relatively wide band gap semiconductors, one can write, $a \to 0, b \to 0, c \to 0$ and $I_{11}(E) \to E$.

Thus, from (6.42), we get,

$$\frac{\hbar^2k^2}{2m_c} = U_\lambda I_{11}(E) - P_\lambda \tag{6.44}$$

$$\frac{\hbar^2k^2}{2m_c} = t_{1\lambda}E + t_{2\lambda}E^2 - \delta_\lambda \tag{6.45}$$

and

$$\frac{\hbar^2k^2}{2m_c} = t_{1\lambda}E - \delta_\lambda \tag{6.46a}$$

where

$$U_\lambda = (1+\theta_\lambda), \quad \theta_\lambda = \frac{C_0}{A}\left(t_\lambda + \frac{BJ_\lambda}{A}\right),$$

$$C_0 = \left[\frac{|e|^2}{96\pi c^3 m_r}\frac{I\lambda^2 E_{g_0}(E_{g_0}+\Delta)}{\sqrt{\varepsilon_{sc}\varepsilon_0}\left(E_{g_0}+\frac{2}{3}\Delta\right)}\frac{\beta^2}{4}\left(1+\frac{\rho}{\sqrt{2}}\right)^2\right],$$

$$A = E_{g_0}, \quad B = \left[1+\frac{m^*}{m_V}\right], \quad G_\lambda = \left[\frac{2B}{(A+\delta')^3} - \frac{BC_\lambda}{(A+\delta')}\right]$$

$$C_\lambda = [(E_{g_0}+\delta')^{-1} + (E_{g_0}+\delta')(E_{g_0}-\delta')^{-2}](A+\delta')^{-1}$$

$$P_\lambda = \frac{C_0}{A}J_\lambda, \quad J_\lambda = (D_\lambda + 2(E_{g_0}-\delta')\sqrt{f_\lambda}),$$

$$D_\lambda = \left(1+\frac{2(E_{g_0}-\delta')}{(A+\delta')}\right), \quad f_\lambda = \left[\frac{1}{(A+\delta')^2} + \frac{1}{(E_{g_0}-\delta')^2} - C_\lambda\right],$$

$$t_{1\lambda} = \left(1+\frac{3m_c}{m_r}\alpha\delta_\lambda\right), \quad \alpha = \frac{1}{E_{g_0}},$$

$$\delta_\lambda = \frac{|e|^2 I\lambda^2}{96 m_r \pi c^3\sqrt{\varepsilon_{sc}\varepsilon_0}} \text{ and } t_{2\lambda} = \alpha t_{1\lambda}$$

Under the condition of heavy doping, following the methods as developed in Chapter 1, the HD dispersion relations in this case in the presence of light waves can be written as

$$\frac{\hbar^2 k^2}{2m_c} = T_1(E, \eta_g, \lambda) \tag{6.46b}$$

$$\frac{\hbar^2 k^2}{2m_c} = T_2(E, \eta_g, \lambda) \tag{6.47}$$

$$\frac{\hbar^2 k^2}{2m_c} = T_3(E, \eta_g, \lambda) \tag{6.48a}$$

where

$T_1(E, \eta_g, \lambda) = [U_\lambda[T_{31}(E, \eta_g) + iT_{32}(E, \eta_g)] - P_\lambda]$,

$T_2(E, \eta_g, \lambda) = [U_{1\lambda}\gamma_3(E, \eta_g) + (t_{2\lambda})2\theta_0(E, \eta_g)[1 + Erf(E/\eta_g)]^{-1} - \delta_\lambda]$

and

$$T_3(E, \eta_g, \lambda) = [t_{1\lambda}\gamma_3(E, \eta_g) - \delta_\lambda].$$

The DOS functions for (6.46b), (6.47) and (6.48) can, respectively, be written as

$$N_L(E) = 4\pi g_v \left(\frac{2m_c}{h^2}\right)^{3/2} \sqrt{[T_1(E, \eta_g, \lambda)]}[T_1(E, \eta_g, \lambda)]' \tag{6.48b}$$

$$N_L(E) = 4\pi g_v \left(\frac{2m_c}{h^2}\right)^{3/2} \sqrt{[T_2(E, \eta_g, \lambda)]}[T_2(E, \eta_g, \lambda)]' \tag{6.48c}$$

$$N_L(E) = 4\pi g_v \left(\frac{2m_c}{h^2}\right)^{3/2} \sqrt{[T_3(E, \eta_g, \lambda)]}[T_3(E, \eta_g, \lambda)]'. \tag{6.48d}$$

The EEM can be expressed in this case by using (6.46), (6.47) and (6.48a).

$$m^*(E_{FHDL}, \eta_g, \lambda) = m_c \text{ Real part of } [T_1 E_{FHDL}, \eta_g, \lambda)]' \tag{6.49}$$

$$m^*(E_{FHDL}, \eta_g, \lambda) = m_c[T_2 E_{FHDL}, \eta_g, \lambda)]' \tag{6.50}$$

$$m^*(E_{FHDL}, \eta_g, \lambda) = m_c[T_3 E_{FHDL}, \eta_g, \lambda)]' \tag{6.51a}$$

$$n_0 = \frac{g_v}{3\pi^2}\left(\frac{2m_c}{\hbar^2}\right)^{3/2} \text{ Real Part of } [[T_1(E_{FHDL}, \eta_g, \lambda)]^{3/2} + \sum_{r=1}^{s} L(r)[T_1(E_{FHDL}, \eta_g, \lambda)]^{3/2} \tag{6.51b}$$

$$n_0 = \frac{g_v}{3\pi^2}\left(\frac{2m_c}{\hbar^2}\right)^{3/2} \text{ Real Part of } [[T_2(E_{FHDL}, \eta_g, \lambda)]^{3/2} + \sum_{r=1}^{s} L(r)[T_2(E_{FHDL}, \eta_g, \lambda)]^{3/2} \tag{6.51c}$$

$$n_0 = \frac{g_v}{3\pi^2}\left(\frac{2m_c}{\hbar^2}\right)^{3/2} \text{ Real Part of } [[T_3(E_{FHDL}, \eta_g, \lambda)]^{3/2} + \sum_{r=1}^{s} L(r)[T_3(E_{FHDL}, \eta_g, \lambda)]^{3/2} \tag{6.51d}$$

6.2.1.1 *Results and discussion*

Using the appropriate equations and the values of the energy band constants from Appendix A, we have plotted in Figs. 6.1a to 6.1d, the EM as functions of electron concentration at T= 4.2 K by taking n-InAs, n-InSb, n-$Hg_{1-x}Cd_xTe$ and n-$In_{1-x}Ga_xAs_yP_{1-y}$ lattice matched to InP as examples of III–V, ternary and quaternary materials which are used for the purpose of numerical computations in accordance with the perturbed three and two band models of Kane and that of perturbed parabolic energy bands respectively. In Figs. 6.2a to 6.2d we have plotted the EM as a function of intensity. In Figs. 6.3a to 6.3d we have plotted the EM as a function of wavelength. In Figs. 6.4 and 6.5, we have plotted the EM as functions of the alloy composition for ternary and quaternary materials respectively.

From Fig. 6.1a it appears that the EM increases with the increasing electron concentration for n-InAs and the numerical values of the EM in the presence of light waves in accordance with all the band models are relatively larger than that of the same in the absence of the external photo-excitation excluding curve (e) where for parabolic energy band the EM is numerically concentration invariant due to large band gap. The reason behind such behavior is the fact that the Fermi energy is the monotonic increasing function of electron concentration and the EM increases monotonically with increasing Fermi energy both in the presence and absence of light waves. By comparing the plot (a) with plot (c) in Fig. 6.1a we observe that the presence of spin-orbit splitting in curve (a), decreases the value of the EM as compared with curve (c) in the whole range of carrier degeneracy as considered here. For relatively low values of n_0, the curves (a),(b),(c) and (d) exhibit converging tendency where as the differ with each other for relatively higher values of carrier degeneracy. The curve (e) of Fig. 6.1a represents the EM both in the presence and absence of external light waves for the relatively wide band-gap model which is independent of doping. Plots (a) and (c) of Fig. 6.1b diverge for relatively low values of n_0, intersect each other for a particular zone of concentration and then exhibit small difference although both of them increase with increasing degeneracy. The numerical values of the EM for the curves (a),(b),(c) and (d) of n-InSb as given in Fig. 6.1b are greater as compared with the same for n-InAs as given in Fig. 6.1a, although the nature of the curve (e) is same for both Figs. 6.1a and 6.1b respectively.

The curves (a) and (c) explore the fact that the influence of the spin-orbit splitting on the EM for n-$Hg_{1-x}Cd_xTe$ in the presence of light waves decreases significantly the same mass as compared with the perturbed two band model of Kane and the two curves exhibit wide difference with each other for relatively low values of doping. In the absence of light wave, the effect of Δ on the EM is much less .It appears by comparing the Figs. 6.1a, 6.1b and 6.1c that the EM for n-$Hg_{1-x}Cd_xTe$ in the presence of external photo-excitation, is much more as that of n-InSb and n-InAs respectively. From plots (a) and (c) of Fig. 6.1d, it appears for $In_{1-x}Ga_xAs_yP_{1-y}$ lattice matched to InP that both the curves maintain constant wide difference under photo-excitation with respect to electron concentration in the whole range of electron degeneracy as considered here .The influence of Δ on the EM in the absence of external light waves for three and two band models of Kane is very small as evident from the curves (b) and (d), although the EM increases with n_0 as usual. The influence of the

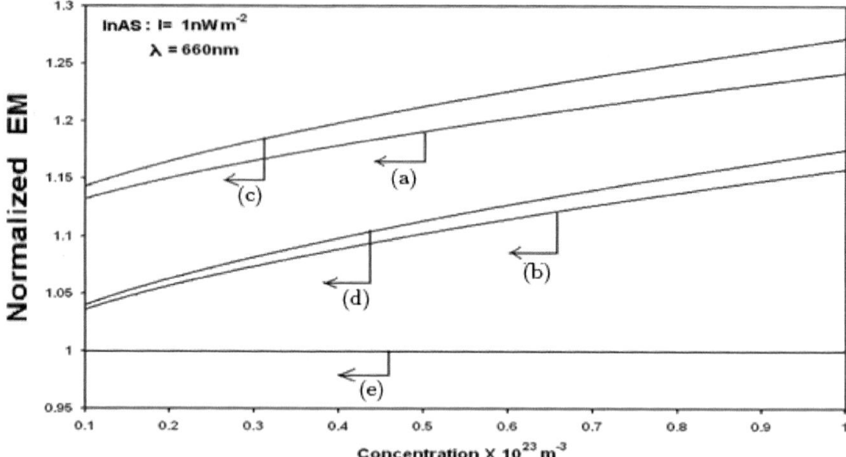

Fig. 6.1a. Plot of the normalized EM as a function of electron concentration for n-InAs in the presence of light waves in which the curves (a) and (c) represent the three and two band models of Kane respectively. The curves (b) and (d) exhibit the same variation in the absence of external photo-excitation. The curve (e) represents the parabolic energy band model both in the presence and in the absence of the external photo-excitation.

energy band constants on the EM is apparent from all the plots of Figs. 6.1a to 6.1d and the numerical values of the EM is greatest for ternary alloys and the least for quaternary systems under light waves. The curves (a) and (c) of Fig. 6.2a explore that the EM increases with increasing light intensity for n-InAs in the presence of light waves for both perturbed three and two band models of Kane whereas for perturbed parabolic energy bands the EM is intensity invariant. The reason behind such behavior is that the Fermi energy increases with increasing light intensity and the EM is the function of Fermi energy. For perturbed parabolic energy bands, due to very large band gap the EM is numerically independent of light intensity. In the absence of light waves the EM is naturally independent of I which is apparent from the curves (b) and (d) of Fig. 6.2a. The curves(a) and (c) of Fig. 6.2b for n-InSb reflects the fact that the influence of spin-orbit splitting increases rapidly with increasing intensity and for low values of light intensity the EM decreases for both perturbed three and two

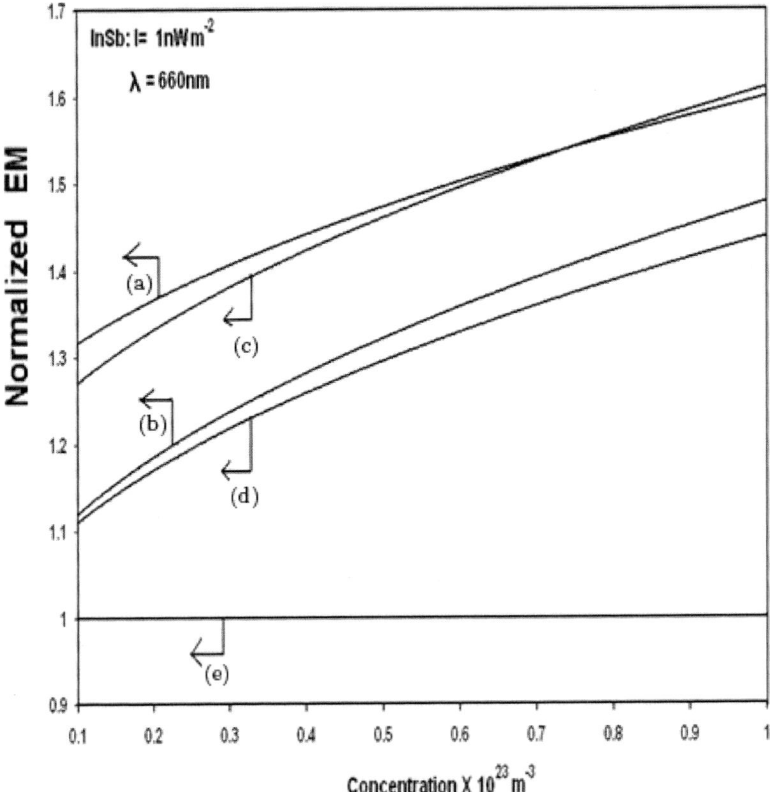

Fig. 6.1b. Plot of the normalized EM as a function of electron concentration for n-InSb in the presence of light waves in which the curves (a) and (c) represent the three and two band models of Kane respectively. The curves (b) and (d) exhibit the same variation in the absence of external photo-excitation. The curve (e) represents the parabolic energy band model both in the presence and in the absence of the external photo-excitation.

band models of Kane, whereas for higher values of I the EM increases significantly.

From the plot (a) of Fig. 6.2c one can infer that the EM for n-$Hg_{1-x}Cd_xTe$ in the presence of light waves increases with increasing I, in a more or less linear fashion in accordance with the perturbed three band model of Kane whereas from plot(c), one observes that the EM on the basis of the perturbed two band model of Kane is greater as compared with plot (a) in the whole range of I. For low values of I,

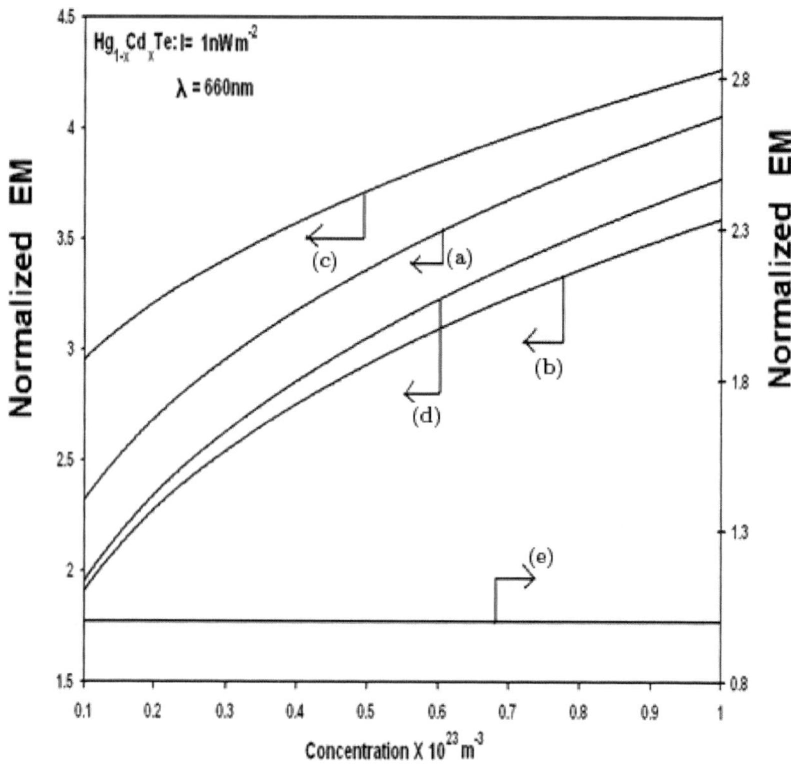

Fig. 6.1c. Plot of the normalized EM as a function of electron concentration for n-$Hg_{1-x}Cd_xTe$ in the presence of light waves in which the curves (a) and (c) represent the three and two band models of Kane respectively. The curves (b) and (d) exhibit the same variation in the absence of external photo-excitation. The curve (e) represents the parabolic energy band model both in the presence and in the absence of the external photo-excitation.

the curves (a) and (c) exhibits converging tendency .It is important to note that with respect to light intensity, the numerical values of the EM in the absence of light waves ,in the case of n-$Hg_{1-x}Cd_xTe$ are greater for both types of band models (curves (b) and (d)) as compared with that of (a) and (c) when I≠0. It appears from the plots (a) and (c) of Fig. 6.2d that the EM increases with I for both types of perturbed band models with different small curvatures and the difference in numerical values of EM increases with increasing in I, although they converge to a particular value for I=10^{-6} nWm^{-2}.

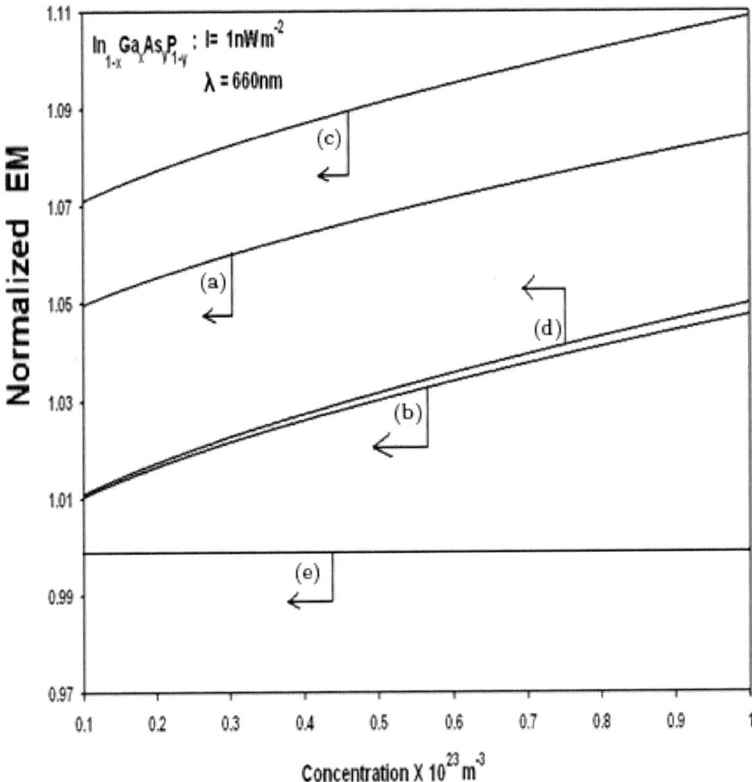

Fig. 6.1d. Plot of the normalized EM as a function of electron concentration for $In_{1-x}Ga_xAs_yP_{1-y}$ lattice matched to InP in the presence of light waves in which the curves (a) and (c) represent the three and two band models of Kane respectively. The curves (b) and (d) exhibit the same variation in the absence of external photo-excitation. The curve (e) represents the parabolic energy band model both in the presence and in the absence of the external photo-excitation.

The curves (a) and (c) of Fig. 6.3a exhibits the fact that the EM for n-InAs increases with increasing wave length in the presence of light waves for both perturbed three and two band models of Kane since the Fermi energy increases with increasing wave length and the EM is the function of Fermi energy. For perturbed parabolic energy bands, the EM is numerically independent of wave length due to large band gap. The curves (a) and (c) maintain wide difference with increasing wave length and the spin-orbit splitting decreases the EM

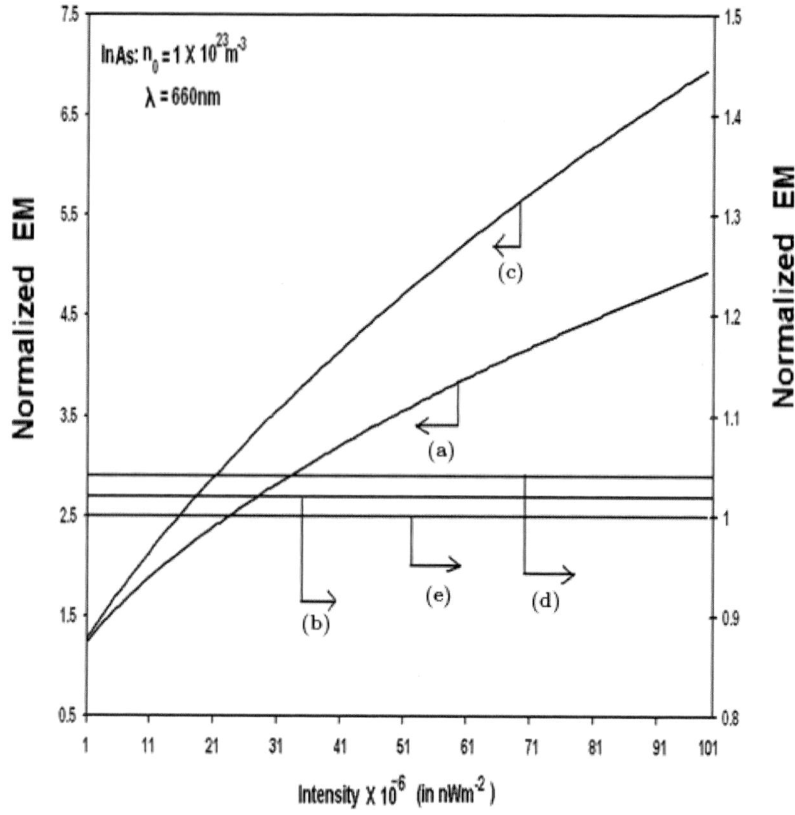

Fig. 6.2a. Plot of the normalized EM as a function of light intensity for n-InAs in the presence of light waves in which the curves (a) and (c) represent the three and two band models of Kane respectively. The curves (b) and (d) exhibit the same variation in the absence of external photo-excitation. The curve (e) represents the parabolic energy band model both in the presence and in the absence of the external photo-excitation.

in the whole range of λ. The curves (b) and (d) exhibit the same variation in the absence of photo excitation and is independent of wave length of the incident light.

From the plots (a) and (c) of Fig. 6.3b for n-InAs, one can infer that the EM increases with increasing wave length in the presence of light waves for both the cases and the exhibit the diverging tendency for relatively low values of λ where as for higher values of the wave length exhibit the converging nature. From plots (b) and (d), we

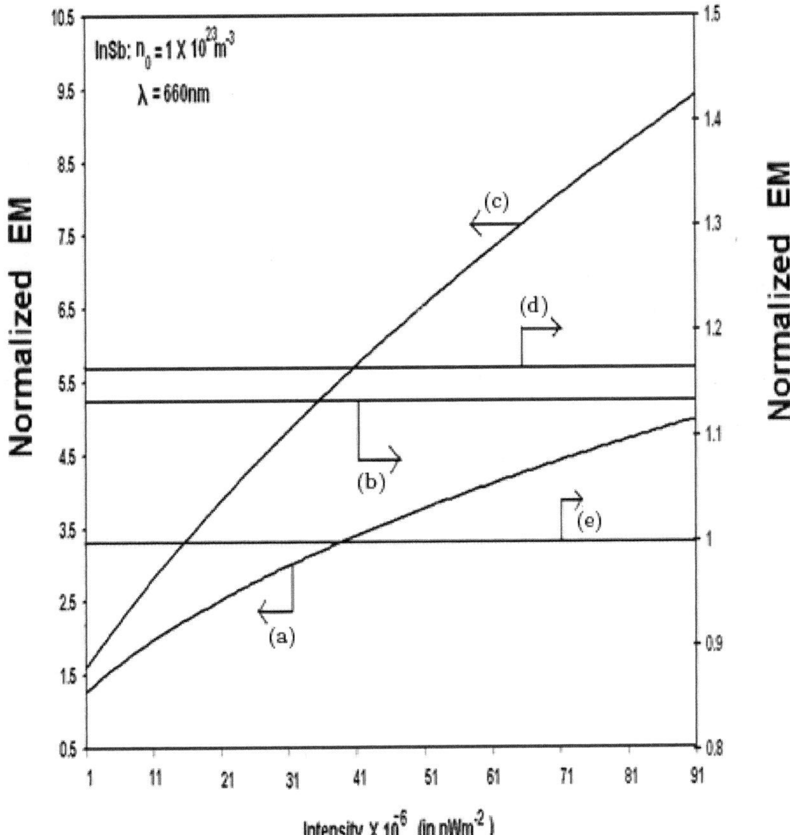

Fig. 6.2b. Plot of the normalized EM as a function of light intensity for n-InSb in the presence of light waves in which the curves (a) and (c) represent the three and two band models of Kane respectively. The curves (b) and (d) exhibit the same variation in the absence of external photo-excitation. The curve (e) represents the parabolic energy band model both in the presence and in the absence of the external photo-excitation.

observe that the numerical values of the EM for both three and two band model of Kane are much larger as compared with the same under photo-excitation for relatively low values of λ. The influence of Δ on the EM for perturbed three band model as observed in Fig 6.3b is less as compared with the same as given in plot (a) of Fig 6.3a.

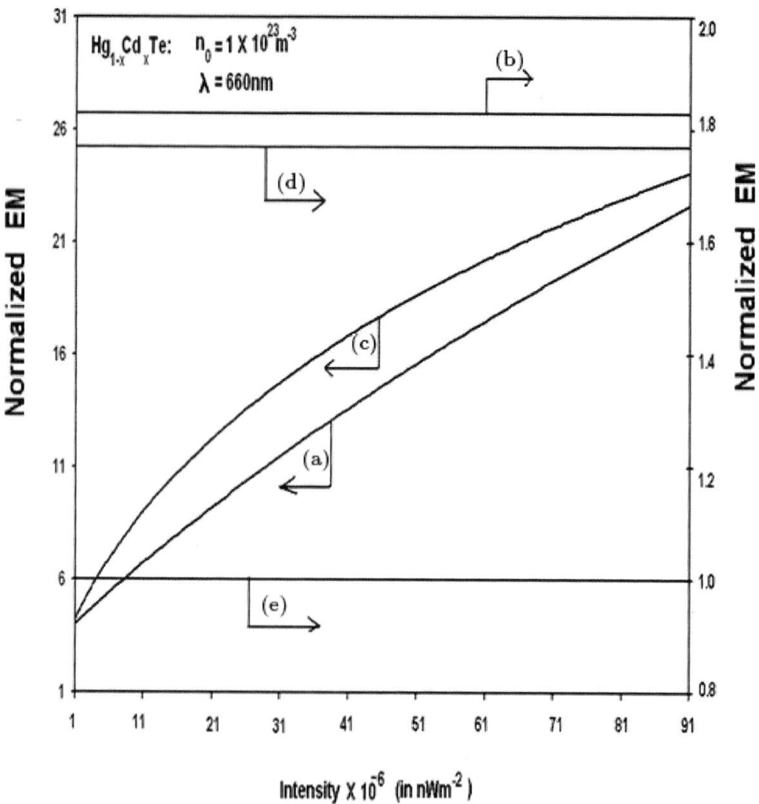

Fig. 6.2c. Plot of the normalized EM as a function of light intensity for n-$Hg_{1-x}Cd_xTe$ in the presence of light waves in which the curves (a) and (c) represent the three and two band models of Kane respectively. The curves (b) and (d) exhibit the same variation in the absence of external photo-excitation. The curve (e) represents the parabolic energy band model both in the presence and in the absence of the external photo-excitation.

The curves (b) and (c)for n- $Hg_{1-x}Cd_xTe$ of Fig 6.3c increase with increasing λ and after intersection they exhibit the wide difference with each other. The spin-orbit splitting decreases the EM remarkably for relatively higher values of λ.

From Fig 6.3d we can write that the EM in this case is much less when compared with the same as given I, the Figs. 6.3a, 6.3b and 6.3c. in the presence of external photo-excitation .From plots (a) and (c) of Fig. 6.3d, it appears for $In_{1-x}Ga_xAs_yP_{1-y}$ lattice matched

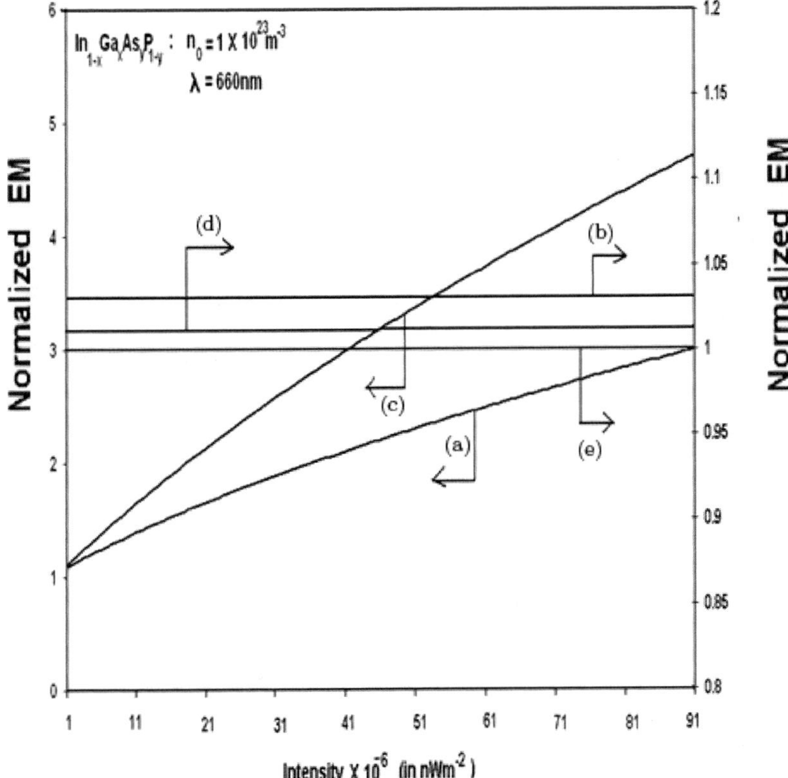

Fig. 6.2d. Plot of the normalized EM as a function of light intensity for $In_{1-x}Ga_xAs_yP_{1-y}$ lattice matched to InP in the presence of light waves in which the curves (a) and (c) represent the three and two band models of Kane respectively. The curves (b) and (d) exhibit the same variation in the absence of external photo-excitation. The curve (e) represents the parabolic energy band model both in the presence and in the absence of the external photo-excitation.

to InP that both the curves maintain wide difference under photo-excitation with respect to λ in the whole range of wave length as considered here. As the wave length increases, from plots (a) and (c) of Fig. 6.3d we infer that the difference also increases. The influence of Δ on the EM in the absence of external light waves for three and two band models of Kane is very small as evident from the curves (b) and (d).

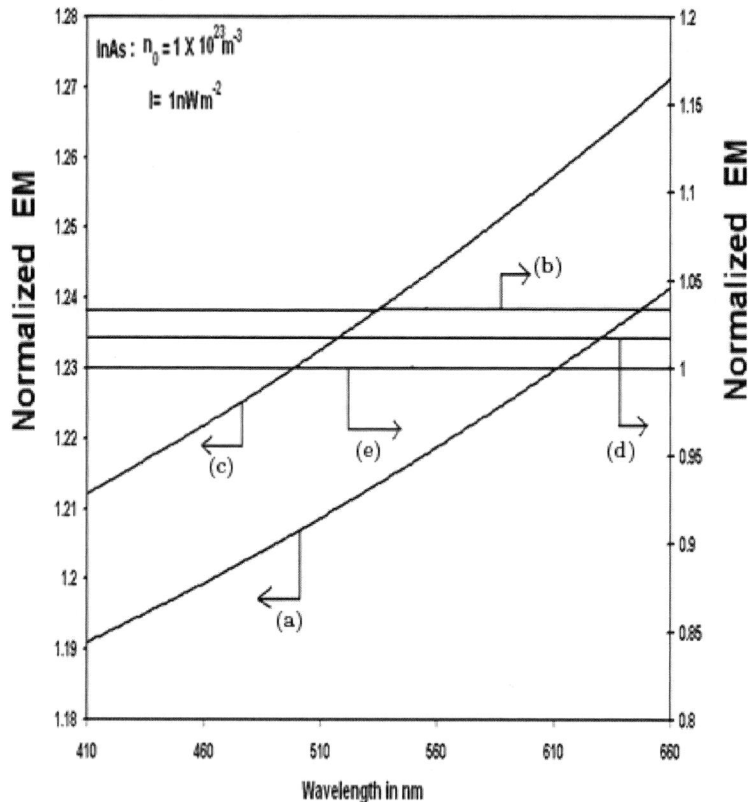

Fig. 6.3a. Plot of the normalized EM as a function of wavelength for n-InAs in the presence of light waves in which the curves (a) and (c) represent the three and two band models of Kane respectively. The curves (b) and (d) exhibit the same variation in the absence of external photo-excitation. The curve (e) represents the parabolic energy band model both in the presence and in the absence of the external photo-excitation.

It appears that the EM increases as the wavelength shifts from violet to red. The influence of light is immediately apparent from the plots in the Figs. 6.2a to 6.3d since the EM depends strongly on I and λ for the three and the two band model of Kane which is in direct contrast with that for the bulk specimens of the said compounds in the absence of external photo-excitation. The variations of the EM in the Figs. 6.2a to 6.3d reflect the direct signature of the light wave on the band structure dependent physical properties of semiconductors

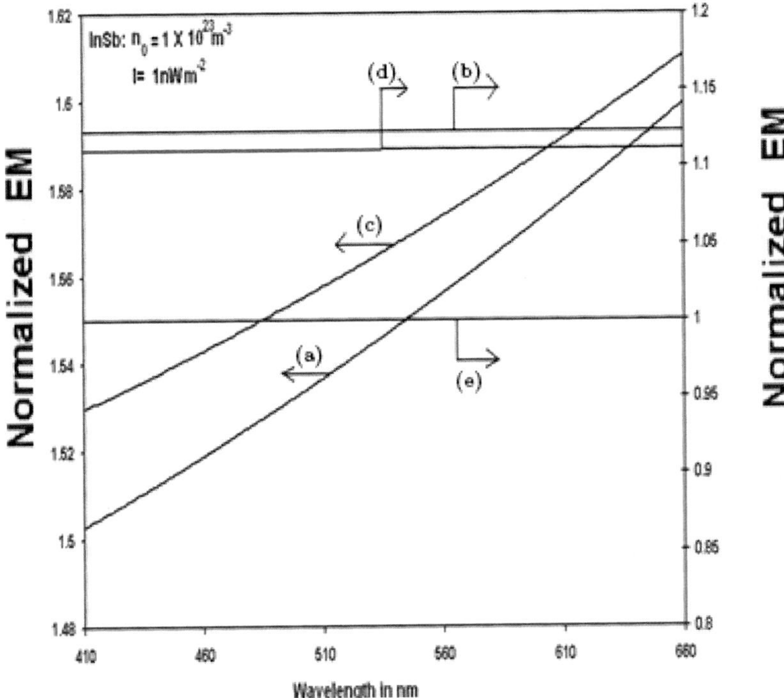

Fig. 6.3b. Plot of the normalized EM as a function of wavelength for n-InSb in the presence of light waves in which the curves (a) and (c) represent the three and two band models of Kane respectively. The curves (b) and (d) exhibit the same variation in the absence of external photo-excitation. The curve (e) represents the parabolic energy band model both in the presence and in the absence of the external photo-excitation.

in general in the presence of external photo-excitation and the photon assisted transport for the corresponding photonic devices. The numerical values of the EM in the presence of the light waves are larger than that of the same in the absence of light wave for both the three and the two band model of Kane. Although, the EM tends to increase with the intensity and the wavelength but the rate of increase is totally band structure dependent. It appears that the numerical values of the EM are greatest for ternary materials and least for quaternary compounds.

In Figs. 6.4 and 6.5, the EM has been plotted as a function of alloy composition for n-$Hg_{1-x}Cd_xTe$ and n-$In_{1-x}Ga_xAs_yP_{1-y}$ lattice

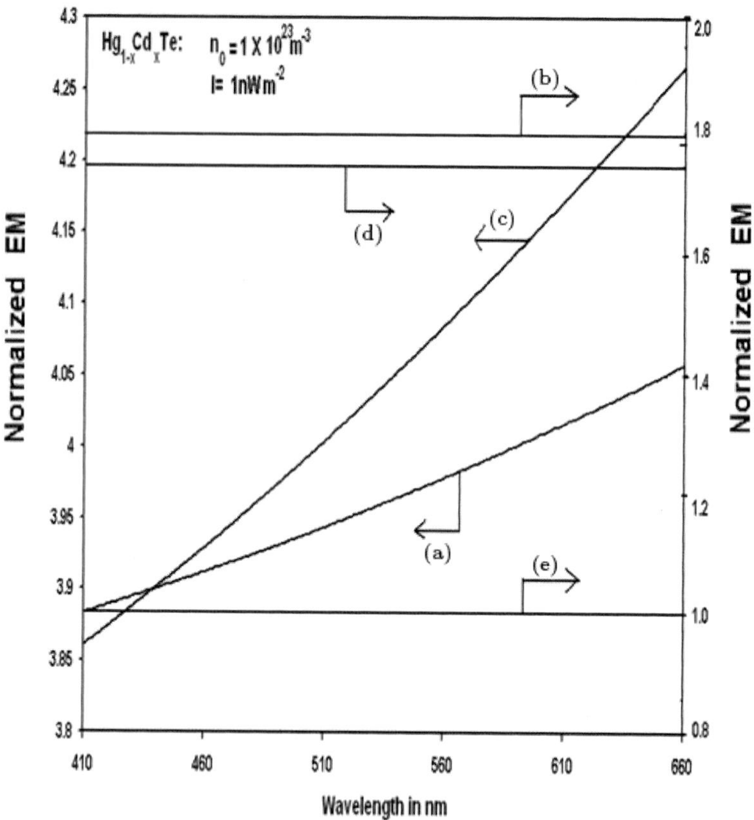

Fig. 6.3c. Plot of the normalized EM as a function of wavelength for n- $Hg_{1-x}Cd_xTe$ in the presence of light waves in which the curves (a) and (c) represent the three and two band models of Kane respectively. The curves (b) and (d) exhibit the same variation in the absence of external photo-excitation. The curve (e) represents the parabolic energy band model both in the presence and in the absence of the external photo-excitation.

matched to InP respectively in which all the cases of Fig. 6.1 have further been plotted for the purpose of relative comparison. The Fermi energy decreases with increasing alloy composition and the EM is a function of Fermi energy. From Fig. 6.4, we can write that the EM in ternary compounds decreases with increasing alloy composition.

The numerical values of EM in the presence of light waves are greater for both the models as appears from the plots (a), (b), (c)

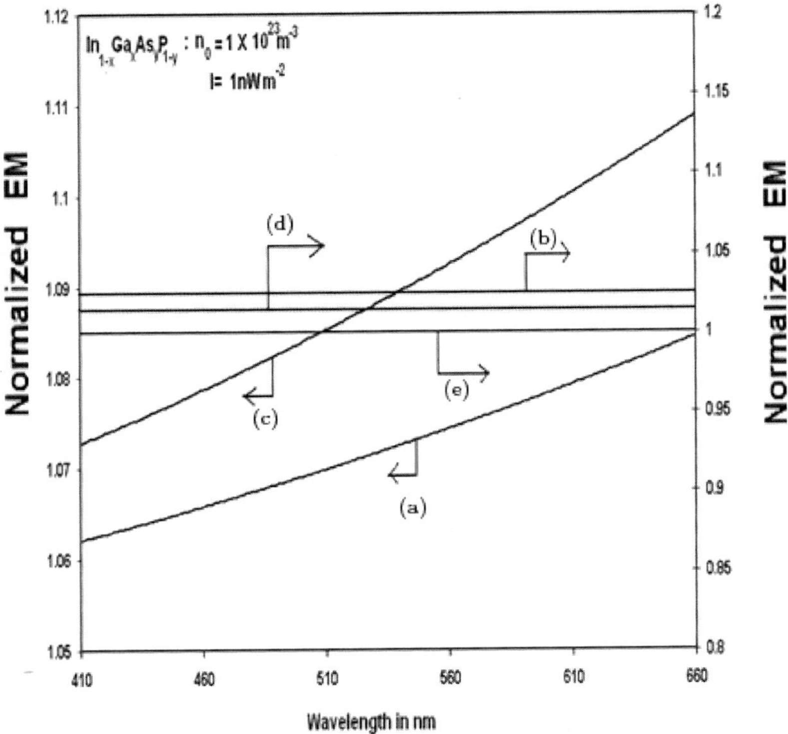

Fig. 6.3d. Plot of the normalized EM as a function of wavelength for $In_{1-x}Ga_xAs_yP_{1-y}$ lattice matched to InP in the presence of light waves in which the curves (a) and (c) represent the three and two band models of Kane respectively. The curves (b) and (d) exhibit the same variation in the absence of external photo-excitation. The curve (e) represents the parabolic energy band model both in the presence and in the absence of the external photo-excitation.

and (d). As alloy composition increases, the EM for all the cases exhibit the converging tendency. The plots of the Fig. 6.4 are valid for x>0.17, since for x<0.17, the band gap becomes negative in n-$Hg_{1-x}Cd_xTe$ leading to semi-metallic state. The plots of the Fig. 6.5 exhibit the variation of the EM with y for n-$In_{1-x}Ga_xAs_yP_{1-y}$ lattice matched to InP. As the Fermi energy increases with the y, from the curves (a), (b),(c) and (d) of Fig 6.5 we observe that the EM increases with increasing y. These four plots also exhibit the fact that the influence of the spin-orbit splitting constant in the presence of light

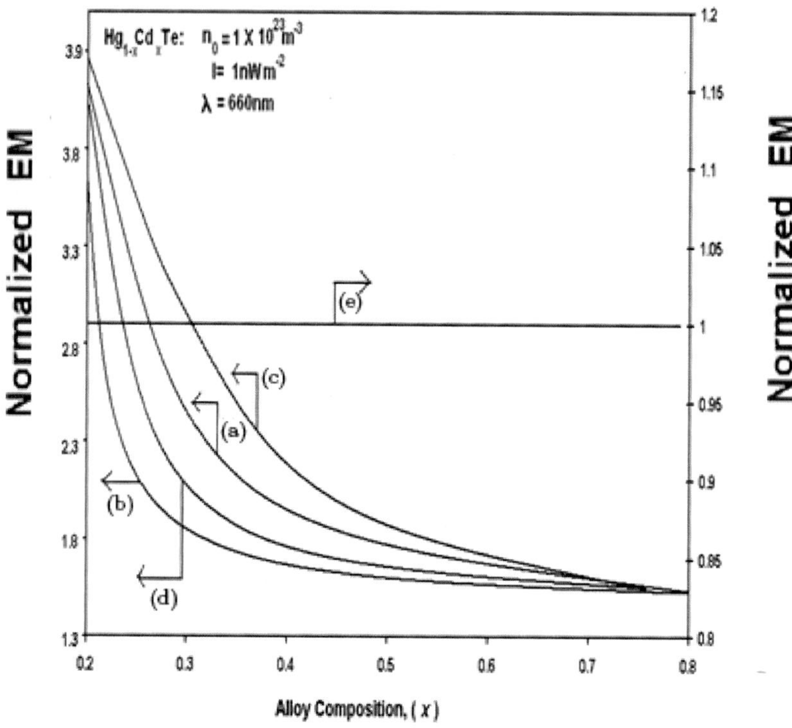

Fig. 6.4. Plot of the normalized EM as a function of alloy composition for $Hg_{1-x}Cd_xTe$ in the presence of light waves in which the curves (a) and (c) represent the three and two band models of Kane respectively. The curves (b) and (d) exhibit the same variation in the absence of external photo-excitation. The curve (e) represents the parabolic energy band model both in the presence and in the absence of the external photo-excitation.

waves is much greater as compared with the same in the absence of photo excitation.

The theoretical results as presented here will be useful in determining the mobility even for relatively wide gap compounds whose energy band structures can be approximated by the parabolic energy bands both in the presence and absence of light waves. It is worth remarking that our basic Equation (6.41) covers various materials having different energy band structures. In this section, the concentration, alloy composition, light intensity and the wavelength dependences of EM in bulk specimens of n-InAs, n-InSb,

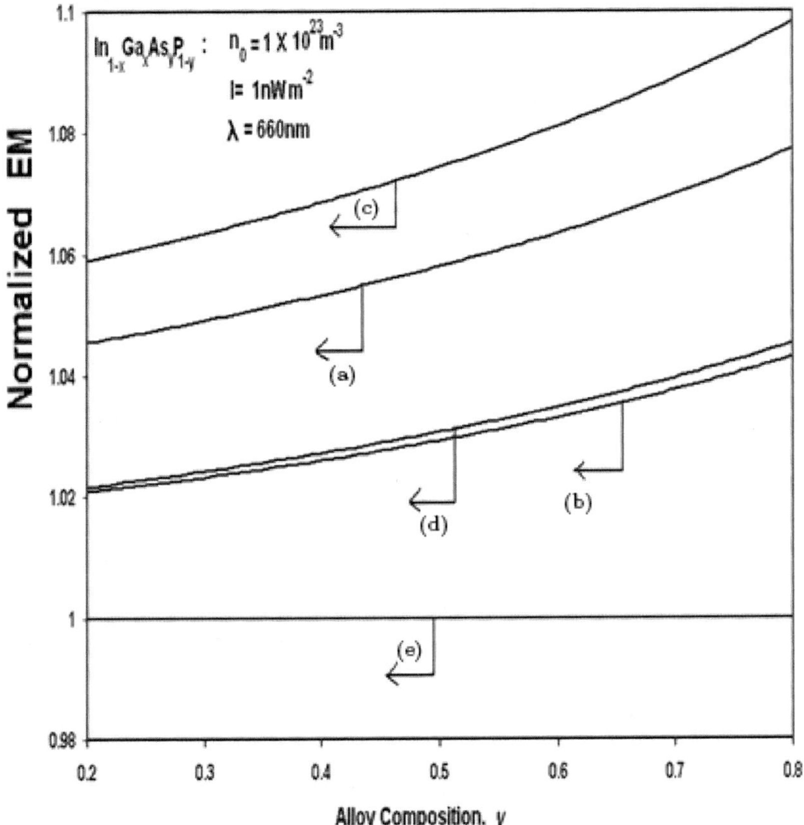

Fig. 6.5. Plot of the normalized EM as a function of alloy composition for $In_{1-x}Ga_xAs_{1-y}P_y$ lattice matched to InPin the presence of light waves in which the curves (a) and (c) represent the three and two band models of Kane respectively. The curves (b) and (d) exhibit the same variation in the absence of external photo-excitation. The curve (e) represents the parabolic energy band model both in the presence and in the absence of the external photo-excitation.

n-$Hg_{1-x}Cd_xTe$ and n-$In_{1-x}Ga_xAs_yP_{1-y}$ lattice matched to InP have been studied. Thus, we have covered a wide class of optoelectronic and allied compounds whose energy band structures are defined by the three and two band models of Kane in the absence of photon field. Under certain limiting conditions, all the results of the EM for different materials having various band structures lead to the well-known expression of the EM for degenerate compounds having parabolic energy band. This indirect test not only exhibits the mathematical compatibility of our formulation but also shows the fact that our simple analysis is a more generalized one, since one can obtain the corresponding results for the relatively wide gap materials having parabolic energy bands under certain limiting conditions from our present derivation.

It is worth remarking that the influence of an external photo-excitation is to change radically the original band structure of the material. Because of this change, the photon field causes to increase the band gap of semiconductor. Our method is not at all related to the DOS technique as used in the literature [7]. From the **E-k** dispersion relation, we can obtain the DOS, but the DOS technique as used in the literature [7] cannot provide the **E-k** dispersion relation. Therefore, our study is more fundamental than those of the existing literature because the Boltzman transport equation, which controls the study of the charge transport properties of semiconductor devices, can be solved if and only if the **E-k** dispersion relation is known. We wish to note that we have not considered the many body effects in this simplified theoretical formalism due to the lack of availability in the literature of proper analytical techniques for including them for the generalized systems as considered in this book. Our simplified approach will be useful for the purpose of comparison when methods of tackling the formidable problem after inclusion of the many body effects for the present generalized systems appear. The inclusion of the said effects would certainly increase the accuracy of the results, although the qualitative features of the EM discussed in this book would not change in the presence of the aforementioned effects. Since the experimental results in the present case are not available in the literature to the best of our knowledge, we cannot compare our generalized theoretical analysis with the corresponding experimental data. Our formalism will be useful in probing the band structure when the experimental results for our generalized systems would appear. It is worth remarking in this context that from our simplified theory, under certain limiting conditions, gets transformed to the well-known result of the EM for wide gap materials having parabolic energy bands. We have not considered other types of optoelectronic materials and other external variables in order to keep the presentation brief. Besides, the influence of energy band models and the various band constants on the EM for different materials can also be studied from all the figures of this book. The numerical results presented in this book would be different for other materials but the nature of variation would be unaltered. The theoretical results as given here would be useful in analyzing various other experimental data related to this phenomenon. Finally, we can write that this theory can be used to investigate the gate capacitance of nano-scale transistors, the carrier contribution to the elastic constants, the Debye screening length, the magnetic susceptibilities, the Burstien Moss shift, plasma frequency, the Hall coefficient, the specific heat and other different transport coefficients of modern semiconductor devices operated in the presence of light waves.

6.2.2 *The EM under magnetic quantization in HD kane type semiconductors in the presence of light waves*

(i) Using (6.46b), the magneto-dispersion law, in the absence of spin, for HD III–V, ternary and quaternary semiconductors, in the presence of photo-excitation, whose unperturbed conduction electrons obey the three band model of Kane, is given by

$$T_1(E,\eta_g) = \left(n+\frac{1}{2}\right)\hbar\omega_0 + \frac{\hbar^2 k_z^2}{2m_c}. \tag{6.52}$$

Using (6.52), the DOS function in the present case can be expressed as

$$D_g(E,\eta_g,\lambda) = \frac{g_v|e|\sqrt{2m_c}}{2\pi^2\hbar^2}\sum_{n=0}^{n_{\max}} \cdot\left[\{T_1(E,\eta_g,\lambda\}'\left\{T_1(E,\eta_g,\lambda) - \left(n+\frac{1}{2}\right)\hbar\omega_0\right\}^{-1/2} H(E-E_{i1})\right] \tag{6.53}$$

where, $E_{n_{l1}}$ is the Landau sub-band energies in this case and is given as

$$T_1(E_{n_{i1}},\eta_g,\lambda) = \left(n+\frac{1}{2}\right)\hbar\omega_0 \tag{6.54}$$

The EEM in this case assumes the form

$$m^*(E_{FHDLB},\eta_g,\lambda) = m_c \text{ Real part of } \{T_1(E_{FHDLB},\eta_g,\lambda)\}' \tag{6.55a}$$

where, E_{FHDLB} is the Fermi energy under quantizing magnetic field in the presence of light waves as measured from the edge of the conduction band in the vertically upward direction in the absence of any quantization.

The electron concentration can be written as

$$n_0 = \frac{g_v|e|B\sqrt{2m}}{\pi^2\hbar^2} \text{ Real Part of } \sum_{n=0}^{n_{\max}} \cdot\left[\left\{T_1(E_{FHDLB},\eta_g,\lambda) - \left(n+\frac{1}{2}\right)\hbar\omega_0\right\}^{1/2} + \sum_{r=1}^{s} L(r)\left[\left\{T_1(E_{FHDLB},\eta_g,\lambda) - \left(n+\frac{1}{2}\right)\hbar\omega_0\right\}^{1/2}\right]\right] \tag{6.55b}$$

(ii) *Using (6.47), the magneto-dispersion law, in the absence of spin, for HD III–V, ternary and quaternary semiconductors, in the presence of photo-excitation, whose unperturbed conduction electrons obey the two band model of Kane, is given by*

$$T_2(E,\eta_g) = \left(n+\frac{1}{2}\right)\hbar\omega_0 + \frac{\hbar^2 k_z^2}{2m_c} \tag{6.56}$$

Using (6.56), the DOS function in the present case can be expressed as

$$D_B(E,\eta_g,\lambda) = \frac{g_v|e|\sqrt{2m_c}}{2\pi^2\hbar^2}\sum_{n=0}^{n_{\max}}\left[\{T_2(E,\eta_g,\lambda)\}' \cdot\left\{T_2(E,\eta_g,\lambda) - \left(n+\frac{1}{2}\right)\hbar\omega_0\right\}^{-1/2} - H(E-E_{n_{i2}})\right] \tag{6.57}$$

where, $E_{\eta_{l2}}$ is the Landau sub-band energies in this case and is given as

$$T_2(E_{\eta_{l2}}, \eta_g, \lambda) = \left(n + \frac{1}{2}\right)\hbar\omega_0 \tag{6.58}$$

The EEM in this case assumes the form

$$m^*(E_{FHDLB}, \eta_g, \lambda) = m_c\{(E_{FHDLB}, \eta_g, \lambda)\}' \tag{6.59a}$$

The electron concentration can be written as

$$n_0 = \frac{g_v |e| B\sqrt{2m}}{\pi^2\hbar^2} \text{ Real Part of } \sum_{n=0}^{n_{\max}} \cdot \left[\left\{T_2(E_{FHDLB}, \eta_g, \lambda) - \left(n + \frac{1}{2}\right)\hbar\omega_0\right\}^{1/2} + \sum_{r=1}^{s} L(r)\left[\left\{T_2(E_{FHDLB}, \eta_g, \lambda) - \left(n + \frac{1}{2}\right)\hbar\omega_0\right\}^{1/2}\right]\right] \tag{6.59b}$$

(iii) *Using (6.47), the magneto-dispersion law, in the absence of spin, for HD III–V, ternary and quaternary semiconductors, in the presence of photo-excitation, whose unperturbed conduction electrons obey the parabolic energy bands, is given by*

$$T_2(E, \eta_g) = \left(n + \frac{1}{2}\right)\hbar\omega_0 + \frac{\hbar^2 k_z^2}{2m_c} \tag{6.60}$$

Using (6.60), the DOS function in the present case can be expressed as

$$D_B(\mathrm{E}, \eta_g, \lambda) = \frac{g_v|e|\sqrt{2m_c}}{2\pi^2\hbar^2} \sum_{n=0}^{n_{\max}} \left[\{T_3(E, \eta_g, \lambda)\}' \cdot \left\{T_3(E, \eta_g, \lambda) - \left(n + \frac{1}{2}\right)\hbar\omega_0\right\}^{-1/2} - H(E - E_{n_{i2}})\right] \tag{6.61}$$

where, $E_{n_i 2}$ is the Landau sub-band energies in this case and is given as

$$T_3(E_{n_{l2}}, \eta_g, \lambda) = \left(n + \frac{1}{2}\right) \hbar\omega_0 \tag{6.62}$$

The EEM in this case assumes the form

$$m^*(E_{FHDLB}, \eta_g, \lambda) = m_c\{T_3(E_{FHDLB}, \eta_g, \lambda)\}' \tag{6.63a}$$

The electron concentration can be written as

$$n_0 = \frac{g_v |e| B\sqrt{2m}}{\pi^2\hbar^2} \text{ Real Part of } \sum_{n=0}^{n_{\max}} \cdot \left[\left\{T_3(E_{FHDLB}, \eta_g, \lambda) - \left(n + \frac{1}{2}\right)\hbar\omega_0\right\}^{1/2} + \sum_{r=1}^{s} L(r)\left[\left\{T_3(E_{FHDLB}, \eta_g, \lambda) - \left(n + \frac{1}{2}\right)\hbar\omega_0\right\}^{1/2}\right]\right. \tag{6.63b}$$

6.2.2.1 *Results and discussion*

Using the values of the energy band constants from Appendix A, we have plotted the EM along the direction of z as functions of 1/B, electron concentration, intensity and wave length (as shown in Figs. 6.6–6.9) at $T = 4.2\,\text{K}$ by taking n-InSb and n-InAs which are used for the purpose of numerical computations in accordance with the perturbed three (using (6.63) and (6.64)) and two (using (6.68) and (6.69)) band models of Kane and that of perturbed parabolic (using (6.73) and (6.74)) energy bands respectively. It appears from Fig. 6.6 that the EM is an oscillatory function of inverse quantizing magnetic field. The oscillatory dependence is due to the crossing over of the Fermi level by the Landau sub-bands in steps resulting in successive reduction the number of occupied Landau levels as the magnetic field is increased. For each coincidence of a Landau level, with the Fermi level, there would be a discontinuity in the density-of-states function resulting in a peak of oscillation.

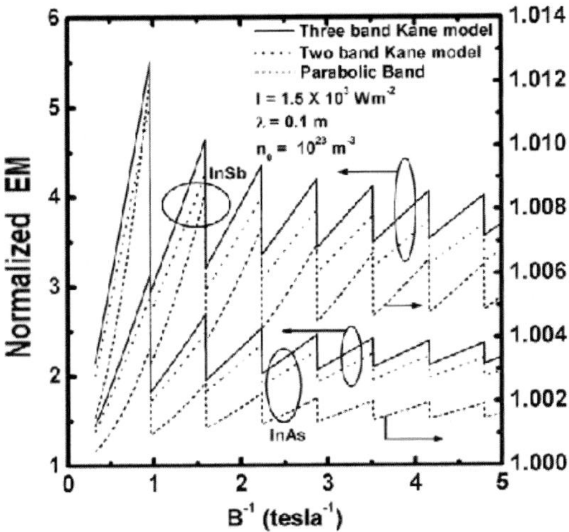

Fig. 6.6. Plot of the normalized EM as a function of inverse magnetic field for n-InSb and n-InAsin the presence of light waves in accordance with the three, the two band models of Kane and the parabolic energy band model in the presence of external photo-excitation.

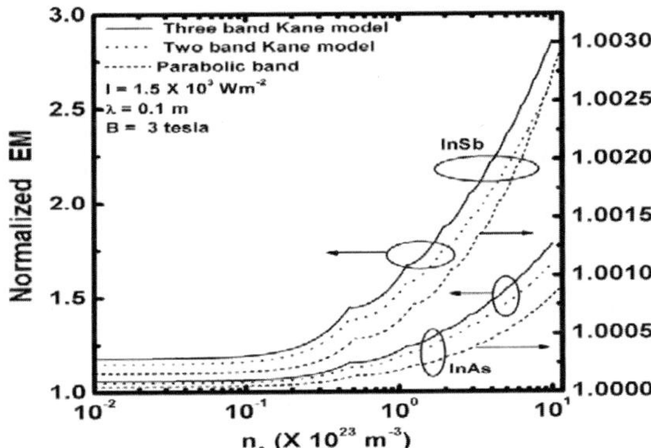

Fig. 6.7. Plot of the normalized EM as a function of carrier concentration for n-InSb and n-InAsin the presence of light waves in accordance with the three, the two band models of Kane and the parabolic energy band model in the presence of external photo-excitation.

Thus the peaks should occur whenever the Fermi energy is a multiple of energy separation between the two consecutive Landau levels and it may be noted that the origin of oscillations in the EM is the same as that of the Shubnikov-de Hass oscillations. With increase

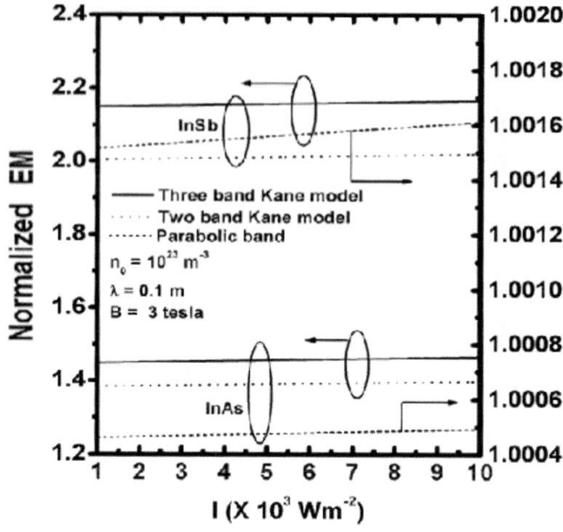

Fig. 6.8. Plot of the normalized EM as a function of light intensity for n-InSb and n-InAsin the presence of light waves in accordance with the three, the two band models of Kane and the parabolic energy band model in the presence of external photo-excitation.

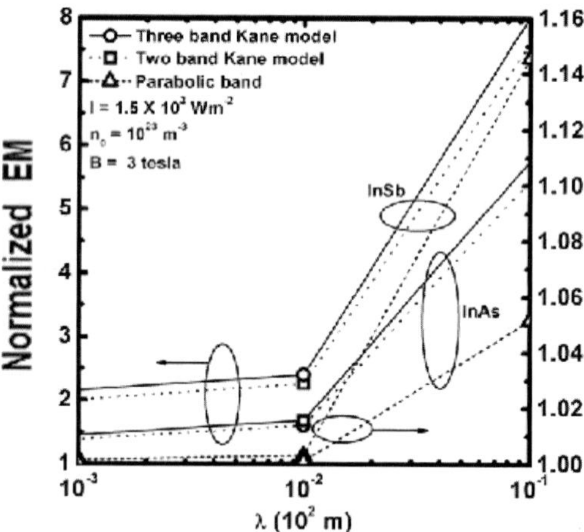

Fig. 6.9. Plot of the normalized EM as a function of light wavelength for n-InSb and n-InAsin the presence of light waves in accordance with the three, the two band models of Kane and the parabolic energy band model in the presence of external photo-excitation.

in magnetic field, the amplitude of the oscillation will increase and, ultimately, at very large values of the magnetic field, the conditions for the quantum limit will be reached (neglecting magnetic freeze out) when the EM will be found to decrease monotonically with increase in magnetic field. In Fig. 6.7, the concentration dependence of the magneto-EM has been plotted for all the cases of Fig. 6.6 for both n-InSb and n-InAs. The EM again shows oscillatory dependence with different numerical values exhibiting the signature of the SdH effect. Although the rate of variations are different, the influence of the energy band constants in accordance with all the type of the band models is apparent from the Figs. One can observe from Fig. 6.8 that the EM has a steady increase with the increase of the light intensity although the same EM increases sharply with the increase in wavelength in different ways, as appears from Figs. 6.8 and 6.9 respectively. The nature of variations in all the cases depends strongly on the energy spectrum constants of the respective materials and the external physical conditions. It should be noted that the numerical value of the EM in the presence of light waves is relatively much higher even at smaller value of the magnetic field, than that in the absence of the magnetic field. Such a high value in the EM can cause a EMastic effect by reducing the electron mobility under the application of a quantized magnetic field and the contribution of the oscillatory mass or the oscillatory mobility would be is more important.

6.2.3 *The EM under crossed electric and quantizing magnetic fields in HD Kane Type Semiconductors in the Presence of Light Waves*

(i) The electron dispersion law in the present case is given by

$$T_1(E,\eta_g,\lambda) = \left(n+\frac{1}{2}\right)\hbar\omega_0 + \frac{[\hbar k_z(E)]^2}{2m_c} - \frac{E_0}{B}\hbar k_y\{T_1(E,\eta_g,\lambda)\}' - \left\{\frac{m_c E_0^2\left[\{T_1(E,\eta_g,\lambda)\}'\right]^2}{2B^2}\right\} \tag{6.64}$$

The use of (6.64) leads to the expressions of the EEM s' along z and y directions as

$$m_z^*(E_{BLHDC}, n, E_0, B, \lambda) = \text{Real part of } m_c \Bigg[\{T_1(E_{F_{BLHDC}}, \eta_g, \lambda)\}'' + \frac{m_c E_0^2\{T_1(E_{FBLHDC}, \eta_g, \lambda)\}'\{T_1(E_{F_{BLHDC}}, \eta_g, \lambda)\}''}{B^2}\Bigg] \quad (6.65)$$

$$m_y^*(E_{BLHDC}, n, E_0, B, \lambda) = \text{Real part of } \left(\frac{B}{E_0}\right)^2 \cdot[\{T_1(E_{F_{BLHDC}}, \eta_g, \lambda)\}']^{-1} \left[T_1(E_{F_{BLHDC}}, \eta_g, \lambda) - \left(n+\frac{1}{2}\right)\hbar\omega_0 + \frac{m_c E_0^2\{T_1(E_{F_{BLHDC}}, \eta_g, \lambda)\}'}{2B^2}\right] \cdot \left[\frac{\{T_1(E_{F_{BLHDC}}, \eta_g, \lambda)\}''}{[\{T_1(E_{F_{BLHDC}}, \eta_g, \lambda)\}']^2} \left[T_1(E_{F_{BLHDC}}, \eta_g, \lambda) - \left(n+\frac{1}{2}\right)\hbar\omega_0 + \frac{m_c E_0^2[\{T_1(E_{F_{BLHDC}}, \eta_g, \lambda)\}']^2}{2B^2}\right] + 1 + \frac{m_c E_0^2\{T_1(E_{F_{BLHDC}}, \eta_g, \lambda)\}''}{B^2}\right. \quad (6.66)$$

where, $E_{F_{BLHDC}}$ is the Fermi energy in this case.

The Landau energy ($E_{n_{l41}}$) can be written as

$$T_1(E_{n_{l41}}, \eta_g, \lambda) = \left(n+\frac{1}{2}\right)\hbar\omega_0 - \left\{\frac{m_c E_0^2\left[\{T_1(\mathrm{E}_{n_{l41}}, \eta_g, \lambda\}'\right]^2}{2B^2}\right\} \quad (6.67a)$$

The electron concentration and the DMR in this case assume the forms

$$n_0 = \frac{2g_v B\sqrt{2m_c}}{3L_x \pi^2 \hbar^2 E_0} \sum_{n=0}^{n_{\max}} [T_{43}(n, E_{F_{BLHDC}}, \eta_g, \lambda) + T_{44}(n, E_{F_{BLHDC}}, \eta_g, \lambda)]$$

where,

$$T_{43}(n, E_{F_{BLHDC}},, \eta_g, \lambda) \equiv \Bigg[\Bigg[T_1(n, E_{F_{BLHDC}}, \eta_g, \lambda) - \left(n+\frac{1}{2}\right)\hbar\omega_0 - \frac{m_2 E_0^2}{2B^2}\left[\{T_{41}(n, E_{F_{BLHDC}}, \eta_g, \lambda)\}'\right]^2$$

$$+ |e|E_0L_x\left[\{T_1(n, E_{F_{BLHDC}}, \eta_g, \lambda)\}'\right]\Bigg]^{3/2}$$

$$-\Bigg[T_1(n, E_{F_{BLHDC}}, \eta_g, \lambda) - \left(n+\frac{1}{2}\right)\hbar\omega_0$$

$$-\frac{m_c E_0^2}{2B^2}[T_1(n, E_{F_{BLHDC}}, \eta_g, \lambda)\}'\rbrack^2\Bigg]\Bigg]^{3/2}$$

$$\cdot\frac{1}{[\{T_1(n, E_{F_{BLHDC}}, \eta_g, \lambda)\}']}$$

and

$$T_{44}(n, E_{F_{BLHDC}}, \eta_g, \lambda) \equiv \sum_{r=1}^{s} L(r)[T_{44}(n, E_{F_{BLHDC}}, \eta_g, \lambda)] \quad (6.67b)$$

(ii) Similarly, the electron dispersion law in this case is given by

$$T_2(E, \eta_g, \lambda) = \left(n+\frac{1}{2}\right)\hbar\omega_0 - \frac{E_0}{B}\hbar k_y\{T_2(E, \eta_g, \lambda)\}' - \frac{m_c E_0^2}{2B^2}[\{T_2(E, \eta_g, \lambda)\}']^2 + \frac{[\hbar k_z(E)]^2}{2m_c} \quad (6.68)$$

The use of (6.68) leads to the expressions of the EEM s' along z and y directions as

$$m_z^*(E_{F_{BLHDC}}, n, E_0, B, \lambda) = m_c\Bigg[\{T_2(E_{F_{BLHDC}}, \eta_g, \lambda)\}'' + \frac{m_c E_0^2\{T_2(E_{F_{BLHDC}}, \eta_g, \lambda)'\}\{T_2(E_{F_{BLHDC}}, \eta_g, \lambda)\}''}{B_2}\Bigg] \quad (6.69)$$

$$m_y^*(E_{F_{BLHDC}}, n, E_0, B, \lambda) = \left(\frac{B}{E_0}\right)^2[\{T_2(E_{F_{BLHDC}}, \eta_g, \lambda)\}']^{-1}$$

$$\cdot\Bigg[T_2(E_{F_{BLHDC}}, \eta_g, \lambda) - \left(n+\frac{1}{2}\right)\hbar\omega_0 + \frac{m_c E_0^2[\{T_2(E_{F_{BLHDC}}, \eta_g, \lambda)\}']^2}{2B^2}\Bigg]\Bigg[\frac{\{T_2(E_{F_{BLHDC}}, \eta_g, \lambda)\}''}{[\{T_2(E_{F_{BLHDC}}, \eta_g, \lambda)\}']^2}$$

$$\cdot\left[T_2\left(E_{F_{BLHDC}},\eta_g,\lambda\right)-\left(n+\frac{1}{2}\right)\hbar\omega_0\right.$$

$$\left.+\frac{m_c E_0^2\left[\{T_2\left(E_{F_{BLHDC}},\eta_g,\lambda\right)\}'\right]^2}{2B^2}\right]+1$$

$$\left.+\frac{m_c E_0^2\left[\{T_2\left(E_{F_{BLHDC}},\eta_g,\lambda\right)\}''\right]^2}{B^2}\right] \tag{6.70}$$

The Landau energy ($E_{n_{l42}}$) can be written as

$$T_2\left(E_{n_{l42}},\eta_g,\lambda\right)=\left(n+\frac{1}{2}\right)\hbar\omega_0-\left\{\frac{m_c E_0^2\left[\{T_2\left(E_{n_{l42}},\eta_g,\lambda\right)\}'\right]^2}{2B^2}\right\} \tag{6.71}$$

(iii) Similarly, the electron dispersion law in this case is given by

$$T_3(E,\eta_g,\lambda)=\left(n+\frac{1}{2}\right)\hbar\omega_0-\frac{E_0}{B}\hbar k_y\left\{T_3\left(E,\eta_g,\lambda\right)\right\}'$$

$$-\frac{m_c E_0^2}{2B^2}\left[\{T_3\left(E,\eta_g,\lambda\right)\}'\right]^2+\frac{\left[\hbar k_z\left(E\right)\right]^2}{2m_c} \tag{6.72}$$

The use of (18.72) leads to the expressions of the EEM s' along z and y directions as

$$m_z^*(E_{F_{BLHDC}},n,E_0,B,\lambda)=m_c\left[\{T_3\left(E_{F_{BLHDC}},\eta_g,\lambda\right)\}''\right.$$

$$\left.+\frac{m_c E_0^2\{T_3\left(E_{F_{BLHDC}},\eta_g,\lambda\right)'\}\{T_3\left(E_{F_{BLHDC}},\eta_g,\lambda\right)\}''}{B^2}\right] \tag{6.73}$$

$$m_y^*\left(E_{F_{BLHDC}},n,E_0,B,\lambda\right)$$

$$=\left(\frac{B}{E_0}\right)^2\left[\{T_3(E_{F_{BLHDC}},\eta_g,\lambda)\}'\right]^{-1}$$

$$\cdot\left[T_3\left(E_{F_{BLHDC}},\eta_g,\lambda\right)-\left(n+\frac{1}{2}\right)\hbar\omega_0\right.$$

$$\left.+\frac{m_c E_0^2[\{T_3(E_{F_{BLHDC}},\eta_g,\lambda)\}']^2}{2B^2}\right]$$

$$\cdot\left[\frac{\{T_3(E_{F_{BLHDC}},\eta_g,\lambda)\}''}{\left[\{T_2(E_{F_{BLHDC}},\eta_g,\lambda)\}'\right]^2}\right.$$

$$\cdot\left[T_3(E_{F_{BLHDC}},\eta_g,\lambda)-\left(n+\frac{1}{2}\right)\hbar\omega_0\right.$$

$$\left.+\ \frac{m_c E_0^2\left[\{T_3(E_{F_{BLHDC}},\eta_g,\lambda)\}'\right]^2}{2B^2}\right]+1$$

$$\left.+\ \frac{m_c E_0^2\left[\{T_3(E_{F_{BLHDC}},\eta_g,\lambda)\}''\right]^2}{B^2}\right] \tag{6.74}$$

The Landau energy ($E_{n_{l43}}$) can be written as

$$T_3(E_{n_{l42}},\eta_g,\lambda)=\left(n+\frac{1}{2}\right)\hbar\omega_0-\left\{\frac{m_c E_0^2\left[\{T_3(E_{n_{l42}},\eta_g,\lambda)\}'\right]^2}{2B^2}\right\} \tag{6.75a}$$

The electron concentration in this case assume the forms

$$n_0=\frac{2g_v B\sqrt{2m_c}}{3L_x\pi^2\hbar^2 E_0}\sum_{n=0}^{n_{\max}}[T_{433}(n,E_{F_{BLHDC}},\eta_g,\lambda)$$
$$+T_{443}(n,E_{F_{BLHDC}},\eta_g,\lambda)] \tag{6.75b}$$

where,

$$T_{433}(n,E_{F_{BLHDC}},\eta_g,\lambda)$$
$$\equiv\left[\left[T_3(E_{F_{BLHDC}},\eta_g,\lambda)-\left(n+\frac{1}{2}\right)\hbar\omega_0\right.\right.$$
$$-\frac{m_c E_0^2}{2B^2}[\{T_3(E_{F_{BLHDC}},\eta_g,\lambda)\}']^2$$
$$\left.+\ |e|E_0L_x[\{T_3(E_{F_{BLHDC}},\eta_g,\lambda)\}']\right]^{3/2}$$
$$-\left[T_3(E_{F_{BLHDC}},\eta_g,\lambda)-\left(n+\frac{1}{2}\right)\hbar\omega_0\right.$$
$$\left.\left.-\ \frac{m_c E_0^2}{2B^2}[\{T_3(E_{F_{BLHDC}},\eta_g,\lambda)\}']^2\right]^{3/2}\right]\frac{1}{[\{T_3(E_{F_{BLHDC}},\eta_g,\lambda)\}']}$$

and

$$T_{443}(n, E_{F_{BLHDC}}, \eta_g, \lambda) \equiv \sum_{r=1}^{s} [L(r) T_{433}(n, E_{F_{BLHDC}}, \eta_g, \lambda)].$$

6.2.4 *The EM in QWs of HD Kane type semiconductors in the presence of light waves*

(i) *The 2D DR in QWs of HD III–V, ternary and quaternary materials, whose unperturbed band structure is defined by the three band model of Kane, in the presence of light waves can be expressed following (6.46b) as*

$$\frac{\hbar^2 k_s^2}{2m_c} + \frac{\hbar^2}{2m_c}\left(\frac{n_z \pi}{d_z}\right)^2 = T_1(E, \eta_g, \lambda) \tag{6.76}$$

The sub band energies ($E_{n_{l7HD}}$) can be written as

$$T_1(E_{n_{l7HD}}, \eta_g, \lambda) = \frac{\hbar^2}{2m_c}(n_z \pi / d_z)^2 \tag{6.77}$$

The expression of the EEM in this case is given by

$$m^*(E_{F2DLHD}, n_z, \lambda) = m_c \text{ Real part of } \{T_1(E_{F2DLHD}, \eta_g, \lambda)\}' \tag{6.78}$$

where, E_{F2DLHD} is the Fermi energy in the present case as measured from the edge of the conduction band in the vertically upward direction in absence of any quantization.

The DOS function can be written as

$$N_{2D}(E, \eta_g, \lambda) = \left(\frac{m_c g_v}{\pi \hbar^2}\right) \sum_{n_z=1}^{n_{z\max}} [T_1(E, \eta_g, \lambda)]' H(E - E_{n_{l7HD}}) \tag{6.79a}$$

The surface electron concentration can be expressed as

$$n_0 = \frac{m_c g_v}{\pi \hbar^2} \text{ Real Part of } \sum_{n_z=1}^{n_{z\max}} (t_{61} + t_{62}) \tag{6.79b}$$

where

$$t_{61} = \left[T_1(E_{F2DLHD}, \eta_g, \lambda) - \frac{\hbar^2}{2m_c}\left(\frac{n_z \pi}{d_z}\right)^2 \right]$$

and

$$t_{62} = \sum_{r=1}^{s} L(r)[t_{61}]$$

(ii) *The 2D DR in QWs of HD III–V, ternary and quaternary materials, whose unperturbed band structure is defined by the two band model of Kane, in the presence of light waves can be expressed following* (6.47) *as*

$$\frac{\hbar^2 k_s^2}{2m_c} + \frac{\hbar^2}{2m_c}\left(\frac{n_z\pi}{d_z}\right)^2 = T_2(E,\eta_g,\lambda) \tag{6.80}$$

The sub band energies $(E_{n_{18HD}})$ can be written as

$$T_2(E_{n_{l8HD}},\eta_g,\lambda) = \frac{\hbar^2}{2m_c}(n_z\pi/d_z)^2 \tag{6.81}$$

The expression of the EEM in this case is given by

$$m^*(E_{F2DLHD},n_z,\lambda) = m_c\{T_2(E_{F2DLHD},\eta_g,\lambda)\}' \tag{6.82}$$

The DOS function can be written as

$$N_{2D}(E,\eta_g,\lambda) = \left(\frac{m_c g_v}{\pi\hbar^2}\right)\sum_{n_z=1}^{n_{z\max}} [T_2(E,\eta_g,\lambda)]' H(E - E_{n_{l8HD}}) \tag{6.83}$$

(iii) *The 2D DR in QWs of HD III–V, ternary and quaternary materials, whose unperturbed band structure is defined by the parabolic energy bands, in the presence of light waves can be expressed following* (6.48) *as*

$$\frac{\hbar^2 k_s^2}{2m_c} + \frac{\hbar^2}{2m_c}\left(\frac{n_z\pi}{d_z}\right)^2 = T_3(E,\eta_g,\lambda) \tag{6.84}$$

The sub band energies $(E_{n_{l9HD}})$ can be written as

$$T_2(E_{n_{l9HD}},\eta_g,\lambda) = \frac{\hbar^2}{2m_c}(n_z\pi/d_z)^2 \tag{6.85}$$

The expression of the EEM in this case is given by

$$m^*(E_{F2DLHD},n_z,\lambda) = m_c\{T_3(E_{F2DLHD},\eta_g,\lambda)\}' \tag{6.86}$$

The DOS function can be written as

$$N_{2D}(E,\eta_g,\lambda) = \left(\frac{m_c g_v}{\pi\hbar^2}\right)\sum_{n_z=1}^{n_{z\max}}[T_3(E,\eta_g,\lambda)]' H(E - E_{n_{l9HD}}) \tag{6.87a}$$

The surface electron concentration can be expressed as

$$n_0 = \frac{m_c g_v}{\pi\hbar^2}\sum_{n_z=1}^{n_{z\max}}(t_{65}+t_{66}) \tag{6.87b}$$

where

$$t_{65} = \left[T_3(E_{F2DLHD},\eta_g,\lambda) - \frac{\hbar^2}{2m_c}\left(\frac{n_z\pi}{d_z}\right)^2\right]$$

and

$$t_{66} = \sum_{r=1}^{s} L(r)[t_{65}]$$

6.2.4.1 *Result and discussions*

Using the values of the energy band constants from Appendix A, we have plotted the EM in the k_s plane as functions of film thickness, surface electron concentration, intensity and wavelength at $T = 4.2\,\mathrm{K}$ by taking ultra thin films of ternary materials which are used for the purpose of numerical computations in accordance with the perturbed three (using (6.77) and (6.79)), two (using (6.82) and (6.84)),band models of Kane and that of perturbed parabolic energy bands (using (6.87) and (6.89)), as shown in Figs. 6.10, 6.11, 6.12 and 6.13 respectively. The influence of carrier confinement in 2D under the presence of an external photo-excitation on the behavior of EM can be understood from the Figs. 6.10 to 6.13. The effect of quantum confinement is immediately apparent from all the curves of Fig. 6.10, since, the 2D EM depend strongly on the nano-thickness, which is in direct contrast with the corresponding bulk specimens which is also the direct signature of quantum confinement. It appears from the said figures that the EM in this case decreases with the increasing film thickness in a step like manner as considered here although the

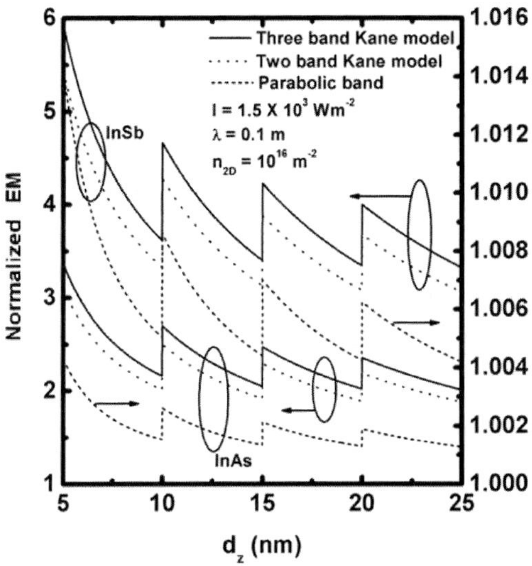

Fig. 6.10. Plot of the normalized EM as a function of film thickness for n-InSb and n-InAsin the presence of light waves in accordance with the three, the two band models of Kane and the parabolic energy band model in the presence of external photo-excitation.

numerical values vary widely and determined by the constants of the energy pectra. The oscillatory dependence is due to the crossing over of the Fermi level by the size quantized levels. For each coincidence of a size quantized level with the Fermi level, there would be a discontinuity in the density-of-states function resulting in a peak of oscillations. With large values of film thickness, the height of the steps decreases and the EM decreases with increasing film thickness in non-oscillatory manner and exhibit monotonic decreasing dependence. The height of step size and the rate of decrement are totally dependent on the band structure. The influence of energy band non-parabolicity is immediately apparent by the comparing the curves of the said figures. The energy band non-parabolicity and the spin orbit splitting constant significantly enhances the numerical values of the EM both the cases of the materials. The numerical values of the EM in accordance with the three band model of Kane are different as compared with the corresponding two band model, which reflects that fact that the presence of the spin orbit splitting constant changes

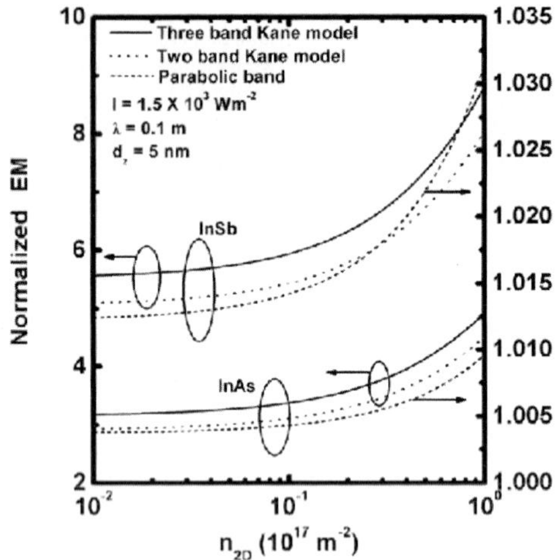

Fig. 6.11. Plot of the normalized EM as a function of surface electron concentration for n-InSb and n-InAsin the presence of light waves in accordance with the three, the two band models of Kane and the parabolic energy band model in the presence of external photo-excitation.

the magnitude of the EM. It may be noted that the presence of the band non-parabolicity in accordance with the two-band model of Kane further changes the peaks of the oscillatory EM for all cases of quantum confinements.

In Fig. 6.11, we have plotted the EM as a function of surface electron concentration per unit area for all cases of Fig. 6.10. It appears that the EM increases with increasing carrier degeneracy and also reflects the signature of the 1D confinement through the non-linear dependence with the 2D electron statistics. Since, most of the electrons at low temperatures occupies the lowest sub-band level, we have plotted the EM by considering the lowest sub-band energy in Fig. 6.11 to 6.13. If more sub-bands were considered, the oscillatory dependence will be less and less prominent with increasing carrier concentration and ultimately, for bulk specimens of the same material, the EM will be found to increase continuously with increasing electron concentration in a non-oscillatory manner. The effects of the light intensity and wavelength on the EM has been exhibited in Figs. 6.12 and 6.13 respectively in the regime of

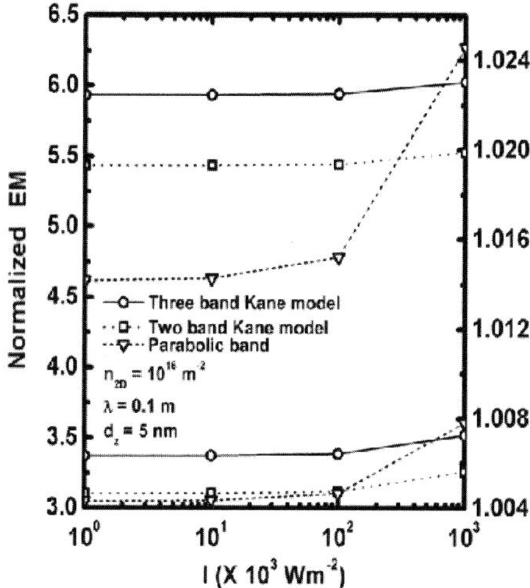

Fig. 6.12. Plot of the normalized EM as a function of light intensity for n-InSb and n-InAsin the presence of light waves in accordance with the three, the two band models of Kane and the parabolic energy band model in the presence of external photo-excitation.

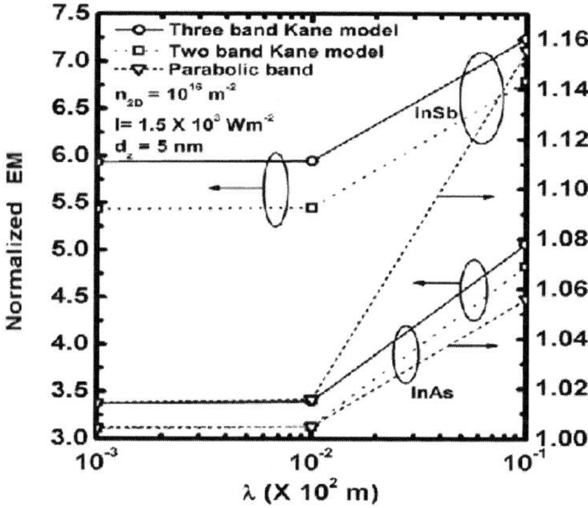

Fig. 6.13. Plot of the normalized EM as a function of wavelength for n-InSb and n-InAsin the presence of light waves in accordance with the three, the two band models of Kane and the parabolic energy band model in the presence of external photo-excitation.

very low temperatures. The EM increases with both the variables with different slopes. It appears that the EM in the case of parabolic energy band varies extremely slowly both with the light intensity and wavelength, although the, sharp and significant variations are exhibited for both three and the two band energy models.

6.2.5 *The EM in doping superlattices of HD kane type semiconductors in the presence of light waves*

(i) The DR in doping superlattices of HD III–V, ternary and quaternary materials in the presence of external photo-excitation whose unperturbed electrons are defined by the three band model of Kane can be expressed following (6.46b) as

$$T_1(E,\eta_g,\lambda) = \left(n_i + \frac{1}{2}\right)\hbar\omega_{91HD}\left(E,\eta_g,\lambda\right) + \frac{\hbar^2 k_s^2}{2m_c} \tag{6.88}$$

where

$$\omega_{91HD}\left(E,\eta_g,\lambda\right) \equiv \left(\frac{n_0\left|e\right|^2}{d_0\varepsilon_{sc}T_1'\left(E,\eta_g,\lambda\right)m_c}\right)^{1/2}.$$

The sub band energies $(E_{n_{i10HD}})$ can be written as

$$T_1(E_{n_{i10HD}},\eta_g,\lambda) = \left(n_i + \frac{1}{2}\right)\hbar\omega_{91HD}\left(E_{n_{i10HD}},\eta_g,\lambda\right) \tag{6.89}$$

The expression of the EEM in this case is given by

$$m^*(E_{F2DLHDD},\eta_g,\lambda,n_i) = m_c\{M_{40HD}(E_{F2DLHDD},\eta_g,\lambda,n_i)\}' \tag{6.90}$$

where,

$$M_{40HD}(E_{F2DLHDD},\eta_g,\lambda,n_i) = \text{Real part of}$$

$$\left\{T_1(E_{F2DLHDD},\eta_g,\lambda) - \left(n_i + \frac{1}{2}\right)\hbar\omega_{91HD}(E_{F2DLHDD},\eta_g,\lambda)\right\}$$

and $E_{F2DLHDD}$ is the Fermi energy in the present case as measured from the edge of the conduction band in the vertically upward direction in absence of any quantization.

The DOS function can be written as

$$N_{2D}(E,\eta_g,\lambda) = \left(\frac{m_c g_v}{\pi\hbar^2}\right)\sum_{n_z=1}^{n_{z\max}}[M_{40HD}(E,\eta_g,\lambda)]' H(E - E_{n_{l10HD}}) \tag{6.91a}$$

The surface electron concentration is given by

$$n_s = \frac{m_c g_v}{\pi\hbar^2} \text{ Real Part of } \left[\sum_{n_i=1}^{n_{i\max}} M_{40HD}(E_{F2DLHDD},\eta_g,\lambda,n_i) + \sum_{r=1}^{s} L(r)[M_{40HD}(E_{F2DLHDD},\eta_g,\lambda,n_i)]\right] \tag{6.91b}$$

(ii) The DR in doping superlattices of HD III–V, ternary and quaternary materials in the presence of external photo-excitation whose unperturbed electrons are defined by the two band model of Kane can be expressed following (6.47) as

$$T_2(E,\eta_g,\lambda) = \left(n_i + \frac{1}{2}\right)\hbar\omega_{92HD}(E,\eta_g,\lambda) + \frac{\hbar^2 k_s^2}{2m_c} \tag{6.92}$$

where

$$\omega_{92HD}(E,\eta_g,\lambda) \equiv \left(\frac{n_0|e|^2}{d_0\varepsilon_{sc}T_2'(E,\eta_g,\lambda)m_c}\right)^{1/2}.$$

The sub band energies ($E_{n_{i11HD}}$) can be written as

$$T_2(E_{n_{i11HD}},\eta_g,\lambda) = \left(n_i + \frac{1}{2}\right)\hbar\omega_{92HD}(E_{n_{i11HD}},\eta_g,\lambda) \tag{6.93}$$

The expression of the EEM in this case is given by

$$m^*(E_{F2DLHDD},\eta_g,\lambda,n_i) = m_c\{M_{41HD}(E_{F2DLHDD},\eta_g,\lambda,n_i)\}' \tag{6.94a}$$

where

$$M_{41HD}(E_{F2DLHDD},\eta_g,\lambda,n_i) = \left\{T_2(E_{F2DLHDD},\eta_g,\lambda) - \left(n_i + \frac{1}{2}\right)\hbar\omega_{92HD}(E_{F2DLHDD},\eta_g,\lambda)\right\} \quad \text{and}$$

The surface electron concentration is given by

$$n_s = \frac{m_c g_v}{\pi\hbar^2}\left[\sum_{n_i=1}^{n_{i_{\max}}} M_{41HD}(E_{F2DLHDD}, \eta_g, \lambda, n_i) + \sum_{r=1}^{s} L(r)[M_{41HD}(E_{F2DLHDD}, \eta_g, \lambda, n_i)]\right] \quad (6.94b)$$

The DOS function can be written as

$$N_{2D}(E, \eta_g, \lambda) = \left(\frac{m_c g_v}{\pi\hbar^2}\right)\sum_{n_z=1}^{n_{z\max}} [M_{41HD}(E, \eta_g, \lambda)]' H(E - E_{n_{l11HD}}) \quad (6.95)$$

(iii) The DR in doping superlattices of HD III–V, ternary and quaternary materials in the presence of external photo-excitation whose unperturbed electrons are defined by the two band model of Kane can be expressed following (6.48) as

$$T_3(E, \eta_g, \lambda) = \left(n_i + \frac{1}{2}\right)\hbar\omega_{93HD}(E, \eta_g, \lambda) + \frac{\hbar^2 k_s^2}{2m_c} \quad (6.96)$$

where

$$\omega_{93HD}(E, \eta_g, \lambda) \equiv \left(\frac{n_0 |e|^2}{d_0 \varepsilon_{sc} T_3'(E, \eta_g, \lambda) m_c}\right)^{1/2}.$$

The sub band energies $(E_{n_{i12HD}})$ can be written as

$$T_3(E_{n_{i12HD}}, \eta_g, \lambda) = \left(n_i + \frac{1}{2}\right)\hbar\omega_{93HD}(E_{n_{i12HD}}, \eta_g, \lambda) \quad (6.97)$$

The expression of the EEM in this case is given by

$$m^*(E_{F2DLHDD}, \eta_g, \lambda, n_i) = m_c\{M_{42HD}(E_{F2DLHDD}, \eta_g, \lambda, n_i)\}' \quad (6.98)$$

where

$$M_{42HD}(E_{F2DLHDD}, \eta_g, \lambda, n_i) = \left\{T_3(E_{F2DLHDD}, \eta_g, \lambda) - \left(n_i + \frac{1}{2}\right)\hbar\omega_{93HD}(E_{F2DLHDD}, \eta_g, \lambda)\right\} \quad \text{and}$$

The DOS function can be written as

$$N_{2D}(E,\eta_g,\lambda) = \left(\frac{m_c g_v}{\pi\hbar^2}\right)\sum_{n_z=1}^{n_{z\max}}[M_{42HD}(E,\eta_g,\lambda)]' H(E - E_{n_{l12HD}}) \tag{6.99a}$$

The surface electron concentration is given by

$$n_s = \frac{m_c g_v}{\pi\hbar^2}\left[\sum_{n_i=1}^{n_{i\max}} M_{42HD}(E_{F2DLHDD},\eta_g,\lambda,n_i) + \sum_{r=1}^{s} L(r)[M_{42HD}(E_{F2DLHDD},\eta_g,\lambda,n_i)]\right] \tag{6.99b}$$

6.2.5.1 *Result and discussions*

Using the values of the energy band constants from Appendix A, we have plotted the EM for the first two sub-bands as functions of wave length, intensity, thickness and electron concentration at $T = 4.2\,\mathrm{K}$ by taking nipi structures of ternary materials which are used for the purpose of numerical computations in accordance with the perturbed three (using (6.119) and (6.122)), two (using (6.124) and (6.127)) band models of Kane and that of perturbed parabolic energy bands (using (6.129) and (6.132)) respectively.

Using the multiple sub-bands, one can numerically evaluate the EM as a function of electron concentration and wavelength in nipi structures of III–V compounds by using the nipi structures of InSb and InAs as shown in Figs. 6.17 and 6.18 respectively, in accordance with three and two band models of Kane. The occurrence of the humps in Fig. 6.17 has been explained earlier in the context of ultrathin films. The effect of nipi structure tailoring increases the EM to an extremely high value which severely affects the electron mobility in such structures.

The effect of increasing wavelength aids to increase the EM in a linearly way, however in case of increasing the light intensity, the tendency of increase in EM is extremely slow. We have not considered the effect of the light intensity and wavelength on the EM governed by the parabolic energy band due to its slow variation from the value 1.

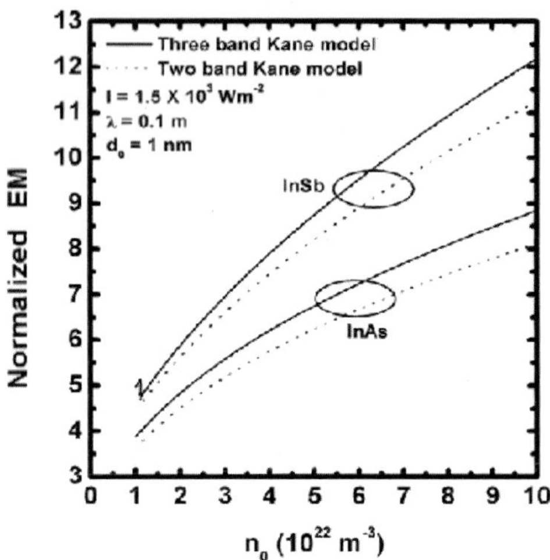

Fig. 6.17. Plot of the normalized EM as a function of surface electron concentration for n-channel inversion layers of n-InSb and n-InAsin the presence of light waves in accordance with the three, the two band models of Kane and the parabolic energy band model in the presence of external photo-excitation.

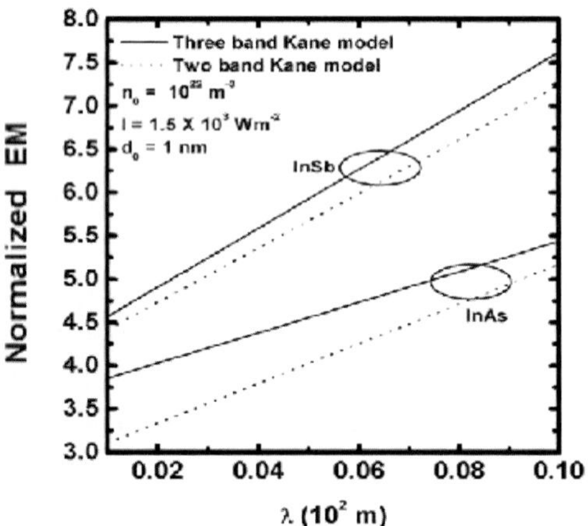

Fig. 6.18. Plot of the normalized EM as a function of electron wavelength for n-channel inversion layers of n-InSb and n-InAs in the presence of light waves in accordance with the three, the two band models of Kane and the parabolic energy band model in the presence of external photo-excitation.

6.2.6 *The EM in accumulation and inversion layers of kane type semiconductors in the presence of light waves*

(a) The 2D DR in accumulation layers of HD III–V, ternary and quaternary materials, whose unperturbed band structure is defined by the two band model of Kane, in the presence of light waves can be expressed following (6.46b) as

$$T_1(E,\eta_g,\lambda) = \frac{\hbar^2 k_s^2}{2m_c} + S_i\left[\frac{\hbar|e|F_2[T_1(E,\eta_g,\lambda)]'}{\sqrt{2m_c}}\right]^{2/3} \tag{6.100}$$

(6.100) represents the EM of the 2D electrons in accumulation layers of HD III–V, ternary and quaternary materials under the weak electric field limit in the presence of light waves whose bulk electrons obey the HD three band model of Kane. Since the EM in accordance with the HD three-band model of Kane is complex in nature, the (6.100) will also be complex. The both complexities occur due to the presence of poles in the finite complex plane of the dispersion relation of the materials in the absence of band tails.

The EM can be expressed as

$$m_L^*(E'_{fL},i,\eta_g) = m_c \text{ Real part of } P'_{3HDL}\left(E'_{fL},i,\eta_g,\lambda\right) \tag{6.101}$$

where,

$$P_{3HDL}\left(E'_{fL},i,\eta_g,\lambda\right) = \left[T_1(E,\eta_g,\lambda) - S_i\left[\frac{\hbar|e|F_2[T_1(E,\eta_g,\lambda]'}{\sqrt{2m_c}}\right]^{2/3}\right]$$

E'_{fL} is the Fermi energy in this case.

Thus, one can observe that the EM is a function of light intensity, scattering potential, the sub-band index, surface electric field, the Fermi energy and the other spectrum constants due to the combined influence of E_g and Δ.

The sub-band energy E_{i1L} is given by

$$0 = \text{ Real part of }\left[T_1(E_{i1L},\eta_g,\lambda) - S_i\left[\frac{\hbar|e|F_s[T_1(E_{i1L},\eta_g,\lambda]'}{\sqrt{2m_c}}\right]^{2/3}\right] \tag{6.102}$$

The DOS function can be written as

$$N_{2DiL}(E) = \frac{m_c g_v}{\pi\hbar^2} \sum_{i=0}^{i_{\max}} \left[P_{3HDL}\left(E, i, \eta_g, \lambda\right) H\left(E - E_{i1L}\right)\right] \quad (6.103)$$

Thus the DOS function is complex in nature.

In the absence of band-tails and under the condition of weak electric field limit, (6.100) assumes the form

$$\beta_0\left(E, \lambda\right) = \frac{\hbar^2 k_s^2}{2m_c} + S_i \left[\frac{\hbar|e|F_s[\beta_0\left(E, \lambda\right)]'}{\sqrt{2m_c}}\right]^{2/3} \quad (6.104)$$

(6.104) represents the EM of the 2D electrons in inversion layers of III–V, ternary and quaternary materials under the weak electric field limit in the present of light waves whose bulk electrons in the absence of any perturbation obey the three band model of Kane.

The EEM can be expressed as

$$m_L^*\left(E_{FiwL}, I, \lambda\right) = m_c\left[P_{3L}\left(E, i, \lambda\right)\right]' |_{E=E_{FiwL}} \quad (6.105)$$

where,

$$P_{3L} + \left(E, i, \lambda\right) = \left[\beta_0\left(E, \lambda\right) - S_i\left[\frac{\hbar|e|F_s\left[\beta_0\left(E, \lambda\right)\right]'}{\sqrt{2m_c}}\right]^{2/3}\right]$$

Thus, one can observe that the EEM is a function of the sub-band index, the light intensity, surface electric field, the Fermi energy and the other spectrum constants due to the combined influence of E_g and Δ.

The sub-band energy $E_{n_{iw2L}}$ in this case can be obtained from the (6.105) as

$$0 = \left[\beta_0\left(E_{n_{iw2L}}, \lambda\right) - S_i\left[\frac{\hbar|e|F_s\left[\beta_0\left(E_{n_{iw2L}}, \lambda\right)\right]'}{\sqrt{2m_c}}\right]^{2/3}\right] \quad (6.106)$$

Thus the 2D total DOS function in weak electric field limit can be expressed as

$$N_{2DiL}\left(E\right) = \frac{m_c g_v}{\pi\hbar^2} \sum_{i=0}^{i_{\max}} \left[P_{3L}\left(E, i, \lambda\right)\right] H\left(E - E_{n_{iwL}}\right) \quad (6.107)$$

(b) The 2D DR in accumulation layers of HD III–V, ternary and quaternary materials, whose unperturbed band structure is defined by the two band model of Kane in the absence of any field, in the presence of light waves can be expressed following (6.47) as

$$T_2(E,\eta_g,\lambda) = \frac{\hbar^2 k_s^2}{2m_c} + S_i\left[\frac{\hbar|e|F_s\left[T_2(E,\eta_g,\lambda)\right]'}{\sqrt{2m_c}}\right]^{\frac{2}{3}} \tag{6.108}$$

(6.108) represents the DR of the 2D electrons in accumulation layers of HD III–V, ternary and quaternary materials under the weak electric field limit in the presence of light waves whose bulk electrons obey the HD two band model of Kane. Since the electron energy spectrum in accordance with the HD two-band model of Kane is real in nature, the (6.108) will also be real.

The EEM can be expressed as

$$m_L^*\left(E'_{fl},i,\eta_g\right) = m_c P'_{3HDL1}\left(E'_{fl},i,\eta_g,\lambda\right) \tag{6.109}$$

where

$$P_{3HDL1}(E'_{fl},i,\eta_g,\lambda) = \left[T_2(E,\eta_g,\lambda) - S_i\left[\frac{\hbar|e|F_s[T_2(E,\eta_g,\lambda)]'}{\sqrt{2m_c}}\right]^{\frac{2}{3}}\right],$$

Thus, one can observe that the EEM is a function of light intensity, scattering potential, the sub-band index, surface electric field, the Fermi energy and the other spectrum constants due to the combined influence of E_g and Δ.

The sub-band energy E_{i1L1} is given by

$$0 = \left[T_2(E_{i1L1},\eta_g,\lambda) - S_i\left[\frac{\hbar|e|F_s\left[T_2(E_{i1L1},\eta_g,\lambda)\right]'}{\sqrt{2m_c}}\right]^{\frac{2}{3}}\right] \tag{6.110}$$

The DOS function can be written as

$$N_{2DiL}(E) = \frac{m_c g_v}{\pi\hbar^2}\sum_{i=0}^{i_{\max}}\left[P_{3HDL1}(E,i,\eta_g,\lambda)\,H(E-E_{i1L1})\right] \tag{6.111}$$

In the absence of band-tails and under the condition of weak electric field limit, (6.126) assumes the form

$$\tau_0(E,\lambda) = \frac{\hbar^2 k_s^2}{2m_c} + S_i\left[\frac{\hbar|e|F_s\left[\tau_0(E,\lambda)\right]'}{\sqrt{2m_c}}\right]^{2/3} \tag{6.112}$$

(6.112) represents the EM of the 2D electrons in inversion layers of III–V, ternary and quaternary materials under the weak electric field limit in the present of light waves whose bulk electrons in the absence of any perturbation obey the two band model of Kane.

The EEM can be expressed as

$$m_L^*\left(E_{FiwL}, i, \lambda\right) = m_c\left[P_{3L2}\left(E, i, \lambda\right)\right]'|_{E=E_{FiwL}} \tag{6.113}$$

where,

$$P_{3L2}\left(E, i, \lambda\right) = \left[\tau_0\left(E, \lambda\right) - S_i\left[\frac{\hbar|e|F_s\left[\tau_0\left(E, \lambda\right)\right]'}{\sqrt{2m_c}}\right]^{2/3}\right]$$

Thus, one can observe that the EM is a function of the sub-band index, the light intensity, surface electric field, the Fermi energy and the other spectrum constants due to the combined influence of E_g and Δ.

The sub-band energy $E_{n_{iw2L2}}$ in this case can be obtained from the (6.112) as

$$0 = \left[\tau_0\left(E_{n_{iw2L2}}, \lambda\right) - S_i\left[\frac{\hbar|e|F_s\left[\tau_0\left(E_{n_{iw2L2}}, \lambda\right)\right]'}{\sqrt{2m_c}}\right]^{2/3}\right] \tag{6.114}$$

Thus the 2D total DOS function in weak electric field limit can be expressed as

$$N_{2D_{iL}}\left(E\right) = \frac{m_c g_v}{\pi\hbar^2}\sum_{i=0}^{i_{\max}}\left[P_{3L2}\left(E, i, \lambda\right) H\left(E - E_{n_{iw2L2}}\right)\right] \tag{6.115a}$$

The surface electron concentration is given by

$$n_0 = \frac{m_c g_v}{\pi\hbar^2}\sum_{i=0}^{i_{\max}}\left[P_{3L2}(E_{FiwL}, i, \lambda) + \sum_{r=1}^{s} L(r)[P_{3L2}(E_{FiwL}, i, \lambda)]\right] \tag{6.115b}$$

The 2D DR in accumulation layers of HD III–V, ternary and quaternary materials, whose unperturbed band structure is defined by the two band model of Kane in the absence of any field, in the presence of light waves can be expressed following (18.48) as

$$T_3\left(E, \eta_g, \lambda\right) = \frac{\hbar^2 k_s^2}{2m_c} + S_i\left[\frac{\hbar|e|F_s\left[T_3\left(E, \eta_g, \lambda\right)\right]'}{\sqrt{2m_c}}\right]^{\frac{2}{3}} \tag{6.116}$$

(6.116) represents the DR of the 2D electrons in accumulation layers of HD III–V, ternary and quaternary materials under the weak electric field limit in the presence of light waves whose bulk electrons obey the HD model of parabolic energy bands. Since the electron energy spectrum in this case is real in nature, the (18.116) will also be real.

The EEM can be expressed as

$$m_L^*\left(E'_{fl}, i, \eta_g\right) = m_c P'_{3HDL2}\left(E'_{fl}, i, \eta_g, \lambda\right) \tag{6.117}$$

where,

$$P_{3HDL2}\left(E'_{fl}, i, \eta_g, \lambda\right) = \left[T_3\left(E, \eta_g, \lambda\right) - S_i\left[\frac{\hbar|e|F_s\left[T_3\left(E, \eta_g, \lambda\right)\right]'}{\sqrt{2m_c}}\right]^{\frac{2}{3}}\right]$$

Thus, one can observe that the EM is a function of light intensity, scattering potential, the sub-band index, surface electric field, the Fermi energy and the other spectrum constants due to the combined influence of E_g and Δ.

The sub-band energy E_{i1L2} is given by

$$0 = \left[T_3\left(E'_{i1L2}, i, \eta_g, \lambda\right) - S_i\left[\frac{\hbar|e|F_s\left[T_3\left(E'_{i1L2}, i, \eta_g, \lambda\right)\right]'}{\sqrt{2m_c}}\right]^{\frac{2}{3}}\right] \tag{6.118}$$

The DOS function can be written as

$$N_{2D_{iL}}(E) = \frac{m_c g_v}{\pi\hbar^2}\sum_{i=0}^{i_{\max}}\left[P_{3HDL2}\left(E, i, \eta_g, \lambda\right) H\left(E - E_{i1L2}\right)\right] \tag{6.119}$$

In the absence of band-tails and under the condition of weak electric field limit, (6.134) assumes the form

$$\rho_0(E, \lambda) = \frac{\hbar^2 k_s^2}{2m_c} + S_i\left[\frac{\hbar|e|F_s[\rho_0(E, \lambda)]'}{\sqrt{2m_c}}\right]^{2/3} \tag{6.120}$$

(6.120) represents the DR of the 2D electrons in inversion layers of III–V, ternary and quaternary materials under the weak electric field limit in the present of light waves whose bulk electrons in the absence of any perturbation obey the model of isotropic parabolic energy bands.

The EM can be expressed as

$$m_L^*(E_{FiwL}, i, \lambda) = m_c [P_{3L3}(E, i, \lambda)]' |_{E=E_{FiwL}} \tag{6.121}$$

where,

$$P_{3L3}(E, i, \lambda) = \left[\rho_0(E, \lambda) - S_i \left[\frac{\hbar|e|F_s [\rho_0(E, \lambda)]'}{\sqrt{2m_c}}\right]^{2/3}\right]$$

Thus, one can observe that the EEM is a function of the sub-band index, the light intensity, surface electric field, the Fermi energy and the other spectrum constants due to the combined influence of E_g and Δ.

The sub-band energy $E_{n_{iw2L3}}$ in this case can be obtained from the (18.120) as

$$0 = \left[\rho_0(E_{n_{iw2L3}}, \lambda) - S_i \left[\frac{\hbar|e|F_s [\rho_0(E_{n_{iw2L3}}, \lambda)]'}{\sqrt{2m_c}}\right]^{2/3}\right] \tag{6.122}$$

Thus the 2D total DOS function in weak electric field limit can be expressed as

$$N_{2D_{iL}}(E) = \frac{m_c g_v}{\pi\hbar^2} \sum_{i=0}^{i_{\max}} [P_{3L3}(E, i, \lambda) H(E - E_{n_{iwL3}})] \tag{6.123}$$

The surface electron concentration is given by

$$n_0 = \frac{m_c g_v}{\pi\hbar^2} \left[\sum_{i=1}^{i_{\max}} P_{3L3}(E_{FiwL}, i, \lambda) + \sum_{r=1}^{s} L(r)[P_{3L3}(E_{FiwL}, i, \lambda)]\right] \tag{6.141b}$$

6.2.6.1 *Result and discussions*

Using the values of the energy band constants from Appendix A, we have plotted the EM in the k_s plane for the first two sub-bands as functions of surface electric field, wave length and intensity at $T = 4.2\,K$ by taking n-channel inversion layers of n-InSb and n-InAs semiconductors which are used for the purpose of numerical computations in accordance with the perturbed three (using (6.92) and (6.95) and two (using (6.103) and (6.106) band models of Kane and that of perturbed parabolic energy bands (using (6.114)

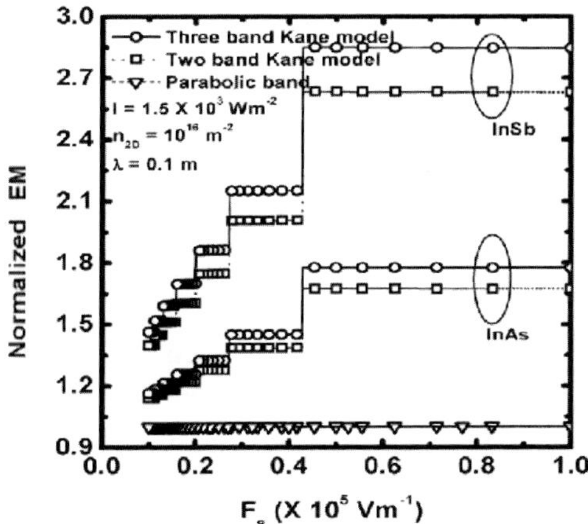

Fig. 6.14. Plot of the normalized EM as a function of electron concentration for nipi structures of n-InSb and n-InAs in the presence of light waves in accordance with the three and the two band models of Kane in the presence of external photo-excitation.

and (6.117) for weak electric field limit respectively. In Fig. 6.14, we have presented the variation of the EM in the n-channel inversion layers of n-InSb and n-InAs as function of surface electric field in accordance with the three band model of Kane, the two band model of Kane and the parabolic energy bands respectively under weak electric field by considering the effect of electric sub-bands.

It appears from Fig. 6.14 that the EM in the n-channel inversion layers increases with increase in surface electric field for weak electric field in a step-like manner with different numerical values and the influence of the energy band constants can also be assessed from the said figures. The EM depends on the electric sub-band index, surface electric field, the Fermi energy and the other spectrum constants due to the combined influence of E_g, Δ and λ which is the characterstic feature of such 2D systems under radiation. In Figs. 6.15 and 6.16, the effect of surface concentration and wavelength on the EM for all the cases of Fig. 6.14 has been considered under the quantum limit approximation. It appears from the Fig 6.15 and 6.16 that the EM increases with both the variables in the straight lines segmentation fashion with different increasing slopes at particular values of surface electron concentration and light wave length. The

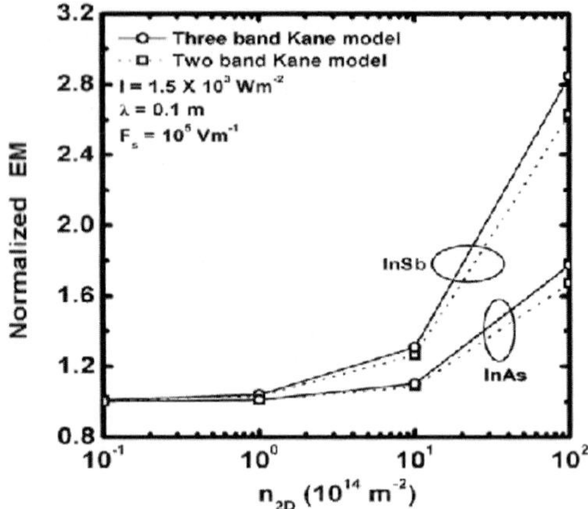

Fig. 6.15. Plot of the normalized EM as a function of wavelength for nipi structures of n-InSb and n-InAs in the presence of light waves in accordance with the three and the two band models of Kane in the presence of external photo-excitation.

values of EM for InSb are greater thant of InAs. The slopes are entirely determined by the spectra constants of n channel InSb and InAs respectively. It may be noted that if the direction of application of the surface electric field applied perpendicular to the surface be taken in an arbitrary direction and not as k_z as assumed in the present work, the EM would be different analytically for both the limits. Nevertheless, the arbitrary choice of the direction normal to the surface would not result in a change of the basic qualitative feature of the EM in n-channel inversion layers of semiconductors in the presence of photo excitation. The approximation of the potential well at the surface by a triangular well introduces some errors, as for instance the omission of the free charge contribution to the potential. This kind of approach is reasonable if there are only few charge carriers in the inversion layer, but is responsible for an overestimation of the splitting when the inversion carrier density exceeds that of the depletion layer. It has been observed that the maximum error due to the triangular potential well is tolerable in the practical sense because for actual calculations, one need a self consistent solution which is a formidable problem, for the present generalized systems due to the non availability of the proper analytical techniques, without exhibiting a widely different qualitative behavior. The second assumption of using only two subbands in the numerical calculations is valid in the range of low temperatures, where the quantum effects become prominent. The errors which are being introduced for these assumptions are found not to be serious enough at low temperatures. We wish to note that the many body effects, the hot electron effects, the formation of band tails, arbitrary orientation of the direction of the electric quantization and the effects of surface of states have been neglected in our simplified theoretical formalism due to the lack of availability in the literature of the proper analytical techniques for including them for the generalized systems as considered in this section. The numerical computation for the corresponding cases of EM under the strong electric field has been leftover for the readers so that they can enjoy the intricate internal computer analysis behind plots in this context. Our simplified approach will be useful for the purpose of comparison, when, the methods of tackling of the aforementioned formidable problems for the present generalized system appear and we admit the fact that the inclusion of the said effects would certainly increase the accuracy of our results.

6.2.7 *The EM in NWs of HD kane type semiconductors in the presence of light waves*

(a) The 1D DR in NWs of HD III–V, ternary and quaternary materials, whose unperturbed band structure is defined by the three band model of Kane in the absence of any field, in the presence of light waves can be expressed following (6.46b) as

$$\frac{\hbar^2 (n_z\pi/d_z)^2}{2m_c} + \frac{\hbar^2 (n_y\pi/d_y)^2}{2m_c} + \frac{\hbar^2 k_x^2}{2m_c} = T_1 (E, \eta_g, \lambda) \qquad (6.124)$$

The DOS function in this case assumes the form 6

$$N_{1DHD\Gamma L} (E, \eta_g)$$

$$= \frac{2g_v}{\pi} \sum_{n_y=1}^{n_{y\max}} \sum_{n_z=1}^{n_{z\max}} T'_{3L1}(E, n_y, n_z, \eta_g, \lambda) H(E - E'_{3HDNWL1}) \quad (6.125)$$

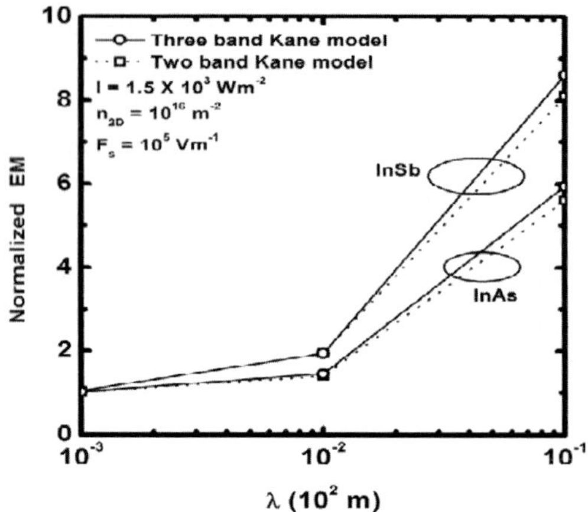

Fig. 6.16. Plot of the normalized EM as a function of surface electric field for n-channel inversion layers of n-InSb and n-InAsin the presence of light waves in accordance with the three, the two band models of Kane and the parabolic energy band model in the presence of external photo-excitation.

where

$$T_{3L1}(E, n_y, n_z, \eta_g, \lambda)$$
$$= \left[\left[T_1(E, \eta_g, \lambda) - \left[\frac{\hbar^2(n_z\pi/d_z)^2}{2m_c} + \frac{\hbar^2(n_y\pi/d_y)^2}{2m_c}\right]\right]\frac{2m_c}{\hbar^2}\right]^{\frac{1}{2}}$$

In (6.125), $E'_{2HDNWL1}$ is the sub-band energy in this case which can be expressed as

$$\frac{\hbar^2\,(n_z\pi/d_z)^2}{2m_c} + \frac{\hbar^2\,(n_y\pi/d_y)^2}{2m_c} = T_1\left(E'_{3HDNWL1,}\eta_g\lambda\right) \tag{6.126}$$

The EM in this case is given by

$$m^*(E_{F1HDNWL1}, \eta_g, \lambda) = m_c \text{ Real part of } [T_1(E_{F1HDNWL1}, \eta_g, \lambda)] \tag{6.127a}$$

where $E_{F1HDNWL1}$ is the Fermi energy in this case.

The electron concentration per unit length is given by

$$n_{1D} = \frac{2g_v}{\pi} \sum_{n_y=1}^{n_{y\max}} \sum_{n_z=1}^{n_{z\max}} \left[T_{3L2}(E_{F1HDNWL2}, n_y, n_z, \eta_g, \lambda) \right.$$

$$\left. + \sum_{r=1}^{s} L(r)[T_{3L2}(E_{F1HDNWL2}, n_y, n_z, \eta_g, \lambda)] \right] \quad (6.127\text{b})$$

The 1D DR, for NWs of III–V materials whose energy band structures are defined by the three-band model of Kane in the absence of band tailing assumes the form

$$\frac{\hbar^2 (n_z\pi/d_z)^2}{2m_c} + \frac{\hbar^2 (n_y\pi/d_y)^2}{2m_c} + \frac{\hbar^2 k_x^2}{2m_c} = \beta_0 (E, \lambda) \quad (6.128)$$

The DOS function can be expressed as

$$N_{1D\Gamma L}(E) = \frac{2g_v}{\pi} \sum_{n_y=1}^{n_{y\max}} \sum_{n_z=1}^{n_{z\max}} T'_{12L1}(E, n_y, n_z, \lambda) H\left(E - E'_{3L1}\right) \quad (6.129)$$

where

$$f_{12L1}(E, n_y, n_z, \lambda) = \left[\left[\beta_0 (E, \lambda) - \left[\frac{\hbar^2 (n_z\pi/d_z)^2}{2m_c} + \frac{\hbar^2 (n_y\pi/d_y)^2}{2m_c}\right]\right] \frac{2m_c}{\hbar^2}\right]^{\frac{1}{2}}$$

and E'_{3L1} is the sub-band energy in this case which can be obtained from the following equation

$$\frac{\hbar^2 (n_z\pi/d_z)^2}{2m_c} + \frac{\hbar^2 (n_y\pi/d_y)^2}{2m_c} + \frac{\hbar^2 k_x^2}{2m_c} = \beta_0 \left(E'_{3L1}, \lambda\right) \quad (6.130)$$

The EEM in this case is given by

$$m^*(E_{F1HDNWL1}, \eta_g, \lambda) = m_c[\beta_0'(E_{F1HDNWL2}, \lambda)] \quad (6.131\text{a})$$

where E_{F1NWL2} is the Fermi energy in this case.

The electron concentration per unit length is given by

$$n_{1D} = \frac{2g_v}{\pi}\sum_{n_y=1}^{n_{y\max}}\sum_{n_z=1}^{n_{z\max}}\left[f_{12L2}(E_{F1NWL2}, n_y, n_z, \eta_g, \lambda) + \sum_{r=1}^{s} L(r)[f_{12L2}(E_{F1NWL2}, n_y, n_z, \eta_g, \lambda)]\right] \quad (6.131b)$$

(b) The 1D EM in NWs of HD III–V, ternary and quaternary materials, whose unperturbed band structure is defined by the two band model of Kane in the absence of any field, in the presence of light waves can be expressed following (6.47) as

$$\frac{\hbar^2 (n_z\pi/d_z)^2}{2m_c} + \frac{\hbar^2 (n_y\pi/d_y)^2}{2m_c} + \frac{\hbar^2 k_x^2}{2m_c} = T_2(E, \eta_g, \lambda) \quad (6.132)$$

The DOS function in this case assumes the form

$$N_{1DHD\Gamma L}(E, \eta_g) = \frac{2g_v}{\pi}\sum_{n_y=1}^{n_{y\max}}\sum_{n_z=1}^{n_{z\max}} \cdot T'_{3L2}(E, n_y, n_z, \eta_g, \lambda) H(E - E'_{3HDNWL2}) \quad (6.133)$$

where

$$T_{3L2}(E, n_y, n_z, \eta_g, \lambda) = \left[\left[T_2(E, \eta_g, \lambda) - \left[\frac{\hbar^2 (n_z\pi/d_z)^2}{2m_c} + \frac{\hbar^2 (n_y\pi/d_y)^2}{2m_c}\right]\right]\frac{2m_c}{\hbar^2}\right]^{\frac{1}{2}}$$

In (6.133), $E'_{3HDNWL2}$ is the sub-band energy in this case which can be expressed as

$$\frac{\hbar^2 (n_z\pi/d_z)^2}{2m_c} + \frac{\hbar^2 (n_y\pi/d_y)^2}{2m_c} = T_2\left(E'_{3HDNWL2,}\eta_g\lambda\right) \quad (6.134)$$

The EEM in this case is given by

$$m^*(E_{F1HDNWL2}, \eta_g, \lambda) = m_c[T_2'(E_{F1HDNWL2}, \eta_g, \lambda)] \quad (6.135a)$$

$$n_{1D} = \frac{2g_v}{\pi} \sum_{n_y=1}^{n_{y\max}} \sum_{n_z=1}^{n_{z\max}} [T_{3L2}(E_{F1HDNWL2}, n_y, n_z, \eta_g, \lambda) + \sum_{r=1}^{s} L(r)[T_{3L2}(E_{F1HDNWL2}, n_y, n_z, \eta_g, \lambda)]] \tag{6.135b}$$

The 1D DR, for NWs of III–V materials whose energy band structures are defined by the two-band model of Kane in the absence of band tailing assumes the form

$$\frac{\hbar^2 (n_z\pi/d_z)^2}{2m_c} + \frac{\hbar^2 (n_y\pi/d_y)^2}{2m_c} + \frac{\hbar^2 k_x^2}{2m_c} = \tau_0(E, \lambda) \tag{6.136}$$

The DOS function can be expressed as

$$N_{1D\Gamma L}(E) = \frac{2g_v}{\pi} \sum_{n_y=1}^{n_{y\max}} \sum_{n_z=1}^{n_{z\max}} f'_{12L2}(E, n_y, n_z, \lambda)\, H(E - E'_{3L2}) \tag{6.137}$$

where

$$f_{12L2}(E, n_y, n_z, \lambda) = \left[\left[\tau_0(E, \lambda) - \left[\frac{\hbar^2 (n_z\pi/d_z)^2}{2m_c} + \frac{\hbar^2 (n_y\pi/d_y)^2}{2m_c}\right]\right] \frac{2m_c}{\hbar^2}\right]^{\frac{1}{2}}$$

and E'_{3L2} is the sub-band energy in this case which can be obtained from the following equation

$$\frac{\hbar^2 (n_z\pi/d_z)^2}{2m_c} + \frac{\hbar^2 (n_y\pi/d_y)^2}{2m_c} + \frac{\hbar^2 k_x^2}{2m_c} = \tau_0(E'_{3L2}, \lambda) \tag{6.138}$$

The EEM in this case is given by

$$m^*(E_{F1HDNWL2}, \eta_g, \lambda) = m_c[\tau'_0(E_{F1HDNWL2}, \lambda)] \tag{6.139a}$$

where $E_{F1HDNWL2}$ is the Fermi energy in this case.

The electron concentration per unit length is given by

$$n_{1D} = \frac{2g_v}{\pi} \sum_{n_y=1}^{n_{y\max}} \sum_{n_z=1}^{n_{z\max}} \left[f_{12L2}(E_{F1NWL2}, n_y, n_z, \eta_g, \lambda) + \sum_{r=1}^{s} L(r)[f_{12L2}(E_{F1NWL2}, n_y, n_z, \eta_g, \lambda)] \right] \quad (6.139b)$$

(c) The 1D DR in NWs of HD III–V, ternary and quaternary materials, whose unperturbed band structure is defined by the parabolic energy bands in the absence of any field, in the presence of light waves can be expressed following (6.48) as

$$\frac{\hbar^2 (n_z\pi/d_z)^2}{2m_c} + \frac{\hbar^2 (n_y\pi/d_y)^2}{2m_c} + \frac{\hbar^2 k_x^2}{2m_c} = T_3 (E, \eta_g, \lambda) \quad (6.140)$$

The DOS function in this case assumes the form

$$N_{1DHD\Gamma L}(E, \eta_g) = \frac{2g_v}{\pi} \sum_{n_y=1}^{n_{y\max}} \sum_{n_z=1}^{n_{z\max}} \cdot T'_{3L3}(E, n_y, n_z, \eta_g, \lambda) H(E - E'_{3HDNWL3}) \quad (6.141)$$

where

$$T_{3L3} (E, n_y, n_z, \eta_g, \lambda) = \left[\left[T_3 (E, \eta_g, \lambda) - \left[\frac{\hbar^2 (n_z\pi/d_z)^2}{2m_c} + \frac{\hbar^2 (n_y\pi/d_y)^2}{2m_c} \right] \right] \frac{2m_c}{\hbar^2} \right]^{\frac{1}{2}}$$

In (6.141), $E'_{3HDNWL3}$ is the sub-band energy in this case which can be expressed as

$$\frac{\hbar^2 (n_z\pi/d_z)^2}{2m_c} + \frac{\hbar^2 (n_y\pi/d_y)^2}{2m_c} = T_3 \left(E'_{3HDNWL3,} \eta_g \lambda\right) \quad (6.142)$$

The EEM in this case is given by

$$m^* (E_{F1HDNWL2}, \eta_g, \lambda) = m_c \left[T'_3 (E_{F1HDNWL2}, \eta_g, \lambda)\right] \quad (6.143a)$$

The electron concentration per unit length

$$n_{1D} = \frac{2g_v}{\pi} \sum_{n_y=1}^{n_{y\max}} \sum_{n_z=1}^{n_{z\max}} \left[T_{3L3}(E_{F1NWL2}, n_y, n_z, \eta_g, \lambda) + \sum_{r=1}^{s} L(r)[T_{3L3}(E_{F1NWL2}, n_y, n_z, \eta_g, \lambda)] \right] \quad (6.143b)$$

The 1D DR, for NWs of III–V materials whose energy band structures are defined by the parabolic energy bands in the absence of band tailing assumes the form

$$\frac{\hbar^2 (n_z\pi/d_z)^2}{2m_c} + \frac{\hbar^2 (n_y\pi/d_y)^2}{2m_c} + \frac{\hbar^2 k_x^2}{2m_c} = \rho_0 (E, \lambda) \quad (6.144)$$

The DOS function can be expressed as

$$N_{1D\Gamma L}(E) = \frac{2g_v}{\pi} \sum_{n_y=1}^{n_{y\max}} \sum_{n_z=1}^{n_{z\max}} f'_{12L3}(E, n_y, n_z, \lambda) H\left(E - E'_{3L3}\right) \quad (6.145)$$

where

$$f_{12L3}(E, n_y, n_z, \lambda) = \left[\left[\rho_0 (E, \lambda) - \left[\frac{\hbar^2 (n_z\pi/d_z)^2}{2m_c} + \frac{\hbar^2 (n_y\pi/d_y)^2}{2m_c}\right]\right] \frac{2m_c}{\hbar^2}\right]^{\frac{1}{2}}$$

and E'_{3L3} is the sub-band energy in this case which can be obtained from the following equation

$$\frac{\hbar^2 (n_z\pi/d_z)^2}{2m_c} + \frac{\hbar^2 (n_y\pi/d_y)^2}{2m_c} + \frac{\hbar^2 k_x^2}{2m_c} = \rho_0 \left(E'_{3L3}, \lambda\right) \quad (6.146)$$

The EM in this case is given by

$$m^* (E_{F1HDNWL2}, \eta_g, \lambda) = m_c \left[\rho'_0 (E_{F1HDNWL2}, \lambda)\right] \quad (6.147a)$$

The electron concentration per unit length

$$n_{1D} = \frac{2g_v}{\pi} \sum_{n_y=1}^{n_{y\max}} \sum_{n_z=1}^{n_{z\max}} \left[f_{12L3}(E_{F1NWL2}, n_y, n_z, \eta_g, \lambda) + \sum_{r=1}^{s} L(r)[f_{12L3}(E_{F1NWL2}, n_y, n_z, \eta_g, \lambda)] \right] \quad (6.147b)$$

6.2.7.1 *Result and discussions*

Using the values of the energy band constants from Appendix A, we have plotted the EM as functions of thickness and electron concentration per unit length at $T = 4.2\,\text{K}$ by taking quantum wires of ternary materials which are used for the purpose of numerical computations in accordance with the perturbed three (using (6.135) and (6.137)) and two (using (6.140) and (6.142)) band models of Kane respectively. In Figs. 6.19 and 6.20, we have plotted the EM in quantum wires of III–V materials as function of lateral dimension and electron concentration respectively. The effect of external photo-excitation increases the EM significantly. The influence of quantum confinement is immediately apparent from all the curves of Figures 6.19 since, the 1D EM depends strongly on the nano-thickness, which is in direct contrast with the corresponding bulk specimens. It appears from the said figures that the 1D EM decreases with the increasing film thickness in a step like manner as considered here although the numerical values vary widely and determined by the constants of the energy spectra. The oscillatory dependence is

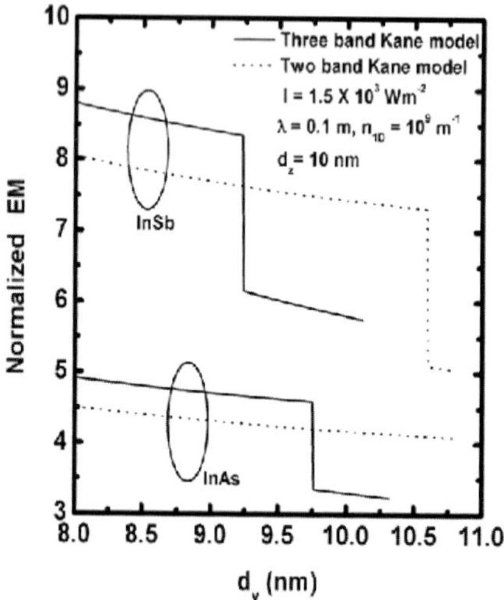

Fig. 6.19. Plot of the normalized EM as a function of lateral film thickness field for quantum wires of n-InSb and n-InAs in the presence of light waves in accordance with the three and the two band models of Kane in the presence of external photo-excitation.

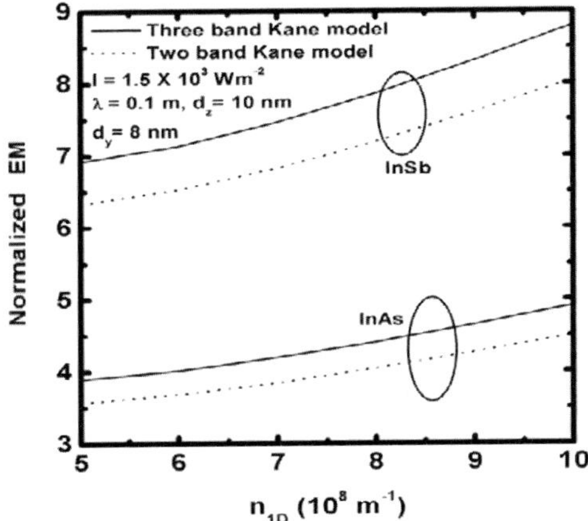

Fig. 6.20. Plot of the normalized EM as a function of linear electron concentration for quantum wires of n-InSb and n-InAs in the presence of light waves in accordance with the three and the two band models of Kane in the presence of external photo-excitation.

due to the crossing over of the Fermi level by the quantized levels. For each coincidence of a quantized level with the Fermi level, there would be a discontinuity in the density-of-states function resulting in a peak of oscillations. With large values of film thickness, the height of the steps decreases and the EM decreases with increasing film thickness in non-oscillatory manner and exhibit monotonic decreasing dependence. The height of step size and the rate of decrement are totally dependent on the band structure. In Fig. 6.20, we have exhibited the effect of the lowest sub-band on the EM when the electron concentration is varied. A direct assessment of the effect of light intensity and electron wavelength can be procured form the said figures. The effect of these two quantities on the parabolic energy bands are very less, for which we have not shown it in this case.

6.2.8 *The EM in QWHD effective mass superlattices of kane type semiconductors in the presence of light waves*

(a) Following Sasaki [9.15], the electron dispersion law in HD III–V effective mass superlattices (EMSLs) in the presence of light waves, the dispersion relations of whose constituent materials in the absence of any perturbation are defined by the three band model of Kane can be written as

$$k_x^2 = \left[\frac{1}{L_0^2}\{\cos^{-1}(f_{HD1}(E, k_y, k_z\lambda))\}^{-2} - k_\perp^2\right] \tag{6.148}$$

in which,

$$f_{HD1}(E, k_y, k_z\lambda) = a_{1HD1}\cos[a_0 C_{1HD1}(E, k_\perp) + b_0 D_{1HD1}(E, k_\perp)]$$
$$-a_{2HD1}\cos[a_0 C_{1HD1}(E, k_\perp) - b_0 D_{1HD1}(E, k_\perp)],$$
$$k_\perp^2 = k_y^2 + k_z^2, \mathrm{L}_0 = a_0 + b_0,$$
$$a_{1HD1} = \left[\sqrt{\frac{m_{c2}\text{Real part of } [T_1(0, \eta_{g2}, \lambda)]}{m_{c1}\text{Real part of } [T_1(0, \eta_{g1}, \lambda)]}} + 1\right]^2$$
$$\cdot\left[4\left(\frac{m_{c2}\text{Real part of } [T_1(0, \eta_{g2}, \lambda)]}{m_{c1}\text{Real part of } [T_1(0, \eta_{g1}, \lambda)]}\right)^{1/2}\right]^{-1}$$
$$a_{2HD1} = \left[-1 + \sqrt{\frac{m_{c2}\text{Real part of } [T_1(0, \eta_{g2}, \lambda)]}{m_{c1}\text{Real part of } [T_1(0, \eta_{g1}, \lambda)]}}\right]^2$$
$$\cdot\left[4\left(\frac{m_{c2}\text{Real part of } [T_1(0, \eta_{g2}, \lambda)]}{m_{c1}\text{Real part of } [T_1(0, \eta_{g1}, \lambda)]}\right)^{1/2}\right]^{-1}$$
$$C_{1HD1}(E, k_\perp, \lambda) = \left[\left(\frac{2m_{c1}}{\hbar^2}\right)T_1(E, \eta_{g_1}, \lambda) - k_\perp^2\right]^{1/2}$$

and

$$D_{1HD1}(E, k_\perp, \lambda) = \left[\left(\frac{2m_{c2}}{\hbar^2}\right)T_1(E, \eta_{g_1}, \lambda) - k_\perp^2\right]^{1/2}.$$

The DR in in QWHD effective mass superlattices of Kane type semiconductors in the presence of light waves, the dispersion relations of whose constituent materials in the absence of any perturbation are defined by the three band model of Kane can be written a

$$\left(\frac{n_x\pi}{d_x}\right) = \left[\frac{1}{L_0^2}\{\cos^{-1}(f_{HDL}(E, k_y, k_z, \lambda))\}^2 - k_1^2\right] \quad (6.149)$$

The EM in this case assumes the form

$$m^*(k_\perp, E, \lambda) = \frac{\hbar^2}{L_0^2}\left|\left[\frac{\cos^{-1}[f_{HD1}(E, k_y, k_z)f_{HD1}(E, k_y, k_z)}{\sqrt{1 - f_{HD1}^2(E, k_y, k_{z)]}}}\right]\right| \quad (6.150)$$

The sub-band energies ($E_{nSL5HDI}$) can be written as

$$\left(\frac{n_z\pi}{d_x}\right)^2 = \left[\frac{1}{L_0^2}\{\cos^{-1}(f_{HD1}(E_{nSLSHD1}, 0, 0, \lambda))\}^2\right] \quad (6.151)$$

The DOS and electron concentration should be calculated numerically

(b) Following Sasaki [9.15], the electron dispersion law in HD III–V effective mass superlattices (EMSLs) in the presence of light waves, the dispersion relations of whose constituent materials in the absence of any perturbation are defined by the two band model of Kane can be written as.

$$k_x^2 = \left[\frac{1}{L_0^2}\left\{\cos^{-1}\left(f_{HD2}\left(E, k_y, k_z\lambda\right)\right)\right\}^{-2} - k_\perp^2\right] \quad (6.152)$$

in which

$$f_{HD2}(E, k_y, k_z, \lambda) = a_{1HD2}\cos[a_0C_{1HD2}(E, k_\perp) + b_0D_{1HD2}(E, k_\perp)]$$
$$-a_{2HD2}\cos[a_0C_{1HD2}(E, k_\perp) + b_0D_{1HD2}(E, k_\perp)],$$

$$a_{1HD1} = \left[\sqrt{\frac{m_{c2}\left[T_2\left(0, \eta_{g2}, \lambda\right)\right]}{m_{c1}\left[T_2\left(0, \eta_{g1}, \lambda\right)\right]}} + 1\right]^2$$
$$\cdot\left[4\left(\frac{m_{c2}\left[T_2\left(0, \eta_{g2}, \lambda\right)\right]}{m_{c1}\left[T_2\left(0, \eta_{g1}, \lambda\right)\right]}\right)^{1/2}\right]^{-1}$$

$$a_{2HD1} = \left[-1 + \sqrt{\frac{m_{c2}\left[T_2\left(0, \eta_{g2}, \lambda\right)\right]}{m_{c1}\left[T_2\left(0, \eta_{g1}, \lambda\right)\right]}}\right]^2$$

$$\cdot \left[4\left(\frac{m_{c2}\left[T_2\left(0, \eta_{g2}, \lambda\right)\right]}{m_{c1}\left[T_2\left(0, \eta_{g1}, \lambda\right)\right]}\right)^{1/2}\right]^{-1}$$

$$C_{1HD1}\left(E, k_{\perp}, \lambda\right) = \left[\left(\frac{2m_{c1}}{\hbar^2}\right) T_2\left(E, \eta_{g_1}, \lambda\right) - k_{\perp}^2\right]^{1/2}$$

and

$$D_{1HD1}\left(E, k_{\perp}, \lambda\right) = \left[\left(\frac{2m_{c2}}{\hbar^2}\right) T_2\left(E, \eta_{g_1}, \lambda\right) - k_{\perp}^2\right]^{1/2}$$

The DR in in QWHD effective mass superlattices of Kane type semiconductors in the presence of light waves, the dispersion relations of whose constituent materials in the absence of any perturbation are defined by the two band model of Kane can be written as

$$\left(\frac{n_x \pi}{d_x}\right) = \left[\frac{1}{L_0^2}\{\cos^{-1}(f_{HD2}(E, k_y, k_z, \lambda))\}^2 - k_{\perp}^2\right] \qquad (6.153)$$

The EM in this case assumes the form

$$m^*(k_{\perp}, E, \lambda) = \frac{\hbar^2}{L_0^2}\left|\left[\frac{\cos^{-1}[f_{HD2}(E, k_y, k_z) f_{HD2}(E, k_y, k_z)}{\sqrt{1 - f_{HD2}^2(E, k_y, k_z)}}\right]\right| \qquad (6.154)$$

The sub-band energies ($E_{nSL5HD1}$) can be written as

$$\left(\frac{n_z \pi}{d_x}\right)^2 = \left[\frac{1}{L_0^2}\{\cos^{-1}(f_{HD1}(E_{nSL5HD2}, 0, 0, \lambda))\}^2\right] \qquad (6.155)$$

The DOS and electron concentration should be calculated numerically

(c) Following Sasaki [9.15], the electron dispersion law in HD III–V effective mass superlattices (EMSLs) in the presence of light waves, the dispersion relations of whose constituent materials in the absence of any perturbation are defined by the parabolic energy bands can be written as.

$$k_x^2 = \left[\frac{1}{L_0^2}\left\{\cos^{-1}\left(f_{HD3}\left(E, k_y, k_z\lambda\right)\right)\right\}^{-2} - k_\perp^2\right] \tag{6.156}$$

in which

$$f_{HD3}(E, k_y, k_z\lambda) = a_{1HD3}\cos[a_0 C_{1HD3}(E, k_\perp) + b_0 D_{1HD3}(E, k_\perp)] - a_{2HD3}\cos[a_0 C_{1HD3}(E, k_\perp) - b_0 D_{1HD3}(E, k_\perp)],$$

$$a_{1HD3} = \left[\sqrt{\frac{m_{c2}\left[T_3\left(0, \eta_{g2}, \lambda\right)\right]}{m_{c1}\left[T_3\left(0, \eta_{g1}, \lambda\right)\right]}} + 1\right]^2 \cdot \left[4\left(\frac{m_{c2}\left[T_3\left(0, \eta_{g2}, \lambda\right)\right]}{m_{c1}\left[T_3\left(0, \eta_{g1}, \lambda\right)\right]}\right)^{1/2}\right]^{-1},$$

$$a_{2HD3} = \left[-1 + \sqrt{\frac{m_{c2}\left[T_3\left(0, \eta_{g2}, \lambda\right)\right]}{m_{c1}\left[T_3\left(0, \eta_{g1}, \lambda\right)\right]}}\right]^2 \cdot \left[4\left(\frac{m_{c2}\left[T_3\left(0, \eta_{g2}, \lambda\right)\right]}{m_{c1}\left[T_3\left(0, \eta_{g1}, \lambda\right)\right]}\right)^{1/2}\right]^{-1},$$

$$C_{1HD3}\left(E, k_\perp, \lambda\right) = \left[\left(\frac{2m_{c1}}{\hbar^2}\right)T_3\left(E, \eta_{g_1}, \lambda\right) - k_\perp^2\right]^{1/2}$$

and

$$D_{1HD3}\left(E, k_\perp, \lambda\right) = \left[\left(\frac{2m_{c2}}{\hbar^2}\right)T_3\left(E, \eta_{g_2}, \lambda\right) - k_\perp^2\right]^{1/2}$$

The DR in in QWHD effective mass superlattices of Kane type semiconductors in the presence of light waves, the dispersion relations of whose constituent materials in the absence of any perturbation are defined by the parabolic energy bands can be written as

$$\left(\frac{n_x\pi}{d_x}\right) = \left[\frac{1}{L_0^2}\{\cos^{-1}(f_{HD3}(E, k_y, k_z, \lambda))\}^2 - k_\perp^2\right] \tag{6.157}$$

The EM in this case assumes the form

$$m^*(k_\perp, E, \lambda) = \frac{\hbar^2}{L_0^2}\left|\left[\frac{\cos^{-1}[f_{HD3}(E,k_y,k_z)f_{HD3}(E,k_y,k_z)}{\sqrt{1-f_{HD3}^2(E,k_y,k_z)}}\right]\right| \tag{6.158}$$

The sub-band energies ($E_{nSL5HD1}$) can be written as

$$\left(\frac{n_z\pi}{d_x}\right)^2 = \left[\frac{1}{L_0^2}\{\cos^{-1}(f_{HD2}(E_{nSL5HD3},0,0,\lambda))\}^2\right] \tag{6.159}$$

The DOS and electron concentration should be calculated numerically.

6.2.8.1 *Result and discussions*

Using the values of the energy band constants from Appendix A, we have plotted the EM as functions of 1/B and electron concentration at $T = 4.2\,\text{K}$ by taking effective mass super lattices of optoelectronic materials under magnetic quantization in accordance with the perturbed three (using (6.150) and (6.151)), two (using (6.152) and (6.153)) band models of Kane and that of perturbed parabolic energy bands (using (6.154) and (6.155)) respectively .In Figs. 6.21 and 6.22, we have plotted the effect of magnetic field and carrier concentration on the EM of effective mass super-lattices in GaAs/AlGaAs structures. The effect of SdH oscillations has been exhibited in this case for multi sub-band generation. In this case, the EM is a sub-band index dependent and we have plotted the EM by considering the lowest sub-band index. In both the figures we see that the effect of the external photo-excitation on the EM dominated by the parabolic energy law does not tend to modulate with either of the variable compared to the bulk value of the EM for InAs and InSb, we find that the EM in GaAs/AlGaAs structures are extremely low and therefore the mobility in superlattices are very large as compared with the value of the mobility of the constituent materials which is very important from the application point of view for modern devices made of superlattices.

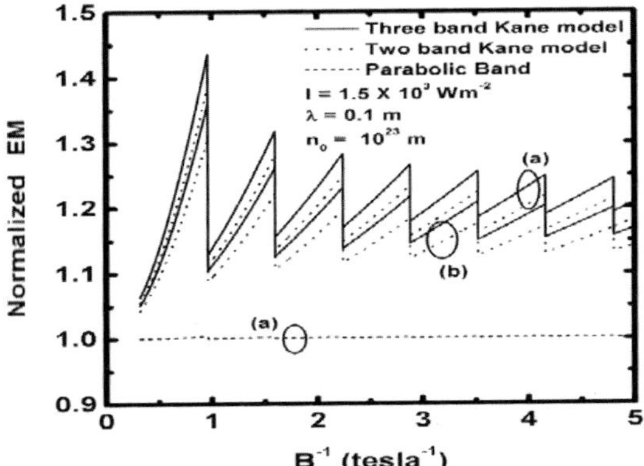

Fig. 6.21. Plot of the normalized EM for the lowest sub-band index as a function of inverse magnetic field of effective mass super-lattices of GaAs/AlGaAs in the presence of light waves in accordance with the three, the two band models of Kane and parabolic energy band model in the presence of external photo-excitation.

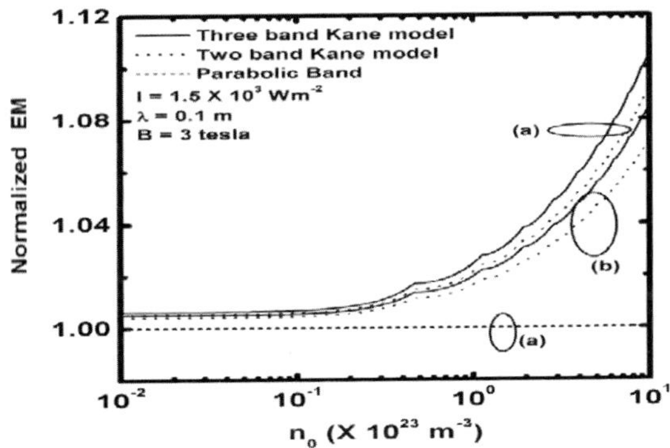

Fig. 6.22. Plot of the normalized EM for the lowest sub-band index as a function of carrier concentration of effective mass super-lattices of GaAs/AlGaAs in the presence of light waves in accordance with the three, the two band models of Kane and parabolic energy band model in the presence of external photo-excitation.

6.2.9 *The EM in NWHD effective mass superlattices of kane type semiconductors in the presence of light waves*

(a) The DR in NWHD effective mass superlattices of Kane type semiconductors in the presence of light waves, the dispersion relations of whose constituent materials in the absence of any perturbation are defined by the three band model of Kane can be written as

$$k_x^2 = \left[\frac{1}{L_0^2}\left\{\cos^{-1}\left(f_{HD1}\left(E, \frac{n_y\pi}{d_y}, \frac{n_z\pi}{d_z}, \lambda\right)\right)\right\}^{-2} \cdot \left[\left(\frac{n_y\pi}{d_y}\right)^2 + \left(\frac{n_z\pi}{d_z}\right)^2\right]\right] \quad (6.160)$$

The EM in this case assumes the form

$$m^*\left(\frac{n_y\pi}{d_y}, \frac{n_z\pi}{d_z}, E, \lambda\right) = \frac{\hbar^2}{L_0^2}\left|\left[\frac{\cos^{-1}[f_{HD1}(E, \frac{n_y\pi}{d_y}, \frac{n_z\pi}{d_z}) f_{HD1}(E, \frac{n_y\pi}{d_y}, \frac{n_z\pi}{d_z})}{\sqrt{1 - f_{HD1}^2(E, \frac{n_y\pi}{d_y}, \frac{n_z\pi}{d_z})}}\right]\right| \quad (6.161)$$

The sub-band energies ($E_{nSL5HD4}$) can be written as

$$0 = \left[\frac{1}{L_0^2}\left\{\cos^{-1}\left(f_{HD1}\left(E_{nSL5HD4}, \frac{n_y\pi}{d_y}, \frac{n_z\pi}{d_z}, \lambda\right)\right)\right\}^{-2} \cdot \left[\left(\frac{n_y\pi}{d_y}\right)^2 + \left(\frac{n_z\pi}{d_z}\right)^2\right]\right] \quad (6.162a)$$

The electron concentration per unit length is given by

$$n_0 = \frac{2g_v}{\pi}\sum_{n_y=1}^{n_{y\max}}\sum_{n_z=1}^{n_{z\max}}\left[P_{629} + \sum_{r=1}^{s} L(r)[P_{629}]\right. \quad (6.162b)$$

where

$$P_{629} = \left[\frac{1}{L_0^2}\left\{\cos^{-1}\left(f_{HD1}\left(E_{F629}, \frac{n_x\pi}{d_x}, \frac{n_y\pi}{d_y}, \lambda\right)\right)\right\}^{2} - \left[\left(\frac{n_x\pi}{d_x}\right)^2 + \left(\frac{n_y\pi}{d_y}\right)^2\right]\right]^{1/2}$$

and E_{F629} is the Fermi energy in this case.

(b) The DR in in NWHD effective mass superlattices of Kane type semiconductors in the presence of light waves, the dispersion relations of whose constituent materials in the absence of any perturbation are defined by the two band model of Kane can be written as

$$k_x^2 = \left[\frac{1}{L_0^2}\left\{\cos^{-1}\left(f_{HD2}\left(E, \frac{n_y\pi}{d_y}, \frac{n_z\pi}{d_z}, \lambda\right)\right)\right\}^{-2} \cdot \left[\left(\frac{n_y\pi}{d_y}\right)^2 + \left(\frac{n_z\pi}{d_z}\right)^2\right]\right] \quad (6.163)$$

The EM in this case assumes the form

$$m^*\left(\frac{n_y\pi}{d_y}, \frac{n_z\pi}{d_z}, E, \lambda\right) = \frac{\hbar^2}{L_0^2}\left|\left[\frac{\cos^{-1}[f_{HD2}(E, \frac{n_y\pi}{d_y}, \frac{n_z\pi}{d_z})f_{HD2}(E, \frac{n_y\pi}{d_y}, \frac{n_z\pi}{d_z})}{\sqrt{1 - f_{HD2}^2(E, \frac{n_y\pi}{d_y}, \frac{n_z\pi}{d_z})}}\right]\right| \quad (6.163)$$

The sub-band energies ($E_{nSL5HD4}$) can be written as

$$0 = \left[\frac{1}{L_0^2}\left\{\cos^{-1}\left(f_{HD2}\left(E_{nSL5HD4}, \frac{n_y\pi}{d_y}, \frac{n_z\pi}{d_z}, \lambda\right)\right)\right\}^{-2} \cdot \left[\left(\frac{n_y\pi}{d_y}\right)^2 + \left(\frac{n_z\pi}{d_z}\right)^2\right]\right] \quad (6.164a)$$

The electron concentration per unit length is given by

$$n_0 = \frac{2g_v}{\pi}\sum_{n_y=1}^{n_{y\max}}\sum_{n_z=1}^{n_{z\max}}\left[P_{630} + \sum_{r=1}^{s} L(r)[P_{630}]\right. \quad (6.165b)$$

where

$$P_{630} = \left[\frac{1}{L_0^2}\left\{\cos^{-1}\left(f_{HD2}\left(E_{F630}, \frac{n_x\pi}{d_x}, \frac{n_y\pi}{d_y}, \lambda\right)\right)\right\}^{2} - \left[\left(\frac{n_x\pi}{d_x}\right)^2 + \left(\frac{n_y\pi}{d_y}\right)^2\right]\right]^{1/2}$$

and E_{F630} is the Fermi energy in this case

(C) (b) The DR in in NWHD effective mass superlattices of Kane type semiconductors in the presence of light waves, the dispersion relations of whose constituent materials in the absence of any perturbation are defined by the parabolic energy bands can be written as

$$k_x^2 = \left[\frac{1}{L_0^2}\left\{\cos^{-1}\left(f_{HD3}\left(E, \frac{n_y\pi}{d_y}, \frac{n_z\pi}{d_z}, \lambda\right)\right)\right\}^{-2} \cdot \left[\left(\frac{n_y\pi}{d_y}\right)^2 + \left(\frac{n_z\pi}{d_z}\right)^2\right]\right] \tag{6.166}$$

The EM in this case assumes the form

$$m^*\left(\frac{n_y\pi}{d_y}, \frac{n_z\pi}{d_z}, E, \lambda\right) = \frac{\hbar^2}{L_0^2}\left|\left[\frac{\cos^{-1}[f_{HD3}(E, \frac{n_y\pi}{d_y}, \frac{n_z\pi}{d_z}) f_{HD3}(E, \frac{n_y\pi}{d_y}, \frac{n_z\pi}{d_z})}{\sqrt{1 - f_{HD3}^2(E, \frac{n_y\pi}{d_y}, \frac{n_z\pi}{d_z})}}\right]\right| \tag{6.167}$$

The sub-band energies ($E_{nSL5HD4}$) can be written as

$$0 = \left[\frac{1}{L_0^2}\left\{\cos^{-1}\left(f_{HD3}\left(E_{nSL5HD4}, \frac{n_y\pi}{d_y}, \frac{n_z\pi}{d_z}, \lambda\right)\right)\right\}^{-2} \cdot \left[\left(\frac{n_y\pi}{d_y}\right)^2 + \left(\frac{n_z\pi}{d_z}\right)^2\right]\right] \tag{6.168a}$$

The electron concentration per unit length is given by

$$n_0 = \frac{2g_v}{\pi}\sum_{n_y=1}^{n_{y\max}}\sum_{n_z=1}^{n_{z\max}}\left[P_{631} + \sum_{r=1}^{s} L(r)[P_{631}]\right] \tag{6.168b}$$

where

$$P_{631} = \left[\frac{1}{L_0^2}\left\{\cos^{-1}\left(f_{HD3}\left(E_{F629}, \frac{n_x\pi}{d_x}, \frac{n_y\pi}{d_y}, \lambda\right)\right)\right\}^2 - \left[\left(\frac{n_x\pi}{d_x}\right)^2 + \left(\frac{n_y\pi}{d_y}\right)^2\right]\right]^{1/2}$$

and E_{F631} is the Fermi energy in this case.

6.2.9.1 *Result and discussions*

Using the values of the energy band constants from Appendix A, we have plotted the EM for the $n_y = 1$ and $n_z = 1$ as a function of the film thickness at $T = 4.2\,K$ by taking quantum wire effective mass super lattices of optoelectronic materials in accordance with the perturbed three (using (6.143) and (6.144)), two (using (6.145) and (6.146)) band models of Kane and that of perturbed parabolic energy bands (using (6.161) and (6.162)) respectively in Fig 6.23.

Fig. 6.23 exhibits the variation of EM in the quantum wire effective mass super-lattices of GaAs/AlGaAs by considering the quantum limit approximation. The EM is greatest for the lowest sub-bands and for higher sub-bands the numerical values of the EMs will be less. It appears that the EM in such structure decreases with the increase in the film thickness in a non-linear way for the three and the two band energy model. Although, it appears that the EM in case of parabolic energy band is linear, however, depends on the photo-excitation factor, which makes the slow variation of EM with both intensity and wavelength.

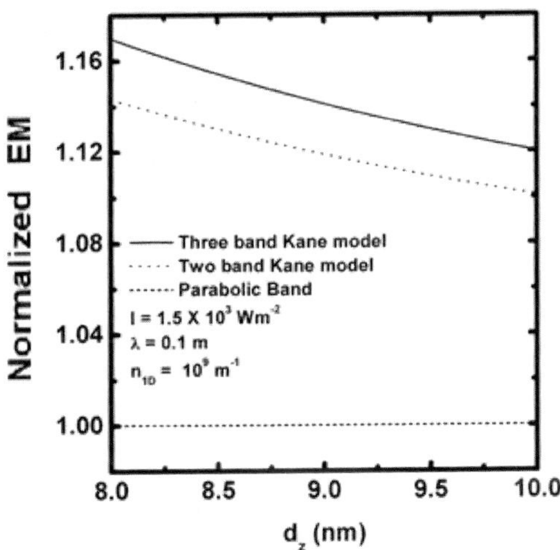

Fig. 6.23. Plot of the normalized EM ($n_y = 1$ and $n_z = 1$) as a function of lateral film thickness of quantum wire effective mass super-lattices of GaAs/AlGaAs in the presence of light waves in accordance with the three, the two band models of Kane and parabolic energy band model in the presence of external photo-excitation.

6.2.10 *The magneto EM in HD effective mass superlattices of kane type semiconductors in the presence of light waves*

(a) In the presence of an external magnetic field along x-direction, the simplified magneto dispersion law in this case can be written as

$$k_x^2 = [\rho_{4HD1}(n, E, \lambda)] \tag{6.169}$$

in which,

$$\rho_{4HD1}(n, E, \lambda) = \frac{1}{L_0^2}\left[\cos^{-1}(f_{HD1}(n, E, \lambda))\right]^2 - \left\{\frac{2|e|B}{\hbar}\left(n + \frac{1}{2}\right)\right\},$$

$$f_{HD1}(E, n, \lambda) = a_{1HD1}\cos[a_0 C_{1HD1}(E, n, \lambda) + b_0 D_{1HD1}(E, n, \lambda)] - a_{2HD1}\cos[a_0 C_{1HD1}(E, n, \lambda) - b_0 D_{1HD1}(E, n, \lambda)]$$

$$C_{1HD1}(E, n, \lambda) \equiv \left[\left(\frac{2m_{c1}}{\hbar^2}\right) T_1(E, \eta_{g_1}, \lambda) - \left\{\frac{2|e|B}{\hbar}\left(n + \frac{1}{2}\right)\right\}\right]^{1/2}$$

and

$$D_{1HD1}(E, n, \lambda) \equiv \left[\left(\frac{2m_{c2}}{\hbar^2}\right) T_1(E, \eta_{g_2}, \lambda) - \left\{\frac{2|e|B}{\hbar}\left(n + \frac{1}{2}\right)\right\}\right]^{1/2}.$$

The EM in this case assumes the form

$$m^*(n, E_{fSLHDB}, \lambda) = \text{Real part of } \frac{\hbar^2}{2}[p_{4HD}(n, E_{fSLHDB}, \lambda)]' \tag{6.170}$$

where the Landau sub-band energies (E_{fSLHDB}) can be written as

The EM in III–V EMSLs under magnetic quantization depends on both, the Fermi energy, magnetic quantum number and wavelength which is the intrinsic property of such SLs.

The DOS function is given by

$$N(E) = \frac{eB}{2\pi^2\hbar}\sum_{n=0}^{n_{\max}} \frac{p'_{4HD1}(n, E, \lambda) H(E - E_{nSL5HD})}{\sqrt{P_{4HD1}(n, E, \lambda)}} \tag{6.171}$$

where the Landau sub-band energies (E_{nSL5HD}) can be written a

$$\rho_{4HD1}(n, E_{fSL5HD}, \lambda) = 0 \tag{6.172}$$

The electron concentration is given by

$$n_0 = \frac{eB}{\pi^2 \hbar} \text{ Real part of } \sum_{n=0}^{n_{\max}} [\sqrt{P_{4HD1}(n, E_{fSlHDB}, \lambda)} + \sum_{r=1}^{s} L(r)[\sqrt{P_{4HD1}(n, E_{fSlHDB}, \lambda)}]] \tag{6.173}$$

(b) In the presence of an external magnetic field along x-direction, the simplified magneto dispersion law in this case can be written as

$$k_x^2 = [\rho_{4HD2}(n, E, \lambda)] \tag{6.174}$$

in which,

$$\rho_{4HD2}(n, E, \lambda) = \frac{1}{L_0^2}\left[\cos^{-1}\left(f_{HD2}(n, E, \lambda)\right)\right]^2 - \left\{\frac{2|e|B}{\hbar}\left(n + \frac{1}{2}\right)\right\},$$

$$f_{HD2}(E, n, \lambda) = a_{1HD2}\cos[a_0 C_{1HD2}(E, n, \lambda) + b_0 D_{1HD2}(E, n, \lambda)] - a_{2HD2}\cos[a_0 C_{1HD2}(E, n, \lambda) - b_0 D_{1HD2}(E, n, \lambda)]$$

$$C_{1HD2}(E, n, \lambda) \equiv \left[\left(\frac{2m_{c1}}{\hbar^2}\right) T_2(E, \eta_{g_1}, \lambda) - \left\{\frac{2|e|B}{\hbar}\left(n + \frac{1}{2}\right)\right\}\right]^{1/2}$$

and

$$D_{1HD2}(E, n, \lambda) = \left[\left(\frac{2m_{c2}}{\hbar^2}\right) T_2(E, \eta_{g_2}, \lambda) - \left\{\frac{2|e|B}{\hbar}\left(n + \frac{1}{2}\right)\right\}\right]^{1/2}.$$

The EM in this case assumes the form

$$m^*(n, E_{fSLHDB}, \lambda) = \frac{\hbar^2}{2}[p_{4HD2}(n, E_{fSLHDB}, \lambda)]' \tag{6.175}$$

where (E_{fSLHDB}) is the Fermi energy in this case.

The EM in III–V EMSLs under magnetic quantization depends on both the Fermi energy, magnetic quantum number and wavelength which is the intrinsic property of such SLs.

The DOS function is given by

$$N(E) = \frac{eB}{2\pi^2\hbar} \sum_{n=0}^{n_{\max}} \frac{p'_{4HD1}(n,E,\lambda)H(E-E_{nSL5HD2})}{\sqrt{P_{4HD2}(n,E,\lambda)}} \tag{6.176}$$

where the Landau sub-band energies $(E_{fSL5HD2})$can be written as

$$\rho_{4HD2}(n, E_{fSL5HD2}, \lambda) = 0 \tag{6.177}$$

The electron concentration is given by

$$n_0 = \frac{eB}{\pi^2\hbar} \text{ Real part of } \sum_{n=0}^{n_{\max}} [\sqrt{P_{4HD2}(n, E_{fSlHDB}, \lambda)}$$

$$+ \sum_{r=1}^{s} L(r)[\sqrt{P_{4HD2}(n, E_{fSlHDB}, \lambda)}]] \tag{6.178}$$

(C) In the presence of an external magnetic field along x-direction, the simplified magneto dispersion law in this case can be written as

$$k_x^2 = [\rho_{4HD3}(n,E,\lambda)] \tag{6.179}$$

in which,

$$\rho_{4HD3}(n,E,\lambda) = \frac{1}{L_0^2}[\cos^{-1}(f_{HD3}(n,E,\lambda))]^2 - \left\{\frac{2|e|B}{\hbar}\left(n+\frac{1}{2}\right)\right\},$$

$$f_{HD3}(E,n,\lambda) = a_{1HD23}\cos[a_0C_{1HD3}(E,n,\lambda) + b_0D_{1HD3}(E,n,\lambda)]$$
$$-a_{2HD3}\cos[a_0C_{1HD3}(E,n,\lambda) - b_0D_{1HD3}(E,n,\lambda)]$$

$$C_{1HD3}(E,n,\lambda) \equiv \left[\left(\frac{2m_{c1}}{\hbar^2}\right) T_3(E,\eta_{g_1},\lambda) - \left\{\frac{2|e|B}{\hbar}\left(n+\frac{1}{2}\right)\right\}\right]^{1/2}$$

and

$$D_{1HD3}(E,n,\lambda) \equiv \left[\left(\frac{2m_{c2}}{\hbar^2}\right) T_3(E,\eta_{g_2},\lambda) - \left\{\frac{2|e|B}{\hbar}\left(n+\frac{1}{2}\right)\right\}\right]^{1/2}.$$

The EM in this case assumes the form

$$m^*(n, E_{fSLHDB}, \lambda) = \frac{\hbar^2}{2}[p_{4HD3}(n, E_{fSLHDB}, \lambda)]' \tag{6.180}$$

where E_{fSLHDB} is the Fermi energy in this case.

The EM in III–V EMSLs under magnetic quantization depends on both the Fermi energy, magnetic quantum number and wavelength which is the intrinsic property of such SLs.

The DOS function is given by

$$N(E) = \frac{eB}{2\pi^2\hbar}\sum_{n=0}^{n_{\max}}\frac{p'_{4HD3}(n,E,\lambda)H(E-E_{nSL5HD3})}{\sqrt{P_{4HD3}(n,E,\lambda)}} \tag{6.181}$$

where the Landau sub-band energies ($E_{fSL5HD3}$)can be written as

$$\rho_{4HD3}(n,E_{fSL5HD3},\lambda) = 0 \tag{6.182}$$

The electron concentration is given by

$$n_0 = \frac{eB}{\pi^2\hbar}\text{ Real part of }\sum_{n=0}^{n_{\max}}[\sqrt{P_{4HD3}(n,E_{fSlHDB},\lambda)} + \sum_{r=1}^{s}L(r)[\sqrt{P_{4HD3}(n,E_{fSlHDB},\lambda)}]] \tag{6.183}$$

6.2.11 *The EM in QWHD superlattices of kane type semiconductors with graded interfaces in the presence of light waves*

The electron dispersion law in bulk specimens of the heavily doped constituent materials of III–V SLs whose energy band structures are defined by (17.46b) be expressed as

$$\frac{\hbar^2k^2}{2m_{cj}} = V_{1j}(E,\eta_{gj},\lambda,\Delta_j,E_{g0j}) + iV_{2j}(E,\eta_{gj},\lambda,\Delta_j,E_{g0j}) \tag{6.184}$$

where

$$j=1,2,\quad V_{1j}(E,\eta_{gj},\lambda,\Delta_j,E_{g0j}) = [U_{\lambda j}T_{1j}(E,\Delta_j,E_{gj},\eta_{gj}) - P_{\lambda j}]$$

$$U_{\lambda j} = (1+\theta_{\lambda j}),\quad \theta_{\lambda j} = \frac{C_{0j}}{A_j}\left(t_{\lambda j} + \frac{B_jJ_{\lambda j}}{A_j}\right),$$

$$C_{0j} = \left[\frac{|e|^2I\lambda^2E_{g0j}(E_{gj}+\Delta_j)\beta_j^2\left(1+\frac{\rho_j}{\sqrt{2}}\right)^2}{384m_{rj}\pi c^3\sqrt{\varepsilon_{scj}\varepsilon_0}\left(E_{g0j}+\frac{2}{3}\Delta_j\right)}\right],$$

$$\beta_j = \left[6\left(E_{g0j} + \frac{2}{3}\Delta_j\right)\frac{(E_{g0j} + \Delta_j)}{\chi_j}\right]^{1/2},$$

$$\chi_j = (6E_{g0}^2 + 9E_{g0j}\Delta_j + 4\Delta_j^2), \quad \rho_j = \left[\frac{4\Delta_j^2}{3\chi_j}\right]^{1/2}, \quad A_j = E_{g0j},$$

$$t_{\lambda j} = [E_{\lambda j} - \frac{G_{\lambda j}(E_{g0j} - \delta'_j)}{\sqrt{t_{\lambda j}}}], \quad E_{\lambda j} = \frac{2B_j(E_{g0j} - \delta'_j)}{\left(A_j + \delta'_j\right)^2},$$

$$B_j = \left[1 + \frac{m_{cj}}{m_{vj}}\right], \quad G_{\lambda j} = \left[\frac{2B_j}{(A_j + \delta'_j)^3} - \frac{B_j C_{\lambda j}}{(A_j + \delta'_j)}\right],$$

$$C_{\lambda j} = [(E_{g0j} + \delta'_j)^{-1} + (E_{g0j} + \delta'_j)(E_{g0j} - \delta'_j)^{-2}](A_j + \delta'_j)^{-1},$$

$$P_{\lambda j} = \frac{C_{0j}}{A_j} J_{\lambda j}, \quad J_{\lambda j} = (D_{\lambda j} + 2(E_{g0j} - \delta'_j)\sqrt{f_{\lambda j}}),$$

$$D_{\lambda j} = \left(1 + \frac{2(E_{g0j} - \delta'_j)}{(A_j + \delta'_j)}\right),$$

$$f_{\lambda j} = \left[\frac{1}{(A_j + \delta'_j)^2} + \frac{1}{(E_{g0j} - \delta'_j)^2} - C_{\lambda j}\right],$$

$$T_{1j}(E, \Delta_j, E_{gj}, \eta_{gj}) = (2/(1 + Erf(E/\eta_{gj}))[(\alpha_j b_j/c_j) \cdot \theta_0(E, \eta_{gj}) + [\alpha_j c_j + b_j c_j - \alpha_j b_j/c_j^2].$$

$$\gamma_0(E, \eta_{gj}) + [(1/c_j)(1 - (\alpha_j/c_j))(1 - b_j/c_j))\frac{1}{2}[1 + Erf(E/\eta_{gj})] - (1/c_j)(1 - (\alpha_j/c_j))(1 - b_j/c_j))(2/c_j\eta_{gj}\sqrt{\pi})\exp(-u_j^2) \cdot \left[\sum_{p=1}^{\infty}(\exp(-p^2/4)/p)\sinh(pu_j)\right]\Bigg],$$

$$b_j \equiv (E_{gj} + \Delta_j)^{-1}, \quad c_j \equiv \left(E_{gj} + \frac{2}{3}\Delta_j\right)^{-1}, \quad u_j \equiv (1 + c_j E)(c_j\eta_{gj})^{-1};$$

$$V_{2j}(E, \eta_{gj}, \lambda, \Delta_j, E_{g0j}) = [U_{\lambda j} T_{2j}(E, \Delta_j, E_{gj}, \eta_{gj})]$$

and

$$T_{2j}(E, \Delta_j, E_{gj}, \eta_{gj})$$
$$\equiv \left(\frac{2}{(1 + Erf(E/\eta_{gj})}\right) \frac{1}{c_j} \left(1 - \frac{\alpha_j}{c_j}\right) \left(1 - \frac{b_j}{c_j}\right) \frac{\sqrt{\pi}}{c_j \eta_{gj}} \exp(-u_j^2)$$

Therefore, the DR in HD III–V SLs with graded interfaces in the presence of light waves can be expressed as [25]

$$k_z^2 = G_8 + iH_8 \tag{6.185}$$

where

$$G_8 = \left[\frac{C_7^2 - D_7^2}{L_0^2} - k_s^2\right], \quad C_7 = \cos^{-1}(\bar{\omega}_7),$$

$$\bar{\omega}_7 = (2)^{\frac{-1}{2}}[(1 - G_7^2 - H_7^2) - \sqrt{(1 - G_7^2 - H_7^2)^2 + 4G_7^2}]^{\frac{1}{2}},$$

$$G_7 = [G_1 + (\rho_5 G_2/2) - (\rho_6 H_2/2) + (\Delta_0/2)\{\rho_6 H_2 - \rho_8 H_3 + \rho_9 H_4 - \rho_{10} H_4 + \rho_{11} H_5 - \rho_{12} H_5 + (1/12)(\rho_{12} G_6 - \rho_{14} H_6)\}],$$

$$G_1 = [(\cos(h_1))(\cosh(h_2))(\cosh(g_1))(\cos(g_2)) + (\sin(h_1))(\sinh(h_2))(\sinh(g_1))(\sin(g_2))],$$

$$h_1 = e_1(b_0 - \Delta_0), \quad e_1 = 2^{\frac{-1}{2}}(\sqrt{t_1^2 + t_2^2} + t_1)^{\frac{1}{2}},$$

$$t_1 = [(2m_{c1}/\hbar^2) \cdot V_{11}(E, E_{g1}, \lambda, \Delta_1, \eta_{g1}) - k_s^2],$$

$$t_2 = [(2m_{c1}/\hbar^2)V_{21}(E, E_{g1}, \lambda, \Delta_1, \eta_{g1})],$$

$$h_2 = e_2(b_0 - \Delta_0), \quad e_2 = 2^{\frac{-1}{2}}(\sqrt{t_1^2 + t_2^2} - t_1)^{\frac{1}{2}}, \quad g_1 = d_1(a_0 - \Delta_0),$$

$$d_1 = 2^{\frac{-1}{2}}(\sqrt{x_1^2 + y_1^2} + x_1)^{\frac{1}{2}},$$

$$x_1 = [-(2m_{c2}/\hbar^2) \cdot V_{11}(E - V_0, E_{g2}, \lambda, \Delta_2, \eta_{g2}) + k_s^2],$$

$$y_1 = [(2m_{c2}/\hbar^2) \cdot V_{22}(E - V_0, E_{g2}, \lambda, \Delta_2, \eta_{g2})], \quad g_2 = d_2(a_0 - \Delta_0),$$

$$d_2 = 2^{\frac{-1}{2}}(\sqrt{x_1^2 + y_1^2} - x_1)^{\frac{1}{2}}, \quad \rho_5 = (\rho_3^2 + \rho_4^2)^{-1}[\rho_1\rho_1 - \rho_2\rho_4],$$

$$\rho_1 = [d_1^2 + e_2^2 - d_2^2 - e_1^2], \quad \rho_3 = [d_1 e_1 + d_2 e_2], \quad \rho_2 = 2[d_1 d_2 + e_1 e_2],$$

$$\rho_4 = [d_1 e_2 - e_1 d_2],$$

$$\begin{aligned}
G_2 &= [(\sin(h_1))(\cosh(h_2))(\sinh(g_1))(\cos(g_2)) \\
&\quad +(\cos(h_1))(\sinh(h_2))(\cosh(g_1))(\sin(g_2))], \\
\rho_6 &= (\rho_3^2 + \rho_4^2)^{-1}[\rho_1\rho_4 + \rho_2\rho_3], \\
H_2 &= [(\sin(h_1))(\cos(h_2))(\sin(g_2))(\cosh(g_1)) \\
&\quad -(\cos(h_1))(\sinh(h_2))(\sinh(g_1))(\cos(g_2))], \\
\rho_7 &= [(e_1^2 + e_2^2)^{-1}[e_1(d_1^2 - d_2^2) - 2d_1 d_2 e_2] - 3e_1], \\
G_3 &= [(\sin(h_1))(\cosh(h_2))(\cosh(g_1))(\cos(g_2)) \\
&\quad + (\cos(h_1))(\sinh(h_2))(\sinh(g_1))(\sin(g_2))], \\
\rho_8 &= [(e_1^2 + e_2^2)^{-1}[e_2(d_1^2 - d_2^2) + 2d_1 d_2 e_1] + 3e_2], \\
H_3 &= [(\sin(h_1))(\cosh(h_2))(\sin(g_2))(\sinh(g_1)) \\
&\quad - (\cos(h_1))(\sinh(h_2))(\cosh(g_1))(\cos(g_2))], \\
\rho_9 &= [(d_1^2 + d_2^2)^{-1}[d_1(e_2^2 - e_1^2) + 2e_2 d_2 e_1] + 3d_1], \\
G_4 &= [(\cos(h_1))(\cosh(h_2))(\cos(g_2))(\sinh(g_1)) \\
&\quad - (\sin(h_1))(\sinh(h_2))(\cosh(g_1))(\sin(g_2))], \\
\rho_{10} &= [-(d_1^2 + d_2^2)^{-1}[d_2(-e_2^2 + e_1^2) + 2e_2 d_2 e_1] + 3d_2], \\
H_4 &= [(\cos(h_1))(\cosh(h_2))(\cosh(g_1))(\sin(g_2)) \\
&\quad + (\sin(h_1))(\sinh(h_2))(\sinh(g_1))(\cos(g_2))], \\
\rho_{11} &= 2[d_1^2 + e_2^2 - d_2^2 - e_1^2], \\
G_5 &= [(\cos(h_1))(\cosh(h_2))(\cos(g_2))(\cosh(g_1)) \\
&\quad - (\sin(h_1))(\sinh(h_2))(\sinh(g_1))(\sin(g_2))],
\end{aligned}$$

$$\rho_{12} = 4[d_1 d_2 + e_1 e_2],$$
$$H_5 = [(\cos(h_1))(\cosh(h_2))(\sinh(g_1))(\sin(g_2)) + (\sin(h_1))(\sinh(h_2))(\cosh(g_1))(\cos(g_2))],$$

$$\rho_{13} = [\{5(d_1 e_1^3 - 3e_1 e_2^2 d_1) + 5d_2(e_1^3 - 3e_1^2 e_2)\}(d_1^2 + d_2^2)^{-1} + (e_1^2 + e_2^2)^{-1}\{5(e_1 d_1^3 - 3d_2 e_1^2 d_1) + 5(d_2^3 e_2 - 3d_1^2 d_2 e_2)\} - 34(d_1 e_1 + d_2 e_2)]$$

$$G_6 = [(\sin(h_1))(\cosh(h_2))(\sinh(g_1))(\cos(g_2)) + (\cos(h_1))(\sinh(h_2))(\cosh(g_1))(\sin(g_2))],$$

$$\rho_{14} = [\{5(d_1 e_2^3 - 3e_2 e_1^2 d_1) + 5d_2(-e_1^3 + 3e_2^2 e_1)\}(d_1^2 + d_2^2)^{-1} + (e_1^2 + e_2^2)^{-1}\{5(-e_1 d_2^3 + 3d_1^2 d_2 e_1) + 5(-d_1^3 e_2 + 3d_2^2 d_1 e_2)\} + 34(d_1 e_2 - d_2 e_1)],$$

$$H_6 = [(\sin(h_1))(\cosh(h_2))(\cosh(g_1))(\sin(g_2)) - (\cos(h_1))(\sinh(h_2))(\sinh(g_1))(\cos(g_2))],$$

$$H_7 = [H_1 + (\rho_5 H_2/2) + (\rho_6 G_2/2) + (\Delta_0/2)\{\rho_8 G_3 + \rho_7 H_3 + \rho_{10} G_4 + \rho_9 H_4 + \rho_{12} G_5 + \rho_{11} H_5 + (1/12)(\rho_{14} G_6 + \rho_{13} H_6)\}],$$

$$H_1 = [(\sin(h_1))(\sinh(h_2))(\cosh(g_1))(\cos(g_2)) + (\cos(h_1))(\cosh(h_2))(\sinh(g_1))(\sin(g_2))],$$

$$D_7 = \sinh^{-1}(\bar{\omega}_7),\ H_8 = (2C_7 D_7/L_0^2)$$

The simplified DR of HD QWs of III–V super-lattices with graded interfaces can be expressed as

$$\left(\frac{n_z \pi}{d_z}\right)^2 = G_8 + iH_8 \tag{6.186}$$

The sub-band equation in this case can be expressed as

$$\left(\frac{n_z \pi}{d_z}\right)^2 = |G_8 + iH_8|_{k_s=0 \text{ and } E=E_{17,100}} \tag{6.187}$$

where $E_{17,100}$ is the sub-band energy in this case.

The EM and the DOS function should be obtained numerically in this case.

6.2.12 *The EM in NWHD superlattices of kane type semiconductors with graded interfaces in the presence of light waves*

The DR in NWHD superlattices of Kane type semiconductors with graded interfaces in the presence of light waves can be expressed as

$$k_z^2 = G_{8,17,50} + iH_{8,17,50} \tag{6.188}$$

where

$$G_{8,17,50} = \left[\frac{C_{7,50}^2 - D_{7,50}^2}{L_0^2} - \left[\left(\frac{n_x\pi}{d_x}\right)^2 + \left(\frac{n_y\pi}{d_y}\right)^2\right]\right],$$

$$C_{7,50} = \cos^{-1}(\bar{\omega}_{7,50}),$$

$$\bar{\omega}_{7,50} = (2)^{\frac{-1}{2}}[(1 - G_{7,50}^2 - H_{7,50}^2) - \sqrt{(1 - G_{7,50}^2 - H_{7,50}^2)^2 + 4G_{7,50}^2}]^{\frac{1}{2}},$$

$$\begin{aligned}G_{7,50} = {} & [G_{1,50} + (\rho_{5,,50}G_{2,,50}/2) - (\rho_{6,50}H_{2,50}/2) + (\Delta_0/2) \\ & \{\rho_{6,50}H_{2,50} - \rho_{8,50}H_{3,50} + \rho_{9,50}H_{4,50} - \rho_{10,50}H_{4,50} \\ & + \rho_{11,50}H_{5,50} - \rho_{12,50}H_{5,50} \\ & + (1/12)(\rho_{12,50}G_{6,50} - \rho_{14,50}H_{6,50})\}],\end{aligned}$$

$$\begin{aligned}G_{1,50} = {} & [(\cos(h_{1,50}))(\cosh(h_{2,50}))(\cosh(g_{1,50}))(\cos(g_{2,50})) \\ & +(\sin(h_{1,50}))(\sinh(h_{2,50}))(\sinh(g_{1,50}))(\sin(g_{2,50}))],\end{aligned}$$

$$h_{1,50} = e_{1,50}(b_0 - \Delta_0), \quad e_{1,50} = 2^{\frac{-1}{2}}(\sqrt{t_{1,50}^2 + t_{2,50}^2} + t_{1,50})^{\frac{1}{2}},$$

$$\begin{aligned}t_{1,50} = {} & \Big[(2m_{c1}/\hbar^2)\cdot V_{11}(E, E_{g1}, \lambda, \Delta_1, \eta_{g1}) \\ & - \left[\left(\frac{n_x\pi}{d_x}\right)^2 + \left(\frac{n_y\pi}{d_y}\right)^2\right]\Big],\end{aligned}$$

$$t_{2,50} = [(2m_{c1}/\hbar^2)V_{21}(E, E_{g1}, \lambda, \Delta_1, \eta_{g1})],$$

$$h_{2,50} = e_{2,50}(b_0 - \Delta_0), \quad e_{2,50} = 2^{\frac{-1}{2}}(\sqrt{t_{1,50}^2 + t_{2,50}^2} - t_{1,50})^{\frac{1}{2}},$$

$$g_{1,50} = d_{1,50}(a_0 - \Delta_0), \quad d_{1,50} = 2^{\frac{-1}{2}}(\sqrt{x_{1,50}^2 + y_{1,50}^2} + x_{1,50})^{\frac{1}{2}},$$

$$x_{1,50} = \left[-(2m_{c2}/\hbar^2) \cdot V_{11}(E - V_0, E_{g2}, \lambda, \Delta_2, \eta_{g2}) \right.$$
$$\left. + \left[\left(\frac{n_x\pi}{d_x}\right)^2 + \left(\frac{n_y\pi}{d_y}\right)^2\right]\right],$$

$$y_{1,50} = [(2m_{c2}/\hbar^2) \cdot V_{22}(E - V_0, E_{g2}, \lambda, \Delta_2, \eta_{g2})],$$

$$g_{2,50} = d_{2,50}(a_0 - \Delta_0), \quad d_{2,50} = 2^{\frac{-1}{2}}(\sqrt{x_{1,50}^2 + y_{1,50}^2} - x_{1,50})^{\frac{1}{2}},$$

$$\rho_{5,50} = (\rho_{3,50}^2 + \rho_{4,50}^2)^{-1}[\rho_{1,50}\rho_{1,50} - \rho_{2,50}\rho_{4,50}],$$

$$\rho_{1,50} = [d_{1,50}^2 + e_{2,50}^2 - d_{2,50}^2 - e_{1,50}^2], \quad \rho_{3,50} = [d_{1,50}e_{1,50} + d_{2,50}e_{2,50}],$$

$$\rho_{2,50} = 2[d_{1,50}d_{2,50} + e_{1,50}e_{2,50}], \quad \rho_{4,50} = [d_{1,50}e_{2,50} - e_{1,50}d_{2,50}],$$

$$G_{2,50} = [(\sin(h_{1,50}))(\cosh(h_{2,50}))(\sinh(g_{1,50}))(\cos(g_{2,50}))$$
$$+(\cos(h_{1,50}))(\sinh(h_{2,50}))(\cosh(g_{1,50}))(\sin(g_{2,50}))],$$

$$\rho_{6,50} = (\rho_{3,50}^2 + \rho_{4,50}^2)^{-1}[\rho_{1,50}\rho_{4,50} + \rho_{2,50}\rho_{3,50}],$$

$$H_{2,50} = [(\sin(h_{1,50}))(\cos(h_{2,50}))(\sin(g_{2,50}))(\cosh(g_{1,50}))$$
$$-(\cos(h_{1,50}))(\sinh(h_{2,50}))(\sinh(g_{1,50}))(\cos(g_{2,50}))],$$

$$\rho_{7,50} = [(e_{1,50}^2 + e_{2,50}^2)^{-1}[e_{1,50}(d_{1,50}^2 - d_{2,50}^2) - 2d_{1,50}d_{2,50}e_{2,50}] - 3e_{1,50}],$$

$$G_{3,50} = [(\sin(h_{1,50}))(\cosh(h_{2,50}))(\cosh(g_{1,50}))(\cos(g_{2,50}))$$
$$+(\cos(h_{1,50}))(\sinh(h_{2,50}))(\sinh(g_{1,50}))(\sin(g_{2,50}))],$$

$$\rho_{8,50} = [(e_{1,50}^2 + e_{2,50}^2)^{-1}[e_{2,50}(d_{1,50}^2 - d_{2,50}^2) + 2d_{1,50}d_{2,50}e_{1,50}] + 3e_{2,50}],$$

$$H_{3,50} = [(\sin(h_{1,50}))(\cosh(h_{2,50}))(\sin(g_{2,50}))(\sinh(g_{1,50}))$$
$$-(\cos(h_{1,50}))(\sinh(h_{2,50}))(\cosh(g_{1,50}))(\cos(g_{2,50}))],$$

$$\rho_{9,50} = [(d_{1,50}^2+d_{2,50}^2)^{-1}[d_{1,50}(e_{2,50}^2-e_{1,50}^2)+2e_{2,50}d_{2,50}e_{1,50}]+3d_{1,50}],$$

$$G_{4,50} = [(\cos(h_{1,50}))(\cosh(h_{2,50}))(\cos(g_{2,50}))(\sinh(g_{1,50})) - (\sin(h_{1,50}))(\sinh(h_{2,50}))(\cosh(g_{1,50}))(\sin(g_{2,50}))],$$

$$\rho_{10,50} = [-(d_{1,50}^2 + d_{2,50}^2)^{-1}[d_{2,50}(-e_{2,50}^2 + e_{1,50}^2) + 2e_{2,50}d_{2,50}e_{1,50}] + 3d_{2,50}],$$

$$H_{4,50} = [(\cos(h_{1,50}))(\cosh(h_{2,50}))(\cosh(g_{1,50}))(\sin(g_{2,50})) + (\sin(h_{1,50}))(\sinh(h_{2,50}))(\sinh(g_{1,50}))(\cos(g_{2,50}))],$$

$$\rho_{11,50} = 2[d_{1,50}^2 + e_{2,50}^2 - d_{2,50}^2 - e_{1,50}^2],$$

$$H_{5,50} = [(\cos(h_{1,50}))(\cosh(h_{2,50}))(\sinh(g_{1,50}))(\sin(g_{2,50})) + (\sin(h_{1,50}))(\sinh(h_{2,50}))(\cosh(g_{1,50}))(\cos(g_{2,50}))],$$

$$\rho_{13,50} = [\{5(d_{1,50}e_{1,50}^3 - 3e_{1,50}e_{2,50}^2d_{1,50}) + 5d_{2,50}(e_{1,50}^3 - 3e_{1,50}^2e_{2,50})\} \cdot(d_{1,50}^2 + d_{2,50}^2)^{-1} + (e_{1,50}^2 + e_{2,50}^2)^{-1}\{5(e_{1,50}d_{1,50}^3 - 3d_{2,50}e_{1,50}^2d_{1,50}) + 5(d_{2,50}^3e_{2,50} - 3d_{1,50}^2d_{2,50}e_{2,50})\} - 34(d_{1,50}e_{1,50} + d_{2,50}e_{2,50})],$$

$$G_{6,50} = [(\sin(h_{1,50}))(\cosh(h_{2,50}))(\sinh(g_{1,50}))(\cos(g_{2,50})) + (\cos(h_{1,50}))(\sinh(h_{2,50}))(\cosh(g_{1,50}))(\sin(g_{2,50}))],$$

$$\rho_{14,50} = [\{5(d_{1,50}e_{2,50}^3 - 3e_{2,50}e_{1,50}^2d_{1,50}) + 5d_{2,50}(-e_{1,50}^3 +3e_{2,50}^2e_{1,50})\}(d_{1,50}^2 + d_{2,50}^2)^{-1} + (e_{1,50}^2 + e_{2,50}^2)^{-1} \cdot\{5(-e_{1,50}d_{2,50}^3 + 3d_{1,50}^2d_{2,50}e_{1,50}) + 5(-d_{1,50}^3e_{2,50} +3d_{2,50}^2d_{1,50}e_{2,50})\} + 34(d_{1,50}e_{2,50} - d_{2,50}e_{1,50})],$$

$$H_{6,50} = [(\sin(h_{1,50}))(\cosh(h_{2,50}))(\cosh(g_{1,50}))(\sin(g_{2,50})) - (\cos(h_{1,50}))(\sinh(h_{2,50}))(\sinh(g_{1,50}))(\cos(g_{2,50}))],$$

$$H_{7,50} = [H_{1,50} + (\rho_{5,50}H_{2,50}/2) + (\rho_{6,50}G_{2,50}/2) + (\Delta_0/2) \{\rho_{8,50}G_{3,50} + \rho_{7,50}H_{3,50} + \rho_{10,50}G_{4,50} + \rho_{9,50}H_{4,50}$$

$$+ \rho_{12,50} G_{5,50} + \rho_{11,50} H_{5,50}$$

$$+ (1/12)(\rho_{14,50} G_{6,50} + \rho_{13,50} H_{6,50})\}],$$

$$H_{1,50} = [(\sin(h_{1,50}))(\sinh(h_{2,50}))(\cosh(g_{1,50}))(\cos(g_{2,50})) + (\cos(h_{1,50}))(\cosh(h_{2,50}))(\sinh(g_{1,50}))(\sin(g_{2,50}))],$$

$$D_{7,50} = \sinh^{-1}(\bar{\omega}_{7,50}),\ H_{8,50} = (2C_{7,50} D_{7,50}/L_0^2)$$

The sub-band equation in this case can be expressed as

$$0 = [G_{8,17,50} + iH_{8,17,50}]|_{E=E_{17,51}} \tag{6.189}$$

where $E_{17,51}$ is the sub-band energy in this case.

At low temperatures where the quantum effects become prominent, the DOS function for the lowest SL mini-band is given by

$$N_{HDSL}(E, \eta_g, \lambda) = \frac{g_v}{\pi} \sum_{n_x=1}^{n_{x\max}} \sum_{n_y=1}^{n_{y\max}} \frac{[G'_{8,17,50} + iH'_{8,17,50}]}{\sqrt{G_{8,17,50} + iH_{8,17,50}}} H(E - E_{17,51}) \tag{6.190}$$

The EM can be written as

$$m^*(E, n_x, n_y, \lambda, \eta_g) = \frac{\hbar^2}{2}(G'_{8,17,50}) \tag{6.191a}$$

The electron concentration is given by

$$n_0 = \frac{2g_v}{\pi} \text{ Real Part of } \sum_{n_x=1}^{n_{x\max}} \sum_{n_x=1}^{n_{y\max}} \left[(\sqrt{G_{8.1750} + iH_{8.1750}})\,|_{E=E_{F662}} + \sum_{r=1}^{s} L(r)(\sqrt{G_{8.1750} + iH_{8.1750}})\,|_{E=E_{F662}} \right] \tag{6.192b}$$

where E_{F662} is the Fermi energy in this case.

6.2.12.1 *Result and discussion*

Using the values of the energy band constants from Appendix A, we have plotted the EM for $n_y = 1$ and $n_z = 1$

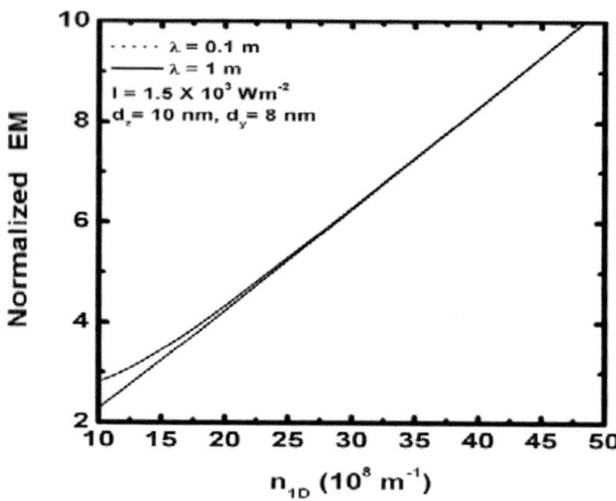

Fig. 6.24. Plot of the normalized EM as a function of linear electron concentration of quantum wire super-lattices of GaAs/AlGaAs with graded interfaces in the presence of light waves in accordance with the two band model of Kane in the presence of external photo-excitation.

as a function of electron concentration at $T = 4.2\,K$ by taking quantum wire super lattices of optoelectronic materials with graded interfaces in accordance with the perturbed two band model of Kane. In Fig. 6.24, we have plotted the EM as function of electron concentration per unit length in quantum wires of GaAs/AlGaAs super-lattices with graded interfaces. It appears from Fig. 6.24 that the effect of a single sub-band linearly increases the EM for low value of wavelength. However, with the increase in the wavelength, the EM tends to coincide with that of the lower wavelength values at higher carrier concentration.

6.2.13 *The magneto EM in HD superlattices of kane type semiconductors with graded interfaces in the presence of light waves*

The magneto EM in HD superlattices of Kane type semiconductors with graded interfaces in the presence of light waves can be expressed as

$$k_z^2 = G_{8,17,54} + iH_{8,17,54} \tag{6.193}$$

where

$$G_{8,17,54} = \left[\frac{C_{7,54}^2 - D_{7,54}^2}{L_0^2} - \left[\frac{2eB}{\hbar}\left(n+\frac{1}{2}\right)\right]\right], \quad \mathrm{C}_{7,54} = \cos^{-1}(\bar{\omega}_{7,54}),$$

$$\bar{\omega}_{7,54} = (2)^{\frac{-1}{2}}[(1 - G_{7,54}^2 - H_{7,54}^2) - \sqrt{(1 - G_{7,54}^2 - H_{7,54}^2)^2 + 4G_{7,54}^2}]^{\frac{1}{2}},$$

$$G_{7,54} = [G_{1,54} + (\rho_{5,,54}G_{2,,54}/2) - (\rho_{6,54}H_{2,54}/2) + (\Delta_0/2) \cdot\{\rho_{6,54}H_{2,55} - \rho_{8,54}H_{3,54} + \rho_{9,54}H_{4,54} - \rho_{10,54}H_{4,54} + \rho_{11,54}H_{5,54} - \rho_{12,54}H_{5,54} + (1/12)(\rho_{12,54}G_{6,54} - \rho_{14,54}H_{6,54})\}],$$

$$G_{1,54} = [(\cos(h_{1,54}))(\cosh(h_{2,54}))(\cosh(g_{1,54}))(\cos(g_{2,54})) +(\sin(h_{1,54}))(\sinh(h_{2,54}))(\sinh(g_{1,54}))(\sin(g_{2,54}))],$$

$$h_{1,54} = e_{1,54}(b_0 - \Delta_0), \quad e_{1,54} = 2^{\frac{-1}{2}}(\sqrt{t_{1,54}^2 + t_{2,54}^2} + t_{1,54})^{\frac{1}{2}},$$

$$t_{1,54} = \left[(2m_{c1}/\hbar^2)\cdot V_{11}(E, E_{g1}, \lambda, \Delta_1, \eta_{g1}) - \left[\frac{2eB}{\hbar}\left(n+\frac{1}{2}\right)\right]\right],$$

$$t_{2,54} = [(2m_{c1}/\hbar^2)V_{21}(E, E_{g1}, \lambda, \Delta_1, \eta_{g1})],$$

$$h_{2,54} = e_{2,54}(b_0 - \Delta_0), e_{2,54} = 2^{\frac{-1}{2}}(\sqrt{t_{1,54}^2 + t_{2,54}^2} - t_{1,54})^{\frac{1}{2}},$$

$$g_{1,54} = d_{1,54}(a_0 - \Delta_0), \quad d_{1,54} = 2^{\frac{-1}{2}}(\sqrt{x_{1,54}^2 + y_{1,54}^2} + x_{1,54})^{\frac{1}{2}},$$

$$x_{1,54} = \left[-(2m_{c2}/\hbar^2)\cdot V_{11}(E - V_0, E_{g2}, \lambda, \Delta_2, \eta_{g2}) + \left[\frac{2eB}{\hbar}\left(n+\frac{1}{2}\right)\right]\right],$$

$$y_{1,54} = [(2m_{c2}/\hbar^2).V_{22}(E - V_0, E_{g2}, \lambda, \Delta_2, \eta_{g2})],$$

$$g_{2,54} = d_{2,54}(a_0 - \Delta_0), \quad d_{2,54} = 2^{\frac{-1}{2}}(\sqrt{x_{1,54}^2 + y_{1,54}^2} - x_{1,54})^{\frac{1}{2}},$$

$$\rho_{5,54} = (\rho_{3,54}^2 + \rho_{4,54}^2)^{-1}[\rho_{1,54}\rho_{1,54} - \rho_{2,54}\rho_{4,54}],$$

$$\rho_{1,54} = [d_{1,54}^2 + e_{2,54}^2 - d_{2,54}^2 - e_{1,54}^2], \quad \rho_{3,54} = [d_{1,54}e_{1,54} + d_{2,54}e_{2,54}],$$

$$\rho_{2,54} = 2[d_{1,54}d_{2,54} + e_{1,54}e_{2,54}], \quad \rho_{4,54} = [d_{1,54}e_{2,54} - e_{1,54}d_{2,54}],$$

$$G_{2,54} = [(\sin(h_{1,54}))(\cosh(h_{2,54}))(\sinh(g_{1,54}))(\cos(g_{2,54})) + (\cos(h_{1,54}))(\sinh(h_{2,54}))(\cosh(g_{1,54}))(\sin(g_{2,54}))],$$

$$\rho_{6,54} = (\rho_{3,54}^2 + \rho_{4,54}^2)^{-1}[\rho_{1,54}\rho_{4,54} + \rho_{2,54}\rho_{3,54}],$$

$$H_{2,54} = [(\sin(h_{1,54}))(\cos(h_{2,54}))(\sin(g_{2,54}))(\cosh(g_{1,54})) - (\cos(h_{1,54}))(\sinh(h_{2,54}))(\sinh(g_{1,54}))(\cos(g_{2,54}))],$$

$$\rho_{7,54} = [(e_{1,54}^2 + e_{2,54}^2)^{-1}[e_{1,54}(d_{1,54}^2 - d_{2,54}^2) - 2d_{1,54}d_{2,54}e_{2,54}] - 3e_{1,54}],$$

$$G_{3,54} = [(\sin(h_{1,54}))(\cosh(h_{2,54}))(\cosh(g_{1,54}))(\cos(g_{2,54})) + (\cos(h_{1,54}))(\sinh(h_{2,54}))(\sinh(g_{1,54}))(\sin(g_{2,54}))],$$

$$\rho_{8,54} = [(e_{1,54}^2 + e_{2,54}^2)^{-1}[e_{2,54}(d_{1,54}^2 - d_{2,54}^2) + 2d_{1,54}d_{2,54}e_{1,54}] + 3e_{2,54}],$$

$$H_{3,54} = [(\sin(h_{1,54}))(\cosh(h_{2,54}))(\sin(g_{2,54}))(\sinh(g_{1,54})) - (\cos(h_{1,54}))(\sinh(h_{2,54}))(\cosh(g_{1,54}))(\cos(g_{2,54}))],$$

$$\rho_{9,54} = [(d_{1,54}^2 + d_{2,54}^2)^{-1}[d_{1,54}(e_{2,54}^2 - e_{1,54}^2) + 2e_{2,54}d_{2,54}e_{1,54}] + 3d_{1,54}],$$

$$G_{4,54} = [(\cos(h_{1,54}))(\cosh(h_{2,54}))(\cos(g_{2,54}))(\sinh(g_{1,54})) - (\sin(h_{1,54}))(\sinh(h_{2,54}))(\cosh(g_{1,54}))(\sin(g_{2,54}))],$$

$$\rho_{10,54} = [-(d_{1,54}^2 + d_{2,54}^2)^{-1}[d_{2,54}(-e_{2,54}^2 + e_{1,54}^2) + 2e_{2,54}d_{2,54}e_{1,54}] + 3d_{2,54}],$$

$$H_{4,54} = [(\cos(h_{1,54}))(\cosh(h_{2,54}))(\cosh(g_{1,54}))(\sin(g_{2,54})) + (\sin(h_{1,54}))(\sinh(h_{2,54}))(\sinh(g_{1,54}))(\cos(g_{2,54}))],$$

$$\rho_{11,54} = 2[d_{1,54}^2 + e_{2,54}^2 - d_{2,54}^2 - e_{1,54}^2],$$

$$G_{5,54} = [(\cos(h_{1,54}))(\cosh(h_{2,54}))(\cos(g_{2,54}))(\cosh(g_{1,54})) - (\sin(h_{1,54}))(\sinh(h_{2,54}))(\sinh(g_{1,54}))(\sin(g_{2,54}))],$$

$$\rho_{12,54} = 4[d_{1,54}d_{2,54} + e_{1,54}e_{2,54}],$$

$$
\begin{aligned}
H_{5,54} &= [(\cos(h_{1,54}))(\cosh(h_{2,54}))(\sinh(g_{1,54}))(\sin(g_{2,54})) \\
&\quad +(\sin(h_{1,54}))(\sinh(h_{2,54}))(\cosh(g_{1,54}))(\cos(g_{2,54}))], \\
\rho_{13,54} &= [\{5(d_{1,54}e_{1,54}^3 - 3e_{1,54}e_{2,54}^2 d_{1,54}) + 5d_{2,54}(e_{1,54}^3 - 3e_{1,54}^2 e_{2,54})\} \\
&\quad \cdot(d_{1,54}^2 + d_{2,54}^2)^{-1} + (e_{1,54}^2 + e_{2,54}^2)^{-1}\{5(e_{1,54}d_{1,54}^3 \\
&\quad - 3d_{2,54}e_{1,54}^2 d_{1,54}) + 5(d_{2,54}^3 e_{2,54} - 3d_{1,54}^2 d_{2,54}e_{2,54})\} \\
&\quad - 34(d_{1,54}e_{1,54} + d_{2,54}e_{2,54})], \\
G_{6,54} &= [(\sin(h_{1,54}))(\cosh(h_{2,54}))(\sinh(g_{1,54}))(\cos(g_{2,54})) \\
&\quad +(\cos(h_{1,54}))(\sinh(h_{2,54}))(\cosh(g_{1,54}))(\sin(g_{2,54}))], \\
\rho_{14,54} &= [\{5(d_{1,54}e_{2,54}^3 - 3e_{2,54}e_{1,54}^2 d_{1,54}) + 5d_{2,54}(-e_{1,54}^3 + 3e_{2,54}^2 e_{1,54})\} \\
&\quad \cdot(d_{1,54}^2 + d_{2,54}^2)^{-1} + (e_{1,54}^2 + e_{2,54}^2)^{-1}\{5(-e_{1,54}d_{2,54}^3 \\
&\quad + 3d_{1,54}^2 d_{2,54}e_{1,54}) + 5(-d_{1,54}^3 e_{2,54} + 3d_{2,54}^2 d_{1,54}e_{2,54})\} \\
&\quad + 34(d_{1,54}e_{2,54} - d_{2,54}e_{1,54})], \\
H_{6,54} &= [(\sin(h_{1,54}))(\cosh(h_{2,54}))(\cosh(g_{1,54}))(\sin(g_{2,54})) \\
&\quad - (\cos(h_{1,54}))(\sinh(h_{2,54}))(\sinh(g_{1,54}))(\cos(g_{2,54}))], \\
H_{7,54} &= [H_{1,54} + (\rho_{5,54}H_{2,54}/2) + (\rho_{6,54}G_{2,54}/2) + (\Delta_0/2) \\
&\quad \cdot\{\rho_{8,54}G_{3,54} + \rho_{7,54}H_{3,54} + \rho_{10,54}G_{4,54} + \rho_{9,54}H_{4,54} \\
&\quad + \rho_{12,54}G_{5,54} + \rho_{11,54}H_{5,54} \\
&\quad + (1/12)(\rho_{14,54}G_{6,54} + \rho_{13,54}H_{6,54})\}], \\
H_{1,54} &= [(\sin(h_{1,54}))(\sinh(h_{2,54}))(\cosh(g_{1,54}))(\cos(g_{2,54})) \\
&\quad + (\cos(h_{1,54}))(\cosh(h_{2,54}))(\sinh(g_{1,54}))(\sin(g_{2,54}))], \\
D_{7,54} &= \sinh^{-1}(\bar{\omega}_{7,54}),\ H_{8,17,54} = (2C_{7,54}D_{7,54}/L_0^2)
\end{aligned}
$$

The sub-band equation in this case can be expressed as

$$0 = [G_{8,17,54} + iH_{8,17,54}]|_{E=E_{17,54}} \tag{6.194}$$

where $E_{17,54}$ is the Landau sub-band energy in this case.

At low temperatures where the quantum effects become prominent, the DOS function for the lowest SL mini-band is given by

$$N_{HDSL}(\mathrm{E},\eta_g,\lambda)=\frac{g_v eB}{2\pi^2\hbar}\sum_{n=0}^{n_{\max}}\frac{[G'_{8,17,54}+iH'_{8,17,54}]}{\sqrt{G_{8,17,54}+iH_{8,17,54}}}H(E-E_{17,54}) \tag{6.195}$$

The EEM can be written as

$$m^*(\mathrm{E},n_x,n_y,\lambda,\eta_g)=\frac{\hbar^2}{2}(G'_{8,17,54}) \tag{6.196a}$$

The electron concentration is given by

$$n_0=\frac{2g_v}{\pi}\text{ Real Part of }\sum_{n_x=1}^{n_{x\max}}\sum_{n_x=1}^{n_{y\max}}\left[(\sqrt{G_{8.1754}+iH_{8.1754}})|_{E=E_{F666}}\right.$$
$$\left.+\sum_{r=1}^{s}L(r)(\sqrt{G_{8.1754}+iH_{8.1754}})|_{E=E_{F666}}\right] \tag{6.196b}$$

where E_{F666} is the Fermi energy in this case.

6.3 Open Research Problems

(R.6.1) Investigate the EM in the presence of intense external non-uniform light waves for all the HD superlattices whose respective dispersion relations of the carriers are given in this chapter.

(R.6.2) Investigate the EM for the heavily–doped semiconductors in SLs the presences of Gaussian, exponential, Kane, Halperian, Lax and Bonch-Burevich types of band tails [16] for all SL systems as discussed in this chapter in the presence of external oscillatory and non-uniform light waves.

(R.6.3) Investigate the EM in the presence of external light waves for short period, strained layer, random and Fibonacci HD superlattices in the presence of an arbitrarily oriented alternating electric field.

(R.6.4) Investigate all the appropriate problems of this chapter for a Dirac electron.

(R.6.5) Investigate all the appropriate problems of this chapter by including the many body, broadening and hot carrier effects respectively.

(R.6.6) Investigate all the appropriate problems of this chapter by removing all the mathematical approximations and establishing the respective appropriate uniqueness conditions.

References

1. P.K.Basu, *Theory of Optical Process in Semiconductors, Bulk and Microstructures* (Oxford University Press, Oxford,1997)
2. K. P. Ghatak, S. Bhattacharya, S. Bhowmik, R. Benedictus, S. Chowdhury, J. Appl. Phys. **103**, 094314 (2008); K. P. Ghatak, S. Bhattacharya, J. Appl. Phys. **102**, 073704 (2007); K. P. Ghatak, S. Bhattacharya, S. K. Biswas, A. De, A. K. Dasgupta, Phys. Scr.,**75**, 820 (2007); P. K. Bose, N. Paitya, S. Bhattacharya, D. De, S. Saha, K. M. Chatterjee, S. Pahari, K. P. Ghatak, Quantum Matter, **1**, 89 (2012); K. P. Ghatak, S. Bhattacharya, A. Mondal, S. Debbarma, P. Ghorai, and A. Bhattacharjee, Quantum Matter, **2**, 25 (2013); S. Bhattacharya, D. De, S. Ghosh, J. P. Bannerje, M. Mitra, B. Nag, S. Saha, S. K. Bishwas, M. Paul, Jour. Comp. Theo. Nanoscience, **7**, 1066 (2010); K. P. Ghatak, S. Bhattacharya, S. Pahari, S. N. Mitra, P. K. Bose, Jour. Phys. Chem. Solids, **70**, 122 (2009), S. Bhattacharya, D. De, R. Sarkar, S. Pahari, A. De, A. K. Dasgupta, S. N. Biswas, K. P. Ghatak, Jour. Comp. Theo. Nanoscience, **5**, 1345 (2008); S. Mukherjee, D. De, D. Mukherjee, S. Bhattacharya, A. Sinha, K. P. Ghatak Physica B, **393**, 347 (2007)
3. K. Seeger, Semiconductor Physics (Springer-Verlag, 7th Edition. 2006)
4. B. R. Nag, *Physics of Quantum Well Devices* (Kluwer Academic Publishers, The Netherlands, 2000)
5. R. K. Pathria, *Statistical Mechanics* 2nd Edition (Butterworth-Hei)

Chapter 7

Few Related Applications

7.1 Introduction

In this monograph, we have investigated different EMs of HD semiconductor nanostructures of different technologically important materials and its implications in materials science in general. The concept of EM is one of the main keys for investigating the carrier transport phenomena of HD quantum effect devices. In this chapter, we shall discuss **twenty seven different applications** in this context. Section 7.3 contains the single gigantic open research problem.

7.2 Different Related Applications

The investigations as presented in this monograph find **twenty seven different applications** in the realm of modern quantum effect devices.

7.2.1 *Thermoelectric power*

In recent years, with the advent of Quantum Hall Effect (QHE) [1], there has been considerable interest in studying the thermoelectric power under strong magnetic field (TPSM) in various types of nanostructured materials having quantum confinement of their charge carriers in one, two and three dimensions of the respective wave vector space leading to different carrier energy spectra [2–31]. The classical TPSM equation $G = \pi^2 k_B/3e$ (which is a function of three fundamental constants of nature) is valid only under the condition of carrier non-degeneracy, being independent of carrier concentration and reflects the fact that the signature of the band structure of any material is totally absent in the same.

Zawadzki [7] demonstrated that the TPSM for electronic materials having degenerate electron concentration is essentially determined by their respective energy band structures. It has, therefore, different values in different materials and changes with the doping, magnitude of the reciprocal quantizing magnetic field under magnetic quantization, quantizing electric field as in inversion layers, nanothickness as in quantum wells, wires and dots, with superlattice period as in quantum confined semiconductor superlattices with graded interfaces having various carrier energy spectra and also in other types of field assisted nanostructured materials.

The magnitude of the thermoelectric power G can be written as [8]

$$G = \frac{1}{|e|Tn_0}\int_{-\infty}^{\infty}(E - E_F)R(E)\left[-\frac{\partial f_0}{\partial E}\right]dE \tag{7.2}$$

where $R(E)$ is the total number of states. The (7.2) can be written under the condition of carrier degeneracy [3] as

$$G\left(\frac{\pi k_B^2 T}{2|e|n_0}\right)\left(\frac{\partial n_0}{\partial E_F}\right) \tag{7.3}$$

For inversion layers and NIPI structures, under the condition of electric quantum limit, (7.3) assumes the form

$$G = \frac{\pi^2 k_B^2 T}{3en_{02D}}\left[\frac{\partial n_{02D}}{\partial(E_{F2D} - E_{02D})}\right] \tag{7.4}$$

where n_{02D}, E_{F2D} and E_{02D} are the surface electron concentration, the Fermi energy and the sub band energy for the said 2D systems at the electric quantum limit

For heavily doped semiconductors (7.2) assumes the form

$$G = \frac{\pi^2 k_B^2 T}{3en_{0HD}} \left[\frac{\partial n_{0HD}}{\partial (E_{FHD} - E_{0HD})} \right] \quad (7.5)$$

where n_{0HD}, E_{FHD} and E_{0HD} are the electron concentration, the Fermi energy and the band tail energy for the heavily doped semiconductors where E_{0HD} should be obtained from the heavily doped dispersion relation of the semiconductor under the conditions $E = E_{0HD}$ and $k = 0$.

Thus, we can use the carrier statistics for different low dimensional HD materials to investigate the TPSM in such compounds.

7.2.2 *Debye screening length*

It is well known that the Debye screening Length (DSL) of the carriers in semiconductors is a very important quantity characterizing the screening of the Coulomb field of the ionized impurity centers by the free carriers [32]. It affects many of the special features of modern nano-devices, the carrier motilities under different mechanisms of scattering, and the carrier plasmas in semiconductors [33]. The DSL is a very good approximation to the accurate self-consistent screening in presence of band tails and is also used to illustrate the interaction between the colliding carriers in Auger effect in solids [32]. The classical value of the DSL is equal to $[\varepsilon_{SC} k_B/(e^2 n_0)]^{1/2}$ ($\varepsilon_{\mathrm{sc}}, k_B, T, e$, and n_0 are the semiconductor permittivity, the Boltzmann's constant, the temperature, the magnitude of the carrier charge, and the electron concentration, respectively) which is valid for both the carriers. In this conventional form, the DSL decreases with increasing carrier concentration at a constant temperature and this relation holds only under the condition of carrier non-degeneracy. It is interesting to note that the under the condition of extreme degeneracy, the expression of DSL for materials having parabolic energy bands can be written as $L_D = (\pi^{2/3} \hbar \sqrt{\varepsilon_{sc}})(e g_v^{1/3} 3^{1/6} n_0^{1/6} \sqrt{m_c})^{-1}$ ($\hbar, m_c$ and g_v are Dirac constant, effective electron mass at the edge of the conduction band and valley degeneracy respectively). Thus we observed that in this case the result is independent of temperature, but depends on n_0, g_v

and m_c. Besides, the indices of inverse electron variation changes from half in the former case to one-sixth in the latter case. Since, the performance of the electron devices at the device terminals and the speed of operation of modern switching transistors are significantly influence by the degree of carrier degeneracy present in these devices, the simplest way of analyzing such devices taking into account of the degeneracy of the band is to use the appropriate DSL to express the performance at the device terminal and switching speed in terms of the carrier concentration [34].

The DSL depends on the density-of-states function which, in turn, is significantly affected by the different carrier energy spectra of different semiconductors having various band structures [32, 34]. In recent years, various energy wave vector dispersion relations of the carriers of different materials have been proposed [35] which have created the interest in studying the DSL in such quantized structures under external conditions. It is well known from the fundamental study of Landsberg [32], that the DSL for electronic materials having degenerate electron concentration is essentially determined by their respective energy band structures. It has, therefore, different values in different materials and varies with the electron concentration, with the magnitude of the reciprocal quantizing magnetic field under magnetic quantization, with the quantizing electric field as in inversion layers, with the nano-thickness as in quantum wells, with super-lattice period as in the quantum confined super-lattices of small gap compounds with graded interfaces having various carrier energy spectra. The nature of these variations has been investigated by in the literature [36–44].

The 3D DSL can, in general, be expressed as [34]

$$L_{3D} = \left[\frac{e^2}{\varepsilon_{sc}} \cdot \frac{\partial n_0}{\partial E_F}\right]^{-1/2} \tag{7.6}$$

Using (7.3) and (7.6) we can write

$$L_{3D} = \left[\frac{\pi^2 k_B^2 \varepsilon_{sc}}{3e^3 n_0 G}\right]^{1/2} \tag{7.7}$$

The 3D DSL for the heavily doped systems is given by

$$L_{3D} = \frac{e^2}{\varepsilon_{sc}}\left[\frac{\partial n_{0HD}}{\partial(E_{FHD} - E_{0HD})}\right]^{-1/2} \tag{7.8}$$

Using (7.8) and (7.5) we obtain the same equation as given by (9.7). Thus, the 3D DSL for degenerate materials can be determined by knowing the experimental values of G.

For inversion layers and NIPI structures, the 2D DSL is given by

$$L_{2D} = \left[\frac{e^2}{2\varepsilon_{sc}}\frac{\partial n_{02D}}{\partial(E_{F2D} - E_{02D})}\right]^{-1} \tag{7.9}$$

Using (7.4) and (7.9) we get,

$$L_{2D} = \frac{2\pi^2 k_B^2 T\varepsilon_{sc}}{3e^3 n_{02D} G} \tag{7.10}$$

For QWs, the G assumes the form

$$G = \frac{\pi^2 k_B^2 T}{3en_{2D}} \cdot \frac{\partial n_{2D}}{\partial E_{Fs}} \tag{7.11}$$

where n_{2D} and E_{F_s} are 2D electron concentration and Fermi energy in this case. Using (7.10) and (7.11) we get

$$L_{2D} = \frac{2\pi^2 k_B^2 T\varepsilon_{sc}}{3e^3 n_{2D} G} \tag{7.12}$$

From the suggestion for the experimental determination of the 3D DSL for degenerate materials having arbitrary dispersion laws as given by (7.7), we observe that for a constant T, the DSL varies inversely with the square root of Gn_0. Only the experimental values of G for any material as a function of electron concentration will generate the experimental values of the 3D DSL for that range of n_0 for that material. Since $(Gn_0)^{-1/2}$ decreases with increasing n_0 for constant T, from (7.7) we can conclude that the 3D DSL will decrease with increasing n_0. For 2D DSL L_D at a constant temperature varies inversely with $n_{02D}G$ or $n_{2D}G$ as appear from (7.10) and (7.12) respectively. Since $n_{02D}G$ and $n_{2D}G$ increase with decreasing surface concentration, from (7.10) and (7.12) we can infer that 2D DSL will increase with decreasing n_{02D} or n_{2D} for the appropriate 2D systems. This statement provides a compatibility test of our theoretical analysis. Thus (7.7), (7.10) and (7.12) provide experimental checks of both 3D and 2D DSLs and also a technique for probing the band structures of the degenerate materials having arbitrary band structures.

7.2.3 *Carrier contribution to the elastic constants*

The knowledge of the carrier contribution to the elastic constants is important in studying the mechanical properties of the materials and has been investigated in the literature [45–48]. The electronic contribution to the second- and third- order elastic constants for HD materials can be written as [45–48]

$$\Delta C_{44} = \frac{-G_0^2}{9}\left[\frac{\partial n_{0HD}}{\partial(E_{FHD} - E_{0HD})}\right] \tag{7.13}$$

and

$$\Delta C_{456} = \frac{G_0^3}{27}\left[\frac{\partial^2 n_{0HD}}{\partial(E_{FHD} - E_{0HD})^2}\right] \tag{7.14}$$

where G_0 is the deformation potential constant. Thus, using (7.5), (7.13) and (7.14), we can write

$$\Delta C_{44} = \left[\frac{-n_0(\bar{G}_0)^2|e|G}{(3\pi^2 k_B^2 T)}\right] \tag{7.15}$$

and

$$\Delta C_{456} = \left(\frac{n_0|e|(\bar{G}_0)^3 G^2}{(3\pi^4 k_B^3 T)}\right)\left(1 + \frac{n_0}{G}\frac{\partial G}{\partial n_0}\right) \tag{7.16}$$

Therefore, by using (7.13) and (7.14) we can investigate ΔC_{44} and ΔC_{456} for all the cases of this monograph. Besides, the experimental graph of G versus n_0 allows us to determine the electronic contribution to the elastic constants for materials having arbitrary spectra.

7.2.4 *Diffusivity-mobility ratio*

It is well known that the diffusivity-mobility ratio (DMR) occupies a central position in the whole field of solid state device electronics and the related sciences since the diffusion constant (a quantity very useful for device analysis where exact experimental determination is rather difficult) can be obtained from this ratio by knowing the experimental values of the mobility. The classical value of the DMR is equal to $(k_BT/|e|)$, (k_B, T, and $|e|$ are Boltzmann's constant, temperature and the magnitude of the carrier charge respectively). This relation in this form was first introduced by Einstein to study of the diffusion of gas particles and is known as the Einstein relation [49, 50]. It appears that the DMR increases linearly with increasing T and is independent of electron concentration. This relation is applicable for both types of charge carriers only under non-degenerate carrier concentration although its validity has been suggested erroneously for degenerate materials [51]. Landsberg first pointed out that the DMR for degenerate semiconductors is essentially determined by their energy band structures [52, 53]. This relation is useful for semiconductor homo-structures [52, 53], semiconductor-semiconductor hetero-structures [56, 57], metals-semiconductor hetero-structures [58–66] and insulator-semiconductor hetero-structures [67–70]. The nature of the variations of the DMR under different physical conditions has been studied in the literature [49, 50, 51, 52, 53, 71–94]. Incidentally, A. N. Chakravarti (a recognized leading expert of DMR in general) and his research group are still contributing significantly under his able leadership regarding this pin pointed research topic on DMR from 1972 [71, 49, 72–75, 81, 86–94] and some of the significant features, which have emerged from these studies, are:

(a) The DMR increases monotonically with increasing carrier concentration in bulk semiconductors and the nature of these variations are significantly influenced by the band structures of different materials.
(b) The DMR increases with the increasing quantizing electric field as in inversion layers.
(c) The DMR oscillates with the inverse quantizing magnetic field under magnetic quantization due to the Shubnikov-de Haas effect.
(d) The DMR shows composite oscillations with the various controlled quantities of semiconductor super lattices.

(e) In ultrathin films, quantum wires and other field assisted low-dimensional systems, the value of the DMR changes appreciably with the external variables depending on the nature of quantum confinements of different materials.

The DMR depends on the density-of-states (DOS) function, which, in turn, is significantly affected by the different carrier energy spectra of different semiconductors having various band structures. It can, in general, be proved that for bulk specimens the DMR is given by [71]

$$\frac{D}{\mu} = \frac{n_{0HD}}{e}\left[\frac{\partial n_{0HD}}{\partial(E_{FHD} - E_{0HD})}\right]^{-1} \tag{7.17}$$

The electric quantum limit as in inversion layers and NIPI structures refers to the lowest electric sub-band and (7.17) assumes the form [71]

$$\frac{D}{\mu} = \frac{n_{02D}}{e}\left[\frac{\partial n_{02D}}{\partial(E_{F2D} - E_{02D})}\right]^{-1} \tag{7.18}$$

Using (7.4), (7.5), (7.17) and (7.18) one obtains

$$\frac{D}{\mu} = \left(\frac{\pi^2 k_B^2 T}{3|e|^2 G}\right) \tag{7.19}$$

Thus, the DMR for degenerate materials can be determined by knowing the experimental values of G.

The suggestion for the experimental determination of the DMR for degenerate semiconductors having arbitrary dispersion laws as given by (7.19) does not contain any energy band constants. For a fixed temperature, the DSL varies inversely as G. Only the experimental values of G for any material as a function of electron concentration will generate the experimental values of the DMR for that range of n_0 for that system. Since G decreases with increasing n_0, from (7.19) one can infer that the DMR will increase with increase in n_0. This statement is the compatibility test so far as the suggestion for the experimental determination of DMR for degenerate materials is concerned.

7.2.5 *Measurement of band-gap in the presence of light waves*

Using the appropriate equations, the normalized incremental band gap (ΔE_g) has been plotted as a function of normalized I_0 (for a given wavelength and considering red light for which $\lambda = 660\,\text{nm}$) at $T = 4.2\,K$ in Figs. 7.1 and 7.2 for n-$Hg_{1-x}Cd_xTe$ and n-$In_{1-x}Ga_xAs_yP_{1-y}$ lattice matched to InP in accordance with the perturbed three and two band models of Kane and that of perturbed parabolic energy bands respectively. In Figs. 7.3 and 7.4, the normalized incremental band gap has been plotted for the aforementioned optoelectronic compounds as a function of λ. It is worth remarking that the influence of an external photo-excitation is to change radically the original band structure of the material. Because of this change, the photon field causes to increase the band gap of semiconductors. We propose

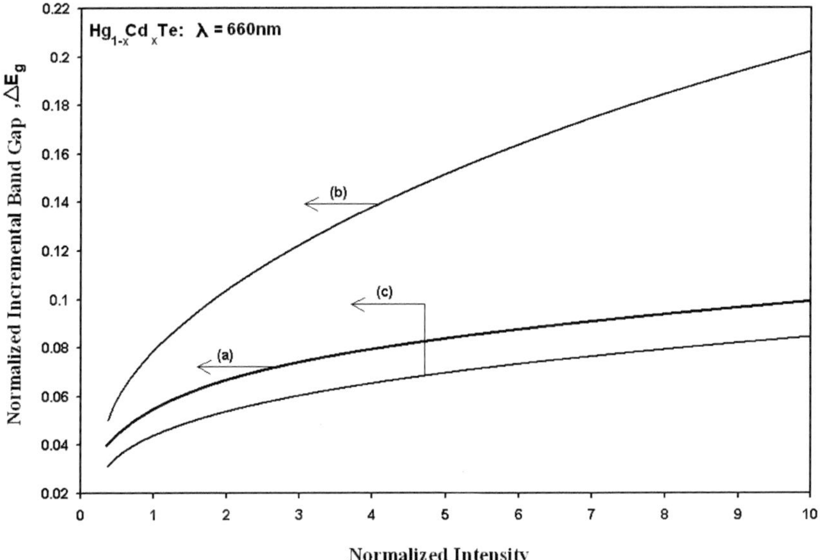

Fig. 7.1. Plots of the normalized incremental band gap (ΔE_g) for n-$Hg_{1-x}Cd_xTe$ as a function of normalized light intensity in which the curves (a) and (b) represent the perturbed three and two band models of Kane respectively. The curve (c) represents the same variation in n-$Hg_{1-x}Cd_xTe$ in accordance with the perturbed parabolic energy bands.

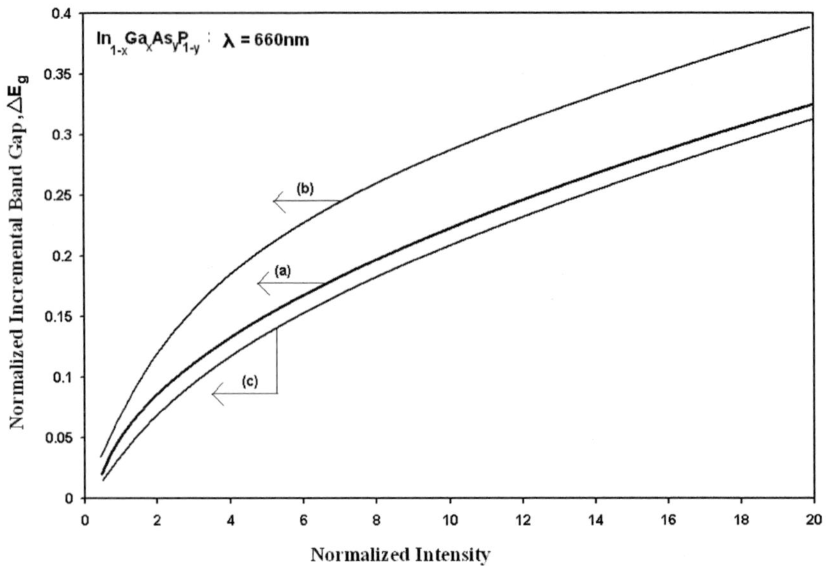

Fig. 7.2. Plots of the normalized incremental band gap (ΔE_g) for $In_{1-x}Ga_xAs_yP_{1-y}$ lattice matched to InP as a function of normalized light intensity for all cases of Fig. 7.1.

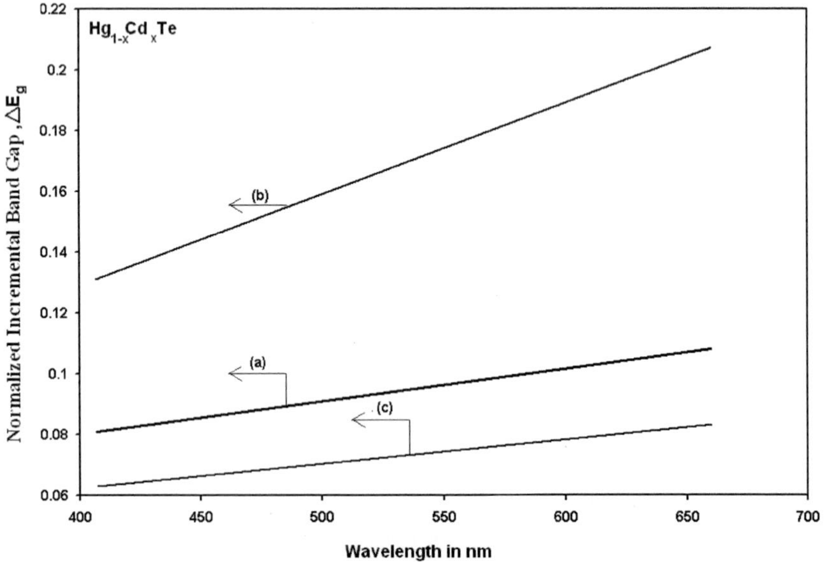

Fig. 7.3. Plots of the normalized incremental band gap (ΔE_g) for $Hg_{1-x}Cd_xTe$ as a function of wavelength for all cases of Fig. 7.2.

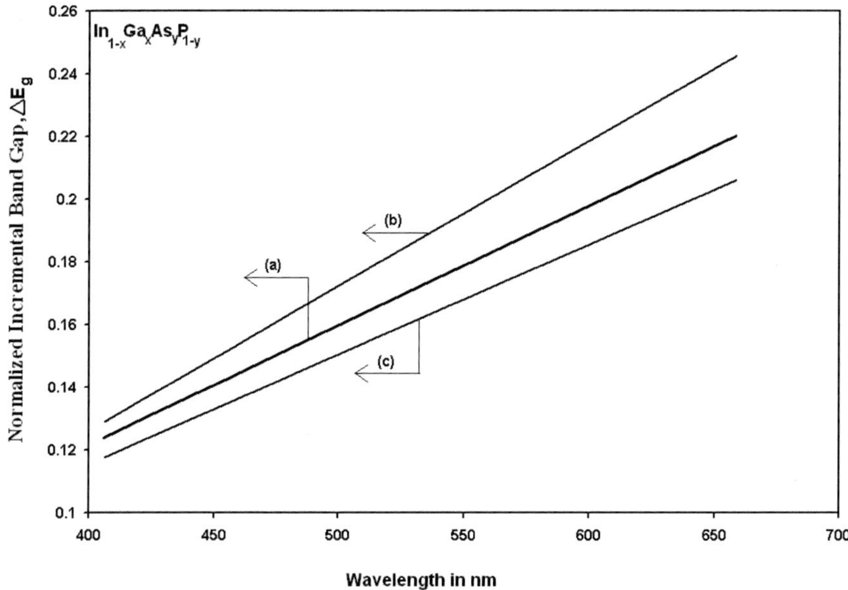

Fig. 7.4. Plots of the normalized incremental band gap (ΔE_g) for $In_{1-x}Gs_xAs_yP_{1-y}$ lattice matched to InP as a function of wavelength for all cases of Fig. 7.1.

the following two experiments for the measurement of band gap of semiconductors under photo-excitation.

(A) A white light with colour filter is allowed to fall on a semiconductor and the optical absorption coefficient ($\bar{\alpha}_0$) is being measured experimentally. For different colours of light, $\bar{\alpha}_0$ is measured and $\bar{\alpha}_0$ versus $\hbar\omega$ (the incident photon energy) is plotted and we extrapolate the curve such that $\bar{\alpha}_0 \to 0$ at a particular value $\hbar\omega_1$. This $\hbar\omega_1$ is the unperturbed band gap of the semiconductor. During this process, we vary the wavelength with fixed I_0. From our present study, we have observed that the band gap of the semiconductor increases for various values of λ when I_0 is fixed (from Figs. 7.3 and 7.4). This implies that the band gap of the semiconductor measured (i.e., $\hbar\omega_1 = E_g$) is not the unperturbed band gap E_{g_0} but the perturbed band gap E_g; where $E_g = E_{g_0} + \Delta E_g, \Delta E_g$ is the increased band gap at $\hbar\omega_1$. Conventionally, we consider this E_g as the unperturbed band gap of the semiconductor and this particular concept needs modification. Furthermore, if we vary I_0 for a monochromatic light (when λ is fixed) the band gap of the semiconductor will also change consequently (Fig. 7.1 and 7.2). Consequently, the absorption coefficient will change with the intensity of light [95]. For the overall understanding, the detailed theoretical and experimental investigations are needed in this context for various materials having different band structures.

(B) The conventional idea for the measurement of the band gap of the semiconductors is the fact that the minimum photon energy $h\nu(\nu$ is the frequency of the monochromatic light) should be equal to the band gap E_{g_0} (unperturbed) of the semiconductor, i.e.,

$$hv = E_{g_0} \tag{7.20}$$

In this case, λ is fixed for a given monochromatic light and the semiconductor is exposed to a light of wavelength λ. Also the intensity of the light is fixed. From Figs. 7.3 and 7.4, we observe that the band gap of the semiconductor is not E_{g_0} (for a minimum value of $h\nu$) but E_g, the perturbed band gap. Thus, we can rewrite the above equality as

$$hv = E_g \tag{7.21}$$

Furthermore, if we vary the intensity of light (Figs. 7.1 and 7.2) for the study of photoemission, the minimum photon energy should be

$$hv_1 = E_{g_1} \tag{7.22}$$

where E_{g_1} is the perturbed band gap of the semiconductor due to various intensity of light when ν and ν_1 are different.

Thus, we arrive at the following conclusions:

(a) Under different intensity of light, keeping λ fixed, the condition of band gap measurement is given by

$$hv_1 = E_{g_1} = E_{g_0} + \Delta E_{g_1} \tag{7.23}$$

(b) Under different colour of light, keeping the intensity fixed, the condition of band gap measurement assumes the form

$$hv = E_g = E_{g0} + \Delta E_g \tag{7.24}$$

and not the conventional result as given by (7.20).

determined by the appropriate carrier statistics. Thus, our present study plays an important role in determining the diffusion coefficients of the minority carriers of quantum-confined lasers with materials having arbitrary band structures. Therefore in the investigation of the optical excitation of the optoelectronic materials which lead to the study of the ambipolar diffusion coefficients the present results contribute significantly.

7.2.6 *Diffusion coefficient of the minority carriers*

This particular coefficient in quantum confined lasers can be expressed as

$$\frac{D_i}{D_0} = \frac{dE_{Fi}}{dE_F} \tag{7.25}$$

where D_i and D_0 are the diffusion coefficients of the minority carriers both in the presence and absence of quantum confinements and E_{Fi} and E_F are the Fermi energies in the respective cases. It appears then that, the formulation of the above ratio requires a relation between E_{Fi} and E_F, which, in turn, can be formed through the respective carrier statistics.

7.2.7 *Nonlinear optical response*

The nonlinear response from the optical excitation of the free carriers is given by [96]

$$Z_0 = \frac{-e^2}{\omega^2\hbar^2}\int_0^\infty \left(k_x\frac{\partial k_x}{\partial E}\right)^{-1} f_0 N(E)dE \tag{7.26}$$

where ω is the optical angular frequency, $N(E)$ is the density-of-states function. From the various E-k relations of different materials under different physical conditions, we can formulate the expression of $N(E)$ and from band structure we can derive the term $(k_x \frac{\partial k_x}{\partial E})$ and thus by using the density-of-states function as formulated, we can study the Z_0 for all types of materials as considered in this monograph.

7.2.8 *Third order nonlinear optical susceptibility*

This particular susceptibility can be written as [107]

$$\chi_{NP}(\omega_1, \omega_2, \omega_3) = \frac{n_0 e^4 \langle \varepsilon^4 \rangle}{24\omega_1\omega_2\omega_3(\omega_1 + \omega_2 + \omega_3)\hbar^4} \tag{7.27}$$

where $n_0\langle\varepsilon^4\rangle = \int_0^\infty \frac{\partial^4 E}{\partial k_z^4} N(E) f_0 dE$ and the other notations are defined in [97]. The term $(\frac{\partial^4 E}{\partial k_z^4})$ can be formulated by using the dispersion relations of different materials as given in appropriate sections of this monograph. Thus one can investigate the $\chi_{NP}(\omega_1, \omega_2, \omega_3)$ for all materials as considered in this monograph.

7.2.9 *Generalized raman gain*

The generalized Raman gain in optoelectronic materials can be expressed as [98]

$$R_G = \bar{I}\left(\frac{16\pi^2 c^2}{\hbar\omega\rho g\omega_s^2 n_s n_p}\right)\left(\frac{\Gamma_\rho}{\Gamma}\right)\left(\left(\frac{e^2}{mc^2}\right)^2 m^2 R^2\right) \tag{7.28}$$

where, $\bar{I} = \sum_{n,t_z}[f_0(n, k_z, \uparrow) - f_0(n, k_z, \downarrow)]$, $f_0(n, k_z, \uparrow)$ is the Fermi factor for spin-up Landau levels, $f_0(n, k_z, \uparrow)$ is the Fermi factor for spin down Landau levels, n is the Landau quantum number and the other notations are defined in [98]. It appears then the formulation of R_G is determined by the appropriate derivation requires the magneto-dispersion relations. By using the different appropriate formulas as formulated in various chapters of this monograph R_G can, in general, be investigated.

7.2.10 *The plasma frequency*

The plasma frequency ω_p can be expressed as[99]

$$\omega_p^2 = \frac{e^2}{\varepsilon_{sc}} \int_{E'}^{\infty} N(E) v_x^2 \frac{\partial f_0(E)}{\partial E} dE \tag{7.29}$$

where E' can be obtained from $N(E') = 0$.

7.2.11 *The activity coefficient*

Dilute solution transport equations introduced by Shockley [100] have been very effective in describing the fluxes of electrons and holes in semiconductors. These expressions, however, are inaccurate at high carrier concentrations due to nonidealities associated with carrier degeneracy and bandgap narrowing. Within the framework of dilute solution theory, these nonidealities can be included by introduction of concentration-dependent activity coefficients. Expressions for activity coefficients should also be useful in the area of chemical kinetics where mass-action principles are applied.Reviews of the literature on activity coefficients are available. Of special interest is a paper by Hwang and Brews [100], where majority carrier activity coefficients were presented that included the effects of Fermistatistics and bandgap narrowing due to exchange interactions and impurity effects. Explicit expressions and plots of electron activity coefficients as a function of concentration for GaAs were included. Activity coefficients were derived under the assumption of complete ionization.No mention was made, however, of minority carrier screening by majority carriers, which are shown here to have a strong influence upon the majority carrier activity coefficients.

The activity coefficient in semiconductors can, in general be expressed as [100]

$$f_e^* = \frac{k_B T}{n_0} \int_0^{\infty} \frac{N(E)}{(1 - f_0)^2} \left[-\frac{\partial f_0}{\partial E} \right] dE \tag{7.30}$$

7.2.12 Magneto-thermal effect in quantized structures

The magneto-thermal effect can be expressed by the dimensionless quantity as [101]

$$\phi = \frac{B}{T}\frac{\partial T}{\partial B} = -\frac{B}{T}\left[\left(\frac{\partial S}{\partial \beta}\right)_{(T,n)}\right]\left[\left(\frac{\partial S}{\partial T}\right)_{(\beta,n)}\right]^{-1} \tag{7.31}$$

where S is the entropy which can be written as [102]

$$S = k_B\int_0^\infty -\left(\frac{\partial f_0}{\partial E}\right)\left(\frac{E-E_F}{k_BT}\right)\left[\int_0^E N(E)dE\right]dE \tag{7.32}$$

Under magnetic quantization we know, the DOS function assumes the form

$$N(E) = \left(\frac{eBg_v}{2\pi^2\hbar}\right)\sum_{n=0}^{n_{\max}}\frac{\partial k_z'(E)}{\partial E}H(E-E') \tag{7.33}$$

where E' is the Landau sub-band energy or Landau level.

Therefore,

$$S = k_B\sum_{n=0}^{n_{\max}}\int_{E'}^{\alpha} -\left(\frac{\partial f_0}{\partial E}\right)\left(\frac{E-E_F}{k_BT}\right)\frac{eBg_v}{2\pi^2\hbar}k_z dE \tag{7.34}$$

Since

$$\int N(E)dE = \frac{eBg_v}{2\pi^2\hbar}\sum_{n=0}^{n_{\max}}k_z(E) \tag{7.35}$$

$$S = c_0\sum_{n=0}^{n_{\max}}\left[\int_{E'}^{\alpha}\left\{k_zE\left(\frac{\partial f_0}{\partial E}\right)dE\right\}\right.$$
$$\left.-E_F\int_{E'}^{\alpha}\left\{k_zE\left(\frac{-\partial f_0}{\partial E}\right)dE\right\}\right] \tag{7.36}$$

where

$$c_0 = \left(\frac{eBg_v}{2\pi^2\hbar T}\right)$$

Using the concept of 3D Quantization (magneto-size quantization, Magneto NIPI, Magneto Inversion layers, Magneto accumulation Layers, magneto quantum dot, Quantum dot superlattices, magneto-size quantized superlattices and the respective heavily doped cases together with influence of light waves and electric fields on the 3D quantized structures), the DOS functions can respectively be written as

$$N(E) = \frac{g_v eB}{2\pi\hbar} \sum_{n=0}^{n_{z\max}} \sum_{n=0}^{n_{\max}} \delta'(E - E'') \tag{7.37}$$

$$N(E) = \frac{g_v eB}{2\pi\hbar} \sum_{i=0}^{i_{\max}} \sum_{n=0}^{n_{\max}} \delta'(E - E'') \tag{7.38}$$

and

$$N(E) = \frac{2g_v}{(d_x d_y d_z)} \sum_{n_x=1}^{n_{x\max}} \sum_{n_y=1}^{n_{y\max}} \sum_{n_z=1}^{n_{z\max}} \delta'(E - E''') \tag{7.39}$$

where E'' is the totally quantized energy in 2D systems in the presence of applied quantizing magnetic field along z-direction and E''' is the totally quantized energy in 0D systems in the absence of quantizing magnetic field along z-direction, (7.37) represents all the cases corresponding to E'', (7.38) represents all the cases corresponding to Magneto Inversion layers and Magneto accumulation Layers and (7.39) represents the rest cases.

In the three cases we can show

$$S = c_0'' \sum \sum E''[F_0(\eta'') - (\eta'')F_{-1}(\eta'')] \tag{7.40}$$

$$S = c_0''' \sum \sum E'''[F_0(\eta''') - (\eta''')F_{-1}(\eta''')] \tag{7.41}$$

where

$$c_0'' = \frac{g_v eBk_B}{2\pi\hbar}, \quad c_0''' = \frac{2g_v}{d_x d_y d_z}, \quad \eta'' = \frac{E_F - E''}{k_B T}, \quad \eta''' = \frac{E_F - E'''}{k_B T}$$

and E_F in this case is the Fermi energy for all the previous respective cases of 3D quantization of wave vector space.

7.2.13 *Normalized hall coefficient*

The normalized Hall coefficient in a semiconductor can be expressed as [103]

$$\frac{R(T)}{R(0)} = 1 + \frac{(\pi k_B T)^2}{3} \left[\frac{1}{N(E)} \frac{\partial N(E)}{\partial E} \right]^2 \Bigg|_{E=E_{FT}} \tag{7.42}$$

where $R(T)$is the Hall coefficient at $T, R(0)$ is the same at T $\to 0$ and E_{FT} is the Fermi energy at T.

7.2.14 *Reflection coefficient*

In the presence of light waves, the reflection coefficient (R) is given by [104]

$$R = \frac{(x-1)^2}{(x+1)^2} \tag{7.43}$$

where

$$x = \sqrt{\varepsilon_{sc} - \frac{n_0 e^2 \lambda^2}{m^* \pi c^2}}, m^*$$

is the effective electron mass and C is the velocity of light.

7.2.15 *Heat capacity*

The heat capacity of the carriers in semiconductors is a very important quantity.

The total heat capacity c_t is given by

$$c_t = c_e + c_l \tag{7.44}$$

where c_e is the electronic heat capacity and is given by [105]

$$c_e = \int_0^\infty -\frac{\partial f_0}{\partial E}\left[\frac{E-E_F}{T} + \frac{\partial E_F}{\partial T}\right](E-E_F)N(E)dE \tag{7.45}$$

Thus (7.44) can be written as

$$c_t = \frac{\pi^2 k_B^2 T N(E_F)}{3}\left[1 + T\left(\frac{\partial E_F}{\partial T}\right)\left(\frac{N'(E_F)}{N(E_F)}\right)\right] + c_l \tag{7.46}$$

7.2.16 *Magnetic susceptibilities*

The Landau dia-magnetic susceptibility (χ_d) can be written as

$$\chi_d = -\mu_0 \frac{\partial^2 F(B)}{\partial B^2} \tag{7.47}$$

where μ_0 is the free space permeability and $F(B)$ is the free energy which can be written as

$$F(B) = n_0 E_{FB} + \int_0^\infty D_2(E) \frac{\partial f_0}{\partial E} dE \tag{7.48}$$

where $D_2(E)$ is twice integration of the DOS function with respect to E.

The Pauli's paramagnetic susceptibility (χ_p) can be expressed as [106]

$$\chi_p = -2\mu_B^2 \int_{E_0}^\infty N(E) \frac{\partial f_0}{\partial E} dE \tag{7.49}$$

where μ_B is the Bohr magnetic field.

7.2.17 *Faraday rotation*

The Faraday rotation of the electrons in semiconductors can in general, be expressed under the condition $\omega\tau >> 1$, as [107]

$$\theta = AI \tag{7.50}$$

where ω is the angular frequency of the incident light, τ is the momentum relaxation time,

$$A \equiv \frac{e^3 l}{8\pi^3 \hbar^4 \omega^2} \left(\frac{1}{2 n_r \varepsilon_0 c} \right),$$

l is the thickness of the semiconductor specimen, n_r is the real part of the refractive index at that frequency and

$$I = \frac{(2\pi)^3}{2g_v} \int_0^\infty \frac{\partial E}{\partial k_x} \left[\frac{\partial E}{\partial k_y} \cdot \frac{\partial^2 E}{\partial k_x \partial k_y} - \frac{\partial E}{\partial k_x} \cdot \frac{\partial^2 E}{\partial k_y^2} \right] \frac{\partial f_0}{\partial E} N(E) dE \tag{7.51}$$

Thus from different EMs as formulated in this book, we can formulate the partial differentials and from the expressions of $N(E)$ as derived in different chapters we can study the Faraday rotations for various HD materials.

7.2.18 *Fowler-nordheim field emission*

The Fowler-Nordheim field emission (FNFE) is a well-known quantum-mechanical phenomenon that involves tunneling of electrons through a surface barrier due to the application of an intense external electric field. Normally, at field strengths of the order of 10^8 V/m (below the electrical breakdown), the potential barriers at the surfaces of metals and semiconductors usually become very thin and result in field emission of electrons due to the tunnel effect [108]. This has been well-investigated with reference to three dimensional electron gases in metals and semiconductors and the FNFE from quantum confined structures has also been studied in this context [109–115].

The current (I) due to Fowler-Nordheim {FN} field emission can be written as

$$I = \frac{1}{2} \sum_{n_x=1}^{n_{x\max}} \sum_{n_y=1}^{n_{y\max}} \left[\int_{E_{11}}^{i} [e \cdot v n_1 t_{11}] \right] \tag{7.52}$$

where is the magnitude of the electron charge, ν is the velocity of the electron and given by

$$v = \frac{1}{\hbar} \frac{\partial E}{\partial k_z} n_1 = N_{1D}(E) \cdot f_0 dE, \quad N_{1D}(E)$$

is the 1D DOS function per sub-bands and can be written as

$$N_{1D}(E) = \frac{2g_v}{\pi} \frac{\partial k_z}{\partial E},$$

t_{11} is the transmission coefficient which can be expressed as

$$t_{11} = \exp \left[-2 \int_i |k_i| di \right]$$

in which $i = x, i, z$

7.2.19 *Optical effective mass*

The investigation of the energy band structure of the noble metals is one of the standard problems in solid state physics. The topology and geometry of the Fermi surfaces of the noble metals are very well known in detail from the de Hass-van Alphen measurements of Shoenberg. Very accurate values for the cyclotron masses of the noble metals were experimentally obtained from de Hass-van Alphen investigations [116]. The optical effective mass being used to study the different optical properties of semiconductors can be expressed following Madelung [116] as

$$m_{opt} = 3\hbar^2 n_0 \left[\int_0^{\infty} N(E) \frac{\partial E}{\partial k} f_0 dE \right]^{-1} \tag{7.53}$$

7.2.20 *Einstein's photoemission*

Einstein's Photoemission (EP) is a physical phenomenon and occupies a singular position in the whole arena of materials science and related disciplines in general.The EP in recent years finds extensive applications in modern optoelectronics, characterization and investigation of condensed matter systems, photoemission spectroscopy and related aspects in connection with the investigations of the optical properties of nanostructures [117–121]. Interest in low dimensional silicon nanostructures also grew up and gained momentum, after the discovery of room temperature photoluminescence and electroluminescence of silicon nano-wires in porous silicon [117]. Work on ultrathin layers of $SiSiO_2$ superlattices resulting into visible light emission at room temperature clearly exhibited low dimensional quantum confinement effect [118] and one of the most popular technique for analyzing the low dimensional structures is to employ photoemission techniques. Recent observation of room temperature photoluminescence and electro luminescence in porous silicon has stimulated vigorous research activities in silicon nanostructures [119].

It is well-known that the classical equation of the photo-emitted current density is [123] $J = [4\pi\alpha_0 e m_c g_v (k_B T)^2/h^3] \exp[(hv - \phi)/(k_B T)]$, where $\alpha_0, e, m^*, g_v, k_B, T, h, hv$ and ϕ are the probability of photoemission, electron charge, effective electron mass at the edge of the conduction band, valley degeneracy, the Boltzmann constant, temperature, the Planck constant, incident photon energy along z-axis and work function respectively. The afore-mentioned equation is valid for both the charge carriers and in this conventional form it appears that, the photoemission changes with the effective mass, temperature, work function and the incident photon energy respectively. This relation holds only under the condition of carrier non-degeneracy.The EP has different values for different materials and varies with doping and with external fields which creates quantization of the wave-vector space of the carriers leading to various types of quantized structures. The nature of these variations has been studied in [117–148].

7.2.21 *Righi-Leduc coefficient*

It is well known that some transport effects under strong magnetic field are independent of relaxation mechanism and are determined by the EMs only. Among them one is thermoelectric power and the other one is Righi-Leduc coefficient, which can in turn, be written as [149]

$$R = \frac{cn_0}{eB^2\chi_{ph}}[\langle (E-E_F)^2\rangle - \langle E-E_F\rangle^2] \tag{7.54}$$

where, χ_{ph} is the phonon thermal conductivity,

$$\langle (E-E_F)^2\rangle = \frac{1}{n}\int (E-E_F)^2 G(E)\left[-\frac{\partial f_0}{\partial E}\right] dE, \quad G(E)$$

is the total number of states and is given by $G(E) = \int_{E_{30}}^{E} N(\zeta)d\zeta$ in which ζ is the variable of integration, E_{30} is the lower limit of integration and

$$\langle E-E_F\rangle = \frac{1}{n}\int (E-E_F)G(E)\left[-\frac{\partial f_0}{\partial E}\right] dE$$

7.2.22 *Electric susceptibility*

In semiconductors the effect of free carriers on the optical properties becomes important at wavelengths longer than the intrinsic absorption edge. Free carriers produce absorption and affect also the dispersion at sufficiently long wavelengths.Electric susceptibility can be written as [150]

$$\chi_c = \frac{-e^2}{3\omega^2\varepsilon_0\hbar^2}\int_0^\infty N(E)\frac{\partial E}{\partial k} f_0 dE \tag{7.55}$$

7.2.23 *Electric susceptibility mass*

In recent years there has been considerable interest in studying the electric susceptibility mass of the carriers in narrow gap materials because of its direct influence in the study of carrier scattering mechanisms in semiconductor. Electric susceptibility mass can be determined from measurements in the infrared region of the frequency dependence of the spectral reflectivity at normal incidence and provides useful information regarding band structures. Electric Susceptibility Mass can be written as [151]

$$m_s^* = 3A_{20}\left[\int_{E_{20}}^{\infty} N(E)\frac{\partial E}{\partial k} f_0 dE\right]^{-1} \tag{7.56}$$

where $A_{20} = n_0 \varepsilon_0 \hbar^2$ and E_{20} is the lower limit of integration.

7.2.24 *Electron diffusion thermo-power*

The electron diffusion thermopower (S_e) is an important property of AlGaN/GaN QLD structures since it yields useful information about the carrier transport mechanism. S_e can be written as [162]

$$S_e = -\frac{1}{n_0 e}\frac{df}{dT} \tag{7.57}$$

where,

$$f = n_0 E_F - \sum \int_{E_{21}}^{\infty} G(E) f_0 dE$$

in which

$$G(E) = \int_{E_{30}}^{E} N(\zeta) d\zeta,$$

E_{21} and E_{30} are the lower limits of integrations of f and G(E) respectively.

7.2.25 *Hydrostatic piezo-resistance coefficient*

The importance of Hydrostatic Piezo-resistance Coefficient ($\mathrm{H_C}$) is well known in the sensor technology and $\mathrm{H_c}$ can be expressed as [153]

$$H_c = \frac{-\partial n_0}{n_0 \partial P} \tag{7.58}$$

where P is the pressure in kilobar.

7.2.26 *Relaxation time for acoustic mode scattering*

The importance of momentum relaxation time for acoustic mode scattering in semiconductors is well known for the evaluation of the electron mobility and in the simplest scattering theory, the relaxation time for acoustic mode scattering ($\tau_{am}(E)$) [154] is directly proportional to the inverse of $N(E)$ and can, in general be expressed as

$$\tau_{am}(E) = \frac{\tau_0}{N(E)} \tag{7.59}$$

where τ_0 is a constant. Thus using various expressions of $N(E)$ in different cases we can study not only ($\tau_{am}(E)$) but also acoustic mode limited mobility in HD semiconductors and their nanostructures.

7.2.27 *Gate capacitance*

In recent years there has been considerable interest in studying the surface capacitance of inversion layers in Si MOS structures. The dependence of the MOS capacitance on gate voltage and on a quantizing magnetic field has been studied both theoretically and experimentally. The fact that the MOS capacitance can be very easily controlled by varying the gate voltage is of much importance from the point of view of technical applications. It may also be stated that, as far as the determination of the effective mass under the degenerate electron distribution at the surface is concerned, measurement of magneto-capacitance as compared to that of conductivity or cyclotron resonance would not be more advantageous regarding the experimental facilities required or accuracies achieved. Nevertheless, it is felt that the theoretical investigation presented here would be of much significance as the interest on gate capacitance has been growing very much in recent years from the point of view of exploration of other fundamental aspects of semiconductor surfaces of MOS structures [155]. The gate capacitance (C_g) can in general be expressed as

$$C_g = \frac{e\partial n_s}{\partial V_g} \tag{7.38}$$

where n_s is the surface electron concentration per unit area and V_g is the gate voltage. It may be noted that although C_g for inversion layers have been studied in literature but investigations of C_g in accumulation layers are rather small. Thus by using the expressions of the DOS functions of inversion and accumulation layers as given in this monograph we can study the gate capacitance for different important cases.

7.3 Open Research Problems

Investigate Carrier Statistics, Thermoelectric Power, Debye Screening Length, Carrier contribution to the elastic constants, Diffusivity-mobility ratio, Measurement of Band-gap in the presence of Light Waves, Diffusion coefficient of the minority carriers, Nonlinear optical response, Third order nonlinear optical susceptibility, Generalized Raman gain, The plasma frequency, The activity coefficient, Magneto-Thermal effect in Quantized Structures, Normalized Hall coefficient, Reflection coefficient, Heat Capacity, Magnetic Susceptibilities, Faraday rotation, Fowler-NordheimFied Emission, Optical Effective Mass, Einstein's Photoemission, Righi-Leduc coefficient, Electric Susceptibility, Electric Susceptibility Mass, Electron Diffusion Thermo-power, Hydrostatic Piezo-resistance Coefficient, Relaxation time for Acoustic Mode Scattering and Gate Capacitance for the appropriate problems of this monograph and perform related experiments for experimental investigations.

References

1. K. v. Klitzing, G. Dorda, M. Pepper, *Phys. Rev. Lett.* **45**, 494 (1980); K. v. Klitzing, *Rev. Mod. Phys.* **58**, 519 (1986)
2. J. Hajdu, G. Landwehr, In: *Strong and Ultrastrong Magnetic Fields and Their Applications*, F. Herlach (ed.), pp. 17 (Springer 1985)
3. I. M. Tsidilkovskii, *Band Structure of Semiconductors* (PergamonPress, Oxford, 1982)
4. E. A. Arushanov, A. F. Knyazev, A. N. Natepov, S. I. Radautsan, *Sov. Phys. Semi.* **15**, 828 (1981)
5. S. P. Zelenim, A. S. Kondratev, A. E. Kuchma, *Sov. Phys. Semi.* **16**, 355 (1982)

6. F. M. Peeters, P. Vasilopoulos, *Phys. Rev. B* **46**, 4667 (1992)
7. W. Zawadzki, In: *Two-Dimensional Systems, Heterostructures and Superlattices*, Springer Series in Solid State Sciences, Vol. 53, edited by G. Bauer, F. Kuchar, H. Heinrich (Springer, Berlin, Heidelberg, 1984).
8. B. M. Askerov, N. F. Gashimzcede, M. M. Panakhov,*Sov. Phys. Sol. State* **29**, 465 (1987).
9. G. P. Chuiko, *Sov. Phys. Semi.* **19**, 1279 (1985).
10. S. Pahari, S. Bhattacharya, K. P. Ghatak, *J. Comput. Theor. Nanosci.*, Invited review article, 2009 [In the press].
11. K. P. Ghatak, S. Bhattacharya, S. Bhowmik, R. Benedictus, S. Choudhury, *J. Appl. Phys.* **103**, 034303 (2008).
12. K. P. Ghatak, S. Bhattacharya, S. Pahari, D. De, S. Ghosh, M. Mitra, *Annalen der Physik* **17**, 195 (2008).
13. K. P. Ghatak, S. N. Biswas, *J. Appl. Phys.* **70**, 299 (1991).
14. K. P. Ghatak, S. N. Biswas, *J. Low Temp. Phys.* **78**, 219 (1990).
15. K. P. Ghatak, M. Mondal, *J. Appl. Phys.* **65**, 3480 (1989).
16. K. P. Ghatak, B. Nag, *Nanostructured Materials* **5**, 769 (1995).
17. K. P. Ghatak, M. Mondal, *Phys. Stat. Sol. (b)* **185**, K5 (1994).
18. K. P. Ghatak, B. Mitra, *IlNuovoCimento D* **15**, 97 (1993).
19. K. P. Ghatak, S. N. Biswas, *Phys. Stat. Solidi (b)* **140**, K107 (1987).
20. K. P. Ghatak, *IlNuovoCimento D* **13**, 1321 (1991).
21. K. P. Ghatak, A. Ghoshal, *Phys. Stat. Sol. (b)* **170**, K27 (1992).
22. K. P. Ghatak, B. De, B. Nag, P. K. Chakraborty, *Molecular Crystals and Liquid Crystals Science and Technology Section B: Nonlinear Optics* **16**, 221 (1996).
23. K. P. Ghatak, M. Mitra, B. Goswami, B. Nag, *Molecular Crystals and Liquid Crystals Science and Technology Section B: Nonlinear Optics* **16**, 167 (1996).
24. K. P. Ghatak, D. K. Basu, D. Basu, B. Nag, *NuovoCimentodellaSocietaItaliana di Fisica D — Condensed Matter, Atomic, Molecular and Chemical Physics, Biophysics* **18**, 947 (1996).
25. B. Mitra, K. P. Ghatak, *Phys. Letts. A* **141**, 81 (1989);S. K. Biswas, A. R. Ghatak, A. Neogi, A. Sharma, S. Bhattacharya, K. P. Ghatak, *Physica E:Low-Dimensional Systems and Nanostructures* **36**, 163 (2007).
26. M. Mondal, A. Ghoshal, K. P. Ghatak, *Il NuovoCimento D* **14**, 63 (1992); K. P. Ghatak, M. Mondal*Phys. Stat. Sol. (b)*, **135**, 819 (1986); L. J. Singh, S. Choudhury, D. Baruah, S. K. Biswas, S. Pahari, K. P. Ghatak, *Phys. B* **368**, 188 (2005).
27. K. P. Ghatak, B. De, M. Mondal, S. N. Biswas, *Materials Research Society Symposium — Proceedings* **198**, 327 (1990); K. P. Ghatak, *Proceedings of SPIE — The International Society for Optical Engineering* **1584**, 435 (1992).
28. K. P. Ghatak, B. De, *Materials Research Society Symposium — Proceedings* **234**, 55 and 59 (1991).
29. K. P. Ghatak, B. De, M. Mondal, S. N. Biswas, *Materials Research Society Symposium — Proceedings* **184**, 261 (1990).

30. M Mondal, KP Ghatak, PhysicaScripta **31**, 613 (1985)
31. K. P. Ghatak, S. N. Biswas, *Materials Research Society Symposium — Proceedings* **216**, 465 (1990).
32. P. T. Landsberg, Eur. J. Phys. **2**, 213(1981).
33. R. B. Dingle, Philos. Mag. **46**, 813(1955); D. Redfield, M. A. Afromowitz, *ibid.* **19**, 831(1969); H. C. Casey, F. Stern, J. Appl. Phys. **47**, 631(1976).
34. S. N. Mohammad, J. Phys. C **13**, 2685(1980).
35. K. P. Ghatak, S. Bhattacharya, S. Pahari, S. N. Mitra, P. K. Bose, D. De, *J. Phys. Chem. Sol.* **70**, 122 (2009); S. Chowdhary, L. J. Singh, K. P. Ghatak, *Physica B: Condensed Matter* **365**, 5 (2005); B. Nag, K. P. Ghatak, *Molecular Crystals and Liquid Crystals Science and Technology Section B: Nonlinear Optics* **19**, 1 (1998); K. P. Ghatak, B. Mitra, *Phys. Scr.* **46**, 182 (1992);
36. K. P. Ghatak, S.Bhattacharya, J. Appl. Phys. **102**, 073704 (2007); M. Mondal, K. P. Ghatak, Phys. Lett. **102A**, 54(1984); S. Bhattacharya, N. C Paul, K. P Ghatak, Physica B, **403,** 4139(2008); K. P Ghatak, S. Bhattacharya, H. Saikia, D. Baruah, A. Saikia, K. M. Singh, A. Ali, S. N. Mitra, P. K. Bose, A. Sinha, Journal of Comp. and Theo. Nanoscience, **3**, 727(2006); P. K. Chakraborty, G. C. Datta, K. P. Ghatak, Phys. Scr. **68**, 368 (2003). E. O. Kane, Solid-State Electron. **28**, 3(1985); W. Zawadzki, in *Handbook on Semiconductors*, edited by W. Paul, North Holland, New York, (1982), **1**, p. 715;
37. A. N. Chakravarti, D. Mukherjee, Phys. Lett. **53A**, 403(1975).
38. B. Mitra, D. K. Basu, B. Nag, K. P. Ghatak, Nonlinear Opt. **17**, 171(1997)
39. M. Mondal, K. P. Ghatak, Phys. Status Solidi B **135**, 239 (1986)
40. A. N. Chakravarti, S. Swaminathan, Phys. Status Solidi A**23**, K191 (1974); A. N. Chakravarti, *ibid.* **25**, K105 (1974).
41. A. N. Chakravarti, K. P. Ghatak, K. K. Ghosh, A. Dhar, Phys. Status Solidi B **103**, K55 (1981)
42. T. Ando, A. H. Fowler, and F. Stern, Rev. Mod. Phys. **54**, 437 (1982); P. K. Basu, *Optical Processes in Semiconductors*; Oxford University Press, New York, 2001.
43. N. Kampf, D. Ben-Yaakov, D. Andelman , S. A. Safran, J. Klein, Phys. Rev. Lett., **103**, 118304 (2009); P. Arnold, L. G. YafFe, Phys. Rev. D. **52**, 7208(1995); G. S. Kulkarni and Z. Zhong, Nano letters, **12**, 719(2012); E. Stern, R. Wagner, F. J. Sigworth, R. Breaker, T. M. Fahmy, M. A. Reed, *Nano Lett.*, **7**, 3405(2007)
44. A. N. Chakravarti, D. Mukherji; Phys. Lett. A, **53**, 57(1975); W. Zawadzki, Adv. Phys. **23**, 435 (1974).
45. A. K. Sreedhar and S. C. Gupta, Phys. Rev. B **5**, 3160 (1972); R. W. Keyes, IBM. J. Res. Devlopm. **5**, 266(1961); Solid State Phys. **20**, 37(1967).
46. S. Bhattacharya, S. Chowdhury and K. P. Ghatak, J. Compu. Theor. Nanosc., **3**, 423(2006); K. P. Ghatak, J. Y. Siddiqui and B. Nag, Phys. Letts. A, **282**, 428 (2001); K. P. Ghatak, J. P. Banerjee and B. Nag, J. Appl. Phys., **83**, 1420 (1998); K. P. Ghatakand B. Nag, Nanostruc. Material., **10**, 923 (1998); B. Nag and K. P. Ghatak, J. Phys. Chem. Sol., **58**, 427 (1997);

47. K. P. Ghatak, J. P. Banerjee, B. Goswami and B. Nag, in Nonlin. Opt. Quant. Opt., **16**, 241 (1996); K. P. Ghatak, J. P. Banerjee, D. Bhattacharyya and B. Nag, N*anotechnology*, **7**, 110(1996); K. P. Ghatak, J. P. Banerjee, M. Mitra and B. Nag, Nonlin. Opt. 17, 193 (1996); K. P. Ghatak, Inter. J. Electron., **71**, 239(1991).
48. K. P. Ghatak, B. De, S. N. Biswas and M. Mondal, Mechanical behavior of materials and structures in microelectronics, MRS Symposium Proceedings, Spring Meeting, **2216**,191, (1991); K. P. Ghatak and B. De, MRS Symposium Proceedings, **226**, 191 (1991); K. P. Ghatak, B. Nag and G. Majumdar, MRS, **379**, 109, (1995); D. Baruah, S. Choudhury, K. M. Singh and K. P. Ghatak, J. Phys., Conf. Series, **61**, 80, (2007).
49. K. P. Ghatak, S. Bhattacharya, D. De, *Einstein Relation in Compound Semiconductors and Their Nanostructures*; Springer Series in Materials Science **116**, (Springer, Heidelberg, 2009)
50. A. Einstein, Ann. der Physik, **17**, 549 (1905); H. Kroemer, IEEE Trans, Electron Devices **25**, 850 (1978); W. Nernst, Z. Phys. Chem., **2**, 613 (1888); J. S. Townsend, Trans. Roy. Soc. **193A**, 129 (1900); C. Wagner, Z. Physik, Chem. **B21**, 24 (1933); C. Herring and M. H. Nichols, Rev. Mod. Phys. **21**, 185 (1949); P. T. Landsberg, *Thermodynamics and Statistical Mechanics*, (Oxford University Press, Oxford 1978); *In Recombination in Semiconductors*, (Cambridge University Press, U.K. 1991);
51. R. W. Lade, Proc. IEEE **52**, (743) (1965)
52. P. T. Landsberg, Proc R. Soc. **A 213**, (226) (1952); Proc. of Phys. Soc. **A62**, (806) (1949);
53. C. H. Wang, A. Neugroschel, IEEE Electron. Dev. Lett.**ED-11**, (576) (1990)
54. I. Y. Leu, A. Neugroschel, IEEE Trans. Electron. Dev. **ED-40**, (1872) (1993)
55. F. Stengel, S. N. Mohammad, H. Morkoç, J. Appl. Phys. **80**, (3031) (1996)
56. H. J. Pan, W. C. Wang, K. B. Thai, C. C. Cheng, K. H. Yu, K. W. Lin, C. Z. Wu, W. C.Liu, Semicond. Sci. Technol. **15**, (1101) (2000)
57. S. N. Mohammad, J. Appl. Phys. **95**, (4856) (2004)
58. V. K. Arora, Appl. Phys. Lett. **80**, (3763) (2002)
59. S. N. Mohammad, J. Appl. Phys. **95**, (7940) (2004)
60. S. N. Mohammad, Philos. Mag. **84**, (2559) (2004)
61. S. N. Mohammad, J. Appl. Phys. **97**, (063703) (2005)
62. K. Suzue, S. N. Mohammad, Z. F. Fan, W. Kim, O. Aktas, A. E. Botchkarev, H. Morkoç, J. Appl. Phys. **80**, (4467) (1996)
63. S. N. Mohammad, Z. F. Fan, W. Kim, O. Aktas, A. E. Botchkarev, A. Salvador, H. Morkoç, Electron. Lett. **32**, (598) (1996)
64. Z. Fan, S. N. Mohammad, W. Kim, O. Aktas, A. E. Botchkarev, K. Suzue, H. Morkoç, J. Electron. Mater. **25**, (1703) (1996)
65. C. Lu, H. Chen, X. Lv, X. Xia, S. N. Mohammad, J. Appl. Phys. **91**, (9216) (2002)
66. S. G. Dmitriev, Yu V. Markin, Semiconductors **34**, (931) (2000)

67. M. Tao, D. Park, S. N. Mohammad, D. Li, A. E. Botchkerav, H. Morkoç, Philos. Mag. B **73**, (723) (1996)
68. D. G. Park, M. Tao, D. Li, A. E. Botchkarev, Z. Fan, S. N. Mohammad, H. Morkoç, J. Vac. Sci. Technol. B **14**, (2674) (1996)
69. Z. Chen, D. G. Park, S. N. Mohammad, H. Morkoç, Appl. Phys. Lett. **69**, (230) (1996)
70. A. N. Chakravarti, D. P. Parui, Phys. Letts **40A**, (113) (1972)
71. A. N. Chakravarti, D. P. Parui, Phys. Letts **43A**(60), (1973); B. R. Nag, A. N. Chakravarti, Solid State Electron. **18**, (109) (1975), Phys. Stal Sol. (a) **22**, (K153) (1974)
72. B. R. Nag, A. N. Chakravarti, P. K. Basu, Phys. Stat. Sol. (a) **68**, (K75) (1981)
73. B. R. Nag, A. N. Chakravarti, Phys. Stal Sol. (a) **67**, (K113) (1981)
74. A. N. Chakravarti, B. R. Nag, Phys. Stat. Sol. (a) **14**, (K55) (1972); Int. J. Elect. **37**, (281) (1974); Phys. Stat. Sol. (a) **14**, (K23) (1972); A. N. Chakravarti, D. P. Parui, Canad. J. Phys. **51**, (451) (1973). D. Mukherjee, A. N. Chakravarti, B. R. Nag, Phys. Stat. Sol (a) **26**, (K27) (1974); S. Ghosh, A. N. Chakravarti, Phys. Stat. Sol. (b), **147**, (355) (1988); A. N. Chakravarti, K. K. Ghosh, K. P. Ghatak, H. M. Mukherjee, Phys. Stat. Sol (b) **118**, (843) (1983); A. N. Chakravarti, K. P. Ghatak, K. K. Ghosh, G. B. Rao, Phys. Stat. Sol (b) **111**, (K61) (1982); A. N. Chakravarti, K. P. Ghatak , K. K. Ghosh, H. M. Mukherjee, S. Ghosh, Phys. Stat. Sol (b) **108**, (609) (1981); A. N. Chakravarti, K. P. Ghatak , S. Ghosh, A. Dhar, Phys. Stat. Sol (b) **105**, (K55) (1981); A. N. Chakravarti, A. K. Choudhury, K. P. Ghatak , D. Roy Choudhury, Phys. Stat. Sol (b) **59**, (K211) (1980); A. N. Chakravarti, A. K. Chowdhury, K. P. Ghatak, D. R. Choudhury, Acta Phys. Polon. A **58**, (251) (1980); K. P. Ghatak, A. K. Chowdhury, S. Ghosh, A. N. Chakravarti, Phys. Stat. Sol (b) **99**, (K55) (1980); A. N. Chakravarti, A. K. Chowdhury, K. P. Ghatak, D. R. Choudhury ,Czech. Jour. of Phys. B **30**, (1161) (1980); A. N. Chakravarti, K. P. Ghatak, A. Dhar, K. K. Ghosh, S. Ghosh, Acta Phys. Polon A **60**, (151) (1981); A. N. Chakravarti, A. K. Chowdhury, K. P. Ghatak, A. Dhar, D. R. Choudhury, Czech. Jour. of Phys. B **31**, (905) (1981); A. N. Chakravarti, K. P. Ghatak, K. K. Ghosh, S. Ghosh, H. M. Mukherjee, Czech. Jour. of Phys. B **31,** (1138) (1981)
75. P. T. Landsberg, J. of Appl. Phys. **56**, (1696) (1984); P. T. Landsberg, Phys. Rev. B **33**, (8321) (1986); Y. Roichman, N. Tessler, Appl. Phys. Lett. **80**, (1948) (2002); J. M. H. Peters, Eur. J. Phys. **3**, (19) (1982); H. Van Cong, S. Brunet, S. Charar, Phys. Stat. Solidi B **109**, (K1) (1982); H. Van Cong, Phys. Stat. Solidi A **56**, (395) (1979); H. Van Cong, Solid State Electron. **24**,(495) (1981)
76. S. N. Mohammad, S. T. H. Abidi, J. Appl. Phys. **61**, (4909) (1987); Solid State Electron. **27**, (1153) (1985); J. Appl. Phys. **56**, (3341) (1984); M.A. Sobhan, S. N. Mohammad, J. Appl. Phys. **58**, (2634) (1985); S. N. Mohammad, A. V. Bemis, IEEE Trans. Electron. Dev. ED-**39**, (282) (1992); S. N. Mohammad, R. L. Carter, Philos. Mag. B **72**, (13) (1995);

S. N. Mohammad, Solid State Electron. **46**, (203) (2002); S. N. Mohammad, J. Chen, J. I. Chyi, H. Morkoç, Appl. Phys. Lett. **56**, (937) (1990)
77. P. T. Landsberg, S. A. Hope Solid State Electronics **20**, (421) (1977); S. A. Hope, G. Feat, P. T. Landsberg, J. Phys. A. math. Gen. **14**, (2377) (1981)
78. W. Elsäber, E. O. Göbel, Electron. Lett. **19**, (335) (1983); R. Hilfer, A. Blumen, Phys. Rev. A **37**, (578) (1988); T. G. Castner, Phys. Rev. B **55**, (4003) (1997); E. Barkai, V. N. Fleurov, Phys. Rev. E., **58**, (1296) (1998); T. H. Nguyen, S. K. O'Leary, Appl. Phys. Lett. **83**, (1998) (2003); T. H. Nguyen, S. K. O'Leary, J. Appl. Phys. **98**, (076102) (2005); C. G. Rodrigues, Á. R. Vasconcellos, R. Luzzi, J. Appl. Phys. **99**, (073701) (2006)
79. R. K. Jain, Phys. Stat. Sol. (a) **42**, (K221) (1977); B. A. Aronzon, E. Z. Meilikhov, Phys. Stat. Sol. (a) **19**, (313) (1973)
80. S. Choudhury, D. De, S. Mukherjee, A. Neogi, A. Sinha, M. Pal, S. K. Biswas, S. Pahari, S. Bhattacharya, K. P. Ghatak, J. Comp. Theo. Nanosc., **5**, (375) (2008); S. M. Adhikari, K. P. Ghatak, Quantum Matter **2**, (296) (2013); S. Pahari, S. Bhattacharya, D. De, S. M. Adhikari, A. Niyogi, A. Dey, N. Paitya, K. P. Ghatak, Physica B: Condensed Matter **405**, (4064) (2010); K. P. Ghatak, S. Bhattacharya, S. Pahari, D. De, R BenedictusSuperlatt. Microst. **46**, (387) (2009)
81. J. P. Bouchaud, A. Georges, Phys. Rep. **195**, (127) (1996); V. Blickle, T. Speck, C. Lutz, U. Seifert, C. Bechinger, Phys. Rev. Lett. **98**, (210601) (2007); Y. Kang, E. Jean, C. M. Fortmonn, Appl. Phys. Lett., **88**, (112110) (2006); F. Neumann, Y. A. Genenko, H. V. Seggern, J. Appl. Phys., **99**, (013704) (2006); J. van de Lagemaat, Phys. Rev. B., **73**, (235319) (2005); Q. Gu, E. A. Schiff, S. Grneber, F. Wang, R. Schwarz, Phys. Rev. Lett., **76**, (3196) (1996); M. Y. Azbel, Phys. Rev. B., **46**, (15004) (1992)
82. A. H. Marshak, Solid State Electron. **30**, (1089) (1987); A. H. Marshak, C. M. V. Vliet, Proc. IEEE **72**, (148) (1984); C. M. V. Vliet, A. van der Zeil, Solid State Electron., **20**, (931) (1977)
83. A. Khan, A. Das, Appl. Phys. A, **89**, (695) (2007)
84. O. Madelung, *Semiconductors: Data handbook*, 3rd Ed. Springer (2004); M. Krieehbaum, P. Kocevar, H. Pascher, G. Bauer, *IEEE QE*, **24** (1727) (1988)
85. K. P. Ghatak, S. N. Biswas, J. of Appl. Phys., **70**, (4309) (1991); K. P. Ghatak, B. Mitra, M. Mondal Ann. der Physik, **48**, (283) (1991); B. Mitra, K. P. Ghatak, in Phys. Letts., **135A**, (397) (1989); K. P. Ghatak, N. Chattopadhyay, M. Mondal, J. Appl. Phys., **63**, (4536) (1988); J. Low Temp. Phys., **73**, (321) (1988)
86. K. P. Ghatak, D. Bhattacharyya, Phys. Letts. **A184**, (366) (1994)
87. K. P. Ghatak, D. Bhattacharyya, PhysicaScripta**52**, (343) (1995); M. Mondal, K. P. Ghatak, J. Mag. Mag. Materials, **62**, (115) (1986)
88. K. P. Ghatak, B. Nag, D. Bhattacharyya, J. of Low Temp. Phys.**14**, (1) (1995)
89. K. P. Ghatak, M. Mondal, Thin Solid Films **148**, (219) (1987)
90. K. P. Ghatak, S. N. Biswas, Nanostructured Materials **2**, (91) (1993)

91. K. P. Ghatak, *Influence of Band Structure on Some Quantum Processes in Nonlinear optical Semiconductors*, D. Eng. Thesis, (1991), Jadavpur University, Kolkata, India.
92. K. P. Ghatak, N. Chattropadhyay, M. Mondal, *Appl. Phys.* A, **44**, (305) (1987)
93. S. N. Biswas , K. P. Ghatak, *Proceedings of the Society of Photo-optical and Instrumentation Engineers (SPIE), Quantum Well and Superlattice Physics*, **792**, (239) (1987); K. P. Ghatak, M. Mondal , S. Bhattacharyya, SPIE, **1284**, (113) (1990); K. P. Ghatak, S. Bhattacharyya, M. Mondal, SPIE, **1307**, (588) (1990); K. P. Ghatak , B. De, *Defect Engineering in Semiconductor Growth, Processing and Device technology Materials Research Society Proceedings*, (MRS) Spring meeting, **262**, (911) (1992); S. Bhattacharya, K. P. Ghatak, S. N. Biswas, *Optoelectronic materials, Devices, Packaging Interconnects*, SPIE, **836**, (72) (1988)
94. M. Mondal, K. P. Ghatak, J. Phys. C, (Sol. State.), **20**, (1671) (1987); M. Mondal. S. N. Banik, K. P. Ghatak, Canad. J. Phys. **67**, (72) (1989); K. P. Ghatak, M. Mondal, J. Appl. Phys. **70**, (1277) (1992)
95. P. K. Chakraborty, S. Bhattacharya, K. P. Ghatak, Jour. Appl. Phys. **98**, 053517(2005)
96. A. S. Filipchenko, I. G. Lang, D. N. Nasledov, S. T. Pavlov, L. N. Radaikine, Phys. Stat. Sol. (b) **66**, 417 (1974)
97. M. Wegener, *Extreme Nonlinear Optics* (Springer-Verlag, Germany,2005)
98. B. S. Wherreff, W. Wolland, C. R. Pidgeon, R. B. Dennis, S. D. Smith, in *Proceedings of the* 12th *International Conference of the Physics of the Semiconductors*, ed. by M. H. Pilkahn, R. G. Tenbner (Staffgard,1978), p. 793
99. C. H. Peschke, Phys. Stat. Sol. (b) **191**, 325 (1995)
100. P. T. Landsberg and H. C. Cheng, Phys. Rev. 32B, 8021 (1985); Semiconductors,"D. van Nostrand Co., Inc., New York (1950); H. Reiss, C. S. Fuller, and F. J. Morin, *Bell Syst. Tech.* J., **35**, 535 (1956); P. T. Landsberg and A. G. Guy, *Phys. Rev.* B, **28**, 1187(1983); "Handbook on Semiconductors," W. Paul, Editor, pp. 359–449, North-Holland Publishing Co. (1982);
101. A. Isihara, Y. Shiwa, J. Phys. C, **18**, 4703 (1985)
102. Semiconducting Lead Chalocogenides, V. G. I. Ravish, B. A. Efimova and I. A. Smirnov, Plenum Press, New York, p. 163 (1970)
103. F. J. Blatt, J. Phys. Chem. Solids **17**, 177 (1961)
104. D. K. Roy, Quantum Mechanical Tunnelling and its applications, p. **309** 1986, World Scientific Publication Co.Ptc.Ltd
105. W. Zawadzki, Physica **127**, 388 (1980)
106. E. V. Rozenfeld, Yu. P. Irkhin and P. G. Guletskii, Sov. Phys. Solid State **29**, 1134(1987)
107. H. Piller, Semiconductors and Semimetals, Ed. by R. K. Willardson and A. C. Beer, **124, 8** (1972)
108. R. H. Fowler , L. Nordheim, Proc.of Royal Society of London,Series-A, **119**, 173 (1928); A. Van Der Ziel, *SolidState Physical Electronics* (Prentice-Hall, Inc., Englewood Cliffs U.S.A. 1957 (p.176)

109. B. Mitra, K. P. Ghatak, Phys. Lett. A, **357**, 146(1990); K. P. Ghatak, M. Mondal, Jour. Mag. Mag. Mat., **74**, 203(1988); K. P. Ghatak, B. Mitra, Phys. Lett., **156A**,. 233, (1991); K.P.Ghatak, A.Ghosal, S.N.Biswas ,M.Mondal, Proc. of SPIE, USA, **1308**, 356 (1990)
110. V. T. Binh, Ch. Adessi, Phys. Rev. Lett. **85**, 864(2000); R. G. Forbes,Ultramicroscopy, **79**, 11(1999); J. W. Gadzuk, E. W. Plummer, Rev. Mod. Phys, **45**, 487(1973); J. M. Beebe, B. Kim, J. W. Gadzuk, C. D. Frisbie, J. G. Kushmerick, Phys. Rev. Lett, **97**, 026801(1999); Y. Feng, J. P. Verboncoeur, Phys. Plasmas, **12**, 103301(2005); W. S. Koh, L. K. Ang, Nanotechnology, **19**, 235402(2008); M. Razavy, *QuantumTheoryof Tunneling* (World Scientific Publishing Co. Pte. Ltd, Singapore2003).
111. S. I. Baranchuk, N. V. Mileshkina, SovietPhys. SolidState, **23** 1715(1981); P. G. Borzyak, A. A. dadykin, SovietPhys. Docklady **27**, 335(1982); S. Bono, R. H. Good, Jr., SurfaceSci.**134**, 272(1983); S. M. Lyth, S. R. P. Silva, Appl. Phys. Letts. **90**, 173124(2007); C. Xu, X. Sun, Int. Jour. of Nanotech. **1**, 452(2004); S. D. Liang, L. Chen, Phys. Rev. Letts. **101**, 027602(2008)
112. E. C. Heeres, E. P. A. M. Bakkers, A. L. Roest, M. Kaiser, T. H. Oosterkamp, N. de Jonge, Nano Lett., **7**, 536 (2007); L. Dong, J. Jiao, D. W. Tuggle, J. M. Petty, S. A. Elliff , M. Coulter, Appl. Phys. Lett., **82**, 1096 (2003); S. Y. Li , P.Lin, C. Y. Lee , T. Y. Tseng, Jour. of Appl. Phys., **95**, 3711(2004); N. N. Kulkarni, J. Bae, C. K. Shih, S. K. Stanley, S. S. Coffee, J. G. Ekerdt ,*Appl. Phys. Lett.*, **87**, 213115 (2005)
113. K. Senthil, K. Yong, Mat. Chem. and Phys., **112**, 88, (2008); R. Zhou, H. C. Chang, V. Protasenko, M. Kuno, A. K. Singh, D. Jena, H. Xing, Jour. of Appl. Phys., **101**, 073704 (2007); K. S. Yeong, J. T. L. Thong, Jour. of Appl. Phys., **100**, 114325 (2006); C. H. Oon, S. H. Khong, C. B. Boothroyd, J. T. L. Thong, Jour.of Appl. Phys., **99**, 064309 (2006)
114. B. H. Kim, M. S. Kim, K. T. Park, J. K. Lee, D. H. Park, J. Joo, S. G. Yu , S. H. Lee, Appl. Phys. Lett., **83**, 539 (2003); Z. S. Wu, S. Z. Deng, N. S. Xu, J. Chen, J. Zhou, J. Chen, Appl. Phys. Lett., **80**, 3829 (2002); Y. W. Zhu, T. Yu, F. C. Cheong, X. J. Xu, C. T. Lim, V. B. C. Tan, J. T. L. Thong, C. H. Sow, Nanotechnology, **16**, 88 (2005); Y. W. Zhu, H. Z. Zhang, X. C. Sun, S. Q. Feng, J. Xu, Q. Zhao, B. Xiang, R. M. Wang, D. P. Yu, Appl. Phys. Lett., **83**, 144(2003); S. Bhattacharjee, T. Chowdhury, Appl. Phys. Lett., **95**, 061501 (2009); S. Kher, A. Dixit, D. N. Rawat, M. S. Sodha, Appl. Phys. Lett. **96**, 044101 (2010)
115. I. Shigeo, W. Teruo, O. Kazuyoshi, T. Masateru, U. Satoshi, N. Norio, Jour. of Vac. Sc. and Tech. B:, **13**, 487(2009); C. A. Spindt, I. Brodie, L. Humphrey, E. R. Westerberg, *Jour. of Appl. Phys.*, **47**, 5248 (2009); Q. Fu, A. V. Nurmikko, L. A. Kolodziejski, R. L. Gunshor, J. W. Wu, *Appl. Phys. Lett.*, **51**, 578 (2009)
116. I. Mertig, E. Mrosan, V. N. Antonov, V.I. Antonov, P. Ziesche, Phys. Stat. sol. (b), **135**, K13 (1986); O. Madelung, Physics of III-V compounds John Wiley and Sons, Inc. New York, 1966, p. 80
117. L. T. Canham, Appl. Phys. Letts. **57**, 1046 (1990)
118. Z. H. Lu, D. J. Lockwood and J. M. Baribeam, Nature **378**, 258 (1995)

119. A. G. Cullis, L. T. Canham, P. D. O. Calocott, J. Appl. Phys. **82**, 909 (1997)
120. M. Cardona, L. Ley, *Photoemission in Solids 1 and 2, Topics in Applied Physics*, vols. **26**, **27**, (Springer-Verlag, Germany, 1978); S. Hüfner, *Photoelectron Spectroscopy*, (Springer, Germany, 2003); S. Hüfner, (Ed.), *Very High Resolution Photoelectron Spectroscopy, Lecture Notes in Physics*, Vol. 715 (Springer-Verlag, Germany, 2007); D. W. Lynch, C. G. Olson, *Photoemission Studies of High-Temperature Superconductors*, (Cambridge University Press, UK, 1999); *Photoemission and the Electronic Properties of Surfaces* ed. B. Feuerbacher, B. Fitton, R. F. Willis (Wiley, New York, 1978); W. Schattke, M. A. V. Hove, *Solid-State Photoemission and Related Methods: Theory and Experiment*, (Wiley, USA, 2003); V. V. Afanas'ev, *Internal Photoemission Spectroscopy: Principles and Applications* (Elsevier, North Holland, 2010); D. J. Lockwood, *LightEmission in Silicon in silicon based materials and devices*, vol 2, ed H. S. Nalwa (Academic Press, San Diego, USA, 2001)
121. K. P. Ghatak, D. De, S. Bhattacharya, *Photoemission from optoelectronic materials and their nanostructures*, Springer Series in Nanostructure Science and Technology, (Springer, 2009)
122. B. Mitra, K. P. Ghatak, Solid-state electronics **32**, 810 (1989); S. Choudhury, L. J. Singh, K. P. Ghatak, Nanotechnology **15**, 180 (2004); K. P. Ghatak, A. K. Chowdhury, S. Ghosh, A. N. Chakravarti, Appl. Phys. **23**, 241 (1980); K. P. Ghatak, D. K. Basu, B. Nag, J. of Phys. and Chem. of Solids **58**, 133 (1997); P. K. Chakraborty, G. C. Datta, K. P. Ghatak, Physica B: Condensed Matter **339**, 198 (2003); K. P. Ghatak, A. Ghoshal, B. Mitra, Il NuovoCimento **D 14**, 903 (1992); B. Mitra, A. Ghoshal, K. P. Ghatak, Phys. Stat. Sol. (b), **154**, K147 (1989); K. P. Ghatak, S. Bhattacharya, S. Bhowmik, R. Benedictus, S. Choudhury J. Appl. Phys., **103**, 094314 (2008); M. Mondal, S. Bhattacharya, K. P. Ghatak Appl. Phys. **A 42**, 331 (1987); A. N. Chakravarti, K. P. Ghatak, A. Dhar, K. K. Ghosh, S. Ghosh, Appl. Phys. A **26**, 165 (1981)
123. R. K. Pathria, *Statistical Mechanics, 2nd ed.* (Butterworth-Heinemann, Oxford, 1996)
124. D. De, S. Bhattacharya, S. M. Adhikari, A. Kumar, P. K. Bose, K. P. Ghatak, Beilstein Jour. Nanotech. **2** , 339 (2012); D. De, Kumar, S. M. Adhikari, S. Pahari, N. Islam, P. Banerjee, S. K. Biswas, S. Bhattacharya, K. P. Ghatak, Superlatti. andMicrostruc. **47**, 377 (2010); K. P. Ghatak, M. Mondal, S. N. Biswas, J. Appl. Phys. **68**, 3032 (1990); M. Mondal, S. Banik, K. P. Ghatak, J. Low Temp. Phys. **74**, 423 (1989); B. Mitra, A. Ghoshal, K. P. Ghatak, Phys. Stat. Sol. (b)**150**, K67 (1988); K. P. Ghatak, D. Bhattacharyya, B. Nag, S. N. Biswas, J. Nonlin. Optics and Quant. Optics **13**, 267 (1995); L. Torres, L. Lopez-Diaz, J. Iniguez, Appl. Phys. Lett. **73**, 3766, (1996); A. Y. Toporov, R. M. Langford, A. K. Petford-Long, Appl. Phys. Lett. **77**, 3063 (2000) ; L. Torres, L. Lopez-Diaz, O. Alejos, J. Iniguez, J. Appl. Phys. **85**, 6208 (1999); S. Tiwari, S. Tiwari, Cryst. Res. Technol. **41**, 78 (2006); R. Houdré, C. Hermann, G. Lampel, P. M. Frijlink,

A. C. Gossard, Phys. Rev. Lett., **55**, 734 (1985)

125. S. M. Adhikari, K. P. Ghatak, Jour. Adv. Phys. **2** , 130 (2013); D. De, S. Bhattacharaya, S. Ghosh, K. P. Ghatak, Adv. Sci., Engi. and Med **4**, 211 (2012); A. Kumar, S. Chowdhury, S. M. Adhikari, S. Ghosh, M. Mitra, D. De, A. Sharma, S. Bhattacharya, A. Dey, K. P. Ghatak, Jour. of Comput. Theo. Nanosc. **7**, 115 (2010); R. Houdré, C. Hermann, G. Lampel, P. M. Frijlink, Surface Sci. **168**, 538 (1986); T. C. Chiang, R. Ludeke, D. E. Eastman, Phys. Rev. **B**. **25**, 6518 (1982); S. P. Svensson, J. Kanski, T. G. Andersson, P. O. Nilsson, J. Vacuum Sci. Technol. **B 2**, 235 (1984); S. F. Alvarado, F. Ciccacci, M. Campagna, Appl. Phys. Letts. **39**, 615 (1981); L. Fleming, M. D. Ulrich, K. Efimenko, J. Genzer, A. S. Y. Chan, T. E. Madey, S. J. Oh, O. Zhou, J. E. Rowe, Vac. Sci. Technol. B **22**, 2000 (2004); H. Shimoda, B. Gao, X. –P. Tang, A. Kleinhammes, L. Fleming, Y. Wu, O. Zhou, Phys. Rev. Lett. **88**, 015502 (2002); M. Maillard, P. Monchicourt, M. P. Pileni, Chem. Phys. Lett., **380**, 704 (2003)
126. B. Mitra, K. P. Ghatak, Phys. Scrip. **40**, 776 (1989); D. De, S. Bhattacharya, K. P. Ghatak, International work shop in Physics of Semiconductor Devices, IEEE 897 (2007); S. Bhattacharya, D. De, S. Chowdhury, S. Karmakar, D. K. Basu, S. Pahari, K. P. Ghatak, Jour. of Comput. Theo. Nanosc**3**, 280 (2006)
127. I. Matsuda, S. Hasegawa, A. Konchenko, Y. Nakayama, Y. Nakamura, M. Ichikawa, Phys. Rev. B, **73**, 113311 (2006); V. L. Colvin, A. P. Alivisatos, J. G. Tobin, Phys. Rev. Lett. **66**, 2786 (1991); B. Schroeter, K. Komlev, W. Richter, Mat. Sc. and Eng., **B88**, 259 (2002); G. F. Bertsch, N. Van Giai, N. Vinh Mau, Phys. Rev. A **61**, 033202 (2000);
128. K. P. Ghatak, S. N. Biswas, Nonlinear Opt. **4**, 39 (1993)
129. K. P. Ghatak, S. N. Biswas, *SPIE, Growth and Characterization of Materials for Infrared Detectors and Nonlinear Optical Switches*, vol. **1484** (USA, 1991) p. 136
130. K. P. Ghatak, B. De, *Polymeric materials for Integrated Optics and Information Storage, Materials Research Society (MRS) Symposium Proceedings*, MRS Spring Meeting, 1991, vol. **228**, p. 237; L. J. Heyderman, F. Nolting, Quitmann, Appl. Phys. Letts. **83**, 1797 (2003); R. P. Cowburn, A. O. Adeyeye, J. A. C. Bland, Appl. Phys. Lett. **70**, 2309 (1997);
131. K. P. Ghatak, B. Nag, G. Majumdar, Strained Layer Expitaxy — Materials, Processing, and Device Applications, MRS Symposium Proceedings, MRS Spring Meeting, 1995, vol. **379**, p. 85.
132. K. P. Ghatak, SPIE, High Speed Phenomena in Photonic Materials and Optical Bistability, USA, 1990, vol. **1280**, p. 53.
133. K. P. Ghatak, Long Wave Length Semiconductor devices, Materials and Processes Symposium Proceedings, MRS Symposium Proceedings, MRS Spring Meeting, 1990, vol. **216**, p. 469.
134. K. P. Ghatak, A. Ghoshal, S. Bhattacharyya, SPIE, Nonlinear Optical Materials and Devices for Photonic Switching, USA, 1990, vol. **1216**, p. 282.
135. K. P. Ghatak, SPIE, Nonlinear Optics III, USA, 1992, vol. **1626**, p. 115.

136. K. P. Ghatak, A. Ghoshal, B. De, SPIE, Optoelectronic Devices and Applications, USA, 1990, vol. **1338**, p. 111.
137. K. P. Ghatak, S. N. Biswas, in *Proceedings of the Society of Photo-optical and Instrumentation Engineers (SPIE)*, Nonlinear Optics II,USA,1991,vol. **1409**, p. 28; K. P. Ghatak, SPIE, Process Module Metrology, USA, 1992, vol. **1594**, p. 110; K. P. Ghatak, SPIE, International Conference on the Application and Theory of Periodic Structures, 1991, vol. **1545**, p. 282
138. K. P. Ghatak, M. Mondal, Solid State Electron. **31**, 1561 (1988)
139. K. P. Ghatak, M. Mondal, J. Appl. Phys. **69**, 1666 (1991)
140. C. Majumdar, A. B. Maity, A. N. Chakravarti, Phys. Stat. Sol. (b) **140,** K7 (1987)
141. C. Majumdar, A. B. Maity, A. N. Chakravarti, Phys. Stat. Sol. (b) **141**, K35 (1987)
142. N. R. Das, K. K. Ghosh, D. Ghoshal, Phys. Stat. Sol. (b) **197**, 97 (1996)
143. C. Majumdar, A. B. Maity, A. N. Chakravarti, Phys. Stat. Sol. (b), **144**, K13 (1987)
144. N. R. Das, A. N. Chakravarti, Phys. Stat. Sol. (b) **176**, 335 (1993)
145. S. Sen, N. R. Das and A. N. Chakravarti, Jour. of Phys: Conden. Mat. **19**, 186205 (2007); N. R. Das, S. Ghosh, A. N. Chakravarti, Phys. Stat. Sol. (b) **174**, 45 (1992)
146. A. B. Maity, C. Majumdar, A. N. Chakravarti, Phys. Stat. Sol. (b) **144**, K93, (1987)
147. A. B. Maity, C. Majumdar, A. N. Chakravarti, Phys. Stat. Sol. (b) **149**, 565 (1988)
148. A. V. D. Ziel, *Solid State Physical Electronics*, (Prentice Hall, Inc. Eaglewood Cliffs, 1957); A. Modinos, *Field, Thermionic and Secondary Electron Emission Spectroscopy*, (Plenum Press, New York, 1984)
149. L. J. Singh, S. Choudhary, A. Mallik, K. P. Ghatak, Journal of Computational and Theoretical Nanoscience **2** (2), 287 (2005)
150. O. Madelung, Physics of III-V compounds John Wiley and Sons, Inc. New York, 1966)
151. B. Mitra A. Ghoshal and K. P. Ghatak, physica status solidi (b), **155**, K23(1989)
152. J. A. Woollam, Phys. Rev., **3**, 1148, (1971)
153. V. V. Kaminoskii,N.NStepanav, L. M. Smirnov, Sov. Phys. Solid state, **27**, 1295(1985)
154. P. I. Baranskii, V. V. Kolomoets and S. S. korolyuk, Phys. Stat Sol.(b), **116**, K109 (1983)
155. K. P Ghatak, S.Biswas, J. Vac. Sci. Technol. B7 **1**, 104 (1989); D. R Choudhury, A. K. Chowdhury , A. N. Chakravarti, PhysicaScripta, 22, 656 (1981); D. R. Choudhury, A. K. Chowdhury , A. N. Chakravarti, Czech. J. of Phys.,B **30** (1980); D. R. Choudhury, A. K. Chowdhury , A. N. Chakravarti, B. R. Nag, Phys. Stst. Sol. (a), **58**, K51 (1980); B. Mitra, A. Ghoshal , K. P. Ghatak, Phys. Stat. Sol. (b), **153**, K209 (1989); M. Mondal, K. P. Ghatak, Phys. Stat. Sol. (a), **93**, 377 (1986); D. R. Choudhury, A. K. Chowdhury , A. N. Chakravarti, Appl. Phys., **22**, 145 (1980)

Chapter 8

Conclusion and Scope for Future Research

This monograph deals with the EMs in various types of HD materials and their quantized counter parts. The external photo excitation, quantization and strong electric field alter profoundly the basic band structures, which, in turn, generate pinpointed knowledge regarding EM in various HDS and their nanostructures. The in-depth experimental investigations covering the whole spectrum of solid state and allied science in general, are extremely important to uncover the underlying physics and the related mathematics in this particular aspect. We have formulated the simplified expressions of EM for few HD quantized structures together with the fact that our investigations are based on the simplified $\boldsymbol{k.p}$ formalism of solid-state science without incorporating the advanced field theoretic techniques. In spite of such constraints, the role of band structure, which generates, in turn, new concepts are truly amazing and discussed throughout the text.

We present the last bouquet of open research problem in this pin pointed topic of research of modern physics.

8.1 Open Research Problems

(R.8.1) Investigate the EM in the presence of a quantizing magnetic field under exponential, Kane, Halperin, Lax and Bonch-Bruevich band tails [1] for all the problems of this monograph of all the HD materials whose unperturbed carrier energy spectra are defined in Chapter 1 by including spin and broadening effects.

(R.8.2) Investigate all the appropriate problems after proper modifications introducing new theoretical formalisms for the problems as defined in (R.8.1) for HD negative refractive index, macro molecular, nitride and organic materials.

(R.8.3) Investigate all the appropriate problems of this monograph for all types of HD quantum confined p-InSb, p-CuCl and semiconductors having diamond structure valence bands whose dispersion relations of the carriers in bulk materials are given by Cunningham [2], Yekimov *et al.* [3] and Roman *et al.* [4] respectively.

(R.8.4) Investigate the influence of defect traps and surface states separately on the EM of the HD materials for all the appropriate problems of all the chapters after proper modifications.

(R.8.5) Investigate the EM of the HD materials under the condition of non-equilibrium of the carrier states for all the appropriate problems of this monograph.

(R.8.6) Investigate the EM for all the appropriate problems of this monograph for the corresponding HD p-type semiconductors and their nanostructures.

(R.8.7) Investigate the EM for all the appropriate problems of this monograph for all types of HD semiconductors and their nanostructures under mixed conduction in the presence of strain.

(R.8.8) Investigate the EM for all the appropriate problems of this monograph for all types of HD semiconductors and their nanostructures in the presence of hot electron effects.

(R.8.9) Investigate the EM for all the appropriate problems of this monograph for all types of HD semiconductors and their nanostructures for nonlinear charge transport.

(R.8.10) Investigate the EM for all the appropriate problems of this monograph for all types of HD semiconductors and their nanostructures in the presence of strain in an arbitrary direction.

(R.8.11) Investigate all the appropriate problems of this monograph for strongly correlated electronic HD systems in the presence of strain.

(R.8.12) Investigate all the appropriate problems of this chapter in the presence of arbitrarily oriented photon field and strain.

(R.8.13) Investigate all the appropriate problems of this monograph for all types of HD nanotubes in the presence of strain.

(R.8.14) Investigate all the appropriate problems of this monograph for HD Bi_2Te_3-Sb_2Te_3 super-lattices in the presence of strain.

(R.8.15) Investigate the influence of the localization of carriers on the EM in HDS for all the appropriate problems of this monograph in the presence of crossed fields.

(R.8.16) Investigate EM for HD p-type SiGe under different appropriate physical conditions as discussed in this monograph in the presence of strain and crossed fields.

(R.8.17) Investigate EM for HD GaN under different appropriate physical conditions as discussed in this monograph in the presence of strain and crossed fields.

(R.8.18) Investigate EM for different disordered HD conductors under different appropriate physical conditions as discussed in this monograph in the presence of strain and crossed fields.

(R.8.19) Investigate all the appropriate problems of this monograph for HD $Bi_2Te_{3-x}Se_x$ and $Bi_{2-x}Sb_xTe_3$ respectively in the presence of strain and crossed fields.

(R.8.20) Investigate all the appropriate problems of this monograph in the presence of crossed electric and alternating quantizing magnetic fields.

(R.8.21) Investigate all the appropriate problems of this monograph in the presence of crossed alternating electric and quantizing magnetic fields.

(R.8.22) Investigate all the appropriate problems of this monograph in the presence of crossed alternating non uniform electric and alternating quantizing magnetic fields.

(R.8.23) Investigate all the appropriate problems of this monograph in the presence of alternating crossed electric and alternating quantizing magnetic fields.

(R.8.24) Investigate all the appropriate problems of this monograph in the presence of arbitrarily oriented pulsed electric and quantizing magnetic fields.

(R.8.25) Investigate all the appropriate problems of this monograph in the presence of arbitrarily oriented alternating electric and quantizing magnetic fields.

(R.8.26) Investigate all the appropriate problems of this monograph in the presence of crossed in homogeneous electric and alternating quantizing magnetic fields.

(R.8.27) Investigate all the appropriate problems of this monograph in the presence of arbitrarily oriented electric and alternating quantizing magnetic fields under strain.

(R.8.28) Investigate all the appropriate problems of this monograph in the presence of arbitrarily oriented electric and alternating quantizing magnetic fields under light waves.

(R.8.29) (a) Investigate the EM for all types of HD materials of this monograph in the presence of many body effects, strain and arbitrarily oriented alternating light waves respectively.

(b) Investigate all the appropriate problems of this chapter for the Dirac electron.

(c) Investigate all the problems of this monograph by removing all the physical and mathematical approximations and establishing the respective appropriate uniqueness conditions.

The formulation of* EM *for all types of HD materials and their quantum confined counter parts considering the influence of all the bands created due to all types of quantizations after removing all the assumptions and establishing the respective appropriate uniqueness conditions is, in general, an extremely difficult problem.* 200 open research problems** have been presented in this monograph and we hope that the readers will not only solve them but also will generate new concepts, both theoretical and experimental. Incidentally, we can easily infer how little is presented and how much more is yet to be investigated in this exciting topic which is the signature of coexistence of new physics, advanced mathematics combined with the inner fire for performing creative researches in this context from the young scientists since like Kikoin [5] we firmly believe that ***"A young scientist is no good if his teacher learns nothing from him and gives his teacher nothing to be proud of". In the mean time our research interest has been shifted and we are leaving this particular beautiful topic with the hope that (R.8.29) alone is sufficient to draw the attention of the researchers from diverse fields and our readers are surely in tune with the fact that ***"Exposition, criticism, appreciation is the work for second-rate minds"*** [6].

References

1. B. R. Nag, *Electron Transport in Compound Semiconductors*, Springer Series in Solid State Sciences, Vol. 11 (Springer-Verlag, Germany, 1980)
2. R. W. Cunningham, Phys. Rev. **167**, 761 (1968)
3. A. I. Yekimov, A. A. Onushchenko, A. G. Plyukhin, Al, L. Efros, J. Expt. Theor. Phys. **88**, 1490 (1985)
4. B. J. Roman, A. W. Ewald, Phys. Rev. B **5**, 3914 (1972)
5. I. K. Kikoin, *Science for Everyone: Encounters with Physicists and Physics* (Mir Publishers, Russia, 1989), p. 154
6. G. H. Hardy, *A mathematician's Apology* (Cambridge University Press, 1990, pp. 61)

Chapter 9

Appendix A: The Numerical Values of the Energy Band Constants of A Few Materials

	Materials	Numerical values of the energy band constants
1	The conduction electrons of n-Cadmium Germanium Arsenide can be described by three types of band models	1. The values of the energy band constants in accordance with the generalized electron dispersion relation of nonlinear optical materials (as given by (2.1)) are as follows $$E_{g_0} = 0.57eV, \quad \Delta_{\parallel} = 0.30eV, \quad \Delta_{\perp} = 0.36eV, \quad m_{\parallel}^* = 0.034m_0, \quad m_{\perp}^* = 0.039m_0,$$ $T = 4K, \delta = -0.21eV, g_v = 1$[1,2], $\varepsilon_{sc} = 18.4\varepsilon_0$ [3] (ε_{sc} and ε_0 are the permittivity of the semiconductor material and free space respectively) and $W = 4eV$[4].

(*Continued*)

(Continued)

	Materials	Numerical values of the energy band constants
		2. In accordance with the three band model of Kane (as given by (2.19)), the spectrum constants are given by $$\Delta = (\Delta_{\parallel} + \Delta_{\perp})/2 = 0.33eV,$$ $$E_{g_0} = 0.57eV,$$ $$m^* = (m^*_{\parallel} + m^*_{\perp})/2 = 0.0365m_0 \quad \text{and}$$ $$\delta = 0eV.$$
		3. In accordance with two band model of Kane, $E_{g_0} = 0.57eV$ and $m^* = 0.0365m_0$.
2	n-Indium Arsenide	The values $E_{g_0} = 0.36eV, \Delta = 0.43eV, m^* = 0.026m_0, g_v = 1, \varepsilon_{sc} = 12.25\varepsilon_0$ [5] and $W = 5.06eV$ [6] are valid for three band model of Kane as given by (2.19).
3	n-Gallium Arsenide	The values $E_{g_0} = 1.55eV, \Delta = 0.35eV, m^* = 0.07m_0, g_v = 1, \varepsilon_{sc} = 12.9\varepsilon_0$ [5] and $W = 4.07eV$ [7] are valid for three band model of Kane as given by (2.19). The values $\alpha_{13} = -1.97 \times 10^{-37}\ eVm^4$ and $a_{15} = -2.3 \times 10^{-34} eVm^4$ [8] are valid for (3.25). The values $\alpha_{11} = -2132 \times 10^{-40}$ eVm^4, $\alpha_{12} = 9030 \times 10^{-50} eVm^5, \beta_{11} = -2493 \times 10^{-40} eVm^4, \beta_{12} = 12594 \times 10^{-50} eVm^5, \gamma_{11} = 30 \times 10^{-30} eVm^3, \gamma_{12} = -154 \times 10^{-42} eVm^4$ [9] are valid for (3.30).

(Continued)

(Continued)

	Materials	Numerical values of the energy band constants
4	n-Gallium Aluminium Arsenide	$E_{g_0} = (1.424 + 1.266x + 0.26x^2)eV, \Delta = (0.34 - 0.5x)$ $eV, m^* = [0.066 + 0.088x]m_0, g_v = 1,$ $\varepsilon_{sc} = [13.18 - 3.12x]\ \varepsilon_0$ [10] and $W = (3.64 - 0.14x)eV$ [11].
5	n-Mercury Cadmium Telluride	$E_{g_0} = (-0.302 + 1.93x + 5.35 \times 10^{-4}(1 - 2x)T - 0.810x^2 + 0.832x^3)eV, \Delta = (0.63 + 0.24x - 0.27x^2)$ $eV, m^* = 0.1m_0\ E_{g_0}(eV)^{-1}, g_v = 1, \varepsilon_{sc} =$ $[20.262 - 14.812x + 5.22795x^2]\ \varepsilon_0$ [12] and $W = (4.23 - 0.813(E_{g_0} - 0.083))eV$ [13].
6	n-Indium Gallium Arsenide Phosphide lattice matched to Indium Phosphide	$E_{g_0} = (1.337 - 0.73y + 0.13y^2)eV, \Delta = (0.114 + 0.26y - 0.22y^2)eV,$ $m^* = (0.08 - 0.039y)m_0, y = (0.1896 - 0.4052x)/(0.1896 - 0.0123x), g_v = 1, \varepsilon_{sc} = [10.65$ $+0.1320y]\varepsilon_0$ and $W(x,y) = [5.06(1-x)y$ $+4.38(1-x)$ $(1-y) + 3.64xy + 3.75\{x(1-y)\}]eV$ [14].
7	n-Indium Antimonide	$E_{g_0} = 0.2352eV, \Delta = 0.81eV, m^* = 0.01359m_0,$ $g_v = 1, \varepsilon_{sc} = 15.56\varepsilon_0$ [5] and $W = 4.72eV$ [6].
8	n-Gallium Antimonide	The values of $E_{g_0} = 0.81eV$, $\Delta = 0.80eV$, $P = 9.48 \times 10^{-10} eVm$, $\bar{\varsigma}_0 = -2.1$, $\bar{\upsilon}_0 = -1.49$, $\bar{\omega}_0 = 0.42$, $g_v = 1$ [15] and $\varepsilon_{sc} = 15.85\varepsilon_0$ [15, 16] are valid for the model of Seiler *et al.* [15] as given by (3.113). The values $\bar{E_1} = 1.024eV$, $\bar{E_2} = 0eV$, $\bar{E_3} = -1.132eV$, $\bar{E_4} = 0.05eV$, $\bar{E_5} = 1.107eV$, $\bar{E_6} = -0.113eV$ and $\bar{E_7} = -0.0072eV$ [16] are valid for the model of Zhang [16] as given by (3.128).

(Continued)

(Continued)

	Materials	Numerical values of the energy band constants
9	n-Cadmium Sulphide	$m_{\parallel}^* = 0.7m_0$, $m_{\perp}^* = 1.5m_0$, $\bar{\lambda}_0 = 1.4 \times 10^{-8}\ eVm$ (value changed), $g_v = 1$ [5], $\varepsilon_{sc} = 15.5\varepsilon_0$ [17] and $W = 4.5eV$ [6].
10	n-Lead Telluride	The values $m_t^- = 0.070m_0$, $m_l^- = 0.54m_0$, $m_t^+ = 0.010m_0$, $m_l^+ = 1.4m_0$, $P_{\parallel} = 141meVnm$, $P_{\perp} = 486meVnm$, $E_{g_0} = 190meV$, $g_v = 4$ [5], $\varepsilon_{sc} = 33\varepsilon_0$ [5, 18] and $W = 4.6eV$ [19] are valid for the Dimmock model [20] as given by (3.165). The values $(\overline{R})^2 = 2.3 \times 10^{-19}(eVm)^2$, $E_{g_0} = 0.16eV$, $(\overline{s})^2 = 4.6(\overline{R})^2$, $\Delta_c' = 3.07eV$, $(\overline{Q})^2 = 1.3(\overline{R})^2$, $\Delta_c'' = 3.28eV$, $(\overline{A})^2 = 0.83 \times 10^{-19}(eVm)^2$ [21] and $W = 4.21eV$ [6] are valid for the model of Bangert and Kastner [21] as given by (3.172). The values $m_{tv} = 0.0965m_0$, $m_{lv} = 1.33m_0$, $m_{tc} = 0.088m_0$, $m_{lc} = 0.83m_0$ [19] are valid for the model of Foley *et al.* [19] as given by (3.179). The values $m_1 = 0.0239m_0$, $m_2 = 0.024m_0$, $m_2' = 0.31m_0$, $m_3 = 0.24m_0$ [22] are valid for the Cohen model [23] as given by (2.144).
11	Stressed n-Indium Antimonide	The values $m^* = 0.048m_o$, $E_{g_0} = 0.081eV$, $B_2 = 9 \times 10^{-10}eVm$, $C_1^c = 3eV$, $C_2^c = 2eV$, $a_0 = -10eV$, $b_0 = -1.7eV$, $d = -4.4eV$, $S_{xx} = 0.6 \times 10^{-3}(kbar)^{-1}$, $S_{yy} = 0.42 \times 10^{-3}(kbar)^{-1}$, $S_{zz} = 0.39 \times 10^{-3}(kbar)^{-1}$, $S_{xy} = 0.5 \times 10^{-3}(kbar)^{-1}$, $\varepsilon_{xx} = \sigma S_{xx}$, $\varepsilon_{yy} = \sigma S_{yy}$, $\varepsilon_{zz} = \sigma S_{zz}$, $\varepsilon_{xy} = \sigma S_{xy}$, σ is the stress in kilobar, $g_v = 1$ [24] are valid for the model of Seiler *et al.* [24] as given by (2.102).

(Continued)

(Continued)

	Materials	Numerical values of the energy band constants
12	Bismuth	$E_{g_0} = 0.0153eV$, $m_1 = 0.00194m_0$, $m_2 = 0.313$ m_0, $m_3 = 0.00246m_0$, $m_2' = 0.36m_0$, $g_v = 3$, $g_s = 2$ [25], $M_2 = 0.128m_0$, $M_2' = 0.80m_0$ [26] and $W = 4.34eV$.
13	Mercury Telluride	$m_v^* = 0.028m_0$, $g_v = 1$, $\varepsilon_\infty = 15.2\varepsilon_0$ [27] and $W = 5.5eV$ [28].
14	Platinum Anti-monide	For valence bands, along <111> direction, $\lambda_1 = 0.33eV$, $l_1 = 1.09eV$, $\nu_1 = 0.17eV$, $\overline{\overline{n}} = 0.22eV$, $\overline{\overline{a}} = 0.643nm$, $I_0 = 0.30(eV)^2$, $\delta_0' = 0.33eV$, $g_v = 8$ [29], $\varepsilon_{sc} = 30\varepsilon_0$ [30] and $\phi \approx 3.0eV$ [31].
15	n-Gallium Phosphide	$m_\parallel^* = 0.92m_0$, $m_\perp^* = 0.25m_0$, $k_0 = 1.7 \times 10^{19}$ m^{-1}, $\lvert V_G \rvert = 0.21eV$, $g_v = 6$, $g_s = 2$ [32] and $W = 3.75eV$ [6].
16	Germanium	$E_{g_0} = 0.785eV$, $m_\parallel^* = 1.57m_0$, $m_\perp^* = 0.0807m_0$ [7] and $W = 4.14eV$ [6].
17	Tellurium	The values $A_6 = 6.7 \times 10^{-16} meVm^2$, $A_7 = 4.2$ $\times 10^{-16} meVm^2$, $A_8 = (6 \times 10^{-8} meVm)^2$ and $A_9 = (3.6 \times 10^{-8} meVm)^2$ [33] are valid for the model of Bouat *et al.* [33] as given by (3.76). The values $t_1 = 0.06315eV$, $t_2 = -10.0\hbar^2/2m_0$, $t_3 = -5.55\hbar^2/2m_0$, $t_4 = 0.3 \times 10^{-36} eVm^4$, $t_5 = 0.3 \times 10^{-36} eVm^4$, $t_6 = -5.55\hbar^2/2m_0$, $t_7 = 6.18 \times 10^{-20}(eVm)^2$ [34] and $W = 1.9708eV$ [35] are valid for the model of Ortenberg and Button [34] as given by (3.82).

(Continued)

(*Continued*)

	Materials	Numerical values of the energy band constants
18	Graphite	The values $\Delta_1 = -0.0002eV$, $\gamma_1 = 0.392eV$, $\gamma_5 = 0.194eV$, $\bar{c} = 0.674nm$, $\gamma_2 = -0.019eV$, $\bar{a} = 0.246nm$, $\gamma_0 = 3eV$, $\gamma_4 = 0.193eV$ [36] and $W = 4.6eV$ [37] are valid for the model of Brandt *et al.* [36] as given by (3.88).
19	Lead Germanium Telluride	The values $\bar{E}_{g0} = 0.21eV$, $g_v = 4$ [38] and $\phi \approx 6eV$ [39] are valid for the model of Vassilev [38] as given by (3.107a).
20	Cadmium Antimonide	The values $A_{10} = -4.65 \times 10^{-19}eVm^2$, $A_{11} = -2.035 \times 10^{-19}eVm^2$, $A_{12} = -5.12 \times 10^{-19}eVm^2$, $A_{13} = -0.25 \times 10^{-10}eVm$, $A_{14} = 1.42 \times 10^{-19}eVm^2$, $A_{15} = 0.405 \times 10^{-19}eVm^2$, $A_{16} = -4.07 \times 10^{-19}eVm^2$, $A_{17} = 3.22 \times 10^{-10}eVm$, $A_{18} = 1.69 \times 10^{-20} (eVm)^2$,$A_{19} = 0.070eV$ [40] and $\varphi \approx 2eV$ [41] are valid for the model of Yamada [40] as given by (3.186).
21	Cadmium Diphosphide	The values $\beta_1 = 8.6 \times 10^{-21}eVm^2$, $\beta_2 = 1.8 \times 10^{-21}(eVm)^2$, $\beta_4 = 0.0825eV$, $\beta_5 = -1.9 \times 10^{-19}eVm^2$ [42] and $\varphi \approx 5eV$ [43] are valid for the model of Chuiko [42] and is given by (3.193).
22	Zinc Diphosphide	The values $\beta_1 = 8.7 \times 10^{-21}eVm^2$, $\beta_2 = 1.9 \times 10^{-21}(eVm)^2$, $\beta_4 = 0.0875eV$, $\beta_5 = -1.9 \times 10^{-19}eVm^2$ [42] and $W \approx 3.9eV$ [43] are valid for the model of Chuiko [42] and is given by (3.193).

(*Continued*)

(*Continued*)

	Materials	Numerical values of the energy band constants
23	Bismuth Telluride	The values $E_{g_0} = 0.145eV$, $\alpha_{11} = 3.25$, $\alpha_{22} = 4.81$, $\alpha_{33} = 9.02$, $\alpha_{23} = 4.15$, $g_s = 2$, $g_v = 6$ [44] and $\varphi = 5.3eV$ [45] are valid for the model of Stordeur *et al.* [44] using (3.200).
24	Carbon Nanotube	The values $a_c = 0.144nm$ [46], $t_c = 2.7eV$ [47], $r_0 = 0.7nm$ [48] and $W = 3.2eV$ [49] are valid for graphene band structure realization of carbon nanotube [48] and are given by (2.167) and (2.168).
25	Antimony	The values $\bar{\alpha}_{11} = 16.7$, $\bar{\alpha}_{22} = 5.98$, $\bar{\alpha}_{33} = 11.61$, $\bar{\alpha}_{23} = 7.54$ [50] and $W = 4.63eV$ [6] are valid for the model of Ketterson [50] and are given by (3.207), (3.208) and (3.209) respectively.
26	Zinc Selenide	$m_2^* = 0.16m_0, \Delta_2 = 0.42eV, E_{g02} = 2.82eV$ [7] and $W = 3.2eV$ [51].
27	Lead Selenide	$m_t^- = 0.23m_0, m_l^- = 0.32m_0, m_t^+ = 0.115m_0,$ $m_l^+ = 0.303m_0, P_{\|} \approx 138meVnm, P_\perp = 471$ $meVnm, E_{g_0} = 0.28eV$ [52], $\varepsilon_{sc} = 21.0\varepsilon_0$ [32] and $W = 4.2eV$ [53].

References

1. J. L. Shay and J. W. Wernik, Ternary Chalcoprite Semiconductors: Growth, Electronic properties and applications, Pergamon Press, London (1975)
2. E. A. Arushanov, A. A. Kaynzev, A. N. Natepov and S. I. Radautsan, Sov. Phys. Semicond. 15, 828 (1981).
3. K. S. Hong, R. F. Speyer and R. A. Condrate, J. Phys. Chem. Solids, 51, 969 (1990).
4. M. Mondal, S. Banik and K. P. Ghatak, J. Low. Temp. Phys. 74, 423 (1989)

5. B. R. Nag, *Electron Transport in Compound Semiconductor*, Springer-Verlag, Germany (1980); M. Krieehbaum, P. Kocevar, H. Pascher and G. Bauer, *IEEE QE*, 24 1727 (1988); J. J. Hopfield, J. Appl. Phys., **32**, 2277 (1961).
6. S. Adachi, Properties of Group-IV, III-V and II-VI Semiconductors, John Wiley and Sons, 2005.
7. O. Madelung, *Semiconductors: Data Handbook*, 3rd ed. (Springer-Verlag, Germany, (2003).
8. D. J. Newson, A. Kurobe, Semicond. Sci. Technol. 3, 786 (1988).
9. U. Rossler, SolidState Commun. 49, 943 (1984).
10. S. Adachi, J. Appl. Phys., 58, R1, (1985).
11. S. Adachi, "*GaAs and Related Materials: Bulk Semiconductors and Superlattice Properties*", (World Scientific, USA, 1994).
12. G. L. Hansen, J. L. Schmit and T. N. Casselman, J. Appl. Phys. 63, 7079 (1982)
13. J. Wenus, J. Rutkowski, and A. Rogalski, IEEE Trans. Elect. Dev. 48, 1326, (2001)
14. S. Adachi, J. Appl. Phys. 53, 8775 (1982).
15. D. G. Seiler, W. M. Beeker and L. M. Roth, *Phys. Rev.* 1, 764 (1970).
16. I. V. Skryabinskii, Yu. I. Ukhanov, Sov. Phys. Solid State, 14, 2838, (1973); H. I. Zhang, Phys. Rev. B, 1, 3450 (1970); P. C. Mathur and S. Jain, Phys. Rev. 19, 1359 (1979).
17. S. Tiwari and S. Tiwari, Cryst. Res. Technol., 41, 78, (2006).
18. J. R. Lowney and S. D. Senturia, J. Appl. Phys., 47, 1771 (1976).
19. W. E. Spicer and G. J. Lapeyre, Phys. Rev. 139, A565–A569 (1965); G. M. T. Foley and P. N. Langenberg, Phys. Rev. B, 15, 4850 (1977).
20. J. O. Dimmock in *"The Physics of semimetals and narrowgap semiconductors"* ed. by D. L. Carter and R. T. Bates, Pergamon Press, Oxford, 319 (1971)
21. E. Bangert and P. Kastner, Phys. Stat. Sol. (b), 61, 503, (1974).
22. D. R. Lovett, in *Semimetals and Narrow Band Gap Semiconductors* (Pion Limited, London, 1977).
23. M. H. Cohen, Phys. Rev. 121, 387 (1961).
24. D. G. Seiler, B. D. Bajaj and A. E. Stephens, Phys. Rev. B, 16, 2822 (1977); A. V. Germaneko and G. M. Minkov, Phys. Stat. Sol. (b) 184, 9 (1994); G. L. Bir and G. E. Pikus, Symmetry and Strain–Induced effects in Semiconductors Nauka, Russia (1972). (in Russian); M. Mondal and K. P. Ghatak, Phys. Stat. Sol. (b) 135, K21 (1986).
25. C. C. Wu and C. J. Lin, J. Low. Temp. Phys. 57, 469 (1984).
26. S. Takaoka, H. Kawamura, K. Murasa, S. Takano, Phys. Rev. B, 13, 1428 (1976).
27. V. I. Ivanov-Omskii, A. Sh. Mekhtisev, S. A. Rustambekova and E. N. Ukraintsev, *Phys. Stat. Sol. (b)* 119, 159 (1983).
28. H. Kim, K. Cho, H. Song, B. Min, J. Lee, G. Kim, S. Kim, S. H. Kim and T. Noh, *Appl. Phys. Lett.* 83, 4619 (2003).
29. P. R. Emtage, *Phys. Rev. A*, 246, 138 (1965).

30. R. A. Reynolds, M. J. Brau, R. A. Chapman, J. Phys. Chem. Solids 29, 755, (1968).
31. J. O'Shaughnessy and C. Smith, Solid State Communications, 8, 481, (1970).
32. G. J. Rees, *Phys. of Compounds*, Proc. of the 13th Inter. Nat. Conf. Ed. F. G. Fumi, pp. 1166, North Holland Company, (1976).
33. J. Bouat and J. C. Thuillier, Surface Sci. 73, 528 (1978).
34. M. V. Ortenberg and K. J. Button, Phys. Rev. B, 16, 2618 (1977).
35. G. Haeffler, A. E. Klinkmüller, J. Rangell, U. Berzinsh and D. Hanstorp, Z. Phys. D, 38, 211, (1996).
36. N. B. Brandt, V. N. Davydov, V. A. Kulbachinskii and O. M. Nikitina, Sov. Phys. Sol. Stat. 29, 1014, (1987).
37. L. M. Viculis, J. J. Mack, O. M. Mayer, H. T. Hahn and R. B. Kaner, J. Mater. Chem., 15, 974 (2005).
38. L. A. Vassilev, Phys. Stat. Sol. (b), 121, 203 (1984); S. Takaoka and K. Murase, Phys. Rev. B, Phys. Rev. B 20, 2823, (1979).
39. D. L. Partin, Superlattices and Microstructures, 1, 131, (1985).
40. Y. Yamada, J. Phys. Japan, 35, 1600 (1973); M. Singh, P. R. Wallace, S. D. Jog and E. Arushanov, J. Phys. Chem. Solids, 45, 409, (1984).
41. W. J. Turner, A. S. Fischler, and W. E. Reese, Phys. Rev. 121, 759 (1961).
42. G. P. Chuiko, Sov. Phys. Semi., 19 (12), 1381, (1985).
43. W. E. Swank, P. G. Le Comber, Phys. Rev. 153, 844, (1967).
44. M. Stordeur and W. Kuhnberger, Phys. Stat. Sol. (b), 69, 377, (1975); D. R. Lovett, Semimetals and Narrow-Bandgap Semiconductors, Pion Limited, 185 (1977); H. Köhler, Phys. Stat. Sol. (b), 74, 591 (1976).
45. D. Haneman, J. Phys. Chem. Solids, 11, 205, (1959).
46. N. Wei, G. Wu and J. Dong, Phys. Lett. A 325, 403 (2004).
47. S. Reich, J. Maultzsch, C. Thomsen and P. Ordejoń, Phys. Rev. B, 66, 035412, (2006).
48. M. S. Lundstrom, J. Guo, "*Nanoscale Transistors: Device Physics, Modeling and Simulation*", (Springer, USA, 2006); J. W. Mintmire and C. T. White, Phys. Rev. Lett., 81, 2506 (1998)
49. F. Buonocore, F. Trani, D. Ninno, A. Di Matteo, G. Cantele and G. Iadonisi, Nanotechnology, 19, 025711, (2008).
50. J. B. Ketterson, Phys. Rev. 129, 18 (1963).
51. R. C. Vilão, J. M. Gil, A. Weidinger, H. V. Alberto, J. Piroto Duarte, N. A. de Campos, R. L. Lichti, K. H. Chow, S. P. Cottrell and S. F. J. Cox, Phys. Rev. B, 77, 235212 , (2008).
52. I. Kang and F. W. Wise, Phys. Rev. B, J. Opt. Soc. Am. B, 14, 1632, (1997).
53. D. Cui, J. Xu, S.-Y. Xu, G. Paradee, B. A. Lewis and M. D. Gerhold, IEEE, Trans. Elect. Dev. 5, 362 (2006).

Chapter 10

Appendix B: The EM in HDs Under Cross-Fields Configuration

10.1 Introduction

The influence of crossed electric and quantizing magnetic fields on the transport properties of semiconductors having various band structures are relatively less investigated as compared with the corresponding magnetic quantization, although, the cross-fields are fundamental with respect to the addition of new physics and the related experimental findings. In 1966, Zawadzki and Lax [1] formulated the electron dispersion law for III–V semiconductors in accordance with the two band model of Kane under cross fields configuration which generates the interest to study this particular topic of semiconductor science in general [2–14].

In Section 10.2.1 of theoretical background, the EM in HD nonlinear optical materials in the presence of crossed electric and quantizing magnetic fields has been investigated by formulating the electron dispersion relation. Section 10.2.2 reflects the study of the EM in HD III–V, ternary and quaternary compounds as a special case of Section 10.2.1. Section 10.2.3 contains the study of the EM for the HD II–VI semiconductors in the present case. In Section 10.2.4, the EM under cross field configuration in HD IV-VI semiconductors has been investigated in accordance with the models of the Cohen, the Lax nonparabolic ellipsoidal and the parabolic ellipsoidal respectively. In Section 10.2.5, the EM for the HD stressed Kane type semiconductors has been investigated. Sections 10.2.6,

10.2.7, 10.2.8, 10.2.9 and 10.2.10 discuss the EMs' in QWs of the above HD semiconductors in the presence of cross-fields configuration respectively. Section 10.3 contains the summary and conclusion of this chapter and lastly Section 10.4 presents three open research problems

10.2 Theoretical Background

10.2.1 *The EM in HD nonlinear optical semiconductors under cross-fields configuration*

The (1.26) of Chapter 1 can be expressed as

$$T_{22}(E,\eta_g) = \frac{p_s^2}{2m_\perp^*} + \frac{p_z^2}{2M_\parallel} T_{22}(E,\eta_g)[T_{21}(E,\eta_g)]^{-1} \tag{10.1}$$

where, $p_s = \hbar k_s$ and $p_z = \hbar k_z$

We know that from electromagnetic theory that,

$$\vec{B} = \nabla \times \vec{A} \tag{10.2}$$

where, $\vec{A}$ is the vector potential. In the presence of quantizing magnetic field B along z direction, the (10.2) assumes the from

$$0\hat{\mathrm{i}} + 0\hat{\mathrm{j}} + B\hat{\mathrm{k}} = \begin{vmatrix} \hat{\mathrm{i}} & \hat{\mathrm{j}} & \hat{\mathrm{k}} \\ \dfrac{\partial}{\partial x} & \dfrac{\partial}{\partial y} & \dfrac{\partial}{\partial z} \\ \mathrm{A_x} & \mathrm{A_y} & \mathrm{A_z} \end{vmatrix} \tag{10.3}$$

where, $\hat{\mathrm{i}}, \hat{\mathrm{j}}$ and $\hat{\mathrm{k}}$ are orthogonal triads. Thus, we can write

$$\begin{aligned} \frac{\partial \mathrm{A_z}}{\partial \mathrm{y}} - \frac{\partial \mathrm{A_y}}{\partial \mathrm{z}} &= 0 \\ \frac{\partial \mathrm{A_x}}{\partial \mathrm{z}} - \frac{\partial \mathrm{A_z}}{\partial \mathrm{x}} &= 0 \\ \frac{\partial \mathrm{A_y}}{\partial \mathrm{x}} - \frac{\partial \mathrm{A_x}}{\partial \mathrm{y}} &= \mathrm{B} \end{aligned} \tag{10.4}$$

This particular set of equations is being satisfied for $\mathrm{A}_x = 0, \mathrm{A}_y = B_x$ and $\mathrm{A}_z = 0$.

Therefore in the presence of the electric field E_o along x axis and the quantizing magnetic field B along z axis for the present case following (10.1) one can approximately write,

$$T_{22}(E,\eta_g) + |e|E_o\hat{x}p(e,\eta_g) = \frac{\hat{p}_x^2}{2m_\perp^*} + \frac{(\hat{p}_x - |e|B\hat{x})^2}{2m_\perp^*} + \frac{\hat{p}_z^2}{2a(E,\eta_g)} \tag{10.5}$$

where

$$\rho(E) \equiv \frac{\partial}{\partial E}[T_{22}(E,\eta_g)] \quad \text{and}$$
$$\alpha(E,\eta_g) \equiv m_\|^*[T_{22}(E,\eta_g)]^{-1}[T_{21}(E,\eta_g)]$$

Let us define the operator $\hat{\theta}$ as

$$\hat{\theta} = -\hat{p}_y + |e|B\hat{x} - \frac{m_\perp^* E_o \rho(E,\eta_g)}{B} \tag{10.6}$$

Eliminating the operator $\hat{x}$, between (10.5) and (10.6) the dispersion relation of the conduction electron in tetragonal semiconductors in the presence of cross fields configuration is given by

$$T_{22}(E,\eta_g) = \left[\left(\left(n+\frac{1}{2}\right)\hbar\omega_{01}\right)\right] + \left(\frac{[\hbar k_z(E)]^2}{2a(E,\eta_g)}\right) - \left(\frac{E_0\hbar k_y \rho(E,\eta_g)}{B}\right) - \left(\frac{M_\perp \rho^2(E,\eta_g)E_o^2}{2B^2}\right) \tag{10.7}$$

where,

$$\omega_{01} = \frac{|e|B}{m_\perp^*}$$

The EEMs along Z and Y directions can, respectively be expressed from (10.7) as

$$m_z^*(\bar{E}_{FBDH},\eta_g, n, E_0)$$
$$= \text{Real part of}\left[a'(\bar{E}_{FBDH},\eta_g)\left[T_{22}(\bar{E}_{FBDH},\eta_g) - \left(n+\frac{1}{2}\right)\hbar\omega_{01}\right.\right.$$

$$+\frac{M_{\perp}\rho^2(\bar{E}_{FBDH},\eta_g)E_0^2}{2B^2}\Bigg]$$

$$+\left[a(\bar{E}_{FBDH},\eta_g)\left[T'_{22}(\bar{E}_{FBDH},\eta_g)\right.\right.$$

$$\left.\left.+\frac{M_{\perp}\rho(\bar{E}_{FBDH},\eta_g)\rho'(\bar{E}_{FBDH},\eta_g)E_0^2}{B^2}\right]\right] \tag{10.8}$$

and

$$m_y^*(\bar{E}_{FBDH},\eta_g,n,E_0)$$

$$=\left(\frac{B}{E_0}\right)^2 \text{Real part of}$$

$$\times\left[\rho(\bar{E}_{FBDH},\eta_g)^{-3}\left[T_{22}(\bar{E}_{FBDH},\eta_g)-\left(n+\frac{1}{2}\right)\hbar\omega_{01}\right.\right.$$

$$\left.+\frac{M_{\perp}\rho^2(\bar{E}_{FBDH},\eta_g)E_0^2}{2B^2}\right]$$

$$\times\left[\rho(\bar{E}_{FBDH},\eta_g)\left[T_{22}(\bar{E}_{FBDH},\eta_g)\right.\right.$$

$$\left.+\frac{M_{\perp}\rho(\bar{E}_{FBDH},\eta_g)\rho'(\bar{E}_{FBDH},\eta_g)E_0^2}{B^2}\right]$$

$$-\left[T_{22}(\bar{E}_{FBDH},\eta_g)-\left(n+\frac{1}{2}\right)\hbar\omega_{01}\frac{M_{\perp}\rho^2(\bar{E}_{FBDH},\eta_g)E_0^2}{2B^2}\right]$$

$$\left.\times\rho'(\bar{E}_{FBDH},\eta_g)\right] \tag{10.9}$$

where $\bar{E}_{FBHD}$ is the Fermi energy in the presence of cross-fields configuration and heavy doping as measured from the edge of the conduction band in the vertically upward direction in the absence of any quantization.

When $E_0 \to 0, m_z^*(\bar{E}_{FBHD},\eta_g,n,E_0) \to \infty$, which is a physically justified result. The dependence of the EM along y direction on the Fermi energy, electric field, magnetic field and the magnetic quantum number is an intrinsic property of cross fields together with the fact in the present case of heavy doping, the EM exists in the band gap. Another characteristic feature of cross field is that various transport coefficients will be sampled dimension dependent. These conclusions are valid for even isotropic parabolic energy bands and cross fields introduce the index dependent anisotropy in the effective mass.

The formulation of DR requires the expression of the electron concentration which can, in general, be written excluding the electron spin as

$$n_o = \frac{-g_v}{L_x\pi^2}\sum_{n=0}^{n_{\max}}\int_{\bar{E}_0}^{\infty} I(E,\eta_g)\frac{\partial f_o}{\partial E}dE \tag{10.10}$$

where L$_x$ is the sample length along x direction, $\bar{E}_0$ is determined by the equation

$$I(\bar{E},\eta_g) = 0$$

where

$$I(E,\eta_g) = \int_{x_l(E,\eta_g)}^{x_h(E,\eta_g)} k_z(E)dk_y \tag{10.11}$$

in which,

$$x_l(E,\eta_g) \equiv \frac{-E_0 M_\perp \rho(E,\eta_g)}{\hbar B} \quad \text{and} \quad x_h(E,\eta_g) \equiv \frac{|e|BL_x}{\hbar} + x_1(E,\eta_g)$$

Thus we get

$$I(E,\eta_g) = \frac{2}{3}\left[\frac{B\sqrt{2a(E,\eta_g)}}{\hbar^2 E_0\rho(E,\eta_g)}\left[\left[T_{22}(E,\eta_g) - \left(n+\frac{1}{2}\right)\frac{\hbar|e|B}{m_\perp^*}\right.\right.\right.$$
$$\left.+|e|E_0L_x\rho(E,\eta_g) - \frac{m_\perp^* E_0^2[\rho(E,\eta_g)]^2}{2B^2}\right]^{\frac{3}{2}}$$
$$-\left[\left[T_{22}(E,\eta_g) - \left(n+\frac{1}{2}\right)\frac{\hbar|e|B}{m_\perp^*}\right.\right.$$
$$\left.\left.\left.\left. - \frac{m_\perp^* E_0^2[\rho(E,\eta_g)]^2}{2B^2}\right]^{\frac{3}{2}}\right]\right]\right] \tag{10.12}$$

The electron concentration is given by

$$n_0 = \left(\frac{2g_v B\sqrt{2}}{3L_x\pi^2\hbar^2 E_0}\right) \text{Real part of} \sum_{n=0}^{n_{\max}} [T_{41HD}(n, \bar{E}_{FBHD}, \eta_g) + T_{42HD}(n, \bar{E}_{FBHD}, \eta_g)] \quad (10.13)$$

where

$$T_{41}(n, \bar{E}_{FBHD}, \eta_g) \equiv \frac{\sqrt{a(\bar{E}_{FBHD}, \eta_g)}}{\rho(\bar{E}_{FBHD}, \eta_g)} \left[\left[T_{22}(\bar{E}_{FBHD}, \eta_g) - \left(n+\frac{1}{2}\right)\frac{\hbar|e|B}{M_\perp} + |e|E_0 L_x \rho(\bar{E}_{FBHD}, \eta_g) - \frac{m_\perp^*[\rho(\bar{E}_{FBHD}, \eta_g)]^2}{2B^2}\right]^{\frac{3}{2}} - \left[T_{22}(\bar{E}_{FBHD}, \eta_g) - \left(n+\frac{1}{2}\right)\frac{\hbar|e|B}{m_\perp^*} - \frac{m_\perp^* E_0^2[\rho(\bar{E}_{FBHD}, \eta_g)]^2}{2B^2}\right]^{\frac{3}{2}}\right]$$

where $\bar{E}_{FBHD}$ is the Fermi energy in this case and

$$T_{42HD}(n, \bar{E}_{FBHD}, \eta_g) \equiv \sum_{r=1}^{s} [L(r) T_{41HD}(n, \bar{E}_{FBHD}, \eta_g)]$$

The DOS function can be written as

$$N(E_0, B) = \left(\frac{2g_v B\sqrt{2}}{3L_x\pi^2\hbar^2 E_0}\right) \sum_{n=0}^{n_{\max}} [T_{41HD}(n, E, \eta_g)]' H(E - \bar{E}_{71}) \quad (10.14)$$

Where $\bar{E}_{71}$ is the complex Landau level energy in this case.

10.2.2 *The EM in HD Kane type III–V semiconductors under cross-fields configuration*

(a) Under the conditions $\delta = 0, \Delta_{\|} = \Delta_{\perp} = \Delta$ and $m_{\|}^* = m_{\perp}^* = m_{c,}$, (10.7) assumes the form

$$T_{33}(E,\eta_g) = \left(n+\frac{1}{2}\right)\hbar\omega_0 + \frac{[\hbar k_z(E)]^2}{2m_c} - \frac{E_0}{B}\hbar k_y\{T_{33}(E,\eta_g)\}' - \frac{m_c E_0^2[\{T_{33}(E,\eta_g)'\}]^2}{} \quad (10.15a)$$

where $T_{33}(E,\eta_g) = T_{31}(E,\eta_g) + iT_{32}(E,\eta_g)$

The use of (10.15a) leads to the expressions of the EEMs' along z and y directions as

$$m_z^*(\bar{E}_{FBHD},\eta_g,n,E_0) = m_c \text{ Real part of } \left[\{T_{33}(\bar{E}_{FBHD},\eta_g)\}'' + \frac{m_c E_0^2\{T_{33}(\bar{E}_{FBHD},\eta_g)'\}\{T_{33}(\bar{E}_{FBHD},\eta_g)\}''}{B^2}\right]$$

$$m_y^*(\bar{E}_{FBHD},\eta_g,n,E_0) = \left(\frac{B}{E_0}\right)^2 \text{ Real part of } \left[\{T_{33}(\bar{E}_{FBHD},\eta_g)\}'\right]^{-1} + \left[T_{33}(\bar{E}_{FBHD},\eta_g) - \left(n+\frac{1}{2}\right)\hbar\omega_0 \right. \quad (10.15b)$$

$$\left. + \frac{m_c E_0^2\{T_{33}(\bar{E}_{FBHD},\eta_g)\}''{]}^2}{2B^2}\right]$$

$$+\left[\frac{-\{T_{33}(\bar{E}_{FBHD},\eta_g)\}''}{[\{T_{33}(\bar{E}_{FBHD},\eta_g)\}']^2}\left[T_{33}(\bar{E}_{FBHD},\eta_g) - \left(n+\frac{1}{2}\right)\hbar\omega_0\right.\right.$$

$$+\frac{m_c E_0^2[\{T_{33}(\bar{E}_{FBHD},\eta_g)\}']}{2B^2}\Bigg]$$

$$+1\frac{m_c E_0^2\{T_{33}(\bar{E}_{FBHD},\eta_g)\}''}{B^2}\Bigg] \tag{10.16}$$

The Landau energy ($\bar{E}_{n_1}$) can be written as

$$T_{33}(\bar{E}_{n_1},\eta_g) = \left(n+\frac{1}{2}\right)\hbar\omega_0 - \frac{m_c E_0^2[\{T_{33}(\bar{E}_{n_1},\eta_g)\}']^2}{2B^2} \tag{10.17}$$

The electron concentration in this case assumes the form

$$n_0 = \frac{2g_v B\sqrt{2m_c}}{3L_x\pi^2\hbar^2 E_0} \text{ Real Part of } \sum_{n=0}^{n_{\max}} [T_{43HD}(n,\bar{E}_{FB},\eta_g)$$
$$+ T_{44HD}(n,\bar{E}_{FB},\eta_g)] \tag{10.18}$$

where,

$$T_{43HD}(n,\bar{E}_{FBHD},\eta_g)$$
$$\equiv \Bigg[\Bigg[T_{33}(\bar{E}_{FB},\eta_g) - \left(n+\frac{1}{2}\right)\hbar\omega_0 - \frac{m_c E_0^2}{2B^2}[\{T_{33}(\bar{E}_{FB},\eta_g)\}']^2$$
$$+ |e|E_0 L_x[\{T_{33}(\bar{E}_{FBHD},\eta_g)\}']\Bigg]^{\frac{3}{2}}$$
$$-\Bigg[T_{33}(\bar{E}_{FB},\eta_g) - \left(n+\frac{1}{2}\right)\hbar\omega_0 - \frac{m_c E_0^2}{2B^2}[\{T_{33}(\bar{E}_{FB},\eta_g)'\}]^2\Bigg]^{\frac{3}{2}}$$
$$\times \frac{1}{[\{T_{33}(\bar{E}_{FB},\eta_g)\}']}$$

and

$$T_{44HD}(n,\bar{E}_{FBHD},\eta_g) \equiv \sum_{r=1}^{s}[L(r)T_{44HD}(n,\bar{E}_{FBHD},\eta_g)].$$

The DOS function can be written as

$$N(E_0,B) = \frac{2g_v B\sqrt{2m_c}}{3L_x\pi^2\hbar^2 E_0}\sum_{n=0}^{n_{\max}}[T_{43HD}(n,E,\eta_g)]H(E-\bar{E}_{n_1}) \tag{10.19}$$

(b) Under the condition $\Delta \gg E_g$, (10.15a) assumes the form

$$\gamma_2(E,\eta_g) = \left(n+\frac{1}{2}\right)\hbar\omega_0 - \frac{E_0}{B}\hbar k_y \gamma_2'(E,\eta_g) - \frac{m_c E_0^2}{2B^2}(\gamma_2'(E,\eta_g) + \frac{[\hbar k_z(E)]^2}{2m_c} \tag{10.20}$$

The use of (10.20) leads to the expressions of the EEMs' along z and y directions as

$$m_z^*(\bar{E}_{FBHD},\eta_g,n,E_0) = m_c\left[\{\gamma_2(\bar{E}_{FBHD},\eta_g)\}'' + \frac{m_c E_o^2\{\gamma_2(\bar{E}_{FBHD},\eta_g)\}'\{\gamma_2(\bar{E}_{FBHD},\eta_g)\}''}{B^2}\right] \tag{10.21}$$

$$\begin{aligned} m_y^*(\bar{E}_{FBHD},\eta_g,n,E_0) &= \left(\frac{B}{E_0}\right)^2 \frac{1}{[\{\gamma_2(\bar{E}_{FBHD},\eta_g)\}']} \\ &\quad\times\left[\gamma_2(\bar{E}_{FBHD},\eta_g) - \left(n+\frac{1}{2}\right)\hbar\omega_0 + \frac{m_c E_o^2[\{\gamma_2(\bar{E}_{FBHD},\eta_g)\}']^2}{2B^2}\right] \\ &\quad\times\left[\frac{-\{\gamma_2(\bar{E}_{FBHD},\eta_g)''}{[\{\gamma_2(\bar{E}_{FBHD},\eta_g)\}']^2}\left[\gamma_2(\bar{E}_{FBHD},\eta_g) - \left(n+\frac{1}{2}\right)\hbar\omega_0 + \frac{m_c E_0^2\left[\{\gamma_2(\bar{E}_{FBHD},\eta_g)\}'\right]^2}{2B^2}\right] + 1 + \frac{m_c E_0^2\left\{\gamma_2(\bar{E}_{FBHD},\eta_g)\right\}''}{B^2}\right] \end{aligned} \tag{10.22}$$

The Landau energy ($\bar{E}_{n_2}$) can be written as

$$\gamma_2(\bar{E},\eta_g) = \left(n+\frac{1}{2}\right)\hbar\omega_0 - \frac{m_c E_0^2}{2B^2}(\gamma_2'(\bar{E}_{n2},\eta_g))^2 \tag{10.23}$$

The expressions for n_0 in this case assume the forms

$$n_0 = \frac{2g_v B\sqrt{2m_c}}{3L_x\pi^2\hbar^2 E_0}\sum_{n=0}^{n_{\max}}[T_{47HD}(n, \bar{E}_{FBHD}, \eta_g) + T_{48HD}(n, \bar{E}_{FBHD}, \eta_g)] \quad (10.24)$$

where

$$T_{47HD}(n, \bar{E}_{FBHD}, \eta_g) \equiv \left[\left[\gamma_2(\bar{E}_{FBHD}, \eta_g) - \left(n+\frac{1}{2}\right)\hbar\omega_0 + |e|E_0 L_x(\gamma_2'(\bar{E}_{FBHD}, \eta_g)) - \frac{m_c E_0^2}{2B^2}(\gamma_2(\bar{E}_{FBHD}, \eta_g))^2\right]^{\frac{3}{2}} - \left[(\gamma_2(\bar{E}_{FBHD}, \eta_g)) - \left(n+\frac{1}{2}\right)\hbar\omega_0 - \frac{m_c E_0^2}{2B^2}(\gamma_2(\bar{E}_{FBHD}, \eta_g))^2\right]^{\frac{3}{2}}\right]\left[\gamma_2'(\bar{E}_{FBHD}, \eta_g)\right]^{-1}$$

and

$$T_{48HD}(n, \bar{E}_{FBHD}, \eta_g) \equiv \sum_{r=0}^{s}[L(r)T_{47HD}(n, \bar{E}_{FBHD}, \eta_g)]$$

Therefore the DOS function can be written as

$$N(E_0, B) = \frac{2g_v B\sqrt{2m_c}}{3L_x\pi^2\hbar^2 E_0}\sum_{n=0}^{n_{\max}}[T_{47HD}(n, E, \eta_g)]H(E - \bar{E}_{n2}) \quad (10.25)$$

(c) For $a \to 0$ and we can write,

$$\gamma_3(E, \eta_g) = \left(n+\frac{1}{2}\right)\hbar\omega_0 - \frac{E_0}{B}\hbar k_y \gamma_3'(E, \eta_g) - \frac{m_c E_0^2}{2B^2}\left(\gamma_3'(E, \eta_g)\right)^2 + \frac{[\hbar k_z(E)]^2}{2m_c} \quad (10.26)$$

The use of (10.25) leads to the expressions of the EEMs' along z and y directions as

$m_z^*(\bar{E}_{FBHD}, \eta_g, n, E_0)$

$$= m_c\left[\left\{\gamma_3(\bar{E}_{FBHD},\eta_g)'' + \frac{m_c E_0^2\{\gamma_3(\bar{E}_{FBHD},\eta_g)\}'\{\gamma_3(\bar{E}_{FBHD},\eta_g)\}''}{B^2}\right\}\right] \tag{10.27}$$

$$m_y^*(\bar{E}_{FBHD},\eta_g,n,E_0)$$

$$= \left(\frac{B}{E_0}\right)^2 \frac{1}{[\{\gamma_3(\bar{E}_{FBHD},\eta_g)\}']} \times\left[\gamma_3(\bar{E}_{FBHD},\eta_g) - \left(n+\frac{1}{2}\right)\hbar\omega_0 + \frac{m_c E_0^2[\{\gamma_3(\bar{E}_{FBHD},\eta_g)\}']^2}{2B^2}\right]$$
$$\times\left[\frac{-\{\gamma_3(\bar{E}_{FBHD},\eta_g)\}''}{[\{\gamma_3(\bar{E}_{FBHD},\eta_g)\}']^2}\left[\gamma_3(\bar{E}_{FBHD},\eta_g) - \left(n+\frac{1}{2}\right)\hbar\omega_0 + \frac{m_c E_0^2[\{\gamma_3(\bar{E}_{FBHD},\eta_g)\}']^2}{2B^2}\right] + 1 + \frac{m_c E_0^2\left\{\gamma_3(\bar{E}_{FBHD},\eta_g)\right\}''}{B^2}\right] \tag{10.28}$$

The Landau energy ($\bar{E}_{n_3}$) can be written as

$$\gamma_3(\bar{E}_{n_3},\eta_g) = \left(n+\frac{1}{2}\right)\hbar\omega_0 - \frac{m_c E_0^2}{2B^2}(\gamma_3'(\bar{E}_{n_3},\eta_g))^2 \tag{10.29}$$

The expressions for n_0 in this case assume the forms

$$n_0 = \frac{2g_v B\sqrt{2m_c}}{3L_x\pi^2\hbar^2 E_0}\sum_{n=0}^{n_{\max}}[T_{49HD}(n,\bar{E}_{FBHD},\eta_g) + T_{50HD}(n,\bar{E}_{FBHD},\eta_g)] \tag{10.30}$$

where

$$T_{49HD}(n,\bar{E}_{FBHD},\eta_g)$$

$$\equiv \left[\left[\gamma_3'(\bar{E}_{FBHD},\eta_g) - \left(n+\frac{1}{2}\right)\hbar\omega_0\right.\right.$$

$$+|e|E_0 L_x(\gamma_3'(\bar{E}_{FBHD},\eta_g)) - \frac{m_c E_0^2}{2B^2}(\gamma_3'(\bar{E}_{FBHD},\eta_g))^2\Bigg]^{3/2}$$

$$-\Bigg[\Bigg[(\gamma_3'(\bar{E}_{FBHD},\eta_g)) - \left(n+\frac{1}{2}\right)\hbar\omega_0$$

$$-\frac{m_c E_0^2}{2B^2}(\gamma_3'(\bar{E}_{FBHD},\eta_g))^2\Bigg]^{3/2}\Bigg][\gamma_3'(\bar{E}_{FBHD},\eta_g)]^{-1}$$

and

$$T_{50HD}(n,\bar{E}_{FBHD},\eta_g) \equiv \sum_{r=0}^{s}[L(r)T_{49HD}(n,\bar{E}_{FBHD},\eta_g)].$$

Therefore the DOS function can be written as

$$N(E_0,B) = \frac{2g_v B\sqrt{2m_c}}{3L_x\pi^2\hbar^2 E_0}\sum_{n=0}^{n_{\max}}[T_{49HD}(n,E,\eta_g)]' H(E-\bar{E}_{n_3}) \tag{10.31}$$

10.2.3 *The EM in HD II–VI semiconductors under cross-fields configuration*

The electron energy spectrum in HD II–VI semiconductors in the presence of electric field E_0 along x direction and quantizing magnetic field B along z direction can approximately be written as

$$\gamma_3(E,\eta_g) = \beta_1(n,E_0) - \frac{E_0}{B}\hbar k_y\gamma_3'(E,\eta_g) - \frac{m_\parallel^* E_0^2}{2B^2}(\gamma_3'(E,\eta_g))^2 + \frac{[\hbar k_z(E)]^2}{2m_\parallel^*} \tag{10.32}$$

where

$$\beta_1(n,E_0) \equiv \left[\left(n+\frac{1}{2}\right)\hbar\omega_0 - \left(\frac{E_0^2 m_\perp^*}{2B^2}\right)\right.$$

$$+D\left\{\left(n+\frac{1}{2}\right)\hbar\omega_{02}-\left(\frac{E_0^2 m_\perp^*}{2B^2}\right)\right\}^{\frac{1}{2}}\Bigg],$$

$$\omega_{02}\equiv\frac{|e|B}{m_\perp^*},\quad\text{and}\quad D\equiv\pm\frac{\bar{\lambda}_0\sqrt{2m_\perp^*}}{\hbar}$$

The use of (10.30) leads to the expressions of the EMs' along z and y directions as

$$m_z^*(\bar{E}_{FBHD},\eta_g,n,E_0)$$
$$=m_\parallel^*\Bigg[\{\gamma_3(\bar{E}_{FBHD},\eta_g)\}''$$
$$+\frac{m_\parallel^* E_0^2\{\gamma_3(\bar{E}_{FBHD},\eta_g)\}'\{\gamma_3(\bar{E}_{FBHD},\eta_g)\}''}{B^2}\Bigg] \tag{10.33}$$

$$m_y^*(\bar{E}_{FBHD},\eta_g,n,E_0)$$
$$=\left(\frac{B}{E_0}\right)^2\frac{1}{[\{\gamma_3(\bar{E}_{FBHD},\eta_g)\}']}\Bigg[\{\gamma_3(\bar{E}_{FBHD},\eta_g)\}-\beta_1(n,E_0)$$
$$+\frac{m_\parallel^* E_0^2[\{\gamma_3(\bar{E}_{FBHD},\eta_g)\}']^2}{2B^2}\Bigg]$$
$$\times\left[\frac{-\{\gamma_3(\bar{E}_{FBHD},\eta_g)\}''}{[-\{\gamma_3(\bar{E}_{FBHD},\eta_g)\}']^2}\left[\gamma_3(\bar{E}_{FBHD},\eta_g)-\beta_1(n,E_0)\right.\right.$$
$$\left.+\frac{m_\parallel^* E_0^2[\{\gamma_3(\bar{E}_{FBHD},\eta_g)\}']^2}{2B^2}\right]$$
$$\left.+1+\frac{m_\parallel^* E_0^2[\{\gamma_3(\bar{E}_{FBHD},\eta_g)\}']}{B^2}\right] \tag{10.34}$$

The Landau energy $(\bar{E}_{n_4})$ can be written as

$$\gamma_3(\bar{E}_{n_4},\eta_g)=\beta_1(n,E_0)-\frac{m_\parallel^* E_0^2}{2B^2}(\gamma_3'(\bar{E}_{n_4},\eta_g))^2 \tag{10.35}$$

The expression for n_0 in this case assumes the form

$$n_0 = \frac{2g_v B\sqrt{2m_c}}{3L_x\pi^2\hbar^2 E_0}\sum_{n=0}^{n_{\max}}[T_{53HD}(n, \bar{E}_{FBHD}, \eta_g) + T_{54HD}(n, \bar{E}_{FBHD}, \eta_g)] \quad (10.36)$$

where

$$T_{53HD}(n, \bar{E}_{FBHD}, \eta_g) \equiv \left[\left[\gamma_3(\bar{E}_{FBHD}, \eta_g) - \beta_1(n, E_0) + |e|E_0 L_x(\gamma_3'(\bar{E}_{FBHD}, \eta_g)) - \frac{m_\parallel^* E_0^2}{2B^2}(\gamma_3'(\bar{E}_{FBHD}, \eta_g))^2\right]^{3/2} - \left[(\gamma_3(\bar{E}_{FBHD}, \eta_g)) - \beta_1(n, E_0) - \frac{m_\parallel^* E_0^2}{2B^2}(\gamma_3'(\bar{E}_{FBHD}, \eta_g))^2\right]^{3/2}\right][\gamma_3'(\bar{E}_{FBHD}, \eta_g)]^{-1}$$

and

$$T_{54HD}(n, \bar{E}_{FBHD}, \eta_g) \equiv \sum_{r=0}^{s}[L(r)T_{53HD}(n, \bar{E}_{FBHD}, \eta_g)].$$

The DOS function can be written as

$$N(E_0, B) = \frac{2g_v B\sqrt{2m_c}}{3L_x\pi^2\hbar^2 E_0}\sum_{n=0}^{n_{\max}}[T_{53HD}(n, E, \eta_g)]' H(E - \bar{E}_{n_4}) \quad (10.37)$$

10.2.4 *The EM in HD IV–VI semiconductors under cross-fields configuration*

(1.185) can be written as

$$\frac{p_s^2}{2M_1^*(E, \eta_g)} + \frac{p_z^2}{2M_3^*(E, \eta_g)} = g(E, \eta_g) \quad (10.38)$$

where

$$M_1^*(E,\eta_g) = \left[\frac{(\bar{R})^2}{E_g}\{c_1(\alpha_1,E,E_g) - iD_1(\alpha_1,E,E_g)\}\right.$$
$$+\frac{(\bar{S})^2}{\Delta_c}\{c_2(\alpha_2,E,E_g) - iD_2(\alpha_2,E,E_g)\}$$
$$\left.\frac{(\bar{Q})^2}{\Delta_c''}\{c_3(\alpha_3,E,E_g) - iD_3(\alpha_3,E,E_g)\}\right]^{-1}$$

$$M_3^*(E,\eta_g) = \left[\frac{2(\bar{A})^2}{E_g}\{c_1(\alpha_1,E,E_g) - iD_1(\alpha_1,E,E_g)\}\right.$$
$$\left.+\frac{(\bar{S}+\bar{Q})^2}{\Delta_c''}\{c_3(\alpha_3,E,E_g) - iD_3(\alpha_3,E,E_g)\}\right]^{-1}$$

and

$$g^*(E,\eta_g) = 2\hbar^2\gamma_0(E,\eta_g)$$

In the presence of quantizing magnetic field B along z direction and the electric field along x-axis, from above equation one obtains

$$\frac{\hat{p}_x^2}{2M_1^*(E,\eta_g)} + \frac{(\hat{p}_y - |e|B\hat{x})^2}{2M_1^*(E,\eta_g)} + \frac{\hat{p}_z^2}{2M_3^*(E,\eta_g)}$$
$$= g^*(E,\eta_g) + |e|E_0\hat{x}\rho_1^*(E,\eta_g) \tag{10.39}$$

where

$$\rho_1^*(E,\eta_g) = \frac{\partial}{\partial E}[g^*(E,\eta_g)]$$

Let us define the operator $\hat{\theta}$ as

$$\hat{\theta} = -\hat{p}_y + |e|B\hat{x} - \frac{\rho_1^*(E,\eta_g)E_0\left[M_1^*(E,\eta_g)\right]}{B} \tag{10.40}$$

Eliminating $\hat{x}$, between the above two equations, the dispersion relation of the conduction electrons in HD stressed Kane type semiconductors in the presence of cross fields configuraration can be expressed as

$$g^*(E,\eta_g) = \left(n+\frac{1}{2}\right)\hbar\bar{\omega}_{i1}(E,\eta_g) + \frac{\hbar^2 k_z^2}{2M_3^*(E,\eta_g)} - \frac{E_0}{B}\rho_1^*(E,\eta_g)\hbar k_y - \frac{E_0^2}{2B^2}[\rho_1^*(E,\eta_g)]^2 M_1^*(E,\eta_g)$$

where

$$\bar{\omega}_{i1}(E,\eta_g) = eB[M_1^*(E,\eta_g)]^{-1} \tag{10.41}$$

The use of (10.41) leads to the expressions of the EMs' along z and y directions as

$$\begin{aligned} m_z^*(\bar{E}_{FBHD},\eta_g,n,E_0) &= \text{Real part of}\Bigg[[M_3^*(\bar{E}_{FBHD},\eta_g)]' \\ &\times\Bigg[g^*(\bar{E}_{FBHD},\eta_g) - \left(n+\frac{1}{2}\right)\hbar\bar{\omega}_i(\bar{E}_{FBHD},\eta_g) \\ &+ \frac{E_0^2}{2B^2}[\rho_1^*(\bar{E}_{FBHD},\eta_g)^2 M_1^*(\bar{E}_{FBHD},\eta_g)] \\ &+ [M_3^*(\bar{E}_{FBHD},\eta_g)]\Bigg[[g^*(\bar{E}_{FBHD},\eta_g)]' \\ &- \left(n+\frac{1}{2}\right)\hbar[\bar{\omega}_i(\bar{E}_{FBHD},\eta_g)]' \\ &+ \frac{E_0^2}{2B^2}[[\rho_1^*(\bar{E}_{FBHD},\eta_g)]^2[M_1^*(\bar{E}_{FBHD},\eta_g)]'] \\ &+2[M_3^*(\bar{E}_{FBHD},\eta_g)] + [\rho_1^*(\bar{E}_{FBHD},\eta_g)][\rho_1^*(\bar{E}_{FBHD},\eta_g)]'\Bigg]\Bigg] \end{aligned} \tag{10.42}$$

and

$$\begin{aligned} m_y^*(\bar{E}_{FBHD},\eta_g,n,E_0) &= (B/E_0)^2 \text{ Real part of } [\rho_1^*(\bar{E}_{FBHD},\eta_g)]^{-3} \end{aligned}$$

$$\times\left[g^*(\bar{E}_{FBHD},\eta_g)-\left(n+\frac{1}{2}\right)\hbar\bar{\omega}_{i1}(\bar{E}_{FBHD},\eta_g)\right.$$

$$+\frac{E_0^2}{2B^2}[\rho^*(\bar{E}_{FBHD},\eta_g)^2M_1^*(\bar{E}_{FBHD},\eta_g)]\Bigg]\Bigg[\Bigg[[\rho_1^*(\bar{E}_{FBHD},\eta_g)]$$

$$\times\left[[g^*(\bar{E}_{FBHD},\eta_g)]'-\left(n+\frac{1}{2}\right)\hbar[\bar{\omega}_i(\bar{E}_{FBHD},\eta_g)]'\right.$$

$$+\frac{E_0^2}{2B^2}\left[[\rho_1^*(\bar{E}_{FBHD},\eta_g)^2[M_1^*(\bar{E}_{FBHD},\eta_g)]'\right]$$

$$-[\rho_1^*(\bar{E}_{FBHD},\eta_g)]'\left[g^*(\bar{E}_{FBHD},\eta_g)-\left(n+\frac{1}{2}\right)\hbar\bar{\omega}_{i1}\right.$$

$$\left.\times(\bar{E}_{FBHD},\eta_g)+\frac{E_0^2}{2B^2}[\rho_1^*(\bar{E}_{FBHD},\eta_g)^2M_1^*(\bar{E}_{FBHD},\eta_g)]\right]\tag{10.43}$$

The Landau level energy (E_{η_g}) in this case can be expressed through the equation

$$g^*(E_{n9},\eta_g)=\left(n+\frac{1}{2}\right)\hbar\bar{\omega}_{i1}(E_{n9},\eta_g)$$
$$-\frac{E_0^2}{2B^2}\left[\rho_1^*(E_{n9},\eta_g)\right]^2M_1^*(E_{n9},\eta_g)\tag{10.44}$$

The electron concentration can be written as

$$n_0=\frac{2B}{3L_x\pi^2\hbar^2E_0}\text{ Real part of }\sum_{n=0}^{n_{\max}}[T_{4131HD}(n,\bar{E}_{FBHD},\eta_g)$$
$$+T_{4141HD}(n,\bar{E}_{FBHD},\eta_g)]\tag{10.45a}$$

where

$$T_{4131HD}(n,\bar{E}_{FBHD},\eta_g)$$
$$=\left[\frac{\sqrt{2M_3^*(\bar{E}_{FBHD},\eta_g)}}{\rho_1^*(\bar{E}_{FBHD},\eta_g)}\right]\Bigg[\Bigg[T_{51}(n,\bar{E}_{FBHD},\eta_g)$$

$$+\frac{E_0}{B}\rho_1^*(\bar{E}_{FBHD},\eta_g)\hbar x_{hHD1}(\bar{E}_{FBHD},\eta_g)\rho_1^*(\bar{E}_{FBHD},\eta_g)\Big]^{3/2}$$

$$-\Big[T_{51}(n,\bar{E}_{FBHD},\eta_g)$$

$$+\frac{E_0}{B}\rho_1^*(\bar{E}_{FBHD},\eta_g)\hbar x_{hHD1}(\bar{E}_{FBHD},\eta_g)\rho_1^*(\bar{E}_{FBHD},\eta_g)\Big]^{3/2}\Big]$$

$$\times T_{51}(n,\bar{E}_{FBHD},\eta_g)$$

$$=\Big[g^*(\bar{E}_{FBHD},\eta_g)-\left(n+\frac{1}{2}\right)\hbar\bar{\omega}_{i1}(\bar{E}_{FBHD},\eta_g)$$

$$+\frac{M_1^*(\bar{E}_{FBHD},\eta_g)E_0^2}{2B^2}[\rho_1^*(\bar{E}_{FBHD},\eta_g)]^2\Big]x_{1HD1}(\bar{E}_{FBHD},\eta_g)$$

$$=\frac{-M_1^*(\bar{E}_{FBHD},\eta_g)E_0\left[\rho_1^*(\bar{E}_{FBHD},\eta_g)\right]}{B},x_{hHD1}(\bar{E}_{FBHD},\eta_g)$$

$$=\frac{|e|BL_x}{\hbar}+x_{1HD1}(\bar{E}_{FBHD},\eta_g)$$

and

$$T_{4141HD}(n,\bar{E}_{FBHD},\eta_g)\equiv\sum_{r=1}^{s}\left[L(r)T_{4131HD}(n,\bar{E}_{FBHD},\eta_g)\right]$$

The DOS in this case is given by

$$N(E_0,B)=\frac{2g_vB\sqrt{2m_c}}{3L_x\pi^2\hbar^2E_0}\sum_{n=0}^{n_{\max}}\left[T_{4131HD}(n,E,\eta_g)\right]'H(E-\bar{E}_{n9})\tag{10.45b}$$

10.2.5 *The EM in HD stressed Kane type semiconductors under cross-fields configuration*

The use of (1.124) can be written as

$$\frac{p_x^2}{2M_1^*(E,\eta_g)}+\frac{p_y^2}{2M_2^*(E,\eta_g)}+\frac{p_z^2}{2M_3^*(E,\eta_g)}=G^*(E,\eta_g)\tag{10.46}$$

where

$$m_1^*(E,\eta_g) = [2\hbar^2[\gamma_0(E,\eta_g) - I(1)T_{17}]]^{-1},$$

$$T_{17} \equiv \left[E_g - C_1\varepsilon - (\bar{a}_0 + C_1)\varepsilon + \frac{3}{2}\bar{b}_0\varepsilon_{xx} - \frac{\bar{b}_0}{2}\varepsilon + \left(\frac{\sqrt{3}}{2}\right)\varepsilon_{xy}\bar{d}_0\right]$$

$$m_2^*(E,\eta_g) = [2\hbar^2[\gamma_0(E,\eta_g) - I(1)T_{27}]]^{-1},$$

$$T_{27} \equiv \left[E_g - C_1\varepsilon - (\bar{a}_0 + C_1)\varepsilon + \frac{3}{2}\bar{b}_0\varepsilon_{xx} - \frac{\bar{b}_0}{2}\varepsilon + \left(\frac{\sqrt{3}}{2}\right)\varepsilon_{xy}\bar{d}_0\right]$$

$$m_3^*(E,\eta_g) = [2\hbar^2[\gamma_0(E,\eta_g) - I(1)T_{37}]]^{-1},$$

$$T_{37} \equiv \left[E_g - C_1\varepsilon - (\bar{a}_0 + C_1)\varepsilon + \frac{3}{2}\bar{b}_0\varepsilon_{xx} - \frac{\bar{b}_0}{2}\varepsilon + \left(\frac{\sqrt{3}}{2}\right)\varepsilon_{xy}\bar{d}_0\right].$$

and the other symbols are written in Chapter 2.

In the presence of quantizing magnetic field B along z direction and the electric field along x-axis, from (10.46) one obtains

$$\frac{\hat{p}_x^2}{2M_1^*(E,\eta_g)} + \frac{(\hat{p}_y - |e|B\hat{x})^2}{2M_2^*(E,\eta_g)} + \frac{\hat{p}_z^2}{2M_3^*(E,\eta_g)}$$
$$= G^*(E,\eta_g) + |e|E_0\hat{x}\left[\frac{m_1^*(E,\eta_g)}{m_2^*(E,\eta_g)}\right]^{\frac{1}{2}}\rho^*(E,\eta_g) \qquad (10.47)$$

where

$$\rho^*(E,\eta_g) = \frac{\partial}{\partial E}[G^*(E,\eta_g)]$$

Let us define the operator $\hat{\theta}$ as

$$\hat{\theta} = -\hat{p}_y + |e|B\hat{x} - \frac{\rho^*(E,\eta_g)E_0[m_1^*(E,\eta_g)m_2^*(E,\eta_g)]^{1/2}}{B} \qquad (10.48a)$$

Eliminating $\hat{x}$, between the above two equations, the dispersion relation of the conduction electrons in HD stressed Kane type semiconductors in the presence of cross fields configuration can be expressed as

$$G^*(E,\eta_g) = \left(n+\frac{1}{2}\right)\hbar\bar{\omega}_i(E,\eta_g) + \frac{\hbar^2 k_z^2}{2m_3^*(E,\eta_g)} - \frac{E_0}{B}\rho^*(E,\eta_g)\left[\frac{m_1^*(E,\eta_g)}{m_2^*(E,\eta_g)}\right]^{1/2}\hbar k_y - \frac{E_0^2}{2B^2}[\rho^*(E,\eta_g)]^2 m_1^*(E,\eta_g) \quad (10.48b)$$

where

$$\bar{\omega}_i(E,\eta_g) = eB[m_1^*(E,\eta_g)m_2^*(E,\eta_g)]^{-\frac{1}{2}}$$

The use of (8.48b) leads to the expressions of the EMs' along z and y directions as

$$m_z^*(\bar{E}_{FBHD},\eta_g,n,E_0) = \Bigg[[m_3^*(\bar{E}_{FBHD},\eta_g)]' \times\Bigg[G^*\left(\bar{E}_{FBHD},\eta_g\right) - \left(n+\frac{1}{2}\right)\hbar\bar{\omega}_i\left(\bar{E}_{FBHD},\eta_g\right) + \frac{E_0^2}{2B^2}[\rho^*(\bar{E}_{FBHD},\eta_g)^2 m_1^*(\bar{E}_{FBHD},\eta_g)] + [m_3^*(\bar{E}_{FBHD},\eta_g)] \times\Bigg[[G^*(\bar{E}_{FBHD},\eta_g)]' - \left(n+\frac{1}{2}\right)\hbar[\bar{\omega}_i(\bar{E}_{FBHD},\eta_g)]' + \frac{E_0^2}{2B^2}\Bigg[2\Bigg[\rho^*(\bar{E}_{FBHD},\eta_g)[\rho^*(\bar{E}_{FBHD},\eta_g)'][m_1^*(\bar{E}_{FBHD},\eta_g)] + [m_1^*(\bar{E}_{FBHD},\eta_g)]'[\rho^*(\bar{E}_{FBHD},\eta_g)]^2\Bigg]\Bigg]\Bigg] \quad (10.49)$$

$$m_y^*(\bar{E}_{FBHD},\eta_g,n,E_0) = (B/E_0)^2[m_4^*(\bar{E}_{FBHD},\eta_g)]^{-3} \Bigg[G^*(\bar{E}_{FBHD},\eta_g) - \left(n+\frac{1}{2}\right)\hbar\bar{\omega}_i(\bar{E}_{FBHD},\eta_g)$$

$$+\frac{E_0^2}{2B^2}[\rho^*(\bar{E}_{FBHD},\eta_g)^2 m_1^*(\bar{E}_{FBHD},\eta_g)]$$

$$\times\left[[m_4^*(\bar{E}_{FBHD},\eta_g)]\left[[G^*(\bar{E}_{FBHD},\eta_g)]' - \left(n+\frac{1}{2}\right)\right.\right.$$

$$\times\,\hbar[\bar{\omega}_i(\bar{E}_{FBHD},\eta_g)]'$$

$$+\frac{E_0^2}{2B^2}[[\rho^*(\bar{E}_{FBHD},\eta_g)^2 m_1^*(\bar{E}_{FBHD},\eta_g)]']$$

$$\times\,[m_4^*(\bar{E}_{FBHD},\eta_g)]'$$

$$\times[G^*(\bar{E}_{FBHD},\eta_g)] - \left(n+\frac{1}{2}\right)\hbar\bar{\omega}_i(\bar{E}_{FBHD},\eta_g)$$

$$\left.+\,\frac{E_0^2}{2B^2}[\rho^*(\bar{E}_{FBHD},\eta_g)^2 m_1^*(\bar{E}_{FBHD},\eta_g)]\right] \tag{10.50}$$

where

$$m_4^*\left(\bar{E}_{FBHD},\eta_g\right) = \left[\left[\rho^*\left(\bar{E}_{FBHD},\eta_g\right)\right]\left[\frac{m_1^*\left(\bar{E}_{FBHD},\eta_g\right)}{m_2^*\left(\bar{E}_{FBHD},\eta_g\right)}\right]^{\frac{1}{2}}\right]$$

The Landau level energy (E_{η_g}) in this case can be expressed through the equation

$$G^*(E_{n_9},\eta_g) = \left(n+\frac{1}{2}\right)\hbar\bar{\omega}_i(E_{n_8},\eta_g) - \frac{E_0^2}{2B^2}[\rho^*(E_{n_8},\eta_g)]^2 m_1^*(E_{n_8},\eta_g) \tag{10.51}$$

The electron concentration can be written as

$$n_0 = \frac{2B}{3L_x\pi^2\hbar^2 E_0}\sum_{n=0}^{n_{\max}}[T_{413HD}(n,\bar{E}_{FBHD},\eta_g) + T_{414HD}(n,\bar{E}_{FBHD},\eta_g)] \tag{10.52}$$

where

$T_{413HD}(n,\bar{E}_{FBHD},\eta_g)$

$$= \left[\frac{\sqrt{2m_3^*(\bar{E}_{FBHD},\eta_g)}}{\rho^*(\bar{E}_{FBHD},\eta_g)}\right]\left[\left[T_5(n,\bar{E}_{FBHD},\eta_g)\right.\right.$$

$$\left. + \frac{E_0}{B}\rho^*(\bar{E}_{FBHD},\eta_g)\hbar x_{hHD}(\bar{E}_{FBHD},\eta_g)\rho^*(\bar{E}_{FBHD},\eta_g)\right]^{\frac{3}{2}}$$

$$- \left[T_5(n,\bar{E}_{FBHD},\eta_g) + \frac{E_0}{B}\rho^*(\bar{E}_{FBHD},\eta_g)\hbar x_{lHD}(\bar{E}_{FBHD},\eta_g)\right.$$

$$\left.\left.\times\,\rho^*(\bar{E}_{FBHD},\eta_g)\right]^{\frac{3}{2}}\right],$$

$$T_5(n,\bar{E}_{FBHD},\eta_g) = \left[G^*(\bar{E}_{FBHD},\eta_g) - \left(n+\frac{1}{2}\right)\hbar\overline{\omega_i}(\bar{E}_{FBHD},\eta_g)\right.$$

$$\left. + \frac{m_1^*(\bar{E}_{FBHD},\eta_g)E_0^2}{2B^2}[\rho^*(\bar{E}_{FBHD},\eta_g)]^2\right]$$

$$x_{lHD}(\bar{E}_{FBHD},\eta_g) = \frac{-m_1^*(\bar{E}_{FBHD},\eta_g)E_0[\rho^*(\bar{E}_{FBHD},\eta_g)]}{B},$$

$$x_{hHD}(\bar{E}_{FBHD},\eta_g) = \frac{|e|BL_x}{\hbar} + x_{lHD}(\bar{E}_{FBHD},\eta_g)$$

and

$$T_{414HD}\left(n,\bar{E}_{FBHD},\eta_g\right) \equiv \sum_{r=1}^{s} L(r)\,T_{413HD}\left(n,\bar{E}_{FBHD},\eta_g\right)$$

The DOS in this case is given by

$$N(E_0,B) = \frac{2B}{3L_x\pi^2\hbar^2E_0}\sum_{n=0}^{n_{\max}}[T_{413HD}(n,E,\eta_g)]'H(E-E_{n8}) \tag{10.53}$$

10.3 Open Research Problems

(R.10.1) Investigate the EM in the presence of an arbitrarily oriented quantizing magnetic and crossed electric fields in HD tetragonal semiconductors by including broadening and the electron spin. Study all the special cases for HD III–V, ternary and quaternary materials in this context.

(R.10.2) Investigate the EMs for all models of HD IV–VI, II–VI and stressed Kane type compounds in the presence of an arbitrarily oriented quantizing magnetic and crossed electric fields by including broadening and electron spin.

(R.10.3) Investigate the EM for all the materials as stated in R.1.1 of chapter 1 in the presence of an arbitrarily oriented quantizing magnetic and crossed electric fields by including broadening and electron spin.

References

1. W. Zawadzki, B. Lax, Phys. Rev. Lett. **16**, 1001 (1966)
2. M. J. Harrison, Phys. Rev. A **29**, 2272 (1984); J. Zak, W. Zawadzki, Phys. Rev. **145**, 536 (1966)
3. W. Zawadzki, Q. H. Vrehen, B. Lax, Phys. Rev. **148**, 849 (1966); Q. H. Vrehen, W. Zawadzki, M. Reine, Phys. Rev. **158**, 702 (1967); M. H. Weiler, W. Zawadzki, B. Lax, Phys. Rev. **163**, 733 (1967)
4. W. Zawadzki and J. Kowalski, Phys. Rev. Lett. **27**, 1713 (1971); C. Chu, M.S Chu, T. Ohkawa, Phys. Rev. Lett. **41**, 653 (1978); P. Hu and C. S. Ting, Phys. Rev. **B36**, 9671 (1987)
5. E.I. Butikov, A.S. Kondratev, A.E. Kuchma, Sov. Phys. Sol. State, **13**, 2594 (1972)
6. K.P. Ghatak, J.P. Banerjee, B. Goswami, B. Nag, Non. Optics and Quantum Optics, **16**, 241 (1996); M. Mondal, K.P. Ghatak, Phys. Stat. Sol. (b), **133**, K67 (1986)
7. M. Mondal, N. Chattopadhyay, K. P. Ghatak, Jour. Low Temp. Phys., **66**, 131 (1987); K.P. Ghatak, M. Mondal, Zeitschrift fur Physik B, **69**, 471 (1988)
8. M. Mondal, K.P. Ghatak, Phys. Lett. A, **131A**, 529 (1988); M. Mondal, K.P. Ghatak, in Phys. Stat. Sol. (b) Germany, **147**, K179 (1988); B. Mitra, K. P. Ghatak,Phys. Lett., **137A**, 413 (1989)
9. B. Mitra, A. Ghoshal, K.P. Ghatak, Phys. Stat. Soli. (b), **154**, K147 (1989)
10. B. Mitra, K.P. Ghatak, Phys. Stat. Sol. (b), **164,** K13 (1991); K.P. Ghatak, B. Mitra, Int. J. Electron., **70**, 345 (1991); S.M. Adhikari, D. De, J.K. Baruah, S. Chowdhury, K.P. Ghatak, Adv. Sci. Focus **1**, 57 (2013); S. Pahari, S. Bhattacharya, D. De, S. M. Adhikari, A. Niyogi, A. Dey, N. Paitya, S. C. Saha, K. P. Ghatak, P. K. Bose, Physica B: Condensed Matter **405**, 4064 (2010); S. Bhattaacharya, S. Choudhary, S. Ghoshal, S.K.Bishwas, D. De, K.P. Ghatak, J. Comp. Theo. Nanoscience. **3**, 423 (2006); M. Mondal, K.P. Ghatak, Ann. der Physik. **46**, 502 (1989); K. P. Ghatak, B. De MRS Proceedings **242**, 377 (1992); K. P. Ghatak, B. Mitra, Internat. Jour.

Electron. Theo. Exp., **70**, 343 (1991); B. Mitra, K. P. Ghatak, Phys. Lett. A **141**, 81 (1989); B. Mitra, K. P. Ghatak, Phys. Lett. **A 137**, 413 (1989); M. Mondal, N. Chattopadhyay, K. P. Ghatak, Jour. of low Temp. Phys. **73**, 321 (1988); K. P. Ghatak, N. Chattopadhyay, S. N. Biswas, International Society for Optics and Photonics, Proc. Soc. Photo-Optical Instru. Engg., USA, 203 (1987)

11. K.P. Ghatak, M. Mitra, B. Goswami and B. Nag, Nonlinear Optics, **16**, 167 (1996); K.P. Ghatak, D.K. Basu, B. Nag, J. Phys. Chem. Sol., **58**, 133 (1997)
12. S. Biswas, N. Chattopadhyay, K. P. Ghatak Internat. Soc. Optics and Photonics, Proc. Soc. Photo-Optical Instru. Engg., USA, **836**, 175 (1987); K.P. Ghatak, M. Mondal and S. Bhattacharyya, *SPIE*, **1284**, 113 (1990); K.P. Ghatak, *SPIE*, **1280**, Photonic materials and optical bistability, 53 (1990)
13. K.P. Ghatak and S.N. Biswas, *SPIE,* Growth and Characterization of Materials for Infrared Detectors and Nonlinear Optical Switches, **1484**, 149 (1991)
14. K.P. Ghatak, SPIE, Fiber Optic and Laser Sensors IX, **1584**, 435 (1992)

Chapter 11

Appendix C: The EM Under Intense Electric Field in HD Kane Type Semiconductors

11.1 Introduction

With the advent of modern nano devices, there has been considerable interest in studying the electric field induced processes in semiconductors having different band structures. It appears from the literature that the studies have been made on the assumption that the carrier dispersion laws are invariant quantities in the presence of intense electric field, which is not fundamentally true. In this chapter, we shall study the EM in quantum confined optoelectronic semiconductors under strong electric field. In Section 11.2.1, and attempt is made to investigate the EM in the presence of intense electric field in HD III-V, ternary and quaternary semiconductors Section 11.2.2 contains the investigation of the EM under magnetic quantization in HD Kane type semiconductors in the presence of intense electric field. In Section 11.2.3, the EM in QWs of HD Kane Type Semiconductors in the presence of intense electric field has been studied. In Section 11.2.4 we have investigated the EM in NWs of HD Kane type semiconductors in the presence of intense electric field. In Section 11.2.5 the magneto EM in QWs in HD Kane type semiconductors in the presence of intense electric field has been studied. In Section 11.2.6 the EM in accumulation and inversion layers of Kane type semiconductors in the presence of intense electric field has been studied. In Section 11.2.7 the magneto EM in accumulation and inversion layers of Ka.ne type semiconductors in the presence of intense electric field has been studied. In Section 11.2.8 the EM in doping superlattices of HD Kane type semiconductors in the presence of intense electric field has been studied. In Section 11.2.9 the magneto EM in doping superlattices of HD Kane type semiconductors in the presence of intense electric field has been investigated. In Section 11.2.10 the EM in QWHD effective mass superlattices of Kane type semiconductors in the presence of intense electric field has been investigated. In Section 11.2.11 the EM in NWHD effective mass superlattices of Kane type semiconductors in the presence of intense electric field has been studied. In Section 11.2.12 the EM in Quantum dot HD superlattices of Kane type semiconductors in the presence of intense electric field has been studied. In Section 11.2.13 the magneto EM in HD effective mass superlattices of Kane type semiconductors in the presence of intense electric field has been studied. In Section 11.2.14 the EM in QWHD superlattices of Kane type semiconductors with graded interfaces in the presence of intense electric field has been studied. In Section 11.2.15 the EM in NWHD superlattices of Kane type semiconductors with graded interfaces in the presence of intense electric field has been investigated. In Section 11.2.16 the magneto EM in HD superlattices of Kane type semiconductors with graded interfaces in the presence of intense electric field has been investigated. In Section 11.2.17 the magneto EM in QWHD superlattices of Kane type semiconductors with graded interfaces in the presence of intense electric field has been investigated. Section 11.3 presents 6 open research problems which challenges the first order creativity from the readers of diverse fields.

11.2 Theoretical Background

11.2.1 *The formulation of the electron dispersion law in the presence of intense electric field in HD III–V, ternary and quaternary semiconductors*

The expression of the inter-band transition matrix element (X_{12}) in this case can be written as

$$X_{12} = i \int u_{\bar{k}1}^{*}(\bar{r}) \cdot \frac{\partial}{\partial k_x} u_{\bar{k}2}(\bar{r}) d^3 r \tag{11.1}$$

where $u_{\bar{k}1}(\bar{r}) \equiv u_1(\bar{k},\bar{r})$ and $u_{\bar{k}2}(\bar{r}) \equiv u_2(\bar{k},\bar{r})$ in which $u_1(\bar{k},\bar{r})$ and $u_2(\bar{k},\bar{r})$ are given by (6.15) and (6.16) respectively.

In the case of the presence of an external electric field, F_s along x axis,the inter-band transition matrix-element, X_{12}, has finite interaction band same band, e.g.,

$$\langle S|S\rangle = \langle X|X\rangle = \langle Y|Y\rangle = \langle Z|Z\rangle = 1$$
$$\langle X|Y\rangle = \langle Y|Z\rangle = \langle Z|X\rangle = 0$$
$$\langle S|X\rangle = \langle X|S\rangle = 0; \ \langle S|Y\rangle = \langle Y|S\rangle = 0 \quad \text{and} \quad \langle S|Z\rangle = \langle Z|S\rangle = 0.$$

Using the appropriate equations we can write

$$\begin{aligned}
X_{12} = i\int d^3r \Bigg\{ & a_{k_+}[(iS)\downarrow'] + b_{k_+}\left[\left(\frac{X'-iY'}{\sqrt{2}}\right)\uparrow'\right] \\
& + c_{k_+}\left[Z'\downarrow'\right]\Bigg\}^* \frac{\partial}{\partial k_x}\left\{a_{k_-}\left[(iS)\uparrow'\right]\right. \\
& \left. - b_{k_-}\left[\left(\frac{X'+iY'}{\sqrt{2}}\downarrow'\right)\right] + c_{k_-}\left[Z'\uparrow'\right]\right\} \\
= i\int d^3r \Bigg[\Bigg\{ & \left(a_{k_+}\frac{\partial}{\partial k_x}a_{k_-}\right)\cdot\left([(iS)\downarrow']^* . [(iS)\uparrow']\right) \\
& + \left(\frac{b_{k_+}}{\sqrt{2}}\frac{\partial}{\partial k_x}a_{k_-}\right)\left[(X'-iY')\uparrow'\right]^*\cdot\left[(iS)\uparrow'\right] \\
& + \left(c_{k_+}\frac{\partial}{\partial k_x}a_{k_-}\right)\left[(Z'\downarrow')\right]^* . \left[(iS)\uparrow'\right]\Bigg\}\Bigg] \\
+ i\int d^3r \Bigg[\Bigg\{ & \left(-\frac{a_{k_+}}{\sqrt{2}}\frac{\partial}{\partial k_x}b_{k_-}\right)\cdot\left[(iS)\downarrow'\right]^* . \left[(X'+iY')\downarrow'\right] \\
& - \left(\frac{b_{k_+}\frac{\partial}{\partial k_x}b_{k_-}}{2}\right)\left[(X'-iY')\uparrow'\right]^*\cdot\left[(X'+iY')\downarrow'\right] \\
& - \left(\frac{c_{k_+}\frac{\partial}{\partial k_x}b_{k_-}}{\sqrt{2}}\right)\left[(Z'\downarrow')\right]^*\cdot\left[(X'+iY')\downarrow'\right]\Bigg\}\Bigg]
\end{aligned}$$

$$
\begin{aligned}
&= i \int d^3r \left[\left\{ \left(a_{k_+} \frac{\partial}{\partial k_x} c_{k_-} \right) \cdot \left([(iS) \downarrow']^* . [Z' \uparrow'] \right) \right.\right. \\
&\quad + \left(\frac{b_{k_+}}{\sqrt{2}} \frac{\partial}{\partial k_x} c_{k_-} \right) [(X' - iY') \uparrow']^* . [Z' \uparrow'] \\
&\quad \left.\left. + \left(c_{k_+} \frac{\partial}{\partial k_x} c_{k_-} \right) [(Z' \downarrow')]^* \cdot [Z' \uparrow'] \right\} \right]
\end{aligned} \tag{11.2}
$$

Therefore,

$$
\begin{aligned}
\frac{X_{12}}{i} &= \left[\left\{ \left[\left(a_{k_+} \frac{\partial}{\partial k_x} a_{k_-} \right) \cdot \langle iS|iS \rangle \langle \downarrow' \mid \uparrow' \rangle \right] \right.\right. \\
&\quad + \left[\left(\frac{b_{k_+}}{\sqrt{2}} \frac{\partial}{\partial k_x} a_{k_-} \right) \langle (X' - iY')|iS \rangle \langle \uparrow' \mid \downarrow' \rangle \right] \\
&\quad \left.\left. + \left[\left(c_{k_+} \frac{\partial}{\partial k_x} a_{k_-} \right) \langle Z'|iS \rangle \langle \downarrow' \mid \uparrow' \rangle \right] \right\} \right] \\
&\quad - \left[\left\{ \left[\left(\frac{a_{k_+}}{\sqrt{2}} \frac{\partial}{\partial k_x} b_{k_-} \right) . \langle iS|(X' + iY') \rangle \langle \downarrow' \mid \downarrow' \rangle \right] \right.\right. \\
&\quad + \left[\left(\frac{b_{k_+}}{2} \frac{\partial}{\partial k_x} b_{k_+} \right) \langle (X' - iY')|(X' + iY') \rangle \langle \uparrow' \mid \downarrow' \rangle \right] \\
&\quad \left.\left. + \left[\left(\frac{c_k}{\sqrt{2}} \frac{\partial}{\partial k_x} b_{k_-} \right) \langle Z'|(X' + iY') \rangle \langle \downarrow' \mid \downarrow' \rangle \right] \right\} \right] \\
&\quad + \left\{ \left[\left(a_k \frac{\partial}{\partial k_x} c_{k_-} \right) . \langle iS|Z' \rangle \langle \downarrow' \mid \uparrow' \rangle \right] \right. \\
&\quad + \left[\left(\frac{b_{k_+}}{\sqrt{2}} \frac{\partial}{\partial k_x} c_{k_-} \right) . \langle (X' - iY')|Z' \rangle \langle \uparrow' \mid \uparrow' \rangle \right] \\
&\quad \left. + \left[\left(c_k \frac{\partial}{\partial k_x} c_{k_-} \right) \cdot \langle Z'|Z' \rangle \langle \downarrow' \mid \uparrow' \rangle \right] \right\}
\end{aligned} \tag{11.3}
$$

Therefore we can write,

$$
X_{12} = i \left\{ - \left(a_{k_+} \frac{\partial}{\partial k_x} a_{k_-} \right) \langle \downarrow' \mid \uparrow' \rangle + \left(c_{k_+} \frac{\partial}{\partial k_x} c_{k_-} \right) \langle \downarrow' \mid \uparrow' \rangle \right\} \tag{11.4}
$$

We can prove that

$$\langle \downarrow' \mid \uparrow' \rangle = \frac{1}{2}(\hat{r}_1 + i\hat{r}_2) \tag{11.5}$$

Therefore (11.4) and (11.5) we get

$$\begin{aligned}
X_{12} &= i\left\{-\left(a_{k+}\frac{\partial}{\partial k_x}a_{k-}\right) + \left(c_{k+}\frac{\partial}{\partial k_x}c_{k-}\right)\right\}\langle \downarrow' \mid \uparrow' \rangle \\
&= -i\left\{\left(a_{k+}\frac{\partial}{\partial k_x}a_{k-}\right) - \left(c_{k+}\frac{\partial}{\partial k_x}c_{k-}\right)\right\}\cdot\frac{1}{2}(\hat{r}_1 + i\hat{r}_2) \\
&= \frac{-iA(k)}{2}(\hat{r}_1 + i\hat{r}_2)
\end{aligned} \tag{11.6}$$

where

$$A(k) = \left(a_{k+}\frac{\partial}{\partial k_x}a_{k-}\right) - \left(c_{k+}\frac{\partial}{dk_x}c_{k-}\right) \tag{11.7}$$

From (11.6), we find,

$$|X_{12}|^2 = \frac{1}{4}A^2(k)(1+1) = \frac{1}{2}A^2(k)[\text{since}, |\hat{r}_1| = |\hat{r}_2| = 1] \tag{11.8}$$

Considering spin-up and spin-down, we have to multiply by 2

$$|X_{12}|^2 = 2 \times \frac{1}{2}A^2(k) = A^2(k) \tag{11.9}$$

We can evaluate X_{11} and X_{22} in the following way:

$$\begin{aligned}
X_{11} &= i\int u^*_{\bar{k}1}(\bar{r})\cdot\frac{\partial}{\partial k_x}u_{\bar{k}1}(\bar{r}).d^3r \\
&= i\int d^3r\left\{\left(a_{k_+}\frac{\partial}{\partial k_x}a_{k_+}\right) + \left(b_{k_+}\frac{\partial}{\partial k_x}b_{k_+}\right) + \left(c_{k_+}\frac{\partial}{\partial k_x}c_{k_+}\right)\right\} \\
&= \frac{1}{2}i\int d^3r\left\{\frac{\partial}{\partial k_x}(a_k^2 + b_k^2 + c_k^2)\right\} = \frac{1}{2}i\int d^3r\left\{\frac{\partial}{\partial k_x}(1)\right\} = 0,
\end{aligned}$$

since

$$a^2_{k+} + b^2_{k+} + c^2_{k+} = 1.$$

Therefore $X_{11} = 0$, and similarly we can prove $X_{22} = 0$. Thus we conclude that intraband momentum matrix element due to external electric field (X_{CC}) is zero. From the expression of $a_{k\pm}$ we can write

$$a_{k_+}^2 = r_0^2 \left[\frac{E_g - \gamma_{k_+}^2 (E_g - \delta')}{E_g + \delta'} \right]^2$$

and

$$a_{k-}^2 = r_0^2 \left[\frac{E_g - \gamma_{k-}^2 (E_g - \delta')}{E_g + \delta'} \right]^2.$$

Therefore,

$$2a_{k-} \frac{\partial}{\partial k_x} a_{k-} = r_0^2 \left[- \left(\frac{E_g - \delta'}{E_g + \delta'} \right) \right] \frac{\partial \gamma_{k-}^2}{\partial k_x}$$

and

$$\frac{\partial a_{k-}}{\partial k_x} = -\frac{r_0^2}{2} \left(\frac{E_g - \delta'}{E_g + \delta'} \right) \frac{1}{a_{k-}} \frac{\partial \gamma_{k-}^2}{\partial k_x}$$

Combining we can write

$$a_{k+} \frac{\partial a_{k-}}{\partial k_x} = -\frac{r_0^2}{2} \left(\frac{E_g - \delta'}{E_g + \delta'} \right) \frac{a_{k+}}{a_{k-}} \frac{\partial \gamma_{k-}^2}{\partial k_x}$$

Similarly, $c_{k_+} = t\gamma_{k_+}$ and $c_{k_-} = t\gamma_{k_-}$. Therefore,

$$c_{k+} \frac{\partial}{\partial k_x} c_{k-} = \frac{t^2}{2} \frac{c_{k+}}{c_{k-}} \frac{\partial \gamma_{k-}^2}{\partial k_x}$$

$$A(k) = \left\{ -\frac{r_0^2}{2} \left[\frac{E_g - \delta'}{E_g + \delta'} \right] \frac{a_{k+}}{a_{k-}} - \frac{t^2}{2} \frac{c_{k+}}{c_{k-}} \right\} \frac{\partial \gamma_{k-}^2}{\partial k_x}$$

Now,

$$\left(\frac{a_{k+}}{a_{k-}} \right)^2 = \frac{E_g - \gamma_{k+}^2 (E_g - \delta')}{E_g - \gamma_{k-}^2 (E_g - \delta')} = \frac{E_g - \frac{\eta - E_g}{2(\eta + \delta')} (E_g - \delta')}{E_g - \frac{\eta + E_g}{2(\eta + \delta')} (E_g - \delta')}$$

$$= \frac{2E_g - (\eta + \delta') - (\eta - E_g)(E_g - \delta')}{2E_g(\eta + \delta') - (\eta + E_g)(E_g - \delta')}$$

$$= \frac{\eta(E_g + \delta') + E_g(E_g + \delta')}{\eta(E_g + \delta') - E_g(E_g - 3\delta')}$$

Therefore,

$$\left(\frac{a_{k+}}{a_{k-}}\right)^2 = \frac{\eta + E_g}{\eta - E_g\left(\frac{E_g - 3\delta'}{E_g + \delta'}\right)} = \frac{\eta + E_g}{\eta - E_g^{'}}$$

where,

$$E_g^{'} = \frac{E_g(E_g - 3\delta')}{E_g + \delta'}.$$

Thus,

$$\frac{a_{k+}}{a_{k-}} = \sqrt{\frac{\eta + E_g}{\eta - E_g^{'}}}$$

Similarly,

$$\frac{c_{k+}}{c_{k-}} = \frac{\gamma_{k+}}{\gamma_{k-}} = \sqrt{\frac{\eta - E_g}{\eta + E_g^{'}}}$$

and thus,

$$A(k) = -\left\{P\left(\frac{\eta + E_g}{\eta - E_g^{'}}\right)^{1/2} + Q\left(\frac{\eta - E_g}{\eta + E_g}\right)^{1/2}\right\}\frac{\partial \gamma_{k-}^2}{\partial k_x} \tag{11.10}$$

where,

$$P = \frac{r_0^2}{2}\left(\frac{E_g - \delta'}{E_g + \delta'}\right)$$

and

$$Q = \frac{t^2}{2}.$$

Now

$$\gamma_{k-}^2 = \frac{\eta + E_g}{2(\eta + \delta')}$$

so that

$$\frac{\partial \gamma_{k-}^2}{\partial k_x} = \frac{1}{2}\left[\frac{\partial \eta/\partial k_x}{(\eta + \delta')} - \frac{\eta + E_g}{(\eta + \delta')^2}\frac{\partial \eta}{\partial k_x}\right]$$

Thus,

$$\frac{\partial \gamma_{k-}^2}{\partial k_x} = \frac{1}{2}\left[\frac{\eta+\delta'-\eta-E_g}{(\eta+\delta')}\right]\frac{\partial \eta}{\partial k_x}$$
$$= -\frac{1}{2}\frac{(E_g-\delta')}{(\eta+\delta')^2}\cdot\frac{\partial \eta}{\partial k_x} \tag{11.11}$$

From (11.10) and (11.11), we get

$$A(k) = \frac{1}{2}\frac{(E_g-\delta')}{(\eta+\delta')^2}\cdot\frac{\partial \eta}{\partial k_u}\cdot\left\{P\left(\frac{\eta+E_g}{\eta-E_g'}\right)^{1/2}+Q\left(\frac{\eta-E_g}{\eta+E_g}\right)^{1/2}\right\} \tag{11.12}$$

This implies

$$\frac{\partial \eta}{\partial k_x} = \frac{E_g\hbar^2}{m_r}\cdot\frac{k_x}{\eta} \tag{11.13}$$

From (11.12) and (11.13), one can write

$$A(k) = \frac{E_g\hbar^2}{2m_r}\cdot\frac{k_u}{\eta}\cdot\frac{(E_g-\delta')}{(\eta+\delta')^2}\cdot\left\{P\left(\frac{\eta+E_g}{\eta-E_g'}\right)^{1/2}+Q\left(\frac{\eta-E_g}{\eta+E_g}\right)^{1/2}\right\} \tag{11.14}$$

Thus,

$$|A(k)|^2 = \frac{E_g^2(E_g-\delta')\hbar^2}{4m_r}\frac{\hbar^2k_x^2}{m_r}\frac{1}{\eta^2}\frac{1}{(\eta+\delta')^4}\cdot$$
$$\times\left\{P\left(\frac{\eta+E_g}{\eta-E_g'}\right)^{1/2}+Q\left(\frac{\eta-E_g}{\eta+E_g}\right)^{1/2}\right\}^2 \tag{11.15}$$

and

$$|X_{12}|^2 = |A(\bar{k})|^2. \tag{11.16}$$

From (11.15) and (11.16) we can write the square of the magnitude of the inter-band transition matrix element due to external electric field ($|X_{CV}|^2$)

It is well-known that the energy Eigen value, $E_n^{(2)}(\bar{k})$ in the presence of a perturbed Hamiltonian, H', is given by [1]

$$E_n^{(2)}(\bar{k}) = E_n(\bar{k}) + \langle n\bar{k}|H'|n\bar{k}\rangle + \left\{ \frac{|\langle n\bar{k}|H'|n\bar{k}\rangle|^2}{[E_n(\bar{k}) - E_m(\bar{k})]} \right\} \tag{11.17}$$

where

$$H\psi_n\left(\bar{k},\bar{r}\right) = E\psi\left(\bar{k},\bar{r}\right) \tag{11.18}$$

$$H = H_0 + H' \tag{11.19}$$

$$H_0 u_n\left(\bar{k},\bar{r}\right) = E_n\left(\bar{k}\right) u_n\left(\bar{k},\bar{r}\right) \tag{11.20}$$

in which, H is the total Hamiltonian, $\psi(\bar{k},\bar{r})$ is the wave function, where $u_n(\bar{k},\bar{r})$ is the periodic function of it, H_0 is the unperturbed Hamiltonian, n is the band index, and $E_n(\bar{k})$ is the energy of an electron in the periodic lattice.

For an external electric field (F_S) applied along the x-axis, the perturbed Hamiltonian (H') can be written as

$$H' = -F \cdot x \tag{11.21}$$

where $F(= eF_S)$

Therefore we get

$$E_n^{(2)}(\bar{k}) = E_n(\bar{k}) - F\langle n\bar{k}|H'|n\bar{k}\rangle + F^2 \left\{ \frac{|\langle n\bar{k}|H'|n\bar{k}\rangle|^2}{[E_n(\bar{k}) - E_m(\bar{k})]} \right\} \tag{11.22}$$

In (11.22), the second and the third terms are due to the perturbation factor.

For

$$X_{nm}\left(\bar{k}\right) = \langle n\bar{k}|x|m\bar{k}\rangle \tag{11.23}$$

we find

$$X_{nm}\left(\bar{k}\right) = i \int u_n^*\left(\bar{k},\bar{r}\right) \left(\frac{\partial}{\partial u}\right) \left[u_m\left(\bar{k},\bar{r}\right)\right] d^3r \tag{11.24}$$

where k_x is the x component of the $\bar{k}$ and the integration in (11.24) extends over the unit cell. From (11.22), (11.23) and (11.24) with the

n corresponds to the conduction band (C) and m corresponds to the valance band (V), we get

$$E_c^{(2)}\left(\bar{k}\right) = E_c\left(\bar{k}\right) - FX_{cc} + \left\{\frac{F^2|X_{CV}|^2}{\left[E_C\left(\bar{k}\right) - E_V\left(\bar{k}\right)\right]}\right\} \tag{11.25}$$

Thus combining the appropriate equations, the dispersion relation of the conduction electrons in the presence of electric field along x-axis can be written as

$$\begin{aligned} I_{11}(E) &= \frac{\hbar^2 k_x^2}{2m_c} + \frac{\hbar^2 k_y^2}{2m_c} + \frac{\hbar^2 k_z^2}{2m_c} + \frac{F^2|X_{12}|^2}{\eta} \\ &= \left[\frac{\hbar^2 k_x^2}{2m_c} + \frac{\hbar^2 k_y^2}{2m_c} + \frac{\hbar^2 k_z^2}{2m_c} + \left\{\frac{\hbar^2 k_x^2}{2m_c} \cdot \frac{2m_c}{m_r} \cdot \frac{F^2\hbar^2 E_g^2(E_g-\delta')^2}{4m_r}\right.\right. \\ &\quad \left.\left. \times \frac{1}{\eta^3}\frac{1}{(\eta+\delta')^4}\left[P\left(\frac{\eta+E_g}{\eta-E_g'}\right)^{1/2} + Q\left(\frac{\eta-E_g}{\eta+E_g}\right)^{1/2}\right]^2\right\}\right] \end{aligned} \tag{11.26}$$

When $F \to 0$, we have from (19.26),

$$k^2 \to \frac{2m_c}{\hbar^2} I_{11}(E)$$

and

$$\eta_1^2 = \left[E_g^2 + E_g \frac{2m_c}{m_r} I_{11}(E)\right].$$

Using the method of successive approximation one can write

$$1 = \frac{\hbar^2 k_x^2}{2m_c I_{11}(E)} + \frac{\hbar^2 k_y^2}{2m_c I_{11}(E)} + \frac{\hbar^2 k_z^2}{2m_c I_{11}(E)} + \frac{\hbar^2 k_x^2}{2m_c I_{11}(E)} . \Phi\left(E, F\right) \tag{11.27}$$

where,

$$\Phi(E,F) = \frac{2m_c}{m_r}\frac{F^2\hbar^2 E_g^2(E_g-\delta')^2}{4m_r}\frac{1}{\eta_1^3}\frac{1}{(\eta_1+\delta')^4} \times \left[P\left(\frac{\eta_1+E_g}{\eta_1-E_g'}\right)^{1/2} + Q\left(\frac{\eta_1-E_g}{\eta_1+E_g}\right)^{1/2}\right]^2$$

Therefore, the $E-k$ dispersion relation in the presence of an external electric field for III–V, ternary and quaternary materials whose unperturbed energy band structures are defined by the three band model of Kane can be expressed as

$$\frac{k_x^2}{\frac{2m_c}{\hbar^2}\left[\frac{I_{11}(E)}{1+\Phi(E,F)}\right]}+\frac{k_y^2}{\frac{2m_c}{\hbar^2}I_{11}\left(E\right)}+\frac{k_z^2}{\frac{2m_c}{\hbar^2}I_{11}\left(E\right)}=1 \tag{11.28}$$

In (11.28), the coefficients of k_x, k_y and k_z are not same and for this reason, this basic equation is "anisotropic" in nature together with the fact that the anisotropic dispersion relation is the ellipsoid of revolution in the k-space.

From (11.28) the expressions of the effective electron masses along x, y, and z directions can, respectively, be written as

$$\begin{aligned} m_x^*(E,F) &= \hbar^2 k_x \left.\frac{\partial k_x}{\partial E}\right|_{k_y=0,k_z=0} \\ &= m_c[1+\Phi(E,F)]^{-2} \\ &\quad \times[[1+\Phi(E,F)]I'_{11}(E)-I_{11}(E)\Phi'(E,F)] \end{aligned} \tag{11.29}$$

$$m_y^*\left(E,F\right) = \hbar^2 k_y \left.\frac{\partial k_y}{\partial E}\right|_{k_x=0,k_z=0} = m_0 I'_{11}\left(E\right) \tag{11.30}$$

$$m_z^*\left(E,F\right) = \hbar^2 k_z \left.\frac{\partial k_z}{\partial E}\right|_{k_x=0,k_y=0} = m_0 I'_{11}\left(E\right) \tag{11.31}$$

where

$$I'_{11}\left(E\right)=\frac{\partial}{\partial E}\left(I_{11}\left(E\right)\right)$$

and

$$\Phi'\left(E,F\right)=\frac{\partial}{\partial E}\left[\Phi\left(E,F\right)\right].$$

It may be noted from (19.29) that the effective mass along x-direction is a function of both electron energy and electric field respectively whereas from (11.30) and (11.31) we can infer the expressions of the effective masses along y and z directions are same and they depend on the electron energy only. Thus in the presence of an electric field, the mass anisotropy for Kane type semiconductors depends both on electron energy and electric field respectively.

The use of the usual approximation [2]

$$k_x^2 \approx \frac{1}{3}k^2 \tag{11.32}$$

in (11.28), leads to the simplified expression of the electron energy spectrum in the present case as

$$\begin{aligned} I_{11}(E) = \frac{\hbar^2 k^2}{2m_c} + \frac{F^2\hbar^2 E_g^2 (E_g - \delta')^2}{12m_r}\frac{2m_c}{m_r} \\ \times I_{11}(E)\frac{1}{\eta_1}\frac{1}{(\eta_1 + \delta')^4}\left\{P\left(\frac{\eta_1 + E_g}{\eta_1 - E_g'}\right)^{1/2} + Q\left(\frac{\eta_1 - E_g}{\eta_1 + E_g}\right)^{1/2}\right\}^2 \end{aligned} \tag{11.33}$$

The (11.33) can approximately be written as

$$\frac{\hbar^2 k^2}{2m_c} = \left[e_1 E^4 + e_2 E^3 + e_3 E^2 + e_4 E + e_5 - \frac{e_6}{1 + CE} + e_7(1 + CE)^{-2}\right] \tag{11.34}$$

where

$$e_1 = Q_f \cdot \omega_1, \quad Q_f = \frac{m_c}{m_r}E_g^{-4}[5e_f E_g^{-2} - 6G_f + 7h_f E_g^{-4}],$$

$$e_f = A_f P_f, \quad A_f = [F\hbar E_g(E_g - \delta')]^2 m_c (6m_r^2(\delta')^4)^{-1},$$

$$F = eF_s, \quad G_f = e_f(4\delta' + C_f),$$

$$C_f = (2E_g Q^2 + PQ(E_g - E_g') - 2P^2 E_g),$$

$$E_g' = \frac{E_g(E_g - 3\delta')}{E_g + \delta'}, \quad P = \frac{r_0^2}{2}\left(\frac{E_g - \delta'}{E_g + \delta'}\right),$$

$$r_0 = \left[\frac{6}{\chi}(E_g + \Delta)(E_g + \frac{2}{3}\Delta)\right]^{1/2}, \quad Q = \frac{t^2}{2},$$

$$t = \left[\frac{6}{\chi}(E_g + \frac{2}{3}\Delta)\right]^{1/2}, \quad h_f = (4\delta' e_f C_f)(B_f)^{-1}, \quad B_f = (P + Q)^2$$

$$P_f = E_g^{-3}(e_f E_g^{-2} - G_f + h_f E_g^{-4}), \quad \omega_1 = a_1^2, \quad a_1 = \frac{ab}{c}, \quad a = \frac{1}{E_g},$$

$$b = \frac{1}{E_g + \Delta}, \quad c = \left(E_g + \frac{2}{3}\Delta\right)^{-1},$$

$$e_2 = Q_f \cdot \omega_2, \quad \omega_2 = 2a_1b_1, \quad b_1 = (c)^{-2}(ac + bc - ab),$$

$$e_3 = (1 - P_f)a_1 + Q_f \cdot \omega_3, \quad \omega_3 = (b_1^2 + 2a_1c_1),$$

$$c_1 = \left[\frac{1}{c}\left(1 - \frac{a}{c}\right)\left(1 - \frac{b}{c}\right)\right], \quad e_4 = [(1 - P_f)b_1 + Q_f\omega_4],$$

$$\omega_4 = 2b_1c_1, \quad e_5 = [(1 - P_f)c_1 + \omega_5 Q_f]$$

$$\omega_5 = (c_1^2 - 2c_1b_1), \quad e_7 = Q_f\omega_7, \omega_7 = c_1^2, \quad e_6 = [(1 - P_f)c_1 - Q_f\omega_6]$$

and

$$\omega_6 = \frac{2c_1b_1}{c}\left(1 - \frac{cc_1}{b_1}\right)$$

Using (11.26b) and (11.34) we get

$$\begin{aligned}
\frac{\hbar^2k^2}{2m_c}\int_{-\infty}^{E} F(V)dV = e_1 \int_{-\infty}^{E} (E - V)^4 F(V)dV \\
+ e_2 \int_{-\infty}^{E} (E - V)^3 F(V)dV \\
+ e_3 \int_{-\infty}^{E} (E - V)^2 F(V)dV \\
+ e_4 \int_{-\infty}^{E} (E - V) F(V)dV + e_5 \int_{-\infty}^{E} F(V)dV \\
- e_6 \int_{-\infty}^{E} \frac{F(V)dV}{[1 + c(E - V)]} \\
+ e_7 \int_{-\infty}^{E} \frac{F(V)dV}{[1 + c(E - V)]^2}
\end{aligned} \tag{11.35}$$

We can prove that

$$\int_{-\infty}^{E} (E - V)^4 F(V)dV = \frac{E^4}{2}\left[1 + Erf\left(\frac{E}{\eta_g}\right)\right]$$

$$
\begin{aligned}
&+\frac{3\eta_g^4}{8\pi}\left[1+Erf\left(\frac{E}{\eta_g}\right)-\frac{2\eta_g}{3E}\exp\left(\frac{-E^2}{\eta_g^2}\right)\right]\\
&+\frac{3}{2}(E\eta_g)^2\left[1+Erf\left(\frac{E}{\eta_g}\right)\right]\\
&+\frac{E^3\eta_g}{\sqrt{\pi}}\exp\left(\frac{-E^2}{\eta_g^2}\right)\\
&+\frac{2}{\sqrt{\pi}}E\eta_g^3\exp\left(\frac{-E^2}{\eta_g^2}\right)\left[1+\frac{E^2}{\eta_g^2}\right]\\
=\;&\psi_0(E,\eta_g) \qquad (11.36)
\end{aligned}
$$

From Chapter 2, we know that

$$
\int_{-\infty}^{E}\frac{F(V)dV}{[1+c(E-V)]}=c_1(c,E,\eta_g)-ic_2(c,E,\eta_g) \qquad (11.37)
$$

where

$$
c_1(c,E,\eta_g)=\left[\frac{2}{c\eta_g\sqrt{\pi}}\exp(-u^2)\left[\sum_{p=1}^{\infty}\exp\left(\frac{-p^2}{4}\right)(p)^{-1}sinh(pu)\right]\right],
$$

$$
c_2(c,E,\eta_g)=\frac{\sqrt{\pi}}{c\eta_g}\exp(-u^2)
$$

and

$$
u=\frac{1+cE}{c\eta_g}
$$

We know that

$$
\begin{aligned}
\frac{\partial}{\partial x}\int_{A(x)}^{B(x)}F(x,\zeta)d\zeta=&\int_{A(x)}^{B(x)}\frac{\partial}{\partial x}[F(x,\zeta)]d\zeta+F(x,B(x))\frac{\partial B(x)}{\partial x}\\
&-F(x,A(x))\frac{\partial A(x)}{\partial x} \qquad (11.38)
\end{aligned}
$$

Using (11.37) and (11.38), we get

$$
\int_{-\infty}^{E}\frac{F(V)dV}{[1+c(E-V)]^2}=c_3(c,E,\eta_g)-ic_4(c,E,\eta_g) \qquad (11.39)
$$

where

$$c_3(c,E,\eta_g) = \left[\frac{-4u\exp(-u^2)}{c^2\eta_g^2\sqrt{\pi}}\left[\sum_{p=1}^{\infty}\exp\left(\frac{-p^2}{4}\right)(p)^{-1}sinh(pu)\right]\right.$$
$$+\frac{1}{\pi c\eta_g}\exp\left(\frac{-E^2}{\eta_g^2}\right)$$
$$\left.-\frac{2\exp(-u^2)}{c^2\eta_g^2\sqrt{\pi}}\sum_{p=1}^{\infty}\exp\left(\frac{-p^2}{4}\right)\cos h(pu)\right] \quad \text{and}$$

$$D_3(c,E,\eta_g) = \frac{2u}{c^2\eta_g^2}\exp(-u^2)$$

Therefore the DR in HD Kane type semiconductors can be written using (11.35), (11.36), (11.37) and (11.39) as

$$\frac{\hbar^2 k^2}{2m_c} = J_1(E,c,\eta_g,F) + iJ_2(E,c,\eta_g,F) \tag{11.40}$$

where,

$$J_1(E,c,\eta_g,F) = 2\left[1+Erf\left(\frac{E}{\eta_g}\right)\right]^{-1}\left[e_1\psi_0(E,\eta_g) + e_2\psi_1(E,\eta_g)\right.$$
$$+ e_3\theta_0(E,\eta_g) + e_4\gamma_0(E,\eta_g) + e_5\frac{1}{2}\left[1+Erf\left(\frac{E}{\eta_g}\right)\right]$$
$$\left.- e_6c_1(E,c,\eta_g) + e_7c_3(E,c,\eta_g)\right],$$

$$\psi_1(E,\eta_g) = \left[\frac{E}{2}\left[1+Erf\left(\frac{E}{\eta_g}\right)\right]\left[E^2+\frac{3}{2}\eta_g^2\right]\right.$$
$$\left.+\frac{\eta_g}{2\sqrt{\pi}}\exp\left(\frac{-E^2}{\eta_g^2}\right)(4E^2+\eta_g^2)\right]$$

and

$$J_2(E,c,\eta_g,F) = 2\left[1+Erf\left(\frac{E}{\eta_g}\right)\right]^{-1}\left[e_6c_2(E,c,\eta_g) + e_7D_3(E,c,\eta_g)\right]$$

The DOS function is given by

$$N_F(E) = 4\pi g_v \left(\frac{2m_c}{h^2}\right)^{\frac{3}{2}} [J_1'(E,c,\eta_g,F) + iJ_2'(E,c,\eta_g,F)] \times \sqrt{[J_1(E,c,\eta_g,F) + iJ_2(E,c,\eta_g,F)]} \quad (11.41)$$

The EEM in this case is given by

$$m^*(E,\eta_g,F) = m_c J_1'(E,c,\eta_g,F) \quad (11.42)$$

11.2.2 *The EM under magnetic quantization in HD Kane Type Semiconductors in the Presence of intense electric field*

The DR of the conduction electrons in HD optoelectronic materials under electric field can be written in presence of quantizing magnetic field B along x-direction whose unperturbed electron energy spectra are defined by the three band model of Kane as

$$\left(n+\frac{1}{2}\right)\hbar\omega_0 + \frac{\hbar^2 k_x^2}{2m_c} = [J_1(E,c,\eta_g,F)+iJ_2(E,c,\eta_g,F)], \quad \omega_0 = \frac{eB}{m_c} \quad (11.43)$$

From (11.43) we get,

$$k_x^2 = w_{11}(E,F,n,\eta_g) \quad (11.44)$$

where,

$$w_{11}(E,F,n,\eta_g) = \frac{2m_c}{\hbar^2}\left[[J_1(E,c,\eta_g,F) + iJ_2(E,c,\eta_g,F)] - \left(n+\frac{1}{2}\right)\hbar\omega_0\right]$$

The density-of-states function for both the cases can, respectively, be expressed as

$$N(E,c,\eta_g,F) = \frac{g_v eB\sqrt{2m_c}}{2\pi^2\hbar^2} \times \sum_{n=0}^{n_{\max}} \frac{[J_1'(E,c,\eta_g,F) + iJ_2'(E,c,\eta_g,F)]H(E-E_{n1HD})}{\sqrt{[J_1(E,c,\eta_g,F) + iJ_2(E,c,\eta_g,F)] - (\mathrm{n}+\frac{1}{2})\hbar\omega_o}} \quad (11.45)$$

where, E_{n1HD} is the Landau level in this case and can be expressed as

$$\left(n+\frac{1}{2}\right)\hbar w_0 = [J_1(E_{n1HD}, c, \eta_g, F) + iJ_2(E_{n1HD}, c, \eta_g, F)] \quad (11.46)$$

The EEM in this case is given by

$$m^*(E, \eta_g, F) = m_c J_1'(E, c, \eta_g, F) \quad (11.47)$$

11.2.3 *The EM in QWs in HD Kane type Semiconductors in the Presence of intense electric field*

The DR in this case is given by

$$\frac{\hbar^2}{2m_c}\left(\frac{n_z\pi}{d_z}\right)^2 + \frac{\hbar^2 k_s^2}{2m_c} = [J_1(E, c, \eta_g, F) + iJ_2(E, c, \eta_g, F)] \quad (11.48)$$

The DOS function in this case is given by

$$N_{2D}(E, c, \eta_g, F) = \frac{m_c g_v}{\pi\hbar^2} \times \sum_{n_z=1}^{n_{z\max}} [J_1'(E, c, \eta_g, F) + iJ_2'(E, c, \eta_g, F)] H(E - E_{n_{z18,1HD}}) \quad (11.49)$$

where, $E_{n_{z18,1HD}}$ is the sub-band energy in this case and can be expressed as

$$\frac{\hbar^2}{2m_c}\left(\frac{n_z\pi}{d_z}\right)^2 = [J_1(E_{n_{z18,1HD}}, c, \eta_g, F) + iJ_2(E_{n_{z18,1HD}}, c, \eta_g, F)] \quad (11.50)$$

The EEM in this case is given by

$$m^*(E, \eta_g, F) = m_c J_1'(E, c, \eta_g, F) \quad (11.51)$$

11.2.4 *The EM in NWs in HD Kane type semiconductors in the presence of intense electric field*

The DR in this case is given by

$$\frac{\hbar^2}{2m_c}\left(\frac{n_z\pi}{d_z}\right)^2+\frac{\hbar^2}{2m_c}\left(\frac{n_y\pi}{d_y}\right)^2+\frac{\hbar^2k_x^2}{2m_c} = [J_1(E,c,\eta_g,F)+iJ_2(E,c,\eta_g,F)] \tag{11.52}$$

The DOS function in this case is given by

$$N_{1D}(E,c,\eta_g,F)=\frac{g_v\sqrt{2m_c}}{\pi\hbar}\sum_{n_y=1}^{n_{y\max}}\sum_{n_z=1}^{n_{z\max}} \times\frac{[J_1'(E,c,\eta_g,F)+iJ_2'(E,c,\eta_g,F)]H(E-E_{n_{z18,2HD}})}{\sqrt{[J_1(E,c,\eta_g,F)+iJ_2(E,c,\eta_g,F)]-G_2(n_y,n_z)}} \tag{11.53}$$

where, $E_{n_{z18,2HD}}$ is the sub-band energy in this case and can be expressed as

$$\frac{\hbar^2}{2m_c}\left(\frac{n_z\pi}{d_z}\right)^2+\frac{\hbar^2}{2m_c}\left(\frac{n_y\pi}{d_y}\right)^2 = [J_1(E_{n_{z18,2HD}},c,\eta_g,F)+iJ_2(E_{n_{z18,2HD}},c,\eta_g,F)] \tag{11.54}$$

and $G_2(n_y,n_z)$, is defined in (3.13) of Chapter 3.

The EEM in this case is given by

$$m^*(E,\eta_g,F)=m_cJ_1'(E,c,\eta_g,F) \tag{11.55}$$

11.2.5 *The magneto EM in QWs of HD Kane type semiconductors in the presence of intense electric field*

The DR in this case is given by

$$\frac{\hbar^2}{2m_c}\left(\frac{n_x\pi}{d_x}\right)^2+\left(n+\frac{1}{2}\right)\hbar\omega_0 = [J_1(E_{n_{z18,4HD}},c,\eta_g,F)+iJ_2(E_{n_{z18,4HD}},c,\eta_g,F)] \tag{11.56}$$

where, $E_{n_{z18,4HD}}$ is the totally quantized energy in this case

The DOS function in this case is given by

$$N_{QWBHD}(E,\eta_g,F) = \frac{g_v eB}{\pi\hbar}\sum_{n_x=1}^{n_{x\max}}\sum_{n=0}^{n_{\max}}\delta'(E - E_{n_{z18,4HD}}) \quad (11.57)$$

11.2.6 *The EM in accumulation and inversion layers of kane type semiconductors in the presence of intense electric field*

(a) The 2D DR in accumulation layers of HD III-V, ternary and quaternary materials, in this case can be expressed as

$$\begin{aligned}&[J_1(E,c,\eta_g,F) + iJ_2(E,c,\eta_g,F)]\\ &= \frac{\hbar^2 k_s^2}{2m_c} + \frac{\hbar|e|F_s}{\sqrt{2m_c}}\left(\frac{2\sqrt{2}S_i^{3/2}}{3}\right)[J_1'(E,c,\eta_g,F) + iJ_2'(E,c,\eta_g,F)]'\end{aligned} \quad (11.58)$$

Since the DR in accordance with the HD three-band model of Kane is complex in nature, the (11.60) will also be complex. The both complexities occur due to the presence of poles in the finite complex plane of the dispersion relation of the materials in the absence of band tails.

The EEM can be expressed as

$$m^*(E,c,\eta_g,F,i) = m_c \text{ Real part of } [P_{3HDL1}(E,c,\eta_g,F,i)]' \quad (11.59)$$

$$P_{3HDL1}(E,c,\eta_g,F,i) = \Bigg[[J_1(E,c,\eta_g,F) + iJ_2(E,c,\eta_g,F)]$$

where,

$$-\frac{\hbar|e|F_s}{\sqrt{2m_c}}\left(\frac{2\sqrt{2}S_i^{3/2}}{3}\right)[J_1'(E,c,\eta_g,F) + iJ_2'(E,c,\eta_g,F)]\Bigg]'$$

Thus, one can observe that the EEM is a function of electric field, scattering potential, the sub-band index, surface electric field, the Fermi energy and the other spectrum constants due to the combined influence of E_g and Δ.

The sub-band energy E_{400HD} is given by

$$[J_1(E_{400HD}, c, \eta_g, F) + iJ_2(E_{400HD}, c, \eta_g, F)] = \frac{\hbar|e|F_s}{\sqrt{2m_c}}\left(\frac{2\sqrt{2}S_i^{3/2}}{3}\right)[J_1'(E_{400HD}, c, \eta_g, F) + iJ_2'(E_{400HD}, c, \eta_g, F)]' \quad (11.60)$$

The DOS function can be written as

$$N_{2D}(E, c, \eta_g, F) = \frac{m_c g_v}{\pi\hbar^2}\sum_{i=0}^{i_{\max}}[P_{3HDL1}(E, c, \eta_g, F, i)]' H(E - E_{400HD}) \quad (11.61)$$

Thus the DOS function is complex in nature.

In the absence of band-tails, the DR in this case assumes the form

$$[J_{11}(E, F)] = \frac{\hbar^2 k_s^2}{2m_c} + \frac{\hbar|e|F_s}{\sqrt{2m_c}}\left(\frac{2\sqrt{2}S_i^{3/2}}{3}\right)[J_{11}'(E, F)]' \quad (11.62)$$

where

$$J_{11}(E, F) = [e_1E^4 + e_2E^3 + e_3E^2 + e_4E + e_5 - \frac{e_6}{1+CE} + e_7(1+CE)^{-2}]$$

(11.63) represents the DR of the 2D electrons in inversion layers of III-V, ternary and quaternary materials under the intense electric field limit whose bulk electrons in the absence of any perturbation obey the three band model of Kane.

The EEM can be expressed as

$$m^*(E, F, i) = m_c[P_{31}(E, F, i)]'$$

where,

$$P_{31}(E, F, i) = \left[[J_{11}(E, F)] - \frac{\hbar|e|F_s}{\sqrt{2m_c}}\left(\frac{2\sqrt{2}S_i^{3/2}}{3}\right)[J_{11}'(E, F)]'\right] \quad (11.63)$$

Thus, one can observe that the EEM is a function of the sub-band index, the light intensity, surface electric field, the Fermi energy and the other spectrum constants due to the combined influence of E_g and Δ.

The sub-band energy E_{401} in this case can be obtained from the (19.64) as

$$0 = \left[[J_{11}(E_{401}, F)] - \frac{\hbar|e|F_s}{\sqrt{2m_c}} \left(\frac{2\sqrt{2}S_i^{3/2}}{3} \right) [J_{11}'(E_{401}, F)]' \right] \tag{11.64}$$

Thus the 2D total DOS function in weak electric field limit can be expressed as

$$N_{2D}(E, F) = \frac{m_c g_v}{\pi\hbar^2} \sum_{i=0}^{i_{\max}} [P_{31}(E, F, i)]' H(E - E_{401}) \tag{11.65}$$

11.2.7 *The EM in doping superlattices of HD kane type semiconductors in the presence of intense electric field*

The DR in doping superlattices of HD III-V, ternary and quaternary materials in the presence of intense electric field whose unperturbed electrons are defined by the three band model of Kane can be expressed as

$$[J_1(E, c, \eta_g, F) + iJ_2(E, c, \eta_g, F)] = \left(\mathrm{n}_i + \frac{1}{2} \right) \hbar\omega_{91HD1}(E, c, \eta_g, F) + \frac{\hbar^2 k_s^2}{2m_c} \tag{11.66}$$

where,

$$\omega_{91HD1}(E, c, \eta_g, F) = \left(\frac{n_s|e|^2}{d_0\varepsilon_{sc}[J_1(E, c, \eta_g, F) + iJ_2(E, c, \eta_g, F)]'\mathrm{m}_c} \right)^{\frac{1}{2}}$$

The sub band energies E_{452} can be written as

$$[J_1(E_{452}, c, \eta_g, F) + iJ_2(E_{452}, c, \eta_g, F)] = \left(n_i + \frac{1}{2} \right) \hbar\omega_{91HD1}(E_{452}, c, \eta_g, F) \tag{11.67}$$

The EEM in this case is given by

$$m^*(E, c, \eta_g, F, n_i) = m_c \text{ Real part of } [P_{3HDL3}(E, c, \eta_g, F, n_i)]' \tag{11.68}$$

where,

$$[P_{3HDL3}(E,c,\eta_g,F,n_i)] = \left[[J_1(E,c,\eta_g,F) + iJ_2(E,c,\eta_g,F)] - \left(n_i + \frac{1}{2}\right)\hbar\omega_{91HD1}(E,c,\eta_g,F)\right]$$

The DOS function in this case is given by

$$N_{2DDSL}(E,c,\eta_g,F) = \frac{m_c g_v}{\pi\hbar^2}\sum_{n_i=0}^{n_{i\max}} \frac{\left[J_1'(E,c,\eta_g,F) + iJ_2'(E,c,\eta_g,F) - \left(\mathrm{n}_i + \frac{1}{2}\right)\hbar\omega_{91HD1}'(E,c,\eta_g,F)\right] H(E - E_{452})}{\sqrt{[J_1(E,c,\eta_g,F) + iJ_2(E,c,\eta_g,F)] - \left(\mathrm{n}_i + \frac{1}{2}\right)\hbar\omega_{91HD1}(E,c,\eta_g,F)}} \quad (11.69)$$

11.2.8 *The EM in QWHD effective mass superlattices of kane type semiconductors in the presence of intense electric field*

Following Sasaki [3], the electron dispersion law in HD III-V effective mass superlattices (EMSLs) in the presence of light waves, the dispersion relations of whose constituent materials in the absence of any perturbation are defined by the three band model of Kane can be written as

$$k_x^2 = \left[\frac{1}{L_0^2}\{\cos^{-1}(f_{18HD1}(E,k_y,k_z,F))\}^2 - k_\perp^2\right] \quad (11.70)$$

In which,

$$f_{18HD1}(E,k_y,k_z,F) = a_{1HD1,18}\cos[a_0 C_{1HD1,18}(E,k_\perp,F) + b_0 D_{1HD1,18}(E,k_\perp,F)] - a_{2HD1,18}\cos[a_0 C_{1HD1,18}(E,k_\perp,F) + b_0 D_{1HD1,18}(E,k_\perp,F)]$$

$$a_{1HD1,18} = \left[\sqrt{\frac{m_{c2}\text{ Real part of }[J_3(0,c,\eta_{g2},F)]}{m_{c1}\text{ Real part of }[J_3(0,c,\eta_{g1},F)]}}+1\right]^2$$

$$\times\left[4\left(\frac{m_{c2}\text{ Real part of }[J_3(0,c,\eta_{g2},F)]}{m_{c1}\text{ Real part of }[J_3(0,c,\eta_{g1},F)]}\right)^{\frac{1}{2}}\right]^{-1},$$

$$a_{2HD1,18} = \left[\sqrt{\frac{m_{c2}\text{ Real part of }[J_3(0,c,\eta_{g2},F)]}{m_{c1}\text{ Real part of }[J_3(0,c,\eta_{g1},F)]}}-1\right]^2$$

$$\times\left[4\left(\frac{m_{c2}\text{ Real part of }[J_3(0,c,\eta_{g2},F)]}{m_{c1}\text{ Real part of }[J_3(0,c,\eta_{g1},F)]}\right)^{\frac{1}{2}}\right]^{-1}$$

$J_1(E,c,\eta_g,F)+iJ_2(E,c,\eta_g,F)=J_3(E,c,\eta_g,F)$,

$$C_{1HD1,18}(E,\mathrm{k}_\perp,F)\equiv\left[\left(\frac{2m_{c1}}{\hbar^2}\right)J_3(E,c,\eta_{g1},F)-\mathrm{k}_\perp^2\right]^{\frac{1}{2}}$$

and

$$D_{1HD1,18}(E,\mathrm{k}_\perp,F)\equiv\left[\left(\frac{2m_{c1}}{\hbar^2}\right)J_3(E,c,\eta_{g2},F)-\mathrm{k}_\perp^2\right]^{\frac{1}{2}}.$$

The DR in QWHD effective mass superlattices of Kane type semiconductors in the presence of intense electric field, the dispersion relations of whose constituent materials in the absence of any perturbation are defined by the three band model of Kane can be written as

$$\left(\frac{n_x\pi}{d_x}\right)^2=\left[\frac{1}{L_0^2}\{\cos^{-1}(f_{18HD1}(\mathrm{E},\mathrm{k_y},\mathrm{k_z},\mathrm{F}))\}^2-\mathrm{k}_\perp^2\right] \quad (11.71)$$

The EEM in this case assumes the form

$$m^*(\mathrm{k}_\perp,\mathrm{E},F)$$
$$=\frac{\hbar^2}{L_0^2}\left|\left[\frac{\cos^{-1}[\mathrm{f}_{18HD1}(\mathrm{E},\mathrm{k_y},\mathrm{k_z},\mathrm{F})]\mathrm{f}'_{18HD1}(\mathrm{E},\mathrm{k_y},\mathrm{k_z},\mathrm{F})}{\sqrt{1-\mathrm{f}^2_{18HD1}(\mathrm{E},\mathrm{k_y},\mathrm{k_z},\mathrm{F})}}\right]\right| \quad (11.72)$$

The sub-band energies E_{600} can be written as

$$\left(\frac{n_x \pi}{d_x}\right) = \left[\frac{1}{L}\{\cos^{-1}(f_{18HD1}(E_{600}, k_y, k_z, F))\}\right] \tag{11.73}$$

11.2.9 *The EM in NWHD effective mass superlattices of kane type semiconductors in the presence of intense electric field*

Following Sasaki [3], the magneto EM in HD III-V effective mass superlattices (EMSLs) in the presence of intense electric field, the dispersion relations of whose constituent materials in the absence of any perturbation are defined by the three band model of Kane can be written as

$$k_x^2 = \left[\frac{1}{L_0^2}\{\cos^{-1}(\mathrm{f}_{18HD2}(\mathrm{E}, n, \mathrm{F}))\}^2 - \frac{2eB}{\hbar}\left(\mathrm{n} + \frac{1}{2}\right)\right] \tag{11.74}$$

In which,

$$\begin{aligned} f_{18HD2}(E, n, F) = a_{1HD1,18} \cos[&a_0 C_{1HD2,18}(E, n, F) \\ &+ b_0 D_{1HD2,18}(E, n, F)] \\ &- a_{2HD1,18} \cos[a_0 C_{1HD2,18}(E, n, F) \\ &+ b_0 D_{1HD2,18}(E, n, F)] \end{aligned}$$

$$\begin{aligned} a_{1HD1,18} = &\left[\sqrt{\frac{m_{c2} \text{ Real part of } [J_3(0, c, \eta_{g2}, F)]}{m_{c1} \text{ Real part of } [J_3(0, c, \eta_{g1}, F)]}} + 1\right]^2 \\ &\times \left[4\left(\frac{m_{c2} \text{ Real part of } [J_3(0, c, \eta_{g2}, F)]}{m_{c1} \text{ Real part of } [J_3(0, c, \eta_{g1}, F)]}\right)^{\frac{1}{2}}\right]^{-1}, \end{aligned}$$

$$\begin{aligned} a_{2HD1,18} = &\left[\sqrt{\frac{m_{c2} \text{ Real part of } [J_3(0, c, \eta_{g2}, F)]}{m_{c1} \text{ Real part of } [J_3(0, c, \eta_{g1}, F)]}} - 1\right]^2 \\ &\times \left[4\left(\frac{m_{c2} \text{ Real part of } [J_3(0, c, \eta_{g2}, F)]}{m_{c1} \text{ Real part of } [J_3(0, c, \eta_{g1}, F)]}\right)^{\frac{1}{2}}\right]^{-1} \end{aligned}$$

$$J_1(E,c,\eta_g,F)+iJ_2(E,c,\eta_g,F)=J_3(E,c,\eta_g,F),$$

$$C_{1HD2,18}(E,n,F)\equiv\left[\left(\frac{2m_{c1}}{\hbar^2}\right)J_3(E,c,\eta_{g1},F)-\frac{2eB}{\hbar}\left(\mathrm{n}+\frac{1}{2}\right)\right]^{\frac{1}{2}}$$

and

$$D_{1HD2,18}(E,n,F)\equiv\left[\left(\frac{2m_{c1}}{\hbar^2}\right)J_3(E,c,\eta_{g2},F)-\frac{2eB}{\hbar}\left(\mathrm{n}+\frac{1}{2}\right)\right]^{\frac{1}{2}}$$

The DOS function in this case is given by

$$N_{100}(\mathrm{E},\mathrm{F})=\frac{g_veB}{\pi^2\hbar}\sum_{n=0}^{n_{\max}}[\omega_{100}(\mathrm{E},n,\mathrm{F})]' \tag{11.75}$$

where

$$\omega_{100}(\mathrm{E},n,\mathrm{F})=\left[\left[\frac{1}{L_0^2}\{\cos^{-1}(f_{18HD2}(\mathrm{E},n,\mathrm{F}))\}^2-\frac{2eB}{\hbar}\left(\mathrm{n}+\frac{1}{2}\right)\right]\right]^{\frac{1}{2}}$$

The EEM in this case assumes the form

$$m^*(n,\mathrm{E},\mathrm{F})=\frac{\hbar^2}{L_0^2}\left|\left[\frac{\cos^{-1}[f_{18HD2}(\mathrm{E},n,\mathrm{F})]f'_{18HD2}(\mathrm{E},n,\mathrm{F})}{\sqrt{1-f_{18HD2}^2(\mathrm{E},n,\mathrm{F})}}\right]\right| \tag{11.76}$$

The Landau sub-band energies E_{602} can be written as

$$0=\left[\frac{1}{L_0^2}\{\cos^{-1}(f_{18HD2}(E_{602},n,F))\}^2-\frac{2eB}{\hbar}\left(\mathrm{n}+\frac{1}{2}\right)\right] \tag{11.77}$$

11.2.10 *The magneto EM in QWHD effective mass superlattices of kane type semiconductors in the presence of intense electric field*

Following Sasaki [3], the electron dispersion law in HD III-V effective mass superlattices (EMSLs) in the presence of light waves, the dispersion relations of whose constituent materials in the absence of any perturbation are defined by the three band model of Kane can be written as

$$\left(\frac{n_x\pi}{d_x}\right)^2 = \left[\frac{1}{L_0^2}\{\cos^{-1}(\mathrm{f}_{18HD2}(\mathrm{E}_{601}, n, F))\}^2 - \frac{2eB}{\hbar}\left(\mathrm{n} + \frac{1}{2}\right)\right] \tag{11.78}$$

where E_{601} is the totally quantized energy in this case.

The DOS function is given by

$$N_{601}(\mathrm{E}, \mathrm{F}, \eta_g) = \frac{g_v eB}{\pi\hbar} \sum_{n_x=1}^{n_{x\max}} \sum_{n=0}^{n_{\max}} \delta'(\mathrm{E} - \mathrm{E}_{601}) \tag{11.79}$$

11.2.11 *The EM in QWHD superlattices of kane type semiconductors with graded interfaces in the presence of intense electric field*

The DR in bulk specimens of the heavily doped constituent materials of III-V SLs in the presence of intense electric field can be expressed as

$$\frac{\hbar^2 k^2}{2m_{cj}} = J_{1j}(\mathrm{E}, \Delta_j, \mathrm{E}_{g0j}, \eta_j, \mathrm{F}) + iJ_{2j}(\mathrm{E}, \Delta_j, \mathrm{E}_{g0j}, \eta_j, \mathrm{F}) \tag{11.80}$$

where $j = 1, 2$,

$$\begin{aligned} J_{1j}(\mathrm{E}, \Delta_j, \mathrm{E}_{g0j}, \eta_j, \mathrm{F}) = 2\left[1 + Erf\left(\frac{E}{\eta_{g_j}}\right)\right]^{-1} & [e_{1j}\psi_{0j}(E, \eta_{g_j}) \\ & + e_{2j}\psi_{1j}(E, \eta_{g_j}) + e_{3j}\theta_{0j}(E, \eta_{g_j}) \\ & + e_{4j}\gamma_0(E, \eta_{g_j}) + e_{5j}\frac{1}{2}\left[1 + Erf\left(\frac{E}{\eta_{g_j}}\right)\right] \\ & - e_{6j}c_{1j}(E, c, \eta_{g_j}) + e_{7j}c_{3j}(E, c, \eta_{g_j})], \end{aligned}$$

$$\begin{aligned} \psi_{1j}(E, \eta_{g_j}) = & \left[\frac{E}{2}\left[1 + Erf\left(\frac{E}{\eta_{g_j}}\right)\right]\left[E^2 + \frac{3}{2}\eta_{g_j}^2\right]\right. \\ & \left. + \frac{\eta_{g_j}}{2\sqrt{\pi}}\exp\left(\frac{-E^2}{\eta_{g_j}^2}\right)\left(4E^2 + \eta_{g_j}^2\right)\right], \end{aligned}$$

$$J_{2j}(\mathrm{E}, \Delta_j, \mathrm{E}_{g0j}, \eta_j, \mathrm{F}) = 2\left[1 + Erf\left(E/\eta_{g_j}\right)\right]^{-1} \times [e_{6j}c_{2j}(E, c, \eta_{g_j}) + e_{7j}D_{3j}(E, c, \eta_{g_j})],$$

$$e_{1j} = Q_{fj} \cdot \omega_{1j},$$

$$Q_{fj} = \frac{m_{cj}}{m_{rj}} E_{g_j}^{-4}[5e_{fj}E_{g_j}^{-2} - 6G_{fj} + 7h_{fj}E_{g_j}^{-4}],$$

$$e_{fj} = A_{fj}P_{fj}, \quad A_{fj} = [F\hbar E_{g_j}(E_{g_j} - \delta_j')]^2 m_{cj}(6m_{rj}^2(\delta_j')^4)^{-1},$$

$$F = eF_s, \quad G_{fj} = e_{fj}(4\delta_j' + C_{fj}),$$

$$C_{fj} = (2E_{g_j}Q_j^2 + P_jQ_j(E_{g_j} - E_{g_j}') - 2P_j^2E_{g_j}),$$

$$E_{g_j}' = \frac{E_{g_j}(E_{g_j} - 3\delta_j')}{E_{g_j} + \delta_j'}, \quad P_j = \frac{r_{0j}^2}{2}\left(\frac{E_{g_j} - \delta_j'}{E_{g_j} + \delta_j'}\right),$$

$$r_{0j} = \left[\frac{6}{\chi_j}(E_{g_j} + \Delta_j)\left(E_{g_j} + \frac{2}{3}\Delta_j\right)\right]^{1/2}, \quad Q_j = \frac{t_j^2}{2},$$

$$t_j = \left[\frac{6}{\chi_j}\left(E_{g_j} + \frac{2}{3}\Delta_j\right)\right]^{1/2}, \quad h_{fj} = (4\delta_j'e_{fj}C_{fj})(B_{fj})^{-1},$$

$$B_{fj} = (P_j + Q_j)^2$$

$$P_{fj} = E_{g_j}^{-3}(e_{fj}E_{g_j}^{-2} - G_{fj} + h_{fj}E_{g_j}^{-4}), \quad \omega_{1j} = a_{1j}^2, \quad a_{1j} = \frac{a_jb_j}{c_j},$$

$$a_j = \frac{1}{E_{g_j}}, \quad b_j = \frac{1}{E_{g_j} + \Delta_j}, \quad c_j = \left(E_{g_j} + \frac{2}{3}\Delta_j\right)^{-1},$$

$$e_{2j} = Q_{fj} \cdot \omega_{2j}, \quad \omega_{2j} = 2a_{1j}b_{1j}, \quad b_{1j} = (c_j)^{-2}(a_jc_j + b_jc_j - a_jb_j),$$

$$e_{3j} = (1 - P_{fj})a_{1j} + Q_{fj} \cdot \omega_{3j}, \quad \omega_{3j} = (b_{1j}^2 + 2a_{1j}c_{1j}),$$

$$c_{1j} = \left[\frac{1}{c_j}\left(1 - \frac{a_j}{c_j}\right)\left(1 - \frac{b_j}{c_j}\right)\right], \quad e_{4j} = [(1 - P_{fj})b_{1j} + Q_{fj}\omega_{4j}],$$

$$\omega_{4j} = 2b_{1j}c_{1j}, \quad e_{5j} = [(1 - P_{fj})c_{1j} + \omega_{5j}Q_{fj}]$$

$$\omega_{5j} = (c_{1j}^2 - 2c_{1j}b_{1j}), \quad e_{7j} = Q_{fj}\omega_{7j}, \quad \omega_{7j} = c_{1j}^2,$$

$$e_{6j} = [(1 - P_{fj})c_{1j} - Q_{fj}\omega_{6j}]$$

and

$$\omega_{6j} = \frac{2c_{1j}b_{1j}}{c_j}\left(1 - \frac{c_j c_{1j}}{b_{1j}}\right)$$

Therefore, the DR in HD III-V SLs with graded interfaces in the presence of intense electric field can be expressed as [4]

$$k_z^2 = G_{8,19} + iH_{8,19} \tag{11.81}$$

where

$$G_{8,19} = \left[\frac{C_{7,19}^2 - D_{7,19}^2}{L_0^2} - k_s^2\right], \quad C_{7,19} = \cos^{-1}(\bar{\omega}_{7,19}),$$

$$\bar{\omega}_{7,19} = (2)^{\frac{-1}{2}}[(1 - G_{7,19}^2 - H_{7,19}^2) - \sqrt{(1 - G_{7,19}^2 - H_{7,19}^2)^2 + 4G_{7,19}^2}]^{\frac{1}{2}}$$

$$\begin{aligned} G_{7,19} = {} & [G_{1,19} + (\rho_{5,19}G_{2,19}/2) - (\rho_{6,19}H_{2,19}/2) + (\Delta_0/2)\{\rho_{6,19}H_{2,19} \\ & - \rho_{8,19}H_{3,19} + \rho_{9,19}H_{4,19} - \rho_{10,19}H_{4,19} \\ & + \rho_{11,19}H_{5,19} - \rho_{12,19}H_{5,19} + (1/12)(\rho_{12,19}G_{6,19} \\ & - \rho_{14,19}H_{6,19})\}], \\ G_{1,19} = {} & [(\cos(h_{1,19}))(\cosh(h_{2,19}))(\cosh(g_{1,19}))(\cos(g_{2,19})) \\ & + (\sin(h_{1,19}))(\sinh(h_{2,19}))(\sinh(g_{1,19}))(\sin(g_{2,19}))],\ h_{1,19} \\ & = e_{1,19}(b_0 - \Delta_0), e_{1,19} = 2^{\frac{-1}{2}}\left(\sqrt{t_{1,19}^2 + t_{2,19}^2} + t_{1,19}\right)^{\frac{1}{2}}, \end{aligned}$$

$$t_{1,19} = [(2m_{c1}/\hbar^2)\cdot J_{11}(\mathrm{E}, \Delta_1, \mathrm{E}_{g01}, \mathrm{F}) - k_s^2],$$

$$t_{2,19} = [(2m_{c1}/\hbar^2)J_{21}(\mathrm{E}, \Delta_1, \mathrm{E}_{g01}, \mathrm{F})],$$

$$h_{2,19} = e_{2,19}(b_0 - \Delta_0), \quad e_{2,19} = 2^{\frac{-1}{2}}\left(\sqrt{t_{1,19}^2 + t_{2,19}^2} - t_{1,19}\right)^{\frac{1}{2}},$$

$$g_{1,19} = d_{1,19}(a_0 - \Delta_0), \quad d_{1,19} = 2^{\frac{-1}{2}}\left(\sqrt{x_{1,19}^2 + y_{1,19}^2} + x_{1,19}\right)^{\frac{1}{2}},$$

$$x_{1,19} = [-(2m_{c2}/\hbar^2)\cdot J_{11}(E - V_0, E_{g2}, \Delta_2, \eta_{g2}, \mathrm{F}) + k_s^2],$$

$$y_1 = [(2m_{c2}/\hbar^2) \cdot J_{22}(E - V_0, E_{g2}, \Delta_2, \eta_{g2}, \mathrm{F})],$$

$$g_{2,19} = d_{2,19}(a_0 - \Delta_0), \quad d_{2,19} = 2^{\frac{-1}{2}}(\sqrt{x_{1,19}^2 + y_{1,19}^2} - x_{1,19})^{\frac{1}{2}},$$

$$\rho_{5,19} = (\rho_{3,19}^2 + \rho_{4,19}^2)^{-1}[\rho_{1,19}\rho_{3,19} - \rho_{2,19}\rho_{4,19}],$$

$$\rho_{1,19} = [d_{1,19}^2 + e_{2,19}^2 - d_{2,19}^2 - e_{1,19}^2], \quad \rho_{3,19} = [d_{1,19}e_{1,19} + d_{2,19}e_{2,19}],$$

$$\rho_{2,19} = 2[d_{1,19}d_{2,19} + e_{1,19}e_{2,19}], \quad \rho_{4,19} = [d_{1,19}e_{2,19} - e_{1,19}d_{2,19}],$$

$$G_{2,19} = [(\sin(h_{1,19}))(\cosh(h_{2,19}))(\sinh(g_{1,19}))(\cos(g_{2,19})) + (\cos(h_{1,19}))(\sinh(h_{2,19}))(\cosh(g_{1,19}))(\sin(g_{2,19}))],$$

$$\rho_{6,19} = (\rho_{3,19}^2 + \rho_{4,19}^2)^{-1}[\rho_{1,19}\rho_{4,19} + \rho_{2,19}\rho_{3,19}],$$

$$H_{2,19} = [(\sin(h_{1,19}))(\cos(h_{2,19}))(\sin(g_{2,19}))(\cosh(g_{1,19})) - (\cos(h_{1,19}))(\sinh(h_{2,19}))(\sinh(g_{1,19}))(\cos(g_{2,19}))],$$

$$\rho_{7,19} = [(e_{1,19}^2 + e_{2,19}^2)^{-1}[e_{1,19}(d_{1,19}^2 - d_{2,19}^2) - 2d_{1,19}d_{2,19}e_{2,19}] - 3e_{1,19}],$$

$$G_{3,19} = [(\sin(h_{1,19}))(\cosh(h_{2,19}))(\cosh(g_{1,19}))(\cos(g_{2,19})) + (\cos(h_{1,19}))(\sinh(h_{2,19}))(\sinh(g_{1,19}))(\sin(g_{2,19}))],$$

$$\rho_{8,19} = [(e_{1,19}^2 + e_{2,19}^2)^{-1}[e_{2,19}(d_{1,19}^2 - d_{2,19}^2) + 2d_{1,19}d_{2,19}e_{1,19}] + 3e_{2,19}],$$

$$H_{3,19} = [(\sin(h_{1,19}))(\cosh(h_{2,19}))(\sin(g_{2,19}))(\sinh(g_{1,19})) - (\cos(h_{1,19}))(\sinh(h_{2,19}))(\cosh(g_{1,19}))(\cos(g_{2,19}))],$$

$$\rho_{9,19} = [(d_{1,19}^2 + d_{2,19}^2)^{-1}[d_{1,19}(e_{2,19}^2 - e_{1,19}^2) + 2e_{2,19}d_{2,19}e_{1,19}] + 3d_{1,19}],$$

$$G_{4,19} = [(\cos(h_{1,19}))(\cosh(h_{2,19}))(\cos(g_{2,19}))(\sinh(g_{1,19})) - (\sin(h_{1,19}))(\sinh(h_{2,19}))(\cosh(g_{1,19}))(\sin(g_{2,19}))],$$

$$\rho_{10,19} = [-(d_{1,19}^2 + d_{2,19}^2)^{-1}[d_{2,19}(-e_{2,19}^2 + e_{1,19}^2) + 2e_{2,19}d_{2,19}e_{1,19}] + 3d_{2,19}],$$

$$H_{4,19} = [(\cos(h_{1,19}))(\cosh(h_{2,19}))(\cosh(g_{1,19}))(\sin(g_{2,19})) + (\sin(h_{1,19}))(\sinh(h_{2,19}))(\sinh(g_{1,19}))(\cos(g_{2,19}))],$$

$$\rho_{11,19} = 2[d_{1,19}^2 + e_{2,19}^2 - d_{2,19}^2 - e_{1,19}^2],$$

$$G_{5,19} = [(\cos(h_{1,19}))(\cosh(h_{2,19}))(\cos(g_{2,19}))(\cosh(g_{1,19})) - (\sin(h_{1,19}))(\sinh(h_{2,19}))(\sinh(g_{1,19}))(\sin(g_{2,19}))],$$

$$\rho_{12,19} = 4[d_{1,19}d_{2,19} + e_{1,19}e_{2,19}],$$

$$H_{5,19} = [(\cos(h_{1,19}))(\cosh(h_{2,19}))(\sinh(g_{1,19}))(\sin(g_{2,19})) + (\sin(h_{1,19}))(\sinh(h_{2,19}))(\cosh(g_{1,19}))(\cos(g_{2,19}))],$$

$$\rho_{13,19} = [\{5(d_{1,19}e_{1,19}^3 - 3e_{1,19}e_{2,19}^2 d_{1,19}) + 5d_{2,19}(e_{1,19}^3 - 3e_{1,19}^2 e_{2,19})\}(d_{1,19}^2 + d_{2,19}^2)^{-1} + (e_{1,19}^2 + e_{2,19}^2)^{-1}\{5(e_{1,19}d_{1,19}^3 - 3d_{2,19}e_{1,19}^2 d_{1,19}) + 5(d_{2,19}^3 e_{2,19} - 3\,d_{1,19}^2 d_{2,19}e_{2,19})\} - 34(d_{1,19}e_{1,19} + d_{2,19}e_{2,19})],$$

$$G_{6,19} = [(\sin(h_{1,19}))(\cosh(h_{2,19}))(\sinh(g_{1,19}))(\cos(g_{2,19})) + (\cos(h_{1,19}))(\sinh(h_{2,19}))(\cosh(g_{1,19}))(\sin(g_{2,19}))],$$

$$\rho_{14,19} = [\{5(d_{1,19}e_{2,19}^3 - 3e_{2,19}e_{1,19}^2 d_{1,19}) + 5d_{2,19}(-e_{1,19}^3 + 3e_{2,19}^2 e_{1,19})\}(d_{1,19}^2 + d_{2,19}^2)^{-1} + (e_{1,19}^2 + e_{2,19}^2)^{-1} \times \{5(-e_{1,19}d_{2,19}^3 + 3d_{1,19}^2 d_{2,19}e_{1,19}) + 5(-d_{1,19}^3 e_{2,19} + 3d_{2,19}^2 d_{1,19}e_{2,19})\} + 34(d_{1,19}e_{2,19} - d_{2,19}e_{1,19})],$$

$$H_{6,19} = [(\sin(h_{1,19}))(\cosh(h_{2,19}))(\cosh(g_{1,19}))(\sin(g_{2,19})) - (\cos(h_{1,19}))(\sinh(h_{2,19}))(\sinh(g_{1,19}))(\cos(g_{2,19}))],$$

$$H_{7,19} = [H_{1,19} + (\rho_{5,19}H_{2,19}/2) + (\rho_{6,19}G_{2,19}/2) + (\Delta_0/2) \{\rho_{8,19}G_{3,19} + \rho_{7,19}H_{3,19} + \rho_{10,19}G_{4,19} + \rho_{9,19}H_{4,19} + \rho_{12,19}G_{5,19} + \rho_{11,19}H_{5,19} + (1/12)(\rho_{14,19}G_{6,19} + \rho_{13,19}H_{6,19})\}],$$

$$H_{1,19} = [(\sin(h_{1,19}))(\cosh(h_{2,19}))(\cosh(g_{1,19}))(\sin(g_{2,19})) + (\cos(h_{1,19}))(\sinh(h_{2,19}))(\sinh(g_{1,19}))(\cos(g_{2,19}))],$$

$$D_{7,19} = \sin^{-1}(\bar{\omega}_{7,19}),\ H_{8,19} = (2C_{7,19}D_{7,19}/L_0^2)$$

The simplified DR of HD QWs of III-V super-lattices with graded interfaces can be expressed as

$$\left(\frac{n_z\pi}{d_z}\right)^2 = G_{8,19} + iH_{8,19} \tag{11.82}$$

The sub-band equation in this case can be expressed as

$$\left(\frac{n_z\pi}{d_z}\right)^2 = |G_{8,19} + iH_{8,19}|_{k_s=0} \quad \text{and} \quad {}_{E=E_{700}} \tag{11.83}$$

where E_{700} is the sub-band energy in this case.

The EEM and the DOS function should be obtained numerically in this case.

11.2.12 *The EM in NWHD superlattices of kane type semiconductors with graded interfaces in the presence of intense electric field*

The EM in NWHD III-V SLs with graded interfaces in the presence of intense electric fieldcan be expressed as [4]

$$k_z^2 = G_{8,20} + iH_{8,20} \tag{11.84}$$

where

$$G_{8,20} = \left[\frac{C_{7,20}^2 - D_{7,20}^2}{L_0^2} - k_s^2\right], \quad C_{7,20} = \cos^{-1}(\bar{\omega}_{7,20}),$$

$$\bar{\omega}_{7,20} = (2)^{-1/2}[(1 - G_{7,20}^2 - H_{7,20}^2) - \sqrt{(1 - G_{7,20}^2 - H_{7,20}^2)^2 + 4G_{7,20}^2}]^{1/2}$$

$$\begin{aligned} G_{7,20} = {} & [G_{1,20} + (\rho_{5,20}G_{2,20}/2) - (\rho_{6,20}H_{2,20}/2) + (\Delta_0/2) \\ & \times \{\rho_{6,20}H_{2,20} - \rho_{8,20}H_{3,20} + \rho_{9,20}H_{4,20} - \rho_{10,20}H_{4,20} \\ & + \rho_{11,20}H_{5,20} - \rho_{12,20}H_{5,20} \\ & +(1/12)(\rho_{12,20}G_{6,20} - \rho_{14,20}H_{6,20})\}] \end{aligned}$$

$$\begin{aligned} G_{1,20} = {} & [(\cos(h_{1,20}))(\cosh(h_{2,20}))(\cosh(g_{1,20}))(\cos(g_{2,20})) \\ & + (\sin(h_{1,20}))(\sinh(h_{2,20}))(\sinh(g_{1,20}))(\sin(g_{2,20}))], \end{aligned}$$

$$h_{1,20} = e_{1,20}(b_0 - \Delta_0), e_{1,20} = 2^{-1/2}(\sqrt{t_{1,20}^2 + t_{2,20}^2} + t_{1,20})^{1/2},$$

$$t_{1,20} = [(2m_{c1}/\hbar^2) \cdot J_{11}(E, \Delta_1, E_{g_{01}}, F) - G_2(n_y, n_z)],$$

$$t_{2,20} = [(2m_{c1}/\hbar^2) J_{21}(E, \Delta_1, E_{g_{01}}, F)],$$

$$h_{2,20} = e_{2,20}(b_0 - \Delta_0), \quad e_{2,20} = 2^{-1/2}(\sqrt{t_{1,20}^2 + t_{2,20}^2} + t_{1,20})^{1/2},$$

$$g_{1,20} = d_{1,20}(a_0 - \Delta_0), \quad d_{1,20} = 2^{-1/2}(\sqrt{x_{1,20}^2 + y_{1,20}^2} + x_{1,20})^{1/2},$$

$$x_{1,20} = [-(2m_{c2}/\hbar^2) \cdot J_{11}(E - V_0, E_{g2}, \Delta_2, \eta_{g2}, F) + G_2(n_y, n_z)],$$

$$y_1 = [(2m_{c2}/\hbar^2) \cdot J_{22}(E - V_0, E_{g2}, \Delta_2, \eta_{g2}, F)],$$

$$g_{2,20} = d_{2,20}(a_0 - \Delta_0), \quad d_{2,20} = 2^{-1/2}(\sqrt{x_{1,20}^2 + y_{1,20}^2} - x_{1,20})^{1/2},$$

$$\rho_{5,20} = (\rho_{3,20}^2 + \rho_{4,20}^2)^{-1}[\rho_{1,20}\rho_{3,20} - \rho_{2,20}\rho_{4,20}],$$

$$\rho_{1,20} = [d_{1,20}^2 + e_{2,20}^2 - d_{2,20}^2 - e_{1,20}^2], \quad \rho_{3,20} = [d_{1,20}e_{1,20} + d_{2,20}e_{2,20}],$$

$$\rho_{2,20} = 2[d_{1,20}d_{2,20} + e_{1,20}e_{2,20}], \quad \rho_{4,20} = [d_{1,20}e_{2,20} - e_{1,20}d_{2,20}],$$

$$G_{2,20} = [(\sin(h_{1,20}))(\cosh(h_{2,20}))(\sinh(g_{1,20}))(\cos(g_{2,20})) + (\cos(h_{1,20}))(\sinh(h_{2,20}))(\cosh(g_{1,20}))(\sin(g_{2,20}))],$$

$$\rho_{6,20} = (\rho_{3,20}^2 + \rho_{4,20}^2)^{-1}[\rho_{1,20}\rho_{4,20} + \rho_{2,20}\rho_{3,20}],$$

$$H_{2,20} = [(\sin(h_{1,20}))(\cos(h_{2,20}))(\sin(g_{1,20}))(\cosh(g_{1,20})) - (\cos(h_{1,20}))(\sinh(h_{2,20}))(\sinh(g_{1,20}))(\cos(g_{2,20}))],$$

$$\rho_{7,20} = (e_{1,20}^2 + e_{2,20}^2)^{-1}[e_{1,20}(d_{1,20}^2 - d_{2,20}^2) - 2d_{1,20}d_{2,20}e_{2,20}] - 3e_{1,20}],$$

$$G_{3,20} = [(\sin(h_{1,20}))(\cosh(h_{2,20}))(\cosh(g_{1,20}))(\cos(g_{2,20})) + (\cos(h_{1,20}))(\sinh(h_{2,20}))(\sinh(g_{1,20}))(\sin(g_{2,20}))],$$

$$\rho_{8,20} = (e_{1,20}^2 + e_{2,20}^2)^{-1}[e_{2,20}(d_{1,20}^2 - d_{2,20}^2) + 2d_{1,20}d_{2,20}e_{1,20}] + 3e_{2,20}],$$

$$H_{3,20} = [(\sin(h_{1,20}))(\cos(h_{2,20}))(\sin(g_{2,20}))(\sinh(g_{1,20})) - (\cos(h_{1,20}))(\sinh(h_{2,20}))(\cosh(g_{1,20}))(\cos(g_{2,20}))],$$

$$\rho_{9,20} = (d_{1,20}^2 + d_{2,20}^2)^{-1}[d_{1,20}(e_{2,20}^2 - e_{1,20}^2) + 2e_{2,20}d_{2,20}e_{1,20}] + 3d_{1,20}],$$

$$\begin{aligned} G_{4,20} &= [(\cos(h_{1,20}))(\cosh(h_{2,20}))(\cos(g_{2,20}))(\sinh(g_{1,20})) \\ &\quad - (\sin(h_{1,20}))(\sinh(h_{2,20}))(\cosh(g_{1,20}))(\sin(g_{2,20}))], \\ \rho_{10,20} &= [-(d_{1,20}^2 + d_{2,20}^2)^{-1}[d_{2,20}(-e_{2,20}^2 + e_{1,20}^2) \\ &\quad + 2e_{2,20}d_{2,20}e_{1,20}] + 3d_{2,20}], \\ H_{4,20} &= [(\cos(h_{1,20}))(\cosh(h_{2,20}))(\cosh(g_{1,20}))(\sin(g_{2,20})) \\ &\quad + (\sin(h_{1,20}))(\sinh(h_{2,20}))(\sinh(g_{1,20}))(\cos(g_{2,20}))], \end{aligned}$$

$$\rho_{11,20} = 2[d_{1,20}^2 + e_{2,20}^2 - d_{2,20}^2 - e_{1,20}^2],$$

$$\begin{aligned} G_{5,20} &= [(\cos(h_{1,20}))(\cosh(h_{2,20}))(\cos(g_{2,20}))(\cosh(g_{1,20})) \\ &\quad - (\sin(h_{1,20}))(\sinh(h_{2,20}))(\sinh(g_{1,20}))(\sin(g_{2,20}))], \end{aligned}$$

$$\rho_{12,20} = 4[d_{1,20}d_{2,20} + e_{1,20}e_{2,20}],$$

$$\begin{aligned} H_{5,20} &= [(\cos(h_{1,20}))(\cosh(h_{2,20}))(\sinh(g_{1,20}))(\sin(g_{2,20})) \\ &\quad + (\sin(h_{1,20}))(\sinh(h_{2,20}))(\cosh(g_{1,20}))(\cos(g_{2,20}))], \\ \rho_{13,20} &= [\{5[d_{1,20}e_{1,20}^3 - 3e_{1,20}e_{2,20}^2d_{1,20}) \\ &\quad + 5d_{2,20}(e_{1,20}^3 - 3e_{1,20}^2e_{2,20})\}(d_{1,20}^2 + d_{2,20}^2)^{-1} \\ &\quad + (e_{1,20}^2 + e_{2,20}^2)^{-1}\{5(e_{1,20}d_{1,20}^3 - 3d_{2,20}e_{1,20}^2d_{1,20}) \\ &\quad + 5(d_{2,20}^3e_{2,20} - 3d_{1,20}^2d_{2,20}e_{2,20})\} \\ &\quad - 34(d_{1,20}e_{1,20} + d_{2,20}e_{2,20})], \\ G_{6,20} &= [(\sin(h_{1,20}))(\cosh(h_{2,20}))(\sinh(g_{1,20}))(\cos(g_{2,20})) \\ &\quad + (\cos(h_{1,20}))(\sinh(h_{2,20}))(\cosh(g_{1,20}))(\sin(g_{2,20})], \\ \rho_{14,20} &= [\{5[d_{1,20}e_{1,20}^3 - 3e_{1,20}e_{1,20}^2d_{1,20}) \\ &\quad + 5d_{2,20}(-e_{1,20}^3 + 3e_{2,20}^2e_{1,20})\}(d_{1,20}^2 + d_{2,20}^2)^{-1} \\ &\quad + (e_{1,20}^2 + e_{2,20}^2)^{-1}\{5(-e_{1,20}d_{1,20}^3 + 3d_{1,20}^2d_{2,20}e_{1,20}) \\ &\quad + 5(-d_{1,20}^3e_{2,20} + 3d_{2,20}^2d_{1,20}e_{2,20})\} \\ &\quad + 34(d_{1,20}e_{2,20} - d_{2,20}e_{1,20})], \\ H_{6,20} &= [(\sin(h_{1,20}))(\cosh(h_{2,20}))(\cosh(g_{1,20}))(\sin(g_{2,20})) \\ &\quad - (\cos(h_{1,20}))(\sinh(h_{2,20}))(\sinh(g_{1,20}))(\cos(g_{2,20}))], \end{aligned}$$

$$H_{7,20} = [H_{1,20} + (\rho_{5,20}H_{2,20}/2) + (\rho_{6,20}G_{2,20}/2) + (\Delta_0/2)\{\rho_{8,20}G_{3,20} + \rho_{7,20}H_{3,20} + \rho_{10,20,}G_{4,20} + \rho_{9,20}H_{4,20} + \rho_{12,20}G_{5,20} + \rho_{11,20}H_{5,20} + (1/12)(\rho_{14,20}G_{6,20} + \rho_{13,20}H_{6,20})\}],$$

$$H_{1,20} = [(\sin(h_{1,20}))(\sinh(h_{2,20}))(\cosh(g_{1,20}))(\cos(g_{2,20})) + (\cos(h_{1,20}))(\cosh(h_{2,20}))(\sinh(g_{1,20}))(\sin(g_{2,20}))],$$

$$D_{7,20} = \sinh^{-1}(\bar{\omega}_{7,20}), \quad H_{8,20} = (2C_{7,19}D_{7,20}/L_0^2)$$

The sub-band equation in this case can be expressed as

$$0 = [G_{8,20} + iH_{8,20}]|_{E=E_{610}} \tag{11.85}$$

where E_{610} is the sub-band energy in this case.

At low temperatures where the quantum effects become prominent, the DOS function for the lowest SL mini-band is given by

$$N_{HDSL}(E, \eta_g, F) = \frac{g_v}{\pi} \sum_{n_x=1}^{n_{x\max}} \sum_{n_y=1}^{n_{y\max}} \frac{[G'_{8,20} + iH'_{8,20}]}{\sqrt{G_{8,20} + iH_{8,20}}} H(E - E_{601}) \tag{11.86a}$$

The EEM can be written as

$$m^*(E, n_y, n_z\eta_g, F) = \frac{\hbar^2}{2}(G'_{8,20}) \tag{11.86b}$$

11.2.13 *The EM in quantum dot HD superlattices of kane type semiconductors with graded interfaces in the presence of intense electric field*

The DR in QDHD superlattices of Kane type semiconductors with graded interfaces in the presence of intense electric field can be expressed as

$$\left(\frac{n_z\pi}{d_z}\right)^2 = [G_{8,20} + iH_{8,20}]|_{E=E_{620}} \tag{11.89}$$

where E_{620} is the totally quantized energy in this case.

The DOS function is given by

$$N_{QDHDSL}(E,\eta_g,F) = \frac{2g_v}{d_x d_y d_z}\sum_{n_x=1}^{n_{x\max}}\sum_{n_y=1}^{n_{y\max}}\sum_{n_z=1}^{n_{z\max}}\delta'(E - E_{620}) \tag{11.90}$$

11.2.14 *The magneto EM in HD superlattices of kane type semiconductors with graded interfaces in the presence of intense electric field*

The magneto EM in HD III–V SLs with graded interfaces in the presence of intense electric field can be expressed as

$$k_z^2 = G_{8,19n} + iH_{8,19n} \tag{11.91}$$

where

$$G_{8,19n} = \left[\frac{C_{7,19n}^2 - D_{7,19n}^2}{L_0^2} = \frac{2eB}{\hbar}\left(n+\frac{1}{2}\right)\right],$$

$$C_{7,19n} = \cos^{-1}(\bar{\omega}_{7,19n}),$$

$$\begin{aligned}
\bar{\omega}_{7,19n} &= (2)^{-1/2}[(1 - G_{7,19n}^2 - H_{7,19n}^2)\\
&\quad - \sqrt{(1 - G_{7,19n}^2 - H_{7,19n}^2)^2 + 4G_{7,19n}^2}]^{1/2},\\
G_{7,19n} &= [G_{7,19n} + (\rho_{5,19n}G_{2,19n}/2) - (\rho_{6,19n}H_{2,19n}/2)\\
&\quad + (\Delta_0/2)\{\rho_{7,19n}H_{2,19n} - \rho_{2,19n}H_{3,19n}\\
&\quad + \rho_{9,19n}H_{4,19n} - \rho_{10,19n}H_{4,19n} + \rho_{11,19n}H_{5,19n}\\
&\quad - \rho_{12,19n}H_{5,19n} + (1/12)(\rho_{12,19n}G_{6,19n} - \rho_{14,19n}H_{6,19})\}],\\
G_{1,19n} &= [(\cos(h_{1,19n}))(\cosh(h_{2,19n}))(\cosh(g_{1,19n}))(\cos(g_{2,19n}))\\
&\quad + (\sin(h_{1,19n}))(\sinh(h_{2,19n}))(\sinh(g_{1,9n}))(\sin(g_{2,19n}))]
\end{aligned}$$

$$h_{1,19n} = e_{1,19n}(b_0 - \Delta_0),$$

$$e_{1,19n} = 2^{-1/2}\left(\sqrt{t_{1,19n}^2 + t_{2,19n}^2} + t_{1,19n}\right)^{1/2},$$

$$t_{1,19n} = \left[(2m_{c1} + \hbar^2)\cdot J_{11}(E,\Delta_1,E_{g01},F) - \frac{2eB}{\hbar}\left(n+\frac{1}{2}\right)\right],$$

$$t_{2,19n} = [(2m_{c1}/\hbar^2)J_{21}(E,\Delta_1,E_{g01},F)],$$

$$h_{2,19n} = e_{2,19n}(b_0 - \Delta_0),$$

$$e_{2,19n} = 2^{-1/2}\left(\sqrt{t_{1,19n}^2 + t_{2,19n}^2} - t_{1,19n}\right)^{1/2},$$

$$g_{1,19n} = d_{1,19n}(a_0 - \Delta_0),$$

$$d_{1,19n} = 2^{-1/2}\left(\sqrt{x_{1,19n}^2 + y_{1,19n}^2} + x_{1,19n}\right)^{1/2},$$

$$x_{1,19n} = \left[-(2m_{c2}+\hbar^2)\cdot J_{11}(E-V_0,E_{g2},\Delta_2,\eta_{g2},F)+\frac{2eB}{\hbar}\left(n+\frac{1}{2}\right)\right],$$

$$y_1 = [(2m_{c2}/\hbar^2)J_{22}(E-V_0,E_{g2},\Delta_2,E_{g21},F)],$$

$$g_{2,19n} = d_{2,19n}(a_0 - \Delta_0),$$

$$d_{2,19n} = 2^{-1/2}\left(\sqrt{x_{1,19n}^2 + y_{1,19n}^2} - x_{1,19n}\right)^{1/2},$$

$$\rho_{5,19n} = (\rho_{3,19n}^2 + \rho_{4,19n}^2)^{-1}[\rho_{1,19n}\rho_{3,19n}^2 - \rho_{2,19n}\rho_{4,19n}],$$

$$\rho_{1,19n} = [d_{1,19n}^2 + e_{2,19n}^2 - d_{2,19n}^2 - e_{1,19n}^2],$$

$$\rho_{3,19n} = [d_{1,19n}e_{1,19n} + d_{2,19n}e_{2,19n}],$$

$$\rho_{2,19n} = 2[d_{1,19n}d_{2,19n} + e_{1,19n}e_{2,19n}],$$

$$\rho_{4,19n} = [d_{1,19n}e_{2,19n} - e_{1,19n}d_{2,19n}],$$

$$\begin{aligned} G_{2,19n} = {} & [(\sin(h_{1,19n}))(\cosh(h_{2,19n}))(\sinh(g_{1,19n}))(\cos(g_{2,19n})) \\ & + (\cos(h_{1,19n}))(\sinh(h_{2,19n}))(\cosh(g_{1,19n}))(\sin(g_{2,19n}))] \end{aligned}$$

$$\rho_{6,19n} = (\rho_{3,19n}^2 + \rho_{4,19n}^2)^{-1}[\rho_{1,19n}\rho_{4,19n} + \rho_{2,19n}\rho_{3,19n}],$$

$$\begin{aligned} H_{2,19n} = {} & [(\sin(h_{1,19n}))(\cos(h_{2,19n}))(\sin(g_{2,19n}))(\cosh(g_{1,19n})) \\ & - (\cos(h_{1,19n}))(\sinh(h_{2,19n}))(\sinh(g_{1,19n}))(\cos(g_{2,19n}))] \end{aligned}$$

$$\begin{aligned} \rho_{7,19n} = {} & [(e_{1,19n}^2 + e_{2,19n}^2)^{-1}[e_{1,19n}(d_{1,19n}^2 - d_{2,19n}^2] \\ & - 2d_{1,19n}d_{2,19n}e_{2,19n}] - 3e_{1,19n}], \end{aligned}$$

$$\begin{aligned} G_{3,19n} = {} & [(\sin(h_{1,19n}))(\cosh(h_{2,19n}))(\cosh(g_{1,19n}))(\cos(g_{2,19n})) \\ & + (\cos(h_{1,19n}))(\sinh(h_{2,19n}))(\sinh(g_{1,19n}))(\sin(g_{2,19n}))], \end{aligned}$$

$$
\begin{aligned}
\rho_{8,19n} &= [(e_{1,19n}^2 + e_{2,19n}^2)^{-1}[e_{2,19n}(d_{1,19n}^2 - d_{2,19n}^2] \\
&\quad + 2d_{1,19n}d_{2,19n}e_{1,19n}] + 3e_{2,19n}], \\
H_{3,19n} &= [(\sin(h_{1,19n}))(\cosh(h_{2,19n}))(\sin(g_{2,19n}))(\sinh(g_{1,19n})) \\
&\quad - (\cos(h_{1,19n}))(\sinh(h_{2,19n}))(\cosh(g_{1,19n}))(\cos(g_{2,19n}))], \\
\rho_{9,19n} &= [(d_{1,19n}^2 + d_{2,19n}^2)^{-1}[d_{1,19n}(e_{2,19n}^2 - e_{1,19n}^2] \\
&\quad + 2e_{2,19n}d_{2,19n}e_{1,19n}] + 3d_{1,19n}], \\
G_{4,19n} &= [(\cos(h_{1,19n}))(\cosh(h_{2,19n}))(\cos(g_{2,19n}))(\sinh(g_{1,19n})) \\
&\quad - (\sin(h_{1,19n}))(\sinh(h_{2,19n}))(\cosh(g_{1,19n}))(\sin(g_{2,19n}))], \\
\rho_{10,19n} &= [-(d_{1,19n}^2 + d_{2,19n}^2)^{-1}[d_{2,19n}(-e_{2,19n}^2 + e_{1,19n}^2] \\
&\quad +2e_{2,19n}d_{2,19n}e_{1,19n}] + 3d_{2,19n}], \\
H_{4,19n} &= [(\cos(h_{1,19n}))(\cosh(h_{2,19n}))(\cosh(g_{1,19n}))(\sin(g_{2,19n})) \\
&\quad + (\sin(h_{1,19n}))(\sinh(h_{2,19n}))(\sinh(g_{1,19n}))(\cos(g_{2,19n}))], \\
&\quad \rho_{11,19n} = 2[(d_{1,19n}^2 + e_{2,19n}^2 - d_{2,19n}^2 - e_{1,19n}^2], \\
G_{5,19n} &= [(\cos(h_{1,19n}))(\cosh(h_{2,19n}))(\cos(g_{2,19n}))(\cosh(g_{1,19n})) \\
&\quad - (\sin(h_{1,19n}))(\sinh(h_{2,19n}))(\sinh(g_{1,19n}))(\sin(g_{2,19n}))], \\
&\quad \rho_{12,19n} = 4[(d_{1,19n}d_{2,19n} + e_{1,19n}e_{2,19n}], \\
H_{5,19n} &= [(\cos(h_{1,19n}))(\cosh(h_{2,19n}))(\sinh(g_{1,19n}))(\sin(g_{2,19n})) \\
&\quad + (\sin(h_{1,19n}))(\sinh(h_{2,19n}))(\cosh(g_{1,19n}))(\cos(g_{2,19n}))], \\
\rho_{13,19n} &= [\{5(d_{1,19n}e_{1,19n}^3 - 3e_{1,19n}e_{2,19n}^2d_{1,19n}) \\
&\quad + 5d_{2,19n}(e_{1,19n}^3 - 3e_{1,19n}^2e_{2,19n})\}(d_{1,19n}^2 + d_{2,19n}^2)^{-1} \\
&\quad +(e_{1,19n}^2 + e_{2,19n}^2)^{-1}\{5(e_{1,19n}d_{1,19n}^3 - 3d_{2,19n}e_{1,19n}^2d_{1,19n}) \\
&\quad + 5(d_{2,19n}^3e_{2,19n} - 3d_{1,19n}^2d_{2,19n}e_{2,19n})\} \\
&\quad - 34(d_{1,19n}e_{1,19n} + d_{2,19n}e_{2,19n})] \\
G_{6,19n} &= [(\sin(h_{1,19n}))(\cosh(h_{2,19n}))(\sinh(g_{1,19n}))(\cos(g_{2,19n})) \\
&\quad + (\cos(h_{1,19n}))(\sinh(h_{2,19n}))(\cosh(g_{1,19n}))(\sin(g_{2,19n}))], \\
\rho_{14,19n} &= [\{5(d_{1,19n}e_{2,19n}^3 - 3e_{2,19n}e_{1,19n}^2d_{1,19n}) \\
&\quad + 5d_{2,19n}(-e_{1,19n}^3 + 3e_{2,19n}^2e_{1,19n})\}(d_{1,19n}^2 + d_{2,19n}^2)^{-1}
\end{aligned}
$$

$$+ (e_{1,19n}^2 + e_{2,19n}^2)^{-1}\{5(-e_{1,19n}d_{2,19n}^3 + 3d_{1,19n}^2 d_{1,19n}e_{1,19n})$$
$$+ 5(-d_{1,19n}^3 e_{2,19n} + 3d_{2,19n}^2 d_{1,19n}e_{2,19n})\}$$
$$+ 34(d_{1,19n}e_{2,19n} - d_{2,19n}e_{1,19n})]$$

$$H_{6,19n} = [(\sin(h_{1,19n}))(\cosh(h_{2,19n}))(\cosh(g_{1,19n}))(\sin(g_{2,19n})) - (\cos(h_{1,19n}))(\sinh(h_{2,19n}))(\sinh(g_{1,19n}))(\cos(g_{2,19n}))],$$

$$H_{7,19n} = [H_{7,19n} + (\rho_{5,19n}H_{2,19n}/2) + (\rho_{6,19n}G_{2,19n}/2) + (\Delta_0/2)\{\rho_{8,19n}G_{3,19n} + \rho_{7,19n}H_{3,19n} + \rho_{10,19n}G_{4,19n} + \rho_{9,19n}H_{4,19n} + \rho_{12,19n}G_{5,19n} + \rho_{11,19n}H_{5,19n} + (1/12)(\rho_{14,19n}G_{6,19n} + \rho_{13,19n}H_{6,19n})\}],$$

$$H_{1,19n} = [(\sin(h_{1,19n}))(\sinh(h_{2,19n}))(\cosh(g_{1,19n}))(\cos(g_{2,19n})) + (\cos(h_{1,19n}))(\cosh(h_{2,19n}))(\sinh(g_{1,19n}))(\sin(g_{2,19n}))],$$

$$D_{7,19n} = \sinh^{-1}(\bar{\omega}_{7,19n}), H_{8,19n} = (2C_{7,19n}D_{7,19n}/L_0^2)$$

The sub-band equation in this case can be expressed as

$$0 = [G_{8,19} + iH_{8,19n}]|_{E=E_{630}} \tag{11.92}$$

where E_{630} is the Landau sub-band energy in this case.

At low temperatures where the quantum effects become prominent, the DOS function for the lowest SL mini-band is given by

$$N_{HDSL}(E, \eta_g, F) = \frac{g_v eB}{2\pi^2\hbar} \sum_{n=0}^{n_{\max}} \frac{[G'_{8,19n} + iH'_{8,19n}]}{\sqrt{G_{8,19n} + iH_{8,10n}}} H(E - E_{630}) \tag{11.93}$$

The EEM can be written as

$$m^*(E, n, \eta_g) = \frac{\hbar^2}{2}(G'_{8,19n}) \tag{11.94}$$

11.2.15 *The magneto EM in QWHD superlattices of kane type semiconductors with graded interfaces in the presence of intense electric field*

The magneto DR in QWHD superlattices of Kane type semiconductors with graded interfaces in the presence of intense electric field can be expressed as

$$\left(\frac{n_z\pi}{d_z}\right)^2 = [G_{8,19n} + iH_{8,19n}]|_{E=E_{650}} \tag{11.95}$$

where E_{650} is the totally quantized energy in this case.

The DOS function is given by

$$N_{QWHDSLB}(E, \eta_g, F) = \frac{g_v}{\pi\hbar} \sum_{n_z=1}^{n_{z\max}} \sum_{n=0}^{n_{\max}} \delta'(E - E_{650}) \tag{11.96}$$

11.3 Open Research Problems

(R.11.1) Investigate the EM in the presence of intense external non-uniform electric field for all the HD superlattices whose respective dispersion relations of the carriers are given in this chapter.

(R.11.2) Investigate the EM for the heavily–doped semiconductors in SLs the presences of Gaussian, exponential, Kane, Halperian, Lax and Bonch-Burevich types of band tails [16] for all SL systems as discussed in this chapter in the presence of external oscillatory and non-uniform electric field.

(R.11.3) Investigate the EM in the presence of external non-uniform electric field for short period, strained layer, random and Fibonacci HD superlattices in the presence of an arbitrarily oriented alternating electric field.

(R.11.4) Investigate all the appropriate problems of this chapter for a Dirac electron.

(R.11.5) Investigate all the appropriate problems of this chapter by including the many body, broadening and hot carrier effects respectively.

(R.11.6) Investigate all the appropriate problems of this chapter by removing all the mathematical approximations and establishing the respective appropriate uniqueness conditions.

References

1. L. Schiff, Quantum Mechanics, McGraw-Hill, USA (1968)
2. B. R. Nag, *Electron Transport in compound Semiconductors* (Springer, Heidelberg, 1980)
3. H. Sasaki, Phys. Rev. B **30**, 7016 (1984)
4. H. X. Jiang, J. Y. Lin, J. Appl. Phys. **61**, 624 (1987)

Chapter 12

Appendix D: The EM in QWHDSLs

12.1 Introduction

In recent years, modern fabrication techniques have generated altogether a new dimension in the arena of quantum effect devices through the experimental realization of an important artificial structure known as semiconductor superlattice (SL) by growing two similar but different semiconducting compounds in alternate layers with finite thicknesses [1]. The materials forming the alternate layers have the same kind of band structure but different energy gaps. The concept of SL was developed for the first time by Keldysh [2] and was successfully fabricated by Esaki and Tsu [2]. The SLs are being extensively used in thermal sensors [3], quantum cascade lasers [4], photodetectors [5], light emitting diodes [6], multiplication [7], frequency multiplication [8], photocathodes [9], thin film transistor [10], solar cells [11], infrared imaging [12], thermal imaging [13], infrared sensing [14] and also in other microelectronic devices.

The most extensively studied III–V SL is the one consisting of alternate layers of GaAs and Ga_{1-x} Al_{x} *As owing to the relative easiness of fabrication. The GaAs and* Ga_{1-x} Al_x *As layers form the quantum wells and the potential barriers respectively. The III–V SL's are attractive for the realization of high speed electronic and optoelectronic devices [15]. In addition to SLs with usual structure, other types of SLs such as II–VI [16], IV–VI [17] and HgTe/CdTe [18] SL's have also been investigated in the literature. The IV-VI SLs exhibit quite different properties as compared to the III–V SL due to the specific band structure of the constituent materials [19]. The epitaxial growth of II–VI SL is a relatively recent development and the primary motivation for studying the mentioned SLs made of materials with the large band gap is in their potential for optoelectronic operation in the blue [19]. HgTe/CdTe SL's have raised a great deal of attention since 1979, when as a promising new materials for long wavelength infrared detectors and other electro-optical applications [20]. Interest in Hg-based SL's has been further increased as new properties with potential device applications were revealed [20, 21]. These features arise from the unique zero band gap material HgTe [22] and the direct band gap semiconductor CdTe which can be described by the three band mode of Kane [23]. The combination of the aforementioned materials with specified dispersion relation makes HgTe/CdTe SL very attractive, especially because of the tailoring of the material properties for various applications by varying the energy band constants of the SLs.*

We note that all the aforementioned SLs have been proposed with the assumption that the interfaces between the layers are sharply defined, of zero thickness, i.e., devoid of any interface effects. The SL potential distribution may be then considered as a one dimensional array of rectangular potential wells. The aforementioned advanced experimental techniques may produce SLs with physical interfaces between the two materials crystallographically abrupt; adjoining their interface will change at least on an atomic scale. As the potential form changes from a well (barrier) to a barrier (well), an intermediate potential region exists for the electrons [24]. The influence of finite thickness of the interfaces on the electron dispersion law is very important, since; the electron energy spectrum governs the electron transport in SLs. In addition to it, for effective mass SLs, the electronic subbands appear continually in real space [25].

In this Chapter, the EM in III-V, II-VI, IV-VI, HgTe/CdTe and strained layer quantum well heavily doped superlattices (QWHDSLs) with graded interfaces has been studied in Sections 12.2.1 to 12.2.5 From Sections 12.2.6 to 12.2.10, the EM from III-V, II-VI, IV-VI, HgTe/CdTe and strained layer quantum well heavily doped effective mass superlattices respectively has been presented. Lastly Section 12.3 presents 1 multi-dimensional open research problem.

12.2 Theoretical Background

12.2.1 *The EM in III–V quantum well HD superlattices with graded interfaces*

The electron dispersion law in bulk specimens of the heavily doped constituent materials of III-V SLs whose un-doped energy band structures are defined by three band model of Kane can be expressed as

$$\frac{\hbar^2 k^2}{2m_{cj}^*} = T_{1j}(E, \Delta_j, E_{gj}, \eta_{gj}) + iT_{2j}(E, \Delta_j, E_{gj}, \eta_{gj}) \qquad (12.1)$$

where

$$\begin{aligned} j &= 1, 2, T_{1j}(E, \Delta_j, E_{gj}, \eta_{gj}) \\ &= (2/(1 - \mathit{Erf}(\mathrm{E}/\eta_{gj}))[(a_j b_{j/c_j}).\theta_0(\mathrm{E}/\eta_{gj}) \\ &\quad + [(a_j c_j + b_j c_j - a_j b_j)/c_j^2 \\ &\quad \cdot\gamma_0(E/\eta_{gj}) + [1/c_j)(1 - (a_j/c_j))(1 - (b_j/c_j))\frac{1}{2}[1 + \mathit{Erf}(E/\eta_{gj})] \\ &\quad - (1/c_j)(1 - (a_j/c_j))(1 - (b_j/c_j))(2/(c_j \eta_{gj}\sqrt{\pi})\exp(-u_j^2) \\ &\quad \cdot \left[\sum_{p=1}^{\infty} \left(\exp\left(-p^2/4\right)/p\right)\sinh\left(pu_j\right)\right]\Bigg], \end{aligned}$$

$$b_j \equiv (E_{gj} + \Delta_j)^{-1}, \quad c_j \equiv (E_{gj} + \frac{2}{3}\Delta_j)^{-1}, \quad u_j \equiv \frac{1 + c_j E}{c_j \eta_{gj}} \quad \text{and}$$

$$T_{2j}(E, \Delta_j, E_{gj}, \eta_{gj}) \equiv \left(\frac{2}{1 + \mathit{Erf}(E/\eta_{gj}}\right) \cdot\frac{1}{c_j}\left(1 - \frac{\alpha_j}{c_j}\right)\left(1 - \frac{b_j}{c_j}\right)\frac{\sqrt{\pi}}{c_j \eta_{gj}}\exp(-u_j^2)$$

Therefore, the dispersion law of the electrons of heavily doped III-V SLs with graded interfacescan be expressed as [25]

$$k_z^2 = G_8 + iH_8 \tag{12.2}$$

where

$$G_8 = \left[\frac{C_7^2 - D_7^2}{L_0^2} - k_s^2\right], \quad C_7 = \cos^{-1}(\bar{\omega}_7),$$

$$\bar{\omega}_7 = (2)^{\frac{-1}{2}}[(1 - G_7^2 - H_7^2) - \sqrt{(1 - G_7^2 - H_7^2)^2 + 4G_7^2}]^{\frac{1}{2}}$$

$$a_{20} = \left[\sqrt{\frac{M_{s2}(0,\eta_{g2})}{M_{s1}(0,\eta_{g1})}} + 1\right]^2 \left[4\left(\frac{M_{s2}(0,\eta_{g2})}{M_{s1}(0,\eta_{g1})}\right)^{\frac{1}{2}}\right]^{-1},$$

$$a_{21} = \left[\sqrt{\frac{M_{s2}(0,\eta_{g2})}{M_{s1}(0,\eta_{g1})}} - 1\right]^2 \left[4\left(\frac{M_{s2}(0,\eta_{g2})}{M_{s1}(0,\eta_{g1})}\right)^{\frac{1}{2}}\right]^{-1}$$

$$C_{40}(E, k_x, k_y, \eta_{g1}) = \left[1 - \bar{P}_1(E,\eta_{g1})k_x^2 - \bar{Q}_2(E,\eta_{g1})k_y^2\right]^{1/2} \cdot \left[\bar{S}_1(E,\eta_{g1})\right]^{-1/2}$$

$$D_{40}(E, k_x, k_y, \eta_{g2}) = [1 - \bar{P}_2(E,\eta_{g2})k_x^2 - \bar{Q}_2(E,\eta_{g2})k_y^2]^{1/2} \cdot [\bar{S}_1(E,\eta_{g2})]^{-1/2}$$

$$G_7 = [G_1 + (\rho_5 G_2/2) - (\rho_6 H_2/2) + (\Delta_0/2) \cdot \{\rho_6 H_2 - \rho_8 H_3 + \rho_9 H_4 - \rho_{10} H_4 + \rho_{11} H_5 - \rho_{12} H_5 + (1/12)(\rho_{12} G_6 - \rho_{14} H_6)\}],$$

$$\mathrm{G}_1 = [(cos(h_1))(cos(h_2))(cos(g_1))(cos(g_2)) + (\sin(h_1))(\sin(h_2))(\sin(g_1))(\sin(g_2))],$$

$$h_1 = e_1(b_0 - \Delta_0), e_1 2^{\frac{-1}{2}}\left(\sqrt{t_1^2 + t_1^2} + t_1\right)^{\frac{1}{2}},$$

$$t_1 = [(2m_{c1}^*/\hbar^2).T_{11}(E, E_{g1}, \Delta_1, \eta_{g1}) - k_s^2],$$

$$t_2 = [(2m_{c1}^*/\hbar^2)T_{21}(E, E_{g1}, \Delta_1, \eta_{g1})]$$

$$h_2 = e_2(b_0 - \Delta_0), \quad e_2 = 2^{\frac{-1}{2}}\left(\sqrt{t_1^2 + t_1^2} - t_1\right)^{\frac{1}{2}},$$

$$g_1 = d_1(a_0 - \Delta_0), \quad \mathrm{d}_1 = 2^{\frac{-1}{2}}\left(\sqrt{x_1^2 + y_1^2} + x_1\right)^{\frac{1}{2}},$$

$$x_1 = [-(2m_{c2}^*/\hbar^2).T_{11}(E - V_0, \mathrm{E}_{g2}, \Delta_2, \eta_{g2}) + k_s^2],$$

$$\mathrm{y}_1 = [(2m_{c2}^*/\hbar^2)T_{22}(E - V_0, \mathrm{E}_{g2}, \Delta_2, \eta_{g2})]$$

$$g_2 = d_2(a_0 - \Delta_0), \quad \mathrm{d}_2 = 2^{\frac{-1}{2}}\left(\sqrt{x_1^2 + y_1^2} - x_1\right)^{\frac{1}{2}},$$

$$\rho_5 = (\rho_3^2 + \rho_4^2)^{-1}[\rho_1\rho_3 - \rho_2\rho_4],$$

$$\rho_1 = [d_1^2 + e_2^2 - d_2^2 - e_1^2], \quad \rho_3 = [d_1e_1 + d_2e_2],$$

$$\rho_2 = 2[d_1d_2 + e_1e_2], \quad \rho_4 = [d_1e_2 - e_1d_2],$$

$$\begin{aligned}\mathrm{G}_2 = {} & [(\sin(h_1))(\cosh(h_2))(\sinh(g_1))(\cos(g_2)) \\ & + (\cos(h_1))(\sinh(h_2))(\cosh(g_1))(\sin(g_2))]\end{aligned}$$

$$\rho_6 = (\rho_3^2 + \rho_4^2)^{-1}[\rho_1\rho_4 - \rho_2\rho_3],$$

$$\begin{aligned}H_2 = {} & [(\sin(h_1))(\cosh(h_2))(\sin(g_2))(\cosh(g_1)) \\ & - (\cos(h_1))(\sinh(h_2))(\sinh(g_1))(\cos(g_2))]\end{aligned}$$

$$\rho_7 = [(e_1^2 + e_2^2)^{-1}[e_1(d_1^2 - d_2^2) - 2d_1d_2e_2] - 3e_1],$$

$$\begin{aligned}\mathrm{G}_3 = {} & [(\sin(h_1))(\cosh(h_2))(\cosh(g_1))(cos(g_2)) \\ & + (\cos(h_1))(\sinh(h_2))(\sinh(g_1))(\sin(g_2))]\end{aligned}$$

$$\rho_8 = [(e_1^2 + e_2^2)^{-1}[e_2(d_1^2 - d_2^2) + 2d_1d_2e_1] + 3e_2],$$

$$\begin{aligned}H_3 = {} & [(\sin(h_1))(\cosh(h_2))(\sin(g_2))(\sinh(g_1)) \\ & - (\cos(h_1))(\sinh(h_2))(\cosh(g_1))(\cos(g_2))]\end{aligned}$$

$$\rho_9 = [(d_1^2 + d_2^2)^{-1}[d_2(e_2^2 - e_1^2) + 2e_2d_2e_1] + 3d_1],$$

$$\begin{aligned}\mathrm{G}_4 = {} & [(\cos(h_1))(\cosh(h_2))(\cos(g_2))(\sinh(g_1)) \\ & - (\sin(h_1))(\sinh(h_2))(\cosh(g_1))(\sin(g_2))]\end{aligned}$$

$$\rho_{10} = [-(d_1^2 + d_2^2)^{-1}[d_2(-e_2^2 + e_1^2) + 2e_2d_2e_1] + 3d_2],$$

$$H_4 = [(\cos(h_1))(\cosh(h_2))(\cosh(g_1))(\sin(g_2))$$
$$+ (\sin(h_1))(\sinh(h_2))(\sinh(g_1))(\cos(g_2))]$$

$$\rho_{11} = 2[d_1^2 + e_2^2 - d_2^2 - e_1^2],$$

$$\mathrm{G}_5 = [(\cos(h_1))(\cosh(h_2))(\cos(g_2))(\cosh(g_1))$$
$$- (\sin(h_1))(\sinh(h_2))(\sinh(g_1))(\sin(g_2))]$$

$$\rho_{12} = 4[d_1d_2 + e_1e_2],$$

$$H_5 = [(\cos(h_1))(\cosh(h_2))(\sinh(g_1))(\sin(g_2))$$
$$+ (\sin(h_1))(\sinh(h_2))(\cosh(g_1))(\cos(g_2))]$$

$$\rho_{13} = [\{5(d_1e_1^3 - 3e_1e_2^2d_1) + 5d_2(e_1^3 - 3e_1^2e_2)\}(d_1^2 + d_2^2)^{-1}$$
$$+ (e_1^2 + e_2^2)^{-1}\{5(e_1d_1^3 - 3d_2e_1^2d_1)$$
$$+ 5(d_2^3e_2 - 3d_1^2d_2e_2)\} - 34(d_1e_1 + d_2e_2)],$$

$$\mathrm{G}_6 = [(\sin(h_1))(\cosh(h_2))(\sinh(g_1))(\cos(g_2))$$
$$+ (\cos(h_1))(\sinh(h_2))(\cosh(g_1))(\sin(g_2))],$$

$$\rho_{14} = [\{5(d_1e_2^3 - 3e_2e_1^2d_1) + 5d_2(-e_1^3 + 3e_2^2e_1)\}(d_1^2 + d_2^2)^{-1}$$
$$+ (e_1^2 + e_2^2)^{-1}\{5(-e_1d_2^3 + 3d_1^2d_2e_1)$$
$$+ 5(-d_1^3e_2 + 3d_2^2d_1e_2)\} + 34(d_1e_2 - d_2e_1)],$$

$$H_6 = [(\sin(h_1))(\cosh(h_2))(\cosh(g_1))(\sin(g_2))$$
$$- (\cos(h_1))(\sinh(h_2))(\sinh(g_1))(\cos(g_2))],$$

$$H_7 = [H_1 + (\rho_5H_2/2) + (\rho_6G_2/2) + (\Delta_0/2)\{\rho_8G_3 + \rho_7H_3$$
$$+ \rho_{10}G_4 + \rho_9H_4 + \rho_{12}G_5 + \rho_{11}H_5 + (1/12)(\rho_{14}G_6 + \rho_{14}H_6)\}],$$

$$H_1 = [(\sin(h_1))(\sinh(h_2))(\cosh(g_1))(\cos(g_2))$$
$$+ (\cos(h_1))(\cosh(h_2))(\sinh(g_1))(\sin(g_2))],$$

$$D_7 = \sinh^{-1}(\bar{\omega}_7), H_8 = (2C_7D_7/L_0^2)$$

The simplified DR of heavily doped quantum well III-V super-lattices with graded interfaces can be expressed as

$$\left(\frac{n_z\pi}{d_z}\right)^2 = G_8 + iH_8 \tag{12.3a}$$

The sub-band equation in this case can be expressed as

$$\left(\frac{n_z\pi}{d_z}\right)^2 = [G_8 + iH_8]|_{k_s=0 \text{ and } \mathrm{E=E_{12,1}}} \tag{12.3b}$$

where $E_{12,1}$ is the sub-band energy in this case.

12.2.2 *The DR in II–VI quantum well HD superlattices with graded interfaces*

The electron energy spectra of the heavily doped constituent materials of II-VI SLs are given by

$$\gamma_3(E, \eta_{g1}) = \frac{\hbar^2 k_s^2}{2m^*_{\perp,1}} + \frac{\hbar^2 k_z^2}{2m^*_{\parallel,1}} \pm C_0 k_s \tag{12.4}$$

and

$$\frac{\hbar^2 k^2}{2m^*_{c2}} = T_{12}(E, \Delta_2, E_{g2}\eta_{g2}) + iT_{22}(E, \Delta_2, E_{g2}, \eta_{g2}) \tag{12.5}$$

where $m^*_{\perp,1}$ and $m^*_{\parallel,1}$ are the transverse and longitudinal effective electron masses respectively at the edge of the conduction band for the first material. The energy-wave vector dispersion relation of the conduction electrons in heavily doped II–VI SLs with graded interfaces can be expressed as

$$k_z^2 = G_{19} + iH_{19} \tag{12.6}$$

where

$$G_{19} = \left[\frac{C_{18}^2 - D_{18}^2}{L_0^2} - k_s^2\right],$$

$$C_{18} = \cos^{-1}(\omega_{18}), \quad \omega_{18} = (2)^{\frac{-1}{2}}\left[(1 - G_{18}^2 - H_{18}^2) - \sqrt{(1 - G_{18}^2 - H_{18}^2)^2 + 4G_{18}^2}\right]^{\frac{1}{2}}$$

$$G_{18} = \frac{1}{2}\left[G_{11} + G_{12} + \Delta_0 \left(G_{13} + G_{14}\right) + \Delta_0 \left(G_{15} + G_{16}\right)\right],$$

$$G_{11} = 2\left(\cos\left(g_1\right)\right)\left(\cos\left(g_2\right)\right)\left(\cos\gamma_{11}\left(E, k_s\right)\right)$$

$$\gamma_{11}\left(E, k_s\right) = k_{21}\left(E, k_s\right)\left(b_0 - \Delta_0\right),$$

$$k_{21}\left(E, k_s\right) = \left\{\left[\gamma_3\left(E, \eta_{g1}\right) - \frac{\hbar^2 k_s^2}{2m_{\perp,1}^*} \pm C_0 k_s\right]\frac{2m_{\parallel,1}^*}{\hbar^2}\right\}^{\frac{1}{2}}$$

$$C_{18} = \cos^{-1}\left(\omega_{18}\right),$$

$$\omega_{18} = (2)^{\frac{-1}{2}}\left[\left(1 - G_{18}^2 - H_{18}^2\right) - \sqrt{\left(1 - G_{18}^2 - H_{18}^2\right)^2 + 4G_{18}^2}\right]^{\frac{1}{2}}$$

$$G_{18} = \frac{1}{2}\left[G_{11} + G_{12} + \Delta_0 \left(G_{13} + G_{14}\right) + \Delta_0 \left(G_{15} + G_{16}\right)\right],$$

$$G_{11} = 2\left(\cos\left(g_1\right)\right)\left(\cos\left(g_2\right)\right)\left(\cos\gamma_{11}\left(E, k_s\right)\right)$$

$$\gamma_{11}\left(E, k_s\right) = k_{21}\left(E, k_s\right)\left(b_0 - \Delta_0\right), \quad k_{21}\left(E, k_s\right)$$

$$= \left\{\left[\gamma_3\left(E, \eta_{g1}\right) - \frac{\hbar^2 k_s^2}{2m_{\perp,1}^*} \pm C_0 k_s\right]\frac{2m_{\parallel,1}^*}{\hbar^2}\right\}^{\frac{1}{2}}$$

$$G_{12}\left(\left[\Omega_1\left(E, k_s\right)\left(\sinh\ g_1\right)\left(\cos g_2\right)\right.\right.$$
$$\left.\left. -\Omega_2\left(E, k_s\right)\left(\sin\ g_2\right)\left(\cosh\ g_1\right)\right]\left(\sin\gamma_{11}\left(E, k_s\right)\right)\right)$$

$$\Omega_1(E, k_s) = \left[\frac{d_1}{k_{21}(E, k_s)} - \frac{k_{21}(E, k_s)d_1}{d_1^2 + d_2^2}\right] \quad \text{and}$$

$$\Omega_2(E, k_s) = \left[\frac{d_2}{k_{21}(E, k_s)} - \frac{k_{21}(E, k_s)d_2}{d_1^2 + d_2^2}\right]$$

$$G_{13}\left(\left[\Omega_3\left(E, k_s\right)\left(\cosh\ g_1\right)\left(\cos\ g_2\right)\right.\right.$$
$$\left.\left. -\Omega_4\left(E, k_s\right)\left(\sinh\ g_1\right)\left(\sin\ g_2\right)\right]\left(\sin\gamma_{11}\left(E, k_s\right)\right)\right)$$

$$\Omega_3(E, k_s) = \left[\frac{d_1^2 - d_2^2}{k_{21}(E, k_s)} - 3k_{21}(E, k_s)\right], \ \Omega_4(E, k_s) = \left[\frac{2d_1 - d_2}{k_{21}(E, k_s)}\right],$$

$$G_{14}\left(\left[\Omega_5\left(E, k_s\right)\left(\sinh\ g_1\right)\left(\cos\ g_2\right)\right.\right.$$
$$\left.\left. -\Omega_6\left(E, k_s\right)\left(\sin\ g_1\right)\left(\cosh\ g_2\right)\right]\left(\cos\gamma_{11}\left(E, k_s\right)\right)\right)$$

$$\Omega_5(E,k_s) = \left[3d_1 - \frac{d_1}{d_1^2+d_2^2}k_{21}(E,k_s)\right],$$

$$\Omega_6(E,k_s) = \left[3d_2 - \frac{d_2}{d_1^2+d_2^2}k_{21}(E,k_s)\right].$$

$$G_{15}\left(\left[\Omega_9\left(E,k_s\right)\left(\cosh\ g_1\right)\left(\cos\ g_2\right)\right.\right.$$
$$\left.\left.-\,\Omega_{10}\left(E,k_s\right)\left(\sinh\ g_1\right)\left(\sin\ g_2\right)\right]\left(\cos\gamma_{11}\left(E,k_s\right)\right)\right)$$

$$\Omega_9(E,k_s) = \left[2d_1^2 - 2d_2^2k_{21}^2(E,k_s)\right], \quad \Omega_{10}(E,k_s) = [2d_1d_2]$$

$$G_{16}\left(\left[\Omega_7\left(E,k_s\right)\left(\sinh\ g_1\right)\left(\cos\ g_2\right)\right.\right.$$
$$\left.\left.-\,\Omega_8\left(E,k_s\right)\left(\sin\ g_1\right)\left(\cosh\ g_2\right)\right]\left(\sin\gamma_{11}\left(E,k_s\right)/12\right)\right)$$

$$\Omega_7(E,k_s) = \left[\frac{5d_2}{d_1^2+d_2^2}k_{21}^3(E,k_s) + \frac{5\left(d_1^3-3d_2^2d_1\right)}{k_{21}(E,k_s)} - 34k_{21}(E,k_s)d_1\right],$$

$$\Omega_8(E,k_s) = \left[\frac{5d_2}{d_1^2+d_2^2}k_{21}^3(E,k_s) + \frac{5\left(d_1^3-3d_2^2d_2\right)}{k_{21}(E,k_s)} + 34k_{21}(E,k_s)d_2\right],$$

$$H_{18} = \frac{1}{2}\left[H_{11}+H_{12}+\Delta_0\left(H_{13}+H_{14}\right)+\Delta_0\left(H_{15}+H_{16}\right)\right],$$

$$H_{11} = 2\left(\sinh\ g_1 \sin g_2 \cos\gamma_{11}(E,k_s)\right),$$

$$H_{12} = \left(\left[\Omega_2\left(E,k_s\right)\left(\sinh g_1\right)\left(\cos g_2\right)\right.\right.$$
$$\left.\left.+\,\Omega_1\left(E,k_s\right)\left(\sin g_2\right)\left(\cosh g_1\right)\right]\left(\sin\gamma_{11}\left(E,k_s\right)\right)\right),$$

$$H_{13} = \left(\left[\Omega_4\left(E,k_s\right)\left(\cosh g_1\right)\left(\cos g_2\right)\right.\right.$$
$$\left.\left.+\,\Omega_3\left(E,k_s\right)\left(\sinh g_1\right)\left(\sin g_2\right)\right]\left(\sin\gamma_{11}\left(E,k_s\right)\right)\right),$$

$$H_{14} = \left(\left[\Omega_6\left(E,k_s\right)\left(\sinh g_1\right)\left(\cos g_2\right)\right.\right.$$
$$\left.\left.+\,\Omega_5\left(E,k_s\right)\left(\sin g_1\right)\left(\cosh g_2\right)\right]\left(\sin\gamma_{11}\left(E,k_s\right)\right)\right),$$

$$H_{15} = \left(\left[\Omega_{10}\left(E,k_s\right)\left(\cosh g_1\right)\left(\cos g_2\right)\right.\right.$$
$$\left.\left.+\,\Omega_9\left(E,k_s\right)\left(\sinh g_1\right)\left(\sin g_2\right)\right]\left(\cos\gamma_{11}\left(E,k_s\right)\right)\right),$$

$$H_{16} = ([\Omega_8(E,k_s)(\sinh g_1)(\cos g_2) + \Omega_7(E,k_s)(\sin g_1)(\cosh g_2)](\sin \gamma_{11}(E,k_s)/12)),$$

$$H_{19} = \left[\frac{2C_{18}D_{18}}{L_0^2}\right] \quad \text{and} \quad D_{18} = \sin^{-1}(\omega_{18})$$

The simplified DR in heavily doped quantum well II-VI super-lattices with graded interfaces can be expressed as

$$\left(\frac{n_z\pi}{d_z}\right)^2 = G_{19} + iH_{19} \tag{12.7a}$$

The sub-band equation in this case can be expressed as

$$\left(\frac{n_z\pi}{d_z}\right)^2 = [G_{19} + iH_{19}]|_{k_s=0} \text{ and } \mathrm{E{=}E_{12.2}} \tag{12.7b}$$

where $E_{12,2}$ is the sub-band energy in this case.

12.2.3 *The DR in IV–VI quantum well HD superlattices with graded interfaces*

The **E-k** DR of the conduction electrons of the heavily doped constituent materials of the IV-VI SLs can be expressed as

$$k_z^2 = [2\bar{p}_{9,i}]^{-1}[-\bar{q}_{9,i}(E,k_s,\eta_{gi}) + [[\bar{q}_{9,i}(E,k_s,\eta_{gi})]^2 + 4\bar{p}_{9,i}\bar{R}_{9,1}(E,k_s,\eta_{gi})]^{\frac{1}{2}}] \tag{12.8}$$

where,

$$\bar{p}_{9,i} = (\alpha_i\hbar^4)/\bar{p}_{9,i} = (\alpha_i\hbar^4)/(4m_{l,i}^- m_{l,i}^+), \quad \mathrm{i} = 1,2,$$

$$\bar{q}_{9,1} = (E,k_s,\eta_{gi}) = [(\hbar^2/2)((1/m_{li}^*) + (1/m_{li}^-)) + \alpha_i(\hbar^4/4)k_s^2((1/m_{li}^+ m_{li}^-) + (1/m_{li}^+ m_{li}^-)) - \alpha_i\gamma_3(E,\eta_{gi})((1/m_{li}^+) - (1/m_{li}^-)) \quad \text{and}$$

$$\bar{R}_{9,i}(E,k_s,\eta_{gi}) = [\gamma_2(E,\eta_{gi}) + \gamma_3(E,\eta_{gi})[(\hbar^2/2)\alpha_i k_s^2((1/m_{ti}^*) - (1/m_{ti}^-))] - [(\hbar^2/2)k_s^2((1/m_{ti}^*) - (1/m_{ti}^-))] = \alpha_i(\hbar^6/4)k_s^4((1/m_{ti}^+ m_{ti}^-))]$$

The electron dispersion law in heavily doped IV-VI SLs with graded interfaces can be expressed as

$$\cos\left(L_0 k\right)\frac{1}{2}\Phi_2\left(E,k_s\right) \tag{12.9}$$

Where

$$\begin{aligned}
\Phi_2\left(E,k_s\right) \equiv {} & [2\cos\left\{\beta_2\left(E,k_s\right)\right\}\cos\left\{\gamma_2\left(E,k_s\right)\right\} \\
& +\varepsilon_2\left(E,k_s\right)\sinh\left\{\beta_2\left(E,k_s\right)\right\}\sin\left\{\gamma_{22}\left(E,k_s\right)\right\} \\
& +\Delta_0[\left(\frac{\left\{K_{112}\left(E,k_s\right)\right\}^2}{K_{212}\left(E,k_s\right)}-3K_{212}\left(E,k_s\right)\right)\cosh \\
& \cdot\left\{\beta_2\left(E,k_s\right)\right\}\sin\left\{\gamma_{22}\left(E,k_s\right)\right\} \\
& +\left(3K_{112}\left(E,k_s\right)-\frac{\left\{K_{212}\left(E,k_s\right)\right\}^2}{K_{112}\left(E,k_s\right)}\right)\sinh \\
& \cdot\left\{\beta_2\left(E,k_s\right)\right\}\cos\left\{\gamma_{22}\left(E,k_s\right)\right\}] \\
& +\Delta_0[2\left(\left\{K_{112}\left(E,k_s\right)\right\}^2-\left\{K_{212}\left(E,k_s\right)\right\}^2\right)\cosh \\
& \cdot\left\{\beta_2\left(E,k_s\right)\right\}\cos\left\{\gamma_{22}\left(E,k_s\right)\right\} \\
& +\frac{1}{12}\left[\frac{5\left\{K_{112}\left(E,k_s\right)\right\}^3}{K_{212}\left(E,k_s\right)}+\frac{5\left\{K_{212}\left(E,k_s\right)\right\}^3}{K_{112}\left(E,k_s\right)}\right. \\
& \left.-34K_{212}\left(E,k_s\right)K_{112}\right]\sinh \\
& \cdot\left\{\beta_2\left(E,k_s\right)\right\}\sin\left\{\gamma_{22}\left(E,k_s\right)\right\}]],
\end{aligned}$$

$$\beta_2\left(E,k_s\right) \equiv K_{112}\left(E,k_s\right)\left[a_0-\Delta_0\right],$$

$$\begin{aligned}
k_{112}^2(E,k_s) = {} & [2\bar{p}_{9,2}]^{-1}[-\bar{q}_{9,2}(E-V_0,k_s\eta_{g2}) \\
& -[[\bar{q}_{9,2}(E-V_0,k_s\eta_{g2})]^2+4\bar{p}_{9,2}\bar{R}_{9,2}(E-V_0,k_s\eta_{g2})]^{\frac{1}{2}}],
\end{aligned}$$

$$\gamma_{22}(E,k_s) \equiv K_{212}(E,k_s)[b_0-\Delta_0],$$

$$k_{212}^2(E,k_s) = [2\bar{p}_{9,1}]^{-1}[-\bar{q}_{9,1}(E,k_s\eta_{g1}) + [[\bar{q}_{9,1}(E,k_s\eta_{g1})]^2 + 4\bar{p}_{9,1}\bar{R}_{9,1}(E,k_s\eta_{g1})]^{\frac{1}{2}}], \quad \text{and}$$

$$\varepsilon_2(E,k_s) \equiv \left[\frac{K_{112}(E,k_s)}{K_{212}(E,k_s)} - \frac{K_{212}(E,k_s)}{K_{112}(E,k_s)}\right].$$

The simplified DR in heavily doped quantum well IV-VI superlattices with graded interfaces can be expressed as

$$\left(\frac{n_z\pi}{d_z}\right)^2 = \frac{1}{L_0^2}\left[\cos^{-1}\left\{\frac{1}{2}\Phi_2(E,k_s)\right\}\right]^2 - k_s^2 \tag{12.10a}$$

The sub-band equation in this case can be expressed as

$$\left(\frac{n_z\pi}{d_z}\right)^2 = \frac{1}{L_0^2}\left[\cos^{-1}\left\{\frac{1}{2}\Phi_2(E_{12,3},0)\right\}\right]^2 \tag{12.10b}$$

where $E_{12,3}$ is the sub-band energy in this case.

12.2.4 *The DR in HgTe/CdTe quantum well HD superlattices with graded interfaces*

The electron energy spectra of the constituent materials of HgTe/CdTe SLs are given by

$$k^2 = \left[\frac{B_{01}^2 + 4A_1E - B_{01}\sqrt{B_{01}^2 + 4A_1E}}{2A_1^2}\right] \tag{12.11}$$

and

$$\frac{\hbar^2k^2}{2m_{c2}^*} = T_{12}(E,\Delta_2,E_{g2}\eta_{g2}) + iT_{22}(E,\Delta_2,E_{g2},\eta_{g2}) \tag{12.12}$$

where $B_{01} = (3|e|^2/128\varepsilon_{sc1})$, $A_1 = (\hbar^2/2m_{c1}^*)$. ε_{sc1} is the semiconductor permittivity of the first material. The energy-wave vector dispersion relation of the conduction electrons in heavily doped

HgTe/CdTe SLs with graded interfaces can be expressed as

$$k_z^2 = G_{192} + iH_{192} \tag{12.13}$$

where

$$G_{192} = \left[\left(\left(C_{182}^2 - D_{182}^2\right)/L_0^2\right) - k_s^2\right],$$

$$C_{182} = \cos^{-1}\left(\omega_{182}\right),$$

$$\omega_{182} = (2)^{\frac{-1}{2}}\left[\left(1 - G_{182}^2 - H_{182}^2\right) - \sqrt{\left(1 - G_{182}^2 - H_{182}^2\right)^2 + 4G_{182}^2}\right]^{\frac{1}{2}},$$

$$G_{182} = \frac{1}{2}\left[G_{112} + G_{122} + \Delta_0\left(G_{132} + G_{142}\right) + \Delta_0\left(G_{152} + G_{162}\right)\right],$$

$$G_{112} = 2\left(\cos\left(g_{12}\right)\right)\left(\cos\left(g_{22}\right)\right)\left(\cos\gamma_8\left(E, k_s\right)\right)$$

$$\gamma_8\left(E, k_s\right) = k_8\left(E, k_s\right)\left(b_0 - \Delta_0\right),$$

$$k_8\left(E, k_s\right) = \left[\frac{B_{01}^2 + 4A_1E - B_{01}\sqrt{B_{01}^2 + 4A_1E}}{2A_1^2}\right]^{1/2}$$

$$G_{122}\left(\left[\Omega_{12}\left(E, k_s\right)\left(\sinh\ g_{12}\right)\left(\cos\ g_{22}\right) - \Omega_{22}\left(E, k_s\right)\left(\sin\ g_{22}\right)\left(\cosh\ g_{12}\right)\right]\left(\sin\gamma_8\left(E, k_s\right)\right)\right)$$

$$\Omega_{12}(E, k_s) = \left[\frac{d_{12}}{k_8(E, k_s)} - \frac{k_8(E, k_s)d_{12}}{d_{12}^2 + d_{22}^2}\right],$$

$$\Omega_{22}(E, k_s) = \left[\frac{d_{22}}{k_8(E, k_s)} - \frac{k_8(E, k_s)d_{22}}{d_{12}^2 + d_{22}^2}\right]$$

$$G_{132} = \left(\left[\Omega_{32}\left(E, k_s\right)\left(\cosh\ g_{12}\right)\left(\cos\ g_{22}\right) - \Omega_{42}\left(E, k_s\right)\left(\sinh\ g_{12}\right)\left(\sin\ g_{22}\right)\right]\left(\sin\gamma_8\left(E, k_s\right)\right)\right),$$

$$\Omega_{32}(E, k_s) = \left[\frac{d_{12}^2 - d_{22}^2}{k_8(E, k_s)} - 3k_8\left(E, k_s\right)\right], \quad \Omega_{42}(E, k_s) = \left[\frac{2d_{12}d_{22}}{k_8\left(E, k_s\right)}\right],$$

$$G_{142} = \left(\left[\Omega_{52}\left(E, k_s\right)\left(\sinh\ g_{12}\right)\left(\cos\ g_{22}\right) - \Omega_{62}\left(E, k_s\right)\left(\sin\ g_{12}\right)\left(\cosh\ g_{22}\right)\right]\left(\cos\gamma_8\left(E, k_s\right)\right)\right),$$

$$\Omega_{52}(E,k_s) = \left[3d_{12} - \frac{d_{12}}{d_{12}^2+d_{22}^2}k_8^2(E,k_s)\right],$$

$$\Omega_{62}(E,k_s) = \left[3d_{22} + \frac{d_{22}}{d_{12}^2+d_{22}^2}k_8^2(E,k_s)\right],$$

$$G_{152} = \left(\left[\Omega_{92}\left(E,k_s\right)\left(\cosh\ g_{12}\right)\left(\cos\ g_{22}\right) - \Omega_{102}\left(E,k_s\right)\left(\sinh\ g_{12}\right)\left(\sin\ g_{22}\right)\right]\left(\cos\gamma_8\left(E,k_s\right)\right)\right),$$

$$\Omega_{92}(E,k_s) = \left[2d_{12}^2 - 2d_{22}^2 - k_8^2(E,k_s)\right], \quad \Omega_{102}(E,k_s) = \left[2d_{12}d_{22}\right],$$

$$G_{162}\left(\left[\Omega_{72}\left(E,k_s\right)\left(\sinh\ g_{12}\right)\left(\cos\ g_{22}\right) - \Omega_{82}\left(E,k_s\right)\left(\sin\ g_{12}\right)\left(\cosh\ g_{22}\right)\right]\left(\sin\gamma_8\left(E,k_s\right)/12\right)\right)$$

$$\Omega_{72}(E,k_s) = \left[\frac{5d_{12}}{d_{12}^2+d_{22}^2}k_8^3(E,k_s) + \frac{5\left(d_{12}^3 - 3d_{22}^2d_{12}\right)}{k_8(E,k_s)} - 34k_8(E,k_s)d_{12}\right],$$

$$\Omega_{82}(E,k_s) = \left[\frac{5d_{12}}{d_{12}^2+d_{22}^2}k_8^3(E,k_s) + \frac{5\left(d_{12}^3 - 3d_{22}^2d_{12}\right)}{k_8(E,k_s)} + 34k_8(E,k_s)d_{12}\right],$$

$$H_{182} = \frac{1}{2}\left[H_{112}+H_{122}+\Delta_0\left(H_{132}+H_{142}\right)+\Delta_0\left(H_{152}+H_{162}\right)\right]$$

$$H_{112} = 2\left(\sinh\ g_{12}\sin g_{22}\cos\gamma_8(E,k_s)\right),$$

$$H_{122} = \left(\left[\Omega_{22}\left(E,k_s\right)\left(\sinh g_{12}\right)\left(\cos g_{22}\right) + \Omega_{12}\left(E,k_s\right)\left(\sin g_{22}\right)\left(\cosh g_{12}\right)\right]\left(\sin\gamma_8\left(E,k_s\right)\right)\right),$$

$$H_{132} = \left(\left[\Omega_{42}\left(E,k_s\right)\left(\cosh g_{12}\right)\left(\cos g_{22}\right) + \Omega_{32}\left(E,k_s\right)\left(\sinh g_{12}\right)\left(\sin g_{22}\right)\right]\left(\sin\gamma_8\left(E,k_s\right)\right)\right),$$

$$H_{142} = \left(\left[\Omega_{62}\left(E,k_s\right)\left(\sinh g_{12}\right)\left(\cos g_{22}\right) + \Omega_{52}\left(E,k_s\right)\left(\sin g_{12}\right)\left(\cosh g_{22}\right)\right]\left(\cos\gamma_8\left(E,k_s\right)\right)\right),$$

$$H_{152} = ([\Omega_{102}(E,k_s)(\cosh g_{12})(\cos g_{22}) + \Omega_{92}(E,k_s)(\sinh g_{12})(\sin g_{22})](\cos\gamma_8(E,k_s))),$$

$$H_{162} = ([\Omega_{82}(E,k_s)(\sinh g_{12})(\cos g_{22}) + \Omega_{72}(E,k_s)(\sin g_{12})(\cosh g_{22})](\sin\gamma_8(E,k_s)/12)),$$

$$H_{192} = \left[\left((2C_{182}D_{182})/L_0^2\right)\right] \quad \text{and} \quad D_{182} = \sinh^{-1}(\omega_{182})$$

The simplified DR in heavily doped quantum well HgTe/CdTe superllatices with graded interfaces can be expressed as

$$\left(\frac{n_z\pi}{d_z}\right)^2 = G_{192} + iH_{192} \tag{12.14a}$$

The sub-band equation in this case can be expressed as

$$\left(\frac{n_z\pi}{d_z}\right)^2 = [G_{192} + iH_{192}]|_{k_s=0} \text{ and } \mathrm{E{=}E_{12.4}} \tag{12.14b}$$

where $E_{12,4}$ is the sub-band energy in this case.

12.2.5 *The DR in strained layer quantum well HD superlattices with graded interfaces*

The DR of the conduction electrons of the constituent materials of the strained layer super lattices can be expressed as

$$[E - T_{1i}]k_x^2 + [E - T_{2i}]k_y^2 + [E - T_{3i}]k_z^2 = q_iE^3 - R_iE^2 + V_iE + \zeta_i \tag{12.15}$$

where

$$T_{1i} = \bar{\theta}_i, \bar{\theta}_i$$

$$= \left[E_{gi} - C_{1i}^c\varepsilon_i - (a_i + C_{1i}^c)\varepsilon_i + \frac{3}{2}b_i\varepsilon_{xxi} - \frac{b_i\varepsilon_i}{2} + \frac{\sqrt{3d_i\varepsilon_{xyi}}}{2}\right],$$

$$T_{2i} = \omega_i, \omega_i$$
$$= \left[E_{gi} - C_{1i}^c \varepsilon_i - (a_i + C_{1i}^c)\,\varepsilon_i + \frac{3}{2} b_i \varepsilon_{xxi} - \frac{b_i \varepsilon_i}{2} - \frac{\sqrt{3d_i \varepsilon_{xyi}}}{2}\right],$$

$$T_{3i} = \delta_i, \delta_i = \left[E_{gi} - C_{1i}^c \varepsilon_i - (a_i + C_{1i}^c)\,\varepsilon_i + \frac{3}{2} b_i \varepsilon_{zzi} - \frac{b_i \varepsilon_i}{2}\right],$$

$$R_i = q_i\left[2A_i + C_{1i}^c \varepsilon_i\right], \qquad \mathrm{q}_i = \frac{3}{2B_{2i}^2}, \quad A_i = E_{gi} - C_{1i}^c \varepsilon_i,$$

$$V_i = q_i\left[A_i^2 - \frac{2C_{1i}^2 \varepsilon_{xyi}}{3} + 2A_i C_{1i}^c \varepsilon_i\right], \quad \zeta_i = q_i\left[\frac{2C_{1i}^2 \varepsilon_{xyi}}{3} - C_{1i}^c \varepsilon_i A_i^2\right]$$

Therefore the electron energy spectrum in HD stressed materials can be written as

$$\bar{P}_i(E, \eta_{gi})\,k_x^2 + \bar{Q}_i(E, \eta_{gi})\,k_y^2 + \bar{S}_i(E, \eta_{gi})\,k_z^2 = 1 \tag{12.16}$$

where

$$\bar{P}_i(E, \eta_{gi}) = \frac{[\gamma_0(E, \eta_{gi}) - I_0 T_{1i}]}{\bar{\Delta}_i(E, \eta_{gi})},$$

$$\bar{\Delta}_i(E, \eta_{gi}) = \left[\frac{-q_i \eta_{gi}^3}{2\sqrt{\pi}} \exp\left(\frac{-E^2}{\eta_{gi}^2}\right)\left[1 + \frac{E^2}{\eta_{gi}^2}\right] - R_i \theta_0(E, \eta_{gi}) + V_i \gamma_0(E, \eta_{gi}) + \frac{\zeta_i}{2}\left[1 + Erf\left(\frac{E}{\eta_{gi}}\right)\right]\right],$$

$$I_0 = \frac{1}{2}\left[1 + Erf\,(E/\eta_{gi})\right],$$

$$\bar{Q}_i(E, \eta_{gi}) = \frac{[\gamma_0(E, \eta_{gi}) - I_0 T_{2i}]}{\bar{\Delta}_i(E, \eta_{gi})} \quad \text{and}$$

$$\bar{\mathrm{S}}_i(E, \eta_{gi}) = \frac{[\gamma_0(E, \eta_{gi}) - I_0 T_{3i}]}{\bar{\Delta}_i(E, \eta_{gi})}$$

The energy-wave vector dispersion relation of the conduction electrons in heavily doped strained layer SLs with graded interfaces can be expressed as

$$\cos(L_0 k) = \frac{1}{2}\bar{\phi}_6(E, k_s) \tag{12.17}$$

where

$$\begin{aligned}
\bar{\phi}_6(E,k_s) = {} & [2\cosh[T_4(E,\eta_{g2})]\cos[T_5(E,\eta_{g1})]] \\
& + [T_6(E,k_s)]\sinh[T_4(E,\eta_{g2})]\sin[T_5(E,\eta_{g1})] \\
& + \Delta_0[\left(\frac{k_0^2(E,\eta_{g2})}{k'(E,\eta_{g1})} - 3k'(E,\eta_{g1})\right)\cosh \\
& \cdot [T_4(E,\eta_{g2})]\sin[T_5(E,\eta_{g1})] \\
& + \left(3k_0(E,\eta_{g2}) - \frac{k'^2(E,\eta_{g1})}{k_0(E,\eta_{g2})}\right)\sinh \\
& \cdot [T_4(E,\eta_{g2})]\cos[T_5(E,\eta_{g1})]] \\
& + \Delta_0\left[2\left(k_0^2(E,\eta_{g2}) - k'^2(E,\eta_{g2})\right)\cosh\right. \\
& \cdot [T_4(E,\eta_{g2})]\sin[T_5(E,\eta_{g1})]] \\
& + \frac{1}{12}\left(\frac{5k_0^3(E,\eta_{g2})}{k'(E,\eta_{g1})} + \frac{5k'^3(E,\eta_{g1})}{k_0(E,\eta_{g2})}\right. \\
& \left. - 34k_0(E,\eta_{g2})k'(E,\eta_{g1})\right) \\
& \cdot \sinh[T_4(E,\eta_{g2})]\sin[T_5(E,\eta_{g1})]
\end{aligned}$$

$$\begin{aligned}
[T_4(E,\eta_{g2})] &= k_0(E,\eta_{g2})[a_0 - \Delta_0], \\
k_0(E,\eta_{g2}) &= \left[\bar{S}_2(E - V_0,\eta_{g2})\right]^{-1/2} \\
& \cdot \left[\bar{P}_2(E - V_0,\eta_{g2})k_x^2 + \bar{Q}_2(E - V_0,\eta_{g2})k_y^2 - 1\right]^{1/2}, \\
T_5(E,\eta_{g1}) &= k'(E,\eta_{g1})[b_0 - \Delta_0], \\
k'(E,\eta_{g1}) &= \left[\bar{S}_1(E,\eta_{g1})\right]^{-1/2} \\
& \cdot \left[1 - \bar{P}_1(E,\eta_{g1})k_x^2 - \bar{Q}_2(E,\eta_{g1})k_y^2\right]^{1/2} \quad \text{and} \\
T_6(E,k_s) &= \left[\frac{k_0(E,\eta_{g1})}{k'(E,\eta_{g1})} - \frac{k'(E,\eta_{g1})}{k_0(E,\eta_{g1})}\right]
\end{aligned}$$

Therefore the DR of the conduction electrons in heavily doped strained layer quantum well SL with graded interfaces can be expressed as

$$\left(\frac{n_z\pi}{d_z}\right)^2 = \frac{1}{L_0^2}\left[\cos^{-1}\left\{\frac{1}{2}\bar{\Phi}_6\left(E,k_s\right)\right\}\right]^2 - k_s^2 \tag{12.18a}$$

The sub-band equation in this case can be expressed as

$$\left(\frac{n_z\pi}{d_z}\right)^2 = \frac{1}{L_0^2}\left[\cos^{-1}\left\{\frac{1}{2}\bar{\Phi}_6\left(E_{12,5},0\right)\right\}\right]^2 \tag{12.18b}$$

where $E_{12,5}$ is the sub-band energy in this case.

12.2.6 *The DR in III–V quantum well HD effective mass super lattices*

Following Sasaki [24], the electron dispersion law in III-V heavily doped effective mass super-lattices (EMSLs) can be written as

$$k_x^2 = \left[\frac{1}{L_0^2}\left\{\cos^{-1}\left(f_{21}\left(E,k_y,k_z\right)\right)\right\}^2 - k_\perp^2\right] \tag{12.19}$$

In which

$$f_{21}\left(E,k_y,k_z\right) = a_1\cos\left[a_0C_{21}\left(E,k_\perp,\eta_{g1}\right)+b_0D_{21}\left(E,k_\perp,\eta_{g2}\right)\right]$$
$$-a_1\cos\left[a_0C_{21}\left(E,k_\perp,\eta_{g1}\right)-b_0D_{21}\left(E,k_\perp,\eta_{g2}\right)\right],$$
$$\mathrm{k}_\perp^2 = k_y^2+k_z^2,$$
$$a_1 = \left[\sqrt{\frac{M_2\left(0,\eta_{g2}\right)}{M_1\left(0,\eta_{g1}\right)}}+1\right]^2\left[4\left(\frac{M_2\left(0,\eta_{g2}\right)}{M_1\left(0,\eta_{g1}\right)}\right)^{1/2}\right]^{-1}$$
$$a_2 = \left[\sqrt{\frac{M_2\left(0,\eta_{g2}\right)}{M_1\left(0,\eta_{g1}\right)}}-1\right]^2\left[4\left(\frac{M_2\left(0,\eta_{g2}\right)}{M_1\left(0,\eta_{g1}\right)}\right)^{1/2}\right]^{-1}$$

$$
\begin{aligned}
M_{iz}(0,\eta_{gi}) = m_{ci}^* \Bigg[\frac{-2}{\sqrt{\pi}} T(0,\eta_{gi}) + 2 \Bigg[&\frac{\alpha_i b_i}{c_i} \frac{\eta_{gi}}{\sqrt{\pi}} \\
&+ \frac{1}{2} \left(\frac{\alpha_i c_i + c_i b_i - \alpha_i b_i}{c_i^2} \right) \\
&+ \frac{1}{\sqrt{\pi c_i}} \left(1 - \frac{\alpha_i}{c_i}\right) \left(1 - \frac{b_i}{c_i}\right) \\
&- \frac{1}{c_i} \left(1 - \frac{\alpha_i}{c_i}\right) \left(1 - \frac{b_i}{c_i}\right) \frac{2}{c_i \eta_{gi} \sqrt{\pi}} \\
&\cdot \left\{ \frac{-2}{c_i \eta_{gi}} \exp\left(\frac{1}{c_i^2 \eta_{gi}^2}\right) \right. \\
&\cdot \left(\sum_{p=1}^{\alpha} \left(\exp\left(\frac{-p^2}{4}\right)\right) \frac{1}{p} \sinh\left(\frac{p}{c_i \eta_{gi}}\right) \right) \\
&+ \exp\left(\frac{-1}{c_i^2 \eta_{gi}^2}\right) \left(\sum_{p=1}^{\alpha} \exp\left(\frac{-p^2}{4}\right) \right. \\
&\left.\left. \cdot \frac{1}{\eta_{gi}} \cosh\left(\frac{p}{c_i \eta_{gi}}\right)\right)\right\} \Bigg] \Bigg],
\end{aligned}
$$

$$
\begin{aligned}
T(0,\eta_{gi}) = 2 \Bigg[&\frac{\alpha_i b_i}{c_i} \frac{\eta_{gi}^2}{4} + \left(\frac{\alpha_i c_i + b_i c_i - \alpha_i b_i}{c_i^2} \right) \frac{\eta_{gi}}{2\sqrt{\pi}} \\
&+ \frac{1}{2c_i} \left(1 - \frac{\alpha_i}{c_i}\right) \left(1 - \frac{b_i}{c_i}\right) \\
&- \frac{1}{c_i} \left(1 - \frac{\alpha_i}{c_i}\right) \left(1 - \frac{b_i}{c_i}\right) \frac{2}{c_i \eta_{gi} \sqrt{\pi}} \exp\left(\frac{1}{c_i^2 \eta_{gi}^2}\right) \\
&\cdot \sum_{p=1}^{\alpha} \frac{\exp(-p^2/4)}{p} \sinh\left(\frac{p}{c_i \eta_{gi}}\right) \Bigg],
\end{aligned}
$$

$$C_{21}(E, k_\perp, \eta_{g1}) = e_1 + ie_2, \quad D_{21}(E, k_\perp, \eta_{g2}) = e_3 + ie_4,$$

$$e_1 = \left[\left(\left(\sqrt{t_1^2 + t_2^2} + t_1\right)/2\right)\right]^{\frac{1}{2}},$$

$$e_2 = \left[\left(\left(\sqrt{t_1^2 + t_2^2} - t_1\right)/2\right)\right]^{\frac{1}{2}}$$

$$t_1 = \left[\frac{2m_{c1}^*}{\hbar^2} T_{11}(E, \Delta_1, \eta_{g1}, E_{g1}) - k_\perp^2\right],$$

$$t_2 = \frac{2m_{c1}^*}{\hbar^2} T_{21}(E, \Delta_1, \eta_{g1}, E_{g1})$$

$$e_3 = \left[\frac{\sqrt{t_3^2 + t_4^2} + t_3}{2}\right]^{1/2}, \quad e_4 = \left[\frac{\sqrt{t_3^2 + t_4^2} - t_3}{2}\right]^{1/2}$$

$$t_3 = \left[\frac{2m_{c2}^*}{\hbar^2} T_{12}(E, \Delta_2, \eta_{g2}, E_{g2}) - k_\perp^2\right],$$

$$t_4 = \frac{2m_{c2}^*}{\hbar^2} T_{22}(E, \Delta_2, \eta_{g2}, E_{g2}),$$

Therefore (12.19) can be expressed as

$$k_z^2 = \delta_7 + i\delta_8 \tag{12.20}$$

where

$$\delta_7 = \left[\frac{1}{L_0^2}\left(\delta_5^2 - \delta_6^2\right) - k_\perp^2\right], \quad \delta_5 = \cos^{-1} p_5,$$

$$p_5 = \left[\frac{1 - \delta_3^2 - \delta_4^2 \sqrt{\left(1 - \delta_3^2 - \delta_4^2\right)^2 + 4\delta_4^2}}{2}\right]^{1/2},$$

$$\delta_3 = (a_1 \cos\Delta_1 \cosh\Delta_2 - a_2 \cos\Delta_3 \cosh\Delta_4),$$

$$\delta_4 = (a_1 \sin\Delta_1 \sinh\Delta_2 - a_2 \sin\Delta_3 \sinh\Delta_4),$$

$$\Delta_1 = (a_0 e_1 + b_0 e_3), \quad \Delta_2 = (a_0 e_2 + b_0 e_4),$$

$$\Delta_3 = (a_0 e_1 - b_0 e_3), \quad \Delta_4 = (a_0 e_1 - b_0 e_4),$$

$$\delta_6 = \sinh^{-1} p_5 \quad \text{and} \quad \delta_8 = \left[2\delta_5\delta_6 / L_0^2\right]$$

The DR in III–V heavily doped effective mass quantum well superlattices can be written as

$$\left(\frac{n_x\pi}{d_x}\right)^2 = \delta_7 + i\delta_8 \tag{12.21a}$$

The sub-band equation in this case can be expressed as

$$\left(\frac{n_x\pi}{d_x}\right)^2 = [\delta_7 + i\delta_8]|_{k_\perp=0} \text{ and } \mathrm{E{=}E_{12.6}} \tag{12.21b}$$

where $E_{12,6}$ is the sub-band energy in this case.

12.2.7 *The DR in II–VI quantum well HD effective mass super lattices*

Following Sasaki [24], the electron dispersion law in heavily doped II-VI EMSLs can be written as

$$k_z^2 = \Delta_{13} + i\Delta_{14}, \tag{12.22}$$

where

$$\Delta_{13} = \left[\frac{1}{L_0^2}\left(\Delta_{11}^2 - \Delta_{12}^2\right) - k_s^2\right],$$

$$\Delta_5 = \cos^{-1} p_6,$$

$$p_5 = \left[\frac{1 - \Delta_9^2 - \Delta_{10}^2\sqrt{\left(1 - \Delta_9^2 - \Delta_{10}^2\right)^2 + 4\Delta_{10}^2}}{2}\right]^{1/2},$$

$$\Delta_9 = (\bar{a}_1 \cos\Delta_6 \cosh\Delta_7 - \bar{a}_2 \cos\Delta_8 \cosh\Delta_7),$$

$$\Delta_{10} = (\bar{a}_1 \sin\Delta_6 \sinh\Delta_7 + \bar{a}_2 \sin\Delta_8 \sinh\Delta_7),$$

$$\Delta_6 = [a_0 C_{22}(E, k_s\eta_{g1}) + b_0 e_3], \quad \Delta_7 = b_0 e_4,$$

$$\Delta_8 = [a_0 C_{22}(E, k_s\eta_{g1}) - b_0 e_3]$$

$$C_{22} = (E, k_s, \eta_{g1}) = \left[\frac{2m_{\parallel,1}^*}{\hbar^2}\left\{\gamma_3(E, \eta_{g1}) - \frac{\hbar^2 k_s^2}{2m_{\perp,1}^*} \mp C_0 k_s\right\}\right]^{1/2},$$

$$\bar{a}_1 = \left[\sqrt{\frac{M_2(0,\eta_{g2})}{\bar{M}_1(0,\eta_{g1})}} + 1\right]^2 \left[4\left(\frac{M_2(0,\eta_{g2})}{\bar{M}_1(0,\eta_{g1})}\right)^{1/2}\right]^{-1},$$

$$\bar{M}_1(0,\eta_{g1}) = m_{c1}^*\left(1 - \frac{2}{\pi}\right),$$

$$\bar{a}_2 = \left[\sqrt{\frac{M_2(0,\eta_{g2})}{\bar{M}_1(0,\eta_{g1})}} - 1\right]^2 \left[4\left(\frac{M_2(0,\eta_{g2})}{\bar{M}_1(0,\eta_{g1})}\right)^{1/2}\right]^{-1}$$

$$\Delta_{12} = \cos^{-1} p_6, \quad \Delta_{14} = \frac{2\Delta_{11}\Delta_{12}}{L_0^2}$$

The DR in III–V heavily doped effective mass quantum well superlattices can be written as

$$\left(\frac{n_z\pi}{d_z}\right)^2 = \delta_{13} + i\delta_{14} \tag{12.23a}$$

The sub-band equation in this case can be expressed as

$$\left(\frac{n_z\pi}{d_z}\right)^2 = [\delta_{13} + i\delta_{14}]|_{k_s=0} \text{ and } E = E_{12.7} \tag{12.23b}$$

where $E_{12,7}$ is the sub-band energy in this case.

12.2.8 *The DR in IV–VI quantum well HD effective mass super lattices*

Following Sasaki [24], the electron dispersion law in IV-VI, EMSLs can be written as

$$k_z^2 = \left[\frac{1}{L_0^2}\left\{\cos^{-1}\left(f_{23}(E,k_x,k_y)\right)\right\}^2 - k_s^2\right] \tag{12.24}$$

where

$$f_{23}(E,k_x,k_y) = a_3 \cos\left[a_0 C_{23}(E,k_x,k_y\eta_{g1}) + b_0 D_{23}(E,k_x,k_y\eta_{g1})\right] - a_4\cos[a_0 C_{23}(E,k_x,k_y,\eta_{g2}) - b_0 D_{23}(E,k_x,k_y\eta_{g2})],$$

$$a_3 = \left[\sqrt{\frac{M_3(0,\eta_{g2})}{M_3(0,\eta_{g1})}} + 1\right]^2 \left[4\left(\frac{M_3(0,\eta_{g2})}{M_3(0,\eta_{g1})}\right)^{1/2}\right]^{-1},$$

$$a_3 = \left[\sqrt{\frac{M_3(0,\eta_{g2})}{M_3(0,\eta_{g1})}} - 1\right]^2 \left[4\left(\frac{M_3(0,\eta_{g2})}{M_3(0,\eta_{g1})}\right)^{1/2}\right]^{-1}$$

$$M_3(0,\eta_{gi}) = (4\bar{p}_{9,1})^{-1}\left[\left\{\alpha_i\left(1-\frac{2}{\pi}\right)\left(\frac{1}{m_{l,i}^+}-\frac{1}{m_{l,i}^-}\right)\right\}\right.$$
$$+\left[[\bar{q}_{9,i}(0,\eta_{gi})]^2 + (4\bar{p}_{9,i})\bar{R}_{9,i}(0,\eta_{gi})\right]^{-1/2}$$
$$\cdot\left[\alpha_i\left(1-\frac{2}{\pi}\right)\left(\frac{1}{m_{l,i}^+}-\frac{1}{m_{l,i}^-}\right)\bar{q}_{9,i}(0,\eta_{gi})\right.$$
$$\left.\left.+\ 2\bar{p}_{9,i}\left(1-\frac{2}{\pi}+\frac{\alpha_i\eta_{gi}}{\sqrt{\pi}}\right)\right]\right],$$

$$\bar{p}_{9,i} = \frac{\alpha_i\hbar^4}{4m_{l,i}^+m_{l,i}^-},$$

$$\bar{q}_{9,i}(0,\eta_{gi}) = \left[\frac{\hbar^2}{2}\left(\frac{1}{m_{l,i}^+}+\frac{1}{m_{l,i}^-}\right) - \frac{\alpha_i\eta_{gi}}{\sqrt{\pi}}\left(\frac{1}{m_{l,i}^+}-\frac{1}{m_{l,i}^-}\right)\right],$$

$$\bar{\mathrm{R}}_{9,1}(0,\eta_{gi}) = \left[\frac{\eta_{gi}}{\sqrt{\pi}}+\frac{\alpha_i\eta_{gi}^2}{2}\right],$$

$$C_{23}(E,k_x,k_y,\eta_{g1}) = [[2\bar{p}_{9,1}]^{-1}[-\bar{q}_{9,1}(E,k_x,k_y\eta_{g1})$$
$$+[\{\bar{q}_{9,1}(E,k_x,k_y,\eta_{g1})\}^2$$
$$+\ (4\bar{p}_{9,1})\bar{R}_{9,1}(E,k_x,k_y,\eta_{g1})]^{1/2}]]^{1/2},$$

$$D_{23}(E,k_x,k_y,\eta_{g2}) = [[2\bar{p}_{9,2}]^{-1}[-\bar{q}_{9,2}(E,k_x,k_y\eta_{g2})$$
$$+[\{\bar{q}_{9,2}(E,k_x,k_y,\eta_{g2})\}^2$$
$$+\ (4\bar{p}_{9,2})\bar{R}_{9,2}(E,k_x,k_y,\eta_{g2})]^{1/2}]]^{1/2},$$

$$\bar{q}_{9,i}(E,k_x,k_y,\eta_{gi}) = \left[\frac{\hbar^2}{2}\left(\frac{1}{m_{l,i}^+}+\frac{1}{m_{l,i}^-}\right)\right.$$
$$+\alpha_i\frac{\hbar^4}{4}k_s^2\left(\frac{1}{m_{l,i}^+m_{l,i}^-}+\frac{1}{m_{l,i}^+m_{l,i}^-}\right)$$

$$- \alpha_i \gamma_3 (E, \eta_{gi}) \left(\frac{1}{m_{l,i}^+} - \frac{1}{m_{l,i}^-} \right) \Bigg],$$

$$\bar{R}_{9,i}(E, k_x, k_y, \eta_{gi}) = \left[\gamma_2 (E, \eta_{gi}) + \gamma_3 (E, \eta_{gi}) \alpha \frac{\hbar^2}{2} k_s^2 \left(\frac{1}{m_{l,i}^+} - \frac{1}{m_{l,i}^-} \right) \right.$$

$$\left. - \frac{\hbar^2}{2} k_s^2 \left(\frac{1}{m_{l,i}^+} - \frac{1}{m_{l,i}^-} \right) - \frac{\alpha \hbar^6}{4} \frac{k_s^4}{m_{t,i}^- m_{t,i}^+} \right],$$

$$a_5 = \left[\sqrt{\frac{m_2^*}{m_1^*}} + 1 \right]^2 \left[4 \left(\frac{m_2^*}{m_1^*} \right)^{1/2} \right]^{-1}$$

Therefore the DR in heavily doped IV-VI, quantum well EMSLs can be written as

$$\left(\frac{n_z \pi}{d_z} \right)^2 = \left[\frac{1}{L_0^2} \left\{ \cos^{-1} \left(f_{23} (E, k_x, k_y) \right) \right\}^2 - k_s^2 \right] \tag{12.25a}$$

The sub-band equation in this case can be expressed as

$$\left(\frac{n_z \pi}{d_z} \right)^2 = \frac{1}{L_0^2} \left[\cos^{-1} \left(f_{23} (E_{12,8}, 0, 0) \right)^2 \right] \tag{12.25b}$$

where $E_{12,8}$ is the sub-band energy in this case.

12.2.9 *The DR in HgTe/CdTe quantum well HD effective mass super lattices*

Following Sasaki [24], the DR in heavily doped HgTe/CdTeEMSLs can be written as

$$k_z^2 = \Delta_{13H} + i\Delta_{14H}, \tag{12.26}$$

where

$$\Delta_{13H} = \left[\frac{1}{L_0^2} \left(\Delta_{11H}^2 - \Delta_{12H}^2 \right) - k_s^2 \right]$$

$$\Delta_{11H} = \cos^{-1} p_{6H},$$

$$p_{6H} = \left[\frac{1 - \Delta_{9H}^2 - \Delta_{10H}^2 - \sqrt{\left(1 - \Delta_{9H}^2 - \Delta_{10H}^2\right)^2 + 4\Delta_{10H}^2}}{2} \right]^{1/2},$$

$$\Delta_{9H} = (\bar{a}_{1H} \cos \Delta_{5H} \cosh \Delta_{6H} - \bar{a}_{2H} \cos \Delta_{7H} \cosh \Delta_{6H}),$$

$$\Delta_{10H} = (\bar{a}_{1H} \sin \Delta_{5H} \sinh \Delta_{6H} + \bar{a}_{2H} \sin \Delta_{7H} \sinh \Delta_{6H}),$$

$$\Delta_{5H} = [a_0 C_{22H}(E, k_s, \eta_{g1}) + b_0 e_3], \quad \Delta_{6H} = b_0 e_4,$$

$$\Delta_{7H} = [a_0 C_{22H}(E, k_s, \eta_{g1}) - b_0 e_3]$$

$$C_{22H}(E, k_s, \eta_{g1}) = \left[\frac{B_{01}^2 + 2A_1 E - B_{01}(B_{01}^2 + 4A_1 E)}{2A_1^2}\right]^{1/2},$$

$$\bar{a}_{1H} = \left[\sqrt{\frac{M_2(0, \eta_{g2})}{m_{c1}^*}} + 1\right]^2, \quad \left[4\left(\frac{M_2(0, \eta_{g2})}{m_{c1}^*}\right)^{1/2}\right]^{-1},$$

$$\bar{a}_{2H} = \left[\sqrt{\frac{M_2(0, \eta_{g2})}{m_{c1}^*}} + 1\right]^2, \quad \left[4\left(\frac{M_2(0, \eta_{g2})}{m_{c1}^*}\right)^{1/2}\right]^{-1}$$

$$\Delta_{12H} = \cos^{-1} p_{6H}, \quad \Delta_{14H} = \frac{2\Delta_{11H}\Delta_{12H}}{L_0^2}$$

The DR in heavily doped HgTe/CdTe QWEMSLs can be written as

$$\left(\frac{n_z \pi}{d_z}\right)^2 = \Delta_{13H} + i\Delta_{14H} \tag{12.27a}$$

The sub-band equation in this case can be expressed as

$$\left(\frac{n_z \pi}{d_z}\right)^2 = [\Delta_{13H} + i\Delta_{14H}]|_{k_s=0} \text{ and } \mathrm{E}=\mathrm{E}_{12,9} \tag{12.27b}$$

where $E_{12,9}$ is the sub-band energy in this case.

12.2.10 *The DR in strained layer quantum well HD effective mass super lattices*

The DR of the constituent materials of heavily doped III–V super lattices can be written as

$$\overline{P}_i(E, \eta_{gi})k_x^2 + \overline{Q}_i(E, \eta_{gi})k_y^2 + \overline{S}_i(E, \eta_{gi})k_z^2 = 1 \tag{12.28}$$

where

$$\overline{P}_i(E,\eta_{gi}) = (\gamma_0(E,\eta_{gi}) - I_0T_{1i})(\overline{\Delta}_i(E,\eta_{gi}))^{-1},$$

$$\mathrm{I}_0 = (1/2)[1 + Erf(E/\eta_{gi})],$$

$$T_{li} = [E_{gi} - C^c_{li}\varepsilon_i - (a_i + C^c_{li})\varepsilon_i(3/2)b_i\varepsilon_{xxi} - (b_i\varepsilon_i/2) + (\sqrt{3d_i\varepsilon_{xyi}}/2)],$$

$$\Delta_i(E,\eta_{gi}) = [(-q_i\eta^3_{gi}/2\sqrt{\pi})\exp(-(E^2/\eta^2_{gi}))[1 + (E^2/\eta^2_{gi})] - R_i\theta_i(E,\eta_{gi}) + V_i\gamma_0(E,\eta_{gi}) + (\zeta_i/2)[1 + Erf(E/\eta_{gi})]], \quad q_i = (3/2B^2_{2i}),$$

$$R_i = q_i[2A_i + C^c_{li}\varepsilon_i], \quad A_i = E_{gi} - C^c_{li}\varepsilon_i,$$

$$V_i = q_i\left[A_i^2 - \left(2C^2_{2i}\varepsilon_{xyi}/3\right) + 2A_iC^c_{li}\varepsilon_i\right],$$

$$\zeta_i = q_i\left[\left(2C^2_{2i}\varepsilon_{xyi}/3\right) - C^c_{li}\varepsilon_iA_i^2\right],$$

$$\bar{Q}_i\left(E,\eta_{gi}\right) = \left(\gamma_0\left(E,\eta_{gi}\right) - I_0T_{2i}\right)\left(\bar{\Delta}_i\left(E,\eta_{gi}\right)\right)^{-1},$$

$$T_{2i} = [E_{gi} - C^c_{li}\varepsilon_i - (a_i + C^c_{li})\varepsilon_i + (3/2)b_i\varepsilon_{xyi} - (b_i\varepsilon_i/2) - (\sqrt{3d_i\varepsilon_{xyi}}/2)],$$

$$\bar{S}_i\left(E,\eta_{gi}\right) = \left(\gamma_0\left(E,\eta_{gi}\right) - I_0T_{3i}\right)\left(\bar{\Delta}_i\left(E,\eta_{gi}\right)\right)^{-1}$$

$$T_{3i} = \left[E_{gi} - C^c_{li}\varepsilon_i + \left(a_i + C^c_{li}\right)\varepsilon_i + (3/2)\,b_i\varepsilon_{zzi} - (b_i\varepsilon_i/2)\right],$$

The electron energy spectrum in heavily doped strained layer effective mass super-lattices can be written as

$$k_z^2 = \left[\frac{1}{L_0^2}\left\{\cos^{-1}\left(f_{40}\left(E,k_x,k_y\right)\right)\right\}^2 - k_s^2\right] \tag{12.29}$$

where

$$f_{40}\left(E,k_x,k_y\right) = a_{20}\cos\left[a_0C_{40}\left(E,k_x,k_y\eta_{g1}\right) + b_0D_{40}\left(E,k_x,k_y\eta_{g1}\right)\right] - a_{21}\cos[a_0C_{40}(E,k_x,k_y,\eta_{g2}) - b_0D_{40}(E,k_x,k_y\eta_{g2})],$$

$$a_{20} = \left[\sqrt{\frac{M_{s2}\left(0,\eta_{g2}\right)}{M_{s1}\left(0,\eta_{g1}\right)}} + 1\right]^2\left[4\left(\frac{M_{s2}\left(0,\eta_{g2}\right)}{M_{s1}\left(0,\eta_{g1}\right)}\right)^{1/2}\right]^{-1}$$

$$M_{si}\left(0,\eta_{gi}\right)=\left(\hbar/2\right)\rho_i\left(\eta_{gi}\right)$$

$$\begin{aligned}\rho_i(\eta_{gi}) &= [(\eta_{gi}/2\sqrt{\pi})-(T_{3i}/2)]^{-2}\\ &\quad\times[\{(\eta_{gi}/2\sqrt{\pi})-(T_{3i}/2)\}\{(V_i/2)-(R_i\eta_{gi}/\sqrt{\pi})\\ &\quad+(\zeta_i/\eta_{gi}\sqrt{\pi})\}-((1/2)-(T_{3i}/\eta_{gi}\sqrt{\pi}))\{(\zeta_i/2)\\ &\quad+(V_i\eta_{gi}/2\sqrt{\pi})-(R_i\eta_{gi}^2/4)-(q_i\eta_{gi}^3/2\sqrt{\pi})\}]\end{aligned}$$

$$a_{20}=\left[\sqrt{\frac{M_{s2}(0,\eta_{g2})}{M_{s1}(0,\eta_{g1})}}+1\right]^2,\quad\left[4\left(\frac{M_{s2}(0,\eta_{g2})}{M_{s1}(0,\eta_{g1})}\right)^{1/2}\right]^{-1},$$

$$a_{21}=\left[\sqrt{\frac{M_{s2}(0,\eta_{g2})}{M_{s1}(0,\eta_{g1})}}+1\right]^2,\quad\left[4\left(\frac{M_{s2}(0,\eta_{g2})}{M_{s1}(0,\eta_{g1})}\right)^{1/2}\right]^{-1}$$

$$\begin{aligned}C_{40}(E,k_x,k_y,\eta_{g1}) &= [1-\bar{P}_1(E,\eta_{g1})k_x^2-\bar{Q}_1(E,\eta_{g1})k_y^2]^{1/2}\\ &\quad\cdot[\bar{S}_1(E,\eta_{g1})]^{-1/2}\end{aligned}$$

$$\begin{aligned}D_{40}(E,k_x,k_y,\eta_{g2}) &= [1-\bar{P}_2(E,\eta_{g2})k_x^2-\bar{Q}_2(E,\eta_{g2})k_y^2]^{1/2}\\ &\quad\cdot[\bar{S}_2(E,\eta_{g2})]^{-1/2}\end{aligned}$$

Therefore, the DR in heavily doped strained layer effective mass quantum well super-lattices can be expressed as

$$\left(\frac{n_z\pi}{d_z}\right)^2=\left[\frac{1}{L_0^2}\{\cos^{-1}(f_{40}(E,k_x,k_y))\}^2-k_s^2\right]\tag{12.30a}$$

The sub-band equation in this case can be expressed as

$$\left(\frac{n_z\pi}{d_z}\right)^2=\frac{1}{L_0^2}[\cos^{-1}(f_{40}(E_{12,10},0,0)\}^2]\tag{12.30b}$$

where $E_{12,10}$ is the sub-band energy in this case.

The DOS, EM and the electron statistics should be calculated numerically.

12.3 Open Research Problem

(R.12.1) investigate the influence of arbitrarily oriented alternating quantizing magnetic field and strain on the EM for all types of HD super-lattices whose carrier energy spectra are described in this book.

References

1. S. Mukherjee, S. N. Mitra, P. K. Bose, A. R. Ghatak, A. Neoigi, J. P. Banerjee, A. Sinha, M. Pal, S. Bhattacharya, K. P. Ghatak, J. Compu. Theor Nanosc, **4**, 550 (2007); N. G. Anderson, W. D. Laidig, R. M. Kolbas, Y. C. Lo, J. Appl. Phys. **60**, (2361) (1986); N. Paitya, K. P. Ghatak, Jour. Adv. Phys. **1**, (161) (2012); N. Paitya, S. Bhattacharya, D. De, K. P. Ghatak Adv. Sci. Engg. Medi. **4** , (96) (2012); S. Bhattacharya, D. De, S. M. Adhikari, K. P. Ghatak Superlatt. Microst. **51**, (203) (2012); D. De, S. Bhattacharya, S. M. Adhikari, A. Kumar, P. K. Bose, K. P. Ghatak, Beilstein Jour. Nanotech. **2** , (339) (2012); D. De, A. Kumar, S. M. Adhikari, S. Pahari, N. Islam, P. Banerjee, S. K. Biswas, S. Bhattacharya, K. P. Ghatak, Superlatt. and Microstruct. 47, (377) (2010); S. Pahari, S. Bhattacharya, S. Roy, A. Saha, D. De, K. P. Ghatak, Superlatt. and Microstruct. **46** , (760) (2009); S. Pahari, S. Bhattacharya, K. P. Ghatak Jour. of Comput. and Theo. Nanosci. **6**, (2088) (2009); S. K. Biswas, A. R. Ghatak, A. Neogi, A. Sharma, S. Bhattacharya, K. P. Ghatak , Phys. E: Low-dimen. Sys. and Nanostruct. **36**, (163) (2007); L. J. Singh, S. Choudhury, D. Baruah, S. K. Biswas, S. Pahari, K. P. Ghatak, Phys. B: Conden. Matter, **368**, (188) (2005); S. Chowdhary, L. J. Singh, K. P. Ghatak, Phys. B: Conden. Matter, **365**, (5) (2005); L. J. Singh, S. Choudhary, A. Mallik, K. P. Ghatak Jour. of Comput. and Theo. Nanosci. **2**, (287) (2005); K. P. Ghatak, J Mukhopadhyay JP Banerjee, SPIE proceedings series, **4746** (1292) (2002); K P Ghatak, S Dutta, DK Basu, B Nag, Il Nuovo Cimento. **D 20**, (227) (1998); K. P. Ghatak, D. K. Basu, B. Nag, Jour. of Phys. and Chem. of Solids. **58**, (133) (1997); K. P. Ghatak, B. De, Mat. Resc. Soc. Proc. **300**, (513) (1993); K. P. Ghatak, B. Mitra, Il Nuovo Cimento. **D 15**, (97) (1993); K. P. Ghatak, Inter. Soci.Opt. and Photon. Proc. Soc. Photo Opt. Instru. Engg., **1626**, (115) (1992); K. P. Ghatak, A. Ghoshal, Phys. Stat. Sol. (b) **170**, (K27) (1992); K. P. Ghatak, S. Bhattacharya, S. N. Biswas, Proc. Soc. Photo opt. instru. Engg., **836**, (72) (1988); K. P. Ghatak, A Ghoshal, S. N. Biswas, M. Mondal, Proc. Soc. Photo Opt. Instru. Engg. **1308**, (356) (1990); K. P. Ghatak, B. De, Proc. Wide bandgap semi. Symp., Matt. Res. Soc. (377) (1992); K. P. Ghatak, B. De, Defect Engg. Semi. Growth, Processing and Device Tech. Symp., Mat. Res. Soc. **262**, (911) (1992); S. N. Biswas, K. P. Ghatak, Internat. Jour. Electronics Theo. Exp. **70**, (125) (1991); B. Mitra, K. P. Ghatak, Phys. Lett. A. **146**, (357) (1990); B. Mitra, K. P. Ghatak, Phys. Lett. A. **142**, (401) (1989); K. P. Ghatak, B Mitra, A Ghoshal, Phy. Stat. Sol. (b) **154**, (K121) (1989); B. Mitra, K. P. Ghatak, Phys. Stat. Sol. (b). **149**, (K117) (1988); K. P. Ghatak, S. N. Biswas, Proc. Soc. Photo Optical Instru. Engg. **792**,(239) (1987); S. Bhattacharyya, K. P. Ghatak, S. Biswas, OE/Fibers' 87,Inter. Soc. Opt. Photon. (73) (1987);

M. Mondal, K. P. Ghatak, Czech. Jour. Phys. B. **36**, (1389) (1986); K. P. Ghatak, A. N. Chakravarti, Phys. Stat. Sol. (b). **117**, (707) (1983)

2. L. V. Keldysh, *Sov. Phys. Solid State* **4**, 1658 (1962); L. Esaki, R. Tsu, *IBM J. Research and Develop.* **14**, 61 (1970); G. Bastard, *Wave mechanics applied to heterostructures*, (Editions de Physique, Les Ulis, France, 1990); E. L. Ivchenko, G. Pikus, *Superlattices and other heterostructures*, (Springer-Berlin, 1995); R. Tsu, *Superlattices to nanoelectronics*, (Elsevier, The Netherlands, 2005)
3. P. Fürjes, Cs. Dücs, M. Ádám, J. Zettner, I. Bársony, *Superlattices and Microstructures* **35**, 455 (2004); T. Borca-Tasciuc, D. Achimov, W. L. Liu, G. Chen, H. W. Ren, C. H. Lin, S. S. Pei, *Microscale Thermophysical Engg.* **5**, 225 (2001)
4. B. S. Williams, *Nat. Photonics* **1**, 517 (2007); A. Kosterev, G. Wysocki, Y. Bakhirkin, S. So, R. Lewicki, F. Tittel, R. F. Curl, *Appl. Phys. B* **90**, 165 (2008); M. A. Belkin, F. Capasso, F. Xie, A. Belyanin, M. Fischer, A. Wittmann, J. Faist, *Appl. Phys. Lett.* **92**, 201101 (2008)
5. G. J. Brown, F. Szmulowicz, R. Linville, A. Saxler, K. Mahalingam, C.-H. Lin, C. H. Kuo, W. Y. Hwang, *IEEE Photonics Technology Letts.* **12**, 684 (2000); H. J. Haugan, G. J. Brown, L. Grazulis, K. Mahalingam, D. H. Tomich, *Physics E: Low-dimensional Systems and Nanostructures* **20**, 527 (2004)
6. S. A. Nikishin, V. V. Kuryatkov, A. Chandolu, B. A. Borisov, G. D. Kipshidze, I. Ahmad, M. Holtz, H. Temkin, *Jpn. J. Appl. Phys.* **42**, L1362 (2003); Y. K. Su, H. C. Wang, C. L. Lin, W. B. Chen, S. M. Chen, *Jpn. J. Appl. Phys.* **42**, L751 (2003); C. H. Liu, Y. K. Su, L. W. Wu, S. J. Chang, R. W. Chuang, *Semicond. Sci. Technol.* **18**, 545 (2003); S. B. Che, I. Nomura, A. Kikuchi, K. Shimomura, K. Kishino, *Phys. Stat. Sol. (b)* **229**, 1001 (2002)
7. C. P. Endres, F. Lewen, T. F. Giesen, S. SchlEEMer, D. G. Paveliev, Y. I. Koschurinov, V. M. Ustinov, A. E. Zhucov, *Rev. Sci. Instrum.* **78**, 043106 (2007)
8. F. Klappenberger, K. F. Renk, P. Renk, B. Rieder, Y. I. Koshurinov, D. G. Pavelev, V. Ustinov, A. Zhukov, N. Maleev, A. Vasilyev, *Appl. Phys. Letts.* **84**, 3924 (2004)
9. X. Jin, Y. Maeda, T. Saka, M. Tanioku, S. Fuchi, T. Ujihara, Y. Takeda, N. Yamamoto, Y. Nakagawa, A. Mano, S. Okumi, M. Yamamoto, T. Nakanishi, H. Horinaka, T. Kato, T. Yasue, T. Koshikawa, *J. of Crystal Growth* **310**, 5039 (2008); X. Jin, N. Yamamoto, Y. Nakagawa, A. Mano, T. Kato, M. Tanioku, T. Ujihara, Y. Takeda, S. Okumi, M. Yamamoto, T. Nakanishi, T. Saka, H. Horinaka, T. Kato, T. Yasue, T. Koshikawa, *Appl. Phys. Express* **1**, 045002 (2008)
10. B. H. Lee, K. H. Lee, S. Im, M. M. Sung, *Organic Electronics* **9**, 1146 (2008)
11. P. H. Wu, Y. K. Su, I. L. Chen, C. H. Chiou, J. T. Hsu, W. R. Chen, *Jpn. J. Appl. Phys.* **45**, L647 (2006); A. C. Varonides, *Renewable Energy* **33**, 273 (2008)
12. M. Walther, G. Weimann, *Phys. Stat. Sol. (b)* **203**, 3545 (2006)

13. R. Rehm, M. Walther, J. Schmitz, J. Fleißner, F. Fuchs, J. Ziegler, W. Cabanski, *Opto-Electronics Rev.* **14**, 19 (2006); R. Rehm, M. Walther, J. Scmitz, J. Fleissner, J. Ziegler, W. Cabanski, R. Breiter, *Electronics Letts.* **42**, 577 (2006)
14. G. J. Brown, F. Szmulowicz, H. Haugan, K. Mahalingam, S. Houston, *Microelectronics Jour.* **36**, 256 (2005)
15. K. V. Vaidyanathan, R. A. Jullens, C. L. Anderson, H. L. Dunlap, *Solid State Electron.* **26**, 717 (1983)
16. B. A. Wilson, IEEE, *J. Quantum Electron.* **24**, 1763 (1988)
17. M. Krichbaum, P. Kocevar, H. Pascher, G. Bauer, *IEEE, J. Quantum Electron.* **24**, 717 (1988)
18. J. N. Schulman, T. C. McGill, *Appl. Phys. Letts.* **34**, 663 (1979)
19. H. Kinoshita, T. Sakashita, H. Fajiyasu, *J. Appl. Phys.* **52**, 2869 (1981)
20. L. Ghenin, R. G. Mani, J. R. Anderson, J. T. Cheung, *Phys. Rev. B* **39**, 1419 (1989)
21. C. A. Hoffman, J. R. Mayer, F. J. Bartoli, J. W. Han, J. W. Cook, J. F. Schetzina, J. M. Schubman, *Phys. Rev. B.* **39**, 5208 (1989)
22. V. A. Yakovlev, *Sov. Phys. Semicon.* **13**, 692 (1979)
23. E. O. Kane., *J. Phys. Chem. Solids* **1**, 249 (1957)
24. H. X. Jiang, J. Y. Lin, *J. Appl. Phys.* **61**, 624 (1987)
25. H. Sasaki, *Phys. Rev. B* **30**, 7016 (1984)

Chapter 13

Appendix E: The EM in Quantum Wire HDSLs

13.1 Introduction

In this chapter, the EM from III-V, II-VI, IV-VI, HgTe/CdTe and strained layer quantum wire heavily doped superlattices (QWHDSLs) with graded interfaces has been studied in Sections 13.2.1 *to* 13.2.5. *From Sections* 13.2.6 *to* 13.2.10, *the EM from III-V, II-VI, IV-VI, HgTe/CdTe and strained layer quantum wire heavily doped effective mass superlattices respectively has been presented. Lastly Section* 13.3 *presents single open research problem.*

13.2 Theoretical Background

13.2.1 *The EM in III-V quantum wire HD superlattices with graded interfaces*

The simplified DR of heavily doped quantum wire III-V super-lattices with graded interfaces can be expressed as [1,2]

$$k_z^2 = [G_8 + iH_8]|_{k_z - \frac{n_z\pi}{d_x} \text{ and } k_y - \frac{n_z\pi}{d_y}} \tag{13.1a}$$

The DOS function can be written as

$$N(E) = \frac{g_v}{\pi}\sum_{n_x=1}^{n_{x\max}}\sum_{n_y=1}^{n_{y\max}} \frac{H(E - E_{13,1})[G_8' + iH_8']}{\sqrt{G_8 + iH_8}} \tag{13.1b}$$

where $E_{13,1}$ is the sub-band energy and the sub-band equation in this case can be expressed as

$$0 = [G_8 + iH_8]|_{k_x} = \frac{n_x \pi}{d_x}, k_y - \frac{n_z \pi}{d_y} \quad \text{and} \quad E = E_{13,1} \qquad (13.1c)$$

The EEM in this case is given by

$$m^*(E, \eta_g, n_x, n_y) = \frac{\hbar^2}{2} G_8' \qquad (13.2)$$

13.2.2 *The DR in II–VI quantum wire HD superlattices with graded interfaces*

The simplified DR of heavily doped quantum wire III-V super-lattices with graded interfaces can be expressed as

$$k_z^2 = [G_{19} + iH_{19}]|k_x - \frac{n_x \pi}{d_x} \text{ and } k_y - \frac{n_y \pi}{d_y} \qquad (13.3a)$$

The DOS function can be written as

$$N(E) = \frac{g_v}{\pi} \sum_{n_x=1}^{n_{x\max}} \sum_{n_y=1}^{n_{y\max}} \frac{H(E - E_{13,2})[G_{19}' + iH_{19}']}{\sqrt{G_{19} + iH_{19}}} \qquad (13.3b)$$

where $E_{13,2}$ is the sub-band energy and the sub-band equation in this case can be expressed as

$$0 = [G_{19} + iH_{19}]|k_x = \frac{n_x \pi}{d_x}, k_y - \frac{n_y \pi}{d_y} \text{ and } E = E_{13,2} \qquad (13.3c)$$

The EEM in this case is given by

$$m^*(E, \eta_g, n_x, n_y) = \frac{\hbar^2}{2} G_{19}' \qquad (13.4)$$

13.2.3 *The DR in IV–VI quantum wire HD superlattices with graded interfaces*

The simplified DR in heavily doped quantum wire IV–VI superlattices with graded interfaces can be expressed as

$$k_z^2 = \left[\frac{1}{L_0^2}\left[\cos^{-1}\left\{\frac{1}{2}\Phi_2(E,k_s)\right\}\right]^2 - k_s^2\right]\Bigg|_{k_x=\frac{n_x\pi}{d_x}\ \text{and}\ k_y-\frac{n_y\pi}{d_y}} \tag{13.5a}$$

The DOS function can be written as

$$N(E) = \frac{g_v}{\pi L_0}\sum_{n_x=1}^{n_{x\max}}\sum_{n_y=1}^{n_{y\max}} \times \frac{\cos^{-1}\left\{\frac{1}{2}\Phi_2(E,k_s)\right\}\Phi_2(E,k_s)\Phi_2'(E,k_s)H(E-E_{13,3})}{\left(\sqrt{\cos^{-1}\left\{\frac{1}{2}\Phi_2(E,k_s)\right\}^2 - L_0^2\left\{\left(\frac{n_x\pi}{d_x}\right)^2+\left(\frac{n_x\pi}{d_x}\right)^2\right\}}\right) \times\sqrt{1-\frac{1}{4}\Phi_2^2(E,k_s)}} \tag{13.5b}$$

where $E_{13,3}$ is the sub-band energy and the sub-band equation in this case can be expressed as

$$0 = \left[\frac{1}{L_0^2}\left[\cos^{-1}\left\{\frac{1}{2}\Phi_2(E_{13,3},k_s)\right\}\right]^2 - k_s^2\right]|k_x = \frac{n_x\pi}{d_x}\ \text{and}\ k_y - \frac{n_y\pi}{d_y} \tag{13.5c}$$

The EEM in this case is given by

$$m^*(E,\eta_g,n_x,n_y) = \frac{\hbar^2}{2L_0^2}\cos^{-1}\left[\frac{1}{2}\Phi_2(E,k_s)\right]\Phi_2'(E,k_s)\left[1-\frac{1}{4}\Phi_2^2(E,k_s)\right]^{-1/2} \tag{13.6}$$

13.2.4 *The DR in HgTe/CdTe quantum wire HD superlattices with graded interfaces*

The simplified DR in heavily doped quantum wire HgTe/CdTe superllatices with graded interfaces can be expressed as

$$k_z^2 = [G_{192} + iH_{192}]|k_x - \frac{n_x\pi}{d_x} \text{ and } k_y - \frac{n_y\pi}{d_y} \tag{13.7a}$$

The DOS function can be written as

$$N(E) = \frac{g_v}{\pi} \sum_{n_x=1}^{n_{x\max}} \sum_{n_y=1}^{n_{y\max}} \frac{H(E - E_{13,4})[G'_{192} + iH'_{192}]}{\sqrt{G_{192} + iH_{192}}} \tag{13.7b}$$

where $E_{13,4}$ is the sub-band energy and the sub-band equation in this case can be expressed as

$$0 = [G_{192} + iH_{192}]|k_x = \frac{n_x\pi}{d_x}, k_y = \frac{n_y\pi}{d_y} \text{ and } E = E_{13,4} \tag{13.7c}$$

The EEM in this case is given by

$$m^*(E, \eta_g, n_x, n_y) = \frac{\hbar^2}{2} G'_{192} \tag{13.8}$$

13.2.5 *The DR in strained layer quantum wire HD superlattices with graded interfaces*

Therefore the DR of the conduction electrons in heavily doped strained layer quantum well SL with graded interfaces can be expressed as

$$k_z^2 = \left[\frac{1}{L_0^2}\left[\cos^{-1}\left\{\frac{1}{2}\Phi_6(E, k_s)\right\}\right]^2 - k_s^2\right]\Bigg| k_x = \frac{n_x\pi}{d_x} \text{ and } k_y - \frac{n_y\pi}{d_y} \tag{13.9a}$$

The DOS function can be written as

$$N(E) = \frac{g_v}{\pi L_0} \sum_{n_x=1}^{n_{x\max}} \sum_{n_y=1}^{n_{y\max}}$$

$$\times \frac{\cos^{-1}\left\{\frac{1}{2}\bar{\Phi}_6(E,k_s)\right\}\bar{\Phi}_6(E,k_s)[\bar{\Phi}_6(E,k_s)]'H(E-E_{13,5})}{\left(\sqrt{\cos^{-1}\left\{\frac{1}{2}\bar{\Phi}_6(E,k_s)\right\}^2 - L_0^2\left\{\left(\frac{n_x\pi}{d_x}\right)^2+\left(\frac{n_x\pi}{d_x}\right)^2\right\}}\right) \times \sqrt{1-\frac{1}{4}[\bar{\Phi}_6(E,k_s)]^2}} \tag{13.9b}$$

where $E_{13,5}$ is the sub-band energy and the sub-band equation in this case can be expressed as

$$0 = \left[\frac{1}{L_0^2}\left[\cos^{-1}\left\{\frac{1}{2}\Phi_6(E_{13,5},k_s)\right\}\right]^2 - k_s^2\right]\Bigg| k_x = \frac{n_x\pi}{d_x} \text{ and } k_y - \frac{n_y\pi}{d_y} \tag{13.9c}$$

The EEM in this case is given by

$$m^*(E,\eta_g,n_x,n_y) = \frac{\hbar^2}{2L_0^2}\cos^{-1}\left[\frac{1}{2}\bar{\Phi}_6(E,k_s)\right][\bar{\Phi}_6(E,k_s)]'\left[1-\frac{1}{4}[\bar{\Phi}_6(E,k_s)]^2\right]^{-1/2} \tag{13.10}$$

13.2.6 *The DR in III–V quantum wire HD effective mass super lattices*

The DR in III-V heavily doped effective mass quantum wire superlattices can be written as

$$k_z^2 = [\delta_7 + i\delta_8]|k_x - \frac{n_x\pi}{d_x} \text{ and } k_y - \frac{n_y\pi}{d_y} \tag{13.11a}$$

The DOS function can be written as

$$N(E) = \frac{g_v}{\pi}\sum_{n_x=1}^{n_{x\max}}\sum_{n_y=1}^{n_{y\max}}\frac{H(E-E_{13,6})[\delta_7' + i\delta_8']}{\sqrt{\delta_7 + i\delta_8}} \tag{13.11b}$$

where $E_{13,6}$ is the sub-band energy and the sub-band equation in this case can be expressed as

$$0 = [\delta_7 + i\delta_8]|k_x - \frac{n_x \pi}{d_x}, k_y - \frac{n_y \pi}{d_y} \text{ and } E = E_{13,6} \tag{13.11c}$$

The EEM in this case is given by

$$m^*(E, \eta_g, n_x, n_y) = \frac{\hbar^2}{2} \delta'_7 \tag{13.12}$$

13.2.7 *The DR in II–VI quantum wire HD effective mass super lattices*

The DR in III–V heavily doped effective mass quantum wire super-lattices can be written as

$$k_z^2 = [\delta_{13} + i\delta_{14}]|k_x - \frac{n_x \pi}{d_x} \text{ and } k_y - \frac{n_y \pi}{d_y} \tag{13.13a}$$

The DOS function can be written as

$$N(E) = \frac{g_v}{\pi} \sum_{n_x=1}^{n_{x\max}} \sum_{n_y=1}^{n_{y\max}} \frac{H(E - E_{13,6})[\delta'_{13} + i\delta'_{14}]}{\sqrt{\delta_{13} + i\delta_{14}}} \tag{13.13b}$$

where $E_{13,7}$ is the sub-band energy and the sub-band equation in this case can be expressed as

$$0 = [\delta_{13} + i\delta_{14}]|k_x - \frac{n_x \pi}{d_x}, k_y - \frac{n_y \pi}{d_y} \text{ and } E = E_{13,7} \tag{13.13c}$$

The EEM in this case is given by

$$m^*(E, \eta_g, n_x, n_y) = \frac{\hbar^2}{2} \delta'_{13} \tag{13.14}$$

13.2.8 *The DR in IV–VI quantum wire HD effective mass super lattices*

Therefore the DR in heavily doped IV-VI, quantum wire EMSLs can be written as

$$k_z^2 = \left[\frac{1}{L_0^2}\{\cos^{-1}(f_{23}(E, k_x, k_y))\}^2 - k_s^2\right] \Bigg| k_x = \frac{n_x \pi}{d_x} \text{ and } k_y - \frac{n_y \pi}{d_y} \tag{13.15a}$$

The DOS function can be written as

$$N(E) = \frac{g_v}{\pi L_0} \sum_{n_x=1}^{n_{x\max}} \sum_{n_y=1}^{n_{y\max}}$$

$$\times \frac{\cos^{-1}\left\{\frac{1}{2}f_{23}\left(E, \frac{n_x\pi}{d_x}, \frac{n_y\pi}{d_y}\right)\right\} f_{23}\left(E, \frac{n_x\pi}{d_x}, \frac{n_y\pi}{d_y}\right) \times \left[f_{23}\left(E, \frac{n_x\pi}{d_x}, \frac{n_y\pi}{d_y}\right)\right]' H(E - E_{13,8})}{\left(\sqrt{\cos^{-1}\left\{\frac{1}{2}f_{23}\left(E, \frac{n_x\pi}{d_x}, \frac{n_y\pi}{d_y}\right)\right\}^2 - L_0^2\left\{\left(\frac{n_x\pi}{d_x}\right)^2 + \left(\frac{n_x\pi}{d_x}\right)^2\right\}}\right) \times \sqrt{1 - \frac{1}{4}\left[f_{23}\left(E, \frac{n_x\pi}{d_x}, \frac{n_y\pi}{d_y}\right)\right]^2}} \tag{13.15b}$$

where $E_{13,8}$ is the sub-band energy and the sub-band equation in this case can be expressed as

$$0 = \left[\frac{1}{L_0^2}\{\cos^{-1}(f_{23}(E_{13,8}, k_x, k_y))\}^2 - k_s^2\right]\Bigg|\, k_x = \frac{n_x\pi}{d_x} \text{ and } k_y = \frac{n_y\pi}{d_y} \tag{13.15c}$$

The EEM in this case is given by

$$m^*(E, \eta_g, n_x, n_y) = \frac{\hbar^2}{2L_0^2}\cos^{-1}\left[\frac{1}{2}f_{23}\left(E, \frac{n_x\pi}{d_x}, \frac{n_y\pi}{d_y}\right)\right] \times \left[f_{23}\left(E, \frac{n_x\pi}{d_x}, \frac{n_y\pi}{d_y}\right)\right]' \times \left[1 - \frac{1}{4}\left[f_{23}(E, \frac{n_x\pi}{d_x}, \frac{n_y\pi}{d_y})\right]^2\right]^{-1/2} \tag{13.16}$$

13.2.9 *The DR in HgTe/CdTe quantum wire HD effective mass super lattices*

The DR in heavily doped HgTe/CdTe QWEMSLs can be written as

$$k_z^2 = [\Delta_{13H} + i\Delta_{14H}]|k_x = \frac{n_x\pi}{d_x} \text{ and } k_y = \frac{n_y\pi}{d_y} \tag{13.17a}$$

The DOS function can be written as

$$N(E) = \frac{g_v}{\pi} \sum_{n_x=1}^{n_{x\max}} \sum_{n_y=1}^{n_{y\max}} \frac{H(E - E_{13,9})[\Delta'_{13H} + i\Delta'_{14H}]}{\sqrt{\Delta_{13H} + i\Delta_{14H}}} \tag{13.17b}$$

where $E_{13,9}$ is the sub-band energy and the sub-band equation in this case can be expressed as

$$0 = [\Delta_{13H} + i\Delta_{14H}]|_{k_x=\frac{n_x\pi}{d_x}, k_y=\frac{n_y\pi}{d_y} \text{ and } E=E_{13,9}} \tag{13.17c}$$

The EEM in this case is given by

$$m^*(E, \eta_g, n_x, n_y) = \frac{\hbar^2}{2}\Delta'_{13H} \tag{13.18}$$

13.2.10 *The DR in strained layer quantum wire HD effective mass super lattices*

Therefore the DR in heavily doped IV-VI, quantum wire EMSLs can be written as

$$k_z^2 = \left[\frac{1}{L_0^2}\{\cos^{-1}(f_{40}(E, k_x, k_y))\}^2 - k_s^2\right] \Bigg| k_x = \frac{n_x\pi}{d_x} \text{ and } k_y = \frac{n_y\pi}{d_y} \tag{13.19a}$$

The DOS function can be written as

$$N(E) = \frac{g_v}{\pi L_0} \sum_{n_x=1}^{n_{x\max}} \sum_{n_y=1}^{n_{y\max}} \times \frac{\cos^{-1}\left\{\frac{1}{2}f_{40}\left(E, \frac{n_x\pi}{d_x}, \frac{n_y\pi}{d_y}\right)\right\} f_{40}\left(E, \frac{n_x\pi}{d_x}, \frac{n_y\pi}{d_y}\right) \times \left[f_{40}\left(E, \frac{n_x\pi}{d_x}, \frac{n_y\pi}{d_y}\right)\right]' H(E - E_{13,10})}{\left(\sqrt{\cos^{-1}\left\{\frac{1}{2}f_{40}\left(E, \frac{n_x\pi}{d_x}, \frac{n_y\pi}{d_y}\right)\right\}^2 - L_0^2\left\{\left(\frac{n_x\pi}{d_x}\right)^2 + \left(\frac{n_x\pi}{d_x}\right)^2\right\}}\right) \times \sqrt{1 - \frac{1}{4}\left[f_{40}\left(E, \frac{n_x\pi}{d_x}, \frac{n_y\pi}{d_y}\right)\right]^2}} \tag{13.19b}$$

where $E_{13,10}$ is the sub-band energy and the sub-band equation in this case can be expressed as

$$0 = \left[\frac{1}{L_0^2}\{\cos^{-1}(f_{40}(E_{13,10}, k_x, k_y))\}^2 - k_s^2\right]\Bigg|_{k_x=\frac{n_x\pi}{d_x} \text{ and } k_y=\frac{n_y\pi}{d_y}} \tag{13.19c}$$

The EEM in this case is given by

$$m^*(E, \eta_g, n_x, n_y) = \frac{\hbar^2}{2L_0^2}\cos^{-1}\left[\frac{1}{2}f_{40}\left(E, \frac{n_x\pi}{d_x}, \frac{n_y\pi}{d_y}\right)\right] \times \left[f_{40}\left(E, \frac{n_x\pi}{d_x}, \frac{n_y\pi}{d_y}\right)\right]' \times \left[1 - \frac{1}{4}\left[f_{40}\left(E, \frac{n_x\pi}{d_x}, \frac{n_y\pi}{d_y}\right)\right]^2\right]^{-1/2} \tag{13.20}$$

13.3 Open Research Problem

(R.13.1) Investigate the influence of arbitrarily oriented alternating quantizing magnetic field and strain on the EM for all types of HD super-lattices whose carrier energy spectra are described in this book.

References

1. S. Chakrabarti, S. K. Sen, S. Chakraborty, L. S. Singh, K. P. Ghatak *Advanced Science, Engineering and Medicine* **6**, 1042 (2014)
2. K. P. Ghatak, S. Dutta, D. K. Basu, B. Nag *Il Nuovo Cimento* D **20**, 227–240 (1998)
3. K. P. Ghatak, S. Karmakar, D. De, S. Pahari, S. K. Charaborty, S. K. Biswas, S. Chowdhury Journal of Computational and Theoretical Nanoscience, **3**, 153 (2006)

Chapter 14

Appendix F: The EM in HDSLs Under Magnetic Quantization

14.1 Introduction

In recent years the influence of magnetic quantization on the various electronic properties of semiconductor supperlattices having various band structures has been investigated in the literature [1–10]. In this Chapter, the magneto EM in III–V, II–VI, IV–VI, HgTe/CdTe and strained layer heavily doped superlattices (HDSLs) with graded interfaces has been studied in Sections 14.2.1 to 14.2.5 From Sections 14.2.6 to 14.2.10, the magneto EM in III-V, II-VI, IV-VI, HgTe/CdTe and strained layer heavily doped effective mass superlattices respectively has been presented. Lastly Section 14.3 presents 4 open research problems.

14.2 Theoretical Background

14.2.1 *The EM in III–V HD superlattices with graded interfaces under magnetic quantization*

The simplified DR of heavily doped quantum well III–V superlattices with graded interfaces under magnetic quantization can be expressed as

$$k_z^2 = G_{8,E,n} + iH_{8E,n} \tag{14.1a}$$

where

$$G_{8E,n} = \left[\frac{C_{7E,n}^2 - D_{7E,n}^2}{L_0^2} - \left\{\frac{2eB}{\hbar}\left(n+\frac{1}{2}\right)\right\}\right],$$

$$C_{7E,n} = \cos^{-1}\left(\overline{\omega_{7E,n}}\right),$$

$$\overline{\omega_{7E,n}} = (2)^{\frac{-1}{2}}\left[\left(1 - G_{7E,n}^2 - H_{7E,n}^2\right) - \sqrt{\left(1 - G_{7E,n}^2 - H_{7E,n}^2\right)^2 + 4G_{7E,n}^2}\right]^{\frac{1}{2}}$$

$$\begin{aligned} G_{7E,n} = {} & [G_{1E,n} + (\rho_{5E,n}G_{2E,n}/2) - (\rho_{6E,n}H_{2E,n}/2) \\ & + (\Delta_0/2)\{\rho_{6E,n}H_{2E,n} - \rho_{8E,n}H_{3E,n} + \rho_{9E,n}H_{4E,n} \\ & - \rho_{10E,n}H_{4E,n}\rho_{11E,n}H_{5E,n} - \rho_{12E,n}H_{5E,n} \\ & + (1/12)(\rho_{12E,n}G_{6E,n} - \rho_{14E,n}H_{6E,n})\}], \end{aligned}$$

$$\begin{aligned} G_{1E,n} = {} & [(\cos(h_{1E,n}))(\cosh(h_{2E,n}))(\cosh(g_{1E,n}))(\cos(g_{2E,n})) \\ & + (\sin(h_{1E,n}))(\sinh(h_{2E,n}))(\sinh(g_{1E,n}))(\sin(g_{2E,n}))], \end{aligned}$$

$$h_{1E,n} = e_{1E,n}(b_0 - \Delta_0), e_{1E,n=}2^{\frac{-1}{2}}\left(\sqrt{t_{1E,n}^2 + t_2^2} + t_{1E,n}\right)^{\frac{1}{2}}$$

$$t_{1E,n} = \left[\left(2m_{c1}^*/\hbar^2\right)T_{11}\left(E, E_{g1}, \Delta_1\eta_{g1}\right) - \left\{\frac{2eB}{\hbar}\left(n+\frac{1}{2}\right)\right\}\right],$$

$$t_2 = \left(2m_{c1}^*/\hbar^2\right)T_{21}\left(E, E_{g1}, \Delta_1\eta_{g1}\right)$$

$$h_{2E,n} = e_{2E,n}(b_0 - \Delta_0), \quad e_{2E,n} = 2^{\frac{-1}{2}}\left(\sqrt{t_{1E,n}^2 + t_2^2} - t_{1E,n}\right)^{\frac{1}{2}},$$

$$g_{1E,n} = d_{1E,n}(b_0 - \Delta_0), \quad d_{1E,n} = 2^{\frac{-1}{2}}\left(\sqrt{x_{1E,n}^2 + y_1^2} - x_{1E,n}\right)^{\frac{1}{2}}$$

$$x_{1E,n} = \left[-\left(2m_{c2}^*/\hbar^2\right)T_{11}\left(E - V_0, E_{g0}, \Delta_0\eta_{g2}\right) - \left\{\frac{2eB}{\hbar}\left(n+\frac{1}{2}\right)\right\}\right],$$

$$y_2 = \left(2m_{c2}^*/\hbar^2\right) T_{22}\left(E - V_0, E_{g2}, \Delta_2 \eta_{g2}\right)$$

$$g_{2E,n} = d_{2E,n}(\alpha_0 - \Delta_0), d_{2E,n} = 2^{\frac{-1}{2}}\left(\sqrt{x_{1E,n}^2 + y_1^2} - x_{1E,n}\right)^{\frac{1}{2}},$$

$$\rho_{5E,n} = \left(\rho_{3E,n}^2 + \rho_{4E,n}^2\right)^{-1}\left[\rho_{1E,n}\rho_{3E,n} - \rho_{2E,n}\rho_{4E,n}\right],$$

$$\rho_{1E,n} = \left[d_{1E,n}^2 + e_{2E,n}^2 - d_{2E,n}^2 - e_{1E,n}^2\right],$$

$$\rho_{3E,n} = \left[d_{1E,n}e_{1E,n} + d_{2E,n}e_{2E,n}\right],$$

$$\rho_{2E,n} = 2\left[d_{1E,n}d_{2E,n} + e_{1E,n}e_{2E,n}\right],$$

$$\rho_{4E,n} = \left[d_{1E,n}e_{2E,n} - e_{1E,n}d_{2E,n}\right],$$

$$G_{2E,n} = [(\sin(h_{1E,n}))(\cosh(h_{2E,n}))(\sinh(g_{1E,n}))(\cos(g_{2E,n})) + (\cos(h_{1E,n}))(\sinh(h_{2E,n}))(\cosh(g_{1E,n}))(\sin(g_{2E,n}))],$$

$$\rho_{6E,n} = (\rho_{3E,n}^2 + \rho_{4E,n}^2)^{-1}[\rho_{1E,n}\rho_{4E,n} + \rho_{2E,n}\rho_{3E,n}],$$

$$H_{2E,n} = [(\sin(h_{1E,n}))(\cosh(h_{2E,n}))(\sin(g_{2E,n}))(\cosh(g_{1E,n})) - (\cos(h_{1E,n}))(\sinh(h_{2E,n}))(\sinh(g_{1E,n}))(\cos(g_{2E,n}))],$$

$$\rho_{7E,n} = [(e_{1E,n}^2 + e_{2E,n}^2)^{-1}[e_{1E,n}(d_{1E,n}^2 - d_{2E,n}^2) - 2d_{1E,n}d_{2E,n}e_{2E,n}] - 3e_{1E,n}],$$

$$G_{3E,n} = [(\sin(h_{1E,n}))(\cosh(h_{2E,n}))(\cosh(g_{1E,n}))(\cos(g_{2E,n})) + (\cos(h_{1E,n}))(\sinh(h_{2E,n}))(\sinh(g_{1E,n}))(\sin(g_{2E,n}))],$$

$$\rho_{8E,n} = [(e_{1E,n}^2 + e_{2E,n}^2)^{-1}[e_{2E,n}(d_{1E,n}^2 - d_{2E,n}^2) - 2d_{1E,n}d_{2E,n}e_{1E,n}] + 3e_{2E,n}],$$

$$H_{3E,n} = [(\sin(h_{1E,n}))(\cosh(h_{2E,n}))(\sin(g_{2E,n}))(\sinh(g_{1E,n})) - (\cos(h_{1E,n}))(\sinh(h_{2E,n}))(\cosh(g_{1E,n}))(\cos(g_{2E,n}))],$$

$$\rho_{9E,n} = [(d_{1E,n}^2 + d_{2E,n}^2)^{-1}[d_{2E,n}(e_{1E,n}^2 - e_{2E,n}^2) + 2e_{1E,n}d_{2E,n}e_{1E,n}] + 3d_{1E,n}],$$

$$G_{4E,n} = [(\cos(h_{1E,n}))(\cosh(h_{2E,n}))(\cos(g_{2E,n}))(\sinh(g_{1E,n})) - (\sin(h_{1E,n}))(\sinh(h_{2E,n}))(\cosh(g_{1E,n}))(\sin(g_{2E,n}))],$$

$$\rho_{10E,n} = [-(d_{1E,n}^2 + d_{2E,n}^2)^{-1}[d_{2E,n}(-e_{1E,n}^2 + e_{2E,n}^2)$$
$$+2e_{2E,n}d_{2E,n}e_{1E,n}] + 3d_{2E,n}],$$

$$H_{4E,n} = [(\cos(h_{1E,n}))(\cosh(h_{2E,n}))(\cosh(g_{1E,n}))(\sin(g_{2E,n}))$$
$$+ (\sin(h_{1E,n}))(\sinh(h_{2E,n}))(\sinh(g_{1E,n}))(\cos(g_{2E,n}))],$$

$$\rho_{11E,n} = 2[d_{1E,n}^2 + e_{2E,n}^2 - e_{2E,n}^2 - d_{1E,n}^2],$$

$$G_{5E,n} = [(\cos(h_{1E,n}))(\cosh(h_{2E,n}))(\cos(g_{2E,n}))(\cosh(g_{1E,n}))$$
$$- (\sin(h_{1E,n}))(\sinh(h_{2E,n}))(\sinh(g_{1E,n}))(\sin(g_{2E,n}))],$$

$$\rho_{12E,n} = 4[d_{1E,n} + d_{2E,n} - e_{1E,n} + e_{2E,n}],$$

$$H_{5E,n} = [(\cos(h_{1E,n}))(\cosh(h_{2E,n}))(\sinh(g_{1E,n}))(\sin(g_{2E,n}))$$
$$+ (\sin(h_{1E,n}))(\sinh(h_{2E,n}))(\cosh(g_1))(\cos(g_2))],$$

$$\rho_{13E,n} = [\{5(d_{1E,n} + e_{1E,n}^3 - 3e_{1E,n}e_{2E,n}^2 + d_{2E,n}) + 5d_{2E,n},$$
$$\cdot(e_{2E,n}^3 - 3e_{1E,n}^2e_{2E,n})\}(d_{1E,n}^2 + d_{2E,n}^2)^{-1}$$
$$+ (e_{1E,n}^2 + e_{2E,n}^2)^{-1}\{5(-e_{1E,n} + d_{2E,n}^3 - 3e_{2E,n}e_{1E,n}^2 + d_{1E,n})$$
$$+ 5d_{2E,n}, (-e_{1E,n}^3 - 3e_{2E,n}^2e_{1E,n})\}(d_{1E,n}^2 + d_{2E,n}^2)^{-1}$$
$$+ 34(d_{1E,n}e_{2E,n} - d_{2E,n}e_{1E,n})]$$

$$H_{6E,n} = [(\sin(h_{1E,n}))(\cosh(h_{2E,n}))(\cosh(g_{1E,n}))(\sin(g_{2E,n}))$$
$$- (\cos(h_{1E,n}))(\sinh(h_{2E,n}))(\sinh(g_{1E,n}))(\cos(g_{2E,n}))],$$

$$H_{7E,n} = [H_{1E,n} + (\rho_{5E,n}H_{2E,n}/2) + (\rho_{6E,n}G_{2E,n}/2)$$
$$+ (\Delta_0/2)\{\rho_{8E,n}G_{3E,n} + \rho_{7E,n}H_{3E,n} + \rho_{10E,n}G_{4E,n}$$
$$+ \rho_{9E,n}H_{4E,n} + \rho_{12E,n}G_{5E,n} + \rho_{11E,n}H_{5E,n}$$
$$+ (1/12)(\rho_{14E,n}G_{6E,n} + \rho_{13E,n}H_{6E,n})\}]$$

$$H_{1E,n} = [(\sin(h_{1E,n}))(\sinh(h_{2E,n}))(\cosh(g_{1E,n}))(\cos(g_{2E,n}))$$
$$+ (\cos(h_{1E,n}))(\cosh(h_{2E,n}))(\sinh(g_{1E,n}))(\sin(g_{2E,n}))],$$

$$D_{7E,n} = \sinh^{-1}(\overline{\omega_{7E,n}}), H_{8E,n} = (2C_{7E,n}D_{7E,n}/L_0^2)$$

The DOS function can be written as

$$N(E) = \frac{eB}{2\pi^2\hbar}\sum_{n=0}^{n_{\max}} \frac{(G'_{8E,n} + iH'_{8E,n})H(E - E_{15,1})}{\sqrt{G_{8E,n} + iH_{8E,n}}} \tag{14.1b}$$

where $E_{15,1}$ is the sub-band energy in this case and is given by

$$0 = [G_{8E,n} + iH_{8E,n}]|_{E=E_{15,1}} \tag{14.1c}$$

The EEM can be written as

$$m^*(E, \eta_g, n) = \frac{\hbar^2}{2} G'_{8E,n} \tag{14.2a}$$

The electron statistics can be expressed as

$$n_0 = \frac{g_v eB}{\pi^2\hbar} \text{ Real part of } \sum_{n=0}^{n_{\max}} \left[(G_{8,E,n} + iH_{8,E,n})^{1/2} \Big|_{E_{F1321}} + \sum_{r=1}^{s} L(r) \left[(G_{8,E,n} + iH_{8,E,n})^{1/2} \Big|_{E_{F1321}} \right] \right] \tag{14.2b}$$

where E_{F1321} is the fermi energy in this case.

14.2.2 *The EM in II–VI HD superlattices with graded interfaces under magnetic quantization*

The simplified DR in heavily doped II–VI super-lattices with graded interfaces under magnetic quantization can be expressed as

$$k_z^2 = G_{19E,n} + iH_{19E,n} \tag{14.3a}$$

$$G_{19E,n} = \left[\frac{C_{18E,n}^2 - D_{18E,n}^2}{L_0^2} - \left(\frac{2eB}{\hbar}\left(n + \frac{1}{2}\right) \right) \right],$$

where

$$C_{18E,n} = \cos^{-1}(\omega_{18E,n}),$$

$$\omega_{18E,n} = (2)^{\frac{-1}{2}} \left[(1 - G_{18E,n}^2 - H_{18E,n}^2) - \sqrt{(1 - G_{18E,n}^2 - H_{18E,n}^2)^2 + 4G_{180D}^2} \right]^{\frac{1}{2}},$$

$$G_{18E,n} = \frac{1}{2}\left[G_{11E,n} + G_{12E,n} + \Delta_0\left(G_{13E,n} + G_{14E,n}\right)\right.$$
$$\left. + \Delta_0\left(G_{15E,n} + G_{16E,n}\right)\right],$$

$$G_{11E,n} = 2(\cos(g_{1E,n}))(\cos(g_{2E,n}))(\cos\gamma_{11}(E,n)),$$

$$\gamma_{11}(E,n) = k_{21}(E,n)(b_0 - \Delta_0)$$

$$k_{21}(E,n) = \left\{\left[\gamma_3(E,\eta_{g1}) - \frac{\hbar^2}{2m^*_{\perp,1}}\left\{\frac{2eB}{\hbar}\left(n+\frac{1}{2}\right)\right\}\right.\right.$$
$$\left.\left. \pm\ C_0\left\{\frac{2eB}{\hbar}\left(n+\frac{1}{2}\right)\right\}^{\frac{1}{2}}\right]\frac{2m^*_{\parallel,1}}{\hbar^2}\right\}^{1/2},$$

$$G_{12E,n} = ([\Omega_1(E,n)(\sinh g_{1E,n})(\cos g_{2E,n})$$
$$- \Omega_2(E,n)(\sin g_{2E,n})(\cosh g_{1E,n})])(\sin\gamma_{11}(E,n))),$$

$$\Omega_1(E,n) = \left[\frac{d_{1E,n}}{k_{21}(E,n)} - \frac{k_{21}(E,n)d_{1E,n}}{d^2_{1E,n} + d^2_{2E,n}}\right],$$

$$\Omega_2(E,n) = \left[\frac{d_{1E,n}}{k_{21}(E,n)} - \frac{k_{21}(E,n)d_{1E,n}}{d^2_{1E,n} + d^2_{2E,n}}\right]$$

$$G_{13E,n} = ([\Omega_3(E,n)(\cosh g_{1E,n})(\cos g_{2E,n})$$
$$- \Omega_4(E,n)(\sin g_{2E,n})(\cosh g_{1E,n})])(\sin\gamma_{11}(E,n)))$$

$$\Omega_3(E,n) = \left[\frac{d^2_{1E,n} - d^2_{2E,n}}{k_{21}(E,n)} - 3k_{21}(E,n)\right],$$

$$\Omega_4(E,n) = \left[\frac{2d_{1E,n}d_{2E,n}}{k_{21}(E,n)}\right]$$

$$G_{14E,n} = ([\Omega_5(E,n)(\sinh g_{1E,n})(\cos g_{2E,n})$$
$$- \Omega_6(E,n)(\sin g_{2E,n})(\cosh g_{1E,n})])(\cos\gamma_{11}(E,n)))$$

$$\Omega_5(E,n) = \left[3d_{1E,n}\frac{d_{1E,n}}{d^2_{1E,n} - d^2_{2E,n}} - k^2_{21}(E,n)\right],$$

$$\Omega_6(E,n) = \left[3d_{2E,n}\frac{d_{2E,n}}{d^2_{1E,n} - d^2_{2E,n}} - k^2_{21}(E,n)\right]$$

$$G_{15E,n} = ([\Omega_9(E,n)(\cosh g_{1E,n})(\cos g_{2E,n})$$
$$- \Omega_{10}(E,n)(\sinh g_{1E,n})(\sinh g_{2E,n})])(\cos\gamma_{11}(E,n)))$$

$$\Omega_9(E,n) = [2d^2_{1E,n} - 2d^2_{2E,n} - k_{21}(E,n)],\ \ \Omega_{10}(E,n) = [2d_{1E,n}d_{2E,n}]$$

$$G_{16E,n} = ([\Omega_7(E,n)(\sinh g_{1E,n})(\cos g_{2E,n})$$
$$- \Omega_8(E,n)(\sin g_{1E,n})(\cosh g_{2E,n})])(\sin\gamma_{11}(E,n)/12))$$

$$\Omega_7(E,n) = \left[\frac{5d_{1E,n}}{d^2_{1E,n} + d^2_{2E,n}}k^3_{21}(E,n) + \frac{5(d^3_{1E,n} - 3d^2_{2E,n}d_{1E,n}}{k_{21}(E,n)}\right.$$
$$\left. -\ 34k_{21}(E,n)d_{1E,n}\right],$$

$$\Omega_8(E,n) = \left[\frac{5d_{2E,n}}{d^2_{1E,n} + d^2_{2E,n}}k^3_{21}(E,n) + \frac{5(d^3_{2E,n} - 3d^2_{2E,n}d_{1E,n}}{k_{21}(E,n)}\right.$$
$$\left. +\ 34k_{21}(E,n)d_{1E,n}\right],$$

$$H_{18,E,n} = \frac{1}{2}[H_{11,E,n} + H_{12E,n} + \Delta_0(H_{13,E,n} + H_{14E,n})$$
$$+ \Delta_0(H_{15E,n} + H_{16E,n})]$$

$$H_{11E,n} = 2(\sinh g_{1E,n})(\sin g_{2E,n})(\cos\gamma_{11}(E,n)))$$

$$H_{12E,n} = ([\Omega_2(E,n)(\sinh\ g_{1E,n})(\cos\ g_{2E,n})$$
$$+ \Omega_1(E,n)(\sin\ g_{2E,n})(\cosh\ g_{1E,n})](\sin\gamma_{11}(E,n))),$$

$$H_{13E,n} = ([\Omega_4(E,n)(\cosh\ g_{1E,n})(\cos\ g_{2E,n})$$
$$+ \Omega_3(E,n)(\sinh\ g_{1E,n})(\sin\ g_{2E,n})](\sin\gamma_{11}(E,n))),$$

$$H_{14E,n} = ([\Omega_6(E,n)(\sinh\ g_{1E,n})(\cos\ g_{2E,n})$$
$$+ \Omega_5(E,n)(\sin\ g_{1E,n})(\cosh\ g_{2E,n})](\cos\gamma_{11}(E,n)))$$

$$H_{15E,n} = ([\Omega_{10}(E,n)(\cosh\ g_{1E,n})(\cos\ g_{2E,n}) + \Omega_9(E,n)(\sin\ g_{1E,n})(\sin\ g_{2E,n})](\cos\gamma_{11}(E,n))),$$

$$H_{19E,n} = ([\Omega_8(E,n)(\sinh\ g_{1E,n})(\cos\ g_{2E,n}) + \Omega_7(E,n)(\sin\ g_{1E,n})(\cosh\ g_{2E,n})](\sin\gamma_{11}(E,n)/2))$$

$$H_{19E,n} = \left[\frac{2C_{18E,n}D_{18E,n}}{L_0^2}\right] \quad \text{and} \quad D_{18E,n} = \sinh^{-1}(\omega_{18E,n})$$

The DOS function can be written as

$$N(E) = \frac{eB}{2\pi^2\hbar}\sum_{n=0}^{n_{\max}}\frac{(G'_{19E,n} + iH'_{19E,n})H(E - E_{15,2})}{\sqrt{G_{19E,n} + iH_{19E,n}}} \tag{14.3b}$$

where $E_{15,2}$ is the sub-band energy in this case and is given by

$$0 = [G_{19E,n} + H_{19E,n}]|_{E=E_{15,2}} \tag{14.3c}$$

The EEM can be written as

$$m^*(\mathrm{E},\eta_g,n) = \frac{\hbar^2}{2}G'_{19E,n} \tag{14.4a}$$

The electron statistics can be expressed as

$$n_0 = \frac{g_v eB}{\pi^2\hbar}\ \text{Real part of}\ \sum_{n=0}^{n_{\max}}\left[(G_{19,E,n} + +iH_{19,E,n})^{1/2}\Big|_{E_{F1322}} + \sum_{r=1}^{s} L(r)\left[(G_{19,E,n} + +iH_{19,E,n})^{1/2}\Big|_{E_{F1322}}\right]\right] \tag{14.4b}$$

where E_{F1322} is the fermi energy in this case.

14.2.3 *The EM in IV–VI HD superlattices with graded interfaces under magnetic quantization*

The simplified DR in heavily doped IV–VI super-lattices with graded interfaces under magnetic quantization can be expressed as

$$k_z^2 = \frac{1}{L_0^2}\left[\cos^{-1}\left\{\frac{1}{2}\Phi_2(E,n)\right\}\right]^2 - \frac{2eB}{\hbar}\left(n + \frac{1}{2}\right) \tag{14.5a}$$

where

$$
\begin{aligned}
\Phi_2(E,n) \equiv \Big[& 2\cosh\{\beta_2(E,n)\}\cos\{\gamma_2(E,n)\} \\
& + \varepsilon_2(E,n)\sinh\{\beta_2(E,n)\}\sin\{\gamma_{22}(E,n)\} \\
& + \Delta_0 \Big[((\{K_{112}(E,n)\}^2/K_{212}(E,n)) \\
& - 3K_{212}(E,n))\cosh\{\beta_2(E,n)\}\sin\{\gamma_{22}(E,n)\} \\
& + \left(3K_{112}(E,n) - \frac{\{K_{212}(E,n)\}^2}{K_{112}(E,n)}\right) \\
& \cdot \sinh\{\beta_2(E,n)\}\cos\{\gamma_{22}(E,n)\}\Big] + \Delta_0 \Big[2(\{K_{112}(E,n)\}^2 \\
& - \{K_{212}(E,n)\}^2)\cosh\{\beta_2(E,n)\}\cos\{\gamma_{22}(E,n)\} \\
& + \frac{1}{12}\left[\frac{5\{K_{112}(E,n)\}^3}{K_{212}(E,n)} + \frac{5\{K_{212}(E,n)\}^3}{K_{112}(E,n)}\right. \\
& \left. - 34K_{212}(E,n)K_{112}(E,n)\right] \\
& \cdot \sinh\{\beta_2(E,n)\}\sin\{\gamma_{22}(E,n)\}\Big]\Big]
\end{aligned}
$$

$$
\begin{aligned}
\beta_2(E,n) &\equiv K_{112}(E,n)[a_0 - \Delta_0], \\
k_{112}^2(E,n) &= [2\bar{p}_{9,2n}]^{-1}[-\bar{q}_{9,2n}(E - V_0\eta_{g2}) \\
&\quad - [[\bar{q}_{9,2n}(E - V_0\eta_{g2})]^2 + 4\bar{p}_{9,2n}\bar{R}_{9,2n}(E - V_0\eta_{g2})]^{\frac{1}{2}}] \\
&\quad \bar{q}_{9,2n}(E - V_0\eta_{g2})\Big[(\hbar^2/2)((1/\mathrm{m}_{l2}^-)) \\
&\quad\quad + a_2(\hbar^2/4)\frac{2eB}{\hbar}\left(n + \frac{1}{2}\right)((1/m_{l2}^+ m_{t2}^-)(1/m_{t2}^+ m_{l2}^-)) \\
&\quad\quad - \alpha_2\gamma_3(E - V_0\eta_{g2})(1/m_{l2}^+) - (1/m_{t2}^-)\Big]
\end{aligned}
$$

$$\bar{\mathrm{R}}_{9,2\mathrm{n}}\left(\mathrm{E},\eta_{\mathrm{g}^2}\right)+\Bigg[\gamma_2\left(\mathrm{E}-\mathrm{V}_0\eta_{\mathrm{g}^2}\right)$$

$$+\gamma_3\left(\mathrm{E}-\mathrm{V}_0\eta_{\mathrm{g}^2}\right)\left[\left(\hbar^2/2\right)\alpha_2\frac{2\mathrm{eB}}{\hbar}\left(\mathrm{n}+\frac{1}{2}\right)((1/\mathrm{m}_{\mathrm{t}2}^*)\right.$$

$$\left.-(1/\mathrm{m}_{\mathrm{t}2}^-))\right]-\left[\left(\mathrm{h}^2/2\right)\mathrm{k}_{\mathrm{s}0}^2\left((1/\mathrm{m}_{\mathrm{t}2}^*)+(1/\mathrm{m}_{\mathrm{t}2}^-)\right)\right]$$

$$\left.-\alpha_2\left(\hbar^6/4\right)\left[\frac{2\mathrm{eB}}{\hbar}\left(\mathrm{n}+\frac{1}{2}\right)\right]^2((1/\mathrm{m}_{\mathrm{t}2}^+\mathrm{m}_{\mathrm{t}2}^-))\right],$$

$$\begin{aligned}
\gamma_2(E,n) &= K_{212}(E,n)[b_0-\Delta_0], \quad K_{212}^2(E,n) \\
&= [2\bar{p}_{9,1n}]^{-1}[-\bar{q}_{9,1n}(E,\eta_{g1})+[[\bar{q}_{9,1n}(E,\eta_{g1})]^2 \\
&\quad + 4\bar{p}_{9,1n}\bar{R}_{9,1n}(E,\eta_{g1})]^{\frac{1}{2}}] \\
\bar{q}_{9,1n}(E,\eta_{g1}) &= [(\hbar^2/2)((1/m_{l1}^*)+(1/m_{l1}^-)) \\
&\quad + \alpha_1(\hbar^6/4)\frac{2eb}{\hbar}\left(n+\frac{1}{2}\right)((1/m_{l1}^+m_{t1}^-) \\
&\quad + (1/m_{l1}^+m_{t1}^-)) - \alpha_1\gamma_3\left(E,\eta_{g1}\right)((1/m_{l1}^+)-(1/m_{t1}^-)],
\end{aligned}$$

$$\bar{\mathrm{R}}_{9,1\mathrm{n}}\left(\mathrm{E},\eta_{\mathrm{g}^1}\right)=\Bigg[\gamma_2\left(\mathrm{E},\eta_{\mathrm{g}^1}\right)+\gamma_2\left(\mathrm{E},\eta_{\mathrm{g}^1}\right)\Bigg[\left(\hbar^2/2\right)$$

$$\cdot\alpha_1(2\mathrm{eB}/\hbar)\left(\mathrm{n}+\frac{1}{2}\right)((1/\mathrm{m}_{\mathrm{t}1}^*)$$

$$\left.-(1/\mathrm{m}_{t1}^-))\right]-[(\mathrm{h}^2/2)\mathrm{k}_{\mathrm{s}0}^2((1/\mathrm{m}_{\mathrm{t}2}^*)+(1/\mathrm{m}_{\mathrm{t}2}^-))]$$

$$\left.-\alpha_1(\hbar^6/4)\left(2\mathrm{eB}/\hbar\left(\mathrm{n}+\frac{1}{2}\right)\right)^2((1/\mathrm{m}_{\mathrm{t}2}^+\mathrm{m}_{\mathrm{t}2}^-))\right] \quad \text{and}$$

$$\varepsilon_2(E,n)\equiv\left[\frac{K_{112}(E,n)}{K_{212}(E,n)}-\frac{K_{212}(E,n)}{K_{112}(E,n)}\right].$$

The DOS function can be written as

$$N(E) = \frac{eBg_v}{2\pi^2 \hbar L_0} \sum_{n=0}^{n_{\max}} \cdot \frac{\cos^{-1}[\frac{1}{2}\varphi_2(E,n)(1-\frac{1}{4}\varphi_2^2(E,n)]^{-1/2}\varphi_2'(E,n)H(E-E_{15,3})}{[[\cos^{-1}[\frac{1}{2}\varphi_2(E,n)]]^2 - L_0^2\frac{2eB}{\hbar}(1+\frac{1}{2})]^{1/2}} \tag{14.5b}$$

where $E_{15,3}$ is the sub-band energy in this case and is given by

$$0 = \frac{1}{L_0^2}\left[\cos^{-1}\left\{\frac{1}{2}\Phi_2\left(E_{15,3}, n\right)\right\}\right]^2 - \frac{2eB}{\hbar}\left(n+\frac{1}{2}\right) \tag{14.5c}$$

The EEM can be written as

$$m^*(E,\eta_g,n) = \frac{\hbar^2}{2L_0^2}\cos^{-1}\left[\frac{1}{2}\varphi_2(E,n)\left(1-\frac{1}{4}\varphi_2^2(E,n)\right)\right]^{-1/2}\varphi_2'(E,n) \tag{14.6a}$$

The electron concentration can be written as

$$n_0 = \frac{g_v eB}{\pi^2\hbar}\sum_{n=0}^{n_{\max}}\left[\left[\frac{1}{L_0^2}\left[\cos^{-1}\left(\frac{1}{2}\phi_2(E,n)\right)\right]^2 - \frac{2eB}{\hbar}\left(n+\frac{1}{2}\right)\right]^{1/2}\Bigg|_{E_{F1323}} \cdot \sum_{r=1}^{s} L(r)\left[\sum_{n=0}^{n_{\max}}\left[\left[\frac{1}{L_0^2}\left[\cos^{-1}\left(\frac{1}{2}\phi_2(E,n)\right)\right]^2 - \frac{2eB}{\hbar}\left(n+\frac{1}{2}\right)\right]^{1/2}\Bigg|_{E_{F1323}}\right]\right] \tag{14.6b}$$

where E_{F1323} is the fermi energy in this case.

14.2.4 *The EM in HgTe/CdTe HD superlattices with graded interfaces under magnetic quantization*

The simplified DR in heavily doped HgTe/CdTe superlattices with graded interfaces under magnetic quantization can be expressed as

$$(k_z)^2 = G_{192E,n} + iH_{192E,n}$$

where

$$G_{192E,n} = \left[\frac{C_{182E,n}^2 - D_{182E,n}^2}{L_0^2} - (2eB/\hbar)\left(n + \frac{1}{2}\right)\right], \tag{14.7a}$$

$$C_{1820D} = \cos^{-1}(\omega_{182E,n}),$$

$$\omega_{182E,n} = (2)^{\frac{-1}{2}}[(1 - G_{182E,n}^2 - H_{182E,n}^2)$$
$$- \sqrt{\left(1 - G_{182E,n}^2 - H_{182E,n}^2\right)^2 + 4G_{182E,n}^2}]^{\frac{1}{2}}$$

$$G_{112E,n} = 2(\cos(g_{12}))(\cos(g_{22}))(\cos\gamma_8(E,n)),$$

$$\gamma_8(E,n) = k_8(E,n)(b_0 - \Delta_0),$$

$$k_8(E,n) = \left[\frac{B_{01}^2 + 4A_1E - B_{01}\sqrt{B_{01}^2 + 4A_1E}}{2A_1^2}\right.$$
$$\left. - (2eB/\hbar)\left(n + \frac{1}{2}\right)\right]^{1/2},$$

$$G_{120D} = ([\Omega_{12}(E,n)(\sinh g_{12E,n})(\cos g_{12E,n})$$
$$-\Omega_{22}(E,n)(\sinh g_{12E,n})(\cos g_{12E,n})](\sin\gamma_8(E,n))),$$

$$\Omega_{12}(E,n) = \left[\frac{d_{12E,n}}{k_8(E,n)} - \frac{k_8(E,n)\,d_{12E,n}}{d_{12E,n} + d_{22E,n}}\right],$$

$$\Omega_{22}(E,n) = \left[\frac{d_{22E,n}}{k_8(E,n)} - \frac{k_8(E,n)\,d_{22E,n}}{d_{12E,n} + d_{22E,n}}\right],$$

$$G_{1320D} = ([\Omega_{32}(E,n)(\cosh g_{12E,n})(\cos g_{22E,n})$$
$$- \Omega_{42}(E,n)(\sinh g_{12E,n})(\cos g_{12E,n})](\sin\gamma_8(E,n))),$$

$$\Omega_{32}(E,n) = \left[\frac{d_{12E,n}^2 - d_{2E,n}^2}{k_8(E,n)} - 3k_8(E,n)\right],$$

$$\Omega_{42}(E,n) = \left[\frac{2d_{12E,n}d_{22E,n}}{k_8(E,n)}\right]$$

$$G_{1420D} = ([\Omega_{52}(E,n)(\sinh g_{12E,n})(\cos g_{22E,n}) - \Omega_{62}(E,n)(\sinh g_{12E,n})(\cosh g_{22E,n})](\cos\gamma_8(E,n))),$$

$$\Omega_{52}(E,n) = \left[3d_{12E,n} - \frac{d_{12E,n}}{d_{12E,n}^2 + d_{22E,n}^2}k_8^2(E,n)\right],$$

$$\Omega_{62}(E,n) = \left[3d_{22E,n} + \frac{d_{22E,n}}{d_{12E,n}^2 + d_{22E,n}^2}k_8^2(E,n)\right],$$

$$G_{1520D} = ([\Omega_{72}(E,n)(\cosh g_{12E,n})(\cos g_{22E,n}) - \Omega_{102}(E,n)(\sinh g_{12E,n})(\cosh g_{22E,n})] \times (\sin\gamma_{80D}(E,n)/12)),$$

$$\Omega_{92}(E,n) = \left[2d_{12E,n}^2 - 2d_{22E,n}^2 - k_8^2(E,n)\right],$$

$$\Omega_{102}(E,n) = [2d_{12E,n}d_{22E,n}]$$

$$G_{162E,n} = ([\Omega_{72}(E,n)(\sinh g_{12E,n})(\cos g_{22E,n}) - \Omega_{82}(E,n)(\sinh g_{12E,n})(\cosh g_{22E,n})] \times (\sin\gamma_{80D}(E,n)/12)),$$

$$\Omega_{72}(E,n) = \left[\frac{5d_{12E,n}}{d_{12E,n}^2 + d_{22E,n}^2}k_8^3(E,n) + \frac{5\left(d_{12E,n}^3 - 3d_{22E,n}^2 d_{12E,n}\right)}{k_8(E,n)} - 34k_8(E,n)\,d_{12E,n}\right]$$

$$\Omega_{82}(E,n) = \left[\frac{5d_{22E,n}}{d_{12E,n}^2 + d_{22E,n}^2}k_8^3(E,n) + \frac{5\left(d_{22E,n}^3 - 3d_{22E,n}^2 d_{12E,n}\right)}{k_8(E,n)} + 34k_8(E,n)\,d_{12E,n}\right]$$

$$H_{182E,n} = \frac{1}{2}[H_{112E,n} + H_{122E,n} + \Delta_0\,(H_{132E,n} + H_{142E,n})$$
$$+ \Delta_0\,(H_{152E,n} + H_{162E,n})],$$
$$H_{112E,n} = 2\left(\sinh\ \ \mathrm{g}_{12E,n}\right)(\sinh g_{22E,n})\,(\cos\gamma_8\,(E,n))),$$

$$H_{1220D} = ([\Omega_{22}(E,n)(\sinh g_{12E,n})(\cos g_{22E,n})$$
$$+ \Omega_{12}(E,n)(\sinh g_{22E,n})(\cosh g_{12E,n})](\sin\gamma_8(E,n))),$$
$$H_{132E,n} = ([\Omega_{42}(E,n)(\cosh g_{12E,n})(\cos g_{22E,n})$$
$$+ \Omega_{32}(E,n)(\sinh g_{12E,n})(\sinh g_{22E,n})](\sin\gamma_8(E,n))),$$
$$H_{132E,n} = ([\Omega_{42}(E,n)(\cosh g_{12E,n})(\cos g_{22E,n})$$
$$+ \Omega_{32}(E,n)(\sinh g_{12E,n})(\sinh g_{22E,n})](\sin\gamma_8(E,n))),$$
$$H_{142E,n} = ([\Omega_{62}(E,n)(\sinh g_{12E,n})(\cos g_{22E,n})$$
$$+ \Omega_{52}(E,n)(\sinh g_{12E,n})(\cosh g_{22E,n})](\cos\gamma_8(E,n))),$$
$$H_{142E,n} = ([\Omega_{62}(E,n)(\sinh g_{12E,n})(\cos g_{22E,n})$$
$$+ \Omega_{52}(E,n)(\sinh g_{12E,n})(\cosh g_{22E,n})](\cos\gamma_8(E,n))),$$
$$H_{1520D} = ([\Omega_{102}(E,n)(\cosh g_{12E,n})(\cos g_{22E,n})$$
$$+ \Omega_{92}(E,n)(\sinh g_{22E,n})(c\sinh g_{12E,n})](\sin\gamma_8(E,n))),$$
$$H_{162E,n} = ([\Omega_{82}(E,n)(\sinh g_{12E,n})(\cos g_{22E,n})$$
$$+ \Omega_{82}(E,n)(\sinh g_{12E,n})(\cosh g_{22E,n})](\sin\gamma_{80D}(E,n)/2)),$$
$$H_{192E,n} = [((2C_{182E,n}d_{182E,n})/L_0^2)] \text{ and } D_{182E,n} = \sinh^{-1}(\omega_{182E,n})$$

The DOS function can be written as

$$N(E) = \frac{eB}{2\pi^2\hbar}\sum_{n=0}^{n_{\max}}\frac{(G'_{192E,n} + iH'_{192E,n})H(E - E_{15,4})}{\sqrt{G_{192E,n} + iH_{192E,n}}} \tag{14.7b}$$

where $E_{15,4}$ is the sub-band energy in this case and is given by

$$0 = [\mathrm{G}_{192E,n} + \mathrm{H}_{192E,n}]|_{E=E_{15,4}} \tag{14.7c}$$

The EEM can be written as

$$m^*(\mathrm{E}, \eta_g, n) = \frac{\hbar^2}{2} G'_{192E,n} \tag{14.8a}$$

The electron statistics can be expressed as

$$n_0 = \frac{g_v eB}{\pi^2 \hbar} \text{ Real part of } \sum_{n=0}^{n_{\max}} \Bigg[(G_{192,E,n} + +iH_{192,E,n})^{1/2} |_{E_{F1324}}$$

$$\times \sum_{r=1}^{s} L(r) \left[(G_{192,E,n} + +iH_{192,E,n})^{1/2} |_{E_{F1324}} \right] \Bigg] \tag{14.8b}$$

where E_{F1324} is the fermi energy in this case.

14.2.5 *The EM in strained layer HD superlattices with graded interfaces under magnetic quantization*

The DR of the conduction electrons in heavily doped strained layer SL with graded interfaces can be expressed as

$$k_z^2 = \frac{1}{L_0^2} \left[\cos^{-1} \left\{ \frac{1}{2} \bar{\varphi}_6 (E, n) \right\} \right]^2 - \frac{2|e|B}{\hbar} \left(n + \frac{1}{2} \right) \tag{14.9a}$$

where

$$\begin{aligned}
\bar{\phi}_6 (E, n) = & \left[2 \cosh \left[T_4 (E, n, \eta_{g2}) \right] \cos \left[T_5 (E, n, \eta_{g1}) \right] \right] \\
& + \left[T_6 (E, n) \right] \sinh \left[T_4 (E, n, \eta_{g2}) \right] \sin \left[T_5 (E, n, \eta_{g1}) \right] \\
& + \Delta_0 \left[\left(\frac{k_0^2 (E, n, \eta_{g2})}{k_0'^2 (E, n, \eta_{g1})} - 3k_0' (E, n, \eta_{g1}) \right) \right. \\
& \left. \times \cos \left[T_4 (E, n, \eta_{g2}) \right] \sin \left[T_1 (E, n, \eta_{g1}) \right] \right] \\
& + \left(3k_0 (E, n, \eta_{g2}) - \frac{k_0'^2 (E, n, \eta_{g1})}{k_0 (E, n, \eta_{g2})} \right) \\
& \cdot \sin \left[T_4 (E, n, \eta_{g2}) \right] \cos \left[T_5 (E, n, \eta_{g1}) \right] \\
& + \Delta_0 [2 (k_0^2 (E, n, \eta_{g2}) - k_{0D}'^2 (E, n, \eta_{g1})) \\
& \cdot \cos [T_4 (E, n, \eta_{g2})] \cos [T_5 (E, n, \eta_{g1})]]
\end{aligned}$$

$$+\frac{1}{12}\left(\frac{5k_0^3(E,n,\eta_{g2})}{k_0'(E,n,\eta_{g1})}+\frac{5k_0'^3(E,n,\eta_{g1})}{k_0(E,n,\eta_{g2})}\right.$$

$$\left.-34k_0(E,n,\eta_{g2})\,k_0'(E,n,\eta_{g1})\right)$$

$$\cdot\sinh\left[T_4(E,n,\eta_{g2})\sin(E,n,\eta_{g1})\right]$$

$$[T_4(E,n,\eta_{g2})]=k_0(E,n,\eta_{g2})\left[a_0-\Delta_0\right],$$

$$k_0(E,n,\eta_{g2})=\left[\bar{S}_2(E-V_0,\eta_{g2})\right]^{-1/2}$$

$$\cdot\left[\left[\left(n+\frac{1}{2}\right)\hbar eb/\left(\sqrt{\rho_1}(E)\,\rho_2(E)\right)\right]-1\right]^{1/2},$$

$$\rho_1(E)=\hbar^2/\left(2\bar{p}_2(E-V_0,\eta_{g2})\right),$$

$$\rho_2(E)=\hbar^2/\left(2\bar{Q}_2(E-V_0,\eta_{g2})\right)$$

$$T_5(E,n,\eta_{g1})=k_0'(E,n,\eta_{g1})\left[b_0-\Delta_0\right],$$

$$k_0'(E,n,\eta_{g1})=\left[\bar{S}_1(E,n,\eta_{g1})\right]^{-1/2}$$

$$\cdot\left[1-\left[(n+1/2)\,\hbar eB/\left(\sqrt{\rho_3(E)\,\rho_4(E)}\right)\right]\right]^{1/2}$$

$$\rho_3(E)=\hbar^2/\left(2\bar{p}_1(E,\eta_{g2})\right),\quad \rho_4(E)=\hbar^2/\left(2\bar{Q}_1(E,\eta_{g1})\right)$$

$$T_6(E,n)=\left[\frac{k_0(E,n,\eta_{g2})}{k_0'(E,n,\eta_{g1})}-\frac{k_0'(E,n,\eta_{g1})}{k_0(E,n,\eta_{g2})}\right]$$

The DOS function can be written as

$$N(E)=\frac{eBg_v}{2\pi^2\hbar L_0}\sum_{n=0}^{n_{\max}}$$

$$\cdot\frac{\cos^{-1}[\frac{1}{2}\overline{\varphi}_6(E,n)(1-\frac{1}{4}[\overline{\varphi}_6(E,n)]^2]^{-1/2}[\overline{\varphi}_6(E,n)]'H(E-E_{15,5})}{[[\cos^{-1}[\frac{1}{2}\overline{\varphi}_6(E,n)]]^2-L_0^2\frac{2eB}{\hbar}(1+\frac{1}{2})]^{1/2}} \tag{14.9b}$$

where $E_{15,5}$ is the sub-band energy in this case and is given by

$$0=\frac{1}{L_0^2}\left[\cos^{-1}\left\{\frac{1}{2}\bar{\varphi}_6(E_{15,5},n)\right\}\right]^2-\frac{2|e|B}{\hbar}\left(n+\frac{1}{2}\right) \tag{14.9c}$$

The EEM can be written as

$$m^*(E,\eta_g,n) = \frac{\hbar^2}{2L_0^2}\cos^{-1}\left[\frac{1}{2}\overline{\varphi}_6(E,n)1 - \frac{1}{4}[\overline{\varphi}_6(E,n)]^2\right]^{-1/2}[\overline{\varphi}_6(E,n)]' \tag{14.10a}$$

The electron concentration can be written as

$$n_0 = \frac{g_v eB}{\pi^2\hbar}\sum_{n=0}^{n_{\max}} \cdot\left[\left[\frac{1}{L_0^2}\left[\cos^{-1}\left(\frac{1}{2}\phi_2(E,n)\right]^2 - \frac{2eB}{\hbar}\left(n+\frac{1}{2}\right)\right]^{1/2}\Big|_{E_{F1325}} \cdot\sum_{r=1}^{s} L(r)\left[\sum_{n=0}^{n_{\max}}\left[\left[\frac{1}{L_0^2}\left[\cos^{-1}\left(\frac{1}{2}\phi_2(E,n)\right]^2 - \frac{2eB}{\hbar}\left(n+\frac{1}{2}\right)\right]^{1/2}\Big|_{E_{F1325}}\right]\right] \tag{14.10b}$$

where E_{F1325} is the fermi energy in this case.

14.2.6 *The EM in III–V HD effective mass super lattices under magnetic quantization*

The magneto electron dispersion relation in this case assumes the form

$$-(k_x)^2 = \delta_{7E,n} + i\delta_{8E,n} \tag{14.11a}$$

where

$$\delta_{7E,n} = \left[\frac{1}{2}\left(\delta_{5E,n}^2 - \delta_{6E,n}^2\right) - \left\{\frac{2eB}{\hbar}\left(n+\frac{1}{2}\right)\right\}\right],$$

$$\delta_{5E,n} = \cos^{-1} p_{5E,n},$$

$$p_{5E,n} = \left[\frac{1-\delta_{3E,n}^2-\delta_{4E,n}^2-\sqrt{\left(1-\delta_{3E,n}^2-\delta_{4E,n}^2\right)+4\delta_{4E,n}^2}}{2}\right]^{1/2},$$

$$\delta_{3E,n} = (a_1\cos\Delta_{1E,n}\cos\Delta_{2E,n} - a_2\cos\Delta_{3E,n}\cos\Delta_{4E,n})$$

$$\delta_{4E,n} = (a_2 \sin \Delta_{1E,n} \sin \Delta_{2E,n} - a_2 \sin \Delta_{3E,n} \sin \Delta_{4E,n})$$

$$\Delta_{1E,n} = (a_0 e_{1E,n} + b_0 e_{3E,n}), \quad \Delta_{2E,n} = (a_0 e_{2E,n} + b_0 e_{4E,n}),$$

$$\Delta_{3E,n} = (a_0 e_{1E,n} - b_0 e_{3E,n}), \quad \Delta_{4E,n} = (a_0 e_{2E,n} - b_0 e_{4E,n}),$$

$$\delta_{6E,n} = \sinh^{-1} p_{5E,n} \quad \text{and} \quad \delta_{8E,n} = \left[2\delta_{5E,n}\delta_{6E,n}/L_0^2\right],$$

$$e_{1E,n} = \left[\left(\left(\sqrt{t_{1E,n}^2 + t_2^2} + t_{1E,n}\right)/2\right)\right]^{\frac{1}{2}},$$

$$e_{2E,n} = \left[\left(\left(\sqrt{t_{1E,n}^2 + t_2^2} - t_{1E,n}\right)/2\right)\right]^{\frac{1}{2}},$$

$$e_{3E,n} = \left[\frac{\sqrt{t_{3E,n}^2 + t_4^2} + t_{3E,n}}{2}\right]^{1/2},$$

$$e_{4E,n} = \left[\frac{\sqrt{t_{3E,n}^2 + t_4^2} - t_{3E,n}}{2}\right]^{1/2},$$

$$t_{1E,n}\left[\frac{2m_{c1}^*}{\hbar^2} T_{11}(E, \Delta_1, \eta_{g1}, E_{g1}) - \frac{2eB}{\hbar}\left(n + \frac{1}{2}\right)\right],$$

$$t_{3E,n}\left[\frac{2m_{c2}^*}{\hbar^2} T_{12}(E, \Delta_2, \eta_{g2}, E_{g2}) - \frac{2eB}{\hbar}\left(n + \frac{1}{2}\right)\right]$$

The DOS function can be written as

$$N(E) = \frac{eB}{2\pi^2\hbar} \sum_{n=0}^{n_{\max}} \frac{(\delta'_{7E,n} + i\delta'_{8E,n}) H(E - E_{15,6})}{\sqrt{\delta_{7E,n} + i\delta_{8E,n}}} \tag{14.11b}$$

where $E_{15,6}$ is the sub-band energy in this case and is given by

$$0 = \delta_{7E_{15,6},n} + i\delta_{8E_{15,6},n} \tag{14.11c}$$

The EEM can be written as

$$m^*(E, \eta_g, n) = \frac{\hbar^2}{2} \delta'_{7E,n} \tag{14.12a}$$

The electron statistics can be expressed as

$$n_0 = \frac{g_v eB}{\pi^2 \hbar} \text{ Real part of } \sum_{n=0}^{n_{\max}} \left[(\delta_{8,E,n} + +\delta_{8,E,n})^{1/2} |_{E_{F1326}} \right.$$

$$\left. \cdot \sum_{r=1}^{s} L(r) \left[(\delta_{8,E,n} + \delta_{8,E,n})^{1/2} |_{E_{F1326}} \right] \right] \tag{14.12b}$$

where E_{F1326} is the fermi energy in this case.

14.2.7 *The EM in II–VI HD effective mass super lattices under magnetic quantization*

The DR in heavily doped II–VI EMSL can be written as

$$(k_z)^2 = \Delta_{13E,n} + i\Delta_{14E,n}, \tag{14.13a}$$

where

$$\Delta_{13E,n} = \left[\frac{1}{L_0^2} \left(\Delta_{11E,n}^2 - \Delta_{12E,n}^2 \right) - \left\{ \frac{2eB}{\hbar} \left(n + \frac{1}{2} \right) \right\} \right]$$

$$\Delta_{11E,n} = \cos^{-1} p_{6E,n},$$

$$p_{6E,n} = \left[\frac{1 - \Delta_{9E,n}^2 - \Delta_{10E,n}^2 - \sqrt{(1 - \Delta_{9E,n}^2 - \Delta_{10E,n}^2)^2 + 4\Delta_{10E,n}^2}}{2} \right]^{1/2},$$

$$\Delta_{9E,n} = \left(\bar{a}_1 \cos \Delta_{6E,n} \cosh \Delta_{7E,n} - \bar{a}_2 \cos \Delta_{8E,n} \cosh \Delta_{7E,n} \right),$$

$$\Delta_{10E,n} = \left(\bar{a}_1 \sin \Delta_{6E,n} \sinh \Delta_{7E,n} + \bar{a}_2 \sin \Delta_{8E,n} \sinh \Delta_{7E,n} \right),$$

$$\Delta_{6E,n} = \left[a_0 C_{22E,n} \left(E_{E,n}, \eta_{g1} \right) + b_0 e_{3E,n} \right],$$

$$\Delta_{7E,n} = b_0 e_{4E,n},$$

$$\Delta_{8E,n} = \left[a_0 C_{22E,n} \left(E_{E,n}, \eta_{g1} \right) - b_0 e_{3E,n} \right],$$

$$C_{22E,n}\left(E_{E,n,\eta_{g1}}\right) = \left[\frac{2m^*_{\|,1}}{\hbar^2}\left\{\gamma_3\left(E_{E,n,\eta_{g1}}\right) - \frac{\hbar^2}{2m^*_{\perp,1}}\left\{\frac{2eB}{\hbar}\left(n+\frac{1}{2}\right)\right\}\right.\right.$$
$$\left.\left.\mp C_0\left[\left\{\frac{2eB}{\hbar}\left(n+\frac{1}{2}\right)\right\}\right]^{1/2}\right\}\right]^{1/2},$$
$$\Delta_{12E,n} = \cos^{-1} p_{6E,n}, \quad \Delta_{14E,n} = \frac{2\Delta_{11E,n}\Delta_{12E,n}}{L_0^2},$$

The DOS function can be written as

$$N(E) = \frac{eB}{2\pi^2\hbar}\sum_{n=0}^{n_{\max}}\frac{(\Delta'_{13E,n} + i\Delta'_{14E,n})H(E - E_{15,7})}{\sqrt{\Delta_{13E,n} + i\Delta_{14E,n}}} \tag{14.13b}$$

where $E_{15,7}$ is the sub-band energy in this case and is given by

$$0 = \Delta_{13E_{15,7,n}} + i\Delta_{14E_{15,7,n}} \tag{14.13c}$$

The EEM can be written as

$$m^*(E,\eta_g,n) = \frac{\hbar^2}{2}\Delta'_{13E,n} \tag{14.14a}$$

The electron statistics can be expressed as

$$n_0 = \frac{g_v eB}{\pi^2\hbar}\text{ Real part of }\sum_{n=0}^{n_{\max}}\left[(\Delta_{13,E,n} + +\Delta_{14,E,n})^{1/2}|_{E_{F1327}}\right.$$
$$\left.\cdot\sum_{r=1}^{s}L(r)\left[(\Delta_{13,E,n} + +\Delta_{14,E,n})^{1/2}|_{E_{F1327}}\right]\right] \tag{14.14b}$$

where E_{F1327} is the fermi energy in this case.

14.2.8 *The EM in IV–VI HD effective mass super lattices under magnetic quantization*

The DR in heavily doped IV–VI, EMSLs under magnetic quantization can be written as

$$(k_z)^2 = \left[[1/L_0^2]\{\cos^{-1}(f_{23}(E,n))\}^2 - \left(\frac{2eB}{\hbar}\left(n+\frac{1}{2}\right)\right)\right] \tag{14.15a}$$

where,

$$f_{23}(E,n) = a_3 \cos[a_0 C_{23E,n}(E,n,\eta_{g1}) + b_0 D_{23E,n}(E,n,\eta_{g1})] \\ - a_4 \cos[a_0 C_{23E,n}(E,n,\eta_{g2}) - b_0 D_{23E,n}(E,n,\eta_{g2})],$$

$$C_{23}(E,n,\eta_{g1}) = \left[[2\overline{p_{9,1}}]^{-1}\left[-\overline{q_{9,1}}(E,n,\eta_{g1}) + \left[\{\overline{q_{9,1}}(E,n,\eta_{g1})\}^2 \right.\right.\right. \\ \left.\left.\left. + (4\overline{p_{9,1}})\overline{R_{9,1}}(E,n,\eta_{g1})\right]^{1/2}\right]\right]^{1/2},$$

$$D_{23}(E,n,\eta_{g2}) = \left[[2\overline{p_{9,2}}]^{-1}\left[-\overline{q_{9,2}}(E,n,\eta_{g2}) + \left[\{\overline{q_{9,2}}(E,n,\eta_{g2})\}^2 \right.\right.\right. \\ \left.\left.\left. + (4\overline{p_{9,2}})\overline{R_{9,2}}(E,n,\eta_{g2})\right]^{1/2}\right]\right]^{1/2},$$

$$\overline{q_{9,i}}(E,n,\eta_{gi}) = \left[\frac{\hbar^2}{2}\left(\frac{1}{m_{l,i}^+} - \frac{1}{m_{l,i}^-}\right)\right. \\ + \alpha_i \frac{\hbar^4}{4}\left(\frac{2eB}{\hbar}\left(n+\frac{1}{2}\right)\right)\left(\frac{1}{m_{l,i}^+ m_{t,i}^-} + \frac{1}{m_{t,i}^+ m_{l,i}^-}\right) \\ \left. - \alpha_i \gamma_3(E,\eta_{gi})\left(\frac{1}{m_{l,i}^+} - \frac{1}{m_{l,i}^-}\right)\right],$$

$$\overline{R_{9,i}}(E,n,\eta_{gi}) = \left[\gamma_2(E,\eta_{gi}) + \gamma_3(E,\eta_{gi})\right. \\ + \alpha_i \frac{\hbar^2}{2}\left(\frac{2eB}{\hbar}\left(n+\frac{1}{2}\right)\right)\left(\frac{1}{m_{t,i}^+} - \frac{1}{m_{t,i}^+}\right) \\ - \frac{\hbar^2}{2}\left(\frac{2eB}{\hbar}\left(n+\frac{1}{2}\right)\right)\left(\frac{1}{m_{t,i}^+} - \frac{1}{m_{t,i}^+}\right) \\ \left. - \frac{\alpha\hbar^6}{4}\frac{\left(\frac{2eB}{\hbar}\left(n+\frac{1}{2}\right)\right)^2}{m_{t,i}^- m_{t,i}^+}\right]$$

The DOS function can be written as

$$N(E) = \frac{eBg_v}{2\pi^2 \hbar L_0} \sum_{n=0}^{n_{\max}} \cdot \frac{\cos^{-1}[\frac{1}{2} f_{23}(E,n)(1 - \frac{1}{4} f_{23}^2(E,n)]^{-1/2} f'_{23}(E,n) \mathrm{H}(E - E_{15,8})}{[[\cos^{-1}[\frac{1}{2} f_{23}(E,n)]]^2 - L_0^2 \frac{2eB}{\hbar}(1 + \frac{1}{2})]^{1/2}} \tag{14.15b}$$

where $E_{15,8}$ is the sub-band energy in this case and is given by

$$0 = \left[[1/L_0^2]\{\cos^{-1}(f_{23}(E_{15,8}, n))\}^2 - \left(\frac{2eB}{\hbar} \left(n + \frac{1}{2} \right) \right) \right] \tag{14.15c}$$

The EEM can be written as

$$m^*(E, \eta_g, n) = \frac{\hbar^2}{2L_0^2} \cos^{-1} \left[\frac{1}{2} f_{23}(E,n) \left(1 - \frac{1}{4} f_{23}^2(E,n) \right]^{-1/2} f'_{23}(E,n) \tag{14.16a}$$

The electron concentration can be written as

$$n_0 = \frac{g_v eB}{\pi^2 \hbar} \sum_{n=0}^{n_{\max}} \left[\left[\frac{1}{L_0^2} \left[\cos^{-1}(f_{23}(E,n)]^2 - \frac{2eB}{\hbar} \left(n + \frac{1}{2} \right) \right]^{1/2} \right|_{E_{F1328}} \cdot \sum_{r=1}^{s} L(r) \left[\sum_{n=0}^{n_{\max}} \left[\left[\frac{1}{L_0^2} \left[\cos^{-1}(f_{23}(E,n)]^2 - \frac{2eB}{\hbar} \left(n + \frac{1}{2} \right) \right]^{1/2} \right|_{E_{F1328}} \right] \right] \tag{14.16b}$$

where E_{F1328} is the fermi energy in this case.

14.2.9 *The EM in HgTe/CdTe HD effective mass super lattices under magnetic quantization*

The DR in heavily doped HgTe/CdTe EMSLs under magnetic quantization can be written as

$$(k_z)^2 = \Delta_{13HE,n} + i\Delta_{14HE,n} \tag{14.17a}$$

$$\Delta_{13HE,n} = \left[\frac{1}{L_0^2}\left(\Delta_{11HE,n}^2 - \Delta_{12HE,n}^2\right) - \left\{\frac{2eB}{\hbar}\left(n+\frac{1}{2}\right)\right\}\right]$$

$$\Delta_{11HE,n} = \cos^{-1} p_{6HE,n},$$

$$p_{6HE,n} = \left[\frac{\begin{array}{l}1 - \Delta_{9HE,n}^2 - \Delta_{10HE,n}^2 \\ \quad -\sqrt{(1-\Delta_{9HE,n}^2 - \Delta_{10HE,n}^2)^2 + 4\Delta_{10HE,n}^2}\end{array}}{2}\right]^{1/2},$$

$$\Delta_{9HE,n} = (\bar{a}_{1H}\cos\Delta_{5HE,n}\cosh\Delta_{6HE,n} - \bar{a}_{2H}\cos\Delta_{7HE,n}\cosh\Delta_{6HE,n}),$$

$$\Delta_{10HE,n} = (\bar{a}_{1H}\sin\Delta_{5HE,n}\sinh\Delta_{6HE,n} + \bar{a}_{2H}\sin\Delta_{7HE,n}\sinh\Delta_{6HE,n}),$$

$$\Delta_{5HE,n} = \left[a_0 C_{22HE,n}\left(E_{E,n},\eta_{g1}\right) + b_0 e_3\right], \quad \Delta_{6HE,n} = b_0 e_4,$$

$$\Delta_{7HE,n} = \left[a_0 C_{22HE,n}\left(E_{E,n},\eta_{g1}\right) - b_0 e_3\right],$$

$$C_{22HE,n}\left(E_{E,n,\eta_{g1}}\right) = \left[\frac{B_{01}^2 + 2A_1 E_{E,n} - B_{01}(B_{01}^2 + 4A_1 E_{E,n})}{2A_1^2} - \left[\frac{2eB}{\hbar}\left(n+\frac{1}{2}\right)\right]\right]^{1/2},$$

$$\Delta_{12HE,n} = \cos^{-1} p_{6HE,n}, \quad \Delta_{14HE,n} = \frac{2\Delta_{11HE,n}\Delta_{12HE,n}}{L_0^2},$$

The DOS function can be written as

$$N(E) = \frac{eB}{2\pi^2\hbar}\sum_{n=0}^{n_{\max}} \frac{(\Delta'_{13HE,n} + i\Delta'_{14HE,n})H(E - E_{15,9})}{\sqrt{\Delta_{13HE,n} + i\Delta_{14HE,n}}} \tag{14.17b}$$

where $E_{15,9}$ is the sub-band energy in this case and is given by

$$0 = \Delta_{13HE_{15,9},n} + i\Delta_{14HE_{15,9},n} \tag{14.17c}$$

The EEM can be written as

$$m^*(E,\eta_g,n) = \frac{\hbar^2}{2}\Delta'_{13HE,n} \tag{14.18a}$$

The electron statistics can be expressed as

$$n_0 = \frac{g_v eB}{\pi^2 \hbar} \text{ Real part of } \sum_{n=0}^{n_{\max}} \left[(\Delta_{13,HE,n} + +\Delta_{14,HE,n})^{1/2} |_{E_{F1327}} \right.$$

$$\left. \cdot \sum_{r=1}^{s} L(r)[(\Delta_{13,HE,n} + +\Delta_{14,HE,n})^{1/2} |_{E_{F1329}}] \right] \tag{14.18b}$$

where E_{F1329} is the fermi energy in this case.

14.2.10 *The EM in strained layer HD effective mass super lattices under magnetic quantization*

The DR in heavily doped strained layer effective mass super-lattices under magnetic quantization can be expressed as

$$(k_z)^2 = \left[\frac{1}{L_0^2} \left\{ \cos^{-1}(f_{40}(E,n)) \right\}^2 - \left(\frac{2eB}{\hbar} \left(n + \frac{1}{2} \right) \right) \right] \tag{14.19a}$$

where,

$$f_{40}(E,n) = a_{20} \cos[a_{20} C_{40}(E,n,\eta_{g1}) + b_0 D_{40}(E,n,\eta_{g1})]$$
$$-a_{21} \cos[a_0 C_{40}(E,n,\eta_{g2}) - b_0 D_{40}(E,n,\eta_{g2})],$$

$$C_{40}(E,n,\eta_{g1}) = \left[1 - \frac{\hbar eB}{\phi_{50}(E,\eta_{g1})} \left(n + \frac{1}{2} \right) \right]^{1/2} [\bar{S}_1(E,\eta_{g1})]^{=1/2},$$

$$\phi_{50}(E,\eta_{g1}) = \sqrt{\psi_{50}(E,\eta_{g1}) \psi_{51} \eta_{g1}},$$

$$\psi_{50}(E,\eta_{g1}) = \frac{\hbar^2}{2\bar{P}_1(E,\eta_{g1})}, \quad \psi_{51}(E,\eta_{g1}) = \frac{\hbar^2}{2\bar{Q}_1(E,\eta_{g1})}$$

$$D_{40}(E,n,\eta_{g2}) = \left[1 - \frac{\hbar eB}{\phi_{50}(E,\eta_{g2})} \left(n + \frac{1}{2} \right) \right]^{1/2} [\bar{S}_2(E,\eta_{g2})]^{=1/2},$$

$$\phi_{501}(E,\eta_{g2}) = \sqrt{\psi_{501}(E,\eta_{g2}) \psi_{511} \eta_{g2}}$$

$$\psi_{501}(E,\eta_{g2}) = \frac{\hbar^2}{2\bar{P}_1(E,\eta_{g2})}, \quad \psi_{511}(E,\eta_{g2}) = \frac{\hbar^2}{2\bar{Q}_2(E,\eta_{g2})}$$

The DOS function can be written as

$$N(E) = \frac{eBg_v}{2\pi^2\hbar L_0}\sum_{n=0}^{n_{\max}}$$

$$\cdot\frac{\cos^{-1}[\frac{1}{2}f_{40}(E,n)(1-\frac{1}{4}f_{40}^2(E,n)]^{-1/2}f'_{40}(E,n)H(E-E_{15,10})}{[[\cos^{-1}[\frac{1}{2}f_{40}(E,n)]]^2 - L_0^2\frac{2eB}{\hbar}(1+\frac{1}{2})]^{1/2}} \tag{14.19b}$$

where $E_{15,10}$ is the sub-band energy in this case and is given by

$$0 = \left[[1/L_0^2]\{\cos^{-1}(f_{40}(E_{15,10},n))\}^2 - \left(\frac{2eB}{\hbar}\left(n+\frac{1}{2}\right)\right)\right] \tag{14.19c}$$

The EEM can be written as

$$m^*(E,\eta_g,n) = \frac{\hbar^2}{2L_0^2}\cos^{-1}\left[\frac{1}{2}f_{40}(E,n)\left(1-\frac{1}{4}f_{40}^2(E,n)\right]^{-1/2}f'_{40}(E,n) \tag{14.20a}$$

The electron concentration can be written as

$$n_0 = \frac{g_v eB}{\pi^2\hbar}\sum_{n=0}^{n_{\max}}\left[\left[\frac{1}{L_0^2}\left[\cos^{-1}(f_{40}(E,n)]^2 - \frac{2eB}{\hbar}\left(n+\frac{1}{2}\right)\right]^{1/2}\right|_{E_{F13210}}$$

$$\cdot\sum_{r=1}^{s}L(r)\left[\sum_{n=0}^{n_{\max}}\left[\left[\frac{1}{L_0^2}\left[\cos^{-1}(f_{40}(E,n)\right]^2\right.\right.\right.$$

$$\left.\left.\left.-\frac{2eB}{\hbar}\left(n+\frac{1}{2}\right)\right]^{1/2}\right|_{E_{F13210}}\right]\right] \tag{14.20b}$$

where E_{F13210} is the fermi energy in this case.

14.3 Open Research Problems

(R.14.1) Investigate the magneto EM in the presence of an arbitrarily oriented non-quantizing magnetic field in III–V, II–VI, IV–VI, HgTe/CdTe and strained layer heavily doped superlattices (HDSLs) with graded interfaces by including the electron spin.

(R.14.2) Investigate the magneto EM in III–V, II–VI, IV–VI, HgTe/CdTe and strained layer heavily doped effective mass superlattices in the presence of an arbitrarily oriented non-quantizing magnetic field by including the electron spin.

(R.14.3) Investigate the EM for all the problems from R.14.1 to R.14.2 in the presence of an additional arbitrarily oriented electric field.

(R.14.4) Investigate the EM for all the problems from R.14.1 to R.14.3 in the presence of arbitrarily oriented crossed electric and magnetic fields.

References

1. M Mondal, KP Ghatak *Czechoslovak Journal of Physics* B **36**, 1389 (1986)
2. S Bhattacharyya, KP Ghatak, S Biswas — OE/Fibers' 87, 1987, 73, International Society for Optics and Photonics
3. B Mitra, KP Ghatak *physica status solidi (b)* **149**, K117 (1988)
4. KP Ghatak, SN Biswas MRS Proceedings **161**, 371 (1989) (USA)
5. B Mitra, KP Ghatak *Physics Letters A* **142**, 401 (1989)
6. KP Ghatak, SN Biswas in Properties of II–VI semiconductors: Bulk crystals, epitaxial films, quantum well structures, and dilute magnetic systems, 1990 (USA)
7. KA Makhya P Ghatak in Properties of II–VI Semiconductors Bulk Crystals, Epitaxial Films, Quantum Well Structures, and Dilute Magnetic Systems, **161,** 371 (1990) (USA)
8. Kamakhya P Ghatak, Bhaswati Mitra, Shambhu N Biswas in Optics, Electro-Optics, and Laser Applications in Science and Engineering 137, International Society for Optics and Photonics (1991)
9. KP Ghatak *Il Nuovo Cimento D* **13**, 1321 (1991)
10. KP Ghatak, A Ghoshal *physica status solidi (b)* **170**, K27 (1992)

Chapter 15

Appendix G: The EM in Heavily Doped Ultra-thin Films (HDUFs) Under Cross-Fields Configuration

15.1 Introduction

In this chapter in Section 15.2.1 of theoretical background, the EM in ultra thin films of HD nonlinear optical materials in the presence of crossed electric and quantizing magnetic fields has been investigated by formulating the electron dispersion relation. Section 15.2.2 reflects the study of the EM in ultra thin films of HD III–V, ternary and quaternary compounds as a special case of section 15.2.1. Section 15.2.3 contains the study of the EM for the ultra thin films of HD II–VI semiconductors in the present case. In Section 15.2.4, the EM under cross field configuration in ultra thin films of HD IV–VI semiconductors has been investigated in accordance with the models of the Cohen, the Lax non-parabolic ellipsoidal and the parabolic ellipsoidal respectively. In Section 15.2.5, the EM for the ultra thin films of HD stressed Kane type semiconductors has been investigated. The last Section 15.3 eleven open research problems.

15.2 Theoretical Background

15.2.1 *The EM in Heavily Doped Ultra-thin Films (HDUFs) of nonlinear optical semiconductors under cross-fields configuration*

The DR of the conduction electrons in HD ultrathin films of nonlinear optical material in the presence of cross-fields configuration can be written as

$$T_{22}(E,\eta_g) = \left[\left(\left(n+\frac{1}{2}\right)\hbar\omega_{01}\right) + \left(\frac{[\hbar]^2}{2a(E,\eta_g)}\right)\left(\frac{\pi n_z}{d_z}\right) - \left(\frac{E_0\hbar k_y \rho(E,\eta_g)}{B}\right) - \left(\frac{M_\perp \rho^2(E,\eta_g)E_0^2}{2B^2}\right)\right] \quad (15.1)$$

The use of (15.1) leads to the expression of EEM along y direction as

$$m_y^*(e_{fA1},\eta_g,n,E_0,n_z) = \text{Real Part of } (B/E_0)T_{49}(e_{fA1},\eta_g,n_z)\,[T_{49}(e_{fA1},\eta_g,n_z)]' \quad (15.2a)$$

where e_{fA1} is the Fermi energy in this case and

$$T_{49}(e_{fA1},\eta_g,n_z) = \left[T_{22}(e_{fA1},\eta_g) = \left[\left(\left(n+\frac{1}{2}\right)\hbar\omega_{01}\right) + \left(\frac{[\hbar]^2}{2a(e_{fA1},\eta_g)}\right)\left(\frac{\pi n_z}{d_z}\right) - \left(\frac{M_\perp \rho^2(e_{fA1},\eta_g)E_0^2}{2B^2}\right)\right]\rho(e_{fA1},\eta_g)^{-1}\right]$$

The sub band energy $E_{9,1}$ in this case assumes the from

$$T_{22}(E_{9,1},\eta_g) = \left[\left(\left(n+\frac{1}{2}\right)\hbar\omega_{01}\right) + \left(\frac{[\hbar]^2}{2a(E_{9,1},\eta_g)}\right)\left(\frac{\pi n_z}{d_z}\right) - \left(\frac{M_\perp \rho^2(E_{9,1},\eta_g)E_0^2}{2B^2}\right)\right] \quad (15.2b)$$

The surface electron statistics can be written as

$$n_s = \frac{g_v eB}{\pi\hbar} \text{ Real part of } \sum_{n=0}^{n_{\max}} \sum_{n_z=0}^{n_{z\max}} F_{-1}(n_{15,2,1}) \quad (15.2c)$$

Where

$$n_{15,2,1} = \frac{e_{fA1} - E_{9.1}}{k_B T}$$

15.2.2 The EM in Heavily Doped Ultra-thin Films (HDUFs) of type III–V semiconductors under cross-fields configuration

(a) Under the conditions $\delta = 0, \Delta_{\|} = \Delta_{\perp} = \Delta$ and $m_{\|}^* = m_{\perp}^* m_c$ (9.1) assumes the form

(b) $$T_{33}(E, \eta_g) = \left(n + \frac{1}{2}\right)\hbar\omega_{01} + \frac{[\hbar\frac{\pi n_z}{d_z}]^2}{2m_c} - \frac{E_0}{B}\hbar k_y \{T_{33}(E, \eta_g)\}' - \frac{m_c E_0^2[\{T_{33}(E, \eta_g)\}']^2}{2B^2} \quad (15.3)$$

The use of (15.3) leads to the expression of EEM along y direction as

$$m_y^*(e_{fA2}, \eta_g, n, E_0, n_z) = \text{Real Part of } (B/E_0)^2 T_{50}(e_{fA2}, \eta_g, n_z)\,[T_{50}(e_{fA2}, \eta_g, n_z)]' \quad (15.4a)$$

where e_{fA2} is the Fermi energy in this case and

$$T_{50}(e_{fA2}, \eta_g, n_z) = \left[T_{30}(e_{fA2}, \eta_g) - \left(n + \frac{1}{2}\right)\hbar\omega_0 - \frac{[\hbar\frac{\pi n_z}{d_z}]^2}{2m_c} + \frac{m_{\perp} E_0^2[\{T_{33}(e_{fA2}, \eta_g)\}']^2}{2B^2}\right][\{T_{33}(e_{fA2}, \eta_g)\}']^{-1}$$

The sub-band energy $E_{9,2}$ in this case assumes the from

$$T_{33}(E_{9,2}, \eta_g) = \left(n + \frac{1}{2}\right)\hbar\omega_0 + \frac{[\hbar\frac{\pi n_z}{d_z}]^2}{2m_c} - \frac{m_c E_0^2[\{T_{33}(E_{9,2}, \eta_g)\}']^2}{2B^2} \quad (15.4b)$$

(b) HD two band model of Kane

Under the condition $\Delta >> E_g$, (15.3) assumes the form

$$\gamma_2(E,\eta_g) = \left(n+\frac{1}{2}\right)\hbar\omega_0 - \frac{E_0}{B}\hbar k_y \gamma_2'(E,\eta_g) - \frac{m_c E_0^2}{2B^2}\left(\gamma_2'(E,\eta_g)\right) + \frac{[\hbar\frac{n_z\pi}{d_z}]^2}{2m_c} \tag{15.5}$$

The use of (15.5) leads to the expression of EEM along y direction as

$$m_y^*(e_{fA3},\eta_g,n,E_0,n_z) = (B/E_0)^2 T_{51}(e_{fA3},\eta_g,n_z)\,[T_{51}(e_{fA3},\eta_g,n_z)]' \tag{15.6a}$$

where e_{fA3} is the Fermi energy in this case and

$$T_{51}(e_{fA3},\eta_g,n_z) = \left[\gamma_2(e_{fA3},\eta_g) - \left(n+\frac{1}{2}\right)\hbar\omega_0 - \frac{[\hbar\frac{\pi n_z}{d_z}]^2}{2m_c} + \frac{m_c E_0^2[\{\gamma_2(e_{fA3},\eta_g)\}']^2}{2B^2}\right][\{\gamma_2(e_{fA3},\eta_g)\}']^{-1}$$

The sub-band energy $E_{9,3}$ in this case assumes the from

$$\gamma_2(E_{9,3},\eta_g) = \left(n+\frac{1}{2}\right)\hbar\omega_0 - \frac{m_c E_0^2}{2B^2}\left(\gamma_2'(E_{9,3},\eta_g)\right)^2 + \frac{[\hbar\frac{n_z\pi}{d_z}]^2}{2m_c} \tag{15.6b}$$

(c) HD Parabolic energy bands

The DR and for this model under this condition $\alpha \to 0$ can be written as

$$\gamma_3(E,\eta_g) = \left(n+\frac{1}{2}\right)\hbar\omega_0 - \frac{E_0}{B}\hbar k_y \gamma_3'(E,\eta_g) - \frac{m_c E_0^2}{2B^2}\left(\gamma_3'(E,\eta_g)\right)^2 + \frac{[\hbar\frac{n_z\pi}{d_z}]^2}{2m_c} \tag{15.7}$$

The use of (15.7) leads to the expression of EEM along y direction as

$$m_y^*(e_{fA4},\eta_g,n,E_0,n_z) = (B/E_0)^2 T_{52}(e_{fA4},\eta_g,n_z)\,[T_{52}(e_{fA4},\eta_g,n_z)]' \tag{15.8a}$$

where e_{fA4} is the Fermi energy in this case and

$$T_{52}(e_{fA4},\eta_g,n_z) = \left[\gamma_2(e_{fA4},\eta_g) - \left(n+\frac{1}{2}\right)\hbar\omega_0 - \frac{[\hbar\frac{\pi n_z}{d_z}]^2}{2m_c} + \frac{m_c E_0^2[\{\gamma_3(e_{fA4},\eta_g)\}']^2}{2B^2}\right][\{\gamma_3(e_{fA4},\eta_g)\}']^{-1}$$

The sub-band energy $E_{9,4}$ in this case assumes the from

$$\gamma_3(E_{9,4},\eta_g) = \left(n+\frac{1}{2}\right)\hbar\omega_0 - \frac{m_c E_0^2}{2B^2}\left(\gamma_3'(E_{9,4},\eta_g)\right)^2 + \frac{[\hbar\frac{n_z\pi}{d_z}]^2}{2m_c} \quad (15.8b)$$

The surface electron statistics can be written as

$$n_s = \frac{g_v eB}{\pi\hbar}\sum_{n=0}^{n_{max}}\sum_{n_z=0}^{n_{zmax}} F_{-1}(n_{15,2,2c})$$

Where

$$n_{15,2,2c} = \frac{e_{fA4} - E_{9.4}}{k_B T} \quad (15.8c)$$

15.2.3 *The EM in Heavily Doped Ultra-thin Films (HDUFs) of II–VI semiconductors under cross-fields configuration*

TheDRin this case in ultrathin films of HD II–VI semiconductors can be written as

$$\gamma_3(E,\eta_g) = \beta_1(n,E_0) - \frac{E_0}{B}\hbar k_y\gamma_3'(E,\eta_g) - \frac{m_\parallel^* E_0^2}{2B^2}\left(\gamma_3'(E,\eta_g)\right)^2 + \frac{[\hbar\frac{n_z\pi}{d_z}]^2}{2m_\parallel} \quad (15.9)$$

The use of (15.9) leads to the expression of EEM along y direction as

$$m_y^*(e_{fA4},\eta_g,n,E_0,n_z) = (B/E_0)^2 T_{53}(e_{fA5},\eta_g,n_z)\,[T_{53}(e_{fA5},\eta_g,n_z)]' \quad (15.10a)$$

where e_{fA5} is the Fermi energy in this case and

$$T_{53}(e_{fA5},\eta_g,n_z) = \left[\gamma_3(e_{fA5},\eta_g) - \beta_1(n,E_0) - \frac{[\hbar\frac{\pi n_z}{d_z}]^2}{2m_\parallel^*} + \frac{m_\parallel^* E_0^2[\{\gamma_3(e_{fA5},\eta_g)\}']^2}{2B^2}\right][\{\gamma_3(e_{fA5},\eta_g)\}']^{-1}$$

The sub-band energy $E_{9,5}$ in this case assumes the from

$$\gamma_3(\mathrm{E}_{9,5},\eta_\mathrm{g}) = \beta_1(\mathrm{n},\mathrm{E}_0) - \frac{\mathrm{m}_\|^*\mathrm{E}_0^2}{2\mathrm{B}^2}\left(\gamma_3'(\mathrm{E}_{9,5},\eta_\mathrm{g})\right)^2 + \frac{[\hbar\frac{\mathrm{n_z}\pi}{\mathrm{d_z}}]^2}{2\mathrm{m}_\|^*} \quad (15.10\mathrm{b})$$

The surface electron statistics can be written as

$$n_s = \frac{g_v eB}{\pi\hbar}\sum_{n=0}^{n_{\max}}\sum_{n_z=0}^{n_{z\max}} F_{-1}(n_{15,2,3})$$

Where

$$n_{15,2,3} = \frac{e_{fA5} - E_{9.5}}{k_B T} \quad (15.10\mathrm{c})$$

15.2.4 *The EM in Heavily Doped Ultra-thin Films (HDUFs) of IV–VI semiconductors under cross-fields configuration*

The DR in this case is given by

$$g^*(E,\eta_g) = \left(n+\frac{1}{2}\right)\hbar\overline{\omega_{i1}}(E,\eta_g) + \frac{\hbar^2(\frac{n_z\pi}{d_z})^2}{2M_3^*(E,\eta_g)} - \frac{E_0}{B}\rho_1^*(E,\eta_g)\hbar k_y$$
$$-\frac{E_0^2}{2B^2}[\rho_1^*(E,\eta_g)]^2 M_1^*(E,\eta_g) \quad (15.11)$$

The use of (15.11) leads to the expression of EM along y direction as

$$\mathrm{m}_\mathrm{y}^*(\mathrm{e_{fA6}},\eta_\mathrm{g},\mathrm{n},\mathrm{E}_0,\mathrm{n_z})$$
$$= \text{Real Part of } (\mathrm{B}/\mathrm{E}_0)^2\mathrm{T}_{54}(\mathrm{e_{fA6}},\eta_\mathrm{g},\mathrm{n_z})\,[\mathrm{T}_{54}(\mathrm{e_{fA6}},\eta_\mathrm{g},\mathrm{n_z})]'$$

where $\mathrm{e_{fA6}}$ is the Fermi energy in this case and

$$\mathrm{T}_{49}(\mathrm{e_{fA6}},\eta_\mathrm{g},\mathrm{n_z}) = \left[g^*(\mathrm{e}_{fA6},\eta_g) = \left(n+\frac{1}{2}\right)\hbar\overline{\omega_{i1}}(\mathrm{e}_{fA6},\eta_g)\right.$$
$$\left. + \frac{\hbar^2(\frac{n_z\pi}{d_z})^2}{2M_3^*(\mathrm{e}_{fA6},\eta_g)} - \frac{E_0^2}{2B^2}[\rho_1^*(\mathrm{e}_{fA6},\eta_g)]^2 M_1^*(\mathrm{e}_{fA6},\eta_g)\right][\rho_1^*(\mathrm{e}_{fA6},\eta_g)]^{-1}$$
$$(15.12\mathrm{a})$$

The sub-band energy $E_{9,6}$ in this case assumes the from

$$g^*(E_{9,6},\eta_g) = \left(n+\frac{1}{2}\right)\hbar\overline{\omega_{i1}}(E_{9,6},\eta_g) + \frac{\hbar^2(\frac{n_z\pi}{d_z})^2}{2M_3^*(E_{9,6},\eta_g)} - \frac{E_0^2}{2B^2}[\rho_1^*(E_{9,6},\eta_g)]^2 M_1^*(E_{9,6},\eta_g) \tag{15.12b}$$

The surface electron statistics can be written as

$$n_s = \frac{g_v eB}{\pi\hbar}\ \text{Real part of} \sum_{n=0}^{n_{\max}}\sum_{n_z=0}^{n_{z\max}} F_{-1}(n_{15,2,4})$$

Where

$$n_{15,2,4} = \frac{e_{fA6} - E_{9.6}}{k_B T} \tag{15.12c}$$

15.2.5 *The EM in Heavily Doped Ultra-thin Films (HDUFs) of stressed semiconductors under cross-fields configuration*

The DR in this case assumes the form

$$G^*(E,\eta_g) = \left(n+\frac{1}{2}\right)\hbar\overline{\omega_i}(E,\eta_g) + \frac{\hbar^2(\frac{n_z\pi}{d_z})^2}{2m_3^*(E,\eta_g)} - \frac{E_0}{B}\rho^*(E,\eta_g)\left[\frac{m_1^*(E,\eta_g)}{m_2^*(E,\eta_g)}\right]^{\frac{1}{2}}\hbar k_y - \frac{E_0^2}{2B^2}[\rho^*(E,\eta_g)]^2 m_1^*(E,\eta_g) \tag{15.13}$$

The use of (15.13) leads to the expression of EEM along y direction as

$$\begin{aligned}&\mathrm{m_y^*(e_{fA7},\eta_g,n,E_0,n_z)}\\ &\quad= \text{Real Part of } \mathrm{(B/E_0)^2 T_{55}(e_{fA7},\eta_g,n_z)[T_{55}(e_{fA7},\eta_g,n_z)]'}\end{aligned}$$

where e_{fA7} is the Fermi energy in this case and

$$T_{55}(e_{fA7},\eta_g,n_z) = \left[G^*(e_{fA7},\eta_g) = \left(n+\frac{1}{2}\right)\hbar\overline{\omega_i}(e_{fA7},\eta_g) + \frac{\hbar^2(\frac{n_z\pi}{d_z})^2}{2m_3^*(e_{fA7},\eta_g)} - \frac{E_0^2}{2B^2}[\rho^*(e_{fA7},\eta_g)]m_1^*(e_{fA7},\eta_g) - \frac{E_0^2}{2B^2}[\rho^*(E,\eta_g)]^2 m_1^*(E,\eta_g)\right][m_4^*(e_{fA7},\eta_g)]^{-1} \tag{15.14a}$$

The sub-band energy $E_{9,7}$ in this case assumes the from

$$G^*(E_{9,7},\eta_g) = \left(n+\frac{1}{2}\right)\hbar\overline{\omega_i}(E_{9,7},\eta_g) + \frac{\hbar^2(\frac{n_z\pi}{d_z})^2}{2m_3^*(E_{9,7},\eta_g)} - \frac{E_0^2}{2B^2}[\rho^*(E_{9,7},\eta_g)]^2 m_1^*(E_{9,7},\eta_g) \tag{15.14b}$$

The surface electron statistics can be written as

$$n_s = \frac{g_v eB}{\pi\hbar}\sum_{n=0}^{n_{\max}}\sum_{n_z=0}^{n_{z\max}} F_{-1}(n_{15,2,5})$$

where

$$n_{15,2,5} = \frac{e_{fA4} - E_{9.7}}{k_B T} \tag{15.14c}$$

15.3 Open Research Problems

(R.15.1) Investigate the 2D EM in the presence of cross fields for all the HD UFs of the materials whose bulk EMs are given in problems in R. 1.1 of chapter 1.

(R.15.2) Investigate the 2D EM in the presence of cross fields in ultra-thin films of HD nonlinear optical semiconductors by including broadening and the electron spin. Study all the special cases for HD III–V, ternary and quaternary materials in this context.

(R.15.3) Investigate the 2D EMs for HD UFs of IV–VI, II–VI and stressed Kane type compounds in the presence of cross fields by including broadening and electron spin.

(R.15.4) Investigate the 2D EM in the presence of cross fields in HD UFs of nonlinear optical semiconductors by including broadening and the electron spin. Study all the special cases for HD III–V, ternary and quaternary materials in this context.

(R.15.5) Investigate the 2D EMs for HD UFs of IV–VI, II–VI and stressed Kane type compounds in the presence of cross fields by including broadening and electron spin.

(R.15.6) Investigate the 2D EM for all the UFs of materials whose bulk EMs are stated in R.1.1 of chapter 1 in the presence of an arbitrarily oriented quantizing magnetic field and crossed electric field by including broadening and electron spin under the condition of heavy doping.

(R.15.7) Investigate the 2D EM in HD UFs of nonlinear optical, III–V, II-VI, IV-VI and stressed Kane type semiconductors in the presence of non-uniform strain.

(R.15.8) Investigate the 2D EM in the presence of an arbitrarily oriented non-quantizing magnetic field for UFs of HD nonlinear optical semiconductors by including the electron spin. Study all the special cases for III–V, ternary and quaternary materials in this context.

(R.15.9) Investigate the EMs in QWs of HD IV–VI, II–VI and stressed Kane type compounds in the presence of an arbitrarily oriented non-quantizing magnetic field by including the electron spin.

(R.15.10) Investigate the 2D EM for HD UFs of all the materials as stated in R.1.1 of chapter 1 in the presence of an arbitrarily oriented magnetic field by including electron spin and broadening.

(R.15.11) Investigate all the problems of this section by removing all the mathematical approximations and establishing the respective appropriate uniqueness conditions.

Chapter 16

Appendix H: The EMs in Low Dimensional HD Systems in the Presence of Magnetic Field

16.1 Introduction

In this chapter in Sections 16.2.1 and 16.2.2 of the theoretical background 16.2 we study the EM in QWs and NWs of HD III–V semiconductors in the presence of magnetic field respectively. In Sections 16.2.3 and 16.2.4 of the theoretical background 16.2 we study the same for QWs and NWs of HD III–V semiconductors in the presence of crossed fields respectively. Besides Sections 16.2.5 and 16.2.6 explore the EM in QWs and NWs of HD IV–VI semiconductors in the presence of magnetic field. Section 16.2.7 contains the study of EM in QWs of HD III–V materials in the presence of an arbitrarily oriented magnetic field. Section 16.3 quentents 12 open research problems.

16.2 Theoretical Background

16.2.1 *The EM in Quantum Wells of HD III–V, ternary and quaternary materials in the presence of magnetic field*

The DR of the conduction electrons in bulk samples of III–V ternary and quaternary materials can be written as

$$E(1 + \alpha E) = \frac{\hbar^2 k^2}{2m_c} \tag{16.1a}$$

where E is the electron energy as measured from the edge of the conduction band in the vertically upward direction in the absence of any quantization, $\alpha = \frac{1}{E_g} E_g$ is the band gap and m_c is the effective electron mass at the edge of the conduction band. Let us consider a thin layer III–V semiconductor of rectangular cross-section. In such a structure the electron motion is quantized along the z-direction, (a being the nano thickness along z-direction) resulting in formation of electric sub-bands corresponding to different quantum numbers. The electrons motion is free along in the x-y plane and the magnetic field B is applied along the y-direction.

In the absence of magnetic field, $E(k)$ dispersion relation of the electrons in quasi 2D structures can be expressed as

$$E(1+\alpha E) = \frac{\hbar^2}{2m_c}\left(\frac{n\pi}{\alpha}\right)^2 + \frac{\hbar^2 k_x^2}{2m_c} + \frac{\hbar^2 k_y^2}{2m_c} \tag{16.1b}$$

In this case, the potential function assumes the form

$$V(z) = 0, \quad 0 < z < a$$

and $V(z) = \alpha\ 0 < z$ and for $z > a$

The corresponding Eigen function in such structure can be written as

$$\Psi_n(x,y,z) = \Psi_n^0 = u_n = \sqrt{\frac{2}{aL_xL_y}}\sin\left(\frac{n\pi}{a}z\right)\exp\left(ik_x x + ik_y y\right)$$

$$= \sqrt{\frac{2}{aL_xL_y}}\psi_n(z)\exp\left(ik_x x + ik_y y\right),$$

$$\left[\psi_n(z) = \sin\left(\frac{n\pi}{a}z\right)\right] \tag{16.2}$$

The (16.1a) can be expressed as

$$\hat{E} = a\hat{p}^2 - b\hat{p}^4$$

where, $a = \frac{1}{2m_c} b = \alpha a^2$ and $\vec{p} = -i\hbar\vec{\nabla}$

Therefore the Hamiltonian can be expressed as

$$H = a(-i\hbar\vec{\nabla} + e\vec{A})^2 - b(-i\hbar\vec{\nabla} + e\vec{A})^4 \tag{16.3}$$

where $\vec{A}$ is the vector potential

We know that $\vec{B} = \vec{\nabla} X \vec{A}$ and let us choose Coulomb gauge here as $\vec{\nabla} \cdot \vec{A} = 0$

Since magnetic field is along y-direction, so

$$\vec{B} = \vec{j}B, \quad \vec{B} = (0, B, 0)$$

Let $\vec{A} = \vec{i}Bz$ so that it satisfies the above two relations i.e.

$$\vec{\nabla} \cdot \vec{A} = \frac{\partial}{\partial x}(Bz) + 0 + 0 = 0$$

and

$$\vec{B} = \vec{\nabla} \cdot \vec{A} = \begin{vmatrix} \vec{i} & \vec{j} & \vec{k} \\ \frac{\partial}{\partial x} & \frac{\partial}{\partial y} & \frac{\partial}{\partial z} \\ Bz & 0 & 0 \end{vmatrix} = \vec{i} \cdot 0 + \vec{j}\frac{\partial}{\partial z}(Bz) + \vec{k}\left[-\frac{\partial}{\partial y}(Bz)\right] = \vec{j}B$$

Applying this in (16.3), we have the Hamiltonian for the system as

$$\begin{aligned} H &= a(-i\hbar\vec{\nabla} + \vec{i}eBz)^2 - b(-i\hbar\vec{\nabla} + \vec{i}eBz)^4 \\ &= a\left[\left(-i\hbar\frac{\partial}{\partial x} + eBz\right)\vec{i} + \vec{j}\left(-i\hbar\frac{\partial}{\partial y}\right) + \vec{k}\left(-i\hbar\frac{\partial}{\partial z}\right)\right]^2 \\ &\quad - b\left[\left(-i\hbar\frac{\partial}{\partial x} + eBz\right)\vec{i} + \vec{j}\left(-i\hbar\frac{\partial}{\partial y}\right) + \vec{k}\left(-i\hbar\frac{\partial}{\partial z}\right)\right]^4 \\ &= a(\hat{H}_1)^2 - b(\hat{H}_1)^4 \end{aligned} \tag{16.4}$$

Again,

$$\hat{H} = \hat{H}_0 + \hat{H}' \tag{16.5}$$

where $\hat{H}_0$ is referred to as unperturbed Hamiltonian.

$\hat{H}'$ is the perturbation which has to be calculated and the effect of H' is assumed to be small.

$\hat{H}_0 u_n = E_n^0 u_n$ is the Schrodinger equation of the system. (16.6)

$\hat{H}_0 u_n = E_n^0 u_n$ is the Eigen value equation for the unperturbed Hamiltonian (16.7) so that

$$\psi_u^2 = u_n = \sqrt{\frac{2}{aL_xL_y}} \sin\left(\frac{n\pi}{a}z\right) \exp\left(ik_xx + ik_yy\right)$$
$$= \psi_n\left(z\right) \exp\left(ik_xx + ik_yy\right)$$

Now,

$$\hat{H}_1\psi_n^0 = \left[\left(-i\hbar\frac{\partial}{\partial x} + eBz\right)\vec{i} + \vec{j}\left(-i\hbar\frac{\partial}{\partial y}\right) + \vec{k}\left(-i\hbar\frac{\partial}{\partial z}\right)\right] u_n$$
$$= \vec{i}\left(-i\hbar \cdot ik_x + eBz\right) u_n \tag{16.8}$$

The term $(\hat{H}_1)^2\psi_n^0$ is being calculated in the following way

$$(\hat{H}_1)^2\psi_n^0 = \hat{H}_1(\hat{H}_1\psi_n^0)$$
$$= \left[\left(-i\hbar\frac{\partial}{\partial x} + eBz\right)\vec{i} + \vec{j}\left(-i\hbar\frac{\partial}{\partial y}\right)\right.$$
$$\left. + \vec{k}\left(-i\hbar\frac{\partial}{\partial z}\right)\right] \cdot \left(\hat{H}_1\psi_n^0\right)$$
$$= \left(-i\hbar \cdot ik_x + eBz\right)\left(-i\hbar \cdot ik_x + eBz\right) u_n$$
$$+ \left(-i\hbar \cdot ik_y + eBz\right)\left(-i\hbar \cdot ik_y + eBz\right) + u_n$$
$$+ \left(-i\hbar\frac{n\pi}{a}\right)\left(-i\hbar\frac{n\pi}{a}\right)\left(-u_n\right)$$
$$(\hat{H}_1)^2\psi_n^0 = \left(\hbar k_x + eBz\right)^2 u_n + \hbar^2k_y^2u_n + \hbar^2\left(\frac{n\pi}{a}\right)^2 u_n \tag{16.9}$$

Therefore

$$(\hat{H}_1)^2\psi_n^0 = \left[\left\{\hbar^2k_x^2 + \hbar^2\left(\frac{n\pi}{a}\right)^2 + \hbar^2k_y^2\right\} + e^2B^2z^2 + 2\hbar k_xeBz\right] u_n \tag{16.10}$$

$$(\hat{H}_1)^3\psi_n^0 = \left[\left(-i\hbar\frac{\partial}{\partial x} + eBz\right)\vec{i} + \vec{j}\left(-i\hbar\frac{\partial}{\partial y}\right)\right.$$
$$\left. + \vec{k}\left(-i\hbar\frac{\partial}{\partial z}\right)\right] (\hat{H}_1)^2\psi_n^0$$

Therefore

$$(\hat{H}_1)^3\psi_n^0 = \vec{i}\left[(-i\hbar\cdot ik_x + eBz)\left\{\hbar^2k_x^2 + \hbar^2k_y^2 + \hbar^2\left(\frac{n\pi}{a}\right)^2\right.\right.$$
$$\left.\left. + e^2B^2z^2 + 2\hbar k_x eBz\right\}\right]u_n$$
$$+\vec{j}\left[(-i\hbar.ik_y)\left\{\hbar^2k_x^2 + \hbar^2k_y^2 + \hbar^2\left(\frac{n\pi}{a}\right)^2 + e^2B^2z^2\right.\right.$$
$$\left.\left. + 2\hbar k_x eBz\right\}\right]u_n + \vec{k}\left[\left(-i\hbar.\frac{n\pi}{a}\right)\{\hbar^2k_x^2\right.$$
$$+\hbar^2k_y^2 + \hbar^2\left(\frac{n\pi}{a}\right)^2 \tag{16.11}$$

$$(\hat{H}_1)^3\psi_n^0 = \vec{i}\left[(\hbar k_x + eBz)\left\{\hbar^2k_x^2 + \hbar^2k_y^2 + \hbar^2\left(\frac{n\pi}{a}\right)^2\right.\right.$$
$$\left.\left. + e^2B^2z^2 + 2\hbar k_x eBz\right\}\right]u_n + +\vec{j}$$
$$\cdot\left[(\hbar k_y)\left\{\hbar^2k_x^2 + \hbar^2k_y^2 + \hbar^2\left(\frac{n\pi}{a}\right)^2 + e^2B^2z^2 + 2\hbar k_x eBz\right\}\right]$$
$$\cdot u_n + \vec{k}[\left(-i\hbar.\frac{n\pi}{a}\right)\{\hbar^2k_x^2 + \hbar^2k_y^2 + \hbar^2\left(\frac{n\pi}{a}\right)^2 + e^2B^2z^2$$
$$+ 2\hbar k_x eBz\}]u_n' + \vec{k}\left[-2i\hbar e^2B^2z - 2i\hbar^2k_x eB\right]u_n \tag{16.12}$$

where

$$\left[u_n' = \sqrt{\frac{2}{a}}i_x i_y \cos\left(\frac{n\pi}{a}\right)z\ \exp\left(ik_x x + ik_y y\right)\right]$$

Again

$$\left(\hat{H}_1\right)^4\psi_n^0 = \left[\left(-i\hbar\frac{\partial}{\partial x} + eBz\right)\vec{i} + \vec{j}\left(-i\hbar\frac{\partial}{\partial y}\right)\right.$$
$$\left. + \vec{k}\left(-i\hbar\frac{\partial}{\partial z}\right)\right]\left(\hat{H}_1\right)^3\psi_n^0$$

$$
\begin{aligned}
&= (\hbar k_x + eBz)^2 \left\{ \hbar^2 k_x^2 + \hbar^2 k_y^2 + \hbar^2 \left(\frac{n\pi}{a}\right)^2 + e^2 B^2 z^2 \right. \\
&\quad \left. + \ 2\hbar k_x eBz \right\} u_n + \left(\hbar^2 k_y^2\right) \left\{ \hbar^2 k_x^2 + \hbar^2 k_y^2 + \hbar^2 \left(\frac{n\pi}{a}\right)^2 \right. \\
&\quad \left. + e^2 B^2 z^2 + 2\hbar k_x eBz \right\} + u_n \\
&\quad + \left(-i\hbar\frac{n\pi}{a}\right)\left(-i\hbar\frac{n\pi}{a}\right) \left\{ \hbar^2 k_x^2 + \hbar^2 k_y^2 + \hbar^2 \left(\frac{n\pi}{a}\right) \right. \\
&\quad \left. + e^2 B^2 z^2 + 2\hbar k_x + eBz \right\} (-u_n) \\
&\quad + \left(i\hbar\frac{n\pi}{a}\right)(-i\hbar)\left(e^2 B^2 2z + 2\hbar k_x eBz\right) u_n' \\
&\quad + (-i\hbar)\left(-2i\hbar^2 e^2 B^2 z - 2i\hbar^2 k_x eB\right)\left(\frac{n\pi}{a}\right) u_n' \\
&\quad + (-i\hbar)\left(-2i\hbar e^2 B^2\right) u_n \\
&= \left(\hbar^2 k_x^2 + 2\hbar k_x eBz + e^2 B^2 z^2\right) \left\{ \left\{ \hbar^2 k_x^2 \right.\right. \\
&\quad \left.\left. + \ \hbar^2 k_y^2 + \hbar^2 \left(\frac{n\pi}{a}\right)^2 + e^2 B^2 z^2 + 2\hbar k_x eBz \right\} + u_n \right\} + \hbar^2 k_y^2 \\
&\quad \cdot \left\{ \hbar^2 k_x^2 + \hbar^2 k_y^2 + \hbar^2 \left(\frac{n\pi}{a}\right)^2 e^2 B^2 z^2 + 2\hbar k_x eBz \right\} \\
&\quad + u_n + \hbar^2 \left(\frac{n\pi}{a}\right)^2 \left\{ \hbar^2 k_x^2 + \hbar^2 k_y^2 + \hbar^2 \left(\frac{n\pi}{a}\right)^2 \right. \\
&\quad \left. + e^2 B^2 z^2 + 2\hbar k_x eBz \right\} + u_n - 2\hbar^2 e^2 B^2 u_n \\
&\quad - \hbar^2 \left(\frac{n\pi}{a}\right)\left(4e^2 B^2 z + 4\hbar k_x eB\right) \\
&\quad \cdot \sqrt{\frac{2}{a}} i_x i_y \cos\left(\frac{n\pi}{a} z\right) \exp\left(ik_x x + ik_y y\right) \qquad (16.13)
\end{aligned}
$$

Thus finally we get

$$(\hat{H}_1)^4\psi_n^0 = \left\{\hbar^2 k_x^2 + \hbar^2 k_y^2 + \hbar^2\left(\frac{n\pi}{a}\right)^2\right\}^2 u_n$$
$$+2\left\{\hbar^2 k_x^2 + \hbar^2 k_y^2 + \hbar^2\left(\frac{n\pi}{a}\right)^2\right\}(e^2B^2z^2 + 2\hbar k_x eBz)u_n$$
$$+(e^2B^2z^2 + 2\hbar k_x eBz)u_n - 2\hbar^2 e^2 B^2 u_n - \hbar^2\left(\frac{n\pi}{a}\right)$$
$$\cdot(4e^2B^2z + 4\hbar k_x eBz)u_n' \tag{16.14}$$

Thus the total Hamiltonian of the system is

$$\hat{H} = a\left(\hat{H}_1\right)^2 - b\left(\hat{H}_1\right)^4$$
$$= \frac{\hbar^2 k_x^2}{2m_c} + \frac{\hbar^2 k_y^2}{2m_c} + \frac{\hbar^2}{2m_c}\left(\frac{n\pi}{a}\right)^2 + \frac{e^2B^2z^2}{2m_c} + \frac{2\hbar k_x eBz}{2m_c}$$
$$-\alpha\left[\frac{\hbar^2 k_x^2}{2m_c} + \frac{\hbar^2 k_y^2}{2m_c} + \frac{\hbar^2}{2m_c}\left(\frac{n\pi}{a}\right)^2\right]^2$$
$$-2\alpha\left\{\frac{\hbar^2 k_x^2}{2m_c} + \frac{\hbar^2 k_y^2}{2m_c} + \frac{\hbar^2}{2m_c}\left(\frac{n\pi}{a}\right)^2\right\}\cdot\left(\frac{e^2B^2z^2}{2m_c} + \frac{2\hbar k_x eBz}{2m_c}\right)$$
$$-\alpha\left(\frac{e^2B^2z^2}{2m_c} + \frac{2\hbar k_x eBz}{2m_c}\right)^2 + 2\alpha\left(\frac{\hbar eB}{2m_c}\right)^2$$
$$+\frac{\alpha\hbar^2}{2m_c}\left(\frac{n\pi}{a}\right)\left(\frac{4e^2B^2z^2}{2m_c} + \frac{4\hbar k_x eBz}{2m_c}\right)\cot\left(\frac{n\pi}{a}z\right) \tag{16.15}$$

Therefore the perturbed Hamiltonian is given by

$$\hat{H}' = \frac{e^2B^2z^2}{2m_c} + \frac{\hbar k_x eBz}{m_c} + 2\alpha\left(\frac{\hbar eB}{2m_c}\right)^2$$
$$+\frac{\alpha\hbar^2}{2m_c}\left(\frac{n\pi}{a}\right)\left(\frac{2e^2B^2z^2}{m_c} + \frac{2\hbar k_x eBz}{m_c}\right)\cot\left(\frac{n\pi}{a}z\right)$$

$$- 2\alpha \left\{ \frac{\hbar^2 k_x^2}{2m_c} + \frac{\hbar^2 k_y^2}{2m_c} + \frac{\hbar^2}{2m_c} \left(\frac{n\pi}{a} \right)^2 \right\} \cdot \left(\frac{e^2 B^2 z^2}{2m_c} + \frac{2\hbar k_x e B z}{2m_c} \right)$$

$$- \alpha \left(\frac{e^2 B^2 z^2}{2m_c} + \frac{2\hbar k_x e B z}{2m_c} \right)^2 \tag{16.16}$$

So the first order perturbation to the energy is

$$E_n^{(1)} = \hat{H}'_{nn} = \int_\tau u_n^* \hat{H}' u_n d\tau = \langle u_n | \hat{H}' | u_n \rangle$$

$$= \frac{\hbar k_x e B}{m_c} \langle z \rangle + \frac{e^2 B^2}{2m_c} \langle z^2 \rangle + 2\alpha \left(\frac{\hbar e B}{2m_c} \right)^2 + \frac{\alpha \hbar^2}{2m_c} \left(\frac{n\pi}{a} \right) \frac{2e^2 B^2}{m_c}$$

$$\cdot \int_0^a \left(\frac{2}{a} \right) z \sin\left(\frac{n\pi}{a} z \right) \cos\left(\frac{n\pi}{a} z \right) dz + \frac{\alpha \hbar^2}{2m_c} \left(\frac{n\pi}{a} \right) \frac{2\hbar k_x e B}{m_c}$$

$$\cdot \int_0^a \left(\frac{2}{a} \right) z \sin\left(\frac{n\pi}{a} z \right) \cos\left(\frac{n\pi}{a} z \right) dz$$

$$- 2\alpha \left\{ \frac{\hbar^2 k_x^2}{2m_c} + \frac{\hbar^2 k_y^2}{2m_c} + \frac{\hbar^2}{2m_c} \left(\frac{n\pi}{a} \right)^2 \right\} \frac{e^2 B^2}{2m_c} \langle z^2 \rangle$$

$$- 2\alpha \left\{ \frac{\hbar^2 k_x^2}{2m_c} + \frac{\hbar^2 k_y^2}{2m_c} + \frac{\hbar^2}{2m_c} \left(\frac{n\pi}{a} \right)^2 \right\} \frac{\hbar k_x e B z}{m_c} \langle z \rangle$$

$$- \alpha \frac{e^4 B^4}{4m_c^2} \langle z^4 \rangle - 2\alpha \frac{e^2 B^2}{2m_c} \frac{\hbar k_x e B}{m_c} \langle z^3 \rangle - \alpha \frac{\hbar^2 k_x^2 e^2 B^2}{m_c^2} \langle z^2 \rangle \tag{16.17}$$

in which

$$\langle z \rangle = \int_\tau u_n^* z u_n d\tau = \left(\frac{2}{a} \right) \int_0^a \sin^2\left(\frac{n\pi}{a} z \right) dz \tag{16.18}$$

$$\langle z^2 \rangle = \int_\tau u_n^* z^2 u_n d\tau = \left(\frac{2}{a} \right) \int_0^a z^2 \sin^2\left(\frac{n\pi}{a} z \right) dz \tag{16.19}$$

$$\langle z^3 \rangle = \int_\tau u_n^* z^3 u_n d\tau \tag{16.20}$$

$$\langle z^4 \rangle = \int_\tau u_n^* z^4 u_n d\tau \tag{16.21}$$

Now

$$\int_0^a z\sin^2\left(\frac{n\pi}{a}z\right)dz$$

$$= \frac{1}{2}\int_0^a z\left(1-\cos\frac{2n\pi}{a}z\right)dz = \frac{1}{2}\int_0^a \frac{1}{2}\int_0^a z\cos\left(\frac{2n\pi}{a}\right)z\;dz$$

$$= \frac{a^2}{4} - \frac{1}{2}\left[z\int\cos\left(\frac{2n\pi}{a}z\right)dz - \int\left\{\cos\frac{2n\pi}{a}z\;dz\right\}\right]_0^a$$

$$= \frac{a^2}{4} - \frac{1}{2}\left[\frac{az}{2n\pi}\sin\frac{2n\pi}{a}z - \frac{a}{2n\pi}\int\sin\left(\frac{2n\pi}{a}\right)z\;dz\right]_0^a$$

$$= \frac{a^2}{4} - \frac{1}{2}\left[\frac{az}{2n\pi}\sin\left(\frac{2n\pi z}{a}\right) + \left(\frac{a}{2n\pi}\right)^2\cos\left(\frac{2n\pi z}{a}\right)\right]_0^a$$

$$= \frac{a^2}{4} - \frac{1}{2}\left[\frac{a^2}{2n\pi}\sin 2n\pi + \frac{a^2}{4n^2\pi^2}\cos 2n\pi - 0 - \frac{a^2}{4n^2\pi^2}\right]_0^a$$

$$= \frac{a^2}{4} - \frac{1}{2}[0+1-1] = \frac{a^2}{4}$$

Therefore from (16.8) we can write,

$$\langle z\rangle = \frac{2}{a}\frac{a^2}{4} = \frac{a}{2} \tag{16.22}$$

Again

$$\int_0^a z^2\sin^2\left(\frac{n\pi}{a}z\right)dz = \frac{1}{2}\int_0^a z^2\left(1-\cos\frac{2n\pi}{a}z\right)dz$$

$$= \frac{1}{2}\int_0^a z^2dz\frac{1}{2}\int_0^a z^2\cos\left(\frac{2n\pi}{a}\right)zdz$$

$$= \frac{a^3}{6} - \frac{1}{2}\left[z^2\int\cos\left(\frac{2n\pi}{a}z\right)dz - \int\left\{2z\left(\cos\frac{2n\pi}{a}z\right)dz\right\}\right]_0^a$$

$$= \frac{a^3}{6} - \frac{1}{2}\left[\frac{az^2}{2n\pi}\sin\frac{2n\pi}{a}z - \frac{2a}{2n\pi}\int z\sin\left(\frac{2n\pi}{a}\right)z\;dz\right]_0^a$$

$$= \frac{a^3}{6} + \left(\frac{a}{2n\pi}\right)\int_0^a z\sin\left(\frac{2n\pi z}{a}\right)dz$$

$$= \frac{a^3}{6} + \left(\frac{a}{2n\pi}\right)\left[\int_0^a z\ \sin\left(\frac{2n\pi z}{a}\right) dz\right.$$

$$\left.- \int_0^a 1\left\{\sin\left(\frac{2n\pi z}{a}\right) dz\right\} dz\right]$$

$$= \frac{a^3}{6} + \frac{a}{2n\pi}\left[\left\{\frac{az}{2n\pi} - \cos\left(\frac{2n\pi z}{a}\right)\right\}\right]_0^a + \frac{a}{2n\pi}\int_0^a \cos\left(\frac{2n\pi z}{a}\right) dz$$

$$= \frac{a^3}{6} + \frac{a}{2n\pi}\left[-\frac{a^2}{2n\pi}\cos 2n\pi + 0 + \left(\frac{a}{2n\pi}\right)^2 \sin\left(\frac{2n\pi z}{a}\right)\right]_0^a$$

$$= \frac{a^3}{6} + \left(\frac{a}{2n\pi}\right)\left[-\frac{a^2}{2n\pi} + \left(\frac{a}{2n\pi}\right)^2 \sin 2n\pi - 0\right] = \left(\frac{a^3}{6} - \frac{a^3}{4n^2\pi^2}\right)$$

Therefore, from (16.19) we can write,

$$\langle z^2 \rangle = \frac{2}{a}\left(\frac{a^3}{6} - \frac{a^3}{4n^2\pi^2}\right) = \left(\frac{a^2}{3} - \frac{a^2}{2n^2\pi^2}\right) \tag{16.23}$$

Since

$$\int_0^a z^3 \sin^2\left(\frac{n\pi}{a}z\right) dz = \frac{1}{2}\int_0^a z^3\left(1 - \cos\frac{2n\pi}{a}z\right) dz$$

$$= \frac{1}{2}\int_0^a z^3 dz - \frac{1}{2}\int_0^a z^3 \cos\left(\frac{2n\pi}{a}z\right) dz$$

$$= \frac{a^4}{8} - \frac{1}{2}\left[z^3\int_0^a \cos\left(\frac{2n\pi}{a}z\right) dz\right.$$

$$\left.- 3\int_0^a z^2\left\{\int \cos\left(\frac{2n\pi}{a}z\right) dz\right\} dz\right]$$

$$= \frac{a^4}{8} - \frac{1}{2}\left[\frac{z^3 a}{2n\pi}\sin\frac{2n\pi}{a}z \int_0^a z^2 \sin\left(\frac{2n\pi z}{a}\right) dz\right]$$

$$= \frac{a^4}{8} + \frac{3}{2}\left(\frac{a}{2n\pi}\right)\int_0^a z^2 \sin\left(\frac{2n\pi z}{a}\right) dz$$

$$= \frac{a^4}{8} + \frac{3}{2}\left(\frac{a}{2n\pi}\right)\left[z^2 \int_0^a \sin\left(\frac{2n\pi z}{a}\right) dz\right.$$

$$\left. - 2\int_0^a z \int \sin\left(\frac{2n\pi}{a} z\right) dz\right]$$

$$= \frac{a^4}{8} + \frac{3}{2}\left(\frac{a}{2n\pi}\right)\left[-\frac{z^2 a}{2n\pi} \cos\left(\frac{2n\pi z}{a}\right) \int_0^a\right.$$

$$\left. + \frac{a}{n\pi}\int_0^a z \cos\left(\frac{2n\pi z}{a}\right) dz\right]$$

$$= \frac{a^4}{8} + \frac{3}{2}\left(\frac{a}{2n\pi}\right)\left[-\frac{a^3}{2n\pi} \cos(2n\pi) - 0\right.$$

$$\left. + \left(\frac{a}{n\pi}\right)\int_0^a z \cos\left(\frac{2n\pi}{a} z\right) dz\right]$$

$$= \frac{a^4}{8} - \frac{3a^4}{4n^2\pi^2} + \frac{3a^2}{4n^2\pi^2}\int_0^a z \cos\left(\frac{2n\pi}{a} z\right) dz$$

$$= \left(\frac{a^4}{8} - \frac{3a^4}{8n^2\pi^2}\right)$$

Therefore from (16.20) we can write,

$$\langle z^3 \rangle = \frac{2}{a}\left(\frac{a^4}{8} - \frac{3a^4}{8n^2\pi^2}\right) = \frac{a^3}{4} - \frac{3a^3}{4n^2\pi^2} \tag{16.24}$$

$$\langle z^4 \rangle = \frac{2}{a}\int_0^a z^4 \sin^2\left(\frac{n\pi}{a}\right) z dz = \frac{1}{a}\int_0^a z^4\left(1 - \cos\frac{2n\pi}{a} z\right) dz$$

$$= \frac{1}{a}\int_0^a z^4 dz - \frac{1}{a}\int_0^a z^4 \cos\frac{2n\pi}{a} z dz$$

$$= \frac{a^4}{5} - \frac{1}{a}\left[z^4 \frac{a}{2n\pi} \sin\frac{2n\pi}{a} z \int_0^a z^3 \sin \int_0^a z^3 \sin\left(\frac{2n\pi}{a} z\right) dz\right]$$

$$= \frac{a^4}{5} + \frac{2}{n\pi}\int_0^a z^3 \sin\left(\frac{2n\pi}{a} z\right) dz$$

$$= \frac{a^4}{5} + \frac{2}{n\pi}\left[-z^3 \frac{a}{2n\pi} \cos\left(\frac{2n\pi}{a} z\right)\right.$$

$$+\int_0^a 3\frac{a}{2n\pi}\int_0^a z^2\cos\left(\frac{2n\pi}{a}z\right)dz\Bigg]$$

$$=\frac{a^4}{5}-\frac{a^4}{n^2\pi^2}+\frac{3a}{n^2\pi^2}\int_0^a z^2\cos\left(\frac{2n\pi}{a}z\right)dz$$

$$=\frac{a^4}{5}-\frac{a^4}{n^2\pi^2}+\frac{3a}{n^2\pi^2}\left[-\frac{a}{n\pi}\int_0^a z^2\sin\left(\frac{2n\pi}{a}z\right)dz\right]$$

$$=\frac{a^4}{5}-\frac{a^4}{n^2\pi^2}+\frac{3a}{n^2\pi^2}\left[-\frac{a}{n\pi}\left(-\frac{a^2}{2n\pi}\right)\right]$$

$$=\frac{a^4}{5}-\frac{a^4}{n^2\pi^2}+\frac{3a}{n^2\pi^2}\left(\frac{a^3}{2n^2\pi^2}\right)$$

$$\langle z^4\rangle=\frac{a^4}{5}-\frac{a^4}{n^2\pi^2}+\frac{3a^2}{2n^4\pi^4}$$

Since the basic wave function is normalized, the first order contribution of the Eigen energy value can be written as

$$E_n^{(1)}=\frac{\hbar k_x eB}{m_c}\langle z\rangle+\frac{e^2B^2}{2m_c}\langle z^2\rangle+2\alpha\left(\frac{\hbar eB}{2m_c}\right)^2-\frac{\alpha\hbar^2}{2m_c}\frac{e^2B^2}{m_c}$$
$$-2\alpha\left\{\frac{\hbar^2k_x^2}{2m_c}+\frac{\hbar^2k_y^2}{2m_c}+\frac{\hbar^2}{2m_c}\left(\frac{n\pi}{a}\right)^2\right\}\frac{\hbar k_x eB}{m_c}\langle z\rangle$$
$$-2\alpha\left\{\frac{\hbar^2k_x^2}{2m_c}+\frac{\hbar^2k_y^2}{2m_c}+\frac{\hbar^2}{2m_c}\left(\frac{n\pi}{a}\right)^2\right\}\frac{e^2B^2}{2m_c}\langle z^2\rangle$$
$$-\alpha\frac{\hbar^2k_x^2e^2B^2}{m_c^2}\langle z^2\rangle-\frac{\alpha\hbar k_x e^3B^3}{m_c^2}\langle z^3\rangle-\alpha\left(\frac{e^4B^4}{4m_c^2}\right)\langle z^4\rangle \tag{16.25}$$

So the total energy is

$$E=E_n^0+E_n^{(1)}=\frac{\hbar^2k_x^2}{2m_c}+\frac{\hbar^2k_y^2}{2m_c}+\frac{\hbar^2}{2m_c}\left(\frac{n\pi}{a}\right)^2$$
$$-\alpha\left[\frac{\hbar^2k_x^2}{2m_c}+\frac{\hbar^2k_y^2}{2m_c}+\frac{\hbar^2}{2m_c}\left(\frac{n\pi}{a}\right)^2\right]^2$$

$$
\begin{aligned}
&+\frac{\hbar k_x eB}{m_c}\langle z\rangle+\frac{e^2B^2}{2m_c}\langle z^2\rangle+2\alpha\left(\frac{\hbar eB}{2m_c}\right)^2\\
&-2\alpha\left(\frac{\hbar eB}{2m_c}\right)^2-\frac{2\alpha\hbar k_x eB}{m_c}\left\{\frac{\hbar^2k_x^2}{2m_c}+\frac{\hbar^2k_y^2}{2m_c}+\frac{\hbar^2}{2m_c}\left(\frac{n\pi}{a}\right)^2\right\}\langle z\rangle\\
&-\alpha\left(\frac{e^2B^2}{m_c}\right)\left\{\frac{\hbar^2k_x^2}{2m_c}+\frac{\hbar^2k_y^2}{2m_c}+\frac{\hbar^2}{2m_c}\left(\frac{n\pi}{a}\right)^2\right\}\langle z^2\rangle\\
&-\alpha\left(\frac{\hbar k_x eB}{m_c}\right)^2\langle z^2\rangle-\alpha\left(\frac{e^2B^2}{2m_c}\right)^2\langle z^4\rangle-\alpha\left(\frac{\hbar k_x e^3B^3}{m_c^2}\right)^2\langle z^3\rangle\\
=&\frac{\hbar^2}{2m_c}\left(\frac{n\pi}{a}\right)^2+\frac{\hbar^2k_y^2}{2m_c}+\frac{1}{2m_c}[\hbar k_x+eB\langle z\rangle]^2\\
&+\frac{e^2B^2}{2m_c}[\langle z^2\rangle-\langle z\rangle^2]-\alpha\left[\frac{\hbar^2k_x^2}{2m_c}+\frac{\hbar^2k_y^2}{2m_c}\right.\\
&\left.+\frac{\hbar^2}{2m_c}\left(\frac{n\pi}{a}\right)^2+\frac{\hbar k_x eB}{m_c}\langle z\rangle\right]^2+\alpha\left(\frac{\hbar k_x eB}{m_c}\right)^2\langle z^2\rangle\\
&\left.-\alpha\left(\frac{e^2B^2}{2m_c}\right)\left\{\frac{\hbar^2k_x^2}{2m_c}+\frac{\hbar^2k_y^2}{2m_c}+\frac{\hbar^2}{2m_c}\left(\frac{n\pi}{a}\right)^2\right\}\right]\\
&-\alpha\left(\frac{\hbar k_x eB}{m_c}\right)^2\langle z^2\rangle-\alpha\left(\frac{\hbar k_x e^3B^3}{m_c^2}\right)\langle z^3\rangle
\end{aligned}
$$

Therefore,

$$
\begin{aligned}
E=&\frac{\hbar^2}{2m_c}\left(\frac{n\pi}{a}\right)^2+\frac{\hbar^2k_y^2}{2m_c}+\frac{1}{2m_c}(\hbar k_x+eB\langle z\rangle)^2+\frac{e^2B^2}{2m_c}[\langle z^2\rangle-\langle z\rangle^2]\\
&-\alpha\left(\frac{\hbar k_x eB}{m_c}\right)^2[\langle z^2\rangle-\langle z\rangle^2]-\alpha\left(\frac{e^2B^2}{2m_c}\right)^2\langle z\rangle^4-\left(\frac{\hbar k_x e^3B^3}{m_c^2}\right)\langle z^3\rangle\\
&-\alpha\left(\frac{e^2B^3}{m_c}\right)\left\{\frac{\hbar^2k_x^2}{2m_c}+\frac{\hbar^2k_y^2}{2m_c}+\frac{\hbar^2}{2m_c}\left(\frac{n\pi}{a}\right)^2\right\}\langle z\rangle^2
\end{aligned}
$$

$$
\begin{aligned}
&-\alpha\left[\frac{\hbar^2}{2m_c}\left(\frac{n\pi}{a}\right)^2+\frac{\hbar^2k_y^2}{2m_c}\left(\hbar k_x+eB\langle z\rangle\right)^2\right]^2 \\
&+\alpha\left(\frac{e^2B^2}{m_c}\right)\left[\frac{\hbar}{2m_c}+\left(\frac{n\pi}{a}\right)^2+\frac{\hbar^2k_y^2}{2m_c}+\frac{\hbar^2k_x^2}{2m_c}\right]\langle z\rangle^2 \\
&+\alpha\left(\frac{\hbar k_x e^3B^3}{m_c^2}\right)\langle z^3\rangle+\alpha\left(\frac{e^2B^2}{2m_c}\right)^2\langle z\rangle^4 \qquad (16.26)
\end{aligned}
$$

We have to average the kinetic energy in the order to obtain the DR in this case in the presence of band tails. Using the (1.4) and (15.26), we can write that the DR in HDQW of III–V semiconductors (whose unperturbed conduction electrons obey the two band model of Kane) in the presence of a parallel magnetic field B along y-direction can be written

$$
\begin{aligned}
\gamma_3(E,\eta_g)=&\left[\frac{\hbar^2}{2m_c}\left(\frac{n\pi}{a}\right)^2+\frac{\hbar^2k_y^2}{2m_c}+\frac{1}{2m_c}\left(\hbar k_x+eB\langle z\rangle\right)^2\right. \\
&-\alpha\left[\frac{\hbar^2}{2m_c}\left(\frac{n\pi}{a}\right)^2+\frac{\hbar^2k_y^2}{2m_c}+\frac{1}{2m_c}\left(\hbar k_x+eB\langle z\rangle\right)^2\right]^2 \\
&+\frac{e^2B^2}{2m_c}\left[\langle z^2\rangle-\langle z\rangle^2\right]-\alpha\left(\frac{e^2B^2}{m_c}\right) \\
&\cdot\left\{\frac{\hbar^2k_x^2}{2m_c}+\frac{\hbar^2k_y^2}{2m_c}+\frac{\hbar^2}{2m_c}\left(\frac{n\pi}{a}\right)^2\right\}\left[\{\langle z^2\rangle-\langle z\rangle^2\}\right] \\
&-\alpha\left(\frac{\hbar k_x eB}{m_c}\right)^2\left[\langle z^2\rangle-\langle z\rangle^2\right]-\alpha\left(\frac{\hbar k_x e^3B^3}{m_c^2}\right)^2\left[\langle z^3\rangle-\langle z\rangle^3\right] \\
&\left.-\alpha\left(\frac{e^2B^2}{2m_c}\right)^2\left[\langle z^4\rangle-\langle z\rangle^4\right]\right] \qquad (16.27)
\end{aligned}
$$

Putting $k_n = k_y = 0$ and $E = E_n$ in (16.27) we get

$$
\begin{aligned}
\gamma_3(E,\eta_g)=&\frac{\hbar^2}{2m_c}\left(\frac{n\pi}{a}\right)^2+\frac{e^2B^2\langle z\rangle^2}{2m_c} \\
&-\alpha\left[\frac{\hbar}{2m_c}\left(\frac{n\pi}{a}\right)^2+\frac{e^2B^2\langle z\rangle^2}{2m_c}\right]^2+\frac{e^2B^2}{2m_c}\left[\langle z^2\rangle-\langle z\rangle^2\right]
\end{aligned}
$$

$$- \alpha \frac{e^2 B^2}{m_c} \left[\frac{\hbar^2}{2m_c} \left(\frac{n\pi}{a} \right)^2 \right] [\langle z^2 \rangle - \langle z \rangle^2]$$
$$- \alpha \left(\frac{e^2 B^2}{2m_c} \right)^2 [\langle z^4 \rangle - \langle z \rangle^4] \tag{16.28}$$

From (16.27) and (16.28) we get

$$\gamma_3 (E, \eta_g) = \gamma_3 (E_n, \eta_g) + \frac{\hbar^2 k_y^2}{2m_c} + \left(\frac{\hbar^2 k_x^2 + 2\hbar k_x eB \langle z \rangle}{2m_c} \right)$$
$$- \alpha \left[\left(\frac{\hbar^2 k_x^2}{2m_c} + \frac{\hbar^2 k_y^2}{2m_c} + \frac{\hbar k_x eB \langle z \rangle}{m_c} \right) \right.$$
$$\left. + \left\{ \frac{\hbar^2}{2m_c} \left(\frac{n\pi}{a} \right)^2 + \frac{e^2 B^2 \langle z \rangle^2}{2m_c} \right\} \right]^2$$
$$+\alpha \left[\frac{\hbar^2}{2m_c} \left(\frac{n\pi}{a} \right)^2 + \frac{e^2 B^2 \langle z \rangle^2}{2m_c} \right]^2$$
$$-\alpha \frac{e^2 B^2}{m_c} \left(\frac{\hbar^2 k_x^2}{2m_c} + \frac{\hbar^2 k_y^2}{2m_c} \right) [\langle z^2 \rangle - \langle z \rangle^2] - \alpha \frac{\hbar^2 k_x^2}{m_c}$$
$$\cdot \frac{e^2 B^2}{m_c} [\langle z^2 \rangle - \langle z \rangle^2] - \alpha \left(\frac{\hbar k_x e^3 B^3}{m_c^2} \right) [\langle z^3 \rangle - \langle z \rangle^3] \tag{16.29}$$

Putting $\alpha = 0$ we get

$$\gamma_3 (E, \eta_g) = f_n + \frac{\hbar^2 k_x^2}{2m_c} + \frac{\hbar^2 k_y^2}{2m_c} + \frac{\hbar k_x eB \langle z \rangle}{m_c} \tag{16.30}$$

where

$$f_n = \frac{\hbar}{2m_c} \left(\frac{n\pi}{a} \right)^2 + \frac{e^2 B^2 \langle z^2 \rangle}{2m_c}$$

putting the value $\frac{\hbar^2 k_x^2}{2m_c} + \frac{\hbar^2 k_y^2}{2m_c} + \frac{\hbar k_x eB\langle z \rangle}{m_c} = \gamma_3(E, \eta_g) - f_n$ in the perturbed term $\alpha[(\frac{\hbar^2 k_x^2}{2m_c} + \frac{\hbar^2 k_y^2}{2m_c} + \frac{\hbar k_x eB\langle z \rangle}{m_c}) + \{\frac{\hbar}{2m_c}(\frac{n\pi}{a})^2 + \frac{e^2 B^2 \langle z^2 \rangle}{2m_c}\}]^2$

of (1.29) we get

$$\bar{\gamma}(E,\eta_g) = A(n)\,k_x^2 + 2b(n)\,k_x + c(n)\,k_y^2 \tag{16.31}$$

where

$$\begin{aligned}\bar{\gamma}(E,\eta_g) = \Bigg[\Bigg[&\gamma_3(E,\eta_g) - \gamma_3(E_n,\eta_g) + \alpha\Bigg[\gamma_3(E,\eta_g) - f_n \\ &+ \left\{\frac{\hbar}{2m_c}\left(\frac{n\pi}{a}\right)^2 + \frac{e^2B^2\langle z\rangle^2}{2m_c}\right\}\Bigg]^2 \\ &- \alpha\left[\frac{\hbar}{2m_c}\left(\frac{n\pi}{a}\right)^2 + \frac{e^2B^2\langle z\rangle^2}{2m_c}\right]^2\Bigg],\end{aligned}$$

$$\begin{aligned}A(n) = \Bigg[&\frac{\hbar^2}{2m_c} - \alpha\frac{e^2B^2}{m_c}\cdot\frac{\hbar^2}{2m_c}\{\langle z^2\rangle - \langle z\rangle^2\} \\ &- \alpha^2\frac{\hbar^2e^2B^2}{2m_c}\{\langle z^2\rangle - \langle z\rangle^2\}\Bigg],\end{aligned}$$

$$2b(n) = \left[\frac{\hbar eB\langle z\rangle}{m_c} - \frac{\alpha\hbar e^3B^3}{m_c^2}\{\langle z^3\rangle - \langle z\rangle^3\}\right]$$

and

$$c(n) = \left[\frac{\hbar^2}{2m_c} - \alpha\frac{e^2B^2}{m_c}\cdot\frac{\hbar^2}{2m_c}[\langle z^2\rangle - \langle z\rangle^2]\right]$$

(16.31) can be written as

$$\frac{\left[k_x + \frac{b(n)}{A(n)}\right]^2}{\frac{1}{A(n)}\left[\bar{\gamma}(E,\eta_g) - \frac{b^2(n)}{A(n)}\right]} + \frac{k_y^2}{\frac{1}{c(n)}\left[\bar{\gamma}(E,\eta_g) - \frac{b^2(n)}{A(n)}\right]} = 1 \tag{16.32}$$

(16.32) indicates that the DR in this case is quantized ellipses for constant electron energy and scattering potential in the $k_x - k_y$ plane. The area of the ellipse can be written as

$$\tau(E,n,\eta_g) = \pi\tau_1(E,n,\eta_g) \tag{16.33}$$

where

$$\tau_1(E,n,\eta_g) = \frac{1}{\sqrt{A(n)\,c(n)}}\left[\bar{\gamma}(E,\eta_g) - \frac{b^2(n)}{A(n)}\right]$$

The EM in this case is given by

$$m^*(E,n,\eta_g) = \frac{\hbar^2}{2}\tau_1'(E,n,\eta_g) \tag{16.34}$$

Thus magnetic field makes the mass quantum number dependent.

It is important to note that in any semiconductor the EM at the Fermi level is a function of electron concentration. In the case of HDS, the EM at any energy is a function of electron concentration due to the presence of η_g, a concept impossible without band tailing.

The DOS function is given by

$$N_{QWHDB}(E,B,\eta_g) = \frac{g_v}{2\pi}\sum_{n=1}^{n_{\max}}\tau_1'(E,n,\eta_g)H(E-E_{19,1}) \tag{16.35}$$

where $E_{19,1}$ is the sub-band energy and can be expressed through the equation

$$\bar{\gamma}(E_{19,1},\eta_g) = 0 \tag{16.36a}$$

The surface electron concentration is given by

$$n_0 = \frac{g_v}{2\pi}\sum_{n=1}^{n_{\max}}\left[\tau_1(E_{F1536,n,\eta_g}) + \sum_{r=1}^{s}L(r)[\tau_1(E_{F1536,n,\eta_g})]\right] \tag{16.36b}$$

where E_{F1536} is the Fermi energy in this case.

16.2.2 *The EM in nano wires of HD III–V semiconductors in the presence of magnetic field*

The eigen function and the energy eigen values in this case are given by

$$\begin{aligned}\psi_{nl}^0 &= \left(\frac{4}{ab}\right)^{\frac{1}{2}}\sin\left(\frac{n\pi z}{a}\right)\sin\left(\frac{l\pi x}{b}\right)\exp(iyk_y)\\ &= c_1\psi_{nl}(x,z)\exp(iyk_y)\end{aligned} \tag{16.37}$$

and

$$E_{nl}^0 = \frac{\hbar^2}{2m_c}\left[\left(\frac{n\pi}{a}\right)^2 + \left(\frac{l\pi}{b}\right)^2\right] + \frac{\hbar^2 k_y^2}{2m_c}$$
$$-\alpha\left[\frac{\hbar^2}{2m_c}\left\{\left(\frac{n\pi}{a}\right)^2 + \left(\frac{l\pi}{b}\right)^2 + \frac{\hbar^2 k_y^2}{2m_c}\right\}\right]^2 \quad (16.38)$$

where a and b are the nano thickness along z and y-directions respectively, $n(= 1, 2, 3, \ldots\ldots\ldots\ldots)$ is the size quantum number along z-direction and $l(= 1, 2, 3, \ldots\ldots\ldots\ldots)$ is the size quantum number along x-direction, $c_1 = (\frac{4}{ab})^{\frac{1}{2}}$ and $\psi_{nl}(x, z) = \sin(\frac{n\pi z}{a})\sin(\frac{l\pi x}{b})$

In the presence of a magnetic field B along y-direction the term $\hat{H}_1\psi_{nl}^0$ assumes the from

$$\hat{H}_1\psi_{nl}^0 = \left[\left(-i\hbar\frac{\partial}{\partial x}eBz\right)\vec{i} + \vec{j}\left(-i\hbar\frac{\partial}{\partial y}\right) + \vec{k}\left(-i\hbar\frac{\partial}{\partial z}\right)\right]\psi_{nl}^0$$
$$= \left[-i\hbar c\left(\frac{l\pi}{b}\right)\cos\left(\frac{l\pi}{b}x\right)\sin\left(\frac{n\pi}{a}z\right)\exp\left(ik_y y\right) + eBz\psi_{nl}^0\right]\vec{i}$$
$$+\vec{j}\left(-i\hbar.ik_y\right)\psi_{nl}^0$$
$$+\vec{k}\left[-i\hbar c_1\left(\frac{n\pi}{a}\right)\sin\left(\frac{l\pi}{b}x\right)\cos\left(\frac{n\pi}{a}z\right)\exp\left(ik_y y\right)\right] \quad (16.39)$$

Again

$$(\hat{H}_1)^2\psi_{nl}^0 = \left(-i\hbar\frac{\partial}{\partial x} + eBz\right)$$
$$\cdot\left[-i\hbar c_1\left(\frac{n\pi}{a}\right)\sin\left(\frac{l\pi}{b}x\right)\cos\left(\frac{n\pi}{a}z\right)\exp(ik_y y) + eBz\psi_{nl}^0\right]$$
$$+\left(-i\hbar.ik_y\right)^2\psi_{nl}^0 + \left(-i\hbar\frac{\partial}{\partial x} + eBz\right)$$
$$\cdot\left[-i\hbar c_1\left(\frac{n\pi}{a}\right)\sin\left(\frac{l\pi}{b}x\right)\cos\left(\frac{n\pi}{a}z\right)\exp\left(ik_y y\right)\right]$$

$$= \left[-i\hbar - i\hbar c_1 \left(\frac{l\pi}{b}\right)^2 - \sin\left(\frac{l\pi}{b}x\right) \sin\left(\frac{n\pi}{a}z\right) \exp\left(ik_y y\right) \right.$$

$$- 2i\hbar eBzc_1 \left(\frac{l\pi}{b}\right) \cos\left(\frac{l\pi}{b}x\right) \sin\left(\frac{n\pi}{a}z\right) \exp\left(ik_y y\right)$$

$$\left. + \; e^2B^2z^2\psi^0_{nl} \right] + \hbar^2 k_y^2 \psi^0_{nl} + \left(\frac{n\pi}{a}\right)^2 \psi^0_{nl}$$

$$= \left[\hbar^2 \left(\frac{l\pi}{b}\right)^2 \psi^0_{nl} + e^2B^2z^2\psi^0_{nl} - 2i\hbar eBzc_1 \left(\frac{l\pi}{b}\right) \right.$$

$$\left. \cdot \cos\left(\frac{l\pi}{b}x\right) \sin\left(\frac{n\pi}{a}z\right) \exp\left(ik_y y\right) \right]$$

$$+ \hbar^2 k_y^2 \psi^0_{nl} + \hbar^2 \left(\frac{n\pi}{a}\right)^2 \psi^0_{nl}$$

$$= \left[\hbar^2 \left(\frac{l\pi}{b}\right)^2 + \hbar^2 \left(\frac{n\pi}{a}\right)^2 + \hbar^2 k_y^2 + e^2B^2z^2 \right] \psi^0_{nl}$$

$$- 2i\hbar eBz \sqrt{\frac{4}{ab}} \left(\frac{l\pi}{b}\right) \cos\left(\frac{l\pi}{b}x\right) \sin\left(\frac{n\pi}{a}z\right) \exp\left(ik_y y\right)$$

Therefore

$$(\hat{H}_1)^2 \psi^0_{nl} = \left[\hbar^2 \left(\frac{l\pi}{b}\right)^2 + \hbar^2 \left(\frac{n\pi}{a}\right)^2 + \hbar^2 k_y^2 + e^2B^2z^2 \right] \psi^0_{nl}$$

$$- 2i\hbar eBz \left(\frac{l\pi}{b}\right) \cot\left(\frac{l\pi}{b}x\right) \psi^0_{nl} \qquad (16.40)$$

$$(\hat{H}_1)^3 \psi^0_{nl} = \left[\left(-i\hbar\frac{\partial}{\partial x} + eBz\right) \vec{i} + \vec{j} \left(-i\hbar\frac{\partial}{\partial y}\right) \right.$$

$$\left. + \vec{k} \left(-i\hbar\frac{\partial}{\partial z}\right) \right] \cdot \left(\hat{H}_1^2\right) \psi^0_{nl}$$

$$= \vec{i} \left[\left\{ \hbar^2 \left(\frac{l\pi}{b}\right)^2 + \hbar^2 \left(\frac{n\pi}{a}\right)^2 + \hbar^2 k_y^2 + e^2B^2z^2 \right\} \right.$$

$$
\begin{aligned}
&\cdot (-i\hbar)\, c_1 \left(\frac{l\pi}{b}\right) \cos\left(\frac{l\pi}{b}x\right) \sin\left(\frac{n\pi}{a}z\right) \exp\left(ik_y y\right)\\
&+ 2(i\hbar eBz)(-i\hbar)c_1\left(\frac{l\pi}{b}\right)^2 \sin\left(\frac{l\pi}{b}x\right) \sin\left(\frac{n\pi}{a}z\right) \exp(ik_y y)\\
&\cdot eBz \left\{\hbar^2 \left(\frac{l\pi}{b}\right)^2 + \hbar^2 \left(\frac{n\pi}{a}\right)^2 + \hbar^2 k_y^2 + e^2B^2z^2\right\} \psi_{nl}^0\\
&- 2i\hbar e^2B^2z^2c_1 \left(\frac{l\pi}{b}\right) \cos\left(\frac{l\pi}{b}x\right) \sin\left(\frac{n\pi}{a}z\right) \exp\left(ik_y y\right)\Big]\\
&+ \vec{j}[-i\hbar \left[\hbar^2 \left(\frac{l\pi}{b}\right)^2 + \hbar^2 \left(\frac{n\pi}{a}\right)^2 + \hbar^2 k_y^2 + e^2B^2z^2\right]\\
&\cdot (ik_y)\, \psi_{nl}^0 + (i\hbar)\, 2i\hbar eBz\, (ik_y)\, c_1 \left(\frac{l\pi}{b}\right) \cos\left(\frac{l\pi}{b}x\right)\\
&\cdot \sin\left(\frac{n\pi}{a}z\right) \exp\left(iyk_y\right)] + \vec{k}[(-i\hbar)\\
&\cdot \left[\hbar^2 \left(\frac{l\pi}{b}\right)^2 + \hbar^2 \left(\frac{n\pi}{a}\right)^2 + \hbar^2 k_y^2 + e^2B^2z^2\right]
\end{aligned}
\tag{16.41}
$$

$$
\begin{aligned}
(\hat{H}_1)^4 \psi_{nl}^{`0} = &\left[\left\{\hbar^2 \left(\frac{l\pi}{b}\right)^2 + \hbar^2 \left(\frac{n\pi}{a}\right)^2 + \hbar^2 k_y^2 + e^2B^2z^2\right\}\right.\\
&\cdot (-i\hbar)^2\, c_1 \left(\frac{l\pi}{b}\right)^2 . \left(-\sin\left(\frac{l\pi}{b}x\right)\right) \sin\left(\frac{n\pi}{a}z\right)\\
&\cdot \exp\left(iyk_y\right) + 2\,(-i\hbar)^2\,(i\hbar eBz)\, c_1 \left(\frac{l\pi}{b}\right)^3 \cos\left(\frac{l\pi}{b}x\right)\\
&\cdot \sin\left(\frac{n\pi}{a}z\right) \exp\left(iyk_y\right)\\
&+ eBz \left\{\hbar^2 \left(\frac{l\pi}{b}\right)^2 + \hbar^2 \left(\frac{n\pi}{a}\right)^2 + \hbar^2 k_y^2 + e^2B^2z^2\right\}\\
&\cdot (-i\hbar)\, c_1 \left(\frac{l\pi}{b}\right) \cos\left(\frac{l\pi}{b}x\right) \sin\left(\frac{n\pi}{a}z\right) \exp\left(iyk_y\right)
\end{aligned}
$$

$$- 2i\hbar e^2 B^2 z^2 . (-i\hbar)\, c_1 \left(\frac{l\pi}{b}\right)^2 . \left(-\sin\left(\frac{l\pi}{b}x\right)\right)$$

$$\cdot \sin\left(\frac{n\pi}{a}z\right) \exp\left(iyk_y\right)$$

$$+ eBz\left\{\hbar^2\left(\frac{l\pi}{b}\right)^2 + \hbar^2\left(\frac{n\pi}{a}\right)^2 + \hbar^2 k_y^2 + e^2B^2z^2\right\}$$

$$\cdot (-i\hbar)\left(\frac{l\pi}{b}\right)\cot\left(\frac{l\pi}{b}x\right)\psi_{nl}^0 + 2e^2B^2z^2\left(\frac{l\pi}{b}\right)^2\psi_{nl}^0$$

$$+ \left\{\hbar^2\left(\frac{l\pi}{b}\right)^2 + \hbar^2\left(\frac{n\pi}{a}\right)^2 + \hbar^2 k_y^2 + e^2B^2z^2\right\} e^2B^2z^2\psi_{nl}^0$$

$$\left. - 2i\hbar e^3B^3z^3\left(\frac{l\pi}{b}\right)\cot\left(\frac{l\pi}{b}x\right)\psi_{nl}^0\right]$$

$$+ \left[\hbar^2\left(\frac{l\pi}{b}\right)^2 + \hbar^2\left(\frac{n\pi}{a}\right)^2 + \hbar^2 k_y^2 + e^2B^2z^2\right]\hbar^2 k_y^2\psi_{nl}^0$$

$$+ \hbar^2 k_y^2\left(-2i\hbar eBz\right)\left(\frac{l\pi}{b}\right)\cot\left(\frac{l\pi}{b}x\right)\psi_{nl}^0 + \hbar^2\left(\frac{n\pi}{a}\right)^2$$

$$\cdot \left[\hbar^2\left(\frac{l\pi}{b}\right)^2 + \hbar^2\left(\frac{n\pi}{a}\right)^2 + \hbar^2 k_y^2 + e^2B^2z^2\right]\psi_{nl}^0$$

$$- 4\hbar^2e^2B^2z^2\left(\frac{n\pi}{a}\right)\cot\left(\frac{n\pi}{a}z\right)\psi_{nl}^0 - 2\hbar^2e^2B^2\psi_{nl}^0$$

$$+ 4i\hbar^3 eB\left(\frac{l\pi}{b}\right)\left(\frac{n\pi}{a}\right)\cot\left(\frac{l\pi}{b}x\right)\cot\left(\frac{n\pi}{a}z\right)\psi_{nl}^0$$

$$- 2i\hbar^3 eBz\left(\frac{l\pi}{b}\right)\left(\frac{n\pi}{a}\right)^2\cot\left(\frac{l\pi}{b}x\right)\psi_{nl}^0 \tag{16.42}$$

Thus the total Hamiltonian of the system is

$$\hat{H} = a\left(\hat{H}_1\right)^2 - b\left(\hat{H}_1\right)^4$$

Therefore

$$\begin{aligned}
\hat{H} = &\left[\frac{\hbar^2}{2m_c}\left(\frac{l\pi}{b}\right)^2 + \frac{\hbar^2}{2m_c}\left(\frac{n\pi}{a}\right)^2 + \frac{\hbar^2 k_y^2}{2m_c}\right] + \frac{e^2B^2z^2}{2m_c} \\
&- \frac{2i\hbar eBz}{2m_c}\left(\frac{l\pi}{b}\right)\cot\left(\frac{l\pi}{b}x\right) - \alpha a^2\left[\hbar^2\left(\frac{l\pi}{b}\right)^2\right. \\
&\cdot\left\{\hbar^2\left(\frac{l\pi}{b}\right)^2 + \hbar^2\left(\frac{n\pi}{a}\right)^2 + \hbar^2k_y^2 + e^2B^2z^2\right\} \\
&- 2i\hbar^3 eBz\left(\frac{l\pi}{b}\right)^3\cot\left(\frac{l\pi}{b}x\right) - 2i\hbar eBz\left\{\hbar^2\left(\frac{l\pi}{b}\right)^2 + \hbar^2\left(\frac{n\pi}{a}\right)^2\right. \\
&\left. + \hbar^2k_y^2 + e^2B^2z^2\right\}\left(\frac{l\pi}{b}\right)\cot\left(\frac{l\pi}{b}x\right) + 4\hbar^2e^2B^2z^2\left(\frac{l\pi}{b}\right)^2 \\
&+ e^2B^2z^2\left\{\hbar^2\left(\frac{l\pi}{b}\right)^2 + \hbar^2\left(\frac{n\pi}{a}\right)^2 + \hbar^2k_y^2 + e^2B^2z^2\right\} \\
&- 2i\hbar e^3B^3z^3\left(\frac{l\pi}{b}\right)\cot\left(\frac{l\pi}{b}x\right) \\
&+ \left\{\hbar^2\left(\frac{l\pi}{b}\right)^2 + \hbar^2\left(\frac{n\pi}{a}\right)^2 + \hbar^2k_y^2 + e^2B^2z^2\right\}\hbar^2k_y^2 \\
&- 2i\hbar eBz\left(\hbar^2k_y^2\right)\left(\frac{l\pi}{b}\right)\cot\left(\frac{l\pi}{b}x\right) \\
&+ \hbar^2\left(\frac{n\pi}{a}\right)^2\left\{\hbar^2\left(\frac{l\pi}{b}\right)^2 + \hbar^2\left(\frac{n\pi}{a}\right)^2 + \hbar^2k_y^2 + e^2B^2z^2\right\} \\
&- 4\hbar^2e^2B^2z\left(\frac{n\pi}{a}\right)\cot\left(\frac{n\pi}{a}z\right) - 2\hbar^2e^2B^2 \\
&+ 4i\hbar^3eB\left(\frac{l\pi}{b}\right)\left(\frac{n\pi}{a}\right)\cot\left(\frac{l\pi}{b}x\right)\cot\left(\frac{n\pi}{a}z\right) \\
&\left. - 2i\hbar^3eBz\left(\frac{l\pi}{b}\right)\left(\frac{n\pi}{a}\right)^2\cot\left(\frac{l\pi}{b}x\right)\right]
\end{aligned} \tag{16.43}$$

Thus the perturbed Hamiltonian will be

$$\hat{H}' = \frac{e^2B^2}{2m_c}\left(z^2\right) - \left(\frac{2i\hbar eB}{2m_c}\right)(z)\left(\frac{l\pi}{b}\right)\cot\left(\frac{l\pi}{b}x\right)$$
$$- 2\alpha\left(\frac{e^2B^2}{2m_c}\right)\left[\frac{\hbar^2}{2m_c}\left(\frac{l\pi}{b}\right)^2 + \frac{\hbar^2}{2m_c}\left(\frac{n\pi}{a}\right)^2 + \frac{\hbar^2k_y^2}{2m_c}\right]$$
$$\cdot\left(z^2\right) - \alpha\left(\frac{e^2B^2}{2m_c}\right)^2\left(z^4\right) + \alpha\left(\frac{2i\hbar eB}{2m_c}\right)(z)$$
$$\cdot\left\{\frac{\hbar^2}{2m_c}\left(\frac{l\pi}{b}\right)^2 + \frac{\hbar^2}{2m_c}\left(\frac{n\pi}{a}\right)^2 + \frac{\hbar^2k_y^2}{2m_c}\right\}\left(\frac{l\pi}{b}\right)\cot\left(\frac{l\pi}{b}x\right)$$
$$\cdot\alpha\left(\frac{4i\hbar e^3B^3}{4m_c^2}\right)\left(z^3\right)\left(\frac{l\pi}{b}\right)\cot\left(\frac{l\pi}{b}x\right) + \alpha\left(\frac{2i\hbar eB}{2m_c}\right)\left(\frac{\hbar^2k_y^2}{2m_c}\right)(z)$$
$$\cdot\left(\frac{l\pi}{b}\right)\cot\left(\frac{l\pi}{b}x\right) + \left(\frac{2\hbar eB}{2m_c}\right)^2\left(\frac{n\pi}{a}\right)\cot\left(\frac{n\pi}{a}z\right)$$
$$+\alpha\left(\frac{2i\hbar^3eB}{4m_c^2}\right)\left(\frac{l\pi}{b}\right)^3(z)\cot\left(\frac{l\pi}{b}x\right) + \alpha\left(\frac{2i\hbar^3eB}{4m_c^2}\right)\left(\frac{l\pi}{b}\right)$$
$$\cdot\left(\frac{n\pi}{a}\right)^2(z)\cot\left(\frac{l\pi}{b}x\right) - \alpha\left(\frac{2\hbar eB}{2m_c}\right)^2\left(z^2\right)\left(\frac{l\pi}{b}\right)^2 \tag{16.44}$$

So the first order perturbation to the energy is

$$E_{nl}^{(1)} = \hat{H}'_{nl} = \frac{\int_{-\infty}^{+\infty}\psi_{nl}^*\hat{H}'\psi_{nl}d\psi}{\int_{-\infty}^{+\infty}\psi_{nl}^*\psi_{nl}d\psi} \tag{16.45}$$

Now,

$$\int_{-\infty}^{+\infty}\psi_{nl}^*\psi_{nl}d\psi = \left[\frac{2}{a}\int_0^a \sin^2\left(\frac{n\pi}{a}z\right)dz\right]\left[\frac{2}{b}\int_0^b \sin^2\left(\frac{l\pi}{b}x\right)dx\right] = 1 \tag{16.46}$$

$$\langle z^2\rangle = \left[\frac{2}{b}\int_0^b \sin^2\left(\frac{l\pi}{b}x\right)dx\right]\left[\frac{2}{a}\int_0^a \sin^2\left(\frac{n\pi}{a}z\right)dz\right] \tag{16.47}$$

Now

$$\frac{2}{b}\int_0^b \sin^2\left(\frac{l\pi}{b}x\right) dx = 1$$

Therefore

$$\langle z^2\rangle_n = \frac{2}{a}\int_0^a \sin^2\left(\frac{n\pi}{a}z\right) dz = \left(\frac{a^2}{3} - \frac{a^2}{2n^2\pi^2}\right) \tag{16.48}$$

Again

$$\left\langle z \cot\left(\frac{l\pi}{b}x\right)\right\rangle = \left[\frac{2}{a}\int_0^a \sin^2\left(\frac{n\pi}{a}z\right) dz\right]\cdot\left[\frac{2}{b}\int_0^b \sin^2\left(\frac{l\pi}{b}x\right) dx\right] = 0 \tag{16.49}$$

Now

$$\frac{2}{b}\int_0^b \sin^2\left(\frac{l\pi}{b}x\right)\cos\left(\frac{l\pi}{b}x\right) dx = 0$$

We can write

$$\langle z^4\rangle_n = \frac{2}{a}\int_0^a \sin^2\left(\frac{n\pi}{a}z\right) dz = \left(\frac{a^4}{5} - \frac{a^4}{n^2\pi^2} + \frac{3a^4}{2n^4\pi^4}\right) \tag{16.50}$$

Besides

$$\left\langle z^3 \cot\left(\frac{l\pi}{b}x\right)\right\rangle = \left[\frac{2}{a}\int_0^a \sin^2\left(\frac{n\pi}{a}z\right) dz\right] \cdot\left[\frac{2}{b}\int_0^b \cos\left(\frac{l\pi}{b}x\right)\sin\left(\frac{l\pi}{b}x\right) dx\right] = 0 \tag{16.51}$$

and

$$\left\langle z \cot\left(\frac{n\pi}{a}z\right)\right\rangle = \left[\frac{2}{a}\int_0^a z\cos\left(\frac{n\pi}{a}z\right)\sin\left(\frac{n\pi}{a}z\right) dz = -\left(\frac{a}{2n\pi}\right)\right. \tag{16.52}$$

We also obtain

$$\left\langle \cot\frac{l\pi}{b}x\cot\left(\frac{n\pi}{a}z\right)\right\rangle = \left[\frac{2}{a}\int_0^a \sin\left(\frac{n\pi}{a}z\right)\cos\left(\frac{n\pi}{a}z\right) dz\right] \cdot\left[\frac{2}{b}\int_0^b \sin\left(\frac{l\pi}{b}x\right)\cos\left(\frac{l\pi}{b}x\right) dx\right] = 0 \tag{16.53}$$

The first order correction to the energy is given by

$$E_{nl}^{(1)} = \frac{e^2B^2}{2m_c}\langle z^2\rangle_n - 2\alpha\left(\frac{e^2B^2}{2m_c}\right)\left[\frac{\hbar^2}{2m_c}\left(\frac{l\pi}{b}\right)^2 + \frac{\hbar^2}{2m_c}\left(\frac{n\pi}{a}\right)^2 + \frac{\hbar^2k_y^2}{2m_c}\right]$$
$$\cdot\left\langle z^2\right\rangle_n - \alpha\left(\frac{e^2B^2}{2m_c}\right)^2\left\langle z^4\right\rangle_n + \alpha\left(\frac{2\hbar eB}{2m_c}\right)^2\left(\frac{n\pi}{a}\right)\left(-\frac{a}{2n\pi}\right)$$
$$-\alpha\left(\frac{2\hbar eB}{2m_c}\right)^2\left(\frac{l\pi}{b}\right)^2\left\langle z^2\right\rangle_n + 2\alpha\left(\frac{2\hbar eB}{2m_c}\right)^2 \quad (16.54)$$

$$E_{nl}^{(1)} = \frac{e^2B^2}{2m_c}\langle z^2\rangle_n - 2\alpha\left(\frac{e^2B^2}{2m_c}\right)\left[\frac{\hbar^2}{2m_c}\left(\frac{l\pi}{b}\right)^2 + \frac{\hbar^2}{2m_c}\left(\frac{n\pi}{a}\right)^2 + \frac{\hbar^2k_y^2}{2m_c}\right]$$
$$\cdot\langle z^2\rangle_n - \alpha\left(\frac{2\hbar eB}{2m_c}\right)^2\left(\frac{l\pi}{b}\right)^2\left\langle z^2\right\rangle_n - \alpha\left(\frac{e^2B^2}{2m_c}\right)^2\left\langle z^4\right\rangle_n$$
$$-\alpha\left(\frac{\hbar^2e^2B^2}{2m_c^2}\right)^2 + \alpha\left(\frac{\hbar^2e^2B^2}{2m_c^2}\right)^2 \quad (16.55)$$

The DR in NWs of HD III–V semiconductors in the presence of a parallel magnetic field B along y-direction whose unperturbed electrons obey the two band model of Kane can be written as

$$\gamma_3(E,\eta_g) = \left[\frac{\hbar^2}{2m_c}\left(\frac{l\pi}{b}\right)^2 + \frac{\hbar^2}{2m_c}\left(\frac{n\pi}{a}\right)^2 + \frac{\hbar^2k_y^2}{2m_c}\right]$$
$$-\alpha\left[\frac{\hbar^2}{2m_c}\left(\frac{l\pi}{b}\right)^2 + \frac{\hbar^2}{2m_c}\left(\frac{n\pi}{a}\right)^2 + \frac{\hbar^2k_y^2}{2m_c}\right]^2 + \frac{e^2B^2}{2m_c}\left\langle z^2\right\rangle_n$$
$$-\alpha\left(\frac{2\hbar eB}{2m_c}\right)^2\left(\frac{l\pi}{b}\right)^2\left\langle z^2\right\rangle_n - 2\alpha\left(\frac{e^2B^2}{2m_c}\right)$$
$$\cdot\left[\frac{\hbar^2}{2m_c}\left(\frac{l\pi}{b}\right)^2 + \frac{\hbar^2}{2m_c}\left(\frac{n\pi}{a}\right)^2 + \frac{\hbar^2k_y^2}{2m_c}\right]\left\langle z^2\right\rangle_n$$
$$-\alpha\left(\frac{e^2B^2}{2m_c}\right)^2\left\langle z^4\right\rangle_n \quad (16.56a)$$

(16.56a) can be written as

$$k_y^2 = \gamma_{100}\left(E, \eta_g, l, n, B\right) \tag{16.56b}$$

where

$$\gamma_{100}(E, \eta_g, l, n, B) = \frac{1}{2\alpha a^2}[a - 2\alpha\omega a$$
$$- \sqrt{(2\alpha\omega a - a)^2 - 4\alpha a^2[\alpha\omega_1^2 - \omega_1 + \gamma_3(E, \eta_g)]}]$$

$$\omega_1 = \frac{\hbar^2}{2m_c}\left(\frac{n\pi}{a}\right)^2 + \frac{\hbar^2}{2m_c}\left(\frac{l\pi}{b}\right)^2$$

$$\omega_2 = \frac{e^2B^2}{2m_c}\left\langle z^2\right\rangle_n - \alpha\left(\frac{\hbar eB}{2m_c}\right)^2\left(\frac{l\pi}{b}\right)^2$$
$$- 2\alpha\left(\frac{e^2B^2}{2m_c}\right)[\omega_1]\left\langle z^2\right\rangle_n - \alpha\left(\frac{e^2B^2}{2m_c}\right)\left\langle z^4\right\rangle_n$$

and

$$\omega_3 = 2\alpha\left(\frac{e^2B^2}{2m_c}\right)a\left\langle z^4\right\rangle_n$$

The EEM is given by

$$m^*\left(E, n, l, B\right) = \frac{\hbar^2}{2}\gamma'_{100}\left(E, \eta, l, n, B\right) \tag{16.56c}$$

The DOS function is given by

$$N_{1D}\left(E\right) = \frac{gv}{\pi}\sum_{n=0}^{n_{\max}}\sum_{l=0}^{l_{\max}}\frac{\gamma'_{100}\left(E, \eta_g, l, n, B\right)H\left(E - E_{19,2}\right)}{\sqrt{\gamma_{100}\left(E, \eta_g, l, n, B\right)}} \tag{16.56d}$$

where $E_{19,2}$ is the sub-band energy which can be expressed through the equation

$$\gamma_{100}\left(E_{19,2}, \eta_g, l, n, B\right) = 0 \tag{16.56e}$$

The electron concentration per unit length can be written as

$$n_0 = \frac{g_v}{2\pi}\sum_{n=1}^{n_{\max}}\sum_{i=1}^{i_{\max}}\left[\sqrt{y_{100}(E_{F1522}, \eta_g, i, n, B)}\right.$$
$$\left. + \sum_{r=1}^{s}L(r)\sqrt{y_{100}(E_{F1522}, \eta_g, i, n, B)}\right] \tag{16.56f}$$

where E_{F1522} is the Fermi Energy in this case.

16.2.3 The EM in Quantum Wells of HD III–V semiconductors in the presence of cross fields

The total Hamiltonian and the unperturbed wave function in the present case in the presence of the magnetic field B along y direction and the crossed electric field E_0 along z direction can, respectively, be written as

$$\hat{H} = a\left(-i\hbar\vec{\nabla} + e\vec{\nabla}\right)^2 - b\left(-i\hbar\vec{\nabla} + e\vec{\nabla}\right)^4 - eE_0 z \quad (16.71)$$

$$\psi_n^0 = \sqrt{\frac{2}{aL_xL_y}}\sin\left(\frac{n\pi}{a}z\right)\exp\left(ik_x x + ik_y y\right) \quad (16.72)$$

In this case the total Hamiltonian can be expressed as

$$\begin{aligned}
\hat{H} = {} & \frac{\hbar^2k_x^2}{2m_c} + \frac{\hbar^2k_y^2}{2m_c} + \frac{\hbar^2}{2m_c}\left(\frac{n\pi}{a}\right)^2 + \frac{e^2B^2z^2}{2m_c} + \frac{2\hbar k_x eBz}{2m_c} \\
& - \alpha\left[\frac{\hbar^2}{2m_c}\left(k_x^2 + k_y^2 + \left(\frac{n\pi}{a}\right)^2\right)\right]^2 \\
& - 2\alpha\left\{\frac{\hbar^2k_x^2}{2m_c} + \frac{\hbar^2k_y^2}{2m_c} + \frac{\hbar^2}{2m_c}\left(\frac{n\pi}{a}\right)^2\right\}\left(\frac{e^2B^2z^2}{2m_c} + \frac{2\hbar k_x eBz}{2m_c}\right) \\
& - \alpha\left[\frac{e^2B^2z^2}{2m_c} + \frac{2\hbar k_x eBz}{2m_c}\right]^2 + 2\alpha\left(\frac{\hbar eB}{2m_c}\right)^2 + \alpha\frac{\hbar^2}{2m_c}\left(\frac{n\pi}{a}\right) \\
& \cdot\left(\frac{4e^2B^2z}{2m_c} + \frac{4\hbar k_x eBz}{2m_c}\right)^2\cot\left(\frac{n\pi}{a}z\right) - eE_0 z
\end{aligned} \quad (16.73)$$

Therefore, the DR can be written as

$$\begin{aligned}
E = {} & \frac{\hbar^2}{2m_c}\left(\frac{n\pi}{a}\right)^2 + \frac{\hbar^2k_y^2}{2m_c} + \frac{1}{2m_c}\left(\hbar k_x + eB\left\langle z\right\rangle\right)^2 \\
& - \alpha\left[\frac{\hbar^2}{2m_c}\left(\frac{n\pi}{a}\right)^2 + \frac{\hbar^2k_y^2}{2m_c} + \frac{1}{2m_c}(\hbar k_x + eB\langle z\rangle)^2\right]^2
\end{aligned}$$

$$+\frac{e^2B^2}{2m_c}[\langle z^2\rangle-\langle z\rangle^2]-\alpha\left(\frac{e^2B^2}{2m_c}\right)\left\{\frac{\hbar^2k_x^2}{2m_c}+\frac{\hbar^2k_y^2}{2m_c}+\frac{\hbar^2}{2m_c}\left(\frac{n\pi}{a}\right)^2\right\}$$

$$\cdot[\langle z^2\rangle-\langle z\rangle^2]-\alpha\left(\frac{\hbar k_x eb}{m_c}\right)^2[\langle z^2\rangle-\langle z\rangle^2]-\alpha\left(\frac{\hbar k_x e^3B^3}{m_c^2}\right)^2$$

$$\cdot[\langle z^3\rangle-\langle z\rangle^3]-\alpha\left(\frac{e^2B^2}{2m_c}\right)^2[\langle z^4\rangle-\langle z\rangle^4]-eE_0\langle z\rangle \tag{16.74}$$

If the electric field E_0 is along x direction, the total Hamiltonian can be written as

$$\hat{H}=a\left(-i\hbar\vec{\nabla}+e\vec{\nabla}\right)^2-b\left(-i\hbar\vec{\nabla}+e\vec{\nabla}\right)^4-eE_0x \tag{16.75}$$

(16.74) can be expressed as

$$\hat{H}=\frac{\hbar^2k_x^2}{2m_c}+\frac{\hbar^2k_y^2}{2m_c}+\frac{\hbar^2}{2m_c}\left(\frac{n\pi}{a}\right)^2+\frac{e^2B^2z^2}{2m_c}+\frac{2\hbar k_x eBz}{2m_c}$$

$$-\alpha\left[\frac{\hbar^2}{2m_c}\left(k_x^2+k_y^2+\left(\frac{n\pi}{a}\right)^2\right)\right]^2$$

$$-2\alpha\left\{\frac{\hbar^2k_x^2}{2m_c}+\frac{\hbar^2k_y^2}{2m_c}+\frac{\hbar^2}{2m_c}\left(\frac{n\pi}{a}\right)^2\right\}\left(\frac{e^2B^2z^2}{2m_c}+\frac{2\hbar k_x eBz}{2m_c}\right)$$

$$-\alpha\left[\frac{e^2B^2z^2}{2m_c}+\frac{2\hbar k_x eBz}{2m_c}\right]^2+2\alpha\left(\frac{\hbar eB}{2m_c}\right)^2+\alpha\frac{\hbar^2}{2m_c}\left(\frac{n\pi}{a}\right)$$

$$\cdot\left(\frac{4e^2B^2z}{2m_c}+\frac{4\hbar k_x eBz}{2m_c}\right)^2\cot\left(\frac{n\pi}{a}z\right)-eE_0x \tag{16.76}$$

Thus the perturbed part =

$$\langle eE_0x\rangle=eE_0\frac{2}{aL_xL_y}\int_0^2\sin^2\left(\frac{n\pi}{a}z\right)\int_0^{L_x}xdx\int_0^{L_y}dy$$

$$=eE_0\frac{2}{aL_xL_y}\left(\frac{a}{2}\right)\frac{L_x^2}{2}L_y=eE_0\left(\frac{L_x}{2}\right)$$

The DR in Quantum Wells of HD III–V Semiconductors in the Presence of Cross Fields is given by

$$\gamma_3(E,\eta_g) = \frac{\hbar^2}{2m_c}\left(\frac{n\pi}{a}\right)^2 + \frac{\hbar^2 k_y^2}{2m_c} + \frac{1}{2m_c}(\hbar k_x + eB\langle z\rangle)^2$$
$$-\alpha\left[\frac{\hbar^2}{2m_c}\left(\frac{n\pi}{a}\right)^2 + \frac{\hbar^2 k_y^2}{2m_c} + \frac{1}{2m_c}(\hbar k_x + eB\langle z\rangle)^2\right]^2$$
$$+\frac{e^2B^2}{2m_c}[\langle z^2\rangle - \langle z\rangle^2] - \alpha\left(\frac{e^2B^2}{2m_c}\right)$$
$$\cdot\left\{\frac{\hbar^2 k_x^2}{2m_c} + \frac{\hbar^2 k_y^2}{2m_c} + \frac{\hbar^2}{2m_c}\left(\frac{n\pi}{a}\right)^2\right\}[\langle z^2\rangle - \langle z\rangle^2]$$
$$-\alpha\left(\frac{\hbar k_x eb}{m_c}\right)^2[\langle z^2\rangle - \langle z\rangle^2] \tag{16.77}$$

The use of (16.77) leads to the expressions of EEM and DOS function, which is left to be done by the readers.

16.2.4 *The EM in nano-wires of HD III–V semiconductors in the presence of cross fields*

(a) The electric field E_0 is along z direction and the crossed magnetic field B is along y direction

The DR in this case is given by

$$\gamma_3(E,\eta_g) = \left[\frac{\hbar^2}{2m_c}\left(\frac{n\pi}{a}\right)^2 + \frac{\hbar^2}{2m_c}\left(\frac{l\pi}{b}\right)^2 + \frac{\hbar^2}{2m_c}\left(\frac{r\pi}{c}\right)^2\right]$$
$$-\alpha\left[\frac{\hbar^2}{2m_c}\left(\frac{n\pi}{a}\right)^2 + \frac{\hbar^2}{2m_c}\left(\frac{l\pi}{b}\right)^2 + \frac{\hbar^2}{2m_c}\left(\frac{r\pi}{c}\right)^2\right]^2$$
$$-eE_0\langle z\rangle_n + \frac{e^2B^2}{2m_c}\langle z^2\rangle_n - \alpha\left(\frac{2\hbar eB}{2m_c}\right)^2\left(\frac{l\pi}{b}\right)^2\langle z^2\rangle_n$$

$$- 2\alpha\left(\frac{e^2B^2}{2m_c}\right)\left[\frac{\hbar^2}{2m_c}\left(\frac{n\pi}{a}\right)^2+\frac{\hbar^2}{2m_c}\left(\frac{l\pi}{b}\right)^2+\frac{\hbar^2k_y^2}{2m_c}\right]$$
$$\cdot\left\langle z^2\right\rangle_n-\alpha\left(\frac{e^2B^2}{2m_c}\right)^2\left\langle z^4\right\rangle_n \tag{16.78}$$

(a) The electric field E_0 is along x direction and the crossed magnetic field B is along y direction

In this case we can write

$$E=\left[\frac{\hbar^2}{2m_c}\left(\frac{n\pi}{a}\right)^2+\frac{\hbar^2}{2m_c}\left(\frac{l\pi}{b}\right)^2+\frac{\hbar^2k_y^2}{2m_c}\right]$$
$$-\alpha\left[\frac{\hbar^2}{2m_c}\left(\frac{n\pi}{a}\right)^2+\frac{\hbar^2}{2m_c}\left(\frac{l\pi}{b}\right)^2+\frac{\hbar^2k_y^2}{2m_c}\right]^2-eE_0\left\langle x\right\rangle_n$$
$$+\frac{e^2B^2}{2m_c}\left\langle z^2\right\rangle_n-\alpha\left(\frac{2\hbar eB}{2m_c}\right)^2\left(\frac{l\pi}{b}\right)^2\left\langle z^2\right\rangle_n$$
$$-2\alpha\left(\frac{e^2B^2}{2m_c}\right)\left[\frac{\hbar^2}{2m_c}\left(\frac{n\pi}{a}\right)^2+\frac{\hbar^2}{2m_c}\left(\frac{l\pi}{b}\right)^2+\frac{\hbar^2k_y^2}{2m_c}\right]$$
$$\cdot\left\langle z^2\right\rangle_n-\alpha\left(\frac{e^2B^2}{2m_c}\right)^2\left\langle z^4\right\rangle_n \tag{16.79}$$

The perturbed term =

$$eE_0\left\langle x\right\rangle_n=\frac{eE_0}{L_y}\frac{4}{ab}\int_0^a\sin^2\left(\frac{n\pi}{a}z\right)dz\int_0^b x\,\sin^2\left(\frac{l\pi}{b}x\right)dx\int_0^{L_y}dy$$
$$=\frac{eE_0}{L_y}\left(\frac{2}{a}\right)\frac{a}{2}L_y\left(\frac{2}{b}\right)\int_0^b x\,\sin^2\left(\frac{l\pi}{b}x\right)dx$$
$$=\left(\frac{2}{b}\right)\frac{b^2}{4}\left(\frac{b}{2}\right)eE_0 \tag{16.80}$$

Therefore the DR in this case is given by

$$\gamma_3\left(E,\eta_g\right)=\left[\frac{\hbar^2}{2m_c}\left(\frac{n\pi}{a}\right)^2+\frac{\hbar^2}{2m_c}\left(\frac{l\pi}{b}\right)^2+\frac{\hbar^2k_y^2}{2m_c}\right]$$

$$- \alpha \left[\frac{\hbar^2}{2m_c} \left(\frac{n\pi}{a} \right)^2 + \frac{\hbar^2}{2m_c} \left(\frac{l\pi}{b} \right)^2 + \frac{\hbar^2 k_y^2}{2m_c} \right]^2 - eE_0 \left\langle \frac{b}{2} \right\rangle_n$$

$$+ \frac{e^2 B^2}{2m_c} \left\langle z^2 \right\rangle_n - \alpha \left(\frac{2\hbar eB}{2m_c} \right)^2 \left(\frac{l\pi}{b} \right)^2 \left\langle z^2 \right\rangle_n$$

$$- 2\alpha \left(\frac{e^2 B^2}{2m_c} \right) \left[\frac{\hbar^2}{2m_c} \left(\frac{n\pi}{a} \right)^2 + \frac{\hbar^2}{2m_c} \left(\frac{l\pi}{b} \right)^2 + \frac{\hbar^2 k_y^2}{2m_c} \right]$$

$$\cdot \left\langle z^2 \right\rangle_n - \alpha \left(\frac{e^2 B^2}{2m_c} \right)^2 \left\langle z^4 \right\rangle_n \tag{16.81}$$

The use of (16.78) and (16.81) lead to the expressions of EEM and DOS function, which is left to be done by the readers.

16.2.5 *The EM in Quantum Wells of HD IV–VI semiconductors in the presence of magnetic field*

(a) Cohen Model

In accordance with Cohen model, the DR in bulk specimens of IV–VI semiconductors can be written as

$$E\left(1 + \alpha E\right) \left(1 + E \frac{\alpha \hbar^2 k_y^2}{2m_2'} \right) + \frac{\hbar^2 k_x^2}{2m_1} + \frac{\hbar^2 k_z^2}{2m_3} \tag{16.84}$$

Let us substitute

$$\frac{1}{2m_i} = a_i \quad i = 1, 2, 3 \quad \text{and} \quad \frac{1}{2m_2'} = a_2'$$

Using the method of successive approximation we can write

$$\begin{aligned} E &= a_1 p_x^2 + a_3 p_z^2 + a_2 p_y^2 + \alpha (a_1 p_x^2 + a_3 p_z^2 + a_2 p_y^2)^2 \\ &\quad + \alpha a_2 p_y^2 (a_1 p_x^2 + a_3 p_z^2 + a_2 p_y^2) \\ &\quad - \alpha a_2' p_y^2 (a_1 p_x^2 + a_3 p_z^2 + a_2 p_y^2) + \alpha a_1 a_2' p_y^4 \end{aligned} \tag{16.85}$$

Let us choose the formation of QWs is along z direction in presence of magnetic field (B) along y direction.

Choosing the Coulomb gauge as
$\vec{A} = (0, 0, -Bx)$ and $\vec{\nabla} \cdot \vec{A} = 0$ together with the fact

$$\vec{B} = \vec{\nabla} \times \vec{A} = \begin{vmatrix} i & j & k \\ \partial/\partial x & \partial/\partial y & \partial/\partial z \\ 0 & 0 & -Bx \end{vmatrix}$$

In presence of perturbing magnetic field total Hamiltonian can be written as

$$\begin{aligned}
\hat{H} = a_1 \left(-i\hbar\frac{\partial}{\partial x} + eA_x\right)^2 + a_2 \left(-i\hbar\frac{\partial}{\partial y} + eA_y\right)^2 \\
+ a_3 \left(-i\hbar\frac{\partial}{\partial z} + eA_z\right)^2 - \alpha \left[a_1 \left(-i\hbar\frac{\partial}{\partial x} + eA_x\right)^2\right. \\
\left. + \, a_2 \left(-i\hbar\frac{\partial}{\partial y} + eA_y\right)^2 + a_3 \left(-i\hbar\frac{\partial}{\partial x} + eA_z\right)^2\right]^2 \\
+ \alpha a_2 \left(-i\hbar\frac{\partial}{\partial y} + eA_y\right)^2 \left[a_1 \left(-i\hbar\frac{\partial}{\partial x} + eA_x\right)^2\right. \\
\left. + \, a_2 \left(-i\hbar\frac{\partial}{\partial y} + eA_y\right)^2 + a_3 \left(-i\hbar\frac{\partial}{\partial x} + eA_z\right)^2\right] \\
- \alpha a_2' \left(-i\hbar\frac{\partial}{\partial y} + eA_y\right)^2 \left[a_1 \left(-i\hbar\frac{\partial}{\partial x} + eA_x\right)^2\right. \\
\left. + \, a_2 \left(-i\hbar\frac{\partial}{\partial y} + eA_y\right)^2 + a_3 \left(-i\hbar\frac{\partial}{\partial x} + eA_z\right)^2\right] \quad (16.86)
\end{aligned}$$

$$\begin{aligned}
&\alpha a_2 a_2' \left(-i\hbar\frac{\partial}{\partial y} + eA_y\right)^4 \\
&= -a_1\hbar^2\frac{\partial^2}{\partial x^2} - a_2\hbar^2\frac{\partial^2}{\partial y^2} + a_3\left(-\hbar^2\frac{\partial^2}{\partial z^2} + e^2B^2x^2 + 2i\hbar eBx\frac{\partial}{\partial z}\right) \\
&\quad - \alpha\left[-a_1\hbar^2\frac{\partial^2}{\partial x^2} - a_2\hbar^2\frac{\partial^2}{\partial y^2}\right.
\end{aligned}$$

$$+\ a_3\left(-\hbar^2\frac{\partial^2}{\partial z^2}+e^2B^2x^2+2i\hbar eBx\frac{\partial}{\partial z}\right)\Bigg]^2$$

$$+\alpha a_2\left(\hbar^2\frac{\partial^2}{\partial y^2}\right)\left[-a_1\hbar^2\frac{\partial^2}{\partial x^2}-a_2\hbar^2\frac{\partial^2}{\partial y^2}\right.$$

$$\left.+\ a_3\left(-\hbar^2\frac{\partial^2}{\partial z^2}+e^2B^2x^2+2i\hbar eBx\frac{\partial}{\partial z}\right)\right]$$

$$+\alpha a_2'\left(-\hbar^2\frac{\partial^2}{\partial y^2}\right)\left[-a_1\hbar^2\frac{\partial^2}{\partial x^2}-a_2\hbar^2\frac{\partial^2}{\partial y^2}\right.$$

$$\left.+\ a_3\left(-\hbar^2\frac{\partial^2}{\partial z^2}+e^2B^2x^2+2i\hbar eBx\frac{\partial}{\partial z}\right)\right]$$

$$-\alpha a_2'\left(-\hbar^2\frac{\partial^2}{\partial y^2}\right)\left[-a_1\hbar^2\frac{\partial^2}{\partial x^2}-a_2\hbar^2\frac{\partial^2}{\partial y^2}\right.$$

$$\left.+\ a_3\left(-\hbar^2\frac{\partial^2}{\partial z^2}+e^2B^2x^2+2i\hbar eBx\frac{\partial}{\partial z}\right)\right]$$

$$+\alpha a_2 a_2'\left(-\hbar^4\frac{\partial^4}{\partial y^4}\right) \tag{16.87}$$

Considering the effect of magnetic field and non parabolicity as perturbed parameters, the unperturbed wave equation can be written as

$$-a_1\hbar^2\frac{\partial^2\psi_0}{\partial x^2}-a_2\hbar^2\frac{\partial^2\psi_0}{\partial y^2}-a_3\hbar^2\frac{\partial^2\psi_0}{\partial z^2}=E_0\psi_0 \tag{16.88}$$

Therefore energy eigen value (E_n^0) and the eigen function assume the forms

$$E_0=E_n^0=\frac{\hbar k_x^2}{2m_1}+\frac{\hbar^2 k_y^2}{2m_2}+\frac{\hbar}{2m_3}\left(\frac{n\pi}{a}\right)^2 \tag{16.89}$$

$$\psi_0=\psi_n^0=\sqrt{\frac{2}{aL_xL_y}}\sin\left(\frac{n\pi}{a}z\right)\exp(ik_xx+ik_yy) \tag{16.90}$$

The perturbed Hamiltonian can be calculated in the following way

$$\left[-a_1\hbar^2\frac{\partial^2}{\partial x^2} - a_2\hbar^2\frac{\partial^2}{\partial y^2} + a_3\left(-\hbar^2\frac{\partial^2}{\partial z^2} + e^2B^2x^2 + 2i\hbar eBx\frac{\partial}{\partial z}\right)\right]\psi_n$$

$$= a_1\hbar^2k_x^2\psi_n + a_2\hbar^2k_y^2\psi_n + a_3\hbar^2\left(\frac{n\pi}{a}\right)^2\psi_n + a_3e^2B^2x^2\psi_n$$

$$+ 2a_3i\hbar eBx\left(\frac{n\pi}{a}\right)\cot\left(\frac{n\pi}{a}z\right)\psi_n - a_1\hbar^2\frac{\partial^2}{\partial x^2} - a_2\hbar^2\frac{\partial^2}{\partial y^2}$$

$$+ a_3\left(-\hbar^2\frac{\partial^2}{\partial z^2} + e^2B^2x^2 + 2i\hbar eBx\frac{\partial}{\partial z}\right)\Bigg]$$

$$\Bigg[a_1\hbar^2k_x^2\psi_n + a_2\hbar^2k_y^2\psi_n + a_3\hbar^2\left(\frac{n\pi}{a}\right)^2\psi_n + a_3e^2B^2x^2\psi_n$$

$$+ 2a_3i\hbar eBx\left(\frac{n\pi}{a}\right)x\cot\left(\frac{n\pi}{a}z\right)\psi_n\Bigg]$$

$$= \Bigg[a_1\hbar^4k_x^4\psi_n + a_1a_2\hbar^4k_x^2k_y^2\psi_n + a_1a_3\hbar^4k_x^2\left(\frac{n\pi}{a}\right)^2\psi_n$$

$$+ a_1a_3\hbar^2k_x^2e^2B^2x^2\psi_n - a_1a_3\hbar^22e^2B\psi_n$$

$$+ 2a_1a_3i\hbar^3eB\left(\frac{n\pi}{a}\right)k_x^2x\cot\left(\frac{n\pi}{a}z\right)\psi_n + a_2a_1\hbar^4k_y^2k_x^2\psi_n$$

$$+ a_2^2\hbar^4k_y^4\psi_n + a_2a_3\hbar^4\left(\frac{n\pi}{a}\right)^2k_y^2\psi_n + a_2a_3e^2B^2x^2\hbar^2k_y^2\psi_n$$

$$+ 2a_2a_3i\hbar^3k_y^2eB\left(\frac{n\pi}{a}\right)x\cot\left(\frac{n\pi}{a}z\right)\psi_n + a_3a_1\hbar^4\left(\frac{n\pi}{a}\right)^2k_x^2\psi_n$$

$$+ a_3a_2\hbar^4\left(\frac{n\pi}{a}\right)^2k_y^2\psi_n + a_3\hbar^4\left(\frac{n\pi}{a}\right)^4\psi_n$$

$$+ a_3^2\hbar^2\left(\frac{n\pi}{a}\right)^2e^2B^2x^2\psi_n + 2a_3^2i\hbar^3eB\left(\frac{n\pi}{a}\right)^3x\cot\left(\frac{n\pi}{a}z\right)\psi_n$$

$$+ \left[a_3a_1\hbar^2k_x^2e^2B^2 + a_3a_2\hbar^2k_y^2e^2B^2 + a_3^2\hbar^2\left(\frac{n\pi}{a}\right)^2e^2B^2\right]x^2\psi_n$$

$$+ a_3^2e^4B^4x^4\psi_n + 2a_3^2i\hbar e^3B^3\left(\frac{n\pi}{a}\right)x^3\cot\left(\frac{n\pi}{a}z\right)\psi_n$$

$$+ 2a_3i\hbar eB\left[a_1\hbar^2k_x^2 + a_2\hbar^2k_y^2 + a_3\hbar^2\left(\frac{n\pi}{a}\right)^2\right]x\left(\frac{n\pi}{a}\right)$$

$$\cot\left(\frac{n\pi}{a}z\right)\psi_n + 2a_3^2 i\hbar e^3 B^3 x^3 \left(\frac{n\pi}{a}\right)\cot\left(\frac{n\pi}{a}z\right)\psi_n$$
$$+ 4a_3^2\hbar^2 e^2 B^2 \left(\frac{n\pi}{a}\right)^2 x^2\psi_n$$
$$= \left[a_1\hbar^2 k_x^2 + a_2\hbar^2 k_y^2 + a_3\hbar^2\left(\frac{n\pi}{a}\right)^2\right]^2 \psi_n - 2a_1 a_3\hbar^2 e^2 B^2\psi_n$$
$$+ 4a_3 i\hbar eB\left[a_1\hbar^2 k_x^2 + a_2\hbar^2 k_y^2\right.$$
$$\left. + \ a_3\hbar^2\left(\frac{n\pi}{a}\right)\right] x\left(\frac{n\pi}{a}\right)\cot\left(\frac{n\pi}{a}z\right)\psi_n + \left[2a_1 a_3\hbar^2 k_x^2 e^2 B^2\right.$$
$$\left. + \ 2a_2 a_3\hbar^2 k_y^2 e^2 B^2 + 6a_3^2\hbar^2\left(\frac{n\pi}{a}\right)^2 e^2 B^2\right] x^2\psi_n$$
$$+ 4a_3^2 i\hbar e^3 B^3\left(\frac{n\pi}{a}\right) x^3\cot\left(\frac{n\pi}{a}z\right)\psi_n + a_3^2 e^4 B^4 x^4\psi_n \quad (16.91)$$

$$\hat{H}\psi_n = a_1\hbar^2 k_x^2\psi_n + a_2\hbar^2 k_y^2\psi_n + a^3\hbar^2\left(\frac{n\pi}{a}\right)^2\psi_n + a_3 e^2 B^2 x^2\psi_n$$
$$+ 2a_i i\hbar eB\left(\frac{n\pi}{a}\right) x\cot\left(\frac{n\pi}{a}z\right)\psi_n$$
$$- \alpha\left[a_1\hbar^2 k_x^2 + a_2\hbar^2 k_y^2 + a_3\hbar^2\left(\frac{n\pi}{a}\right)^2\right]^2\psi_n$$
$$+ 2\alpha a_1 a_3\hbar^2 e^2 B^2\psi_n - 4\alpha a_3 i\hbar eB\left[a_1\hbar^2 k_x^2 + a_2\hbar^2 k_y^2\right.$$
$$\left. + a_3\hbar^2\left(\frac{n\pi}{a}\right)^2\right]\left(\frac{n\pi}{a}\right) x\cot\left(\frac{n\pi}{a}z\right)\psi_n - \alpha\left[2a_1 a_3\hbar^2 k_x^2 e^2 B^2\right.$$
$$\left. + \ 2a_2 a_3\hbar^2 k_y^2 e^2 B^2 + 6a_3^2\hbar^2\left(\frac{n\pi}{a}\right)^2 e^2 B^2\right]^2 x^2\psi_n$$
$$+\alpha a_2 a_3\hbar^2 k_y^2 e^2 B^2 x^2\psi_n$$
$$+ 2\alpha a_2 a_3 i\hbar^3 k_y^2 eB\left(\frac{n\pi}{a}\right) x\cot\left(\frac{n\pi}{a}z\right)\psi_n$$
$$- \alpha a_2'\left[a_1\hbar^4 k_x^2 k_y^2 + a_2\hbar^4 k_y^4 + a_3\hbar^4 k_y^2\left(\frac{n\pi}{a}\right)^2\right]\psi_n$$

$$- \alpha a_2' a_3 k h^2 k_y^2 e^2 B^2 x^2 \psi_n$$

$$- 2\alpha a_2' a_3 i\hbar^3 k_y^2 eB \left(\frac{n\pi}{a}\right) x \cot\left(\frac{n\pi}{a} z\right) \psi_n + \alpha a_2 a_2' \hbar^4 k_y^4 \psi_n$$

$$= \hat{H}_0 \psi_n + \hat{H}' \psi_n \tag{16.92}$$

where

$$\hat{H}_0 = a_1 \hbar^2 k_x^2 + a_2 \hbar^2 k_y^2 + a_3 \hbar^2 \left(\frac{n\pi}{a}\right)^2$$

The remaining part correspond to perturbed Hamiltonian.

$$\left\langle x \cot\left(\frac{n\pi}{a} z\right)\right\rangle = \frac{2}{a} \int_0^a \sin\left(\frac{n\pi}{a} z\right) \cos\left(\frac{n\pi}{a} z\right) dz \int_0^{L_x} \frac{1}{L_x} dx = 0,$$

$$\left\langle x^3 \cot\left(\frac{n\pi}{a} z\right)\right\rangle = 0, \quad \langle x^2 \rangle = \frac{1}{L_x} \int_0^{L_x} x^2 dx = \frac{L_x^2}{3},$$

$$\langle x^4 \rangle = \frac{1}{L_x} \int_0^{L_x} x^4 dx = \frac{L_x^4}{5}$$

The DR in this case is given by

$$E = a_1 \hbar^2 k_x^2 + a_2 \hbar^2 k_y^2 + a_3 \hbar^2 \left(\frac{n\pi}{a}\right)^2$$

$$-\alpha \left[a_1 \hbar^2 k_x^2 + a_2 \hbar^2 k_y^2 + a_3 \hbar^2 \left(\frac{n\pi}{a}\right)^2\right]^2$$

$$+ \alpha a_2 \hbar^2 k_y^2 \left[a_1 \hbar^2 k_x^2 + a_2 \hbar^2 k_y^2 + a_3 \hbar^2 \left(\frac{n\pi}{a}\right)^2\right]$$

$$- \alpha a_2' \hbar^2 k_y^2 \left[a_1 \hbar^2 k_x^2 + a_2 \hbar^2 k_y^2 + a_3 \hbar^2 \left(\frac{n\pi}{a}\right)^2\right]$$

$$+\alpha a_2 a_2' \hbar^4 k_y^4 + a_3 e^2 B^2 \langle x^2 \rangle + 2\alpha a_1 a_3 \hbar^2 e^2 B^2$$

$$-\alpha \left[2a_1 a_3 \hbar^2 k_x^2 + a_2 a_3 \hbar^2 k_y^2 + 6a_3^2 \hbar^2 \left(\frac{n\pi}{a}\right)^2\right] e^2 B^2 \langle x^2 \rangle$$

$$-\alpha a_3^2 e^4 B^4 \langle x^4 \rangle - \alpha a_2' a_3 \hbar^2 k_y^2 e^2 B^2 \langle x^2 \rangle \tag{16.93}$$

Thus following the method of Chapter 1, the DR in HDQWs of IV–VI semiconductors can be written as

$$\begin{aligned}
\gamma_3(E,\eta_g) &= a_1\hbar^2 k_x^2 + a_2\hbar^2 k_y^2 + a_3\hbar^2\left(\frac{n\pi}{a}\right)^2 \\
&\quad - \alpha\left[a_1\hbar^2 k_x^2 + a_2\hbar^2 k_y^2 + a_3\hbar^2\left(\frac{n\pi}{a}\right)^2\right]^2 \\
&\quad + \alpha a_2\hbar^2 k_y^2\left[a_1\hbar^2 k_x^2 + a_2\hbar^2 k_y^2 + a_3\hbar^2\left(\frac{n\pi}{a}\right)^2\right] \\
&\quad - \alpha a_2'\hbar^2 k_y^2\left[a_1\hbar^2 k_x^2 + a_2\hbar^2 k_y^2 + a_3\hbar^2\left(\frac{n\pi}{a}\right)^2\right] \\
&\quad + \alpha a_2 a_2'\hbar^4 k_y^4 + a_3 e^2 B^2\left\langle x^2\right\rangle + 2\alpha a_1 a_3\hbar^2 e^2 B^2 \\
&\quad - \alpha\left[2a_1 a_3\hbar^2 k_x^2 + a_2 a_3\hbar^2 k_y^2 + 6a_3^2\hbar^2\left(\frac{n\pi}{a}\right)^2\right] e^2 B^2\left\langle x^2\right\rangle \\
&\quad - \alpha a_3^2 e^4 B^4\left\langle x^4\right\rangle - \alpha a_2' a_3\hbar^2 k_y^2 e^2 B^2\left\langle x^2\right\rangle \qquad (16.94)
\end{aligned}$$

Putting $\alpha = 0$, we get

$$\gamma_3(E,\eta_g) = a_1\hbar^2 k_x^2 + a_2\hbar^2 k_y^2 + a_3\hbar^2\left(\frac{n\pi}{a}\right)^2 + a_3 e^2 B^2\langle x^2\rangle \qquad (16.95)$$

(16.95) can be written as

$$\gamma_3(E,\eta_g) - a_3 e^2 B^2\left\langle x^2\right\rangle - a_3\hbar^2\left(\frac{n\pi}{a}\right)^2 = a_1\hbar^2 k_x^2 + a_2\hbar^2 k_y^2 \qquad (16.96)$$

Using the method of successive approximation we can write

$$\begin{aligned}
\gamma_3(E,\eta_g) &= a_1\hbar^2 k_x^2 + a_2\hbar^2 k_y^2 + a_3\hbar^2\left(\frac{n\pi}{a}\right)^2 - \alpha(\gamma_3(E,\eta_g) - c_0)^2 \\
&\quad + \alpha a_2(\gamma_3(E,\eta_g) - c_0)\hbar^2 k_y^2 - \alpha a_2'(\gamma_3(E,\eta_g) - c_0)\hbar^2 k_y^2 \\
&\quad + \alpha a_2 a_2'\hbar^4 k_y^4 + c_0 + c_2 - 2\alpha a_1 a_3 e^2 B^2\left\langle x^2\right\rangle\hbar^2 k_x^2 \\
&\quad - \alpha a_2 a_3 e^2 B^2\left\langle x^2\right\rangle\hbar^2 k_y^2 - 6\alpha a_3^2\hbar^2\left(\frac{n\pi}{a}\right)^2 e^2 B^2\left\langle x^2\right\rangle \\
&\quad - \alpha a_3^2 e^4 B^4\left\langle x^4\right\rangle - \alpha a_2' a_3 e^2 B^2\left\langle x^2\right\rangle\hbar^2 k_y^2 \qquad (16.97)
\end{aligned}$$

where

$$c_0 = a_3 e^2 B^2\left\langle x^2\right\rangle = \frac{a_3 e^2 B^2 L_x^2}{3}, \quad c_1 = a_3\hbar^2\left(\frac{n\pi}{a}\right)^2$$

The (16.97) can be written as

$$\gamma_3(E,\eta_g) - c_0 - c_1 - c_2 + \alpha(\gamma_3(E,\eta_g) - c_0)^2 + c_3 + c_4$$
$$= [a_1\hbar^2 - 2\alpha\hbar^2 a_1 a_3 e^2 B^2 \langle x^2\rangle]k_x^2 + [a_2\hbar^2 + \alpha a_2(\gamma_3(E,\eta_g) - c_0)\hbar^2$$
$$- \alpha a_2'(\gamma_3(E,\eta_g) - c_0)\hbar^2 - \hbar^2\alpha a_2 a_3 e^2 B^2 \langle x^2\rangle$$
$$- \hbar^2\alpha a_2' a_3 e^2 B^2 \langle x^2\rangle]k_y^2 + \alpha a_2 a_2'^2 \hbar^4 k_y^2 \tag{16.98}$$

The DR can be written as

$$\gamma_{10}(E,\eta_g) = \beta_1 k_x^2 + \beta_2 k_y^2 + \beta_3 k_y^4 \tag{16.99}$$

where,

$$\gamma_{10}(E,\eta_g) = \gamma_3(E,\eta_g) - c_0 - c_1 - c_2 + c_3 + c_4 + \alpha(\gamma_3(E,\eta_g) - c_0)^2$$

$$c_2 = 2\alpha a_1 a_3 \hbar^2 e^2 B^2, \quad c_3 = 6\alpha a_3^2 \hbar^2 \left(\frac{n\pi}{a}\right)^2 e^2 B^2 \left(\frac{L_x^2}{3}\right),$$

$$c_4 = \alpha a_3^2 e^4 B^4 \left(\frac{L_x^4}{5}\right),$$

$$\beta_1 = a_1\hbar^2 - 2\alpha\hbar^2 a_1 a_3 e^2 B^2 \left(\frac{L_x^2}{3}\right)$$
$$= a_1\hbar^2 \left[1 - 2\alpha a_3 e^2 B^2 \left(\frac{L_x^2}{3}\right)\right] = \text{positive quantity},$$

$$\beta_2 = a_2\hbar^2 + \alpha a_2(\gamma_3(E,\eta_g) - c_0)\hbar^2 - \alpha a_2'(\gamma_3(E,\eta_g) - c_0)\hbar^2$$
$$- \hbar^2\alpha a_2 a_3 e^2 B^2 \left(\frac{L_x^2}{3}\right) - \hbar^2\alpha a_2' a_3 e^2 B^2 \left(\frac{L_x^2}{3}\right)$$
$$\simeq a_2\hbar^2 - 2\alpha\hbar^2 a_2 a_3 e^2 B^2 \left(\frac{L_x^2}{3}\right) \simeq a_2\hbar^2 \left(1 - 2\alpha a_3 e^2 B^2 \frac{L_x^2}{3}\right)$$

$\simeq$ positive quantity and $\beta_3 = \alpha a_2 a_2' \hbar^4 =$ positive quantity

Therefore, from (16.99) we can write

$$k_x = \pm\frac{1}{\sqrt{\beta_1}}\sqrt{\gamma_{10}(E,\eta_g) - \beta_2 k_y^2 - \beta_3 k_y^4} \tag{16.100}$$

Since area is a positive quantity, therefore

$$k_x = \frac{1}{\sqrt{\beta_1}} \cdot \sqrt{\gamma_{10}(E,\eta_g) - \beta_2 k_y^2 - \beta_3 k_y^4} \tag{16.101}$$

Similarly we get

$$k_y = \pm\sqrt{\frac{-\beta_2}{2\beta_3} \pm \frac{\sqrt{\beta_2^2 + 4\beta_3\gamma_{10}(E,\eta_g)}}{2\beta_3}}$$

Since k_y is real, therefore

$$k_y = \pm\sqrt{-\frac{1}{2}\cdot\left(\frac{\beta_2}{\beta_3}\right) + \frac{1}{2}\sqrt{\left(\frac{\beta_2}{\beta_3}\right)^2 + \frac{4\gamma_{10}(E,\eta_g)}{\beta_3}}} = \pm u$$

So the area enclosed by the 2D surface as given by (16.99) can be written as

$$A = 2I = 2\int_{-u}^{+u} \frac{1}{\sqrt{\beta_1}}\sqrt{\gamma_{10}(E,\eta_g) - \beta_2 k_y^2 - \beta_3 k_y^4 dk_y} \tag{16.102}$$

where,

$$\begin{aligned} I &= \frac{1}{\sqrt{\beta_1}}\int_{-u}^{+u} \sqrt{\gamma_{10}(E,\eta_g) - \beta_2 k_y^2 - \beta_3 k_y^4 dk_y} \\ &= \frac{2}{\sqrt{\beta_1}}\int_{0}^{+u} \sqrt{\gamma_{10}(E,\eta_g) - \beta_2 k_y^2 - \beta_3 k_y^4 dk_y} \\ &= 2\sqrt{\frac{\beta_3}{\beta_1}}\int_{0}^{+u} \sqrt{\frac{\gamma_{10}(E,\eta_g)}{\beta_3} - \left(\frac{\beta_2}{\beta_3}\right) k_y^2 - k_y^4 dk_y} = 2\sqrt{\frac{\beta_3}{\beta_1}} \cdot I_1 \end{aligned}$$

Therefore

$$A = 2I = 4\sqrt{\frac{\beta_3}{\beta_1}}.I_1 \tag{16.103}$$

where

$$I_1 = \int_{0}^{+u} \sqrt{\frac{\gamma_{10}(E,\eta_g)}{\beta_3} - \left(\frac{\beta_2}{\beta_3}\right) k_y^2 - k_y^4 dk_y} \tag{16.104}$$

Now

$$\frac{\gamma_{10}(E,\eta_g)}{\beta_3} - \left(\frac{\beta_2}{\beta_3}\right)k_y^2 - k_y^4 dk_y = (a^2+x^2)(b^2-x^2)$$
$$= a^2b^2 - a^2x^2 + b^2x^2 - x^4$$
$$= a^2b^2(a^2-b^2)x^2 - x^4$$

So that

$$a^2b^2 = \frac{\gamma_{10}(E,\eta_g)}{\beta_3}, \quad a^2 - b^2 = \frac{\beta_2}{\beta_3} \quad \text{and} \quad x = k_y,$$

i.e., $dx = dk_y$

$$I_1 = \int_0^u \sqrt{(a^2+x^2)(b^2-x^2)}dx \tag{16.105}$$

Now

$$(a^2+b^2)^2 = (a^2-b^2)^2 + 4a^2b^2 = \left(\frac{\beta_2}{\beta_3}\right)^2 + \frac{4\gamma_{10}(E,\eta_g)}{\beta_3} \tag{16.106}$$

$$a^2+b^2 = \sqrt{\left(\frac{\beta_2}{\beta_3}\right)^2 + \frac{4\gamma_{10}(E,\eta_g)}{\beta_3}} \quad \text{for real } b.$$

Therefore,

$$a^2 = \frac{1}{2}\left(\frac{\beta_2}{\beta_3}\right) + \frac{1}{2}\sqrt{\left(\frac{\beta_2}{\beta_3}\right)^2 + \frac{4\gamma_{10}(E,\eta_g)}{\beta_3}}$$
$$b^2 = \frac{1}{2}\sqrt{\left(\frac{\beta_2}{\beta_3}\right)^2 + \frac{4\gamma_{10}(E,\eta_g)}{\beta_3}} - \frac{1}{2}\left(\frac{\beta_2}{\beta_3}\right)$$

So all the coefficients of (1.105) are defined. From the expressions of u and b we note that $u = b$. Since b is real and positive, $u > 0$ and also $a > b$. Therefore $a > u$. Thus we observe that $a > u > 0$. Under this condition (1.105) can be written as

$$I_1 = \frac{1}{3}\sqrt{a^2+b^2} \cdot \{a^2 F(\theta, r) - (a^2 - b^2)E(\theta, r)\}$$
$$+ \frac{u}{3}(u^2 + 2a^2 - b^2)\sqrt{\frac{b^2 - u^2}{a^2 + u^2}} \tag{16.107}$$

where

$$F(\theta, r) = \int_0^\theta \frac{dt}{\sqrt{1 - r^2 \sin^2 t}}, \qquad E(\theta, r) = \int_0^\theta \sqrt{1 - r^2 \sin^2 t} dt,$$

$F(\theta, r)$ and $E_{(\theta,r)}$ are known as incomplete Elliptic integral of first and second kind respectively in which $r^2 < 1$ and is known as modulus of the integral,

$$\theta = \sin^{-1}\left[\frac{u}{b}\sqrt{\frac{a^2+b^2}{a^2+u^2}}\right] \quad \text{and} \quad r = \frac{b}{\sqrt{a^2+b^2}}$$

In this case $b = u$, so that

$$I_1 = \frac{1}{3}\sqrt{a^2+b^2} \cdot \{a^2 F(\theta, r) - (a^2 - b^2)E(\theta, r)\} \tag{16.108}$$

when $b = u$, then

$$\theta = \frac{\pi}{2} \tag{16.109}$$

Therefore from (16.109) we get

$$I_1 = \frac{1}{3}\sqrt{a^2+b^2} \cdot \left\{a^2 F\left(\frac{\pi}{2}, r\right) - (a^2 - b^2)E\left(\frac{\pi}{2}, r\right)\right\}$$
$$= \frac{1}{3}\sqrt{a^2+b^2} \cdot \{a^2 K(r) - (a^2 - b^2)E(r)\} \tag{16.110}$$

where $K(r)$ and $E(r)$ are known as the complete Elliptic integrals of first and second kind respectively.

In our case $r = \frac{b}{\sqrt{a^2+b^2}}$ and since $a > b$ therefore $r^2 < 1$

Thus we write

$$r = \sqrt{\frac{b^2}{a^2+b^2}} = \frac{1}{\sqrt{2}}\left[1 - \frac{\frac{\beta_2}{\beta_3}}{\sqrt{\left(\frac{\beta_2}{\beta_3}\right)^2 + \frac{4\gamma_{10}(E,\eta_g)}{\beta_3}}}\right]^{1/2}$$

Since $r << 1$

Therefore

$$K(r) = \frac{\pi}{2}\left[1 + (1/2)^2 r^2 + \left(\frac{1.3}{2.4}\right)^2 r^4 + \cdots \right.$$
$$\left. + \left\{\frac{(2n-1)!!}{2^n n!}\right\}^2 r^{2n} + \cdots\right]$$
$$E(r) = \frac{\pi}{2}\left[1 - (1/2)^2 r^2 + \left(\frac{1^2.3}{2^2.4^2}\right) r^4 - \cdots \right.$$
$$\left. - \left\{\frac{(2n-1)!!}{2^n n!}\right\}^2 \frac{r^{2n}}{(2n-1)} - \cdots\right]$$

Since $r^2 << 1$, we take only r^2 terms as they provide maximum contribution in comparison with the higher order terms. Therefore

$$I_1 = \frac{1}{3}(\sqrt{a^2+b^2})\{a^2K(r) - (a^2-b^2)E(r)\}$$
$$= \frac{1}{3}(\sqrt{a^2+b^2})\left[a^2\frac{\pi}{2}\left\{1+\frac{1}{2^2}r^2\right\} - (a^2-b^2)\frac{\pi}{2}\left\{1-\frac{1}{2^2}r^2\right\}\right]$$
$$= \frac{1}{3}(\sqrt{a^2+b^2})\left[\frac{a^2\pi}{2} + \frac{a^2r^2}{4}\left(\frac{\pi}{2}\right) - \frac{a^2\pi}{2} + \frac{a^2r^2}{4}\left(\frac{\pi}{2}\right)\right.$$
$$\left. + b^2\left(\frac{\pi}{2}\right) - \frac{b^2r^2}{4}\left(\frac{\pi}{2}\right)\right] = \frac{1}{3}(\sqrt{a^2+b^2})\left(\frac{\pi}{2}\right)\frac{3b^2[2a^2+b^2]}{4(a^2+b^2)}$$
$$= \left(\frac{\pi}{2}\right)\frac{b^2[2a^2+b^2]}{4\sqrt{(a^2+b^2)}} = \left(\frac{\pi}{2}\right)\frac{1}{4}\frac{(2a^2b^2+b^4)}{(a^2+b^2)^{1/2}}$$
$$= \left(\frac{\pi}{8}\right)\frac{\left(\frac{\beta_2}{\beta_3}\right)^2\left[\frac{1}{2} + \frac{3\gamma_{10}(E,\eta_g)\beta_3}{\beta_2^2} - \frac{1}{2}\left\{1+\frac{4\gamma_{10}(E,\eta_g)}{\beta_3}\right\}^{1/2}\right]}{\left(\frac{\beta_2}{\beta_3}\right)^2\left[1+\frac{4\gamma_{10}(E,\eta_g)\beta_3}{\beta_3}\right]^{1/2}}\Bigg]^2$$
$$= \left(\frac{\pi}{8}\right)\frac{\frac{2\gamma_{10}(E,\eta_g)}{\beta_3} + \left[\frac{1}{2}\sqrt{\left(\frac{\beta_2}{\beta_3}\right)^2 + \frac{4\gamma_{10}(E,\eta_g)}{\beta_3}} - \frac{1}{2}\left(\frac{\beta_2}{\beta_3}\right)\right]}{\left[\left(\frac{\beta_2}{\beta_3}\right)^2 + \frac{4\gamma_{10}(E,\eta_g)}{\beta_3^2}\right]^{1/4}}\Bigg]\Bigg]$$

So the general expression of Area (A) is given by

$$A = 4\sqrt{\frac{\beta_3}{\beta_1}}\left(\frac{\pi}{8}\right)\left(\frac{\beta_2}{\beta_3}\right)^{3/2} \cdot \frac{\left[\frac{1}{2} + \frac{3\gamma_{10}(E,\eta_g)\beta_3}{\beta_2^2} - \frac{1}{2}\left\{1 + \frac{4\gamma_{10}(E,\eta_g)\beta_3}{\beta_2^2}\right\}^{1/2}\right]}{\left[1 + \frac{4\gamma_{10}(E,\eta_g)\beta_3}{\beta_2^2}\right]^{1/4}}$$

$$= \left(\frac{\pi}{2}\right)\left(\frac{\beta_2}{\beta_3}\right)\left(\frac{\beta_2}{\beta_1}\right)^{1/2} \cdot \frac{\left[\frac{1}{2} + \frac{3\gamma_{10}(E,\eta_g)\beta_3}{\beta_2^2} - \frac{1}{2}\left\{1 + \frac{4\gamma_{10}(E,\eta_g)\beta_3}{\beta_2^2}\right\}^{1/2}\right]}{\left[1 + \frac{4\gamma_{10}(E,\eta_g)\beta_3}{\beta_2^2}\right]^{1/4}} \quad (16.111)$$

Since $\frac{4\gamma_{10}(E,\eta_g)\beta_3}{\beta_2^2} < 1$

We can write

$$\left[1 + \frac{4\gamma_{10}(E,\eta_g)\beta_3}{\beta_2^2}\right]^{1/2} \cong 1 + \frac{2\gamma_{10}(E,\eta_g)\beta_3}{\beta_2^2} \quad \text{and} \quad \left[1 + \frac{4\gamma_{10}(E,\eta_g)\beta_3}{\beta_2^2}\right]^{1/4} \cong 1 + \frac{\gamma_{10}(E,\eta_g)\beta_3}{\beta_2^2}$$

$$\left[1 + \frac{4\gamma_{10}(\mathrm{E},\eta_g)\beta_3}{\beta_2^2}\right]^{1/2} \cong 1 + \frac{2\gamma_{10}(\mathrm{E},\eta_g)\beta_3}{\beta_2^2} \quad \text{and} \quad \left[1 + \frac{4\gamma_{10}(\mathrm{E},\eta_g)\beta_3}{\beta_2^2}\right]^{1/4} \cong 1 + \frac{\gamma_{10}(\mathrm{E},\eta_g)\beta_3}{\beta_2^2}$$

Therefore we can write

$$A = \left(\frac{\pi}{2}\right)\left(\frac{\beta_2}{\beta_3}\right)\left(\frac{\beta_2}{\beta_1}\right)^{1/2}\left[\frac{3\gamma_{10}(E,\eta_g)\beta_3}{\beta_2^2} - \frac{\gamma_{10}(E,\eta_g)\beta_3}{\beta_2^2}\right]$$

$$\cdot\left[1 + \frac{\gamma_{10}(E,\eta_g)\beta_3}{\beta_2^2}\right]^{-1}$$

$$= \pi\left(\frac{\beta_2}{\beta_3}\right)\left(\frac{\beta_2}{\beta_1}\right)^{1/2}\left[\frac{\gamma_{10}(E,\eta_g)\beta_3}{\beta_2^2}\right]\left[1 + \frac{\gamma_{10}(E,\eta_g)\beta_3}{\beta_2^2}\right]^{-1}$$

$$= \pi\left(\frac{\beta_2}{\beta_1}\right)^{1/2}\left(\frac{1}{\beta_2}\right)\gamma_{10}(E,\eta_g)\left[1 + \frac{\gamma_{10}(E,\eta_g)\beta_3}{\beta_2^2}\right]^{-1}$$

Therefore the simplified expression of area can be written as

$$A = \frac{\pi\gamma_{10}(E,\eta_g)}{\sqrt{\beta_1\beta_2}} \cdot \frac{1}{\left[1 + \frac{\gamma_{10}(E,\eta_g)\beta_3}{\beta_2^2}\right]} = \omega_{19,1}(E,\eta_g) \qquad (16.112)$$

The EEM and the DOS function can respectively be written in this case as

$$m^*(E,\eta_n,n) = \frac{\hbar^2}{2\pi}\omega'_{19,1}(E,\eta_g) \qquad (16.113)$$

$$N_{2DQWHD}(E,\eta_g) = \frac{g_v}{(2\pi)^2}\sum_{n=1}^{n_{\max}}\omega'_{19,1}(E,\eta_g)H(E - E_{19,1}) \qquad (16.114)$$

where $E_{19,1}$ is the sub band energy and can be obtained from the equation

$$\gamma_{10}(E_{19,1},\eta_g) = 0 \qquad (16.115)$$

(b) Model of McClure and Choi

In accordance with this model, the DR in bulk specimens of IV–VI materials can be written as

$$E(1+\alpha E) = \frac{\hbar^2k_x^2}{2m_1} + \frac{\hbar^2k_y^2}{2m_2} + \frac{\hbar^2k_z^2}{2m_3} + \frac{\alpha\hbar^4k_y^4}{4m_2m'_2}$$

$$+\frac{\hbar^2k_y^2}{2m_2}\alpha E\left(1 - \frac{m_2}{m'_2}\right) - \frac{\alpha\hbar^4k_x^2k_y^2}{4m_2m_2} - \frac{\alpha\hbar^4k_y^2k_z^2}{4m_2m_3} \qquad (16.116)$$

Let us substitute

$$\frac{1}{2m_i} = a_i, \quad i = 1, 2, 3, \ldots, \quad \frac{1}{2m_2'} = a_2' \quad \text{and} \quad \left(1 - \frac{m_2}{m_2'}\right) = a_0$$

Therefore (1.116) can be written as

$$E(1 + \alpha E) = a_1 p_x^2 + a_2 p_y^2 + a_3 p_z^2 + \alpha a_2 a_2' p_y^4 + \alpha a_2 a_0 p_y^2 E - \alpha a_1 a_2 p_x^2 p_y^2 - \alpha a_2 a_3 p_y^2 p_z^2 \quad (16.117)$$

Putting $\alpha = 0$, in (1.117) we get

$$E = a_1 p_x^2 + a_2 p_y^2 + a_3 p_z^2 \quad (16.118)$$

So, by the method of successive approximation we write

$$E = a_1 p_x^2 + a_2 p_y^2 + a_3 p_z^2 - \alpha[a_1 p_x^2 + a_2 p_y^2 + a_3 p_z^2]^2 + \alpha a_2 a_2' p_y^4 + \alpha a_2 a_0 p_y^2 (a_1 p_x^2 + a_2 p_y^2 + a_3 p_z^2) - \alpha a_1 a_2 p_x^2 p_y^2 - \alpha a_2 a_3 p_y^2 p_z^2 \quad (16.118)$$

Choosing Coulomb Gauge as

$$\vec{A} = (0, 0, -Bx) \quad \text{and} \quad \vec{B}(0, B, 0)$$

So that $\vec{\nabla}\vec{A} = 0$ and $\vec{B} = \vec{\nabla} \times \vec{A}$

In presence of perturbing magnetic field $\vec{B}$ total Hamiltonian ($\hat{H}$) can be written as

$$\hat{H} = a_1\left(-i\hbar\frac{\partial}{\partial x}\right)^2 + a_2\left(-i\hbar\frac{\partial}{\partial y}\right)^2 + a_3\left(-i\hbar\frac{\partial}{\partial z} - eB_x\right)^2 - \alpha\left[a_1\left(-i\hbar\frac{\partial}{\partial x}\right)^2 + a_2\left(-i\hbar\frac{\partial}{\partial y}\right)^2 + a_3\left(-i\hbar\frac{\partial}{\partial z} - eB_x\right)^2\right]^2 + \alpha a_2 a_2'\left(-i\hbar\frac{\partial}{\partial y}\right)^4 + \alpha a_2 a_0\left(-i\hbar\frac{\partial}{\partial y}\right)^2 \cdot \left[a_1\left(-i\hbar\frac{\partial}{\partial x}\right)^2 + a_2\left(-i\hbar\frac{\partial}{\partial y}\right)^2 + a_3\left(-i\hbar\frac{\partial}{\partial z} - eB_x\right)^2\right] \quad (16.119)$$

Considering non-parabolicity and magnetic field ($\vec{B}$) as perturbations we can write

$$\hat{H}_0\psi_0 = E_0\psi_0 \tag{16.120}$$

Therefore from (1.120) we get

$$-a_1\hbar^2\frac{\partial^2\psi_0}{\partial x^2} - a_2\hbar^2\frac{\partial^2\psi_0}{\partial y^2} - a_3\hbar^2\frac{\partial^2\psi_0}{\partial z^2} = E_0\psi_0 \tag{16.121}$$

So that

$$E_0 = E_n^0 = \frac{\hbar^2 k_x^2}{2m_1} + \frac{\hbar^2 k_y^2}{2m_2} + \frac{\hbar^2}{2m_3}\left(\frac{n\pi}{a}\right)^2 \tag{16.122}$$

and

$$\psi_0 = \psi_n^0 = \sqrt{\frac{2}{aL_xL_y}} = \sin\left(\frac{n\pi}{a}z\right)\exp(ik_xx + ik_yy) \tag{16.123}$$

Considering perturbing effects, the total Hamiltonian can be written as

$$\begin{aligned}
\hat{H} = &-a_1\hbar^2\frac{\partial^2}{\partial x^2} - a_2\hbar^2\frac{\partial^2}{\partial y^2} + a_3\left(-\hbar^2\frac{\partial^2}{\partial z^2} + e^2B^2x^2 + 2i\hbar eBx\frac{\partial}{\partial z}\right) \\
&- \alpha\left[-a_1\hbar^2\frac{\partial^2}{\partial x^2} - a_2\hbar^2\frac{\partial^2}{\partial y^2}\right. \\
&\left. + a_3\left(-\hbar^2\frac{\partial^2}{\partial z^2} + e^2B^2x^2 + 2i\hbar eBx\frac{\partial}{\partial z}\right)\right]^2 \\
&+ \alpha a_2 a_2'\left(\hbar^4\frac{\partial^4}{\partial y^4}\right) + \alpha a_2 a_0\left(-\hbar^2\frac{\partial^2}{\partial y^2}\right) \\
&\cdot\left[-a_1\hbar^2\frac{\partial^2}{\partial x^2} - a_2\hbar^2\frac{\partial^2}{\partial y^2} + a_3\left(-\hbar^2\frac{\partial^2}{\partial z^2} + e^2B^2x^2 + 2i\hbar eBx\frac{\partial}{\partial z}\right)\right] \\
&- \alpha a_2 a_2\hbar^4\left(\frac{\partial^2}{\partial x^2}\right)\left(\frac{\partial^2}{\partial y^2}\right) - \alpha a_2 a_3\left(-\hbar^2\frac{\partial^2}{\partial y^2}\right)\left(-\hbar^2\frac{\partial^2}{\partial y^2}\right) \\
&\cdot\left(-\hbar^2\frac{\partial^2}{\partial z^2} + e^2B^2x^2 + 2i\hbar eBx\frac{\partial}{\partial z}\right)
\end{aligned} \tag{16.124}$$

Now

$$\left[-a_1\hbar^2\frac{\partial^2}{\partial x^2}-a_2\hbar^2\frac{\partial^2}{\partial y^2}+a_3\left(-\hbar^2\frac{\partial^2}{\partial z^2}+e^2B^2x^2+2i\hbar eBx\frac{\partial}{\partial z}\right)\right]\psi_n$$

$$=a_1\hbar^2k_x^2\psi_n+a_2\hbar^2k_y^2\psi_n+a_3\hbar^2\left(\frac{n\pi}{a}\right)^2\psi_n+a_3e^2B^2x^2\psi_n$$

$$+2a_3i\hbar eBx\left(\frac{n\pi}{a}\right)\cot\left(\frac{n\pi}{a}z\right)\psi_n$$

$$\left[-a_1\hbar^2\frac{\partial^2}{\partial x^2}-a_2\hbar^2\frac{\partial^2}{\partial y^2}+a_3\left(-\hbar^2\frac{\partial^2}{\partial z^2}+e^2B^2x^2+2i\hbar eBx\frac{\partial}{\partial z}\right)\right]^2\psi_n$$

$$=\left[a_1\hbar^2k_x^2+a_1\hbar^2k_y^2+a_3\hbar^2\left(\frac{n\pi}{a}\right)^2\right]\psi_n-2a_1a_3\hbar^2e^2B^2\psi_n$$

$$+4a_3i\hbar eB\left[a_1\hbar^2k_x^2+a_1\hbar^2k_y^2+a_3\hbar^2\left(\frac{n\pi}{a}\right)^2\right.$$

$$\left.\left(\frac{n\pi}{a}\right)x\cot\left(\frac{n\pi}{a}z\right)\psi_n+\left[2a_1a_3\hbar^2k_x^2+2a_2a_3\hbar^2k_y^2+6a_3^2\hbar^2\left(\frac{n\pi}{a}\right)^2\right]\right.$$

$$\cdot e^2B^2x^2\psi_n+4a_3^2i\hbar e^3B^3\left(\frac{n\pi}{a}\right)x^3\cot\left(\frac{n\pi}{a}z\right)\psi_n+a_3^2e^4B^4x^4\psi_n$$

$$\hbar^4\frac{\partial^4}{\partial y^4}(\psi_n)=\hbar^4(ik_y)^4\psi_n=\hbar^4k_y^4\psi_n$$

$$\left(-\hbar^2\frac{\partial^2}{\partial y^2}\right)\left[a_1\hbar^2k_x^2\psi_n+a_2\hbar^2k_y^2\psi_n+a_3\hbar^2\left(\frac{n\pi}{a}\right)^2\psi_n\right.$$

$$\left.+\ a_3e^2B^2x^2\psi_n+2a_3i\hbar eBx\left(\frac{n\pi}{a}\right)\cot\left(\frac{n\pi}{a}z\right)\psi_n\right]$$

$$=\hbar^2k_y^2\left[a_1\hbar^2k_x^2+a_2\hbar^2k_y^2+a_3\hbar^2\left(\frac{n\pi}{a}\right)^2\right]\psi_n+a_3\hbar^2k_y^2e^2B^2x^2\psi_n$$

$$+2a_3i\hbar^3k_y^2eBx\left(\frac{n\pi}{a}\right)\cot\left(\frac{n\pi}{a}z\right)\psi_n$$

$$\hbar^4\left(\frac{\partial^2}{\partial x^2}\right)\left(\frac{\partial^2}{\partial y^2}\right)\psi_n=\hbar^4\frac{\partial^2}{\partial x^2}[(ik_y)^2\psi_n]=\hbar^4(ik_y)^2(ik_x)^2\psi_n$$

$$=\hbar^4k_x^2k_y^2\psi_n\left(-\hbar^2\frac{\partial^2}{\partial y^2}\right)\left[-\hbar\frac{\partial^2}{\partial z^2}+e^2B^2x^2+2i\hbar eBx\frac{\partial}{\partial z}\right)\right]\psi_n$$

$$= -\hbar^2 \frac{\partial^2}{\partial y^2} \left[\hbar \left(\frac{n\pi}{a} \right)^2 \psi_n + e^2 B^2 x^2 \psi_n \right.$$
$$\left. + \; 2i\hbar eBx \left(\frac{n\pi}{a} \right) \cot \left(\frac{n\pi}{a} z \right) \psi_n \right]$$
$$= \hbar^2 k_y^2 \left[\hbar^2 \left(\frac{n\pi}{a} \right)^2 + e^2 B^2 x^2 + 2i\hbar eBx \left(\frac{n\pi}{a} \right) \cot \left(\frac{n\pi}{a} z \right) \right] \psi_n \tag{16.125}$$

Therefore

$$\hat{H}\psi_n = a_1 \hbar^2 k_x^2 \psi_n + a_2 \hbar^2 k_y^2 \psi_n + a_3 \hbar^2 \left(\frac{n\pi}{a} \right)^2 \psi_n + a_3 e^2 B^2 x^2 \psi_n$$
$$+ 2a_3 \hbar eB \left(\frac{n\pi}{a} \right) x \cot \left(\frac{n\pi}{a} z \right) \psi_n$$
$$- \alpha \left[a_1 \hbar^2 k_x^2 + a_2 \hbar^2 k_y^2 + a_3 \hbar^2 \left(\frac{n\pi}{a} \right)^2 \right]^2 \psi_n + 2\alpha a_1 a_3 \hbar^2 e^2 B^2 \psi_n$$
$$- 4\alpha a_3 i\hbar eB \left[a_1 \hbar^2 k_x^2 + a_2 \hbar^2 k_y^2 + a_3 \hbar^2 \left(\frac{n\pi}{a} \right)^2 \right]$$
$$\cdot \left(\frac{n\pi}{a} \right) x \cot \left(\frac{n\pi}{a} z \right) \psi_n - \alpha \left[2a_1 a_3 \hbar^2 k_x^2 + 2a_2 a_3 \hbar^2 k_y^2 \right.$$
$$\left. + \; 6a_3^2 \hbar^2 \left(\frac{n\pi}{a} \right)^2 \right] e^2 B^2 x^2 \psi_n - 4\alpha i\hbar a_3^2 e^3 B^3 \left(\frac{n\pi}{a} \right) x^3 \cot$$
$$\cdot \left(\frac{n\pi}{a} z \right) \psi_n - \alpha a_3^2 e^4 B^4 x^4 \psi_n + \alpha a_2 a_2' \hbar^4 k_y^4 \psi_n + \alpha a_2 a_0 \hbar^2 k_y^2$$
$$\cdot \left[a_1 \hbar^2 k_x^2 + a_2 \hbar^2 k_y^2 + a_3 \hbar^2 \left(\frac{n\pi}{a} \right)^2 \right] \psi_n + \alpha a_2 a_0 a_3 \hbar^2 k_y^2 e^2 B^2 x^2 \psi_n$$
$$+ 2\alpha a_2 a_0 a_3 i\hbar^3 k_y^2 eB \left(\frac{n\pi}{a} \right) x \cot \left(\frac{n\pi}{a} z \right) \psi_n$$
$$- \alpha a_1 a_2 \hbar^4 k_x^2 k_y^2 \psi_n - \alpha a_2 a_3 \hbar^4 k_y^2 \left(\frac{n\pi}{a} \right)^2 \psi_n$$
$$- \alpha a_2 a_3 \hbar^2 k_y^2 e^2 B^2 x^2 \psi_n$$
$$- 2\alpha a_2 a_3 i\hbar^3 k_y^2 eB \left(\frac{n\pi}{a} \right) x \cot \left(\frac{n\pi}{a} z \right) \psi_n$$
$$= \hat{H}\psi_n + \hat{H}'\psi_n \tag{16.126}$$

where $\hat{H}_0$ is the unperturbed Hamiltonian $= a_1\hbar^2k_x^2 + a_2\hbar^2k_y^2 + a_3\hbar^2(\frac{n\pi}{a})^2$.

Applying first order perturbation we get

$$\left\langle x\cot\left(\frac{n\pi}{a}z\right)\right\rangle = 0, \quad \left\langle x^3\cot\left(\frac{n\pi}{a}z\right)\right\rangle = 0,$$
$$\langle x^2\rangle = \frac{L_x^2}{3} \quad \text{and} \quad \langle x^4\rangle = \frac{L_x^4}{5}$$

Now,

$$a_2a_0 = \frac{1}{2m_c}\left(1 - \frac{m_2}{m_2'}\right) = \frac{1}{2m_2} - \frac{1}{2m_2'} = (a_2 - a_2')$$

Therefore the DR in QWs of HD IV–VI semiconductors in the presence of magnetic field whose conduction electrons obey the model of McClure and Choi can be written as

$$\begin{aligned}
\gamma_3(E,\eta_g) &= a_1\hbar^2k_x^2 + a_2\hbar^2k_y^2 + a_3\hbar^2\left(\frac{n\pi}{a}\right)^2 \\
&\quad - \alpha\left[a_1\hbar^2k_x^2 + a_2\hbar^2k_y^2 + a_3\hbar^2\left(\frac{n\pi}{a}\right)^2\right]^2 \\
&\quad + \alpha a_2a_2'\hbar^4k_y^4 + \alpha a_2\hbar^2k_y^2\left[a_1\hbar^2k_x^2 + a_2\hbar^2k_y^2 + a_3\hbar^2\left(\frac{n\pi}{a}\right)^2\right] \\
&\quad - \alpha a_2'k_x^2k_y^2\left[a_1\hbar^2k_x^2 + a_2\hbar^2k_y^2 + a_3\hbar^2\left(\frac{n\pi}{a}\right)^2\right] \\
&\quad - \alpha a_1a_2\hbar^4k_x^2k_y^2 - \alpha a_2a_3\hbar^4k_y^2\left(\frac{n\pi}{a}\right)^2 + a_3e^2B^2\langle x^2\rangle \\
&\quad + 2\alpha a_1a_3\hbar^2e^2B^2 - \alpha a_3^2B^4\langle x^4\rangle \\
&\quad - \alpha\left[2a_1a_3\hbar^2k_x^2 + 2a_2a_3\hbar^2k_y^2 + 6a_3^2\hbar^2\left(\frac{n\pi}{a}\right)^2\right]^2 e^2B^2\langle x^2\rangle \\
&\quad - \alpha a_2'a_3\hbar^2k_y^2e^2B^2\langle x^2\rangle
\end{aligned} \tag{16.127}$$

The use of (1.127) leads to the expressions of EEM and DOS, the derivations of which are left for the readers.

16.2.6 *The DR in nano wires of HD IV–VI semiconductors in the presence of magnetic field*

(a) McClure and Choi Model

In nano-wire let us assume the free electron motion is along y-direction and the magnetic field B is also applied along y-direction. Following (16.118) the Hamiltonian can be expressed as

$$\hat{H} = -a_1\hbar^2\frac{\partial^2}{\partial x^2} - a_2\hbar^2\frac{\partial^2}{\partial y^2} + a_3\left(-\hbar^2\frac{\partial^2}{\partial z^2} + e^2B^2x^2 + 2i\hbar eBx\frac{\partial}{\partial z}\right)$$

$$\hat{H} = a_1\left(-i\hbar\frac{\partial}{\partial x}\right)^2 + a_2\left(-i\hbar\frac{\partial}{\partial x}\right)^2 + a_3\left(-i\hbar\frac{\partial}{\partial x} - eBx\right)^2$$

$$- \alpha\left[a_1\left(-i\hbar\frac{\partial}{\partial x}\right)^2 + a_2\left(-i\hbar\frac{\partial}{\partial y}\right)^2 + a_3\left(-i\hbar\frac{\partial}{\partial x} - eBx\right)^2\right]^2$$

$$\cdot\alpha a_2 a_2'\left(-i\hbar\frac{\partial}{\partial x}\right)^4 \alpha a_2 a_0\left(-i\hbar\frac{\partial}{\partial y}\right)^2$$

$$\cdot\left[a_1\left(-i\hbar\frac{\partial}{\partial x}\right)^2 + a_2\left(-i\hbar\frac{\partial}{\partial y}\right)^2 + a_3\left(-i\hbar\frac{\partial}{\partial z} - eBx\right)^2\right]$$

$$- \alpha a_1 a_2'\left(-i\hbar\frac{\partial}{\partial x}\right)^2\left(-i\hbar\frac{\partial}{\partial y}\right)^2$$

$$- \alpha a_2 a_3\left(-i\hbar\frac{\partial}{\partial y}\right)^2\left[-i\hbar\frac{\partial}{\partial z} - eBx\right]^2 \tag{16.128}$$

Considering non-parabolicity and magnetic field as perturbed, unperturbed wave equation is given by

$$\hat{H}_0\psi_0 = E_0\psi_0 \tag{16.129}$$

The (16.129) can be written as

$$-a_1\hbar^2\frac{\partial^2\psi_0}{\partial x^2} - a_2\hbar^2\frac{\partial^2\psi_0}{\partial y^2} - a_3\hbar^2\frac{\partial^2\psi_0}{\partial z^2} = E_0\psi_0 \tag{16.130}$$

Therefore

$$E_0 = E_{nl} = \frac{\hbar^2}{2m_l}\left(\frac{l\pi}{b}\right)^2 + \frac{\hbar^2 k_y^2}{2m_2} + \frac{\hbar^2}{2m_3}\left(\frac{n\pi}{a}\right)^2 \tag{16.131}$$

and

$$\psi^0 = \psi_0 = \psi_{nl} = \sqrt{\frac{4}{abL_y}} \sin\left(\frac{l\pi}{b}x\right) \sin\left(\frac{n\pi}{a}z\right) \exp(ik_y y) \quad (16.132)$$

Including perturbing effect, the total Hamiltonian can be written as

$$\begin{aligned}
\hat{H} = \hat{H}_0 + \hat{H}' = &-a_1\hbar^2\frac{\partial^2}{\partial x^2} - a_2\hbar^2\frac{\partial^2}{\partial y^2} \\
&+ a_3\left(-\hbar^2\frac{\partial^2}{\partial z^2} + e^2B^2x^2 + 2i\hbar eBx\frac{\partial}{\partial z}\right) \\
&- \alpha\left[-a_1\hbar^2\frac{\partial^2}{\partial x^2} - a_2\hbar^2\frac{\partial^2}{\partial y^2}\right. \\
&\left.+ a_3\left(-\hbar^2\frac{\partial^2}{\partial z^2} + e^2B^2x^2 + 2i\hbar eBx\frac{\partial}{\partial z}\right)\right]^2 \\
&+ \alpha a_2 a_2'\left(-\hbar^4\frac{\partial^4}{\partial y^4}\right) + \alpha a_2 a_0\left(-\hbar^2\frac{\partial^2}{\partial y^2}\right) \\
&\cdot\left[-a_1\hbar^2\frac{\partial^2}{\partial x^2} - a_2\hbar^2\frac{\partial^2}{\partial y^2} + a_3\left(-\hbar^2\frac{\partial^2}{\partial z^2} + e^2B^2x^2 + 2i\hbar eBx\frac{\partial}{\partial z}\right)\right] \\
&- \alpha a_1 a_2\hbar^4\left(\frac{\partial^2}{\partial x^2}\right)\left(\frac{\partial^2}{\partial y^2}\right) - \alpha a_2 a_3\left(-\hbar^2\frac{\partial^2}{\partial y^2}\right) \\
&\cdot\left(-\hbar^2\frac{\partial^2}{\partial z^2} + e^2B^2x^2 + 2i\hbar eBx\frac{\partial}{\partial z}\right)
\end{aligned} \quad (16.133)$$

Now

$$\begin{aligned}
&\left[a_1\hbar^2\frac{\partial^2}{\partial x^2} - a_2\hbar^2\frac{\partial^2}{\partial y^2} + a_3\left(-\hbar^2\frac{\partial^2}{\partial z^2} + e^2B^2x^2 + 2i\hbar eBx\frac{\partial}{\partial z}\right)\right]\psi_{nl}^0 \\
&= a_1\hbar^2\left(\frac{l\pi}{b}\right)^2\psi_{nl}^0 + a_2\hbar^2k_y^2\psi^0 + a_3\hbar^2\left(\frac{n\pi}{a}\right)^2\psi^0 \\
&\quad + a_3e^2B^2x^2\psi^0 + 2a_3i\hbar eBx\left(\frac{n\pi}{a}\right)\cot\left(\frac{n\pi}{a}z\right)\psi^0
\end{aligned} \quad (16.134)$$

Therefore

$$\left[-a_1\hbar^2\frac{\partial^2}{\partial x^2} - a_2\hbar^2\frac{\partial^2}{\partial y^2} + a_3\left(-\hbar^2\frac{\partial^2}{\partial z^2} + e^2B^2x^2 + 2i\hbar eBx\frac{\partial}{\partial z}\right)\right]^2\psi_{nl}^0$$

$$= \left[a_1\hbar^2\left(\frac{l\pi}{b}\right)^2 + a_2\hbar^2k_y^2 + a_3\hbar^2\left(\frac{n\pi}{a}\right)^2\right]^2\psi^0 - 2a_1a_3\hbar^2e^2B^2\psi^0$$

$$+ 4a_3i\hbar eB\left[a_1\hbar^2\left(\frac{l\pi}{b}\right)^2 + a^2\hbar^2k_y^2 + a_3\hbar^2\left(\frac{n\pi}{a}\right)^2\right]$$

$$\left(\frac{n\pi}{a}\right)x\cot\left(\frac{n\pi}{a}z\right)\psi^0 + \left[2a_1a_3\hbar^2\left(\frac{l\pi}{b}\right)^2 + 2a_2a_3\hbar^2k_y^2\right.$$

$$\left.+ 6a_3^2\hbar^2\left(\frac{n\pi}{a}\right)^2\right]e^2e^2B^2x^2\psi^0 + 4a_3^2i\hbar e^3B^3$$

$$\cdot\left(\frac{n\pi}{a}\right)x^3\cot\left(\frac{n\pi}{a}z\right)\psi^0 + a_3^2e^4B^4x^4\psi^0 \tag{16.135}$$

$$\hbar^4\frac{\partial^4}{\partial y^4}(\psi^0) = \hbar^4(ik_y)^4\psi^0 = \hbar^4k_y^4\psi^0$$

$$\hbar^4\left(\frac{\partial^2}{\partial x^2}\right)\left(\frac{\partial^2}{\partial y^2}\right)\psi^0 = \hbar^4\frac{\partial^2}{\partial x^2}[(ik_y)^2\psi^0]$$

$$= \hbar^4(-k_y^2)\left[-\left(\frac{l\pi}{b}\right)^2\psi^0\right]$$

$$= \hbar^4k_y^2\left(\frac{l\pi}{b}\right)^2\psi^0$$

So

$$\hat{H}\psi_{nl}^0 = \hat{H}_0\psi_{nl}^0 + \hat{H}'\psi_{nl}^0 \tag{16.136a}$$

Therefore

$$\hat{H}\psi_{nl}^0 = a_1\hbar^2\left(\frac{l\pi}{b}\right)^2\psi^0 + a_2\hbar^2k_y^2\psi^0 + a_3\hbar^2\left(\frac{n\pi}{b}\right)^2\psi^0$$

$$+ a_3e^2B^2x^2\psi^0 + 2a_3i\hbar dB\left(\frac{n\pi}{b}\right)x\cot\left(\frac{n\pi}{b}z\right)\psi^0$$

$$
\begin{aligned}
&-\alpha\left[a_1\hbar^2\left(\frac{l\pi}{b}\right)^2+a_2\hbar^2k_y^2+a_3\hbar^2\left(\frac{n\pi}{b}\right)^2\right]^2\psi^0\\
&+2\alpha a_1a_3\hbar^2e^2B^2\psi^0-4\alpha a_3i\hbar eB\left[a_1\hbar^2\left(\frac{l\pi}{b}\right)^2\right.\\
&\left.+\ a_2\hbar^2k_y^2+a_3\hbar^2\left(\frac{n\pi}{b}\right)^2\right]\left(\frac{n\pi}{b}\right)x\cot\left(\frac{n\pi}{b}z\right)\psi^0\\
&-\alpha\left[2a_1a_3\hbar^2\left(\frac{l\pi}{b}\right)^2+2a_2a_3\hbar^2k_y^2+6a_3^2\hbar^2\left(\frac{n\pi}{b}\right)^2\right]\\
&\cdot e^2B^2x^2\psi^0-4\alpha a_3^2i\hbar e^3B^3\left(\frac{n\pi}{b}\right)x^3\cot\left(\frac{n\pi}{b}z\right)\psi^0\\
&-\alpha a_3^2e^4B^4x^4\psi^0+\alpha a_2a_2'\hbar^4k_y^4\psi^0\\
&+\alpha a_2a_0\hbar^2k_y^2\left[a_1\hbar^2\left(\frac{l\pi}{b}\right)^2+a_2\hbar k_y^2+a_3\hbar^2\left(\frac{n\pi}{b}\right)^2\right]\psi^0\\
&+\alpha a_2a_0a_3\hbar^2k_y^2e^2B^2x^2\psi^0+2\alpha a_2a_0a_3i\hbar^3k_y^2eB\\
&\cdot\left(\frac{n\pi}{b}\right)x\cot\left(\frac{n\pi}{b}z\right)\psi^0-\alpha a_1a_2\hbar^4k_y^2\left(\frac{l\pi}{b}\right)^2\psi^0\\
&-\alpha a_2a_3\hbar^4k_y^2\left(\frac{n\pi}{b}\right)^2\psi^0-\alpha a_2a_3\hbar^2k_y^2e^2B^2x^2\psi^0\\
&-2\alpha a_2a_3i\hbar^3k_y^2eB\left(\frac{n\pi}{b}\right)x\cot\left(\frac{n\pi}{b}z\right)\psi^0 \qquad (16.136\text{b})
\end{aligned}
$$

Applying first order perturbation we can write

$$\left\langle x\cot\left(\frac{n\pi}{b}z\right)\right\rangle=0,\quad \left\langle x^3\cot\left(\frac{n\pi}{b}z\right)\right\rangle=0,$$

$$\langle x^2\rangle=\frac{2}{b}\int_0^b x^2\sin^2\left(\frac{l\pi}{b}\right)xdx=\left(\frac{b^2}{3}-\frac{b^2}{2l^2\pi^2}\right)$$

and

$$\langle x^4\rangle=\frac{2}{b}\int_0^b x^4\sin^2\left(\frac{l\pi}{b}\right)xdx=\left(\frac{b^4}{5}-\frac{b^4}{l^2\pi^2}+\frac{3b^4}{2l^4\pi^4}\right)$$

The DR in this case is given by

$$E = a_1\hbar^2\left(\frac{l\pi}{b}\right)^2 + a_2\hbar^2k_y^2 + a_3\hbar^2\left(\frac{n\pi}{b}\right)^2 - \alpha\left[a_1\hbar^2\left(\frac{l\pi}{b}\right)^2 + a_2\hbar^2k_y^2 + a_3\hbar^2\left(\frac{n\pi}{b}\right)^2\right]^2 + \alpha a_2a_2'\hbar^4k_y^4 + \alpha a_2a_0\hbar^2k_y^2\left[a_1\hbar^2\left(\frac{l\pi}{b}\right)^2 + a_2\hbar^2k_y^2 + a_3^2\hbar^2\left(\frac{n\pi}{b}\right)^2\right] - \alpha a_1a_2\hbar^4\left(\frac{l\pi}{b}\right)^2k_y^2 - \alpha a_2a_3\hbar^4k_y^2\left(\frac{n\pi}{b}\right)^2 + a_3e^2B^2\langle x^2\rangle + 2\alpha a_1a_3\hbar^2e^2B - \alpha a_3^2e^4B^4\langle x^4\rangle - \left[2a_1a_3\hbar^2\left(\frac{l\pi}{b}\right)^2 + 2a_2a_3\hbar^2k_y^2 + 6a_3^2\hbar^2\left(\frac{n\pi}{a}\right)^2\right]e^2B^2\langle x^2\rangle + \alpha a_2a_0a_3\hbar^2k_y^2e^2B^2\langle x^2\rangle - \alpha a_2a_3\hbar^2k_y^2e^2B^2\langle x^2\rangle \quad (16.137)$$

Now

$$a_2a_0 = (a_2 - a_2')$$

Therefore the DR in NWs of HD IV–VI semiconductors in the presence of magnetic field in the present case can be written as

$$\gamma_3(E,\eta_g) = a_1\hbar^2\left(\frac{l\pi}{b}\right)^2 + a_2\hbar^2k_y^2 + a_3\hbar^2\left(\frac{n\pi}{b}\right)^2 - \alpha\left[a_1\hbar^2\left(\frac{l\pi}{b}\right)^2 + a_2\hbar^2k_y^2 + a_3\hbar^2\left(\frac{n\pi}{b}\right)^2\right]^2 + \alpha a_2a_2'\hbar^4k_y^4 + \alpha a_2a_0\hbar^2k_y^2\left[a_1\hbar^2\left(\frac{l\pi}{b}\right)^2 + a_2\hbar^2k_y^2 + a_3^2\hbar^2\left(\frac{n\pi}{b}\right)^2\right] - \alpha a_1a_2\hbar^4\left(\frac{l\pi}{b}\right)^2k_y^2 - \alpha a_2a_3\hbar^4k_y^2\left(\frac{n\pi}{b}\right)^2 + a_3e^2B^2\langle x^2\rangle + 2\alpha a_1a_3\hbar^2e^2B - \alpha a_3^2e^4B^4\langle x^4\rangle$$

$$
\begin{aligned}
&- \left[2a_1a_3\hbar^2\left(\frac{l\pi}{b}\right)^2 + 2a_2a_3\hbar^2k_y^2 + 6a_3^2\hbar^2\left(\frac{n\pi}{a}\right)^2\right] \\
&\cdot e^2B^2\langle x^2\rangle + \alpha a_2a_0a_3\hbar^2k_y^2e^2B^2\langle x^2\rangle - \alpha a_2a_3\hbar^2k_y^2e^2B^2\langle x^2\rangle
\end{aligned}
\tag{16.138}
$$

The use of (16.138) leads to the expressions of EEM and DOS function, the derivations of which are left to the readers.

(a) Cohen model

The total Hamiltonian be

$$
\begin{aligned}
\hat{H} &= a_1\left(-i\hbar\frac{\partial}{\partial x}\right)^2 + a_2\left(-i\hbar\frac{\partial}{\partial x}\right)^2 + a_3\left(-i\hbar\frac{\partial}{\partial x} - eBx\right)^2 \\
&\quad - \alpha\left[a_1\left(-i\hbar\frac{\partial}{\partial x}\right)^2 + a_2\left(-i\hbar\frac{\partial}{\partial y}\right)^2 + a_3\left(-i\hbar\frac{\partial}{\partial x} - eBx\right)^2\right]^2 \\
&\quad + \alpha a_2\left(-i\hbar\frac{\partial}{\partial x}\right)^2\left[a_1\left(-i\hbar\frac{\partial}{\partial x}\right)^2 + a_2\left(-i\hbar\frac{\partial}{\partial y}\right)^2\right. \\
&\quad \left. + \, a_3\left(-i\hbar\frac{\partial}{\partial z} - eBx\right)^2\right] - \alpha a_2'\left(-i\hbar\frac{\partial}{\partial y}\right)^2 \\
&\quad \cdot\left[a_1\left(-i\hbar\frac{\partial}{\partial x}\right)^2 + a_2\left(-i\hbar\frac{\partial}{\partial y}\right)^2 + a_3\left[-i\hbar\frac{\partial}{\partial z} - eBx\right]^2\right. \\
&\quad + \alpha a_2a_2'\left(-i\hbar\frac{\partial}{\partial y}\right)^4
\end{aligned}
\tag{16.139}
$$

Considering non-parabolicity and magnetic field as perturbed, the unperturbed wave equation is

$$
\hat{H}_0\psi_0 = E_0\psi_0 \tag{16.140}
$$

$$
-a_1\hbar^2\frac{\partial^2\psi_0}{\partial x^2} - a_2\hbar^2\frac{\partial^2\psi_0}{\partial y^2} - a_3\hbar^2\frac{\partial^2\psi_0}{\partial z^2} = E_0\psi_0
$$

So that

$$E_0 = E_{nl}^{(0)} = \frac{\hbar^2}{2m_1}\left(\frac{l\pi}{b}\right)^2 + \frac{\hbar^4 k_y^2}{2m_2} + \frac{\hbar^2}{2m_3}\left(\frac{n\pi}{a}\right)^2 \tag{16.141}$$

$$\psi_0 = \psi_{nl}^0 = \sqrt{\frac{4}{abLy}}\sin\left(\frac{l\pi}{b}x\right)\sin\left(\frac{n\pi}{a}z\right)\exp(ik_y y) \tag{16.142}$$

Including perturbing effect, the total Hamiltonian can be expressed as

$$\begin{aligned}\hat{H} = \hat{H}_0 + \hat{H}' &= -a_1\hbar^2\frac{\partial^2}{\partial x^2} - a_2\hbar^2\frac{\partial^2}{\partial y^2} \\ &+a_3\left(-\hbar^2\frac{\partial^2}{\partial z^2} + e^2B^2x^2 + 2i\hbar eBx\frac{\partial}{\partial z}\right) \\ &-\alpha\left[-a_1\hbar^2\frac{\partial^2}{\partial x^2} - a_2\hbar^2\frac{\partial^2}{\partial y^2}\right. \\ &\left.+ a_3\left(-\hbar^2\frac{\partial^2}{\partial z^2} + e^2B^2x^2 + 2i\hbar eBx\frac{\partial}{\partial z}\right)\right]^2 + \alpha a_2\left(-\hbar^2\frac{\partial^2}{\partial y^2}\right) \\ &\cdot\left[-a_1\hbar^2\frac{\partial^2}{\partial x^2} - a_2\hbar^2\frac{\partial^2}{\partial y^2} + a_3\left(-\hbar^2\frac{\partial^2}{\partial z^2} + e^2B^2x^2 + 2i\hbar eBx\frac{\partial}{\partial z}\right)\right] \\ &-\alpha a_2'\left(-\hbar^2\frac{\partial^2}{\partial y^2}\right)\left[-a_1\hbar^2\frac{\partial^2}{\partial x^2} - a_2\hbar^2\frac{\partial^2}{\partial y^2}\right. \\ &\left.+ a_3\left(-\hbar^2\frac{\partial^2}{\partial z^2} + e^2B^2x^2 + 2i\hbar eBx\frac{\partial}{\partial z}\right)\right] + \alpha a_2 a_2'\left(-\hbar^4\frac{\partial^4}{\partial y^4}\right)\end{aligned} \tag{16.143}$$

We can write

$$\begin{aligned}&\left[-a_1\hbar^2\frac{\partial^2}{\partial x^2} - a_2\hbar^2\frac{\partial^2}{\partial y^2} + a_3\left(-\hbar^2\frac{\partial^2}{\partial z^2} + e^2B^2x^2 + 2i\hbar eBx\frac{\partial}{\partial z}\right)\right]\psi_n^0 \\ &= a_1\hbar^2\left(\frac{l\pi}{b}\right)^2\psi_n + a_2\hbar^2k_y^2\psi_n + a_3\hbar^2\left(\frac{n\pi}{a}\right)^2\psi_n + a_3e^2B^2x^2\psi_n \\ &\quad+ 2a_3i\hbar eBx\left(\frac{n\pi}{a}\right)\cot\left(\frac{n\pi}{a}z\right)\psi_n\end{aligned} \tag{16.144}$$

$$\left[-a_1\hbar^2\frac{\partial^2}{\partial x^2} - a_2\hbar^2\frac{\partial^2}{\partial y^2} + a_3\left(-\hbar^2\frac{\partial^2}{\partial z^2} + e^2B^2x^2 + 2i\hbar eBx\frac{\partial}{\partial z}\right)\right]^2\psi_n^0$$

$$= \left[a_1^4\hbar^4\left(\frac{l\pi}{b}\right)^4\psi_n + a_1a_2\hbar^4\left(\frac{l\pi}{b}\right)^2k_y^2\psi_n + a_1a_3\hbar^4\left(\frac{l\pi}{b}\right)^2\left(\frac{n\pi}{a}\right)^2\psi_n\right.$$

$$+ a_1a_3\hbar^2\left(\frac{l\pi}{b}\right)^2e^2B^2x^2\psi_n - 2a_1a_3\hbar^2e^2B^2\psi_n$$

$$+ 2a_1a_3i\hbar^3eBx\left(\frac{l\pi}{b}\right)^2\left(\frac{n\pi}{a}\right)\cot\left(\frac{n\pi}{a}z\right)\psi_n$$

$$+ a_2a_1\hbar^4\left(\frac{l\pi}{b}\right)^2\left(\frac{n\pi}{a}\right)^2k_y^2\psi_n + a_2^2\hbar^4k_y^4\psi_n$$

$$+ a_2a_3kh^4\left(\frac{n\pi}{a}\right)^2k_y^2\psi_n + a_2a_3e^2B^2x^2\hbar^2k_y^2\psi_n$$

$$+ 2a_2a_3i\hbar^3k_y^2eB\left(\frac{n\pi}{a}\right)x\cot\left(\frac{n\pi}{a}z\right)\psi_n$$

$$+ a_3a_1\hbar^4\left(\frac{l\pi}{b}\right)^2\left(\frac{n\pi}{a}\right)^2\psi_n + a_3a_2h^4\left(\frac{n\pi}{a}\right)^2k_y^2\psi_n$$

$$+ a_3^2\hbar^4\left(\frac{n\pi}{a}\right)^4\psi_n + a_3^2\hbar^2\left(\frac{n\pi}{a}\right)^2e^2B^2x^2\psi_n$$

$$+ 2a_3^2i\hbar^3eB\left(\frac{n\pi}{a}\right)^3x\cot\left(\frac{n\pi}{a}z\right)\psi_n$$

$$+ \left[a_3a_1\hbar^2\left(\frac{l\pi}{b}\right)^2 + a_3a_2\hbar^2k_y^2 + a_3^2\hbar^2\left(\frac{n\pi}{a}\right)^2\right]e^2B^2x^2\psi_n$$

$$+ a_3^2e^4B^4x^4\psi_n + 2a_3^2i\hbar e^3B^3\left(\frac{n\pi}{a}\right)x^3\cot\left(\frac{n\pi}{a}z\right)\psi_n$$

$$+ 2a_3i\hbar eB\left[a_1\hbar^2\left(\frac{l\pi}{b}\right)^2 + a_2\hbar^2k_y^2 + a_3\hbar^2\left(\frac{n\pi}{a}\right)^2\right]$$

$$\cdot\left(\frac{n\pi}{a}\right)x\cot\left(\frac{n\pi}{a}z\right)\psi_n + 2a_3^2i\hbar e^3B^3x^3\left(\frac{n\pi}{a}\right)\cot\left(\frac{n\pi}{a}z\right)\psi_n$$

$$+ 4a_3^2\hbar^2e^2B^2\left(\frac{n\pi}{a}\right)^2x^2\psi_n$$

$$= \left[a_1\hbar^2\left(\frac{l\pi}{b}\right)^2 + a_2\hbar^2k_y^2 + a_3\hbar^2\left(\frac{n\pi}{a}\right)^2\right]^2\psi_n - 2a_1a_3\hbar^2e^2B^2\psi_n$$

$$+4a_3i\hbar eB\left[a_1\hbar^2\left(\frac{l\pi}{b}\right)^2 + a_2\hbar^2k_y^2 + a_3\hbar^2\left(\frac{n\pi}{a}\right)^2\right]$$

$$\left(\frac{n\pi}{a}\right)x\cot\left(\frac{n\pi}{a}z\right)\psi_n + \left[2a_1a_3\hbar^2\left(\frac{l\pi}{b}\right)^2 + 2a_2a_3\hbar^2k_y^2 + 6a_3^2\hbar^2\left(\frac{n\pi}{a}\right)^2\right]$$

$$\cdot e^2B^2x^2\psi_n + 4a_3^2i\hbar e^3B^3\left(\frac{n\pi}{a}\right)x^3\cot\left(\frac{n\pi}{a}z\right)\psi_n$$

$$+a_3^2e^4B^4x^4\psi_n - \left(-\hbar^2\frac{\partial^2}{\partial y^2}\right)\left[-a_1\hbar^2\frac{\partial^2}{\partial x^2} - a_2\hbar^2\frac{\partial^2}{\partial y^2}\right.$$

$$\left.+\ a_3\left(-\hbar^2\frac{\partial^2}{\partial z^2} + e^2B^2x^2 + 2i\hbar eBx\frac{\partial}{\partial z}\right)\right]\psi_n$$

$$= a_1\hbar^4\left(\frac{l\pi}{b}\right)^2k_y^2\psi_n + a_2\hbar^4k_y^4\psi_n + a_3\hbar^4\left(\frac{n\pi}{a}\right)^2k_y^2\psi_n$$

$$+a_3\hbar^2e^2B^2k_y^2x^2\psi_n + 2a_3i\hbar^3k_y^2eBx\left(\frac{n\pi}{a}\right)\cot\left(\frac{n\pi}{a}z\right)\psi_n$$

Again

$$\alpha a_2a_2'\hbar^4\frac{\partial^4}{\partial y^4}\psi_n = \alpha a_2{a'}_2\hbar^4k_y^4\psi_n \tag{16.145}$$

So

$$\hat{H}\psi_n = \hat{H}_0\psi_n + \hat{H}'\psi_n \tag{16.146}$$

$$\hat{H}\psi_n = \hat{H}_0\psi_n + \hat{H}'\psi_n \tag{1.146}$$

$$\hat{H}'\psi_n = a_1\hbar^2\left(\frac{l\pi}{b}\right)^2\psi_n + a_2\hbar^2k_y^2\psi_n + a_3\hbar^2\left(\frac{n\pi}{a}\right)^2\psi_n$$

$$+a_3e^2B^2x^2\psi_n + 2a_3i\hbar eB\left(\frac{n\pi}{a}\right)x\cot\left(\frac{n\pi}{a}z\right)\psi_n$$

$$-\alpha\left[a_1\hbar^2\left(\frac{l\pi}{b}\right)^2+a_2\hbar^2k_y^2+a_3\hbar^2\left(\frac{n\pi}{a}\right)^2\right]^2\psi_n$$

$$+2\alpha a_1a_3\hbar^2e^2B^2\psi_n-4\alpha a_3i\hbar eB\left[a_1\hbar^2\left(\frac{l\pi}{b}\right)^2\right.$$

$$\left.+\ a_2\hbar^2k_y^2+a_3\hbar^2\left(\frac{n\pi}{a}\right)^2\right]\left(\frac{n\pi}{a}\right)x\cot\left(\frac{n\pi}{a}z\right)\psi_n$$

$$-\alpha\left[2a_1a_3\hbar^2\left(\frac{l\pi}{b}\right)^2+2a_2a_3\hbar^2k_y^2+6a_3^2\hbar^2\left(\frac{n\pi}{a}\right)^2\right]e^2B^2x^2\psi_n$$

$$-4\alpha a_3^2i\hbar e^3B^3\left(\frac{n\pi}{a}\right)x^3\cot\left(\frac{n\pi}{a}z\right)\psi_n-\alpha a_3^2e^4B^4x^2\psi_n$$

$$+\alpha a_2\hbar^2k_y^2$$

$$\left[a_1\hbar^2\left(\frac{l\pi}{b}\right)^2+a_2\hbar^2k_y^2+a_3\hbar^2\left(\frac{n\pi}{a}\right)^2\right]\psi_n+\alpha a_2a_3\hbar^2e^2B^2k_y^2x^2\psi_n$$

$$+2\alpha a_2a_3i\hbar^3k_y^2eB\left(\frac{n\pi}{a}\right)x\cot\left(\frac{n\pi}{a}z\right)\psi_n$$

$$-\alpha a_2'\hbar^2k_y^2\left[a_1\hbar^2\left(\frac{l\pi}{b}\right)^2+a_2\hbar^2k_y^2+a_3\hbar^2\left(\frac{n\pi}{a}\right)^2\right]\psi_n$$

$$-\alpha a_2'a_3\hbar^2k_y^2e^2B^2x^2\psi_n-2\alpha a_2'a_3i\hbar^3k_y^2eB$$

$$\cdot\left(\frac{n\pi}{a}\right)x\cot\left(\frac{n\pi}{a}z\right)\psi_n+\alpha a_2a_2'\hbar^4k_y^4\psi_n$$

The averages are calculated as follow:

$$\left\langle x\cot\left(\frac{n\pi}{b}z\right)\right\rangle=0,\quad\left\langle x^3\cot(\frac{n\pi}{b}z)\right\rangle=0,$$

$$\langle x^2\rangle=\frac{2}{b}\int_0^b x^2\sin^2\left(\frac{l\pi}{b}\right)xdx=\left(\frac{b^2}{3}-\frac{b^2}{2l^2\pi^2}\right)$$

and

$$\langle x^4\rangle=\frac{2}{b}\int_0^b x^4\sin^2\left(\frac{l\pi}{b}x\right)dx=\left(\frac{b^4}{5}-\frac{b^4}{l^2\pi^2}+\frac{3b^4}{2l^4\pi^4}\right)$$

The DR in this case is given by

$$\begin{aligned}
\gamma_3(E,\eta_g) &= a_1\hbar^2\left(\frac{l\pi}{b}\right)^2 + a_2\hbar^2k_y^2 + a_3\hbar^2\left(\frac{n\pi}{b}\right)^2 \\
&\quad - \alpha\left[a_1\hbar^2\left(\frac{l\pi}{b}\right)^2 + a_2\hbar^2k_y^2 + a_3\hbar^2\left(\frac{n\pi}{b}\right)^2\right]^2 \\
&\quad + \alpha a_2\hbar^2k_y^2 + \left[a_1\hbar^2\left(\frac{l\pi}{b}\right)^2 + a_2\hbar^2k_y^2 + a_3^2\hbar^2\left(\frac{n\pi}{b}\right)^2\right] \\
&\quad - \alpha a_2'\hbar^2k_y^2\left[a_1\hbar^2\left(\frac{l\pi}{b}\right)^2 + a_2\hbar^2k_y^2 + a_3^2\hbar^2\left(\frac{n\pi}{b}\right)^2\right] \\
&\quad + \alpha a_2a_2'\hbar^4k_y^4 + a_3e^2B^2\langle x^2\rangle + 2\alpha a_1a_3\hbar^2e^2B^2 \\
&\quad - \alpha\left[2a_1a_3\hbar^2\left(\frac{l\pi}{b}\right)^2 + a_2a_3\hbar^2k_y^2 + 6a_3^2\hbar^2\left(\frac{n\pi}{a}\right)^2\right] \\
&\quad \cdot e^2B^2\langle x^2\rangle - \alpha a_3^2e^4B^4\langle x^4\rangle - \alpha a_2'a_3\hbar^2k_y^2e^2B^2\langle x^2\rangle
\end{aligned} \tag{16.147}$$

where

$$\langle x^2\rangle = \left(\frac{b^2}{3} - \frac{b^2}{2l^2\pi^2}\right) \quad \text{and} \quad \langle x^4\rangle = \left(\frac{b^4}{5} - \frac{b^4}{l^2\pi^2} + \frac{3b^4}{2l^4\pi^4}\right)$$

The EEM and the DOS function can be calculated from (1.147) and are left for the readers to enjoy the same.

16.2.7 *The EM in Quantum Wells of HD III–V semiconductors in the presence of arbitrarily oriented magnetic field*

In this case the total Hamiltonian ($\hat{H}$) can be written as

$$\hat{H} = a'(-i\hbar\nabla' + eA')^2 - b'(-i\hbar\vec{\nabla} + e\vec{A})^4 \tag{16.148}$$

In which

$$a' = \frac{1}{2m_c}, \quad b' = \alpha(a')^2, \quad \alpha = \frac{1}{E_g}, \quad \vec{A} = \vec{i}zB_y + \vec{j}xB_z + \vec{k}yB_x,$$

$$\vec{B} = \vec{i}B_x + \vec{j}B_y + \vec{k}B_z, \quad B_x = B\sin\theta\cos\phi$$

$B_y = B\sin\theta\sin\phi, B_z = B\cos\theta, (B,\theta,\phi)$ are the spherical polar coordinates, $\vec{\nabla}.\vec{A} = 0$,

$$\vec{\nabla}\times\vec{A} = \vec{B} = \begin{vmatrix} \vec{i} & \vec{j} & \vec{k} \\ \frac{\partial}{\partial x} & \frac{\partial}{\partial y} & \frac{\partial}{\partial z} \\ zB_y & xB_z & yB_x \end{vmatrix}$$ and $(\vec{i},\vec{j},\vec{k})$ are orthogonal triads.

$$\hat{H} = a'\left[\left(-i\hbar\frac{\partial}{\partial x} + eB_y z\right)\vec{i} + \left(-i\hbar\frac{\partial}{\partial y} + exB_z\right)\vec{j} + \vec{k}\left(-i\hbar\frac{\partial}{\partial z} + eyB_z\right)\right]^2 - b'\left[\left(-i\hbar\frac{\partial}{\partial x} + ezB_y\right)\vec{i} + \left(-i\hbar\frac{\partial}{\partial y} + exB_z\right)\vec{j} + \left(-i\hbar\frac{\partial}{\partial z} + eyB_x\right)\vec{k}\right]^4$$

From (16.148) we can write

$$\hat{H} = a'\left[\left(-i\hbar\frac{\partial}{\partial x} + eB_y z\right)\vec{i} + \left(-i\hbar\frac{\partial}{\partial y} + exB_z\right)\vec{j} + \vec{k}\left(-i\hbar\frac{\partial}{\partial z} + eyB_z\right)\right]^2 - b'\left[\left(-i\hbar\frac{\partial}{\partial x} + ezB_y\right)\vec{i} + \left(-i\hbar\frac{\partial}{\partial y} + exB_z\right)\vec{j} + \left(-i\hbar\frac{\partial}{\partial z} + eyB_x\right)\vec{k}\right]^4 = a'\hat{H}_1^2 - b'\hat{H}_1^4 \tag{16.145}$$

The unperturbed wave function and energy Eigen values in this case assume the forms

$$u_n = \psi_n^0 = \sqrt{\frac{2}{aL_xL}}\sin\left(\frac{nz}{a}z\right)\exp(ik_xx + ik_yy) \tag{16.146}$$

$$E_n^0 = \frac{\hbar^2}{2m_c}\left(\frac{n\pi}{a}\right)^2 + \frac{\hbar^2k_y^2}{2m_c} + \frac{\hbar^2k_x^2}{2m_c} - \alpha\left[\frac{\hbar^2}{2m_c}\left(\frac{n\pi}{a}\right)^2 + \frac{\hbar^2k_y^2}{2m_c} + \frac{\hbar^2k_x^2}{2m_c}\right]^2 \tag{16.147}$$

Now

$$\hat{H}_1\psi_n^0 = -i\hbar\frac{\partial}{\partial x}(-i\hbar\cdot ik_x + ezB_y)u_n + eB_yz(-i\hbar\cdot ik_x + ezB_y)u_n + (-i\hbar\cdot ik_y + eB_zx)^2u_n \tag{16.148}$$

$$\hat{H}_1\psi_n^0 = -i\hbar\frac{\partial}{\partial x}(-i\hbar\cdot ik_x + ezB_y)u_n + eB_yz(-i\hbar\cdot ik_x + ezB_y)u_n + (-i\hbar\cdot ik_y + eB_zx)^2u_n + eyB_x\left(-i\hbar\frac{\partial}{\partial z} + eyB_x\right)u_n + \left(-i\hbar\frac{\partial}{\partial z} + eyB_x\right)\left(-i\hbar\frac{n\pi}{a}\cot\left(\frac{n\pi}{a}z\right)u_n\right. \tag{16.149}$$

Therefore (16.197) assumes the form

$$\hat{H}_1^2\psi_n^0 = (\hbar k_x + eB_yz)^2u_n + (\hbar k_y + exB_z)^2u_n + \hbar^2\left(\frac{n\pi}{a}\right)^2u_n - 2i\hbar\left(\frac{n\pi}{a}\right)eyB_x\cot\left(\frac{n\pi}{a}z\right)u_n + e^2y^2B_x^2u_n \tag{16.150}$$

$$\begin{aligned}\hat{H}_1^3\psi_n^0 &= \left[\left(-i\hbar\frac{\partial}{\partial x} + eB_yz\right)\vec{i} + \left(-i\hbar\frac{\partial}{\partial x} + exB_z\right)\vec{j} + \vec{k}\left(-i\hbar\frac{\partial}{\partial z} + eyB_x\right)\right]\hat{H}_1^2\psi_n^0 \\ &= \vec{i} = \Big[i\hbar\cdot ik_x\Big\{(\hbar k_x + eB_yz)^2u_n + (\hbar k_y + exB_z)^2u_n \\ &\quad + \hbar^2\left(\frac{n\pi}{a}\right)^2u_n - 2i\hbar\left(\frac{n\pi}{a}\right)eyB_x\cot\left(\frac{n\pi}{a}z\right)u_n + e^2y^2B_x^2u_n\Big\} \\ &\quad + (-i\hbar)(2\hbar k_yeB_z + 2e^2B_z^2x)u_n\Big] + \vec{i}eB_yz\hat{H}_1^2\psi_n^0 \\ &\quad + \vec{j}eB_zx\hat{H}_1^2\psi_n^0 + \vec{k}eyB_x\hat{H}_1^2\psi_n^0 + \vec{j}\Big[-i\hbar ik_y\hat{H}_1^2\psi_n^0 \\ &\quad + (-i\hbar)(-2i\hbar)\cdot\left(\frac{n\pi}{a}\right)eB_x\cot\left(\frac{n\pi}{a}z\right)u_n + (-i\hbar)e^2B_x^2yu_n\Big] \\ &\quad + \vec{k}\Big[-i\hbar\Big\{\left(\frac{n\pi}{a}\right)\cot\left(\frac{n\pi}{a}z\right)\Big\}\hat{H}_1^2\psi_n^0 + (-i\hbar)2eB_y \\ &\quad \cdot(\hbar k_x + eB_yz)u_n + \left(\frac{n\pi}{a}\right)^2(-i\hbar)(-2i\hbar)ey_yB_x)(-u_n)\Big]\end{aligned}$$

$$
\begin{aligned}
&= e\vec{A}\hat{H}_1^2\psi_n^0 + \vec{i}\hbar k_x \hat{H}_1^2\psi_n^0 + \vec{j}\hbar k_y \hat{H}_1^2\psi_n^0 \\
&\quad + \vec{k}\left(-i\hbar\frac{n\pi}{a}\right)\cot\left(\frac{n\pi}{a}z\right)\hat{H}_1^2\psi_n^0 \\
&\quad + \vec{i}\{-2i\hbar^2 k_y eB_z - 2i\hbar e^2 B_z^2 x)u_n \\
&\quad + \vec{j}\left\{-2\hbar^2\left(\frac{n\pi}{a}\right)\cot\left(\frac{n\pi}{a}z\right)eB_x - 2i\hbar e^2 B_x^2 y\right\}u_n \\
&\quad + \vec{k}\left[\left\{-2i\hbar^2 k_x eB_y - 2i\hbar e^2 B_y^2 z + 2\hbar^2\left(\frac{n\pi}{a}\right)^2 eyB_x\right\}u_n\right] \\
&= \vec{i}\hbar k_x \hat{H}_1^2\psi_n^0 + eB_y z\hat{H}_1^2\psi_n^0 + (-2i\hbar^2 k_y eB_z - 2i\hbar e^2 B_z^2 x)u_n] \\
&\quad + \vec{j}\{\hbar k_y \hat{H}_1^2\psi_n^0 + exB_z\hat{H}_1^2\psi_n^0 \\
&\quad + \left\{-2\hbar^2\left(\frac{n\pi}{a}\right)eB_x\cot\left(\frac{n\pi}{a}z\right) - 2i\hbar e^2 B_x^2 y\right\}u_n\Big] \\
&\quad + \vec{k}\left[-i\hbar\left(\frac{n\pi}{a}\right)\cot\left(\frac{n\pi}{a}z\right)\hat{H}_1^2\psi_n^0 + eyB_x\hat{H}_1^2\psi_n^0\right. \\
&\quad \left. + \left\{2\hbar^2\left(\frac{n\pi}{a}\right)^2 eyB_x - 2i\hbar e^2 B_y^2 z - 2i\hbar^2 k_x eB_y\right\}u_n\right]
\end{aligned} \tag{16.151}
$$

Therefore

$$
\hat{H}_1^4\psi_n^0 = R + S + T \tag{16.152}
$$

where

$$
\begin{aligned}
R &= \left(-i\hbar\frac{\partial}{\partial x} + ezB_y\right)[\hbar k_x \hat{H}_1^2\psi_n^0 + ezB_y\hat{H}_1^2\psi_n^0 \\
&\quad - 2i\hbar^2 k_y eB_z u_u - 2i\hbar e^2 B_z^2 x u_u] \\
&= ezB_y\hbar k_x \hat{H}_1^2\psi_n^0 + e^2 z^2 B_y^2 \hat{H}_1^2\psi_n^0 - 2i\hbar^2 k_y e^2 zB_y B_z u_u \\
&\quad - 2i\hbar e^2 zxB_y B_z^2 u_u + (-i\hbar)(-2i\hbar^2)ek_y B_z(ik_x)u_n \\
&\quad + (-i\hbar)(-2i\hbar)e^2 B_z^2 u_u + (-i\hbar)(-2i\hbar)e^2 B_z^2 x(ik_x)u_n \\
&\quad + (\hbar k_x + ezB_y)[\hbar k_x \hat{H}_1^2\psi_n^0 - 2i\hbar^2 k_y eB_z u_u - 2i\hbar e^2 B_z^2 x u_u]
\end{aligned}
$$

$$= (\hbar k_x + ezB_y)^2 \hat{H}_1^2 \psi_n^0 - 2i\hbar^3 ek_y k_x B_z u_u - 2i\hbar^2 e^2 B_z^2 x k_x u_u$$
$$- 2\hbar^2 e^2 B_z^2 u_u - 2i\hbar^3 k_x k_y e B_z u_u - 2i\hbar^2 k_x e^2 B_z^2 x u_u$$
$$- 4i\hbar^2 k_y e^2 z B_y B_z u_u - 4i\hbar e^3 B_z^2 B_y z x u_u \quad (16.153)$$

$$S = \left(-i\hbar\frac{\partial}{\partial y} + exB_z\right)\Big[(\hbar k_y + exB_z)\,\hat{H}_1^2\psi_n^0$$
$$- 2\hbar^2\left(\frac{n\pi}{a}\right) eB_x \cot\left(\frac{n\pi}{a}z\right) u_u - 2i\hbar e^2 B_x^2 y u_u\Big]$$
$$= (\hbar k_y + exB_z)\Big[\hbar k_y \hat{H}_1^2 \psi_n^0 - 2\hbar^2\left(\frac{n\pi}{a}\right) eB_x \cot\left(\frac{n\pi}{a}z\right) u_u$$
$$- 2i\hbar e^2 B_x^2 y u_u\Big] + (-i\hbar i k_y)(-2\hbar^2)\left(\frac{n\pi}{a}\right) eB_x \cot\left(\frac{n\pi}{a}z\right) u_u$$
$$+ (-i\hbar i k_y)(-2i\hbar e^2 B_x^2 y u_u) + (-i\hbar)(-2i\hbar e^2 B_x^2) u_u$$
$$+ exB_z(\hbar k_y + exB_z)\hat{H}_1^2\psi_n^0$$
$$+ exB_z\left[-2\hbar^2\left(\frac{n\pi}{a}\right) eB_x \cot\left(\frac{n\pi}{a}z\right) u_u - 2i\hbar e^2 B_x^2 y u_u\right] \quad (16.154)$$

$$S = (\hbar k_y + exB_z)^2 \hat{H}_1^2 \psi_n^0 - (\hbar k_y + 2exB_z)$$
$$\cdot \left[-2i\hbar e^2 B_x^2 y u_u - 2\hbar^2\left(\frac{n\pi}{a}\right) eB_x \cot\left(\frac{n\pi}{a}z\right) u_u\right]$$
$$- 2\hbar k_y\left(\frac{n\pi}{a}\right) eB_x \cot\left(\frac{n\pi}{a}z\right) u_u - 2\hbar^2 e^2 B_x^2 u_u - 2i\hbar^2 k_y e^2 B_x^2 y u_u \quad (16.156)$$

$$T = \left(-i\hbar\frac{\partial}{\partial y} + eyB_z\right)\Big[-i\hbar^2\left(\frac{n\pi}{a}\right)\cot\left(\frac{n\pi}{a}z\right)\hat{H}_1^2\psi_n^0$$
$$+ eyB_x \hat{H}_1^2 \psi_n^0 + 2\hbar^2\left(\frac{n\pi}{a}\right)^2 eyB_x u_u$$
$$- 2i\hbar e^2 B_y^2 z u_u - 2i\hbar^2 k_x e B_y u_u\Big] \quad (16.157)$$

Thus (16.204) can be written as

$$T = -i\hbar\left(\frac{n\pi}{a}\right)\cot\left(\frac{n\pi}{a}z\right)(-i\hbar)\Big\{\left(\frac{n\pi}{a}\right)\cot\left(\frac{n\pi}{a}z\right)\hat{H}_1^2\psi_n^0$$
$$+ 2(\hbar k_y + eB_y z)eB_y u_u - 2i\hbar\left(\frac{n\pi}{a}\right)^2 eyB_x(-u_n)\Big\}$$

$$
\begin{aligned}
&+(-i\hbar)\left(\frac{n\pi}{a}\right)\left(-\frac{n\pi}{a}\right)\cos ece^2\left(\frac{n\pi}{a}z\right)(-i\hbar)\hat{H}_1^2\psi_n^0 \\
&+eyB_x(-i\hbar)\left\{\left(\frac{n\pi}{a}\right)\cot\left(\frac{n\pi}{a}z\right)\hat{H}_1^2\psi_n^0-2i\hbar\left(\frac{n\pi}{a}\right)^2\right. \\
&\left.\cdot\, eyB_x(-u_n)+2(\hbar x+eB_yz)eB_yu_n\right\} \\
&+2\hbar^2\left(\frac{n\pi}{a}\right)^2 eyB_x(-i\hbar)\left(\frac{n\pi}{a}\right)\cot\left(\frac{n\pi}{a}z\right)u_n-2i\hbar e^2B_y^2z \\
&\cdot(-i\hbar)\left(\frac{n\pi}{a}\right)\cot\left(\frac{n\pi}{a}z\right)u_n-2i\hbar e^2B_y^2(-i\hbar)u_n \\
&-2i\hbar^2k_xeB_y(-i\hbar)\left(\frac{n\pi}{a}\right)\cot\left(\frac{n\pi}{a}z\right)u_n \\
&+eyB_x(-i\hbar)\left(\frac{n\pi}{a}\right)\cot\left(\frac{n\pi}{a}z\right)\hat{H}_1^2\psi_n^0+e^2y^2B_x^2\hat{H}_1^2\psi_n^0 \\
&+e^2y^2B_x^22\hbar^2\left(\frac{n\pi}{a}\right)^2u_n-2i\hbar e^2B_y^2zeyB_xu_n-2i\hbar^2ek_xB_yeB_xyu_n
\end{aligned}
\tag{16.158}
$$

$$
\begin{aligned}
S=&(\hbar k_y+exB_z)^2\hat{H}_1^2\psi_n^0-2\hbar^3k_y\left(\frac{n\pi}{a}\right)eB_x\cot\left(\frac{n\pi}{a}z\right)u_u \\
&-2\hbar^2e^2B_x^2u_u-2i\hbar^2k_ye^2B_x^2yu_u-2i\hbar^2k_ye^2B_x^2yu_u \\
&-2\hbar^3k_y\left(\frac{n\pi}{a}\right)eB_x\cot\left(\frac{n\pi}{a}z\right)u_u-4i\hbar e^3xyB_x^2B_zu_u \\
&-4\hbar\left(\frac{n\pi}{a}\right)e^2xB_xB_z\cot\left(\frac{n\pi}{a}z\right)u_u
\end{aligned}
\tag{16.159}
$$

$$
\begin{aligned}
T=&-\hbar\left(\frac{n\pi}{a}\right)^2\cot^2\left(\frac{n\pi}{a}z\right)\hat{H}_1^2\psi_n^0 \\
&-2\hbar^2\left(\frac{n\pi}{a}\right)\cot\left(\frac{n\pi}{a}z\right)eB_y(\hbar k_x+eB_yz)u_n-2i\hbar^3\left(\frac{n\pi}{a}\right)^3 \\
&\cdot\cot\left(\frac{n\pi}{a}z\right)eyB_xu_n+\hbar^2\cot\left(\frac{n\pi}{a}\right)^2\operatorname{cosec}^2\left(\frac{n\pi}{a}z\right)\hat{H}_1^2\psi_n^0 \\
&-i\hbar\left(\frac{n\pi}{a}\right)eyB_x\cot\left(\frac{n\pi}{a}z\right)\hat{H}_1^2\psi_n^0+2\hbar^2\left(\frac{n\pi}{a}\right)^2e^2y^2B_x^2u_n \\
&-2i\hbar(\hbar k_x+eB_yz)e^2yB_xB_yu_n-2i\hbar^3\left(\frac{n\pi}{a}\right)^3eyB_x\cot\left(\frac{n\pi}{a}z\right)u_n
\end{aligned}
$$

$$
\begin{aligned}
&-2\hbar^2\left(\frac{n\pi}{a}\right)e^2B_y^2z\cot\left(\frac{n\pi}{a}z\right)u_n-2\hbar^2e^2B_y^2u_n\\
&-2\hbar^3\left(\frac{n\pi}{a}\right)ek_xB_y\cot\left(\frac{n\pi}{a}z\right)u_n\\
&-i\hbar\left(\frac{n\pi}{a}\right)eyB_x\cot\left(\frac{n\pi}{a}z\right)\hat{H}_1^2\psi_n^0+e^2y^2B_x^2\hat{H}_1^2\psi_n^0\\
&+2\hbar^2\left(\frac{n\pi}{a}\right)^2e^2y^2B_x^2u_n-2i\hbar e^3zyB_y^2B_xu_n-2i\hbar^2e^2k_xyB_yB_xu_n
\end{aligned}
\tag{16.159}
$$

$$
\begin{aligned}
T=\;&\hbar^2\left(\frac{n\pi}{a}\right)^2\hat{H}_1^2\psi_n^0-2i\hbar\left(\frac{n\pi}{a}\right)eyB_x\cot\left(\frac{n\pi}{a}z\right)\hat{H}_1^2\psi_n^0\\
&-2\hbar^3\left(\frac{n\pi}{a}\right)eB_yk_x\cot\left(\frac{n\pi}{a}z\right)u_n-2\hbar^2\left(\frac{n\pi}{a}\right)e^2B_y^2z\cot\left(\frac{n\pi}{a}z\right)u_n\\
&-4i\hbar^3\left(\frac{n\pi}{a}\right)^3eyB_x\cot\left(\frac{n\pi}{a}z\right)u_n\\
&+4\hbar^2\left(\frac{n\pi}{a}\right)^2e^2y^2B_x^2u_n-2i\hbar^2k_xe^2yB_xB_yu_n-2i\hbar e^3B_y^2B_xzyu_n\\
&-2\hbar^2\left(\frac{n\pi}{a}\right)e^2B_y^2z\cot\left(\frac{n\pi}{a}z\right)u_n-2\hbar^3\left(\frac{n\pi}{a}\right)ek_xB_y\cot\left(\frac{n\pi}{a}z\right)u_n\\
&-2\hbar^2e^2B_y^2u_n+e^2y^2B_x^2\hat{H}_1^2\psi_n^0-2i\hbar^3zyB_y^2B_xu_n\\
&-2i\hbar^2e^2k_xyB_yB_xu_n
\end{aligned}
\tag{16.160}
$$

$$
\begin{aligned}
T=\;&\hbar^2\left(\frac{n\pi}{a}\right)^2\hat{H}_1^2\psi_n^0+e^2y^2B_x^2\hat{H}_1^2\psi_n^0\\
&-2i\hbar\left(\frac{n\pi}{a}\right)eyB_x\cot\left(\frac{n\pi}{a}z\right)\hat{H}_1^2\psi_n^0-4\hbar^3\left(\frac{n\pi}{a}\right)ek_xB_y\\
&\cdot\cot\left(\frac{n\pi}{a}z\right)u_n-4\hbar^2\left(\frac{n\pi}{a}\right)e^2B_y^2z\cot\left(\frac{n\pi}{a}z\right)u_n\\
&-4i\hbar^3\left(\frac{n\pi}{a}\right)^3eyB_x\cot\left(\frac{n\pi}{a}z\right)u_n+4\hbar^2\left(\frac{n\pi}{a}\right)^2e^2y^2B_x^2u_n\\
&-4i\hbar^2e^2k_xyB_yB_xu-4i\hbar e^3zyB_y^2B_xu-2\hbar^2e^2B_y^2u
\end{aligned}
\tag{16.161}
$$

$$
\begin{aligned}
R=\;&(\hbar k_x+ezB_y)^2\hat{H}_1^2\psi_n^0-4i\hbar^3k_xk_yeB_zu_u-4i\hbar^2k_xe^2xB_z^2u_u\\
&-4i\hbar^2k_ye^2zB_yB_zu_u-4i\hbar e^3zxB_z^2B_yu_u-2\hbar^2e^2B_z^2u_u
\end{aligned}
\tag{16.162}
$$

$$\begin{aligned}S = &(\hbar k_y + exB_z)^2 \hat{H}_1^2 \psi_n^0 - 4\hbar^3 k_y \left(\frac{n\pi}{a}\right) eB_x \cot\left(\frac{n\pi}{a} z\right) u_u \\ &- 4i\hbar^2 k_y e^2 B_x^2 y u_u - 4i\hbar e^3 xy B_x^2 B_z u_u \\ &- 4\hbar^2 \left(\frac{n\pi}{a}\right) e^2 x B_x B_z x \cot\left(\frac{n\pi}{a} z\right) u_u - 2\hbar^2 e^2 B_x^2 u_u\end{aligned} \tag{16.163}$$

Therefore

$$\begin{aligned}\hat{H}_1^4 \psi_n^0 = &(\hbar k_x + ezB_y)^2 \hat{H}_1^2 \psi_n^0 + (\hbar k_y + ezB_z)^2 \hat{H}_1^2 \psi_n^0 \\ &+ \hbar^2 \left(\frac{n\pi}{a}\right)^2 \hat{H}_1^2 \psi_n^0 + e^2 y^2 B_x^2 \hat{H}_1^2 \psi_n^0 \\ &- 2i\hbar \left(\frac{n\pi}{a}\right) eyB_x \cot\left(\frac{n\pi}{a} z\right) \hat{H}_1^2 \psi_n^0 \\ &- 4\hbar^3 \left(\frac{n\pi}{a}\right) ek_x B_y \cot\left(\frac{n\pi}{a} z\right) u_n \\ &- 4\hbar^3 \left(\frac{n\pi}{a}\right) ek_y B_x \cot\left(\frac{n\pi}{a} z\right) u_n \\ &- 4\hbar^2 \left(\frac{n\pi}{a}\right) e^2 B_y^2 z \cot\left(\frac{n\pi}{a} z\right) u_n \\ &- 4\hbar^2 \left(\frac{n\pi}{a}\right) e^2 x B_x B_z \cot\left(\frac{n\pi}{a} z\right) u_n \\ &- 4i\hbar^3 \left(\frac{n\pi}{a}\right)^3 eyB_x \cot\left(\frac{n\pi}{a} z\right) u_n \\ &- 4i\hbar^3 k_x k_y eB_z u_n - 4i\hbar^2 k_x e^2 x B_z^2 u_n - 4i\hbar^2 k_y e^2 z B_y B_z u_n \\ &- 4i\hbar^2 k_y e^2 y B_x^2 u_n - 4i\hbar^2 k_x e^2 y B_y B_x u_n - 4i\hbar e^3 zy B_y^2 B_x u_n \\ &- 4i\hbar e^3 zx B_x^2 B_y u_n - 4i\hbar e^3 xy B_x^2 B_z u_n - 2\hbar^2 e^2 B_x^2 u_n \\ &- 2\hbar^2 e^2 B_z^2 u_n - 2\hbar^2 e^2 B_y^2 u_n + 4\hbar^2 \left(\frac{n\pi}{a}\right) e^2 y^2 B_x^2 u_n \\ &- 2\hbar^2 e^2 B_y^2 u_n + 4\hbar^2 \left(\frac{n\pi}{a}\right) e^2 y^2 B_x^2 u_n\end{aligned} \tag{16.163}$$

$$\begin{aligned}\hat{H}_1^4 \psi_n^0 + &\left[\hbar^2 k_x^2 + \hbar^2 k_y^2 + \hbar^2 \left(\frac{n\pi}{a}\right)^2\right]^2 u_n + B_y^4 e^4 z^4 u_n \\ &+ 4\hbar k_x B_y^3 e^3 z^3 u_n + 4\hbar^2 k_x^2 B_y^2 e^2 z^2 u_n + B_z^4 e^4 x^4 u_n\end{aligned}$$

$$+ 4\hbar k_y B_z^3 e^3 x^3 u_n + 4\hbar^2 k_y^2 B_z^2 e^2 x^2 u_n + B_x^4 e^4 y^4 u_n$$
$$+ 2\left[\hbar^2 k_x^2 + \hbar^2 k_y^2 + \hbar^2 \left(\frac{n\pi}{a}\right)^2\right]$$

$$B_y^2 e^2 z^2 u_n + 2\left[\hbar^2 k_x^2 + \hbar^2 k_y^2 + \hbar^2 \left(\frac{n\pi}{a}\right)^2\right] \hbar k_x B_y e z u_n$$
$$+ 4\left[\hbar^2 k_x^2 + \hbar^2 k_y^2 + \hbar^2 \left(\frac{n\pi}{a}\right)^2\right] \hbar k_y B_z e x u_n$$
$$+ 4\left[\hbar^2 k_x^2 + \hbar^2 k_y^2 + \hbar^2 \left(\frac{n\pi}{a}\right)^2\right] \hbar k_x B_y e z u_n$$
$$+ 4\left[\hbar^2 k_x^2 + \hbar^2 k_y^2 + \hbar^2 \left(\frac{n\pi}{a}\right)^2\right] \hbar k_y B_z e x u_n$$
$$+ 4i\hbar \left(\frac{n\pi}{a}\right)\left[\hbar^2 k_x^2 + \hbar^2 k_y^2 + \hbar^2 \left(\frac{n\pi}{a}\right)^2\right] B_x e y \cot\left(\frac{n\pi}{a} z\right) u_n$$
$$+ 2B_z^2 B_y^2 e^4 x^2 z^2 u_n + 4\hbar k_x B_z^2 B_y e^3 x^2 z u_n + 4\hbar k_y B_y^2 B_z e^3 z^2 x u_n$$
$$+ 8\hbar^2 k_y k_x B_y B_z e^2 x z u_n + 2B_y^2 B_x^2 e^4 y^2 z^2 u_n + 4\hbar k_x B_x^2 B_y e^3 y^2 z u_n$$
$$+ 2B_x^2 B_z^2 e^4 x^2 y^2 u_n + 4\hbar k_y B_x^2 B_z e^3 y^2 x u_n$$
$$- 4i\hbar \left(\frac{n\pi}{a}\right) B_x^3 e^3 y^3 \cot\left(\frac{n\pi}{a} z\right) u_n$$
$$- 4i\hbar \left(\frac{n\pi}{a}\right) B_x B_y^2 e^3 z^2 y \cot\left(\frac{n\pi}{a} z\right) u_n$$
$$- 4i\hbar^2 \left(\frac{n\pi}{a}\right)^2 B_x B_z^2 e^3 x^2 y \cot^2\left(\frac{n\pi}{a} z\right) u_n$$
$$- 8i\hbar^2 k_x \left(\frac{n\pi}{a}\right) B_x B_y e^2 y z \cot\left(\frac{n\pi}{a} z\right) u_n$$
$$- 8i\hbar^2 k_y \left(\frac{n\pi}{a}\right) B_x B_z e^2 x y \cot\left(\frac{n\pi}{a} z\right) u_n$$
$$- 4\hbar^2 \left(\frac{n\pi}{a}\right)^2 B_x^2 e^2 y^2 \cot^2\left(\frac{n\pi}{a} z\right) u_n$$
$$- 4\hbar^2 \left(\frac{n\pi}{a}\right)^2 B_y^2 e^2 z \cot^2\left(\frac{n\pi}{a} z\right) u_n$$

$$
\begin{aligned}
&- 4\hbar^2\left(\frac{n\pi}{a}\right)^2 B_xB_ze^2X\cot\left(\frac{n\pi}{a}z\right)u_n \\
&- 4i\hbar^2k_yB_x^2e^2yu_n - 4i\hbar^2k_yB_ye^2zu_n - 4i\hbar^2k_xB_yB_xe^2yu_n \\
&+ 4\hbar^2\left(\frac{n\pi}{a}\right)^2 B_x^2e^2y^2u_n - 4i\hbar B_x^2B_ze^3xyu_n \\
&- 4i\hbar B_y^2B_xe^3yzu_n - 4i\hbar B_z^2B_ye^3zxu_n - 2\hbar^2B_x^2e^2u_n \\
&- 2\hbar^2B_y^2e^2u_n - 2\hbar^2B_z^2e^2u_n - 4i\hbar^3k_xk_yB_zeu_n
\end{aligned} \tag{1.213}
$$

$$
\begin{aligned}
\hat{H}' = &-b'B_y^4e^4z^4 - 4b'\hbar k_xB_y^3e^3z^3 - 4b'\hbar^2k_x^2B_y^2e^2z^2 \\
&- 2b'\left[\hbar^2k_x^2 + \hbar^2k_y^2 + \hbar^2\left(\frac{n\pi}{a}\right)^2\right]B_y^2e^2z^2 \\
&+ a'B_y^2e^2z^2 - 4b'\left[\hbar^2k_x^2 + \hbar^2k_y^2 + \hbar^2\left(\frac{n\pi}{a}\right)^2\right]\hbar k_xB_yez \\
&+ 4bi\hbar^2k_yB_yB_ze^2z + 2a'\hbar k_xB_yez - b'B_x^4e^4y^4 \\
&- 2b'\left[\hbar^2k_x^2 + \hbar^2k_y^2 + \hbar^2\left(\frac{n\pi}{a}\right)^2\right]B_x^2e^2y^2 - 4b'\hbar^2\left(\frac{n\pi}{a}\right)^2B_x^2e^2y^2 \\
&+ a'B_x^2e^2y^2 + 4b'i\hbar^2\left(k_yB_x + k_xB_y\right)B_xe^2y - b'B_z^4e^4x^4 \\
&- 4b'\hbar^2k_y^2B_z^2e^2x^2 - 2b'\left[\hbar^2k_x^2 + \hbar^2k_y^2 + \hbar^2\left(\frac{n\pi}{a}\right)^2\right]B_z^2e^2x^2 \\
&+ a'B_x^2e^2x^2 - 4b'\left[\hbar^2k_x^2 + \hbar^2k_y^2 + \hbar^2\left(\frac{n\pi}{a}\right)^2\right]\hbar k_yB_zex \\
&+ 4b'i\hbar^2k_xB_x^2e^2x + 2a'\hbar k_yB_zex - 2b'B_z^2B_y^2e^4x^2z^2 \\
&- 2b'B_y^2B_x^2e^4y^4z^2 - 4b'\hbar k_yB_y^2B_ze^3xz^2 - 4b'\hbar k_xB_z^2B_ye^3x^2z \\
&+ 4b'\hbar\left(iB_ze - 2\hbar k_yk_x\right)B_yB_ze^2xz - 4b'\hbar k_xB_x^2B_ye^3y^2z \\
&+ 4b'i\hbar B_y^2B_xe^3yz - 2b'B_x^2B_z^2e^4x^2y^2 - 4b'\hbar k_yB_x^2B_ze^3xy^2 \\
&+ 4b'i\hbar B_x^2B_ze^3xy + 4b'\hbar^2\left(\frac{n\pi}{a}\right)B_y^2e^2z\cot\left(\frac{n\pi}{a}z\right) \\
&+ 4b'i\hbar\left(\frac{n\pi}{a}\right)B_xB_y^2B_ze^3xy^2\cot\left(\frac{n\pi}{a}z\right)
\end{aligned}
$$

$$
\begin{aligned}
&+8b'i\hbar^2 k_x \left(\frac{n\pi}{a}\right) B_x B_y e^2 yz \cot\left(\frac{n\pi}{a}z\right) \\
&+4b'i\hbar \left(\frac{n\pi}{a}\right) B_x^3 e^3 y^3 \cot\left(\frac{n\pi}{a}z\right) + 4b'i\hbar\left(\frac{n\pi}{a}\right) \\
&\cdot\left[\hbar^2 k_x^2 + \hbar^2 k_y^2 + \hbar^2\left(\frac{n\pi}{a}\right)^2\right] B_x ey \cot\left(\frac{n\pi}{a}z\right) \\
&-2a'i\hbar\left(\frac{n\pi}{a}\right) B_x ey \cot\left(\frac{n\pi}{a}z\right) + 4b'\hbar^2\left(\frac{n\pi}{a}\right) \\
&\cdot B_x B_z e^2 x \cot\left(\frac{n\pi}{a}z\right) + 4b'i\hbar\left(\frac{n\pi}{a}\right) B_x B_x^2 e^3 yx^2 \cot\left(\frac{n\pi}{a}z\right) \\
&+8b'i\hbar^2 k_y\left(\frac{n\pi}{a}\right) B_x B_z e^2 yx \cot\left(\frac{n\pi}{a}z\right) \\
&+4b'\hbar^2\left(\frac{n\pi}{a}\right)^2 B_x^2 e^2 y^2 \cot^2\left(\frac{n\pi}{a}z\right) \\
&+4b'\hbar^3\left(\frac{n\pi}{a}\right)(k_y B_x + k_x B_y) e \cot\left(\frac{n\pi}{a}z\right) \\
&+2b'\hbar^2 e^2 (B_x^2 + B_y^2 + B_z^2) + 4b'i\hbar^3 k_x k_y B_z e
\end{aligned}
\tag{16.164}
$$

So the first order perturbation to the energy is

$$
E_n^{(1)} = \hat{H}'_m = \frac{\int u_n^* \hat{H} u_n dy}{\int u_n^* u_n dy} = \langle u_n|\hat{H}'|u_n\rangle \tag{16.165}
$$

Now

$$
\begin{aligned}
\int u_n^* u_n dy &= \frac{2}{L_x L_y a}\int_0^a \sin^2\left(\frac{n\pi}{a}z\right) dz \int_0^{l_x} dx \int_0^{l_y} dy \\
&= \frac{2}{a}\left[\frac{1}{2}\int_0^a \left\{1 - \cos\left(\frac{2n\pi}{a}z\right)\right\} dz\right] \\
&= \frac{2}{a}\cdot\frac{1}{2}\left[a - \frac{a}{2n\pi}\sin\frac{2n\pi}{a}z \int_0^a\right] \\
&= \frac{1}{a}\left[a - \frac{a}{2n\pi}\sin 2n\pi\right] = 1
\end{aligned}
\tag{16.166}
$$

Besides various averages are written below:

$$
\langle z^4\rangle = \frac{a^4}{5} - \frac{a^4}{n^2\pi^2} + \frac{a^4}{2n^4\pi^4}, \quad \langle z^3\rangle = \frac{a^3}{4} - \frac{3a^3}{4n^2\pi^2},
$$

$$\langle z^2\rangle = \frac{a^2}{3} - \frac{a^3}{2n^2\pi^2}, \quad \langle z\rangle = \frac{a}{2},$$

$$\langle y^4\rangle = \frac{1}{L_y}\int_0^{L_y} y^4 dy \int_0^a \left(\frac{2}{a}\right)\sin^2\left(\frac{n\pi}{a}z\right)dz = \frac{L_y^4}{5}$$

$$\langle y^2\rangle = \frac{L_y^2}{3}, \quad \langle y\rangle = \frac{L_y}{2}, \quad \langle x^4\rangle = \frac{L_x^4}{5}, \quad \langle x^3\rangle = \frac{L_x^3}{4},$$

$$\langle x^2\rangle = \frac{L_x^2}{3}, \quad \langle x\rangle = \frac{L_x}{2}$$

$$\langle x^2 z^2\rangle = \frac{1}{L_x}\int_0^{L_x} x^2 dx \int_0^a \left(\frac{2}{a}\right) z^2 \sin^2\left(\frac{n\pi}{a}z\right)dz = \frac{L_x^2}{3}\langle z^2\rangle,$$

$$\langle y^2 z^2\rangle = \frac{L_y^2}{2}\langle z^2\rangle, \quad \langle xz^2\rangle = \frac{L_x}{2}\langle z^2\rangle, \quad \langle x^2 z\rangle = \frac{L_x^2}{3}\langle z\rangle, \quad \langle xz\rangle = \frac{L_x}{2}\langle z\rangle$$

$$\langle y^2 z\rangle = \frac{L_y^2}{3}\langle z\rangle, \quad \langle yz\rangle = \frac{L_y}{2}\langle z\rangle, \quad \langle x^2 y^2\rangle = \frac{L_x^2 L_y^2}{6},$$

$$\langle xy^2\rangle = \frac{L_x L_y^2}{6}, \quad \langle xy\rangle = \frac{L_x L_y}{4},$$

$$\begin{aligned}
\left\langle z\cot\left(\frac{n\pi}{a}z\right)\right\rangle &= \frac{2}{a}\int_0^a z\,\cos\left(\frac{n\pi}{a}z\right)\sin\left(\frac{n\pi}{a}z\right)dz \\
&= \frac{1}{a}\left[\int_0^a z\sin\frac{2n\pi}{a}zdz\right] \\
&= \frac{1}{a}\left[-\frac{za}{2n\pi}\cos\frac{2n\pi}{a}z + \frac{a}{2n\pi}\int\cos\frac{2n\pi}{a}zdz\right]_0^a \\
&= \frac{1}{a}\left[-\frac{za}{2n\pi}\cos\frac{2n\pi}{a}z + \left(\frac{a}{2n\pi}\right)^2\sin\frac{2n\pi}{a}z\right]_0^a \\
&= \frac{1}{a}\left[-\frac{a^2}{2n\pi}\cos 2n\pi + \left(\frac{a}{2n\pi}\right)^2\sin 2n\pi\right] \\
&= -\frac{a}{2n\pi}
\end{aligned}$$

$$\left\langle yz^2\cot\left(\frac{n\pi}{a}z\right)\right\rangle = \frac{L_y}{2}\left\langle z^2\cot\left(\frac{n\pi}{a}z\right)\right\rangle = \frac{L_y a^2\left(1-2n^2\pi^2\right)}{8n^3\pi^3}$$

where,

$$\begin{aligned}
\left\langle z^2 \cot\left(\frac{n\pi}{a}z\right)\right\rangle &= \frac{2}{a}\int_0^a z^2 \cos\left(\frac{n\pi}{a}z\right)\sin\left(\frac{n\pi}{a}z\right)dz \\
&= \frac{1}{a}\left[\int_0^a z^2 \sin\left(\frac{2n\pi}{a}z\right)dz\right] \\
&= \frac{1}{a}\left[-\frac{z^2 a}{2n\pi}\cos\frac{2n\pi}{a}z + \frac{2a}{2n\pi}\int_0^a z\cos\left(\frac{2n\pi}{a}z\right)dz\right]_0^a \\
&= \frac{1}{a}\left[-\frac{z^2 a}{2n\pi}\cos\left(\frac{2n\pi}{a}z\right) + \left(\frac{a}{n\pi}\right)\left\{\frac{za}{2n\pi}\sin\frac{2n\pi}{a}z\right.\right. \\
&\quad \left.\left. - \frac{a}{2n\pi}\int \sin\frac{2n\pi}{a}z dz\right\}\right]_0^a \\
&= \frac{1}{a}\left[-\frac{z^2 a}{2n\pi}\cos\left(\frac{2n\pi}{a}z\right) + \left(\frac{a}{n\pi}\right)^2 \frac{z}{2}\sin\left(\frac{n\pi}{a}z\right)\right. \\
&\quad \left. + \frac{1}{4}\left(\frac{a}{n\pi}\right)^3 \cos\frac{2n\pi}{a}z\right]_0^a \\
&= -\frac{a^2}{2n\pi}\cos 2n\pi + \frac{a^2}{n^2\pi^2}\frac{1}{2}\sin 2n\pi + \frac{1}{4}\frac{a^2}{n^3\pi^3}\cos 2n\pi \\
&= -\frac{a^2}{2n\pi} + \frac{a^2}{4n^3\pi^3} = \frac{a^3(1-2n^2\pi^2)}{4n^3\pi^3},
\end{aligned}$$

$$\left\langle yz\cot\left(\frac{n\pi}{a}z\right)\right\rangle = \frac{L_y}{2}\left\langle z\cot\left(\frac{n\pi}{a}z\right)\right\rangle = \frac{L_y}{2}\left(-\frac{a}{2n\pi}\right) = -\frac{aL_y}{4n\pi},$$

$$\begin{aligned}
\left\langle y^3\cot\left(\frac{n\pi}{a}z\right)\right\rangle &= \frac{L_y^4}{4}\left\langle \cot\left(\frac{n\pi}{a}z\right)\right\rangle \\
&= \frac{L_y^4}{4}\left[\frac{2}{a}\int_0^a \sin\left(\frac{n\pi}{a}z\right)\cos\left(\frac{n\pi}{a}z\right)dz\right] \\
&= \frac{L_y^4}{4}\left[\frac{1}{a}\int_0^a \sin\left(\frac{2n\pi}{a}z\right)dz\right]
\end{aligned}$$

$$= \frac{L_y^4}{4a}\left[-\frac{a}{2n\pi}\cos\frac{2n\pi}{a}z\right]_0^a$$

$$= \frac{L_y^4}{4a}\left[-\frac{a}{2n\pi}\{\cos 2n\pi - 1\}\right] = 0,$$

$$\left\langle y\cot\left(\frac{n\pi}{a}z\right)\right\rangle = 0, \quad \left\langle yx^2\cot\left(\frac{n\pi}{a}z\right)\right\rangle = 0, \quad \left\langle x\cot\left(\frac{n\pi}{a}z\right)\right\rangle = 0,$$

$$\left\langle yx\cot\left(\frac{n\pi}{a}z\right)\right\rangle = 0,$$

$$\left\langle y^2\cot^2\left(\frac{n\pi}{a}z\right)\right\rangle = \frac{L_y^2}{3}\left[\frac{2}{a}\int_0^a \cos^2\left(\frac{n\pi}{a}z\right)dz\right]$$

$$= \frac{L_y^2}{3}\frac{1}{a}\left[\int_0^a\left(1+\cos\frac{2n\pi}{a}z\right)dz\right]$$

$$= \frac{L_y^2}{3}\cdot\frac{1}{a}\left[a + \frac{a}{2n\pi}\sin 2n\pi\right] = \frac{L_y^2}{3}$$

So the first order perturbation to the energy is

$$E_n^{(1)} = -b'B_y^4e^4\langle z^4\rangle - 4b'\hbar k_x B_y^3e^3\langle z^3\rangle - 4b'\hbar^2k_x^2B_y^2e^2\langle z^2\rangle$$

$$+ a'B_y^2e^2\langle z^2\rangle - 2b'\left[\hbar^2k_x^2 + \hbar^2k_y^2 + \hbar^2\left(\frac{n\pi}{a}\right)^2\right]B_y^2e^2\langle z^2\rangle$$

$$- 4b'\left[\hbar^2k_x^2 + \hbar^2k_y^2 + \hbar^2\left(\frac{n\pi}{a}\right)^2\right]\hbar k_x B_y e\langle z\rangle + 2a'\hbar k_x B_y e\langle z\rangle$$

$$+ 4bi\hbar^2k_yB_yB_ze^2\langle z\rangle - b'B_x^4e^4\frac{L_y^4}{5}$$

$$- 2b'\left[\hbar^2k_x^2 + \hbar^2k_y^2 + \hbar^2\left(\frac{n\pi}{a}\right)^2\right]B_x^2e^2\frac{L_y^2}{3}$$

$$- 4b'\hbar^2\left(\frac{n\pi}{a}\right)^2B_x^2e^2\frac{L_y^2}{3}a'B_x^2e^2\frac{L_y^2}{3}$$

$$+ 4b'i\hbar^2(k_yB_x + k_xB_y)B_xe^2\frac{L_y}{2} - b'B_z^4e^4\frac{L_x^4}{5} - 4b'\hbar k_yB_z^3e^3\frac{L_x^3}{4}$$

$$
\begin{aligned}
&- 4b'\hbar^2 k_y^2 B_z^2 e^2 \frac{L_x^2}{3} - 2b' \left[\hbar^2 k_x^2 + \hbar^2 k_y^2 + \hbar^2 \left(\frac{n\pi}{a}\right)^2\right] B_z^2 e^2 \frac{L_x^2}{3} \\
&+ a' B_z^2 e^2 \frac{L_x^2}{3} - 4b' \left[\hbar^2 k_x^2 + \hbar^2 k_y^2 + \hbar^2 \left(\frac{n\pi}{a}\right)^2\right] \hbar k_y B_z e \frac{L_x}{2} \\
&+ 2a'\hbar k_y B_z e \frac{L_x}{2} + 4b' i\hbar^2 k_x B_z^2 e^2 \frac{L_x}{2} - 2b' B_z^2 B_y^2 e^4 \frac{L_x^2}{3} \langle z^3 \rangle \\
&- 2b' B_y^2 B_x^2 e \frac{L_y^2}{3} \langle z^2 \rangle - 4b'\hbar k_y B_y^2 B_z e^3 \frac{L_x}{2} \langle z^2 \rangle \\
&- 4b'\hbar k_x B_z^2 B_y e^3 \frac{L_x^2}{3} \langle z \rangle + 4b'\hbar (iB_z e - 2\hbar k_y k_x) B_y B_z e^2 \frac{L_x}{2} \langle z \rangle \\
&- 4b'\hbar k_x B_x^2 B_y e^3 \frac{L_y^2}{3} \langle z \rangle + 4b' i\hbar B_y^2 B_x e^3 \frac{L_y}{2} \langle z \rangle \\
&- 2b' B_x^2 B_z^2 e^4 \left(\frac{L_x^2 L_y^2}{9}\right) - 4b'\hbar k_y B_x^2 B_z e^3 xy^2 \frac{L_x L_y^2}{6} \\
&+ 4b' i\hbar B_x^2 B_z e^3 \left(\frac{L_x L_y}{4}\right) + 4b'\hbar^2 \left(\frac{n\pi}{a}\right) B_y^2 e^2 \left(-\frac{a}{2n\pi}\right) \\
&+ 4b' i\hbar \left(\frac{n\pi}{a}\right) B_x B_y^2 e^3 \frac{L_y}{2} \left(-\frac{a^2}{2n\pi} + \frac{a^2}{4n^3\pi^3}\right) \\
&+ 8b' i\hbar^2 k_x \left(\frac{n\pi}{a}\right) B_x B_y e^2 \left(-\frac{aL_y}{4n\pi}\right) + 4b'\hbar^2 \left(\frac{n\pi}{a}\right)^2 B_x^2 e^2 \frac{L_x^2}{3} \\
&+ 2b'\hbar^2 e^2 (B_x^2 + B_y^2 + B_z^2) + 4b' i\hbar^3 k_x k_y B_z e
\end{aligned}
$$

$$E_n = E_n^{(0)} + E_n^{(1)}$$

$$a_2 a_0 = (a_2 - a_2')$$

The electron energy spectrum in HD QWs of III–V semiconductors in the presence of an arbitrarily oriented quantizing magnetic field B can be written as

$$\gamma_3(E, \eta_g) = \frac{1}{2m_c} \left[\hbar^2 k_x^2 + \hbar^2 k_y^2 + \hbar^2 \left(\frac{n\pi}{a}\right)^2\right]$$

$$-\alpha\left[\frac{\hbar^2k_x^2}{2m_c}+\frac{\hbar^2k_y^2}{2m_c}+\frac{\hbar^2}{2m_c}\left(\frac{n\pi}{a}\right)^2\right]^2+\frac{e^2B_y^2}{2m_c}\langle z^2\rangle$$

$$+\frac{2\hbar k_x eBy}{2m_c}\langle z\rangle+\frac{1}{3}\frac{e^2B_x^2L_y^2}{2m_c}+\frac{1}{3}\frac{e^2B_z^2L_x^2}{2m_c}+\frac{\hbar k_y eB_zL_x}{2m_c}$$

$$-\alpha\left(\frac{e^2B_y^2}{2m_c}\right)^2\langle z^4\rangle-4\alpha\left(\frac{\hbar k_x e^3B_y^3}{4m_c^2}\right)\langle z^3\rangle$$

$$-4\alpha\left(\frac{\hbar k_x eB_y}{2m_c}\right)^2\langle z^2\rangle-2\alpha\left[\hbar^2k_x^2+\hbar^2k_y^2+\hbar^2\left(\frac{n\pi}{a}\right)^2\right]$$

$$\cdot\left(\frac{eB_y}{2m_c}\right)^2\langle z^2\rangle-4\alpha\left[\hbar^2k_x^2+\hbar^2k_y^2+\hbar^2\left(\frac{n\pi}{a}\right)^2\right]\frac{\hbar k_x eB_y}{4m_c^2}\langle z\rangle$$

$$-\frac{\alpha}{5}\left(\frac{e^2L_y^2B_x^2}{2m_c}\right)^2-\left(\frac{2}{3}\right)\alpha\left[\hbar^2k_x^2+\hbar^2k_y^2+\hbar^2\left(\frac{n\pi}{a}\right)^2\right]$$

$$\cdot\left(\frac{eL_yB_x}{2m_c}\right)^2-\frac{4}{3}\alpha\hbar^2\left(\frac{n\pi}{a}\right)^2\left(\frac{eL_yB_x}{2m_c}\right)^2$$

$$-\frac{\alpha}{5}\left(\frac{e^2L_x^2B_z^2}{2m_c}\right)^2-\frac{\alpha\hbar k_y e^3L_x^3B_z^3}{4m_c^2}-\frac{4}{3}\alpha\left(\frac{\hbar k_y eL_xB_z}{2m_c}\right)^2$$

$$-\frac{2}{3}\alpha\left[\hbar^2k_x^2+\hbar^2k_y^2+\hbar^2\left(\frac{n\pi}{a}\right)^2\right]\left(\frac{eL_xB_z}{2m_c}\right)^2$$

$$-2\alpha\left[\hbar^2k_x^2+\hbar^2k_y^2+\hbar^2\left(\frac{n\pi}{a}\right)^2\right]\frac{\hbar k_y eL_xB_z}{4m_c^2}$$

$$-\frac{2}{3}\alpha\left(\frac{e^2L_xB_yB_z}{2m_c}\right)^2\langle z^2\rangle-\frac{2}{3}\alpha\left(\frac{e^2L_yB_yB_x}{2m_c}\right)^2\langle z^2\rangle$$

$$-2\alpha\left[\frac{\hbar k_y e^3L_xB_y^2B_z}{4m_c^2}\right]\langle z^2\rangle-\frac{4}{3}\alpha\left[\frac{\hbar k_x e^3L_x^2B_z^2B_y}{4m_c^2}\right]\langle z\rangle$$

$$-4\alpha\left[\frac{\hbar k_yk_x e^2L_xB_yB_z}{4m_c^2}\right]\langle z\rangle-\frac{4}{3}\alpha\left[\frac{\hbar k_x e^3L_y^2B_x^2B_y}{4m_c^2}\right]\langle z\rangle$$

$$-\frac{2}{9}\left(\frac{e^2 L_x L_y B_x B_z}{2m_c}\right)^2 - \frac{2}{3}\alpha\hbar k_y e^3 L_x\left(\frac{B_x L_y}{2m_c}\right)^2 B_z$$
$$+\frac{4}{3}\alpha\left[\hbar\left(\frac{n\pi}{a}\right)\frac{eL_y B_x}{2m_c}\right]^2 + 2\alpha\left(\frac{e\hbar}{2m_c}\right)^2[B_x^2 + B_z^2] \tag{16.167}$$

Under the condition $\alpha \to 0$, (1.218) get simplified as

$$\frac{1}{2m_c}\left[\hbar^2 k_x^2 + \hbar^2 k_y^2 + \hbar\left(\frac{n\pi}{a}\right)^2\right] = \gamma_3(E,\eta_g) - a_1 k_x - a_2 k_y - a_3 \tag{16.168}$$

where

$$a_1 = \frac{2\hbar e B_y}{2m_c}\langle z\rangle, \quad a_2 = \frac{\hbar e B_z L_x}{2m_c},$$
$$a_3 = \frac{e^2 B_y^2}{2m_c}\langle z^2\rangle + \frac{1}{3}\frac{e^2 B_x^2}{2m_c}L_y^2 + \frac{1}{3}\frac{e^2 B_z^2}{2m_c}L_x^2$$

Using the method of successive approximation from (1.218) and (1.219) we can write

$$\gamma_3(E,\eta_g) = b_1 k_x^2 + b_2 k_y^2 + b_3 - \alpha[(\gamma_3(E,\eta_g) - a_3) - a_1 k_x - a_2 k_y]^2$$
$$-\alpha(a_7 + a_9 + a_{12})[\gamma_3(E,\eta_g) - a_3 - a_1 k_x - a_2 k_y]$$
$$-\alpha a_8[\gamma_3(E,\eta_g) - a_3 - a_1 k_x - a_2 k_y]k_x$$
$$-\alpha a_{13}[\gamma_3(E,\eta_g) - a_3 - a_1 k_x - a_2 k_y]k_y$$
$$-\alpha a_6 k_x^2 + [a_1 - \alpha(a_5 + a_{15} + a_{17})]k_x - \alpha a_{16} k_x k_y$$
$$-\alpha a_{11} k_y^2 + [a_2 - \alpha(a_{10} + a_{14} + a_{18})]k_y + a_3 + \alpha a_4 \tag{16.169}$$

where

$$b_1 = \frac{\hbar^2}{2m_c}, \quad b_2 = \frac{\hbar^2}{2m_c}, \quad b_3 = \frac{\hbar^2}{2m_c}\left(\frac{n\pi}{a}\right)^2,$$
$$a_4 = 2\left(\frac{e\hbar}{2m_c}\right)^2(B_x^2 + B_z^2) + \frac{4}{3}\left[\frac{\hbar}{2m_c}\left(\frac{n\pi}{a}\right)eL_y B_x\right]^2$$

$$-\frac{2}{9}\left(\frac{e^2 L_x L_y B_x B_z}{2m_c}\right)^2 - \frac{2}{3}\left[\left(\frac{e^2 L_x B_y B_z}{2m_c}\right)^2\right.$$

$$\left.+\left(\frac{e^2 L_y B_y B_x}{2m_c}\right)^2\right]\langle z^2\rangle - \frac{4}{3}\hbar^2\left(\frac{n\pi}{a}\right)^2\left(\frac{eL_y B_x}{2m_c}\right)^2$$

$$-\frac{1}{5}\left[\left(\frac{e^2 L_x^2 B_z^2}{2m_c}\right) + \left(\frac{e^2 L_y^2 B_x^2}{2m_c}\right)^2\right] - \frac{e^2 B_y^2}{2m_c}\langle z^4\rangle$$

$$a_5 = 4\left(\frac{\hbar^2 e^3 B_y^3}{4m_c^2}\right)\langle z^3\rangle, \quad a_6 = 4\left(\frac{\hbar e B_y}{2m_c}\right)^2\langle z^2\rangle,$$

$$a_7 = 2\left(\frac{eB_y}{2m_c}\right)^2\langle z^3\rangle, \quad a_8 = 4\left(\frac{\hbar e B_y}{2m_c^2}\right)\langle z\rangle$$

$$a_9 = \frac{2}{3}\left(\frac{eL_y B_x}{2m_c}\right)^2, \quad a_{10} = \frac{\hbar e^3 L_x^3 B_z^3}{4m_c^2}, \quad a_{11} = \frac{4}{3}\left(\frac{\hbar e L_x B_z}{2m_c}\right)^2,$$

$$a_{12} = \left(\frac{eL_x B_z}{2m_c}\right)^2, \quad a_{13} = \frac{2\hbar e L_x B_z}{4m_c^2},$$

$$a_{14} = 2\left[\frac{\hbar e^3 L_x B_y^2 B_z}{4m_c^2}\right]\langle z^2\rangle, \quad a_{15} = \frac{4}{3}\left[\frac{\hbar e^3 L_x^2 B_z^2 B_y}{4m_c^2}\right]\langle z\rangle,$$

$$a_{16} = 4\left[\frac{\hbar e^2 L_x B_z B_y}{4m_c^2}\right]\langle z\rangle, \quad a_{17} = \frac{4}{3}\left[\frac{\hbar e^3 L_y^2 B_x^2 B_y}{4m_c^2}\right]\langle z\rangle$$

(16.220) can be written as

$$\begin{aligned}
&[b_1 - \alpha a_1^2 + \alpha a_8 a_1 - \alpha a_6]k_x^2 + [b_2 - \alpha a_2^2 + \alpha a_{13} a_2 - \alpha a_{11}]k_y^2\\
&\quad + [a_1 - \alpha(a_5 + a_{15} + a_{17}) - \alpha a_8(\gamma_3(E,\eta_g) - a_3)\\
&\quad + \alpha a_1(a_7 + a_9 + a_{12}) + 2\alpha a_1(\gamma_3(E,\eta_g) - a_3)]k_x\\
&\quad + [a_2 - \alpha(a_{10} + a_{14} + a_{18}) - \alpha a_{13}(\gamma_3(E,\eta_g) - a_3)\\
&\quad + \alpha(a_7 + a_9 + a_{12})a_2 + 2\alpha a_2(\gamma_3(E,\eta_g) - a_3)]k_y\\
&\quad + \alpha(a_{13}a_1 + a_8 a_2 - 2a_1 a_2 - a_{16})k_x k_y
\end{aligned}$$

$$+ [b_3 - (\gamma_3(E, \eta_g) - \alpha_3((\gamma_3(E, \eta_g - a_3)^2$$
$$- \alpha(a_7 + a_9 + a_{12})(\gamma_3(E, \eta_g - a_3) + a_3 + \alpha a_4] = 0 \quad (16.170)$$

(16.170) can be expressed as

$$P_1 k_x^2 + Q_1 k_y^2 + k_x k_y + P_2 k_x + Q_2 k_y + c = 0 \quad (16.171)$$

where

$$P_1 = [b_1 - a a_1^2 + \alpha a_8 a_1 - \alpha a_6] \quad Q_1 = [b_2 - \alpha_2^2 + \alpha a_{13} a_2 - \alpha a_{11}]$$
$$P_2 = [a_1 - \alpha(a_5 + a_{15} + a_{17}) - \alpha a_8(\gamma_3(E, \eta_g) - a_3)$$
$$+ \alpha a_1(a_7 + a_9 + a_{12}) + 2\alpha a_1(\gamma_3(E, \eta_g) - a_3)]$$
$$Q_2 = [a_2 - \alpha(a_{10} + a_{14} + a_{18}) - \alpha a_{13}(\gamma_3(E, \eta_g) - a_3)$$
$$+ \alpha(a_7 + a_9 + a_{12}) + 2\alpha a_2(\gamma_3(E, \eta_g) - a_3)]$$
$$c = [b_3 - (\gamma_3(E, \eta_g) - \alpha((\gamma_3(E, \eta_g) - a_3)^2$$
$$- \alpha(a_7 + a_9 + a_{12})(\gamma_3(E, \eta_g) - a_3) + a_3 + \alpha a_4] = 0$$

Let us substitute

$$k_x = k_x' + \alpha_1, \quad k_y = k_y' + \beta_1$$

where

$$\alpha_1 = \frac{RQ_2 - 2Q_1 P_2}{R^2 - 4Q_1 P_1}, \quad \beta_1 = \frac{2P_1 Q_2 - RP_2}{R^2 - 4Q_1 P_1}$$

(16.171) can be expressed as

$$Ak_x'^2 + Bk_y'^2 + 2Hk_x'^2 k_y'^2 = 1 \quad (16.172)$$

where

$$A = \frac{P_1}{C_1}, \quad B = \frac{Q_1}{C_1}, \quad 2H = \frac{R}{C}$$

and

$$C_1 = C - P_1\alpha_1^2 - Q_1\beta_1^2 + R\alpha_1\beta_1 + P_2\alpha_1 - Q_2\beta_1$$

The area of the ellipse is given by

$$A_{19} = \frac{\pi}{\sqrt{AB - H^2}} \quad (16.173)$$

The EEM can be written as

$$m^*(E, n, B_x, B_y, B_z) = \frac{\hbar^2}{2\pi} A'_{19} \tag{16.174a}$$

The surface electron concentration can be expressed as

$$n_0 = \frac{g_v}{2\pi} \sum_{n=0}^{n_{\max}} \left[\frac{1}{\sqrt{AB - H^2}} |E_{F15225} + \sum_{r=1}^{s} L(r) \sum_{n=0}^{n_{\max}} \left[\frac{1}{\sqrt{AB - H^2}} |E_{F15225} \right] \right] \tag{16.174b}$$

where E_{F15225} is the Fermi energy in this case

16.3 Open Research Problems

(R.16.1) Investigate the DR for the QWs of HD negative refractive index, organic, magnetic and other advanced optical materials in the presence of an arbitrarily oriented alternating electric field.

(R.16.2) Investigate the DR for the multiple QWs of HD materials whose unperturbed carrier energy spectra are defined in R16.1

(R.16.3) Investigate the DR for all the appropriate HD low dimensional systems of this chapter in the presence of finite potential wells.

(R.16.4) Investigate the DR for all the appropriate HD low dimensional systems of this chapter in the presence of parabolic potential wells.

(R.16.5) Investigate the DR for all the appropriate HD systems of this chapter forming quantum rings.

(R.16.6) Investigate the DR for all the above appropriate problems in the presence of elliptical Hill and quantum square rings.

(R.16.7) Investigate the DR for triangular two dimensional systems in the presence of an arbitrarily oriented alternating electric field for all the HD materials whose unperturbed carrier energy spectra are defined in R16.1.

(R.16.8) Investigate the DR for HD two dimensional systems of the negative refractive index and other advanced optical materials in the presence of an arbitrarily oriented alternating electric field and non-uniform light waves.

(R.16.9) Investigate the DR for triangular HD two dimensional systems of the negative refractive index, organic, magnetic and other advanced optical materials in the presence of an arbitrarily oriented alternating electric field in the presence of strain.

(R.16.10) (a) Investigate the DR for HD two dimensional systems of the negative refractive index, organic, magnetic and other advanced optical materials in the presence of many body effects.

(b) Investigate all the appropriate problems of this chapter for a Dirac electron.

(R.16.11) Investigate all the appropriate problems of this chapter by including the many body, image force, broadening and hot carrier effects respectively.

(R.16.12) Investigate all the appropriate problems of this chapter by removing all the mathematical approximations and establishing the respective appropriate uniqueness conditions.

Materials Index

Subject Index

N

O

P

Q

R

S

T

U

V

W

Z